Basics of

Carpentry & Construction

for Certificate II

McGraw Hill

2e

DANIEL BONNICI | ALISTER FORD | PAT AIKEN

Title: Basics of construction and carpentry for Certificate II
Daniel Bonnici, Alister Ford, Pat Aiken
ISBN: 9781743767498

A catalogue record for this book is available from the National Library of Australia

Published in Australia by
McGraw-Hill Education (Australia) Pty Ltd
Level 33, 680 George Street, Sydney NSW 2000
Publisher: Norma Angeloni-Tomaras
Content developer: Martina Vascotto
Production editor: Lara McMurray
Editor: Alison Moore
Proofreader: Meredith Lewin
Indexer: SPi, India
Cover image: Onur ERSIN/Shutterstock
Cover and internal design: Seymour Design
Internal design image: Branislav/Getty Images
Typeset in That 10/13 by SPi, India
Printed in Singapore 90 gsm matt art by Markono Print Media Pte Ltd

Contents

Preface

This book addresses the Certificate II training in the VET and secondary sectors, and has been developed to complement the training of students studying in the trade areas of carpentry, joinery and construction. This publication is also mapped to the National Training Package CPC08 and Victorian Building and Construction qualification 22338VIC and references the Australian Standards and state and national building codes.

The text is designed to teach the underpinning knowledge and skills that are required to work effectively in the construction industry. The student can then apply and adapt these skills to current workplace practices. Out in the workplace, techniques will vary as a result of an employer's preference or building philosophy, however the core skills remain the same. The text aims to describe these skills and support students as they practice and refine their own techniques in an ever-changing workplace.

As a teaching resource, this publication covers the major skills and knowledge required by apprentices and pre-apprentices. It will assist in the teaching of hand and power tools, plan reading, WHS/OHS and sustainability, and covers the skills and knowledge required for building a house. Each chapter is aligned to specific units of competency, so trainers will be able to plan their delivery easily. All chapters have worksheet questions that reinforces the content covered, as well as a student research task. Plan reading is integrated into most chapters and refers to the set of plans located in the appendix of this book.

With clear step-by-step instructions and up-to-date content, the major skills and knowledge required to construct a timber-framed and clad house are covered, as well as workplace health and safety, sustainability, hand and power tools and scaffolding. Addressing the majority of the Year 1 and 2 apprenticeship training, this edition will assist students throughout their Certificate II training.

About the authors

DANIEL BONNICI

Daniel currently teaches carpentry and design technology at Bayside P–12 college (Paisley campus, Newport) in their Technical Trades Centre. Prior to this Daniel was employed at Victoria University (TAFE), where he was the VET in schools' (VETiS) coordinator for carpentry, bricklaying, furnishing, construction trade taster and school-based carpentry apprentices. Daniel taught in the areas of VETiS and apprenticeship carpentry.

Daniel has worked in Australia and the United Kingdom as a carpenter in the domestic, industrial and commercial sectors. His qualifications include:

- Certificate III in Carpentry—industrial broad stream
- Diploma of Vocational Education and Training
- Certificate IV in Workplace Assessment (TAE)
- Graduate Certificate in Leadership in Education and Training
- Certificate IV in Building
- Graduate Diploma in Technology Education.

Daniel currently chairs the Victorian Carpentry Teachers Network, which has representatives from all Victorian TAFEs. He has contributed to the Victorian Certificate II in Building & Construction for the past 15 years as a subject matter expert and project steering committee member.

In 2017 Daniel was appointed to the carpentry Technical Advisory Group (TAG) addressing the redevelopment of the current CPC Certificate III in Carpentry and Certificate I & II in Construction.

In 2007 Daniel was recognised for his work in the VET in schools carpentry area, being awarded:

- Australian Trade Teacher of the Year in General Construction 2007 (Institute for Trade Skills Excellence)
- Victoria University Vice Chancellor Peak Award for Excellence in Teaching and Learning 2007
- Victoria University Vice Chancellor Citation for Excellence in Teaching and Learning for the Faculty of Technical and Trades Innovation 2007.

Daniel is an avid supporter of WorldSkills Australia and has been a convenor and judge for VETiS construction and Open carpentry at regional and state competitions for the past 12 years. He has been chief judge in the category of VETiS construction at the last three national competitions.

ALISTER FORD

Alister commenced his carpentry and joinery apprenticeship in June 1979 at the age of 15 and completed in June 1983.

He then worked on some very large commercial and industrial construction sites around Melbourne before relocating to country Victoria in 1990, and starting his own carpentry and joinery business that operated for 15 years.

In 2002 Alister commenced teaching carpentry Certificate II and III part time at the University of Ballarat TAFE division while still running his own business.

In 2005 Alister commenced full-time teaching in carpentry at the UB TAFE and also held the position of Program Coordinator for the Building & Construction department that oversaw the running of the carpentry, joinery, furniture-making, cabinet-making, painting and decorating, bricklaying/blocklaying and design/drafting programs.

In 2010 Alister transferred his employment as a building and construction teacher to the East Gippsland TAFE, now TAFE Gippsland, based at the Bairnsdale Trade Centre campus to teach Certificate II and III in carpentry.

PAT AIKEN

Pat contributed to Chapters 1, 2, 19 and 20.

A licensed builder, Pat has a long history in the construction industry, and later worked as a teacher of bricklaying, becoming head teacher at Hornsby College of TAFE in 1994. Pat was actively involved in various building program projects during his time at TAFE, including an international benchmarking project in London in 2001 with Construction Training Australia. After an 18-month secondment to the NSW Office of Fair Trading, Pat retired from TAFE in 2007 and started working as a licensed builder and part-time TAFE teacher.

Acknowledgments to:

John Friedman contributed to Chapter 19. He teaches construction at TAFE NSW, Randwick.

Alex Neale (retired TAFE Teacher) has a long history in carpentry and training apprentices in Victoria, QLD and NT. Alex contributed to Chapters 3, 4, 12 and 16.

Jack Barrington, Stuart Arden and **Dieter Mylius**, founding authors with a long history in the construction industry and as teachers of carpentry. All of whom have laid the foundations of this textbook.

We would also like to thank the following people who provided beneficial feedback by reviewing the manuscript:

Adrian Lea—Box Hill Institute
Geoffrey Leng—Sunraysia Institute of TAFE
Marcela Ganicoche—TAFE SA
Mathew Ablett—Sunitafe
Michael Callahan—Previously the head of construction at Melbourne Polytechnic. Michael is now involved with DFAT as part of the 'train the trainer' program in SouthEast Asia and China
Michael Hick—Victoria Polytechnic
Trevor Ingram—VSTP, CCF

Basic competency mapping

Publisher's note: current as of 12 March 2020; please refer to Training.gov for the most up-to-date information.

Chapter	UPC CPC08 Units
1. Work health and safety (Worksheet 1)	• CPCCOHS2001A Apply OHS requirements, policies and procedures in the construction industry • CPCCCM2010B Work safely at heights • CPCCCM1013A Plan and organise work
2. Workplace communication (Worksheet 2)	• CPCCCM1014A Conduct workplace communication • CPCCCM1013A Plan and organise work
3. Trade calculations and measurements (Worksheet 3)	• CPCCCM1015A Carry out measurements and calculations
4. Trade drawings (Worksheet 4)	• CPCCCM2001A Read and interpret plans and specifications
5. Working effectively and sustainably within the construction industry (Worksheet 5)	• CPCCCM1012A Work effectively and sustainably in the construction industry
6. Carpentry hand tools (Worksheet 6)	• CPCCCM2005B Use construction tools and equipment • CPCCCA2002B Use carpentry tools and equipment
7. Timber joints (Worksheet 6)	• This chapter is relevant across several units
8. Construction fastenings and adhesives (Worksheet 6)	• This chapter is relevant across several units
9. Carpentry power tools (Worksheet 6)	• CPCCCM2005B Use construction tools and equipment • CPCCCA2002B Use carpentry tools and equipment
10. Site setting out and basic levelling (Worksheets 7 and 8)	• CPCCCM2006B Apply basic levelling procedures • CPCCCA3002A Carry out setting out • CPCCCA3023A Carry out levelling operations
11. Subfloor and flooring (Worksheet 9)	• CPCCCM2006B Apply basic levelling procedures • CPCCCA3002A Carry out setting out • CPCCCA3023A Carry out levelling operations • CPCCCA3003A Install flooring systems
12. Wall framing (Worksheet 10)	• CPCCCA3004A Construct wall frames
13. Basic roofing (Worksheet 11)	• CPCCCA3005B Construct ceiling framing • CPCCCA3007C Construct a pitched roof • CPCCCA3008B Construct eaves
14. The hip and valley roof (Worksheet 11)	• CPCCCA3007C Construct a pitched roof
15. Restricted-height scaffolding and work platforms (Worksheet 12)	• CPCCCM2008B Erect and dismantle restricted height scaffolding • CPCCCM2010B Work safely at height
16. External cladding (Worksheet 13)	• CPCCCA3017B Install exterior cladding
17. Windows and doors (Worksheet 14)	• CPCCCA3010A Install and replace windows and doors • CPCCCA3013A Install lining, panelling and moulding
18. Internal linings and fixings (Worksheet 15)	• CPCCCA3010A Install and replace windows and doors • CPCCCA3013A Install lining, panelling and moulding
19. Formwork construction and basic concreting (Worksheet 16)	• CPCCCO2013A Carry out concreting to simple forms • CPCCCA2003A Erect and dismantle formwork for footings and slabs on ground • CPCCCM2002A Carry out excavation
20. Carry out demolition (Worksheet 17)	• CPCCCM2009A Carry out basic demolition • CPCCCA3001A Carry out demolition of minor building structures

Chapter	Certificate II Building and Construction—Victorian Units 22338VIC	
1	• CPCCOHS2001A • CPCCWHS1001	Apply OHS requirements, policies and procedures in the construction industry Prepare to work safely in the construction industry
2	• CPCCCM1014A	Conduct workplace communication
3	• CPCCCM1015A	Carry out measurements and calculations
4	• VU22015	Interpret and apply basic plans and drawings
5	• CPCCCM1012A	Work effectively and sustainably in the construction industry
6	• VU22022	Identify and handle carpentry tools and equipment
7	• This chapter is relevant across several units	
8	• This chapter is relevant across several units	
9	• VU22022	Identify and handle carpentry tools and equipment
10	• VU22023 • CPCCCM2006	Perform basic setting out Apply basic levelling procedures
11	• VU22023 • CPCCCM2006 • VU22024	Perform basic setting out Apply basic levelling procedures Construct basic subfloor
12	• VU22025	Construct basic wall frames
13	• VU22026	Construct basic roof frame
14	• VU22026	Construct basic roof frame
15	• VU22016	Erect and safely use working platforms
16	• VU22027	Install basic cladding
17	• VU22029 • VU22028	Install interior fixings Install basic window and door frames
18	• VU22029 • VU22028	Install interior fixings Install basic window and door frames
19	• VU22031	Construct basic formwork for concreting
20	• VU22030	Carry out basic demolition of timber structures

Chapter 1

Work health and safety

Learning Objectives

LO 1.1 Train to survive your first day at work

LO 1.2 Know about duty of care: PPE, safety signage and personal wellbeing

LO 1.3 Avoid construction industry injuries

LO 1.4 Identify major health and safety risks

LO 1.5 Deal with emergencies on construction sites

LO 1.6 Know about Australian work health and safety legislation, regulations and codes of practice

LO 1.7 Analyse workplace hazards

LO 1.8 Know about safe work method statements and job safety analysis

LO 1.9 Understand the work health and safety management plan

Introduction

The building and construction industry covers a wide range of workplaces. These workplaces are among the most dangerous of any Australian working environments–by comparison, others appear far less hazardous. Thousands of Australian construction workers go to worksites each day in the belief that they will return home safely after a day's work. This chapter will focus on construction worksites related to residential construction.

The federal legislation that addresses workplace health and safety in Australia is the *Work Health and Safety Act 2011*, which is underpinned by the Work Health and Safety Regulations of individual Australian states and territories. This legislation has been adopted by the majority of Australian states and territories.

Residential building worksites, such as those for the construction, extension or renovation of a house, are not commonly considered dangerous workplaces, but many workers are exposed to hazards every day without realising it. Many Australian building workers are either seriously injured or die each year because they are involved in work accidents, or have been exposed to hazardous substances on residential building worksites that they believed to be reasonably safe.

Among these workers are young people in their first job, struggling to learn new skills and trying hard to fit into a foreign environment where the majority of people they are working with have much greater work and life experience. All around them, they are exposed to hazards that older and much more experienced workers have learnt to deal with over many years.

In this chapter as a continuation of your previous workplace induction training (Construction Induction White Card), you will learn about all aspects of working safely in the construction industry. You will learn about being aware of your own safety at all times, as well as having a duty of care to your fellow workers, and Australian work health and safety laws, regulations and codes of practice.

You will read about personal protective equipment (PPE), how to select the correct equipment, wear or use it properly and identify situations where other workers, although not involved in the particular task, will need to use PPE as well.

Workplace signage will also be explained in this chapter. It is important to be able to identify the different types of signs and understand their specific meanings.

In addition, you will learn how to identify workplace hazards and risks and how to document them in safe work method statements (SWMS), as well as how to formulate strategies to control or eliminate these hazards and risks. You will also learn about workplace health and safety plans, which you will put into practice almost every day on a construction site.

Additionally, you will learn about the accidents, injuries and incidents that could occur and how best to deal with such situations if they take place.

WORKPLACE SCENARIO

Search the Australian or your state's WorkSafe website and you will find articles and stories documenting incidents that have occurred in the Australian construction industry involving the collapse or failure of scaffolding systems.

At worst, these incidents can be fatal to the workers on the construction site and even members of the public who may be nearby. Such incidents can sometimes thankfully result in only minor injuries, but in most cases the cost can be enormous—with lives lost, people badly injured and an enormous amount of damage to property.

One such incident that I recall happening in the Melbourne suburb of Prahran in 2009 could potentially have claimed many lives, but miraculously no one was killed and only three workers were injured.

A scaffolding was constructed to the height of six stories, across the front facade of a building in Commercial Road, Prahran. Approximately 50 workers were on this construction site and, luckily, at the time of collapse, it was morning teatime and most of the workers were in the site sheds having a work break.

However, three workers were on the scaffolding when the collapse occurred. Emergency workers said they were amazed that no one was killed—neither site workers nor the general public. The scaffolding collapsed onto Commercial Road, crushing seven cars parked along the road outside the construction site. It also brought down powerlines and tram lines, and nearby buildings had to be evacuated. Of the three workers who were on the scaffolding at the time of the collapse, one man escaped with a broken finger and cuts and scrapes, another man suffered broken ribs and back injuries and a third man suffered cuts and scrapes to his arms and legs. These men were treated by emergency personnel and transported to the hospital. One of them had jumped off the scaffolding as it started to collapse, and the other two men had 'ridden' it to the ground. It was extremely fortunate that there were no pedestrians walking along the footpath below at the time, or any people in the parked cars.

After this incident occurred, the worksite was closed by the WorkSafe authority while investigations took place, and the road was closed for some time so that the area could be cleared, and the power and tram lines restored. An incident like this has the potential to claim many lives, but luckily most of the workers were not on the scaffolding at the time, and no members of the public were in the vicinity. The material cost of this incident would have been enormous, but it would have been much higher if lives had been lost.

A simple Google search will allow you to access and read all of the details of this scaffolding collapse incident.

1.1 Train to survive your first day at work

Before starting work in the construction industry all workers must undertake compulsory safety training, but to be successful workers must be prepared for lifelong training. Training and skills development is the best defence against workplace accidents on your first day of work and for every working day that follows.

1.1.1 Mandatory construction induction training

In all Australian states and territories, it is mandatory for all construction workers to undertake general induction training *before they commence work on a construction worksite.* This is particularly important for young construction industry workers, as over recent years there have been cases of young workers who did not undertake this training and were fatally injured on their first day of work.

On completion of the **mandatory construction induction training**, an identification card is issued to the successful participant. This card must be carried by workers at all times while they are undertaking building and construction work. Each state or territory is responsible for both regulating the training and issuing of these cards (see Fig. 1.1).

Regardless of the state or territory the card is issued in, they are **mutually recognised** by all other states and territories. Before working in Victoria, it is recommended that workers from another state or territory confirm that their construction induction card will be recognised.

1.1.2 Site-specific induction training

It is also compulsory for workers to undertake **site-specific induction training** before they enter any new worksite. This is particularly important for young workers. This training identifies all hazards specific to that particular worksite and is normally provided by a representative of the employer or the builder in control of the site (e.g. the site manager or safety officer). For example, if there is an area where excavation is being carried out, workers will be warned of this hazard and, most likely, be told not to enter that area unless authorised.

Site-specific induction training is also an opportunity for the employer or the site manager to:

- sight and confirm that the worker has completed their general induction training and record the details of their general induction card
- explain the safety rules that apply to that worksite
- explain site emergency procedures and the location of the emergency assembly area
- provide the location of first-aid stations and where site amenities are located.

On completion of this training, the site manager will record the workers who undertook the training and, on some projects, issue a card or sticker to indicate their participation and that their site induction training is up to date or current (see Fig. 1.2).

Fig. 1.1 A–D Sample construction induction cards for some states—each state and territory has its own card

(a)

NSW GOVERNMENT | WorkCover

WORK HEALTH AND SAFETY

GENERAL CONSTRUCTION INDUCTION

NEW SOUTH WALES

John James Citizen

Date of Birth 01/01/1987

100001 Card No.

12345 RTO No.

Date of Issue 01/06/2019

(b)

THE GOVERNMENT OF WESTERN AUSTRALIA

WorkSafe WA

OCCUPATIONAL HEALTH AND SAFETY

CONSTRUCTION INDUCTION

WESTERN AUSTRALIA

Issued to

Issue Date

Custom White Signature Panel

Card No 000000

(c)

Tasmanian Government

WORK HEALTH AND SAFETY

GENERAL CONSTRUCTION INDUCTION

TASMANIA

Date of Birth

Issue Date

Card No

(d)

Queensland Government

Department of Employment and Industrial Relations

OCCUPATIONAL HEALTH AND SAFETY

CONSTRUCTION INDUCTION

QUEENSLAND

Cardholder's name

Date of birth / /

RTO No.

Issue date / /

Card No.

Source: (a) © State of New South Wales (SafeWork NSW). For current information go to safework.nsw.gov.au. Creative Commons licence https://creativecommons.org/licenses/by/4.0/legalcode; (b) WorkSafe Western Australia (DMIRS); (c) WorkSafe Tasmania; (d) Office of Industrial Relations, Queensland. For further information regarding construction induction cards, go to www.safework.nsw.gov.au/resource-library/licence-and-registrations/recognition-of-general-construction-induction-training-cards-fact-sheet.

Fig. 1.2 Site induction hard hat sticker

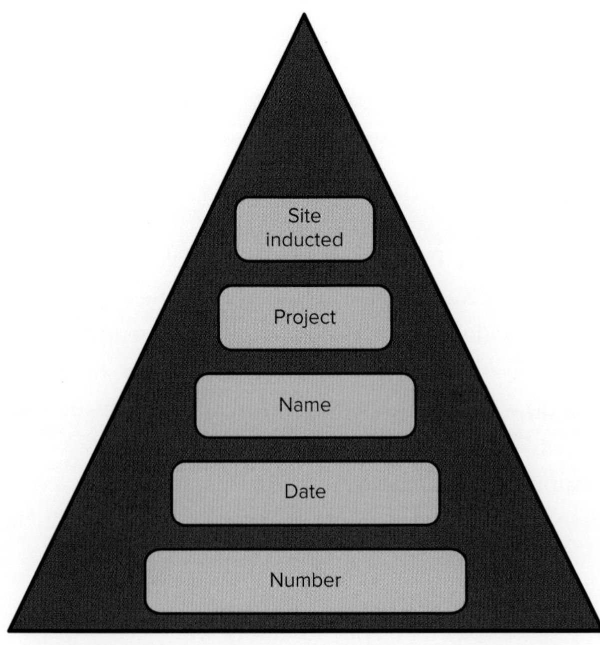

1.1.3 Other training

1.1.3.1 Work activity training

Where there is a new work situation that requires workers to carry out work in which they have no experience, employers must provide appropriate **work activity training** to ensure that those workers can undertake the work safely. Often this type of training is conducted informally on-site as a **toolbox meeting** and workers are given an opportunity to participate in a risk assessment process that identifies any possible risks and how these can be managed or controlled.

If the work is of a high-risk or hazardous nature, the employer will provide workers with a copy of a **safe work method statement**. Work method statements provide a step-by-step approach to a hazardous task where all risks are identified and evaluated, and appropriate controls are put in place. The employer will ensure that all workers have read and understood the information provided before they can commence work.

For young workers and any workers with no previous construction experience, this is especially important. Until a worker can undertake work competently, they must be properly supervised by a person who is authorised by the employer to supervise their work and carry out that same work in a competent manner.

Work classified as high risk includes that which involves:

- a risk of falling more than 2 metres
- work carried out on a telecommunication tower
- the demolition of load-bearing elements of a structure
- the disturbance of asbestos
- work requiring temporary support of structural components
- a confined space
- excavation greater than 1.5 metres deep
- tunnelling
- the use of explosives
- work near pressurised gas mains or pipes
- work on or near chemical, fuel or refrigerant lines
- work near energised electrical installations/services
- a potentially contaminated or flammable atmosphere
- tilt-up or pre-cast concrete construction
- a location on, in or adjacent to a road, railway, shipping lane or other traffic corridor that is in use by traffic
- a location near a powered mobile plant
- artificial extremes of temperature
- a location near water or other liquid that involves a risk of drowning
- diving.

A sample work method statement and risk assessment is provided in the appendix.

1.1.3.2 Nationally accredited training

Formal off-job training in a specific occupation or skill combined with on-job training and supervision by a qualified and competent person is a pathway to a safe workplace for all workers. Some examples of Australian National Construction Qualifications are shown in Table 1.1.

Table 1.1 Australian national construction qualifications

Certificate III in Bricklaying/Blocklaying	Certificate III in Paving
Certificate II in Building and Construction (Victoria)	Certificate III in Post-tensioning
Certificate III in Carpentry	Certificate III in Rigging
Certificate III in Carpentry and Joinery	Certificate III in Roof tiling
Certificate III in Concreting	Certificate III in Scaffolding
Certificate III in Construction waterproofing	Certificate III in Shopfitting
Certificate III in Demolition	Certificate III in Signage
Certificate III in Dogging	Certificate III in Solid plastering
Certificate III in Formwork/Falsework	Certificate III in Steelfixing
Certificate III in Joinery	Certificate III in Stonemasonry (monumental/installation)
Certificate III in Joinery (stairs)	Certificate III in Wall and Ceiling lining
Certificate III in Painting and Decorating	Certificate III in Wall and Floor tiling

More information on these qualifications can be found on www.training.gov.au.

Traditional apprenticeships and entry-level vocational training that result in a national qualification under the Australian Qualifications Framework will provide specific trade training combined with **safe operating procedures** to safely carry out specific trade work.

Fig. 1.3 Safe operating procedures

SAFETY OPERATING PROCEDURES

Power tools

Safety glasses must be worn at all times when using power tools.

Long and loose hair must be contained.

Safety footwear must be worn at all times when using power tools.

Protective clothing must be worn.

Loose jewellery must not be worn.

Hearing protection must be used when using power tools.

Dust mask may be needed for this tool.

Ensure loose clothing is contained or removed.

PRE-OPERATIONAL SAFETY CHECKS

1. Check workspaces and walkways to ensure no slip/trip hazards are present.
2. Read manufacturer's instructions and familiarise yourself with the tool.
3. Ensure all guards and safety shields are in position before starting the power tool. Faulty equipment must not be used. Immediately report any suspect machinery.
4. Ensure the power tool and leads have been checked and the electrical safety tag is current.
5. Check the power tool's lead and plug for cuts or exposed wires.
6. Locate and ensure you are familiar with the operation of the ON/OFF starter.
7. Ensure power supply has an a RCD (safety switch) installed. If in doubt use a portable RCD.

OPERATIONAL SAFETY CHECKS

1. Ensure the work area is stable and level and that the power tool is set up appropriately.
2. Ensure power leads are kept off the ground.
3. Small objects must not be held by hand.
4. Never leave the machine running unattended.
5. Do not bend down near the machine while it is running.
6. Never force the power tool to do its job.
7. Ensure you have a firm grip on the power tool.
8. Slowly move the power tool across the item in a uniform manner.

HOUSEKEEPING

1. Switch off the power tool.
2. Leave the machine in a safe, clean and tidy state.
3. Ensure all off cuts are disposed of in an appropriate receptacle.
4. Ensure any sawdust is swept up.

POTENTIAL POWER TOOL HAZARDS

■ Hot metal ■ Sparks ■ Noise ■ Sharp edges ■ Electrocution

■ Entanglement ■ Dust ■ Eye injuries ■ Splinters ■ Kick back

This SOP does not necessarily cover all possible hazards associated with the machine and should be used in conjunction with other references (SWMS and manufacturer's instructions). It is designed to be used as an adjunct to teaching Safety Procedures and to act as a reminder to users prior to machine use

1.2 Duty of care: PPE, safety signage and personal wellbeing

1.2.1 Personal protective equipment

Together with training, personal protective equipment (PPE) provides the most essential and basic form of protection for all workers. All workers are required by law to use PPE in accordance with:

- safe operating procedures required by their employer and authorities
- any training they receive
- work method statements they must follow.

Where there is a potential risk of injury, employers must supply PPE and ensure that their employees correctly use any PPE they provide.

PPE provided by employers could include such items as:

- safety glasses, goggles or face shields
- overalls or workpants
- high-visibility clothing
- high-visibility reflective vests
- jackets
- steel-capped boots
- earplugs or earmuffs
- gloves–leather or rubber
- respirators or dust masks
- hard hats
- caps or hair nets
- wide-brimmed hats
- UV protective sunscreen or clothing
- knee pads
- arm guards
- aprons.

Fig. 1.4 Personal protective equipment (PPE)

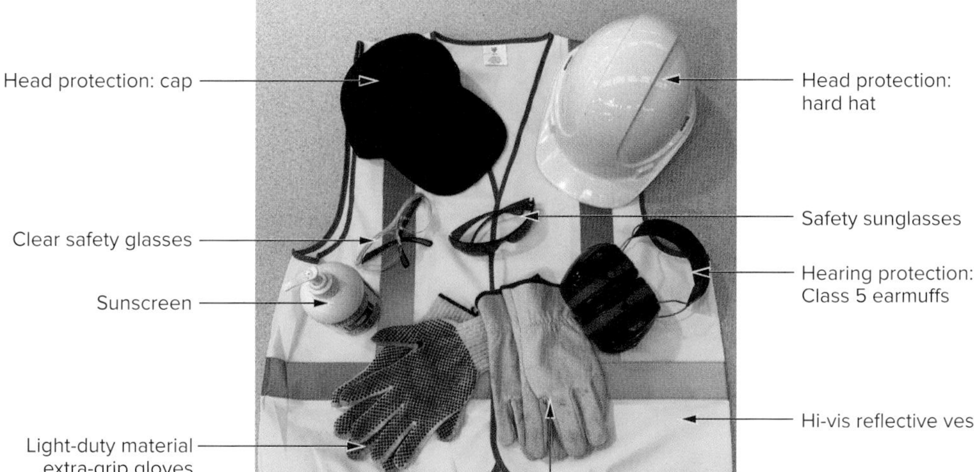

Courtesy of Alister Ford

1.2.2 Safety signage

A range of safety signage is used on construction worksites. For large worksites where there are many workers undertaking a wide range of work activities, there are many different safety signs. On smaller residential building sites, there are fewer and less varied signs.

There can never be 'too many' signs if they prevent injury and save lives!

It is important to take notice of all signage on a worksite and make every effort to read and understand the safety warnings, instructions or information they provide.

Fig. 1.5 Mandatory action signs

Courtesy of Alister Ford

A range of designs and colours are used for signs depending on the information that the sign is conveying.

Blue signs on a white background often with black lettering will tell you that the action you see is mandatory (Fig. 1.5). Generally, these signs will show or give instructions about the type of PPE that you must wear when in the designated area.

A white sign often with black lettering or figures with a red circle and line tell you that the action depicted on the sign is prohibited. This can include such things as No Smoking (Fig. 1.6), No Entry, No Mobile Phones and No Pedestrian Access.

Warning signs are generally yellow with black lettering and a hazard symbol inside a black triangle (Fig. 1.7). The hazard may not be life-threatening but may indicate an activity where extreme caution must still be exercised.

Fig. 1.6 Prohibited action signs

Fig. 1.7 A & B Warning signs

(a)

(b)

Fig. 1.8 A–C Emergency information signs

(a)

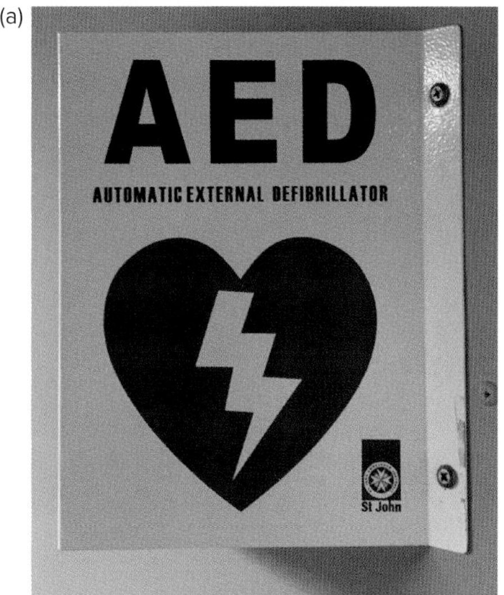

(b)

(c)

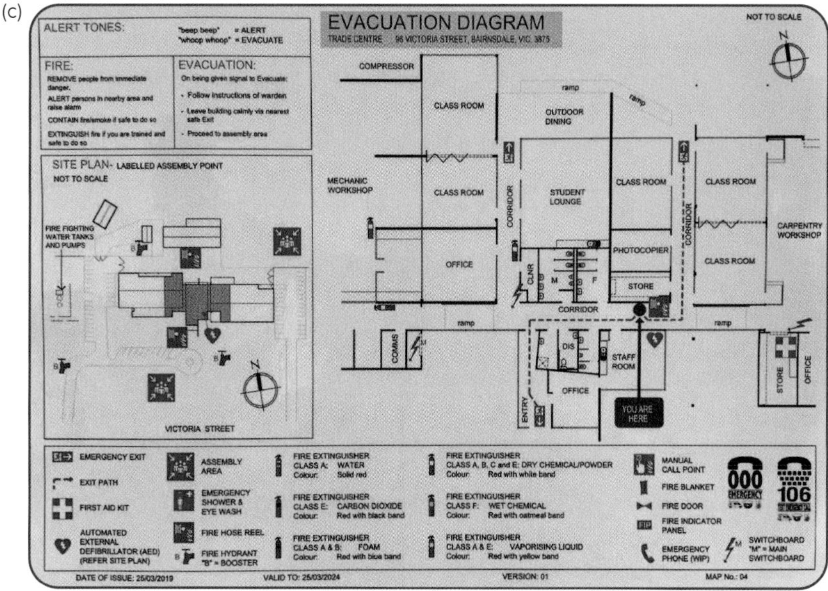

Courtesy of Alister Ford

Emergency information signs consist of a white symbol, lettering or both, often on a green background. Information can include the siting of emergency exits and assembly points, first-aid stations or kits, emergency eye wash or showers and safety equipment (Fig. 1.8).

1.2.3 Personal wellbeing

Every day when you go to work, you and all other workers have the right to a safe environment, both physically and mentally. You have the right to be able to safely carry out all of your tasks for the day, then return home.

1.2.3.1 Drugs and alcohol

Drugs and alcohol should not be used or consumed by anyone on any building site at any time. Anyone presenting for work under the influence of alcohol or drugs should be removed from the workplace immediately and appropriate arrangements made for them to be transported home or to a medical facility if required.

Every worker now also needs to be vigilant for discarded syringes and needles left on worksites. These can be thrown in from the street, over neighbouring fences or left by users after they have entered the worksite illegally for the night seeking shelter.

This is becoming a more common occurrence on building sites. If you do acquire a needlestick injury, remain calm. If there is a first aider on-site, report to them immediately–if not, see your supervisor. Wash the wound thoroughly with soap and water then report it to your doctor as soon as possible for a follow-up consultation.

1.2.3.2 Smoking in the workplace

Smoking is not an illegal activity, but many construction sites are now smoke-free or have designated areas or site sheds where smoking is allowed.

If you are a smoker, you must not smoke anywhere on the building site except in these designated areas. Any no-smoking signage must be noted and adhered to.

1.2.3.3 Bullying, harassment and discrimination

1.2.3.3.1 Bullying

Any form of bullying, be it intentional or unintentional, on the construction site is totally unacceptable. Sometimes a person may unknowingly bully someone, thinking it is just a harmless joke or part of an initiation. Any behaviour of this type needs to cease the moment it is identified and reported if required.

Forms of bullying may include:

- verbal abuse, insulting remarks, teasing, mocking, threats or unwarranted criticism
- physical intimidation
- violence
- deliberate exclusion
- tampering with, or deliberate damage to personal property, tools or work being undertaken.

Fig. 1.9 Site safety signs

1.2.3.3.2 Sexual harassment

Sexual harassment is against the law, meaning it is a criminal offence. No one in the workplace should be subjected to sexual harassment of any type at any time. Sexual harassment is generally looked upon as being a deliberate or premeditated act.

Forms of sexual harassment can include:

- whistling
- comments of a sexual nature
- uninvited physical contact

- offensive messaging or commenting via social media
- subjecting someone to offensive material
- taunting or making comments about a person's presumed sexuality or private life.

1.2.3.3.3 Discrimination

Employers are now subject to all federal anti-discrimination laws, meaning that everyone has an equal opportunity to apply for and be given due consideration for any job that may be available to be filled. In other words, everyone gets a 'fair go'.

Some of the major discrimination acts include:

- *Australian Human Rights Commission Act 1986*
- *Disability Discrimination Act 1992*
- *Racial Discrimination Act 1975*
- *Sex Discrimination Act 1984*
- *Age Discrimination Act 2004*.

If you feel you are being bullied, harassed or discriminated against at any time, do not just try to ignore it, hoping it will go away.

It may be a very awkward or difficult thing to do, but you need to take action by talking to your employer or the correct authority and explaining the situation with a view to having it rectified.

Also document any incidents with the date and time with a description of what has occurred.

CHECK YOUR UNDERSTANDING

1. List five items of personal protective equipment (PPE).
2. Obtain approximate costs for a set of PPE that is suitable for your occupational area of interest.

1.3 Construction industry injuries

1.3.1 Fatal injuries

From 2013 to 2017, a total of 153 workers in the Australian construction industry lost their lives at work.

The construction services sector accounted for 90 of these fatalities, the heavy and civil construction sector for 22 and the building and construction sector for 41.

Of these 41 in the building and construction sector, 21 were from the residential building construction area and 20 from the non-residential building construction area.

1.3.2 Injuries

In 2016–17 the number of serious injury claims (13 280) in the construction industry was at its highest for the past 18 years, rising by 6% in total from 2000 to 2017.

Of the 19 industry divisions, the construction industry is one of only four industries to record a rise in serious claims.

Table 1.2 Worker fatalities: Construction industry sub-divisions by mechanism of incident, 2014 to 2018 (combined)

Construction sub-division and mechanism	No. of fatalities	% of fatalities
Construction services	**92**	**59%**
Falls from a height	31	20%
Vehicle collision*	15	10%
Being hit by falling objects	11	7%
Contact with electricity	10	6%
Being trapped between stationary and moving objects	7	4%
Being hit by moving objects	6	4%
Being trapped by moving machinery	4	3%
Slide or cave-in	3	2%
Rollover of non-road vehicle	2	1%
Other mechanisms	3	2%
Building construction	**42**	**27%**
Falls from a height	18	12%
Being hit by falling objects	9	6%
Being hit by moving objects	5	3%
Contact with electricity	3	2%
Vehicle collision*	2	1%
Being trapped between stationary and moving objects	2	1%
Other mechanisms	3	2%
Heavy and civil engineering construction	**22**	**14%**
Being hit by moving objects	9	6%
Vehicle collision*	3	2%
Being trapped between stationary and moving objects	2	1%
Falls from a height	2	1%
Being hit by falling objects	2	1%
Other mechanisms	4	3%
Construction 5-year total	**156**	**100%**

* Vehicle collisions include fatalities that occurred as a direct result of a vehicle crash. Vehicles include not only road vehicles such as cars and trucks, but also machines such as aircraft, boats, loaders, tractors and quad bikes.

Note: The percentages shown in this table have been rounded to the nearest whole number; therefore the sum of percentage figures for each column may not equal the total.

Table 1.3 Number of serious claims by industry, 2000–01 and 2011–12 to 2016–17

Industry	2000–01	2011–12	2012–13	2013–14	2014–15	2015–16	% chg	2016–17
Health care and social assistance	15 315	19 935	19 240	17 850	17 440	16 705	9%	17 190
Construction	12 295	12 905	12 310	12 400	12 660	13 085	6%	13 280
Manufacturing	27 035	17 735	15 575	14 260	14 160	13 270	−51%	12 860
Retail trade	11 895	10 505	9850	9790	9630	9450	−21%	8490
Transport, postal and warehousing	11 565	11 530	10 385	9755	9225	8615	−25%	8330
Road transport	5260	5100	4800	4555	4380	4230	−20%	4155

Industry	2000–01	2011–12	2012–13	2013–14	2014–15	2015–16	% chg	2016–17
Public administration and safety	8120	9950	10 090	9280	8565	8075	–1%	7710
Education and training	6100	7355	6895	6640	6865	6705	10%	6850
Accommodation and food services	7395	7085	6725	6290	6335	6325	–14%	6175
Administrative and support services	6265	5730	4995	4505	3830	4280	–32%	4730
Wholesale trade	5900	5070	4865	4655	4705	4635	–21%	4500
Agriculture, forestry and fishing	5455	3865	3660	3475	3440	3620	–34%	3620
Agriculture	4155	2970	2875	2650	2640	2825	–32%	2765
Other services	4270	3865	3635	3520	3100	3035	–29%	3065
Arts and recreation services	2320	2620	2125	2260	2245	2200	–5%	2190
Mining	1895	2875	3060	2865	2225	2140	13%	2030
Professional, scientific and technical services	2110	2085	1850	1825	1860	1765	–16%	1795
Electricity, gas, water and waste services	1415	1285	1280	1250	1215	1175	–17%	1210
Rental, hiring and real estate services	1030	1185	1025	1035	1045	985	–4%	970
Information media and telecommunications	1230	755	605	620	600	545	–56%	590
Financial and insurance services	1310	965	835	790	765	680	–48%	585
Total	**133 040**	**127 425**	**119 130**	**113 240**	**110 080**	**107 380**	**–19%**	**106 260**

WORKPLACE SCENARIO

A young construction worker fell backwards into an unguarded drill hole 2.4 m deep. He fell to the bottom of the narrow hole with his feet sticking up above his head, squeezed into a sitting position. He was conscious, his head and hands were free, but he couldn't move. Other workers were prevented from digging him out because of site emergency procedures.

Soil had already fallen on top of him and rescuers couldn't come too close to the edge, concerned with disturbing the loose soil and burying him alive. A few days before the accident another worker had complained that the site was unsafe because of the poor quality of backfill.

An experienced ambulance officer tried to be lowered head-first into the hole to attach ropes to pull him out, but this was considered too dangerous.

Confident that the rescue would be successful, a video was recorded for training purposes. At the Coroner's inquest, the video shows a conversation between the young worker and rescuers in which he expresses his fear of the water rising around his head.

The rescue lasted over two hours. For at least five minutes before he was removed from the drill hole, his head was covered by mud and water. He lived for another five days in hospital before dying. The cause of death was drowning.

1.4 Major health and safety risks

1.4.1 Working at heights

Table 1.4 Worker fatalities: Construction industry, falls from a height fatalities by breakdown agency, 2014 to 2018 (combined)

Falls from a height: breakdown agency	No. of fatalities	% of fatalities
Buildings and other structures	18	35%
Ladders	9	18%
Openings in floors, walls or ceilings	4	8%
Scaffolding	3	6%
Doors and windows	3	6%
Other agencies	14	27%
Construction 5-year total—falls from a height	**51**	**100%**

Note: The percentages shown in this table have been rounded to the nearest whole number; therefore the sum of percentage figures for each column may not equal the total.

1.4.1.1 Ladders

Fig. 1.10 Correct ladder use

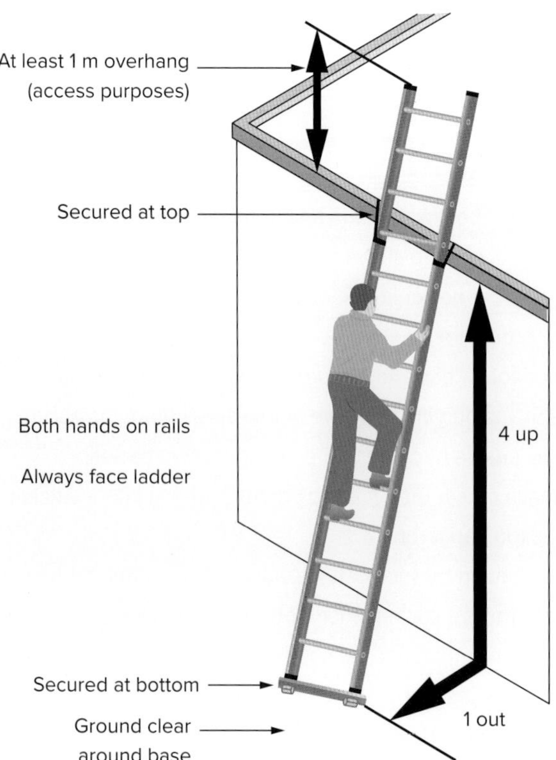

SAFETY TIPS Safe work on ladders

- Ladders should be rated at 150 kg for industrial use.
- Ensure ladder is long enough to extend to a height of 1 metre above a roof.
- Ladders must extend at least 1 metre above the highest rung being stood on or as far as possible above the level where work is being carried out.
- Set up on a solid and stable surface.
- Place ladder at a slope of 4:1.
- Secure top and bottom of ladder to stop slipping.
- Ensure the ladder is in good condition—inspect it.
- Use a tool belt to carry materials and tools.
- Maintain three points of contact when climbing or working from a ladder.
- Wear covered slip-resistant shoes.
- Only one person on a ladder at any time.
- Always face the ladder when ascending or descending.
- Do not climb higher than the third rung from the top of the ladder.

1.4.1.2 Scaffolding

Fig. 1.11 **(a) Perimeter scaffold with a fully decked working platform, guardrails and toeboards; (b) mobile scaffold with an access ladder and trapdoor**

(a)

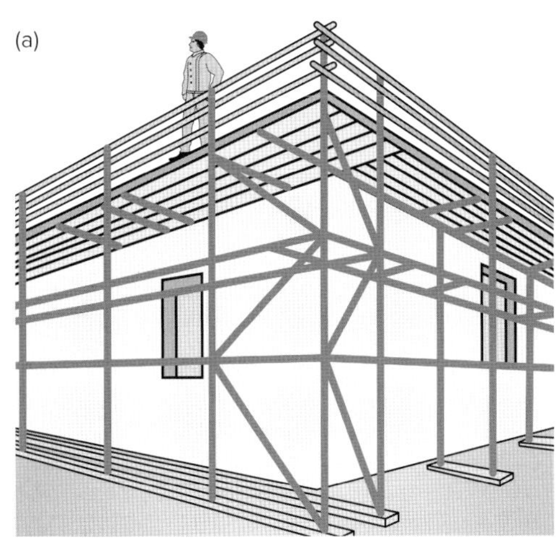

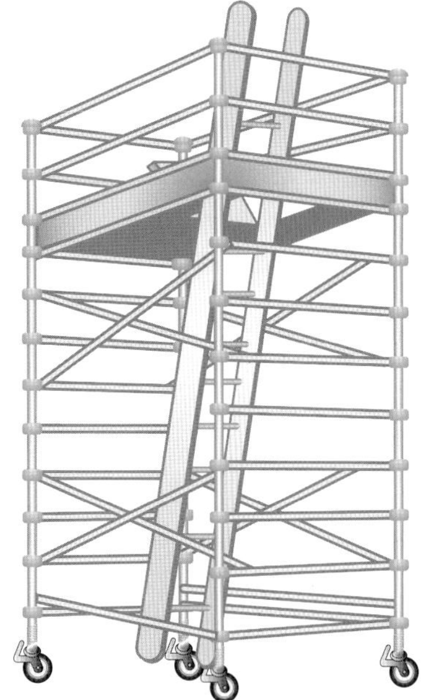

SAFETY TIP Safe work on scaffolding

- Check the scaffolding for faults before using it and when you first get on it.
- Scaffolding should be kept clear of powerlines.
- Don't overload scaffolding.
- Don't alter the scaffolding unless you are authorised and qualified.
- Keep the scaffolding clear of debris and obstructions.

SAFETY TIP Safe work on mobile scaffolding

- Keep mobile scaffolding level and vertical—not leaning.
- Keep clear of powerlines, open floor edges and penetrations.
- Lock castor wheels before using the scaffolding.
- Don't move mobile scaffolding while anyone is still on it.
- Don't climb the scaffolding to access it—an internal ladder must be used.

1.4.1.3 Elevated work platforms

Fig. 1.12 (a) Boom-type elevated work platform; (b) control panel

(a)

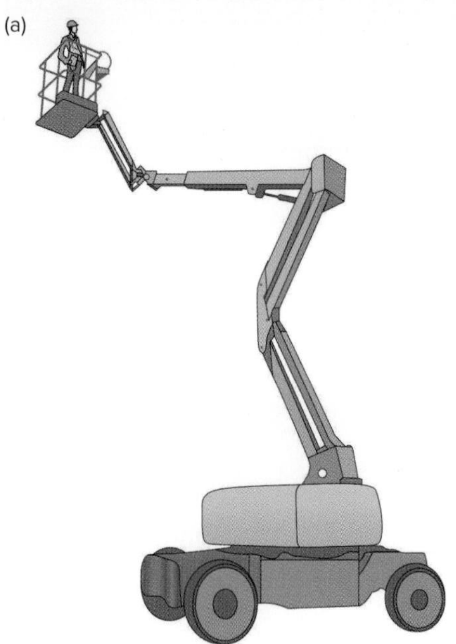

(b)

Courtesy of Alister Ford

SAFETY TIP Safe work on elevated work platforms

- Operators need to be trained to operate an elevated work platform (EWP), including the safe use of fall arrest equipment and emergency rescue procedures.
- Platforms must be used only on a solid, level surface free of obstructions and depressions or holes.
- Safety harnesses must be worn by anyone using an EWP.
- Operators of boom-type EWPs must be licensed if the EWP has a working height of 11 m or more.

1.4.2 Falling objects

Anyone working in the construction industry will at times be faced with the hazard of work being carried out overhead. Normally there will be barriers and signs to prevent anyone from passing through or working in such a hazardous area, or there will be protective hoardings to protect passers-by or anyone working below an active work area.

A major concern, however, is the potential for a falling object to land outside a restricted area or to oversail a protective hoarding. Hard hats must be worn when working in or walking through any areas adjacent to work that is being carried out overhead (Fig. 1.13).

When materials or equipment are being transferred from the ground level to a higher level by cranes or hoists:

- do not work or walk directly under the equipment or materials as they are raised
- stay outside any barricaded area beneath a crane or hoist
- if there are no barricades, don't go near any area under the crane or hoist
- take into account the height the materials or equipment will be raised to and the type of materials or equipment being raised (Fig. 1.14).

Equipment such as steel scaffolding tubes, planks, bricks or bags of materials will be scattered outwards if a chain or wire rope breaks suddenly. The higher the materials or equipment are above the ground, the greater the distance they will be spread across below.

Normally there will be a **crane chaser** or **dogman** on the ground to direct workers and passers-by a safe distance away from the hazardous area, but it is far better to avoid such hazardous areas altogether until the materials and equipment have been moved.

Fig. 1.13 Hard hats must be worn in 'hard hat areas'

Courtesy of Alister Ford

Fig. 1.14 Materials being raised by a crane

© Makeev Petr/Shutterstock

WORKPLACE SCENARIO

A construction worker narrowly avoided death when a stack of concrete sheets hit him after they fell two stories. The man worked as a 'dogman' on the site, a job that requires guiding a crane operator as they move a load. The crane swung into a metal loading platform, which dislodged the concrete.

Paramedics worked to free the man from the rubble before he was taken to hospital with injuries believed to include fractured vertebrae. There were concerns that rules demanding 'exclusion zones' beneath the cranes had not been enforced.

1.4.3 Trucks, excavators, bobcats, backhoes and forklifts

1.4.3.1 Delivery vehicles

Construction sites can receive deliveries of materials at any time. The drivers of large, heavy trucks delivering sand, timber, concrete or bricks often struggle to find a suitable location to unload their materials, or they try to enter or leave a construction site that has restricted access. If they are operating a small hoist, they will be carrying or moving an elevated load of materials that may partly obscure their line of sight.

SAFETY TIP Safe work for deliveries

- Don't assume that the driver can see you.
- Don't walk behind reversing trucks.
- Be alert and make sure you can hear and see delivery vehicles on-site.
- Don't allow yourself to be trapped between a delivery vehicle and any immovable objects such as stockpiled materials, scaffolding or walls.
- Don't stand behind a tipper as it unloads sand or other bulk materials—stand well clear and to one side so that the driver can see you.
- Always have a prepared area where materials or equipment can be safely unloaded and stored to specifications without risk to any workers or the environment.

1.4.3.2 Excavation plant and other vehicles

Excavators, bobcats, backhoes, forklifts, concrete boom pumps and cranes are heavy and noisy machinery. The operators are focused on making sure that they:

- don't damage any services, including underground natural gas, power, sewer and water
- don't hit any overhead electrical or telephone lines, trees, buildings, scaffolding or other stationary items
- are operating their machine effectively.

The operator will not always see or hear anyone beside them or behind them when they are operating an excavator, crane, boom pump or forklift.

Fig. 1.15 Worker in front of an excavator wearing high-visibility clothing

© Bannafarsai_Stock/Shutterstock

If you are close by any heavy mobile plant, ensure the plant operator can see you. When working near any heavy mobile plant, wear high-visibility clothing and work only in an area protected by barricades between the operating plant and workers and where they can see you.

An excavator bucket moving with a sideways swing or the swinging end of a concrete boom pump will cause serious injury.

Fig. 1.16 A & B (a) Concreters working with a concrete boom; (b) pumping concrete

(a)

© Bannafarsai_Stock/Shutterstock

(b)

© Dmitry Kalinovsky/Shutterstock

WORKPLACE SCENARIO

Never underestimate or take anything for granted in the construction industry.

In 40 years of working in this great industry I have seen some accidents that should never have happened and some near misses that potentially could have ended in tragedy.

I've seen things as simple as forgetting to wear the correct PPE, as well as power tool and equipment accidents and structure collapse. I've also seen things on other building sites where I was not working myself that left me wondering, 'What are these workers even thinking?'

It was on one such occasion some 30 years ago that I felt I could not just keep going and ignore what I had seen. I was living and working as a self-employed builder in rural Victoria and was travelling home one evening after my workday had finished.

Driving into the town where I lived, I had to pass the local golf course. The golf course water supply was a series of dams that was an actual water course, sited along the fence line about 20 metres from the road. The golf course had engaged an earth-moving contractor to dig trenches between the roadside table-drain to the top dam and then more trenches between the four dams.

These trenches were for piping to be laid down, and a new pumping station was to be constructed and installed as well.

As I drove past, I noticed that the excavator was operating and had created a deep trench from the roadside to another larger excavation further off the road.

The excavation was so deep, all I could see in there was the top of the excavator's assistant's hard hat. I estimated the trench to be approximately 1.8 metres deep. I couldn't believe what I had just seen. I pulled over and off the road and quickly made my way back to the excavation site, signalling to the operator to cease before assisting the worker in the trench to get clear—there was no way out for him but at the end of the trench some 5 or 6 metres away.

I explained to these guys about the nature of the ground they were working on and that any excavation should be correctly shored up to stop the risk of a collapse and someone being buried alive. They told me that they were ahead of schedule and had started the excavation late that afternoon rather than waiting until the next day. They could get the job finished quickly and would install the trench shoring when it arrived the following day.

That would have been too late if someone had been buried alive that day. The ground was wet, and this could easily have happened.

The excavation workers stopped work for the day and assured me that the shoring would be installed the next day. They were grateful that someone had stopped and called them out. Complacency can be a major contributing factor to workplace accidents or incidents.

So, I went on my way, glad that I had stopped and done something.

The next morning when heading off to work I drove past again. The excavation crew were there on the site, standing outside the barriers they had erected the previous evening before leaving for the day.

They were looking at the trench they had excavated the previous day, but there was a major change: the trench had collapsed and caved in overnight. These workers had had a lucky escape. It was a near miss that could so easily have ended in tragic circumstances.

1.4.3.3 Passing traffic

Often, it may be necessary to leave the confines of the construction site to assist with delivery vehicles delivering such things as pre-fabricated wall frames or roof trusses, timber packs, concrete, plant or machinery. These vehicles often need to reverse into the site.

In this situation, the general public will continue to use the road outside the site. It may be necessary to temporarily stop local traffic so that the delivery vehicle can enter the site safely.

Correct reflective hi-vis clothing must be worn and every precaution taken to alert passing traffic in plenty of time as to whether they need to stop or slow down. A stop/slow bat operated by someone who has been properly trained in traffic control is the most effective method for this. If not, good clear, concise hand signals should be employed.

1.4.4 Manual handling

Manual handling of materials and equipment is still the number one cause of non-fatal work-related injuries in the construction industry. Any manual handling tasks need to be carefully planned and organised to ensure the task is carried out safely. Correct lifting and carrying and setting-down procedures should always be observed.

Different building materials and equipment will all have different characteristics and may even have manual handling instructions included in the manufacturer's instructions or the material safety data sheet.

Injuries that could occur from manual handling can include:

- back injury
- hernia
- muscle tear or sprain
- ligament tear or sprain
- lacerations
- cuts
- puncture wounds
- fractures.

These injuries are usually caused by incorrect lifting, carrying and setting-down procedures, sharp edges or protrusions, incorrect assessment of the load and trying to lift too much.

Preparation for lifting

- Decide how far the load will be moved once it is lifted and how it will be moved.
- Perhaps the load can be rolled rather than lifted?
- Can it be moved onto a trolley?
- Is a crowbar needed to dislodge the load?
- Can it be placed into a wheelbarrow and wheeled to the final location?
- Determine if you have a suitable mechanical device to lift the load.
- Before lifting the load, determine the weight and decide if two people are needed to lift it.

Lifting a load

- Stand as close to the load as you can.
- Place your feet apart with one foot alongside the load to create a stable base.
- Bend your knees and keep your back and body upright and straight as you squat down.
- Adjust your feet to maintain balance.
- Your front foot should still be next to the load and facing the direction you plan to go in.

- Make sure your feet are firmly in position.
- Grasp the load firmly with both arms close to your body and straighten your legs slowly.
- Lift the load with your legs and arms, keeping your body and back straight.
- Keep the load close to your body and don't twist your body on your legs as this twisting movement can damage your knees or cause muscle strains!
- Keep your legs slightly bent and free as you walk. *Don't lock your knees!*

Use the same procedure in reverse when you have moved the load to the new location.

If the load is too heavy, immediately reverse the procedure and place the load back down.

Fig. 1.17 A–D Shovelling loose materials such as sand or dirt: (a) keep the legs apart with one foot in front and close to the end of the shovel; (b) put weight on front foot, place both hands on the shovel with each hand resting against a leg, and bend legs and push forward using body weight to push the shovel; (c) step back onto back foot with loaded shovel and keep the load close to your body; (d) step and turn feet and body in the direction of where the material needs to be placed. Keeping the load close to your body, roll your forearms, wrists and hands to dislodge material

(a)

(b)

(c)

(d)

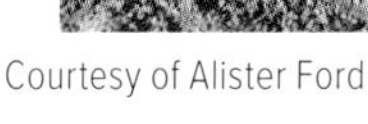

Courtesy of Alister Ford

Moving loads using a wheelbarrow

- Load the wheelbarrow evenly so that it is balanced sideways.
- Place the majority of the load forward and balanced over the wheel.
- Make additional trips rather than overloading.
- Load it facing in the direction you want to wheel it.
- Use your legs, not your back to lift the wheelbarrow.
- Use your normal walking pace.
- Cross all bumps or ridges at 90° to the wheel.
- Keep tyres hard and properly inflated.
- Make sure the pathway is clear of debris.
- Use the legs when travelling downhill to brake the wheelbarrow by lowering them to skid on the ground.

1.4.5 Hazardous substances

1.4.5.1 Asbestos

Asbestos is considered a hazardous substance and is also **carcinogenic**. Diseases such as **mesothelioma** and **asbestosis**, which are associated with asbestos exposure, were responsible for 90 fatalities in the construction industry over the six-year period from 2005–06 to 2010–11.

Asbestos cement, or 'fibro', was used until the late 1980s. It was commonly used in residential construction for wet area linings in bathrooms and laundries, for the eaves lining on brick veneer cottages and for a variety of external cladding profiles, including smooth sheeting on timber-framed cottages.

It is difficult to correctly identify bonded asbestos cement compared to modern fibre cement products. Any material that resembles fibre cement sheeting found when renovating a home built before 1990 should be considered as hazardous asbestos cement until confirmed otherwise.

Immediate exposure to bonded asbestos alone is not a hazard unless there is an attempt to remove it during renovations and demolition.

Bonded asbestos can be damaged by:

- high-pressure water cleaners
- grinding
- saw cutting
- drilling
- snapping or breaking up
- compressed air
- abrasive blasting
- abrasive power tools
- pneumatic tools.

Fig. 1.18 Disposable respirator with valve (note coding below valve)

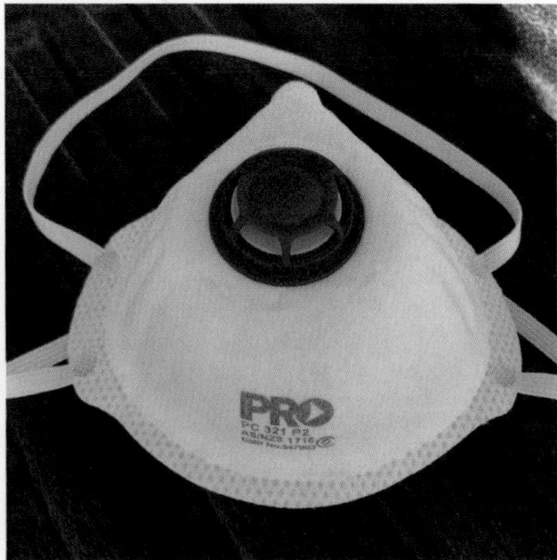

Courtesy of Alister Ford

The asbestos fibres can be exposed and become airborne and also collect among other debris that has to be removed. When undertaking any demolition, it is recommended that a P1 or P2 mask or respirator be worn as these are designed to protect the wearer against a range of fine particles including asbestos (Fig. 1.18). However, they do not provide protection against toxic fumes or gases. Cheap dust masks for protection against 'nuisance dusts' are completely unsuitable for any construction work and offer little protection against any fine particles.

Asbestos is removed by specialist licensed contractors who are trained and qualified.

TIP Normally the builder (the head contractor) who is contracted to renovate a home would investigate the presence of bonded asbestos and arrange for suspect materials to be tested and, if necessary, removed by a specialist asbestos removal contractor before construction work commences.

Fig. 1.19 Post-war fibro home with external asbestos cement sheeting

Courtesy of Alister Ford

Fig. 1.20 Damaged bonded asbestos with asbestos fibres exposed

© Kim Christensen/Shutterstock

1.4.5.2 Dry powder building products

Cement, lime, fireclay, tile adhesives or any combination of these materials in a packaged form are considered to be hazardous substances until they have been mixed and placed.

1.4.5.3 Processed timber products

Processed timber products such as **medium-density fibre board** (MDF), treated pine, plywood and materials made with recycled timber and plastics contain materials that in a dust form can be very harmful. P1 or P2 respirators or dust masks must be worn when cutting these products.

1.4.5.4 Fibre cement board

These products are used for cladding, internal wet area linings and as external and internal wet area flooring. The dust generated by cutting these materials is hazardous.

1.4.5.5 Insulation materials

The most common insulation materials that you will encounter on a domestic building site will be the fibreglass insulation batts that are fitted into wall frames or roof spaces.

The fibres from these batts can cause severe irritation to the skin, eyes and lungs, so the correct clothing, eye protection and respirators should be used to ensure you are well protected.

Alternative insulation materials may be wool, paper or foam-based, and the correct PPE must be used according to the manufacturer's instructions.

1.4.5.6 Protective dust masks or respirators

Any worker using the products listed above must ensure that they do not breathe fine dust particles from these materials into their lungs. They may have to wear protective P1 or P2 dust masks to prevent this occurrence. Such dusts can cause **pneumoconiosis**, a chronic disease of the lungs that can result in tightness of the chest, difficulty in breathing, chronic coughing, bronchitis and emphysema. Although diseases such as pneumoconiosis are not cancers, they can still result in a fatality at worst or at best long-term disability that stops a worker from continuing in their career.

1.4.5.7 Solvents, pesticides, acids, adhesives, gases, alkaline and petroleum products

Any products containing chemicals or gases have the potential to damage your skin, eyes and lungs. Before using any unknown chemical or gaseous product on a construction site, it must be identified and a **materials safety data sheet** must be available to explain what the product contains, what injuries or illnesses it may cause, how these can be treated and what form of PPE must be used by any worker required to use such products on a construction site.

Chemicals must not be mixed under any circumstances unless the directions clearly state that it is a requirement and safe to do so. Risks include causing an explosion or fire, or emitting poisonous fumes. Any mixing of substances must be done strictly according to the product manufacturer's instructions and material safety data sheet, using the correct mixing equipment and PPE.

If you fail to comply, and an incident occurs, many people could be affected–and not just those in the immediate vicinity.

Examples of commonly used products that may contain hazardous chemicals or gases are paints, sealants, adhesives, cleaners, oxyacetylene, waterproofing and fireproofing products.

These may be supplied in many types of tins, drums, plastic containers, glass containers, cylinders or pressurised containers. The material safety data sheet will also give detailed instructions on how any spills or leaks must be contained and cleaned up.

The required equipment detailed must be prepared and be readily available in case such an emergency occurs.

All materials used on any construction site must by law be accompanied by a materials safety data sheet. This requirement includes common construction materials such as cement, bricks, lime, fibre cement board, treated pine and other processed timber products.

1.4.6 Electrical hazards

Electricians are exposed to the danger of electrocution every day. Any building worker may be exposed to an electrical hazard at any time while at work. There are two main controls that are used to protect workers against electrocution.

1.4.6.1 Remove the risk

Electricians do this by turning off the power when they are working with live wiring. The same control can be applied by all workers. If you are in doubt about the condition of any electrical wiring (e.g. in a renovation project), turn the power off and use rechargeable battery-powered tools, a portable generator and other petrol-powered equipment, or use the power available from a neighbour.

TIP Rechargeable battery-powered tools remove the risk of electrocution due to faulty electric tools.

1.4.6.2 Earth leakage circuit breakers and residual current detectors

Both devices switch off power and are extensively used in the construction industry to protect workers. The earth leakage circuit breaker (ELCB) detects any leakage of power into the ground, while the residual current detector (RCD) detects any imbalance of power in the circuit (Fig. 1.21).

These devices are mandatory on all construction sites where electrical power is used. Normally an electrical lead is plugged into a source of power and the ELCB or RCD is connected at the end closest to where electric power tools are being used. The power tools are plugged directly into the ELCB or RCD.

Fig. 1.21 Residual current detector

Courtesy of Alister Ford

1.4.7 Extreme working conditions

If working in extreme conditions, whether it is a hot or cold environment, correct procedures must be taken to ensure the health and safety of yourself and your fellow workmates. Sunburn, dehydration and heat stroke can occur if working outside in the typically hot and sunny Australian conditions.

Every effort must be taken to cover up and protect your skin to avoid sunburn and complications in later years such as skin cancers. Long-sleeved shirts, wide-brimmed hats and sunscreen will give you good protection on hot sunny days. Drink plenty of water, sipping frequently to reduce thirst. A few simple precautions such as these can prevent what may become a traumatic time if ignored.

Extreme conditions may also be encountered if working in environments where humidity levels, radiant heat and temperatures, whether hot or cold, are regulated to a non-natural temperature, or unregulated and at an unknown level.

These could be places such as:

- cool rooms
- cold stores
- boiler houses
- foundries
- industrial kitchens
- factories
- workshops
- subfloor spaces
- roof spaces.

Working conditions in any of these extreme environments must be closely monitored, as well as the physical wellbeing of each worker. Issues such as heat exhaustion and hypothermia can strike quickly and without warning.

Detailed information regarding hot or cold workplace conditions and personal safety can be found in the Safe Work Australia Code of Practice: Managing the work environment and facilities, December 2011, Section 2: The work environment, Part 2.8: Heat and cold.

CHECK YOUR UNDERSTANDING

1. List, in order of risk, the three work activities responsible for the highest numbers of fatalities in the construction industry.
2. Name the two different devices that must be attached to a power source to protect workers from electrocution.
3. Name two classifications of respirators that are suitable for use in asbestos removal.

1.5 Emergencies on construction sites

All construction sites, including residential building sites, must have a site safety plan. The site safety plan has an evacuation plan that must be followed in the event of evacuation of the site.

There should also be designated site personnel (e.g. area chief warden) to take charge of the building site and the evacuation process if such an event occurs. Any instructions issued by these persons must be carried out as directed.

If an emergency occurs, or a situation arises on a construction site, remember to remain calm and work in unison with other workers to carry out the correct procedures as outlined in the site safety and management or evacuation plan. Others may need to rely on you for assistance or guidance at any time, whatever the situation.

There should also be an approved first-aid kit available on any construction site.

1.5.1 Emergency evacuation

The main reasons for evacuation of a residential site could be a bomb hoax, fire, earthquake, flash flooding, gas leak, chemical leak, building or partial building collapse, the threat of violence or any other uncontrolled event or incident. Each site has a designated assembly area that all site workers must go to in the event of an emergency evacuation so that they can be signed off as present. Otherwise, other people may have to return to the site and try to account for anyone who is missing.

1.5.2 Emergency firefighting equipment

With a large fire, the best course of action is to isolate the fire if safe to do so, sound the fire alarm, whether it is a site siren or a break glass alarm, and then evacuate the worksite, moving everyone to the designated emergency assembly area. Call 000 and accurately relay any information for the fire brigade to arrive at the correct place in the minimum amount of time. The fire brigade is best equipped to tackle any large fires with their trucks, hydrants, hoses, reels and breathing apparatus.

For smaller fires, a portable fire extinguisher or fire blanket can be used to prevent the fire spreading. Portable fire extinguishers contain a range of different types of extinguishing agents designed to extinguish specific types of fires (Fig. 1.22).

1.5.3 Medical emergency

A medical emergency on a construction site normally occurs when there has been an accident and a worker is seriously injured. If an accident occurs, other workers have an obligation to respond and assist the injured worker as best they can.

Workers who have been injured may need emergency first aid. On smaller residential construction sites, there may not be a qualified first-aid officer available, but the other workers should have a basic knowledge of first-aid procedures and know how to keep the injured worker safe and comfortable until the trained emergency workers arrive.

Fig. 1.22 Types of fires and suitable extinguishers

Class and type of fire	Colours	A	B	C	D	E	F	
Type of extinguisher		Wood, paper, plastic	Flammable and combustible liquids	Flammable gases	Combustible metals	Electrically-energised equipment	Cooking oils and fats	
Water		Yes	No	No	No	No	No	Dangerous if used on flammable liquid, energised electrical equipment and cooking oil/fat fires.
Carbon dioxide (CO2)		Limited	Limited	No	No	Yes	No	Not suitable for outdoor use or large class A fires.
Dry chemical powder (ABE/BE)		Yes AB(E) No B(E)	Yes	Yes	No	Yes	No AB(E) Yes B(E)	Look carefully at the extinguisher to determine if it is a BE or ABE unit.
Foam		Yes	Yes	No	No	No	Limited	Dangerous if used on energised electrical equipment.
Wet chemical		Yes	No	No	No	No	Yes	Dangerous if used on energised electrical equipment.
Fire blanket		Limited	Limited	No	No	No	Yes	Fire blankets are effective for oil and fat fires within saucepans and for extinguishing clothes that catch on fire. (Ensure you replace after every use.)
Fire hose reel		Yes	No	No	No	No	No	Dangerous if used on flammable liquid energised electrical equipment and cooking oil/fat fires.

The **site safety management plan** should have the contact details for the closest hospital and medical centre or general surgery, including the opening hours.

First-aid kits should be available on-site for the treatment of minor injuries, but they must be properly maintained. Typically, minor injuries on a residential construction site will involve cuts or lacerations. This means there will be blood present. If you are assisting a fellow worker who has had an accident and there is blood present, you should first put on a pair of disposable gloves, which should be found in the first-aid kit. This will give you, as the assisting worker, some protection from any blood-borne diseases that may be present. The main blood-borne infections are hepatitis B, hepatitis C and HIV/AIDS. A simple precaution like this can give you every confidence to assist an injured coworker.

Any injury, no matter how minor, must be correctly documented in case of any later complications arising from the injury, and also to assist in the planning of future tasks to stop any similar incident occurring. This also applies to any near miss, where injury is luckily avoided. If you are working on a residential construction site, it is important to be familiar with the site safety management plan as it may save your life or the life of a fellow worker.

Fig. 1.23 Fire blanket

Courtesy of Alister Ford

1.6 Australian work health and safety legislation, regulations and codes of practice

The model national legislation that addresses workplace health and safety is called the *Work Health and Safety Act 2011* (Cth) (OHS Act). The implementation of this legislation is supported by the Work Health and Safety Regulations 2011 (Cth) (OHS Regulations) together with national standards and codes of practice. Under this model, all workers and employers have shared roles and responsibilities to ensure that construction worksites are safe. These shared responsibilities mean that individual workers must at times participate in the assessment of risks associated with the tasks they undertake.

Most Australian states and territories have now adopted the model safety legislation of the federal government. An immediate benefit has been the mutual recognition of qualifications for mandatory general induction training and the adoption of common codes of practice. The 'harmonisation' of safety requirements across all Australian states and territories not only reduces red tape but also makes the management and implementation of workplace safety more consistent and consequently more effective than it has been in previous years.

There is broad mutual recognition of construction induction White Cards across all state and territory authorities, but interstate workers should check with local authorities to confirm acceptance of the card they hold. A construction induction card will not be accepted if it does not have all of the details required by the local authorities or is out of date.

1.6.1 *Work Health and Safety Act 2011* (Cth)

The following information from the *Guide to the model Work Health and Safety Act* (Safe Work Australia CC BY-NC 4.0 https://creativecommons.org/licenses/by-nc/4.0/) details the main purpose and aim of the *Work Health and Safety Act 2011* (Cth):

The OHS Act provides a framework to protect the health, safety and welfare of all workers at work and of other people who might be affected by the work.

The OHS Act aims to:

- protect the health and safety of workers and other people by eliminating or minimising risks arising from work or workplaces
- ensure fair and effective representation, consultation and cooperation to address and resolve health and safety issues in the workplace
- encourage unions and employer organisations to take a constructive role in improving OHS practices
- assist businesses and workers to achieve a healthier and safer working environment
- promote information, education and training on work health and safety
- provide effective compliance and enforcement measures, and
- deliver continuous improvement and progressively higher standards of work health and safety.

In furthering these aims, there must be regard for the principle that workers and other persons should be given the highest level of protection against harm to their health, safety and welfare from hazards and risks arising from work as is reasonably practicable.

For these purposes 'health' includes psychological health as well as physical health.

A full copy of the Act can be downloaded from: www.comlaw.gov.au/Details/C2011A00137/Download.

The *Guide to the Work Health and Safety Act* can be downloaded from www.safeworkaustralia.gov.au/system/files/documents/1702/guide-to-the-whs-act-at-21-march-2016.pdf (Fig. 1.24).

Fig. 1.24 *Work Health and Safety Act 2011* (Cth)

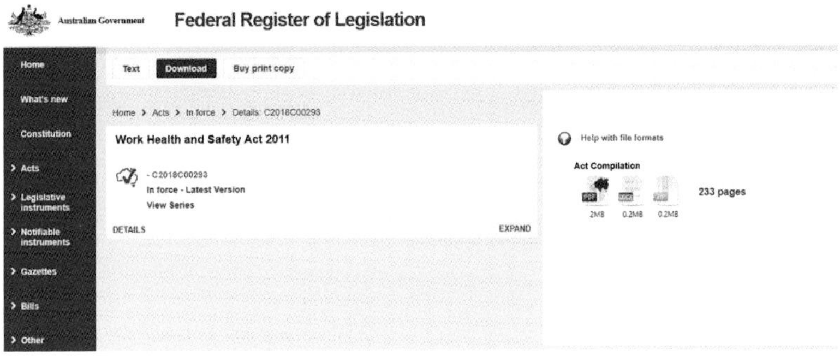

Source: Courtesy of Commonwealth of Australia

1.6.2 Work Health and Safety Regulations 2019 (Cth)

The Work Health and Safety Regulations 2019 (Cth) (OHS Regulations) complement and support the general duties under the OHS Act and support it by setting out mandatory obligations on specific matters. You can access the comprehensive 600-page OHS Regulations document at the Safe Work Australia website (www.safeworkaustralia.gov.au) (Fig. 1.25).

The regulations impose 'duties' (see note on 'duty of care') on people who have a role in undertaking construction work, including:

- the person who commissions the construction work must consult with the designer of the structure on matters relating to health and safety during construction work and provide a health and safety report by the structure designer to the principal contractor
- the designer of the structure must provide a written health and safety report to the person who commissioned the project
- the principal contractor is responsible for the overall management and control of site health and safety, including collecting work method statements for high-risk construction work, providing and maintaining an OHS management plan and informing all workers of the content of the OHS management plan

Fig. 1.25 Work health and safety regulations

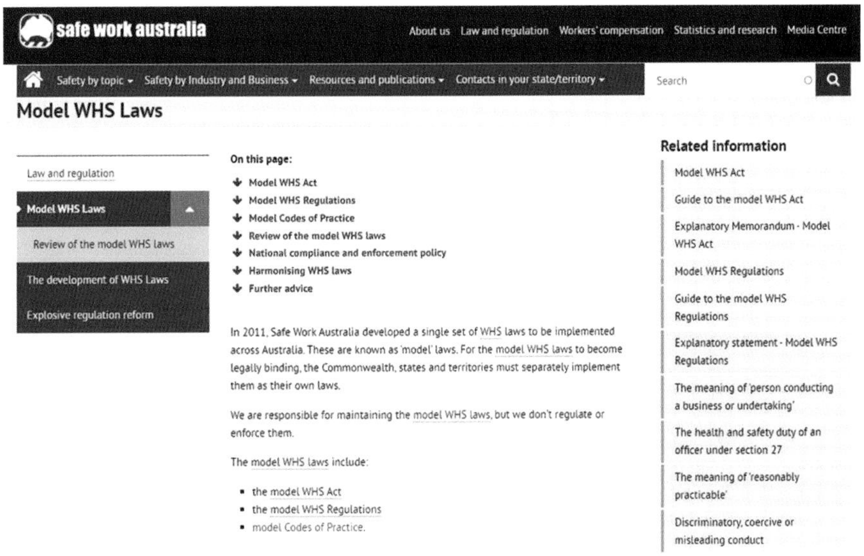

https://www.safeworkaustralia.gov.au/law-and-regulation/model-whs-laws, Safe Work Australia CC BY-NC 4.0
https://creativecommons.org/licenses/by-nc/4.0/

- workers must comply with the OHS management plan and have their general induction training card available.

The term *duty of care* refers to a requirement to exercise due diligence in performing an OHS duty and to act proactively to ensure health and safety at work. It is applied to all people, regardless of the level of responsibility they have in undertaking construction work.

1.6.3 The meaning of construction work

The OHS Regulations (Part 6.1 Preliminary, 289, p. 277) define construction work as follows (Safe Work Australia CC BY-NC 4.0 https://creativecommons.org/licenses/by-nc/4.0/):

(1) . . ., **construction work** means any work carried out in connection with the construction, alteration, conversion, fitting-out, commissioning, renovation, repair, maintenance, refurbishment, demolition, decommissioning or dismantling of a structure.

(2) . . ., **construction work** includes the following:

- **(a)** any installation or testing carried out in connection with an activity referred to in subregulation (1);
- **(b)** the removal from the workplace of any product or waste resulting from demolition;
- **(c)** the prefabrication or testing of elements, at a place specifically established for the construction work, for use in construction work;
- **(d)** the assembly of prefabricated elements to form a structure, or the disassembly of prefabricated elements forming part of a structure;
- **(e)** the installation, testing or maintenance of an essential service in relation to a structure;
- **(f)** any work connected with an excavation;
- **(g)** any work connected with any preparatory work or site preparation (including landscaping as part of site preparation) carried out in connection with an activity referred to in subregulation (1);
- **(h)** an activity referred to in subregulation (1), that is carried out on, under or near water, including work on buoys and obstructions to navigation.

For any construction worker who has responsibility for the actions of other workers in relation to safety, see the comprehensive details in Chapters 18–19 of the Regulations.

1.6.4 Codes of practice

Codes of practice provide guidelines for meeting the requirements of the OHS Act and the OHS Regulations and may be used in a court of law as evidence to prove compliance with the OHS legislation (Fig. 1.26). Construction sites that are not operating safely will often be referred to a copy of a code of practice by OHS inspectors to demonstrate why their work procedures have failed to comply with OHS legislation.

Compliance with codes of practice is not mandatory. In some cases, it may be better to use another method or system of work that results in improved outcomes when compared to the outcomes of a model code of practice. However, all responsibility is on the person using an alternative to a code of practice to prove that it is at least equally safe to use and provides an equivalent or higher standard of OHS than suggested by the code of practice.

A comprehensive register of model codes of practice is available for download at www.safeworkaustralia.gov.au/sites/swa/model-whs-laws/model-cop/a-z-cop/pages/a-z.

Relevant state and territory authorities also have a range of codes of practice related to the construction industry.

Fig. 1.26 Codes of practice for high-risk construction work

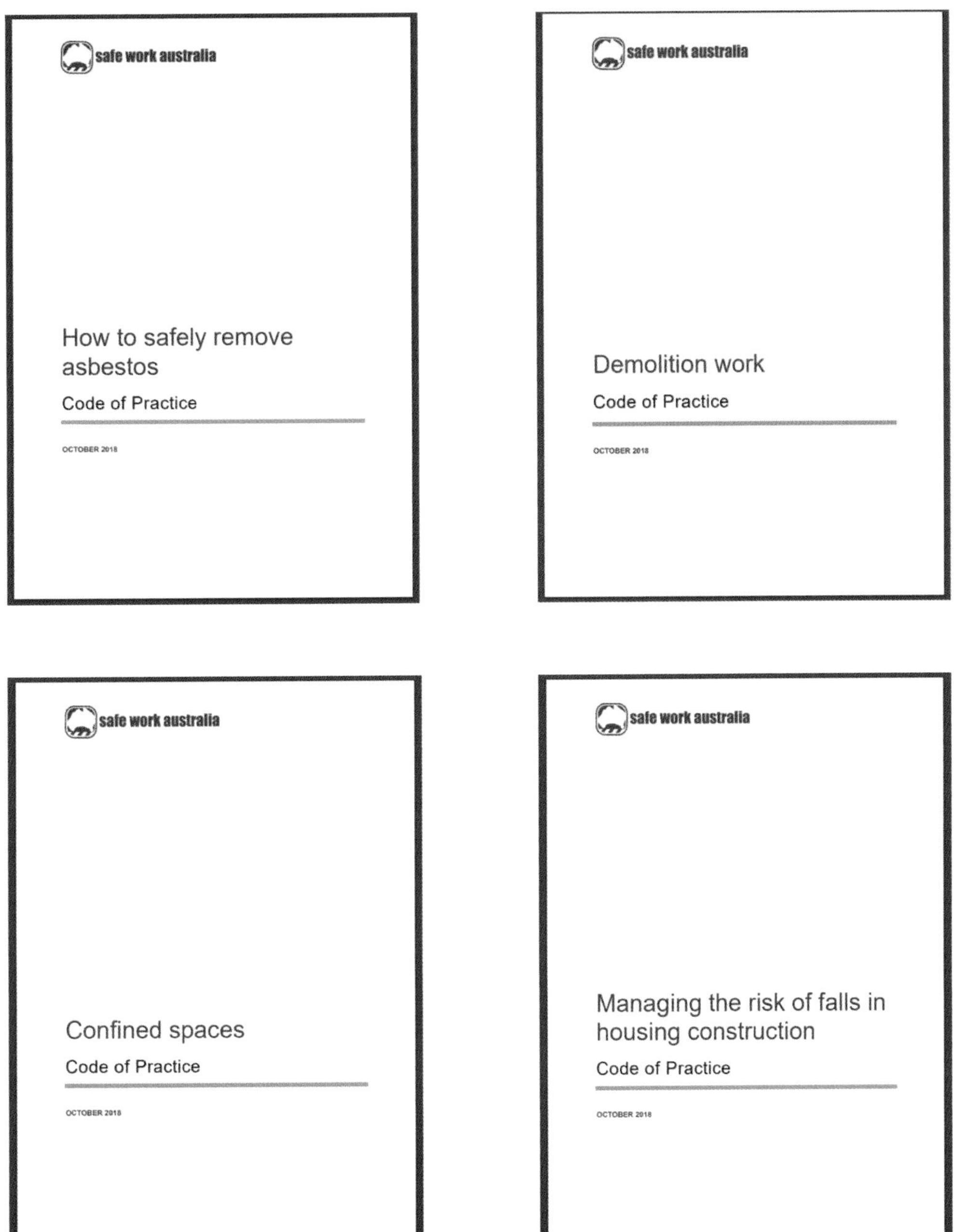

CHECK YOUR UNDERSTANDING

1. Name the two pieces of Australian legislation that address workplace safety.
2. List the three different types of compulsory training that all workers must undertake before commencing any construction work.
3. List your state and name the state authority in it that regulates work health and safety.

1.7 Analysing workplace hazards

The process of analysing workplace hazards would normally commence with a site safety audit or safety inspection. Any worker on site can be called up to assist a safety officer to undertake a safety audit. On larger construction sites, a site safety audit should be undertaken by at least two people to ensure that all hazards are identified. Depending on the level of risk, these may be undertaken every day.

Figure 1.27 shows the front page of a sample site safety checklist to which other items may be added. A very simple handwritten list of risks may also be used in this process provided it is properly signed and dated. Normally the audit report will include recommendations or comments from the person undertaking the safety audit and a date when previous safety issues have been resolved.

Fig. 1.27 Sample front page of a site safety checklist

SITE SAFETY CHECKLIST

Site address	
Contact name	
Contact number	
Date	
Fax number	
Email address	
Prepared by	

SUMMARY OF INSPECTION AND SITE DESCRIPTION Comments, additional safety issues and site-specific hazards List of issues or hazards that require attention or corrective action

REQUIREMENT	YES/NO	COMMENT/ACTION REQUIRED	DATE COMPLETE
Safety management Documented OHS management system Workers inducted to site and record kept			
Site safety signs and notices Construction site safety signs displayed			
Site security/public protection Safeguard for public and after hours (e.g. locked plant and switchboards, barricaded trenches, etc.)			
Site amenities Toilets and handwashing facilities Meals area and drinking water			
First aid and emergencies First-aid kit and contact for first-aid assistance Emergency procedure and contact numbers Register of injuries			
Risk assessments and WMS/JSAs Risks assessed for all activities including work at heights, trenching, working near powerlines, etc.			

The site safety audit can then also be used to assist with writing the following documents:

- safe work method statements
- job safety analysis
- health and safety management plans
- work plans
- work schedules.

After completion, the checklist is filed for future reference as evidence that a risk analysis has been carried out.

1.7.1 Risk categories

Once the risks have been identified, each must then be analysed and a 'risk category' determined. A typical risk assessment matrix used in the construction industry is shown in Figure 1.28. The likelihood of occurrence is first identified in the far left-hand column. The consequence is then estimated and from this the category of risk is classified as follows:

- E–Extreme risk–death or permanent disability
- H–High risk–long-term or serious injury
- M–Moderate risk–lost time
- L–Low risk–first-aid treatment.

Fig. 1.28 Risk assessment matrix

Likelihood	Consequences				
	Insignificant	Minor	Moderate	Major	Severe
Almost certain	M	H	H	E	E
Likely	M	M	H	H	E
Possible	L	M	M	H	E
Unlikely	L	M	M	M	H
Rare	L	L	M	M	H

Where the work undertaken is considered to be a 'high-risk construction activity', a safe work method statement must be developed, identifying an appropriate way–using a hierarchy of hazard control (Fig. 1.29)–to either remove the risk altogether, if feasible, or control the risk to a point where the consequence is either moderate or low.

Fig. 1.29 Hierarchy of hazard control

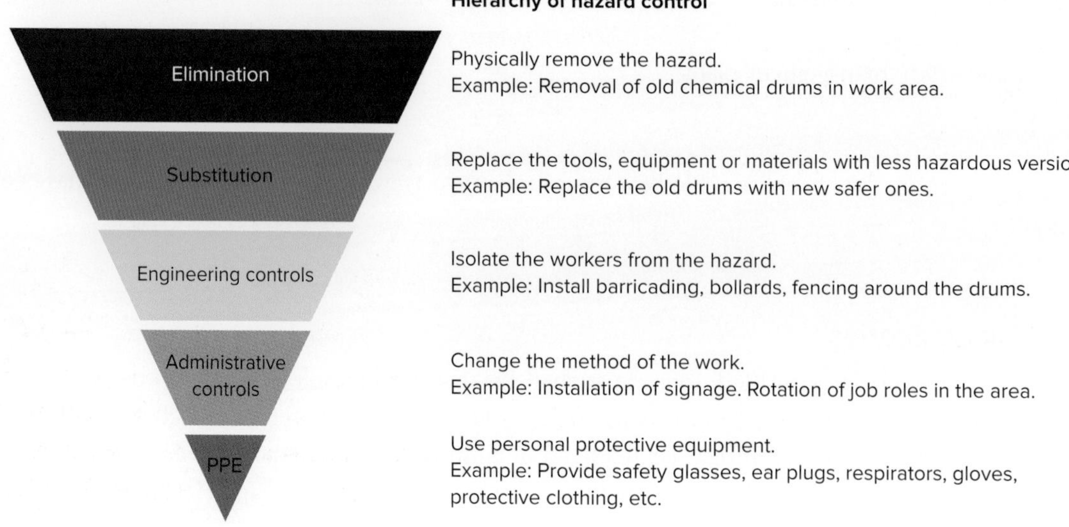
Hierarchy of hazard control
Elimination
Physically remove the hazard.
Example: Removal of old chemical drums in work area.
Substitution
Replace the tools, equipment or materials with less hazardous versions.
Example: Replace the old drums with new safer ones.
Engineering controls
Isolate the workers from the hazard.
Example: Install barricading, bollards, fencing around the drums.
Administrative controls
Change the method of the work.
Example: Installation of signage. Rotation of job roles in the area.
PPE
Use personal protective equipment.
Example: Provide safety glasses, ear plugs, respirators, gloves, protective clothing, etc.

1.8 Safe work method statements and job safety analysis

A safe work method statement (SWMS) is a simple document that outlines the logical steps required to successfully and safely complete a task. When completed, it should be readily accessible and understandable to workers who use it. According to current legislation, an SWMS must be used to remove or minimise risk when people undertake high-risk construction work. On construction sites valued at more than $250 000, it is the responsibility of the principal contractor to ensure that all trade contractors provide an SWMS before they undertake high-risk work. Alternatively, a job safety analysis (JSA) may also be utilised. A JSA is a form of risk assessment that includes step-by-step details of how the task is to be carried out safely. It is broken down into three components–basic steps, hazards and procedure. (See here: https://content.api.worksafe.vic.gov.au/sites/default/files/2018-11/ISBN-Safe-work-method-statements-2018-11.pdf.)

Work Health and Safety Regulations 2011 (Cth) state the following requirements for an SWMS (Safe Work Australia CC BY-NC 4.0 https://creativecommons.org/licenses/by-nc/4.0/):

Regulation 299 Safe work method statement required for high-risk construction work

(2) A safe work method statement must:
- **(a)** identify the work that is high-risk construction work; and
- **(b)** specify hazards relating to the high-risk construction work and risks to health and safety associated with those hazards; and
- **(c)** describe the measures to be implemented to control the risks; and
- **(d)** describe how the control measures are to be implemented, monitored and reviewed.

(3) A safe work method statement must:
- **(a)** be prepared taking into account all relevant matters, including:
  - **(i)** circumstances at the workplace that may affect the way in which the high-risk construction work is carried out; and
  - **(ii)** if the high-risk construction work is carried out in connection with a construction project–the OHS management plan that has been prepared for the workplace; and
- **(b)** be set out and expressed in a way that is readily accessible and understandable to persons who use it.

Safe Work Victoria, the Victorian workplace safety authority, recommends following these five steps when creating a JSA (WorkSafe Victoria, 'Advice for managing major events safely', April 2006, p. 50, https://www.worksafe.vic.gov.au/resources/advice-managing-major-events-safely):

(1) **Document the activity:** Assemble those involved in the activity and then, using the JSA worksheet, write down the tasks that make up the activity, step by step.
(2) **Identify the hazards:** Next to each task, identify what part of the task may cause injury to those doing the work or to anyone else nearby.
(3) **Document the control measures:** For each identified hazard, list the measures that need to be put in place to eliminate or minimise any likely risk of injury to those involved.
(4) **Identify who is responsible:** Document the name of the person responsible for implementing the control measure.
(5) **Monitor and review:** Make sure the activity is supervised to ensure the documented process is being followed. The JSA should be reviewed whenever a documented activity changes, when there is a change of personnel or after an appropriate length of time.

JSAs and SWMSs are identical in concept but there is no prescriptive or mandated layout. Either document can be designed to achieve the outcomes required under national OHS legislation.

CHECK YOUR UNDERSTANDING

1. List three items of information provided in a safe work method statement.

1.9 Work health and safety management plan

The principal contractor must prepare an OHS management plan before construction work commences. Where the value of the construction work is $250 000 or more, an OHS management plan is required by law. The principal contractor has a duty to make the OHS management plan accessible for all workers and to review the plan regularly and advise all workers of any changes to the plan.

The size and complexity of OHS management plans for construction sites varies with the size of the job and the number of workers on-site at any time and the number of high-risk construction activities taking place. So certain jobs will have a highly complex, comprehensive and lengthy document comprising several lever-arch files or volumes; larger projects will also include a separate demolition management plan and traffic control plan. For a small trade contractor or a smaller project, a less complex plan is needed.

The OHS Regulations describe the minimum requirements for an OHS management plan. It must include the following in relation to a construction project:

- the names, positions and responsibilities of all persons at the workplace whose positions or roles involve specific health and safety responsibilities
- how persons conducting a business or undertaking at the workplace will undertake consultation, cooperation and the coordination of activities under their duty to comply with the Act and Regulations
- the arrangements for managing any work health and safety incidents that occur
- any site-specific health and safety rules, and arrangements for ensuring that everyone at the workplace is informed of these rules
- the arrangements for the collection, assessment, monitoring and review of the SWMS.

In practice, an OHS site safety management plan can become a much larger overall safety management system by providing additional guidance and information required under OHS legislation, such as:

- evacuation plan
- contact details for the principal contractor and other key safety personnel such as the first-aid officers and safety officers
- site map of first-aid stations and emergency telephones
- register of all general construction induction card holders working on-site
- register of workers who have undertaken site-specific safety training or induction
- register of major plant maintenance schedules
- schedule of all high-risk construction work, including dates, location and who is undertaking the work
- demolition plan
- traffic control plan
- recycling and waste disposal plan
- asbestos removal plan
- register of incidents/accidents/near misses
- safety committee details
- worker facilities and amenities available on-site
- location and contact details of closest hospital and medical centres
- register of hazardous substances and materials safety data sheets (MSDS)
- register of required signage.

Student research

- Review the codes of practice available at Safe Work Australia (www.safeworkaustralia.gov.au) and determine which of these are specific to the construction industry.
- On the internet, locate a model work method statement that applies to your occupation or vocational area of interest.

End of chapter activities

Activity 1:

Using AS1319 safety signs for the occupational environment as a reference, sketch and label the colouring for two signs from each of the following categories:

- mandatory signs
- prohibition signs
- hazard signs
- emergency information signs.

Activity 2:

Refer to Safe Work Australia, Code of Practice: Managing the work environment and facilities, December 2011, Section 2: The work environment, Part 2.8 Heat and cold and list the following:

- six ways to minimise the risk of heat strain or exhaustion if it is not possible to eliminate the risk
- five ways to minimise the risk of exposure to extreme cold if it is not possible to totally eliminate the risk
- five warning signs of hypothermia to look out for in any of the workers on a particular job.

Activity 3:

Create a safe work method statement for the use of carpentry hand tools required for basic chiselling, sawing, hammering and planing while working at a carpenter's work bench.

AUSTRALIAN STANDARDS AND CODES OF PRACTICE

AS1319-1994
(reconfirmed 2018)
Safety signs for the occupational environment
Safe Work Australia, Code of Practice: Managing the work environment and facilities, December 2011.

Tradie talk—Good enough is never good enough

3D INTERACTIVE

ENSURING FUTURE SKILLS THROUGH SIMULATION TRAINING

Many of the challenges faced in the building and construction industry stem from the complex nature of the structures that its professionals are responsible for designing and constructing. Training and preparing the next generation of building professionals to understand these complexities is essential for the future of the industry.

In 2012, Master Builders Victoria launched the Building Leadership Simulation Centre (BLSC), a realistic, safe learning environment for building and construction professionals that would help accelerate learning by providing training interventions that allowed the immediate application of skills, experience and knowledge. The third such centre in the world and the first in the Southern Hemisphere, the BLSC delivers a realistic, immersive learning experience for its participants.

The realistic, immersive learning approach to training allows participants opportunities to apply, practise and refine new skills almost immediately.

Building Leadership Simulation Centre

© Master Builders Association of Victoria

The BLSC challenges its participants by placing them in real situations, where they are responsible for working through a variety of tasks and activities, are required to report to management and influence team members, all on an immersive, realistic building site. This training compels participants to navigate some tricky situations and deal with typical problems that arise on building sites, while in a safe, immersive learning environment. The immersive learning experience means that no participant flies under the radar, with everyone encouraged at all times by expert behavioural coaches to contribute to and participate in the critical reflection process.

Simulation training encourages participants to think dynamically about the issues and challenges they face, and not just rely on facts alone. More importantly, lateral thinking and intelligence are required for sensible decisions to be made in a manner that achieves the desired outcomes.

Master Builders Victoria has used simulation training to prepare building professionals in a wide range of topics, including OHS, leadership, influential communication, conflict resolution, diplomacy and creative problem-solving.

Chapter 2

Workplace communication

Learning Objectives

LO 2.1 Understand important forms of communication for construction workers

LO 2.2 Communicate in work teams

LO 2.3 Communicate with employers and supervisors

LO 2.4 Resolve disputes

Introduction

Australia's building and construction industry now offers very diverse employment opportunities. People with varying physical and mental abilities and a range of cultural and ethnic backgrounds are employed in many different roles on building sites across the country every day. Ultimately, they all contribute to one common goal: to communicate and work together and get the job done to the required national standards and quality levels.

In the building and construction industry, many different languages are used every day across numerous worksites. The mixture of ethnic backgrounds and different languages combined with a noisy work environment can sometimes make communication difficult. In addition, common construction terms are meaningless to someone who has never worked in the construction industry before, so there is the need to learn a new workplace language, as well as understand the different forms of communication. Plus, not all workers on a construction site will have a complete understanding of the next worker's role. Communication is thus an important factor in achieving a safe work environment and conveying the specific roles of other workers.

When starting a new job, it is important first to introduce yourself and let others know what your role involves, as well as asking questions about what they are doing so you can engage in conversation. By doing so, you will soon gain a better understanding of the many tasks being carried out around you, and you will get to know the other people you are working with, and learn how best to communicate with them.

You will need to understand the different types of communication skills required to work effectively and cohesively with others in the construction industry. Some workers may not speak English very well, others may have hearing or eyesight problems, while some may have a poor grasp of the use of electronic communication devices such as mobile phones, tablets or computers. But effective communication needs to take place nonetheless.

The communication skills that you learn will give you the confidence to conduct face-to-face, visual, electronic and written communication with your employers, supervisors, work team members, other construction site workers and members of the public. But how does a newly recruited construction worker with little industry experience effectively communicate with the other workers that they interact with every day? To be an effective communicator you need to consider some important points to ensure the information you are providing or receiving is understood by the many different people you will be working alongside. With your new skills you will be able to clearly gather, deliver, receive and clarify information using the correct means of communication.

Additionally, workplace meetings occur regularly on construction sites, for various reasons ranging from toolbox meetings to dispute resolution. The ability to participate in and contribute to workplace meetings or to conduct yourself professionally if you are involved in any type of workplace dispute is a very important communication skill to learn.

TIP The mix of different languages, occupations, age groups and levels of work experience on a worksite can create barriers to communication between construction workers. Other barriers include a noisy workplace, misunderstandings that lead to arguments and mistakes, and impoliteness. However, most of these barriers can be overcome by clearly and effectively using the correct communication techniques that you will learn in this chapter.

2.1 Important forms of communication for construction workers

Direct communication with another worker is the main form of communication on construction worksites. However, effective communication involves far more than being able to speak with another person in the same language. There are other equally important visual forms of communication.

2.1.1 Direct communication

A face-to-face discussion, a telephone conversation, radio communication, hand signals and gestures are all direct methods of communication between individuals or a number of workers.

2.1.1.1 Face-to-face verbal communication

Face-to-face verbal communication is probably the most used form of communication on construction worksites. For this to be effective, you must speak clearly and loudly enough to be heard. Remember that older construction workers may have hearing damage. Softly or quietly spoken words are difficult to hear and can be misunderstood.

If you are uncertain about whether your message has been heard clearly, ask if the person you spoke to understood you. For example, you could say, 'Did I make sense?' or 'Did I explain that properly?'. Don't say, 'Didn't you understand me?' or 'Do I need to say that again?' as some workers may interpret this as suggesting they don't understand English or they aren't intelligent.

Pay attention to what others are telling you and face them when they communicate with you. If you can't hear very well because of background noise, read their lips and watch their expressions as well as trying to listen to what they say. Move to a quieter work area if necessary.

Don't pretend to understand what has been said! Ask politely for the information to be repeated and ask any related questions to ensure that you understand what has been said to you. Experienced workers appreciate honesty, and good supervisors will know that you take your job seriously rather than pretending to understand.

SAFETY TIP in noisy work areas

- When wearing hearing protection, don't remove the protection in a noisy work area to speak.
- Stand closely to the worker you wish to speak to and speak loudly or shout and read their lips. They should be able to hear and understand.
- Move to a less noisy work area if necessary.

2.1.1.2 Telephone, radio, mobile phone

If you are using a mobile phone or radio to communicate with another worker, make sure you speak clearly and use these devices only where background noise is low and the person listening can hear you (Fig. 2.1). Alternatively, use text messages and include a related image in the message.

Two-way radios are also commonly used on construction worksites, but you must follow the correct two-way radio procedure to ensure that the communication is understood.

TIP Using a two-way radio

- Turn your microphone on when speaking and off when listening.
- Establish radio contact first by identifying and calling the name or call sign of the person you are communicating with. Then identify yourself.
- Sign off each time you have finished speaking using the term 'over'.
- End the communication by saying 'out'.
- Do not communicate any information that you do not want others to hear. There is no privacy with two-way radio conversations.

Fig. 2.1 Two-way radio communication

Courtesy of Alister Ford

2.1.1.3 Gestures and hand signals

When communicating directly with other workers using gestures such as nodding 'yes', shaking your head 'no' or shrugging that you don't know, make sure the worker you are communicating with is watching you and mouth the words so that they can read your lips if you are in a noisy work area.

Hand signals play an important role in communicating directly with other workers. When using hand signals such as an upright arm with the palm of your hand facing a person to signal 'stop', keep giving that signal until they stop (see Fig. 2.2). Don't assume they have seen your hand signal until they respond by stopping.

2.1.2 Visual communication

Visual communication on construction sites plays an equally important role as direct communication and is a method used to convey official site requirements and rules.

Fig. 2.2 Traffic control: signalling 'Stop!'

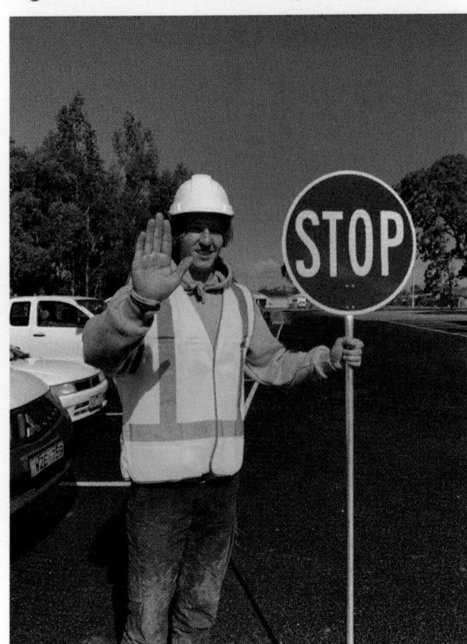

Courtesy of Alister Ford

2.1.2.1 Building plans and specifications

Building drawings, or plans, together with **engineer's detailed drawings** and **specifications** are the principle method of communicating how a building is to be constructed. The ability to read and interpret plans is an essential skill for all construction workers.

Tradespeople are expected to be able to read and interpret the scope and detail of work that they are required to carry out as part of their contribution to the overall construction of a building.

A site manager is expected to be able to read and interpret all aspects of a building plan and to then communicate the requirements of that plan to individual trades. Building plans are the common line of communication between the building designer and those who will construct the building.

An important part of early plan reading is the ability to locate the correct building site if it is in a new estate or a remote location. The site plan will contain details of the suburbs, towns, localities, lot numbers and street names, but the ability to find the correct location via electronic or paper-based maps is very important. No one wants or needs to be driving around looking for the correct building site.

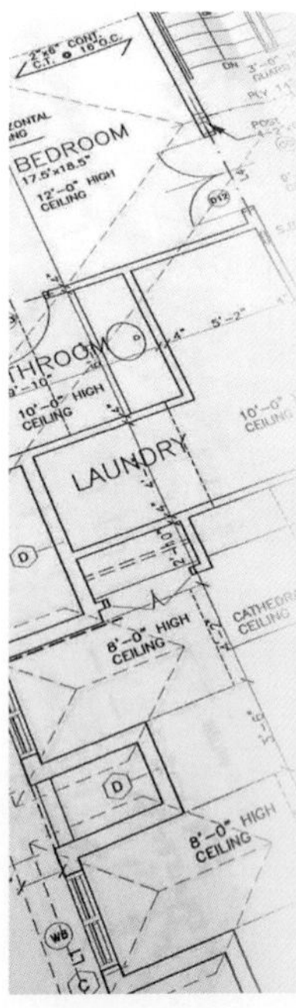

INTERPRETING PLANS

Review the housing plans in the appendix. From Appendix Figure 1 identify and answer the following:

1. What is the address of the building allotment for the proposed new residence?
2. Who is the client?
3. What is the current status of the adjoining lots 8, 10 and 11?
4. What is the difference in height between the floor level and the excavation level?
5. What is the distance between the front and rear boundary of the building site?
6. How wide is the building site?
7. The site has a fall or downward slope. Is the fall from front to back or back to front?
8. What is the soil type (classification) for the building site?
9. What is the wind classification for the building site?
10. What is the BAL (Bushfire Attack Level) rating for the building site?

2.1.2.2 Communication using mobile phones, tablets, personal computers, apps, emails, copier/scanners

Communication using electronic devices has come a long way in recent years and now, at the click of a button or the touch of a screen, you can access any number of installed communication apps and have a video or voice conversation with almost anyone just about anywhere. Again, good communication skills–such as speaking clearly, listening attentively and trying not to speak over the other person–are needed.

This type of communication is very useful on a building site as not only can a conversation take place, but viewing the matter at hand can now be done in real time. So it may be the case that a company supervisor or director, a building materials supplier or a sub-contractor can have a direct view of the situation and be able to make on-the-spot decisions while on another building site or in an off-site office, allowing the job to progress without delay. The time and date of the conversation will also be automatically recorded.

These modern communications devices provide ways to communicate with construction workers that can be included with other forms of communication or used alone. For example:

- construction site issues such as building plan errors, mismatched building finishes, defective building work and unsafe work areas can be photographed and sent via text message or app on a mobile phone or tablet, or as an attachment with an email
- a handwritten order can be photographed and sent as an email attachment via mobile phone or tablet, or alternatively an order can be typed as a text message and forwarded to a supplier
- if a copier/scanner is available, many of these machines now have a scan-and-send email application. You can send a hand-drawn sketch, handwritten message or any other relevant document, which will be delivered immediately to the structural engineer via the scan-and-send email feature.

2.1.2.3 Hand-drawn sketches

The simple sketching of on-site construction details, together with other direct communication, including speech, can overcome many barriers to effective communication. Different languages, occupations and levels of work experience create barriers to communication that can be overcome by including sketches as part of the process of communicating and receiving facts and ideas (see Fig. 2.3).

2.1.2.4 Written communication and site documents

Written communication in the form of handwritten or typed documents is commonly used with construction work. However, to deal with this type of communication you must be able to read or write a range of different documents. The ability to read correctly and understand what is written and the ability to write by hand or type or enter text into a mobile phone are extremely important skills widely used in the construction industry.

Examples of relevant documents include:

- *Site safety plans, safe work method statements, safe operating procedures and accident reports* (refer to Chapter 1).
- *Material labels and material safety data sheets (MSDS).* For safety reasons it is very important that you know how to access, read and understand the MSDS or material labels for any products or materials in the building industry. The information on these documents also needs to be communicated to other workers on the job site so they are aware of any substances or materials that may be harmful to health if they are improperly exposed to the product. Anyone working on the site should be aware of important points such as material handling and working procedures, required personal protective equipment (PPE), spills or leakage

Fig. 2.3 Sketched layout of decking and posts

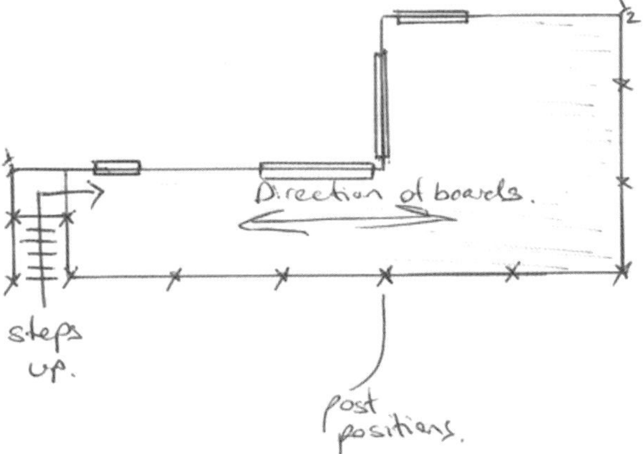

actions and clean-up, emergency medical procedures, and manufacturer and supplier information before the work starts. Information on understanding safety data sheets can be downloaded from www.safeworkaustralia.gov.au/doc/understanding-safety-data-sheets-hazardous-chemicals.

- *Company policies, procedures, regulations and quality requirements.* These items may be written into your contract of employment, giving a clear indication of the company's expectations of you. They can also be used in a broader sense to cover the expectations and quality requirements of all employees in the many job roles over the many different job sites in regard to work practices and finished products. All of these policies, procedures, regulations and requirements must be followed to maintain a safe and harmonious workplace and to uphold the standards of professionalism, workmanship and product quality that the company requires.
- *Work bulletins, site notices, memos, work instructions (including specifications and requirements), work plans and work schedules.* Depending on the worksite, any of these documents could be displayed on a noticeboard in or outside the site office or plan room, or even in the employee break room. These documents may also be sent to individuals or work teams electronically. Another method of distribution is to provide handouts to all team members at toolbox meetings before work commences. However you receive these documents, you must read them carefully and understand what is required of you with regard to the mentioned work. If at any stage you are unsure of what is required, it is important to ask your supervisor to clarify these points to avoid risking any mistakes or accidents.
- *Time sheets, pay dockets (Fig. 2.4), claim forms.* It is very important that you know how to fill out your time sheets clearly and correctly, whether by hand or electronically. Any mistakes on your time

Fig. 2.4 Example of a pay docket

PAY ADVICE Lightning Constructions ABN: 76 872 932 179

Alexander Doe: F/T: National Construction Award: Carpenter

Date of payment: 31 October 2019: Pay period: 14 October to 28 October 2019
Hourly rate: $30.00; Annual salary: $60 000; Bank details: Westpac BSB 1234 A/C 456789
Annual leave accrual: 15 days; Personal/carer's leave accrual: 5 days; As at 31 October 2019

Entitlements	Unit	Rate	Total
Wages for ordinary hours worked	40 hours	$30.00	$1200.00
Travel allowance	5 days	$15.00	$75.00
Minus RTO accrual	2 hours	$30.00	–$60.00
Leave taken during the pay period	00 hours	$00.00	$00.00
TOTAL ORDINARY HOURS PAID = 38 hours			
TOTAL RTO HOURS ACCRUED = 6			
OT × 1.5	00 hours	$00.00	$00.00
OT × 2	00 hours	$00.00	$00.00
		Gross payment	**$1215.00**

Deductions	
Taxation	$300.00-
Credit union	$100.00-
Total deductions	**$400.00-**
Net payment	**$815.00**

Employer superannuation contribution—12%	
BUS	
Contribution	**$145.80**

sheet could mean a delay in your wages being paid. Always check your pay slips. Learn how to read them properly. Don't just take it for granted that your wages will be correct every time. Mistakes do sometimes get made. Your pay slip will contain a lot of information with which you need to keep up to date, such as your pay scale or hourly rate, the amount of tax you pay, how much is being contributed to your superannuation account, how much sick leave you have, how much annual leave you are entitled to and how much long service leave you are entitled to. If at any time you are unsure of the information on your pay slip, talk to your employer. They will explain it to you and the information you take on board from this one simple act of communication will help you in times to come.

- *Orders (Fig. 2.5), delivery dockets and invoices (Fig. 2.6 and 2.7).* It is crucial to understand these very important documents. If you are signing for a delivery, you must know how to check that you are receiving the correct materials, tools or equipment, in terms of both quality and quantity. You must also be able to give clear directions regarding distribution, stacking or storage of the items around the worksite according to the given workplans. Sign for the delivery only when you are sure that all of the items on the delivery docket have been delivered and are undamaged. If you believe that any items are missing or have been damaged during transportation, you should not sign as you may put yourself in a compromised position, and some of the blame may be apportioned back to you. Instead, bring the problem to the attention of your supervisor so that it can be sorted out promptly and effectively with the company providing the goods.

Fig. 2.5 Example of a purchase order

Toby Henderson Licensed Builder 866930 ABN: 93 989 996 992		**Purchase Order**	
224 Jones Road		ORDER NO.	[103-12-19]
Gosford NSW 2250		DATE	30 April 2019
02 4344 2777 or 0411900788		SUPPLIER ID	TJ HARDWARE
TO	Alan Peters		
	TJ Hardware 367 Chamber Road Wyoming 2250		
Quantity	**Description**	**Unit price**	**Total**
415.00	L/M 90 × 20 Merbau Decking	4.50	1867.50
4.80	L/M H3 Treated Pine 290 × 45 F7	14.20	68.16
14.70	L/M H3 Treated Pine 190 × 45 F7	9.50	139.65
39.60	L/M H3 Treated Pine 140 × 45 F7	7.15	283.14
15.00	L/M H3 Treated Pine 90 × 35	4.10	61.50
15.00	L/M H3 Treated Pine 50 × 35	2.50	37.50
18.00	L/M Dressed Hardwood Posts 90 × 90	23.50	423.00
27.00	L/M H3 Treated Pine 140 × 35	5.50	148.50
33.00	L/M 90 × 35 Treated Pine—'Terminator'	4.00	132.00
4.00	Secura FC Interior 2.7 × 0.6 Sheets	100.00	400.00
20.00	Galvanised Post Holders 90 × 90	8.00	160.00
1.00	Box Galvanised 65 mm Coach Screws	25.00	25.00
		Subtotal	**$3745.95**
		GST	**374.60**
		Total	**$4120.55**

Fig. 2.6 Example of a delivery docket

DELIVERY DOCKET			
TJ HARDWARE PTY LTD 367 CHAMBER RD WYOMING			
DELIVERY NO	070513/05	ORDER NO.	[103-12-19]
DATE	7 May 2019	DATE	30 April 2019
DRIVER	Thompson	CUSTOMER ID	Toby Henderson
TO	Toby Henderson		
	224 Jones Road		
	Gosford NSW 2250 PH 02 4344 2777 or 0411900788		

	THE FOLLOWING ITEMS IN GOOD ORDER AND CONDITION
Quantity	**Description**
415.00	L/M 90 × 20 Merbau Decking
4.80	L/M H3 Treated Pine 290 × 45 F7
14.70	L/M H3 Treated Pine 190 × 45 F7
39.60	L/M H3 Treated Pine 140 × 45 F7
15.00	L/M H3 Treated Pine 90 × 35
15.00	L/M H3 Treated Pine 50 × 35
18.00	L/M Dressed Hardwood Posts 90 × 90
27.00	L/M H3 Treated Pine 140 × 35
33.00	L/M 90 × 35 Treated Pine—'Terminator'
4.00	Secura FC Interior 2.7 × 0.6 Sheets
20.00	Galvanised Post Holders 90 × 90
1.00	Box Galvanised 65 mm Coach Screws
SUPPLIER SIGNATURE	RECIPIENT SIGNATURE
TERMS 7 DAYS FOR ANY CLAIMS	

Fig. 2.7 Example of a tax invoice

TJ HARDWARE ABN: 72 534 801 766		**INVOICE**
224 Jones Road	TAX INVOICE	[1021-05-19]
Gosford NSW 2250	DATE	30 May 2019
02 4344 2777 or 0411900788	CUSTOMER ID	Toby Henderson
TO Toby Henderson		
224 Jones Road Gosford NSW 2250		

Quantity	Description	Unit price	Total
415.00	L/M 90 × 20 Merbau Decking	4.50	1867.50
4.80	L/M H3 Treated Pine 290 × 45 F7	14.20	68.16
14.70	L/M H3 Treated Pine 190 × 45 F7	9.50	139.65
39.60	L/M H3 Treated Pine 140 × 45 F7	7.15	283.14
15.00	L/M H3 Treated Pine 90 × 35	4.10	61.50
15.00	L/M H3 Treated Pine 50 × 35	2.50	37.50
18.00	L/M Dressed Hardwood Posts 90 × 90	23.50	423.00
27.00	L/M H3 Treated Pine 140 × 35	5.50	148.50
33.00	L/M 90 × 35 Treated Pine—'Terminator'	4.00	132.00
4.00	Secura FC Interior 2.7 × 0.6 Sheets	100.00	400.00
20.00	Galvanised Post Holders 90 × 90	8.00	160.00
1.00	Box Galvanised 65 mm Coach Screws	25.00	25.00
		SUBTOTAL	**$3745.95**
		GST	**374.60**
		TOTAL	**$4120.55**

TERMS 30 DAYS—THANK YOU FOR YOUR BUSINESS!

2.1.2.5 Mandatory licences and qualifications

Mandatory licences are used to communicate to others that a builder, tradesperson or construction worker has the authority to carry out specific building or construction work. Some tradespeople are licensed to work in just that trade.

Plumbers and electricians must be licensed, whether they are employed or work as contractors.

All construction workers must undertake general induction training before working on a construction site. On completion they are issued with a card to carry at all times as proof that they have undertaken the training.

Hoist operators, crane operators, forklift drivers and workers carrying out scaffolding, dogging and rigging must have a 'National licence to perform high risk work' (Fig. 2.8).

Fig. 2.8 Licence to perform high risk work

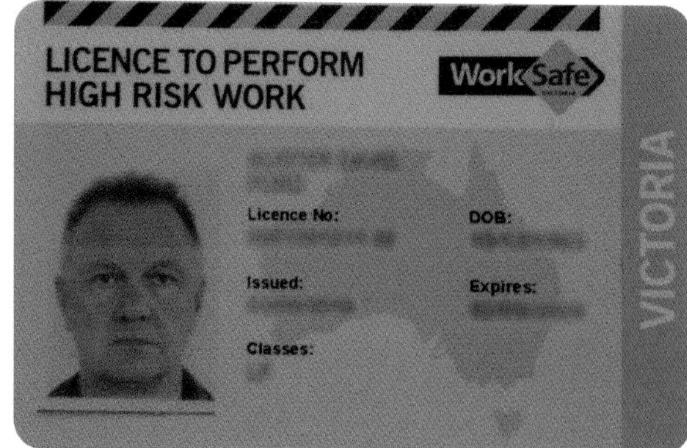

Courtesy of Alister Ford

2.1.2.6 Construction site signs

Construction site signs include safety and warning signs (Fig. 2.9), direction signs (Fig. 2.10) and information signs (Fig. 2.11). Signs play a critical role by:

- reminding workers to use their personal protective equipment (PPE) and avoid unsafe work practices
- warning workers that they must be cautious when entering a dangerous work area
- directing workers to a specific location on-site or directing them away from areas they are unauthorised to enter
- providing advice and information about the worksite, including the safety rules they must obey.

CHECK YOUR UNDERSTANDING

1. List 10 different types of visual communication used in the construction industry.
2. List five points that must be considered when using two-way radio communication.
3. How would you go about communicating with another worker who has a severe hearing impairment?
4. How would you let someone in a noisy work environment know they are doing a good job? Assume you and the other worker are wearing hearing protection and that you can't leave the work area.

Fig. 2.9 A & B Safety and warning signs

(a)

Courtesy of Alister Ford

(b)

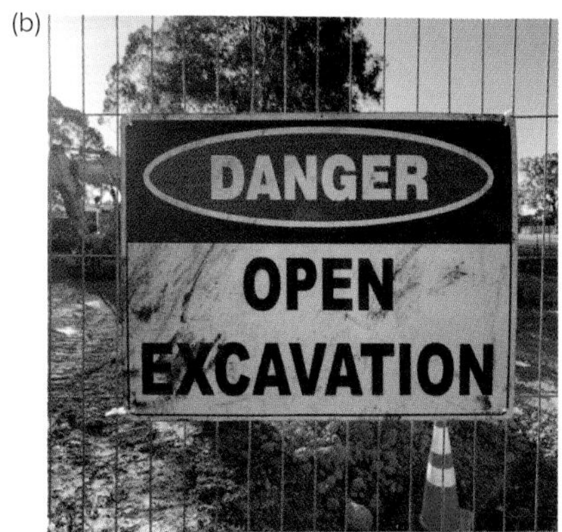

Fig. 2.10 Direction sign

Courtesy of Alister Ford

Fig. 2.11 Information sign

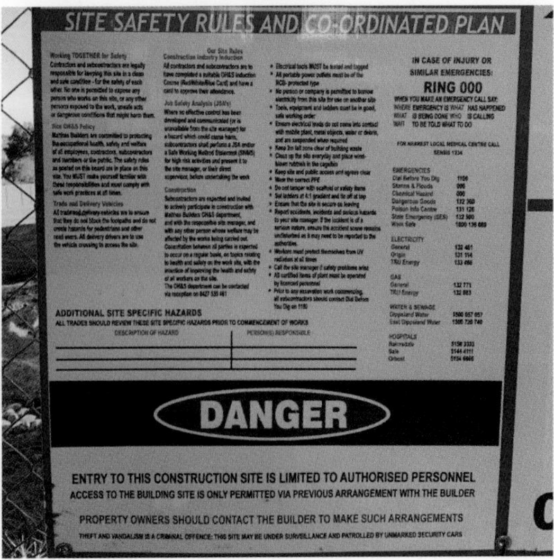

2.2 Communicating in work teams

In today's workplace, all workers have a responsibility to maintain worksite communication with their colleagues. Workers should place a high level of importance on all interactions with their fellow workers, including social aspects.

Once you have a job, how you communicate with your employer, supervisor and your fellow employees will play a major role in whether you have a great job or an ordinary job. You will spend a lot of time during your life at work so make it count!

Good-quality communication between yourself and your colleagues will make your time at work worthwhile, and your working day will go more quickly, regardless of how hard the work may be.

2.2.1 Communicating with other workers

To work effectively with others, you must be able to communicate with them. This means being able to discuss matters not strictly related to work, being honest and friendly, and showing some interest in what they do on-site and off-site.

An ability to interact with other workers is essential. From a safety and productivity perspective, the effectiveness of work teams depends on all members of the team being able to communicate and work together.

For new workers with little industry experience, it is absolutely essential to be able to communicate with more experienced workers. Apprentices and trainees must be able to ask questions and receive and understand both positive and critical feedback in relation to the work they perform.

Older, veteran workers must have the confidence to pass on the knowledge they have acquired over their many years of work. However, they can be confident about passing on their knowledge only if younger, less experienced workers listen and apply that information.

Equally important is non-verbal communication between workers, which plays a vital role in the effectiveness of work teams. Individual team members need to encourage and support other team members by acknowledging good work practices with, say, a 'thumbs up' gesture (Fig. 2.12) or a nod of the head together with a smile.

Fig. 2.12 Thumbs up!

Courtesy of Alister Ford

Through their own actions, every worker conveys information to other workers. Actions that benefit both individuals and the work team include the following:

- Treating other workers politely and greeting them in the morning and saying 'see you later' when leaving work to go home sets the scene for cooperation in the workplace.
- Having skilled workers with high work standards ensures that the work team maintains a good standard of work.
- Ensuring one work member regularly wears PPE signals to all other workers, including those outside of their team, that workplace safety is an important issue.
- Making sure the tools of the trade are well-maintained and cleaned and stored away at the end of each day sends a strong visual message that demonstrates each worker's personal values and responsibility.
- Maintaining a clean and tidy work area is a strong visual message to other workers that encourages safety, high work standards and professionalism.
- Encouraging workers to be diligent and methodical promotes a strong work ethic among all team members.

2.2.2 On-site meetings

On-site meetings may be called for a range of reasons including industrial or safety issues, the use of a new material that requires training, or an employer wishing to make an announcement or issue special work instructions. These meetings are often called **toolbox meetings**.

Meetings on worksites are generally short, as most workers are keen to get back and finish what they were doing before the meeting commenced.

If you participate in an on-site meeting, try to obtain as much information as possible from the meeting.

Participation in meetings

- If you are late, try not to disturb other workers.
- Be quiet when others are speaking.
- If a show of hands is requested to vote at a meeting, make your intention clear.
- If you have a question or would like to make a proposal to the meeting, raise your hand and address your question to the person running the meeting.
- If you would like to make a proposal to the meeting, write down what you would like to propose and read it out, making sure you can be heard and understood.
- If the meeting is important, make brief notes during the meeting or after it finishes.

CHECK YOUR UNDERSTANDING

1. Toolbox meetings can be called at any time on construction sites for all of the workers on the job or just for a specific work group. List three reasons why a toolbox meeting may be called at short notice.
2. Besides providing individual protection for the wearer, how does the use of PPE by one worker benefit other workers?
3. Explain how you would demonstrate to other workers that you have a strong work ethic, without actually telling them so.

2.2.3 Communicating with the public

In the construction industry you will often be required to communicate with the general public. This may be the case if someone enters the building site mistakenly or if they are an expected visitor but are unsure who, or where, to report to. Always greet any visitors politely and ask clearly if you can be of any assistance in directing them to where they need to be.

It may be that the person they wish to see isn't on-site at that time–if so, clearly write down the visitor's details and pass this on to the person concerned so that the visitor may be contacted at a later time.

Alternatively, you may need to communicate with the general public if some of the work takes place outside of the worksite boundaries–for instance, if concrete trucks are arriving for a pour or building material delivery trucks need access to the job site to unload. Note that in such cases, traffic and pedestrians may need to be halted for a period of time. If you are involved, you must ensure that you communicate clearly with any pedestrians, drivers, delivery truck drivers and fellow workers and that they have understood you.

Remember that drivers or pedestrians may not be able to hear you. Make sure that any sign you may be operating, such as a stop/slow bat, is clearly displayed and visible to any members of the public who are affected.

Also ensure that any hand signals you may use in conjunction with the sign are clear and concise.

WORKPLACE SCENARIO

Communication is important at all times. I recall once when I was working as a carpenter for a major Melbourne construction company. We had just commenced work on a vacant site in Dunlop Road, Mulgrave. The datum for the job site needed to be obtained and marked for the building site excavations and setting up for construction to begin.

The datum height was noted on the building plans, but the levels needed to be obtained from a surveying marker located in the middle of the intersection of Springvale Road and Ferntree Gully Road. The company surveyor was contacted and soon arrived on-site to set the correct datum for us. I was assigned to assist, so we introduced ourselves and then the surveyor explained to me how the job would be done. There were no electronic measuring devices at that time, so the levels would be shot using an automatic level and staff. Readings would be transferred to a booking sheet and later the calculations would be made, and the site datum marked for work to proceed further. The surveyor explained the whole process to me, so I understood exactly what we needed to achieve in this job. He explained all of the hand signals that we would use as we would be approximately 80 to 100 m apart at all reading points and change of stations and this would be our only means of communication. We made our way to the 1.5 km intersection and set up the instrument for the first reading.

The surveyor gave me a small orange flag. These were the days before hi-vis work clothes, and it was 8.30 am on a busy Melbourne weekday. There was traffic everywhere. I soon found out what the orange flag was for as I stood in the middle of the intersection holding the staff. It was my only means of communication with the car and truck drivers who passed through that intersection as I stood out there.

I made it safely back to the footpath after the first reading was taken, then, with a series of levels taken using the hand signals as our means of communication, we worked our way back along Springvale Road to the job site. After the calculations were completed, we were able to establish the correct datum for the job and the work on-site was able to proceed. This was the first time I had ever worked with a surveyor, so I made sure I engaged in conversation with him and asked questions about all aspects of the job. I learned a lot that day about surveying and worksite communication that I still put into practice to this day.

2.2.4 Communicating with trainers

1. People often commence employment in the construction industry as an apprentice or trainee.
2. A mandatory component of an apprenticeship or traineeship is to undertake off-site training and even some on-site training, although not under the direction of your immediate supervisor.
3. Whichever training organisation you enrol with will assign a trainer to you, and if you attend off-site training you will most likely be in a class with other students enrolled in the same qualification. Any on-site training would be conducted in a one-on-one situation unless your employer has more than one apprentice or trainee at the same level.
4. Effective communication with your trainer is vitally important. A lack of communication could be the cause of you failing enrolled subjects, which essentially costs your training organisation, your employer and yourself, time and money.
5. All parties involved should discuss and clearly record the details of your training plan requirements, your enrolment, study times and progress.

2.3 Communicate with employers and supervisors

It is extremely important to be able to speak directly to your employer or supervisor about all aspects of your work and your work role.

At your first contact with a potential employer or supervisor, and in any later interactions, make sure they understand you are a productive worker willing to be part of a work team–even if the team is just the two of you!

2.3.1 Face-to-face communication with employers and supervisors

Always greet your supervisor or employer at the start of work and say goodbye when you leave work. This will demonstrate that you enjoy working with them and like your job; it also encourages fair treatment by your employer or supervisor.

Work-related problems are best dealt with by speaking to your employer or supervisor privately. It is best to be upfront and honest about any work issues you may have.

A common problem for many young employees is their understanding of *what* they are actually paid and their conditions of employment. If these conditions were not established at the time of applying for the job, it is appropriate for you to ask what they are. An open and honest employer will appreciate you coming to them directly rather than asking someone else.

TIP As a new worker you must first build trust in other workers before discussing work-related problems.

2.3.2 Non-verbal communication

As discussed earlier, non-verbal communication between you and your employer or supervisor includes a range of documents that you may be required to use, signs that you must comply with, and the use of gestures and sketches.

Other elements of non-verbal communication you must be aware of include:

- *Wearing required PPE.* Employers and supervisors become very agitated when workers fail to wear PPE because if there is an accident, they will share responsibility and may be fined because you didn't comply with work instructions. By not wearing PPE you are demonstrating disregard for your safety and the maintenance of a safe workplace for all workers. Most employers and supervisors are genuinely concerned about your safety and health. Ignoring their concerns is a strong message to them that you don't appreciate their concern for your welfare.
- *Maintaining a good personal appearance, speaking clearly and working diligently.* These all convey the message to your employer that you want to achieve the best possible work performance.
- *Maintaining your tools.* This demonstrates to your employer that you want to achieve good-quality workmanship and that you understand the importance of good workmanship for the success of the business.
- *Constructing or building to standards required by your employer or supervisor.* This conveys your acceptance and adoption of the standard of quality that your employer or supervisor has set and wants to achieve.
- *Using correct trade terminology.* This prevents misunderstandings and demonstrates a desire to be understood and to be acknowledged as a person with a good knowledge of the technical aspects of your trade–in other words, 'Being seen as acting in a professional or tradelike manner'.

2.4 Resolving disputes

Disputes on building sites are generally the result of poor communication. The best way to prevent disputes is to maintain communication at all times. Misunderstandings should be corrected immediately.

If you have an argument with your employer or supervisor or another worker, continue to maintain a polite relationship. Keep saying good morning when you arrive for work and good afternoon when you leave, otherwise you may have to leave a job you really like.

Arguments are generally resolved with time. Just keep talking and keep on smiling! In most cases both parties will eventually apologise to each other, and it is often better to be the first to apologise so that everyone can move on.

If your attempts to resolve a dispute fail, make enquiries outside of your workplace, such as through employer or employee organisations or relevant government agencies. If you discover you have not been treated fairly, discuss the matter again with your employer or your supervisor. If the problem is not resolved as a result of this, you may have to apply for another job or take the matter further.

TIP It is far easier to apply for another job while you have a job. If you plan to resign from a job, get your next job organised first.

CHECK YOUR UNDERSTANDING

1. A worker accuses you of ridiculing them after you attempt to help them complete a task because you believed they needed assistance. How would you remedy this situation?
2. What would you do in the future to prevent such a misunderstanding?
3. You have made a mistake and think you may have to pull down work that you have just completed after working on that job for four hours. Select one of the three options below and explain why you have selected it.
   a. Tell your supervisor immediately.
   b. Not tell your supervisor and instead immediately start to make good the defective work.
   c. Pull the work down and start another job.

Student research

- Do an online search for 'Australian construction terminology' and list and define 10 new construction terms with which you are not familiar.
- Visit Safe Work Australia: www.safeworkaustralia.gov.au. Select 'Model Codes of Practice' on the home page, and then select 'Construction Work' from the list of options. Identify who has health and safety duties relating to construction work.

End of chapter activity

Activity 1:

Look around your worksite or school and see how many signs you can locate; list them and what they are communicating.

Divide them into categories of:

- Safety signs
- Warning signs
- Direction signs
- Information signs.

Note the colours of the signs and the writing.

Can you spot any trends in terms of the colours and the categories in which you have placed them?

AUSTRALIAN BUILDING CODES AND STANDARDS

N/A

Tradie talk—We need an 820 cavity slider

Chapter 3

Trade calculations and measurements

Learning Objectives

LO 3.1 Understand metric units

LO 3.2 Use an electronic calculator

LO 3.3 Apply plane geometry

LO 3.4 Understand solid figures

LO 3.5 Calculate and order materials

Introduction

The carpenter who is active in the trade will need to make calculations daily. These can include simple additions or subtractions of lengths, or they may involve calculations of geometrical areas and volumes. The carpenter may also be required to work out the quantities of material required to complete the job at hand, as well as calculating the costing of these materials.

In this chapter you will learn how to perform some basic calculations and will be shown how to calculate the area, volume and perimeter of various shapes. You will also learn basic calculator procedures and how to tabulate and order materials.

3.1 Metric units

Metric units are used throughout the building industry. They include the units shown in Table 3.1. Note that small letters are used for all symbols except where the unit is derived from a proper name, such as newton (N) and ampere (A), or where the value is over one million, such as the megapascal (MPa), etc.

Table 3.1 Common units in the construction industry

Unit	Physical function	Symbol
Metre	unit of length	m
Millimetre	1000 millimetres = 1 metre	mm
Kilometre	1000 metres = 1 kilometre	km
Square metre	unit of area (m × m)	m^2
Square millimetre	unit of area (mm × mm)	mm^2
Hectare	unit of area (10 000 m^2)	ha
Cubic metre	unit of volume (m × m × m)	m^3
Litre	unit of volume or capacity	L
	(100 mm × 100 mm × 100 mm)	
	1000 litres = 1 m^3	
	1 litre of water = 1 kg	
Kilogram	unit of mass	kg
Gram	1000 grams = 1 kilogram	g
Tonne	1000 kilograms = 1 tonne	t
Newton	unit of force	N
Pascal	unit of pressure = 1 N/m^2	Pa
Kilopascal	1000 pascals	kPa
Megapascal	1 000 000 pascals	MPa
Volt	electrical potential	V
Ampere	electric current	A
Watt	unit of power = V × A	W

It is important that tradespeople working in the construction industry are able to read and record measurements accurately to the nearest millimetre. Measuring equipment used on a building site can include:

- laser measuring equipment
- one-metre folding rulers
- retractable steel tape measures
- trundle wheels
- levelling devices
- straight edges and squares.

Fig. 3.1 Laser measuring devices

© AlexLMX/Shutterstock

Laser measuring equipment is becoming more popular in the building industry, and is used especially when a tradesperson is doing a quote and needs to measure up a job–for example, building partition walls in an open office space. Such tasks would normally take two people to measure up using a long tape. With a laser device, however, the device is set down against one wall, and it calculates the distance by firing a laser beam across the floor until it hits another wall. Then, the electronic display shows the length. The device speeds up the task and is extremely accurate.

It is essential to take care of measuring equipment. A tape should be wiped clean occasionally, especially if it gets wet, as the metal will rust. If it does get wet, a clean rag with some oil on it will help stop any corrosion.

For most other equipment, avoid dropping it and ensure batteries are charged if present. If it comes in a case, keep it in that case, so that the equipment is protected and remains accurate. Calibration of some equipment is required periodically, so please read the manufacturer's instructions prior to its first use.

The metric units used in the building industry for linear measurements are metres and millimetres, where one metre (m) = 1000 millimetres (mm). Usually, dimensions on building plans are given in millimetres, but on some large site plans they may be given in metres. This system eliminates the possibility of making errors in reading any dimension, as the decimal point can go in only one place. Plus, plans can often be kept in a dirty environment and it can be difficult to see if it's a dot or a spec of dirt. If there is no decimal point shown in a figure, it must be in millimetres. For example, a dimension of 3648 is in millimetres and no symbol of 'mm' is necessary to indicate this fact.

If a decimal point is shown, and the figure is written to three decimal places, then all of the figures before the point are metres and all after the point are millimetres. For example, a dimension of 27.632 is 27 m and 632 mm, and again no symbol is necessary to indicate this on the plan.

3.1.1 Converting millimetres to metres

To convert millimetres to metres, divide the figure by 1000 (e.g. 3648 mm divided by 1000 = 3.648 m). The easy way to divide by 1000 is to move the decimal point three places to the left, adding zeros to make up the three places if necessary. For example:

3648 mm = 3.648 m
17846 mm = 17.846 m
5 mm = 0.005 m
75 mm = 0.075 m

Note that a zero is placed before the decimal point if there is no significant full figure. With practice, dividing by 10, 100 or 1000 becomes a procedure of mental arithmetic. For example:

- To divide by 10, move one place to the left.
- To divide by 100, move two places to the left.
- To divide by 1000, move three places to the left.
- To multiply by 10, move one place to the right.
- To multiply by 100, move two places to the right.
- To multiply by 1000, move three places to the right.

If necessary, add zeros to the right; for example, 306 multiplied by 10 equals 3060.

To add metric units

To add metric units, write the quantities in a column with the decimal points aligned, then add up the columns; the decimal point in the answer will be aligned with those above it. For example, to add the dimensions 6.372 m, 4.021 m, 50 mm, 120 mm and 6 mm:

	6.372
	4.021
	0.050
	0.120
	0.006
Total	10.569 m

To subtract metric units

To subtract metric units, write down the quantities–the smaller figure under the larger figure–and, with the decimal points aligned, subtract. For example, subtract 3.645 m from 12.429 m:

```
          12.429 –
           3.645
         --------
Total      8.784 m
```

or subtract 28 mm from 3.620 m:

```
          3.620 –
          0.028
         -------
Total     3.592 m
```

To multiply metric units

To multiply metric units, multiply in the normal manner and, counting from the right-hand side of the final figure, place the same number of decimal points in the answer as there are in total in the two numbers being multiplied together. For example, multiply 2.632 m by 6.3 m:

```
           2.632 ×
             6.3
         --------
            7896
         15 792
         --------
Total    16 5816
```

Then count back four decimal points, so the answer is 16.5816 m^2

To divide metric units

To divide metric units, divide the dividend by the divisor, and the decimal point in the quotient should be over the decimal point in the dividend. For example, divide 39.392 by 8:

```
     4.924
  8 )39.392
    32
     7 3
     7 2
       19
       16
        32
```

or

to divide 39.392 by 2.5, first move the decimal point in both the dividend and the divisor one position to the right:

```
       15.757
  25 )393.920
      25
      143
      125
       18 9
       17 5
        1 42
        1 25
          170
```

Answer = 15.757, to three decimal places

TIP The common metric units used in the building industry for linear measurements are metres and millimetres.

CHECK YOUR UNDERSTANDING

1. Which metric units of length are used in the building industry?

3.2 The calculator

It should be mentioned from the outset that the calculator is regarded as one of the most valuable tools of the trade–and probably one of the cheapest. The keyboard of the basic calculator is shown in Figure. 3.2, with examples of how to use the calculator following. It is assumed that the calculator will be used wherever possible throughout this book, and solutions have been arranged in such a way that the figures can be entered and the problems solved using the calculator.

A useful hint when using a calculator is to make a rough estimate of what the answer should be, then to check the magnitude of the number. This is just in case a mistake was made in entering the number, which often puts the result out by a factor of ten or more.

Fig. 3.2 Basic calculator keyboard

© Ingram Publishing/SuperStock

Most calculators are similar in operation, but the order in which some of the function keys are pressed may vary from brand to brand. Check the operating instructions supplied with the calculator.

3.2.1 Addition

Units must be entered in the same units; for example, either metres or millimetres only, in one calculation.

To add 23.5 m to 87.45 m on a calculator enter:

[C] [2] [3] [.] [5] [+] [8] [7] [.] [4] [5] [=]

This displays the answer 110.95 (m)

Where mixed units are to be added, they must all be converted to the same units. For example, add 9.38 m, 0.09 m, 125 mm and 84 mm.

1. Convert all units to metres:

   $= 9.38 + 0.09 + 0.125 + 0.084$

   On a calculator enter:

   [C] [9] [.] [3] [8] [+] [.] [0] [9] [+]
   [.] [1] [2] [5] [+] [.] [0] [8] [4] [=]

   This displays 9.679 (m)

2. Convert all units to millimetres:

   = 9380 + 90 + 125 + 84

   [C] [9] [3] [8] [0] [+] [9] [0] [+]
   [1] [2] [5] [+] [8] [4] [=]

   This displays 9679 (mm)

3.2.2 Subtraction

Units subtracted must also be in the same units. For example, subtract 23.5 m from 87.45 m:

[C] [8] [7] [.] [4] [5] [−] [2] [3] [.] [5] [=]

This displays the answer 63.95 (m)

3.2.3 Multiplication

Measurements may be multiplied as measurement by measurement for areas, or by a factor to increase the number. For example, multiply 2.4 m by 3.6 m:

[C] [2] [.] [4] [×] [3] [.] [6] [=]

This displays the answer 8.64 (m^2)

Multiply 2.456 m by 6:

[C] [2] [.] [4] [5] [6] [×] [6] [=]

This displays the answer 14.736 (linear metres)

3.2.4 Division

Measurements may be divided by similar units or by a factor to decrease the number. For example, divide 72.8 by 34:

[C] [7] [2] [.] [8] [÷] [3] [4] [=]

This displays the answer 2.141

3.2.5 Squaring a number

To square a number is to multiply it by itself. For example:

$$4^2 = 4 \times 4$$
$$= 16$$

Some calculators have a function key marked x^2 that, when pressed, will square any number that has been entered.

[C] [1] [2] [x^2] displays the result 144

If there is no squaring function on the calculator, the entry should be as follows:

[C] [1] [2] [×] [1] [2] [=] also displays 144

Some calculators may function automatically by multiplying by the last number entered when the × button is pressed, if no further entry is made:

[C] [1] [2] [×] [=] displays 144

3.2.6 The square root of a number

The square root of a number is the reverse of the square of a number. The square root of a number is that figure which, when multiplied by itself, will result in the given number.

For example, $4 \times 4 = 16$, so $\sqrt{16} = 4$.

This can be done on a calculator that has a square root function key. The alternative is to look at a set of tables, or to derive the answer by trial and error:

[C] [1] [6] [√x] [=] displays the answer 4

3.2.7 Percentage calculations

Percentage calculations (symbol %) convert fractions into units per 100 units and are commonly used as a basis for calculating such things as discounts on accounts or allowances for adding waste to various quantities. Percentage calculation can also be used to establish a profit margin when costing a job.

Percentages commonly found and their equivalent fractions are:

5% = 5/100 = 1/20
10% = 10/100 = 1/10
20% = 20/100 = 1/5
50% = 50/100 = 1/2
12.5% = 1/8
33.33% = 1/3
66.66% = 2/3
16.66% = 1/6

For example, 7% of \$36.25 = (36.25 × 7)/100
= \$2.54

Some calculators have a percentage key marked with the symbol %, which carries out this calculation when the numbers are entered and the percentage key pressed. For example:

[C] [3] [6] [.] [2] [5] [×] [7] [%]

This displays the answer 2.5375 (\$2.54)

3.2.8 Memory keys

Memory keys are available on most calculators and enable the operator to memorise several minor calculations and bring them together as a whole. The relevant keys are:

[CM] clear memory
[MR] memory recall
[M−] subtract from memory or memory minus
[M+] add to memory or memory plus.

The result is finally given when the recall button is used.

Calculate 7 articles at \$6.25
plus 4 articles at \$0.62
plus 14 articles at \$13.40

[C] [CM] [7] [×] [6] [.] [2] [5] [=] [M+]....43.75
[4] [×] [.] [6] [2] [=] [M+]................2.48
[1] [4] [×] [1] [3] [.] [4] [=] [M+]......187.60

Then press the memory recall [MR] and the answer 233.83 (\$233.83) will be displayed.

3.2.9 Ratios and proportions

Sometimes problems can be solved quickly by using proportions. For example:

A plot is 8.25 m long by 2.5 m wide. If the width were to be increased to 3.5 m, what will the length be if it is to preserve the same proportions? The ratio of 8.25:2.5 must be maintained, so the new ratio will be *x*:3.5.

$$\text{Therefore } x/3.5 = 8.25/2.5$$
$$x = (8.25 \times 3.5)/2.5$$
$$= 11.55 \text{ m}$$

The following shows an example of how the use of ratios and proportions are put into practice on a building site when designing and constructing an access ramp.

An access ramp is to be built with a maximum gradient of 1:14 as determined by the Building Code of Australia. The height of the ramp is to finish 250 mm above ground level. Therefore, the horizontal length of the ramp will need to be a minimum 3500 mm so that it complies with the design ratio, i.e. 14 × 250 mm = 3500 mm (Fig. 3.3).

Fig. 3.3 Ratios and proportions

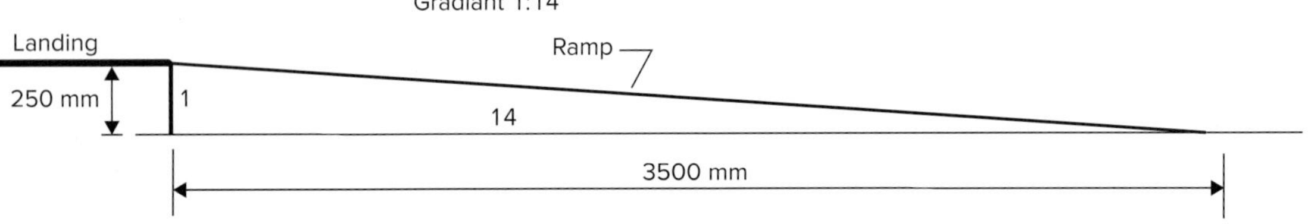

3.2.10 Parentheses

When some quantities are shown in parentheses, it means that this part of the problem must be treated as a complete unit and must be calculated first. For example:

$$(4 \times 2) + (3 \times 8) = 8 + 24$$
$$= 32$$

If no parentheses are present, the calculations may be confusing. Mathematical conventions dictate that the multiplications should be carried out first.

CHECK YOUR UNDERSTANDING

1. Add together the following units: 6.3 m, 2673 mm, 0.050 m, 18 mm and 24.050 m.

3.3 Plane geometry

A *plane surface* is a flat surface and a *plane figure* is a figure that lies wholly on a plane surface. The *perimeter* of a plane figure is the distance around the sides of the figure. A plane figure has dimensions in two directions only and has area only. The measurement of plane figures in various combinations is an important part of the carpentry trade.

3.3.1 Quadrilaterals

Quadrilaterals are plane figures bounded by four straight sides; they contain four angles that together add up to 360°.

A particular group of quadrilaterals are called *parallelograms*, which can be defined as plane figures having two pairs of equal and parallel sides. There are four types of parallelograms: a square, a rectangle, an L-shaped figure and a rhombus.

3.3.1.1 Square

A square is a plane figure having four equal sides and four equal angles (Fig. 3.4). Each angle is 90°.

$$\textbf{Perimeter} = 4 \times l$$
$$= 4 \times 3$$
$$= 12 \text{ m}$$
$$\textbf{Area} = l^2$$
$$= 3^2$$
$$= 9 \text{ m}^2$$

Fig. 3.4 Square

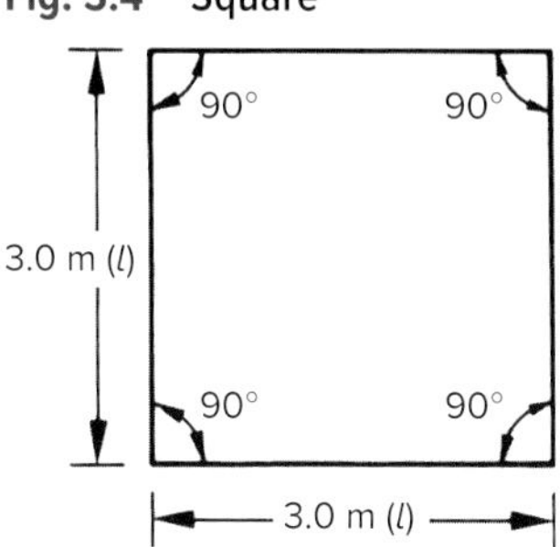

3.3.1.2 Rectangle

A rectangle is a plane figure having two pairs of equal and parallel sides and four equal angles, each of 90° (Fig. 3.5). A rectangle has length (*l*) and breadth (*b*) or width (*w*).

$$\textbf{Perimeter} = (2 \times l) + (2 \times b)$$
$$= (2 \times 4.5) + (2 \times 2.5)$$
$$= 9 + 5$$
$$= 14 \text{ m}$$
$$\textbf{Area} = l \times b$$
$$= 4.5 \times 2.5$$
$$= 11.25 \text{ m}^2$$

Fig. 3.5 Rectangle

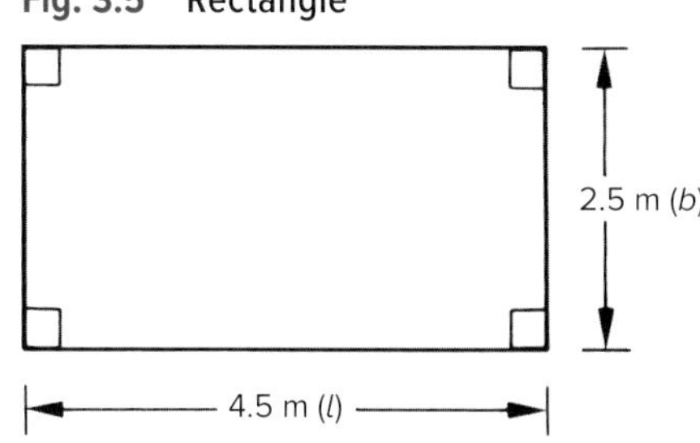

3.3.1.3 L-shaped figure

An L-shaped figure is the extension of a square or rectangular-shaped figure (Fig. 3.6).

$$\text{Perimeter} = 15 \text{ m} + 6 \text{ m} + 6 \text{ m} + 3 \text{ m} + 9 \text{ m} + 9 \text{ m} = 48 \text{ m, or}$$
$$\text{Perimeter} = (2 + \text{L}) + (2 \times \text{W})$$
$$= (2 \times 15) + (2 \times 9)$$
$$= 30 + 18$$
$$= 48 \text{ m.}$$

To calculate the total surface area of the L-shaped figure, you will need to first divide it into two separate areas (note the broken line).

$$\text{Area 1} = 9.0 \times 9.0 = 81 \text{ m}^2$$
$$\text{Area 2} = 6.0 \times 6.0 = 36 \text{ m}^2$$
$$\text{Total area} = 117 \text{ m}^2$$

Fig. 3.6 L-shaped building

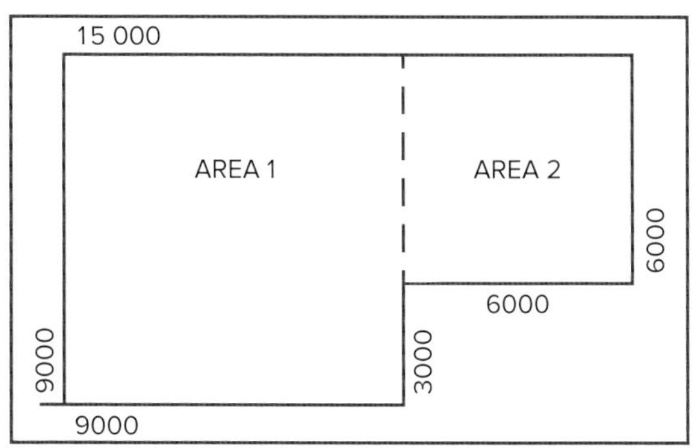

> **TIP** For best practice when working out surface area, it is always easier if you convert the dimensions to metres prior to calculating as area is normally expressed in m^2.

3.3.1.4 Rhombus

A rhombus is a plane figure having four equal sides, with two sets of dissimilar angles (Fig. 3.7). The base of any rhombus figure is the side on which it is shown to be standing–any side can become the base. The perpendicular height (*h*) of a figure is the height of the figure measured at 90° to the base.

$$\begin{aligned}\textbf{Perimeter} &= 4 \times l\\ &= 4 \times 3\\ &= 12\text{ m}\\ \textbf{Area} &= l \times h\\ &= 3.0 \times 2.6\\ &= 7.8\text{ m}^2\end{aligned}$$

Fig. 3.7 Rhombus

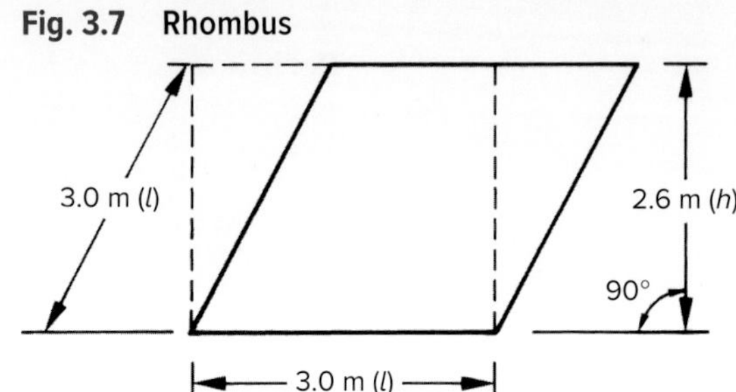

3.3.1.5 Parallelogram (rhomboid)

This is a plane figure having two pairs of equal and parallel sides, with two sets of dissimilar angles (Fig. 3.8).

$$\begin{aligned}\textbf{Perimeter} &= (2 \times l_1) + (2 \times l_2)\\ &= (2 \times 4.5) + (2 \times 2.5)\\ &= 14\\ \textbf{Area} &= l_1 \times h\\ &= 4.5 \times 2.0\\ &= 9\text{ m}^2\end{aligned}$$

Fig. 3.8 Parallelogram

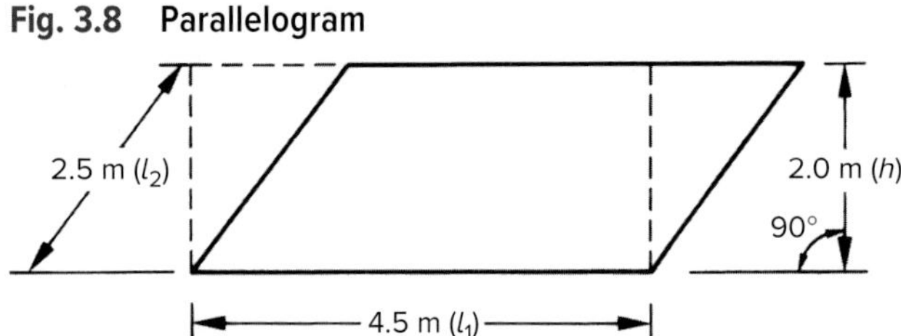

3.3.1.6 Trapezoid or trapezium

This is a plane figure, not a parallelogram, bounded on four sides, two of which are parallel (Fig. 3.9).

$$\begin{aligned}\textbf{Area} &= \frac{l_1 + l_2}{2} \times h\\ &= ((4.6 + 2.8)/2) \times 2.2\\ &= 7.4/2 \times 2.2\\ &= 8.14\text{ m}^2\end{aligned}$$

Fig. 3.9 Trapezium

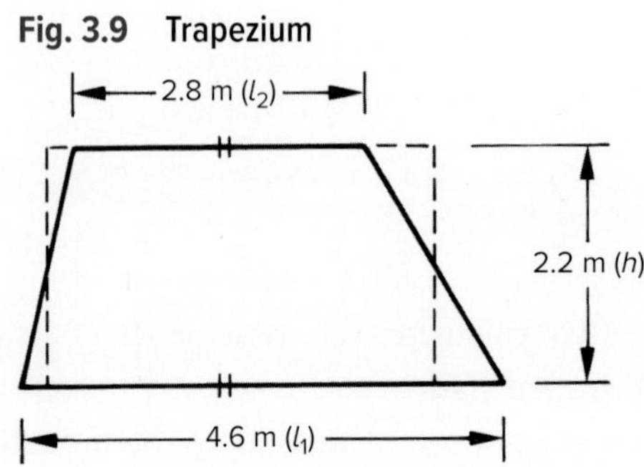

CHECK YOUR UNDERSTANDING

1. Calculate the perimeter of a rectangular block of land that measures 43 m × 26 m.

3.3.2 Triangles

Triangles are plane figures bounded by three straight sides (Fig. 3.10). Triangles have the following properties:

1. Any two sides added together are greater than the third side.
2. The sum of the included angles equals 180°.
3. The exterior angle is equal to the sum of the two opposite interior angles.

Triangles are identified by indicating the corner points with an uppercase letter and the side opposite the corner with the same letter in lowercase. The side on which the triangle stands is the base and the perpendicular height is the distance from the base to the apex, measured at 90°.

$$\textbf{Area} = \frac{\boldsymbol{b \times h}}{\mathbf{2}}$$

$$= \frac{4.6 \times 2.4}{2}$$

$$= 5.52\ \text{m}^2$$

Fig. 3.10 Triangle

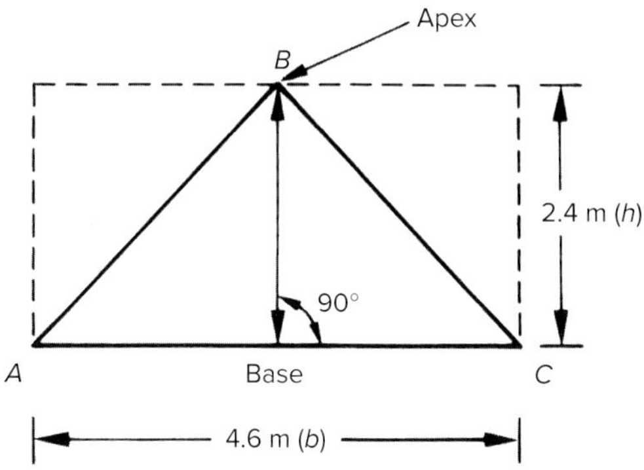

There are six types of triangles, equilateral, isosceles, scalene, right angle, obtuse and acute. In this chapter, we will discuss equilateral, isosceles, scalene and right-angle triangles.

3.3.2.1 Equilateral triangle

An equilateral triangle is a triangle with three equal sides and therefore three equal angles (Fig. 3.11).

Fig. 3.11 Equilateral triangle

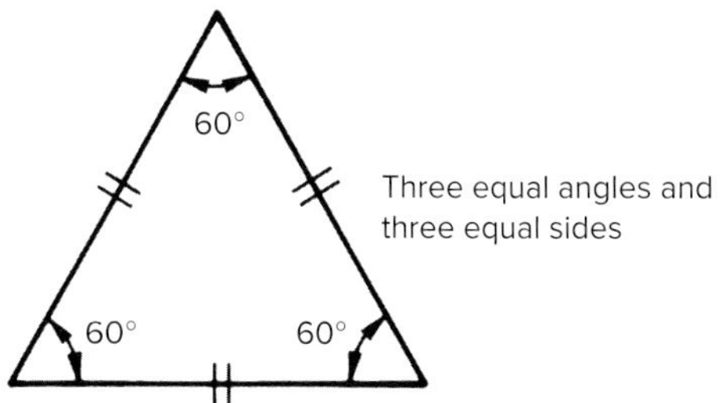

3.3.2.2 Isosceles triangle

An isosceles triangle has two equal sides and therefore two equal angles (Fig. 3.12). A perpendicular line drawn from the apex to the base will bisect the base.

Fig. 3.12 Isosceles triangle

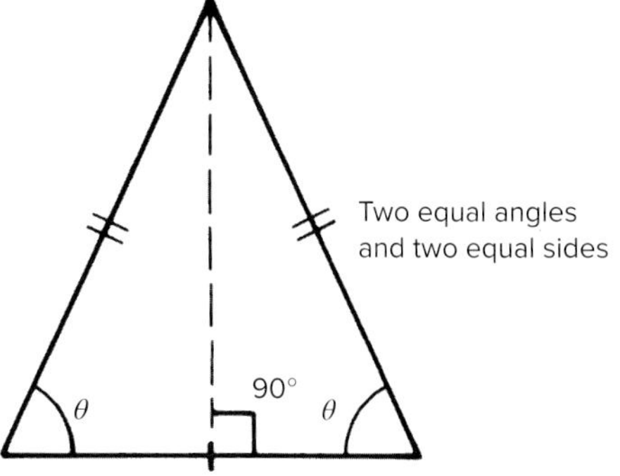

3.3.2.3 Scalene triangle

A scalene triangle has no equal sides and no equal angles, but each single angle is less than 90° (Fig. 3.13).

Fig. 3.13 Scalene triangle

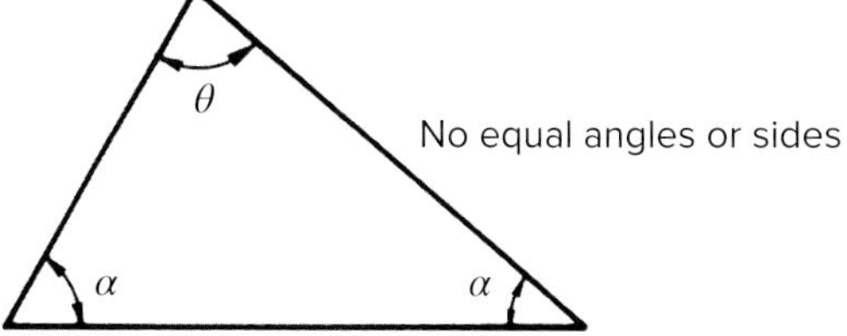

3.3.3 Right-angled triangles

These are triangles in which one of the included angles is a right angle (90°) (Fig. 3.14). The side opposite the right angle is called the *hypotenuse*.

The right-angled triangle is most valuable for solving various problems that confront the carpenter, such as setting out right angles and calculating the lengths of roof rafters using *Pythagoras' theorem*. This theorem states: 'In a right-angled triangle, the square on the hypotenuse equals the sum of the squares on the other two sides.'

The most readily adaptable right-angled triangle is one with sides in the proportions 3 to 4 to 5 (Fig. 3.15). We can prove the theorem in the following way:

$a = 3$

$b = 4$

$c = 5$

$$c^2 = a^2 + b^2$$
$$25 = 9 + 16$$

Using this theorem, if any two sides of a right-angled triangle are known, the third can be calculated (Fig. 3.16):

$$c^2 = a^2 + b^2$$
$$c^2 = 2.3^2 + 6.4^2$$
$$\text{Therefore } c = \sqrt{(5.29 + 40.96)}$$
$$= \sqrt{46.25}$$
$$= 6.8 \text{ m}$$

Fig. 3.14 Right-angled triangle

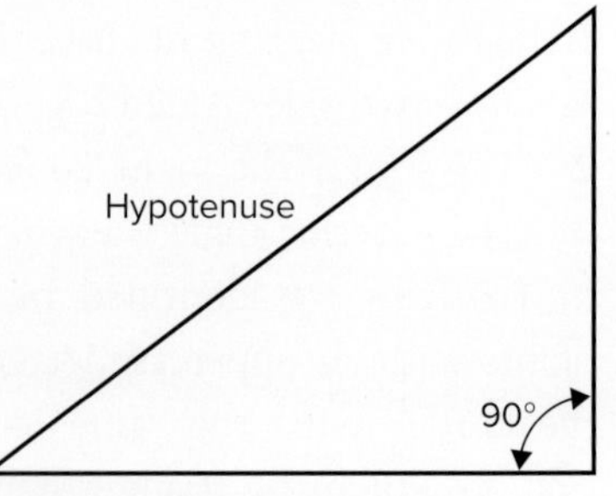

Fig. 3.15 3, 4, 5 triangle

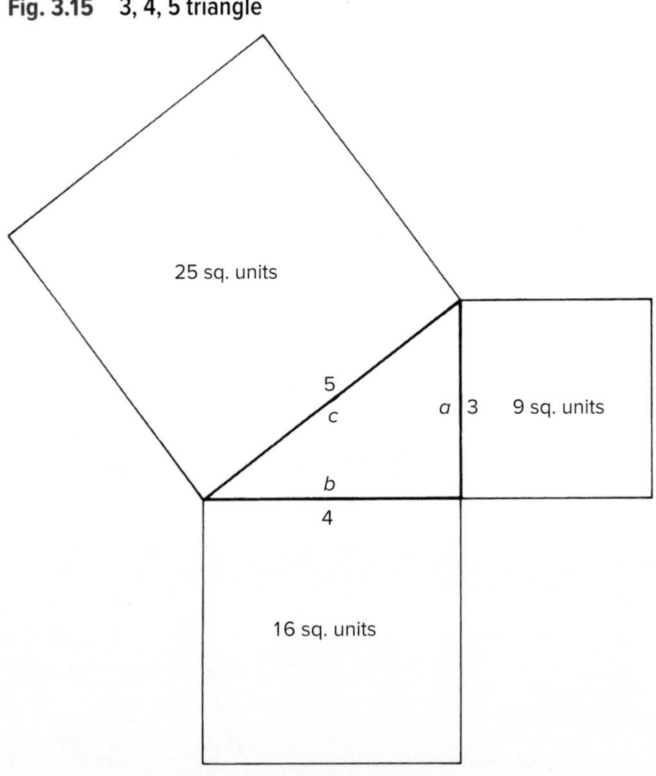

Fig. 3.16 Triangle problem

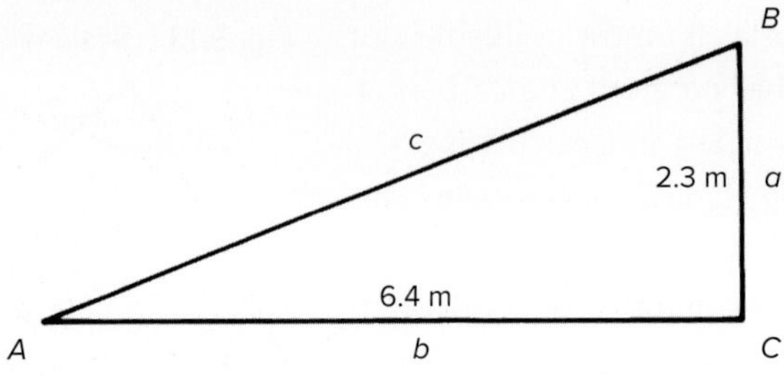

3.3.4 Circles

Circles are plane figures, bounded by a line traced out by a point moving at a fixed constant distance from another given fixed point (Fig. 3.17).

- **Circumference** of the circle is the length of the line traced out by the moving point, to again reach its starting point (the distance around the outside of the circle).
- **Centre** is the fixed point.
- **Diameter (symbol *d*)** is a straight line drawn through the centre and terminated at each end by the circumference.
- **Radius (symbol *r*)** is a straight line drawn from the centre of the circle to the circumference, its length being half the diameter.

Fig. 3.17 Circle

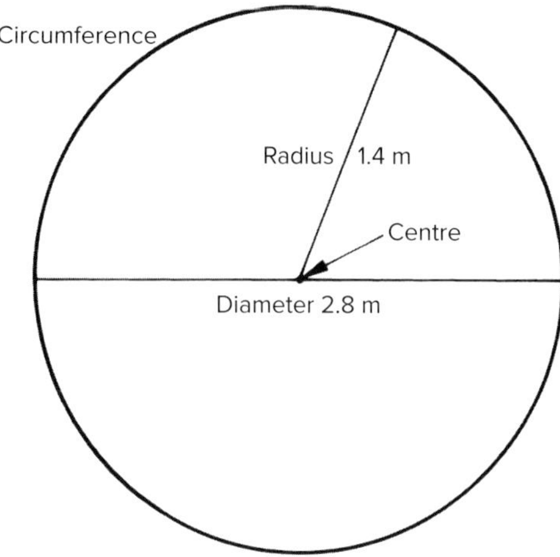

In any circle, there is always a fixed ratio between the length of the circumference and the diameter. This ratio is termed pi (π) and has a value of 3.142 (to three decimal places). That is, the circumference of a circle is always 3.142 times the length of the diameter.

If the diameter is 2.8 m, then:

$$\textbf{Circumference of circle} = \pi \times d$$
$$= 3.142 \times 2.8$$
$$= 8.797 \text{ m}$$

Some calculators have a key π, which when pressed will automatically display the value of π. In this problem, the solution would be:

[C] [π] [×] [2] [.] [8] [=]

This displays 8.7964

If the key π is not available, then the value of π can of course be simply entered in the usual way.

The area of a circle can be calculated as follows:

$$\textbf{Area of circle} = \pi r^2$$
$$r = 1.4 \text{ m}$$
$$\text{Area} = 3.142 \times 1.4 \times 1.4 \text{ m}^2$$
$$= 6.158 \text{ m}^2$$

3.3.4.1 Parts of a circle

As shown in Figure 3.18, the parts of a circle are as follows:

- **Arc** is any part of the circumference.
- **Semi-circle** is half a circle bounded by half the circumference and the diameter (*d*).
- **Quadrant** is one-quarter of a circle bounded by two radii at 90°, and one-quarter of the circumference.
- **Sector** is part of a circle bounded by two radii and the arc between them.

Fig. 3.18 Parts of a circle

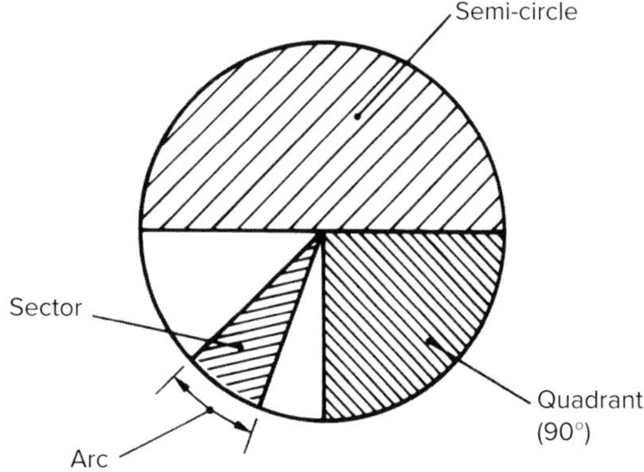

3.3.4.2 Degree

When a radius of a circle rotates through 1/360th part of the circumference it forms an angle and the angle is termed 1°. Therefore:

- ¼ circle = 90°, or one right angle
- ½ circle = 180°, or two right angles
- Full circle = 360°, or four right angles, as shown in Figure. 3.19.

Parts of a degree are called minutes (') and seconds ("), where one degree equals 60' and one minute equals 60". Degrees are the usual way to measure angles.

Fig. 3.19 Degree increments of the circle

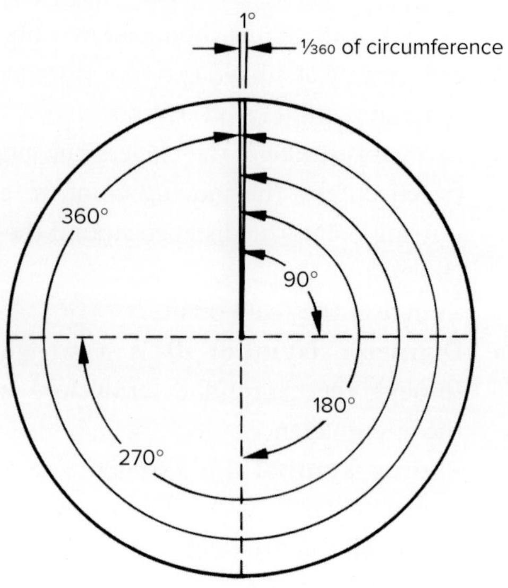

3.3.4.3 Chord

A *chord* is a straight line joining any two points on the circumference other than the diameter; a *segment* is the part of a circle bounded by a chord and its arc (Fig. 3.20).

Fig. 3.20 Chord, segment and radius

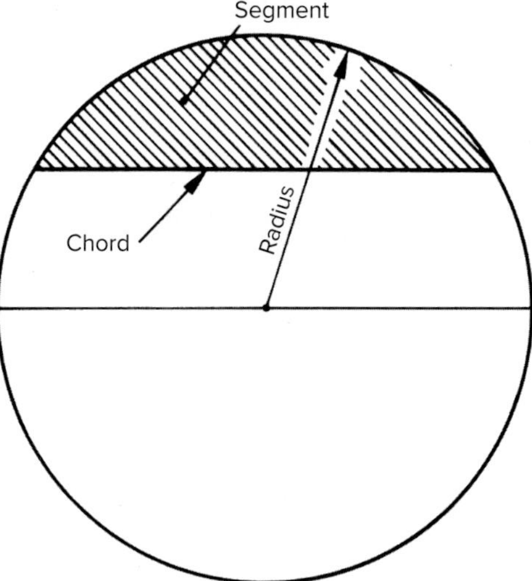

3.3.5 Polygons

There are many polygons (i.e. plane figures having more than four sides), but the two most likely to be encountered are the hexagon and the octagon.

3.3.5.1 Hexagon

The regular hexagon is a plane figure bounded by six equal sides, each side at 120° to the one adjacent (Fig. 3.21). A hexagon can be thought of as six equilateral triangles grouped around a central point. The perimeter of a hexagon is equal to the length of a side multiplied by six.

The area of a hexagon is equal to the area of the equilateral triangle formed by joining two adjacent corners to the centre of the figure, multiplied by six. Therefore, using the formula for the area of triangles, the area of the regular hexagon will equal the shortest distance from the perimeter to the centre, multiplied by half of one side, multiplied by six.

$$\text{Area} = \frac{l}{2} \times \text{shortest distance to centre} \times 6$$
$$= l \times \text{shortest distance to centre} \times 3$$

If the distance to the centre is not known, it can be calculated by trigonometrical methods, and will be found

Fig. 3.21 Hexagon

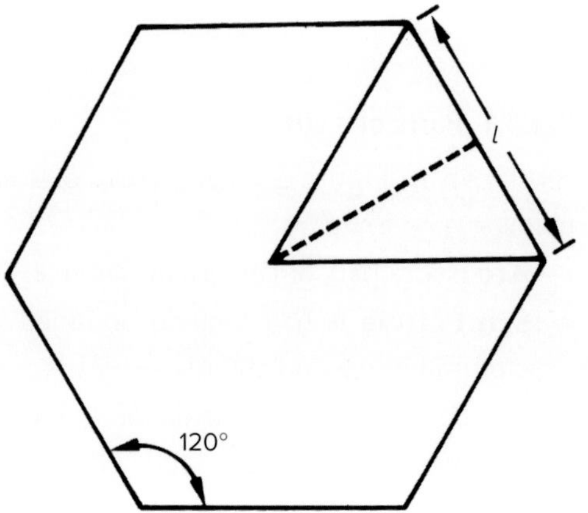

to be a constant ratio with the sides of the regular hexagon. Using this method, the area of a hexagon will be:

$$\textbf{Area} = l^2 \times 2.598$$

3.3.5.2 Octagon

The regular octagon is a plane figure bounded by eight equal sides, each side at 135° to the one adjacent (Fig. 3.22). The perimeter of a regular octagon is calculated by taking the length of one side and multiplying by eight, while the area of a regular octagon can be calculated by two methods:

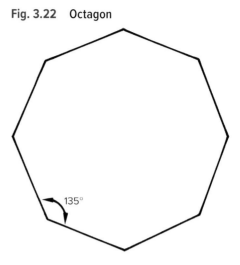

Fig. 3.22 Octagon

135°

1. A similar method to the hexagon can be used, where the area of the isosceles triangle bound by two adjacent corners and the centre of the figure can be calculated and multiplied by eight. This can be represented as:

$$\text{Area} = \frac{l}{2} \times \text{shortest distance to centre} \times 8$$
$$= l \times \text{shortest distance to centre} \times 4$$

   Once again there is a constant relationship between the length of a side and the distance to the centre of a regular octagon, giving the simple formula:

$$\textbf{Area} = l^2 \times 4.8284$$

2. Another method is to take the area of the square that would just encompass the octagon and to subtract the area of the four small triangles at each corner of the square. This will be useful if calculating areas of non-regular octagons.

CHECK YOUR UNDERSTANDING

1. What is the area of a rectangular figure 4.6 m × 450 mm? Show the formula used and all workings.
2. Calculate the area of a circle with a diameter of 8.360 m. Show the formula used and all workings.

3.4 Solid figures

Solid figures are those that occupy space, and the amount of space is called their 'volume'. Solid figures have three dimensions–length, breadth and height.

3.4.1 Cube

This is a solid figure contained by six square faces. It has length, width and height, which are all equal (Fig. 3.23).

$$\textbf{volume} = l \times l \times l, \textbf{ or } l^3$$
$$= 0.5 \times 0.5 \times 0.5$$
$$= 0.125 \text{ m}^3$$

Fig. 3.23 Cube

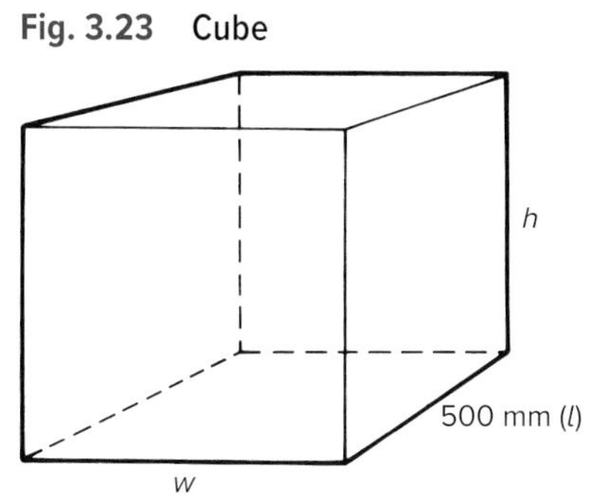

3.4.2 Prism

A prism is a solid figure with two equal and parallel ends and straight parallel sides, which are parallelograms. Prisms are named according to the shape of the end faces.

The *axis* is an imaginary line joining the centres of the end faces. When the axis is at right angles to the end faces, it is termed a *rectangular or right prism* (Fig. 3.24). Otherwise they are called *oblique prisms* (Fig. 3.25).

$$\textbf{volume of rectangular prism} = \boldsymbol{l \times b \times h}$$
$$= 0.7 \times 0.5 \times 2.5$$
$$= 0.875\ \text{m}^3$$

Fig. 3.24 Rectangular or right prism

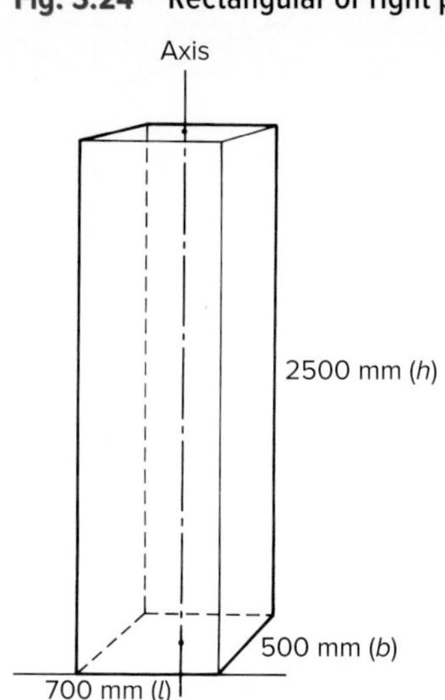

Fig. 3.25 Oblique prism

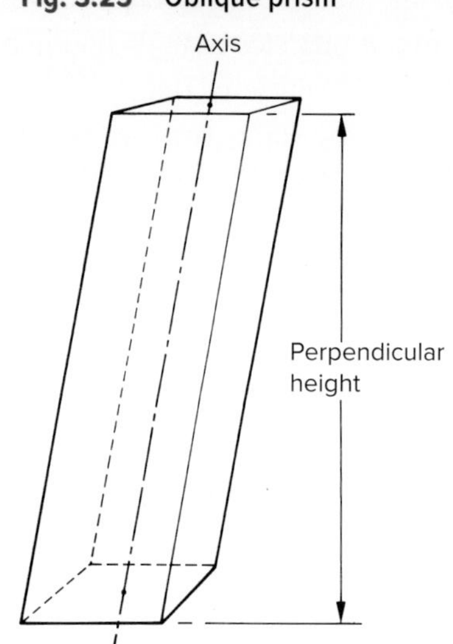

Prisms are often found lying on their side, such as in a concrete beam, etc. (e.g. Fig. 3.26) and the formula for the volume can be modified to:

$$\text{volume} = \text{area of end} \times l$$
$$\textbf{volume of square prism} = \boldsymbol{w^2 \times l}$$
$$= 0.3 \times 0.3 \times 4.6$$
$$= 0.414\ \text{m}^3$$
$$\textbf{volume of triangular prism} = \boldsymbol{(b \times Ph)/2 \times l}$$
$$= (1.5 \times 0.6)/2 \times 3.8$$
$$= 1.71\ \text{m}^3$$

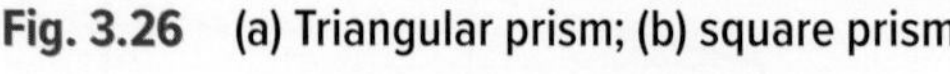

Fig. 3.26 (a) Triangular prism; (b) square prism

(a)

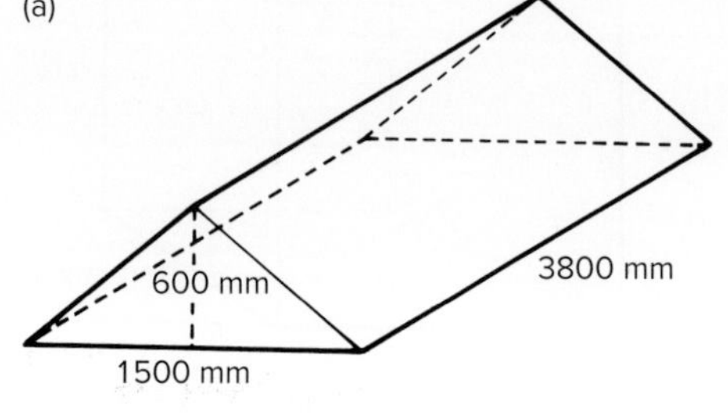

(b)

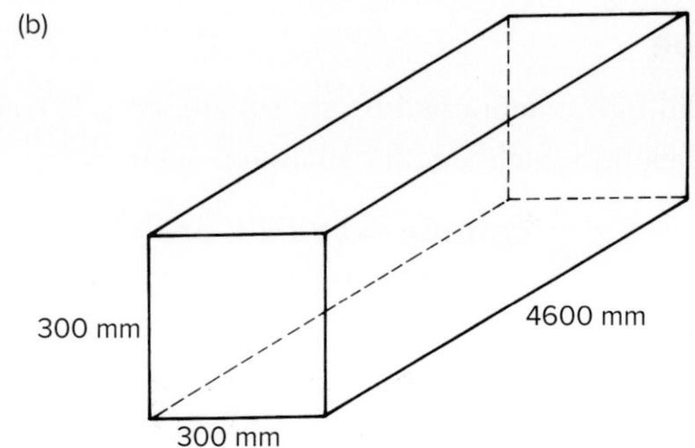

3.4.3 Cylinder

A cylinder is a solid figure, having two circular ends and straight parallel sides (Fig. 3.27).

$$\textbf{volume} = \textbf{area of end} \times \boldsymbol{h}$$
$$= \boldsymbol{\pi r^2 \times h}$$

For example, calculate the volume of a cylinder with a diameter of 800 and a height of 3600.

$$\textbf{volume} = \boldsymbol{\pi r^2 \times h}$$
$$= 3.142 \times 0.4 \times 0.4 \times 3.6$$
$$= 1.81\ \text{m}^3$$

Fig. 3.27 Cylinder

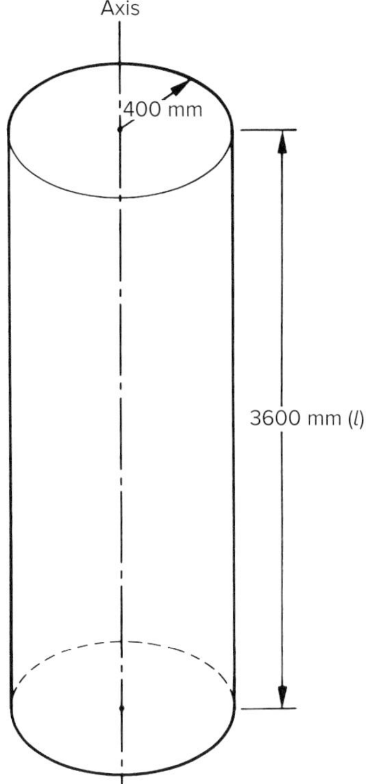

3.4.4 Cone

A cone is a solid figure with a circular base and straight sloping sides meeting at an apex (Fig. 3.28). The *generator* is a straight line joining any point on the circumference with the apex. The *slant height* is the length of the side along the generator, and the *perpendicular height* is the distance from the base to the apex measured along the axis.

The volume of a cone is equal to the area of the base times one-third of the perpendicular height (*Ph*). For example, the volume of a cone with a radius of 200 mm and a perpendicular height of 800 mm is:

$$\textbf{volume} = \boldsymbol{\pi r^2 \times 1/3 \times Ph}$$
$$= 3.142 \times 0.2 \times 0.2 \times 0.8/3$$
$$= 0.0335\ \text{m}^3$$

Fig. 3.28 Cone

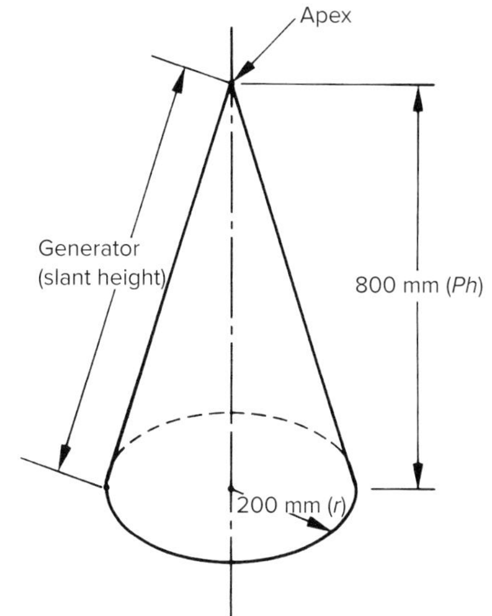

CHECK YOUR UNDERSTANDING

1. A concrete path measures 22 m in length × 1600 mm in width × 100 mm in depth.
   a. Calculate how many cubic metres of concrete will be required to pour that path (show the correct formula and workings).
   b. What would be the total cost of the pour if the cost of ready-mix concrete is $165.00 per cubic metre? Note: when ordering ready-mix concrete, the company will supply only in increments of 0.2 cubic metres. Therefore, when placing an order and costing, you will need to round up your order to the next 0.2 of a cubic metre.

3.4.5 Pyramid

A pyramid is a solid figure standing on a geometrical base with straight sides sloping to an apex. Pyramids are named according to the shape of their base–square (Fig. 3.29), rectangular, hexagonal, etc. The volume of a pyramid is equal to the area of the base × 1/3 × *Ph*. For example:

volume of a pyramid = area of base × 1/3 × *Ph*

$$= (0.3 \times 0.3 \times 1.2)/3$$

$$= 0.036\ \text{m}^3$$

Fig. 3.29 Square pyramid

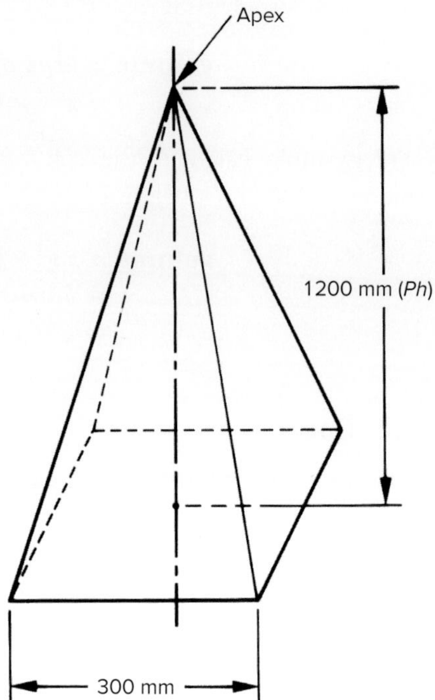

3.4.6 Weight/mass

Weight/mass is important for a number of activities in the building industry. For instance, it needs to be known when calculating roof members for a timber roof, as the mass of the roof covering needs to be determined so that the appropriate size of the member can be calculated.

It would also need to be known when picking up materials such as crushed rock from a garden supplies centre using a trailer. For loose materials such as crushed rock, sand, concrete etc., the volume is generally calculated and expressed in cubic metres (as mentioned earlier in this chapter). However, the weight also needs to be known so that the trailer is not overloaded. All trailers, utes and trucks have a maximum weight rating, and it is against the law to exceed that rating–it is also dangerous. The weight of loose material can vary depending on its water content. For instance, 1 square metre of sand can have different weights depending on how much moisture is present, although the volume doesn't vary. The company selling the product will generally know the weight of a cubic metre of that material, so it is important to ask the question prior to loading.

Alternatively, a weigh bridge could be used to work out how much the loaded material weighs. The trailer would need to be first weighed empty than again once loaded. However, this is not really practical and would be time-consuming and costly. The weight of the products would be stated in either kilograms or fractions of a tonne.

3.4.7 Tolerances

Knowing what tolerance is being worked to will ensure that the work being carried out is completed to the desired standard. For example, when setting out a rectangular house, it is important to check that it is square once set out. So, the two diagonal measurements would be measured to see if they were the same. A difference of 5 mm in the two diagonal measurements would be deemed acceptable. But if the difference was 50 mm, this would be considered to be outside the acceptable tolerance of error. The house would therefore need to be set out again.

Tolerances can also be used when teaching apprentices about the expected quality when completing certain tasks. For example, as an apprentice I was taught the acceptable gap in a mitered joint when installing the architraves around a door, was the thickness of a cigarette paper. This was a way of gauging what standard of work was acceptable rather than providing a measurement.

3.5 Calculating and ordering materials

Table 3.2 is an example of how a carpenter would record what materials are required for a particular project. It's important that the notes on the table are clear and concise so that the company supplying the materials understands what the carpenter wants, but also so that at a later date the carpenter does not have to rely on memory alone for the finer details of the project and does not need to recall what the original thinking was when working out the quantities.

The table below can be adapted to use for ordering any construction materials. The key is to be clear with the information being recorded and to include enough information so that people can make sense of it.

Fig. 3.30 Timber arbour with seat

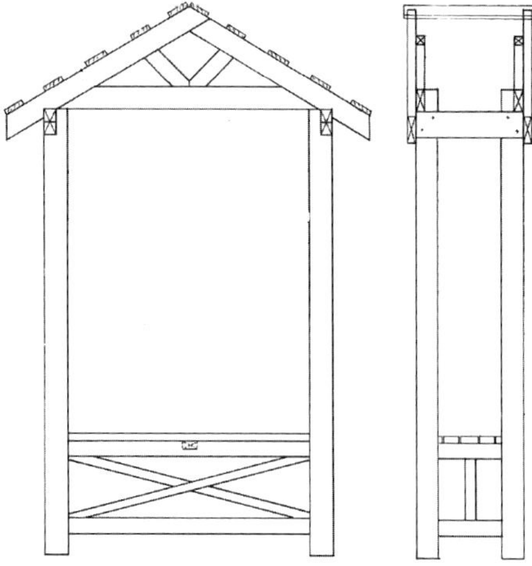

Courtesy of WorldSkills Australia

Table 3.2 Arbour material list

Item	Part	Timber species	Quantity required	Length	Width	Thickness	Remarks
1	Posts	Oregon	4	2.4 m	90 mm	90 mm	
2	Roof members	Hoop pine	2	2.4 m	90 mm	45 mm	Rafters
3			1	2.1 m	70 mm	35 mm	Battens
4			2	1.5 m	90 mm	45 mm	Fascia beam
5	Seat members	Hoop pine	1	1.8 m	70 mm	45 mm	Top and bottom back rail
6			1	2.4 m	70 mm	45 mm	Top and bottom front rail
7			1	2.4 m	70 mm	45 mm	Seat supports
8			1	1.8 m	70 mm	45 mm	Side
9	Seat slats	Jarrah	3	2.7 m	90 mm	19 mm	
10	Seat brace	Hoop pine	1	2.4 m	70 mm	35 mm	

Courtesy of WorldSkills Australia

WORKPLACE SCENARIO

When I was a third-year apprentice, we were working on my boss's house. One day we were installing the stumps for the subfloor. The concrete truck was there and we were barrowing the concrete to each hole ready for the stumps to go in. It was a blistering hot day, about 36 degrees, so the concrete was going off quickly and we were working as fast as possible to get the concrete poured and the stumps in place. We were also looking forward to finishing for the day, but as we finished placing concrete into the last hole, my boss yelled out that the next concrete truck would be arriving in 15 minutes. 'Oh?' we said. 'Another truck?' Guess who had doubled the concrete order by mistake! There was no way of returning it, so needless to say, by 9 pm that night we had also poured blind concrete for the patio, and all of the external steps for the house had been poured, too!

End of chapter activity

Activity 1:

Use the table below to record the timber required to build the walls of the pantry in the plans located in the appendix. Order lengths to the nearest 300 mm.

Wall height: 2.4 m

Timber size: 90 mm x 45 mm

1 row of noggins

450 mm stud spacings

Jamb stud: 90 x 45

Common stud: 90 x 35

Plates: 90 x 45

How many do we need of each?

- Top plates
- Bottom plates
- Jamb studs
- Common studs
- Sill trimmer
- Head trimmer
- Noggins

Table 3.3 Recording the number of timbers required

Item	Part	Timber species	Quantity required	Length	Width	Thickness	Remarks

Tradie talk—Measure twice and cut once

Chapter 4

Trade drawings

Learning Objectives

LO 4.1 Identify types of working drawings and their functions

LO 4.2 Recognise commonly used symbols and abbreviations

LO 4.3 Locate and identify key features on building plans

LO 4.4 Use drawing programs and tools

Introduction

The ideas and construction methods used to produce buildings are set down in the form of drawings and building specifications. These allow different trades to understand the work that needs to be done without referring to one specific person. All architectural drawings and specifications are produced using a specific language. It is necessary to understand how the information is conveyed by architects and draughtspersons in order to read the drawings and specifications and then build the planned structure as intended by the designer.

In this chapter you will learn how to identify and read basic building plans and documents. This will include elements such as the different plans, elevations and sectional views associated with working drawings in the building trade.

You will learn how to identify and read aspects such as title panels, specifications, amendments, abbreviations, symbols and scales. You will then produce some working sketches and drawings of your own using the drawing equipment discussed in the chapter.

4.1 Identifying types of working drawings and their functions

There are basically two levels of drawings:

1. **Design drawings**, where the ideas on relationships, sizes and stylistic treatment of rooms are explored. These drawings allow the architect and the client to agree about what is to be done.

2. **Building or working drawings**, which set down the finalised design. These, together with the job specification document, provide all of the information the builder and tradesperson need for construction work to take place.

The information conveyed in all types of architectural drawings is standardised so that people in different places and from different backgrounds can understand the construction of the building.

The basic form of drawing is called *orthographic* projection and consists of a series of related views of the building, which, when looked at in conjunction with each other, allow a complete understanding of the building (Fig. 4.1–4.4).

The three main types of 'view' of a building are the plan, elevation and section.

4.1.1 The plan view

The plan view shows the layout or arrangement of parts of a building when seen from above (bird's eye view) such as the rooms in the building. There are many types of plan view, for example:

1. site plan
2. floor plan
3. roof plan
4. footing plan.

4.1.2 The elevation view

This is a view of a side of a building, facing it squarely. For a building with four sides, the architectural drawings will show four elevation views. To avoid confusion between the elevation views, we name each elevation according to what direction it faces on the compass, for example, the south elevation.

Fig. 4.1 Naming views on orthographic drawing

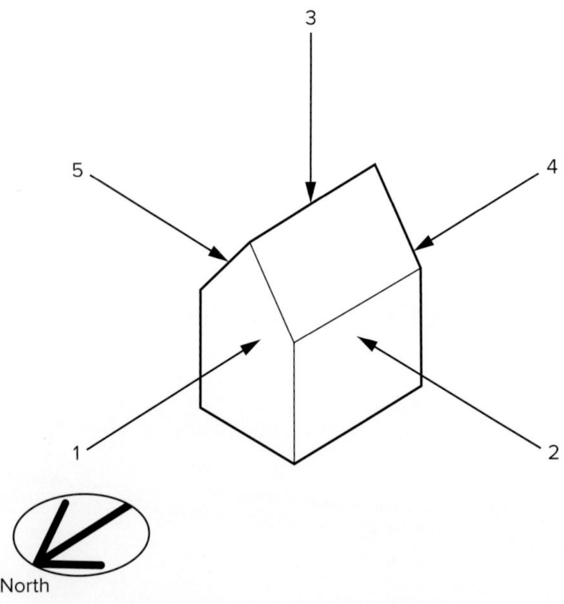

Naming or numbering of views	
View 1 or Elevation 1	North elevation
View 2 or Elevation 2	West elevation
View 3 or Plan	Plan
View 4 or Elevation 3	South elevation
View 5 or Elevation 4	East elevation

4.1.3 The sectional view

This is like a vertical plan view; just as a floor plan is a view of the walls as if the roof of a building had been removed, so a section is a side view as if an exterior wall had been removed. More correctly it is a 'cut' through a building at a certain point. It shows heights and internal structures that are not evident in the plan or elevation views.

4.1.4 Understanding scaling

The architectural information comes in many varied forms, and is set out on single or multiple sheets of paper, often referred to as the 'drawings' or 'plans' of the building.

The process of drawing large objects such as buildings or parts of buildings to a proportionate size that can be fitted on to paper is called *drawing to scale*. The scale is, in effect, a reduction of the full-sized object to a small picture–like a reproduction that

Fig. 4.2 Orthographic drawing—concept

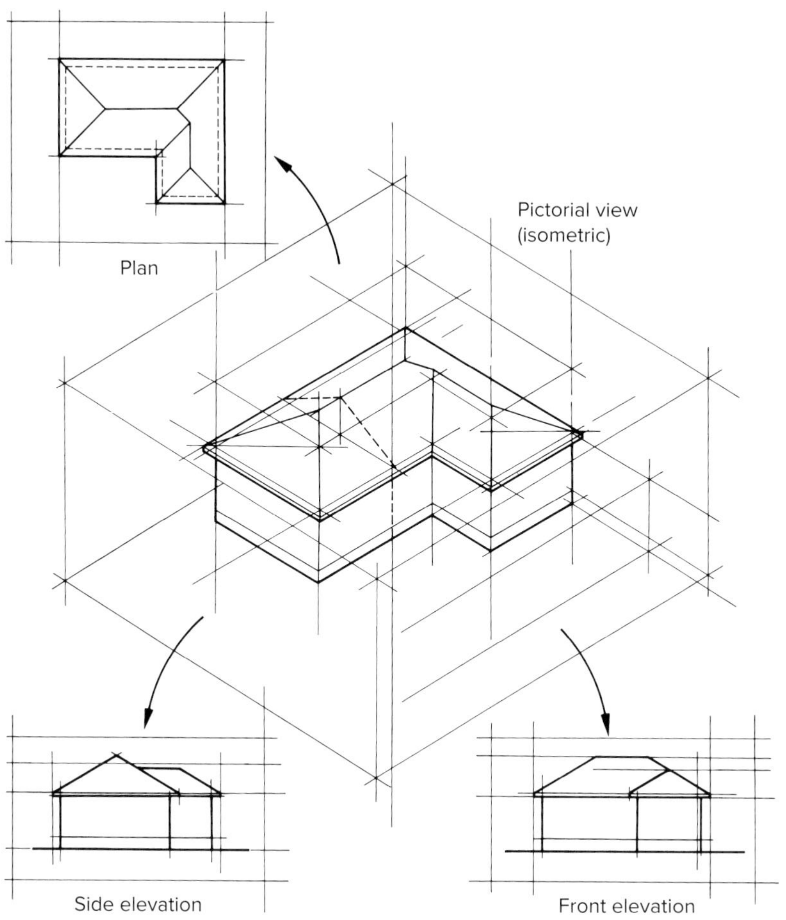

Fig. 4.3 Concept of the section

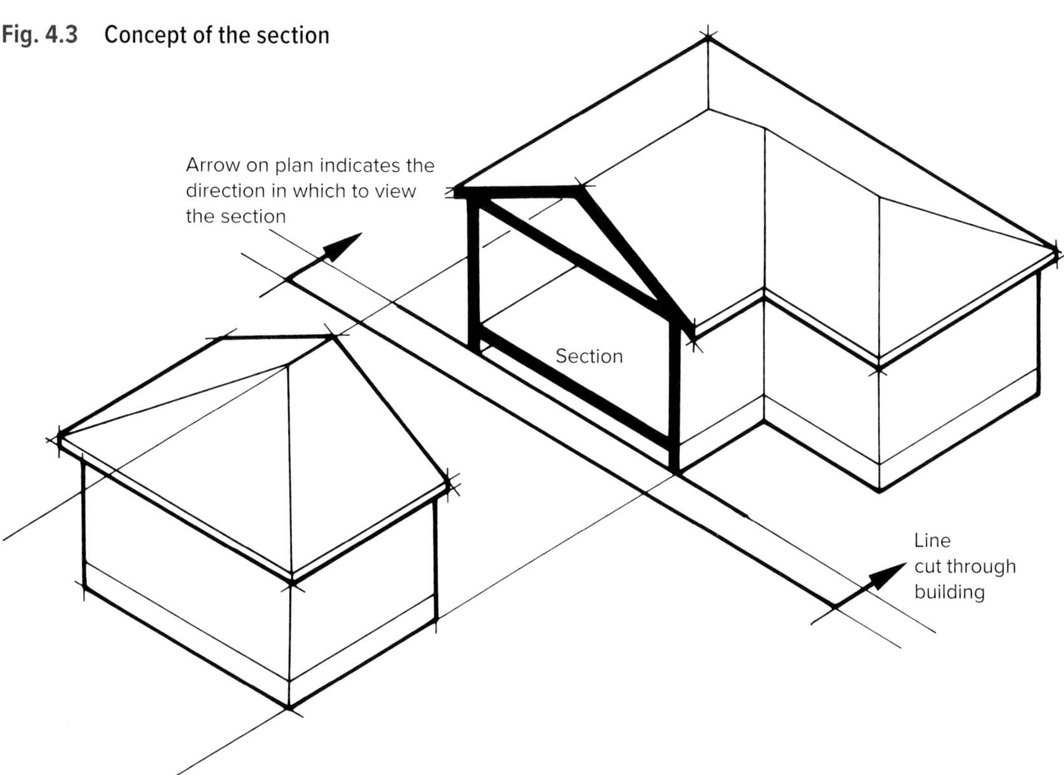

Fig. 4.4 Arrangement of orthographic views on a drawing

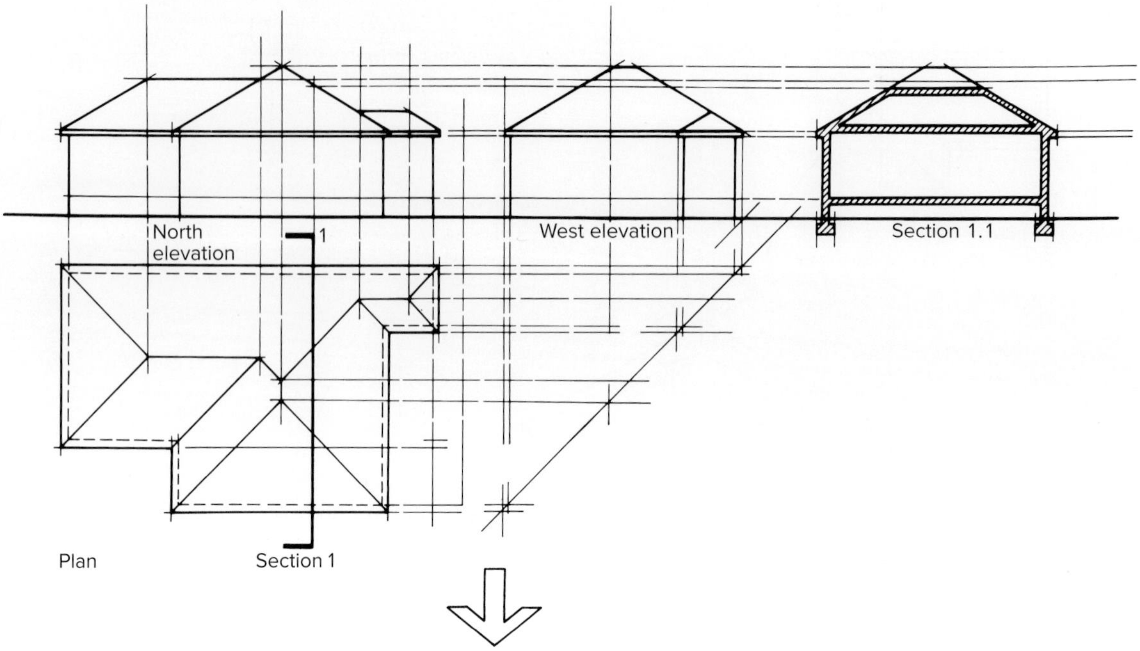

Fig. 4.5 A 350 mm wide strip footing shown at scale 1:20 by scale rule

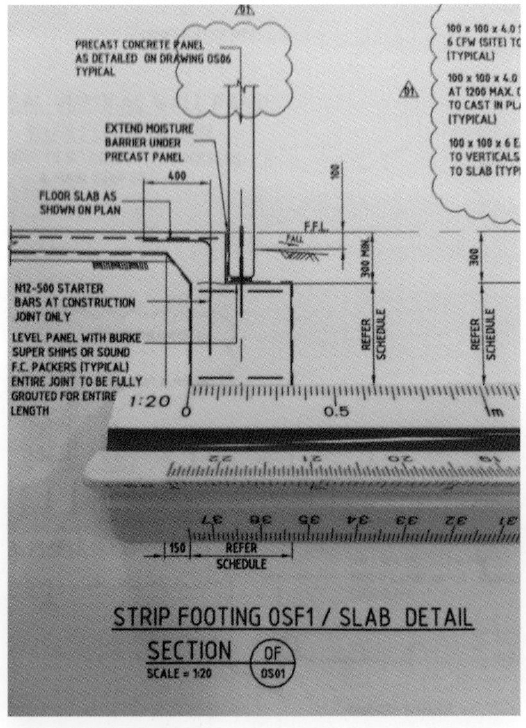

Courtesy of Alister Ford

exactly represents the real thing. For example, instead of drawing a window that is 900 mm long, it is drawn at a reduced scale, say, 1/50th of its full size. This enables all of the orthographic views to be reduced and placed on handy-sized sheets.

The most common scales used in construction practice are:

- site plans–1:500, 1:200, 1:100
- plan views–1:200, 1:100, 1:50, 1:20
- elevations–1:200, 1:100, 1:50, 1:20
- sections–1:200, 1:100, 1:50, 1:20
- construction details–1:10, 1:5, 1:2, 1:1 (full size).

A simple way to understand how scaling works is to measure between the dimension lines and multiply by the scale to which the plan has been drawn (Fig. 4.5). This may need to happen if the dimension has been missed and not recorded.

If plans need to be produced from preliminary sketches that have dimensions jotted in, simply divide the measurement by the scale to which the plan will be drawn to get the correct measurement for drawing.

TIP The measurements on architectural working drawings will always take precedence over a scaled measurement.

Put simply, true length divided by scale = plan length. Or plan length multiplied by scale = true length.

For domestic constructions, the most common scales of drawings are 1:100 and 1:50 (Fig. 4.6). This is because domestic construction practice has become relatively standardised and there is no need to show details that are common knowledge throughout the industry. In order to represent this common construction practice at reduced scales, a symbolised code for building elements (e.g. windows and doors) has evolved. A sample of these conventions displaying the symbols for materials is shown in Figure. 4.7.

Fig. 4.6 Reduction scales

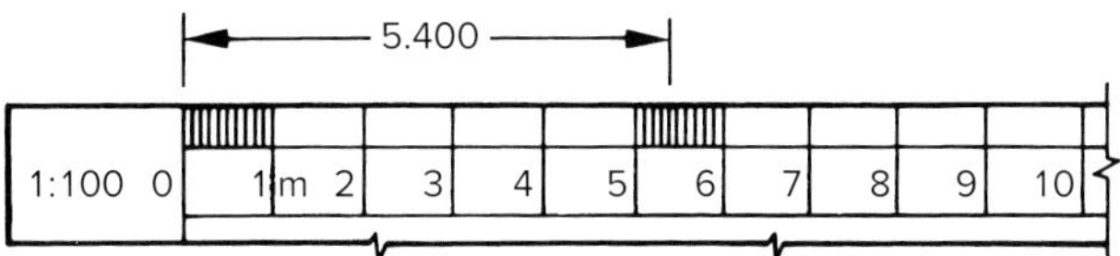

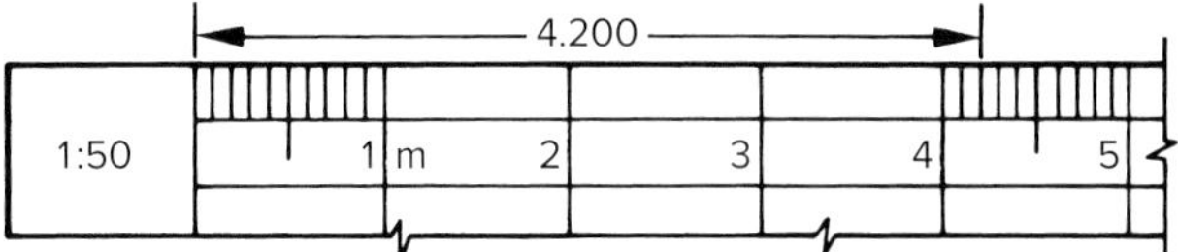

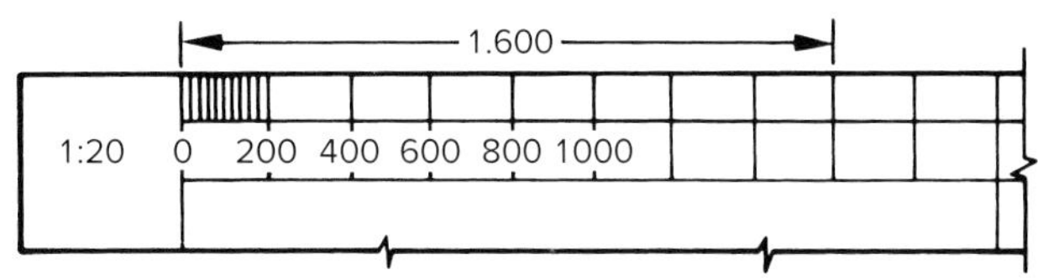

These symbols in fact comprise the 'language' of building drawing and it is necessary for carpenters to familiarise themselves with them in order to interpret construction drawings. The symbols appear on plans, elevations and in sections at various scales. On some types of drawing (notably restorations and extensions to existing buildings), the symbols are coloured in order to distinguish new work from the existing structure.

WORKPLACE SCENARIO

I can recall a day early in my career as a qualified carpenter and joiner. I had finished my apprenticeship in a carpentry/joinery workshop environment and had now commenced a new full-time job with a major construction company in Melbourne.

I was sent to a building site in the city and when I arrived, I reported to my foreman. This site was the construction of a multistory building well in excess of 30 floors high. The foreman showed me around and introduced me to some of my new work colleagues then told me which job he needed me to do.

The work is comprised of some framing and formwork that needed to be set up on one of the floors so that we could then proceed further with that job. I asked for some plans and specifications, so off we went to the shed that was specifically allocated to housing all of the building plans and documents.

When we walked into the shed there were racks and benches along each wall totally covered with folders full of plans and documents. I had no idea where to look. I had never seen anything like this before. I was totally overwhelmed with the enormity of it all. I told the foreman that I had no idea what to do in regards to obtaining plans for the job at hand, so he showed me how to reference all of the plans and specifications.

After that I spent a lot of my spare time in the plan shed teaching myself how everything worked on a major job site such as this one.

4.1.5 Specifications

A specification is the extension of the working drawings and contains additional written instructions that complement the working drawings. While the working drawings contain written details, their main purpose is to show information relating to the shape, location and dimensions of the building.

Fig. 4.7 Architectural information—symbols for materials

Material	General location drawings (section) Scale: 1:50 or less	Large scale drawings (section)
Brickwork	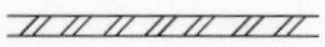	
Cement render, plaster	Too fine to hatch	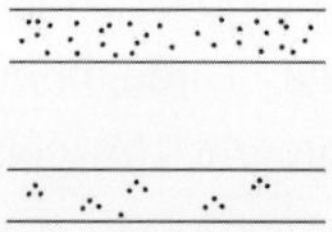
Concrete	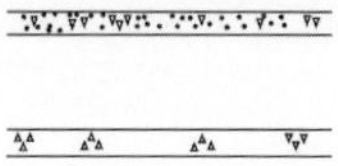	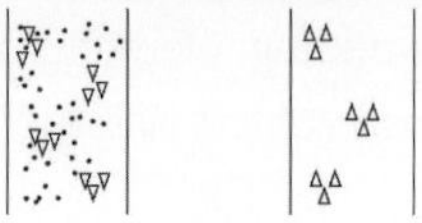
Concrete block		
Cut stone masonry		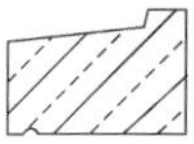
Earth		
Fill	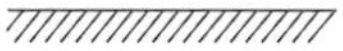	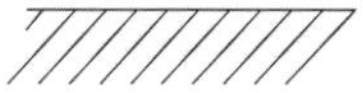
Glass	Too fine to hatch	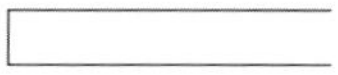
Hardcore		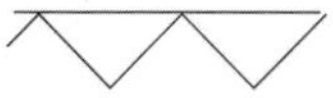
Insulation		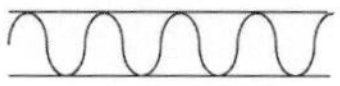
Partition block	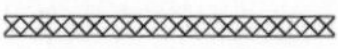	
Rock		
Structural steel	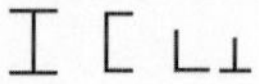	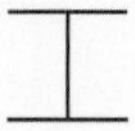
Stud walls	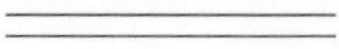 (Grey shading)	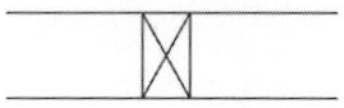
Timber	Usually too fine to hatch	Sawn Dressed 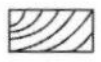

Additional information on the quality of the work, materials, fabrication sizes, colour and finishes of the building project are contained within the written specification document. Information that is not found in one document should in most instances be found within the other.

Specifications for an extremely large construction such as a multistorey building may be provided in book form, or even in multiple volumes. For smaller construction jobs there may only be a smaller booklet. Sometimes the specifications may be included on the working drawings.

A standard specification will include information for the builder and various tradespeople working on the specific project and is provided under headings such as concretor, bricklayer, carpenter, plumber, electrician, etc.

When the specification is read in conjunction with the working drawings, sufficient information will be available for the builder and contractors to submit accurate tenders, sign contracts and engage in the construction of the building.

4.1.6 Amendments

If a change is decided upon during the construction of the building, the plans and specifications will need to be changed and approved. This is called an amendment, or a minor variation, depending on the size and scope of the change (Fig 4.8). Amendments need to be shown on the most recent set of plans,

Fig. 4.8 An amended plan referring to the addendum for more detail

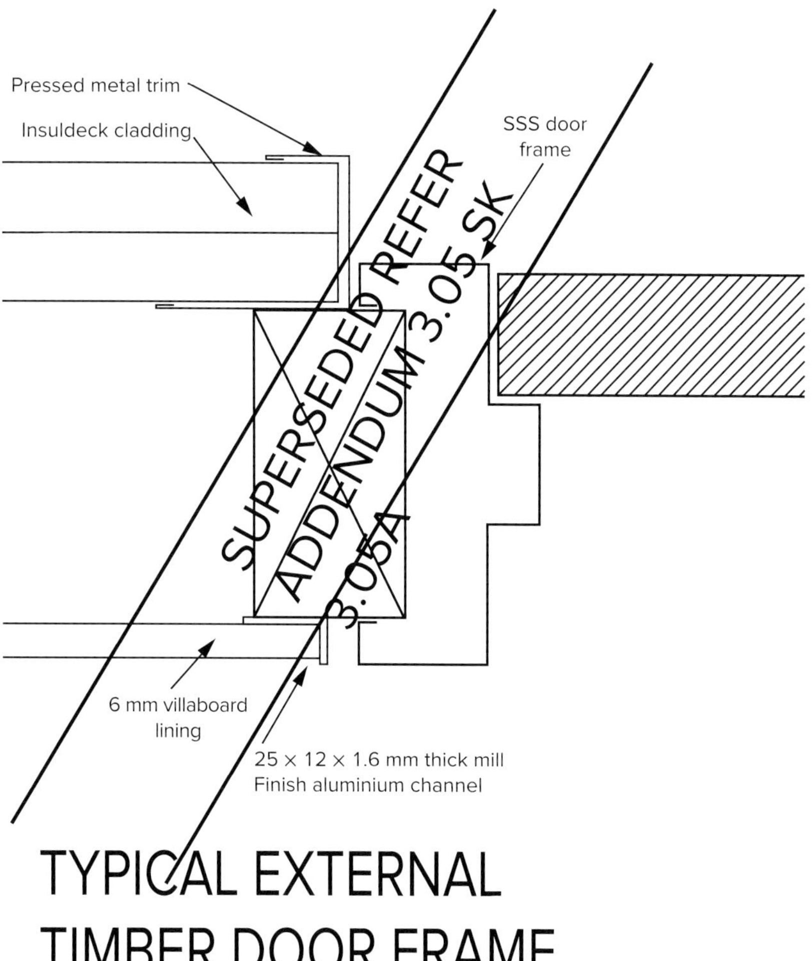

Courtesy of Alister Ford

and must also be signed by the client and builder so there is no chance of confusion between any of the parties involved with the build.

Amendments can be shown on the working drawings and also noted in the plan title panels. If there are changes to the specifications of the finished job, the amendment may need to be noted in the job specifications as well.

Most amendments will occur when the client decides to make some changes or if any specific building materials or items are no longer available after the plans have all been approved and work has commenced.

Some common changes include the resizing of rooms, requirements for walls to be constructed in a different position, a window being repositioned or made bigger, and bathroom or kitchen alterations or redesigns.

Amendments to the plans on large-scale construction jobs will often be marked on the plan with a direction to check the addendum for more information. This will usually be a specific book or document that accompanies the plans and specifications.

The size or scope of the alteration doesn't matter–it is always best to have the amendments approved by the relevant authorities and the plans altered by the architect or draughtsperson.

4.2 Recognising commonly used symbols and abbreviations

4.2.1 Abbreviations

Using abbreviations is a way to show summarised written details–shortened or abbreviated names for appliances, fixtures and other items–on a plan without deliberately overcrowding or cluttering up the document.

The most common abbreviations used are compiled in a list; a small sample of these is shown in Table 4.1. A more comprehensive list is available in Australian Standard AS 1100 Technical drawing, Part 301 Architectural drawing (AS 1100.301–2008).

4.2.2 Contents of drawings

Refer to Figures 4.9-4.11 and the sample working drawings located in the appendix for examples of the following drawing types.

INTERPRETING PLANS

Refer to the electrical plan in Appendix Figure 5.

Identify, sketch and label the symbols for the following items:

- double power point
- light point
- manhole—(ceiling access point)
- exhaust fan
- hot water service
- smoke detector.

Table 4.1 Interpretation of abbreviations used in construction

Abbreviation	Word(s)	Abbreviation	Word(s)
AHD	Australian height datum	FW	Floor waste
AJ	Articulated joint	G	Gas
AL	Aluminium	HW	Hot water unit
AS	Australian Standard	NCC	National Construction Code
B	Basin	P	Pier
BCA	Building Code of Australia	PBD	Plasterboard
BDYL	Boundary line	PCC	Pre-cast concrete
BL	Building line	PF	Pad footing
BT	Brick	PG	Plate glass
BV	Brick veneer	RSJ	Rolled steel joist
BWK	Brick work	S	Sink
C	Cooker	SA	Smoke alarm
CF	Concrete floor	SD	Sewer drain
CL	Centre line	SHR	Shower
COL	Column	SWBD	Switchboard
CONC	Concrete	SWD	Stormwater drain
CR	Cement render	T	Truss
CTR	Contour	TBM	Temporary bench mark
CW	Cavity wall	TM	Trench mesh
D	Door	TRH	Trough
DGE	Drainage	UNO	Unless noted otherwise
DP	Downpipe	V	Vent
FCL	Finished ceiling level	W	Window
FFL	Finished floor level	WC	Water closet
FHT	Floor height		

4.3 Locating and identifying key features on building plans

4.3.1 Site plan

Refer to Figure 4.9 and Appendix Figure 1.

Information found on a site plan can include:

1. site boundaries and dimensions
2. contours of the block of land
3. location of the proposed building
4. roads and pathways
5. soil and surface water drains
6. location of services–electricity, gas, sewer, water
7. service runs from mains to building
8. site datum point

Fig. 4.9 Site plan

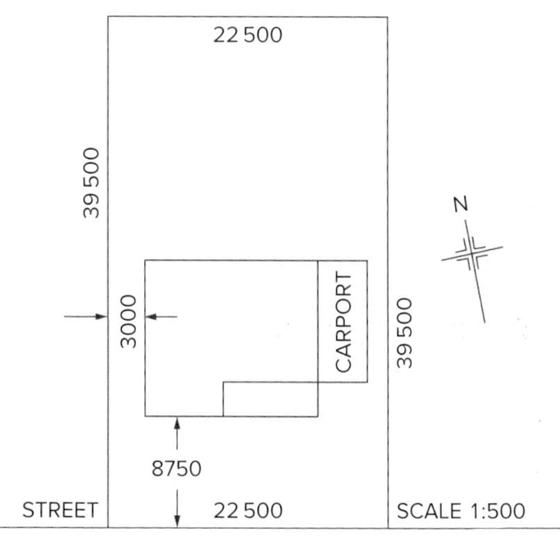

9. earthworks–banks or excavations
10. landscaping
11. existing trees
12. direction of north
13. location of easements.

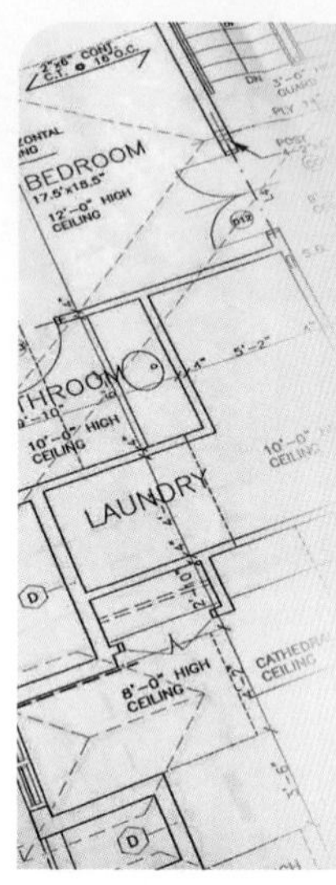

INTERPRETING PLANS

Refer to the site plan view in Appendix Figure 1 and obtain the following information:

- size of the building block
- width of the service easement
- distance set back from the street alignment to the front building line
- scale that the site plan is drawn to
- street name and lot number of the building site
- number of downpipe outlets required for connection to stormwater drainage.

The direction north should always be indicated on the site plan.

4.3.2 Floor plan

Refer to Figure 4.10 and Appendix Figure 2.

Floor plan details can show:

1. internal and external walls
2. overall building dimensions
3. internal dimensions
4. openings in walls–doors and windows
5. wall thicknesses
6. location of fittings and fixtures
7. names of rooms
8. floor finish materials
9. location of stairs and number of treads
10. location of sections.

4.3.3 Footing plans

The footing plan will show the:

1. location of the footing system
2. width and depth of footings
3. location and levels of drains.

Fig. 4.10 Floor plan

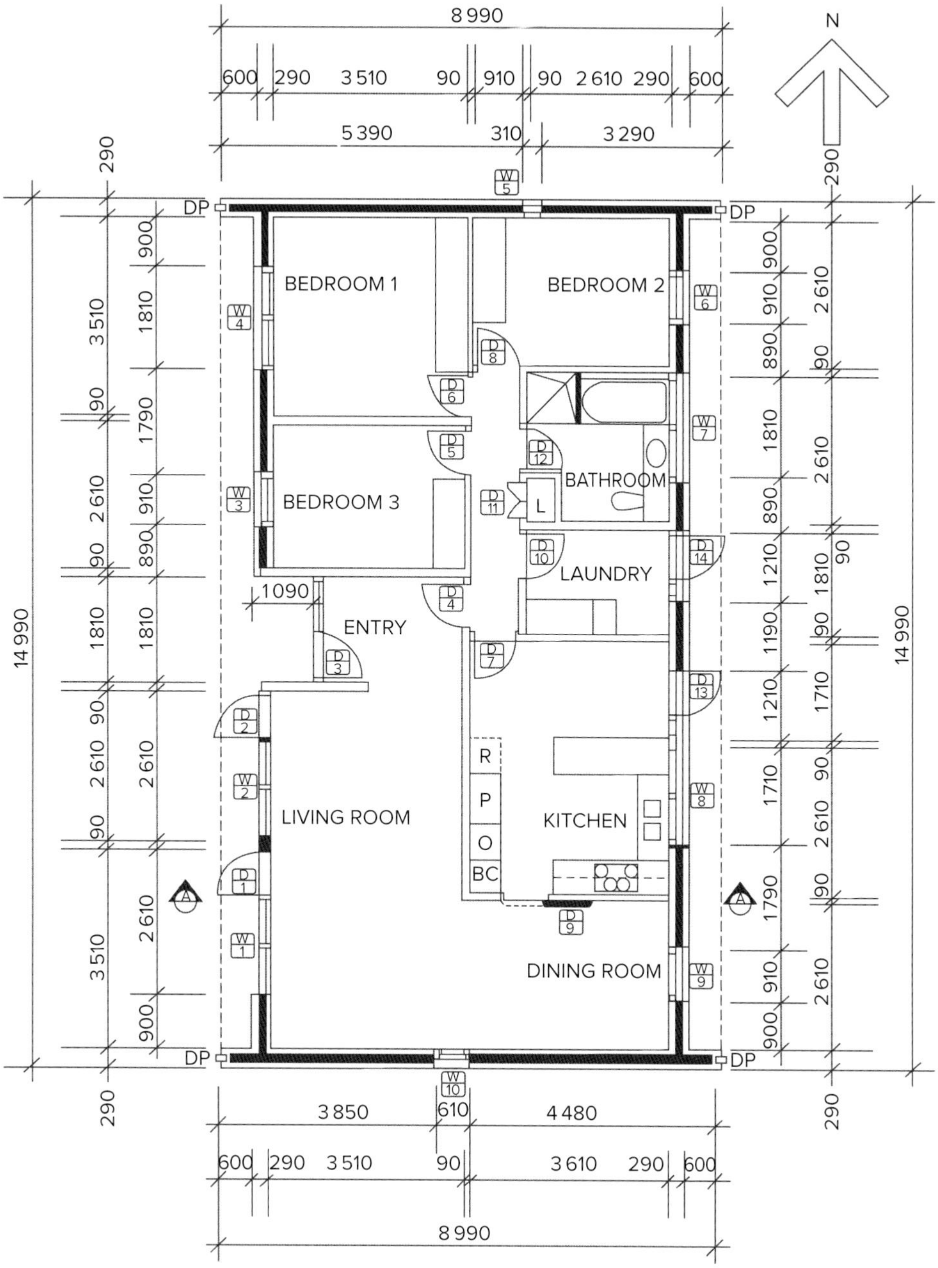

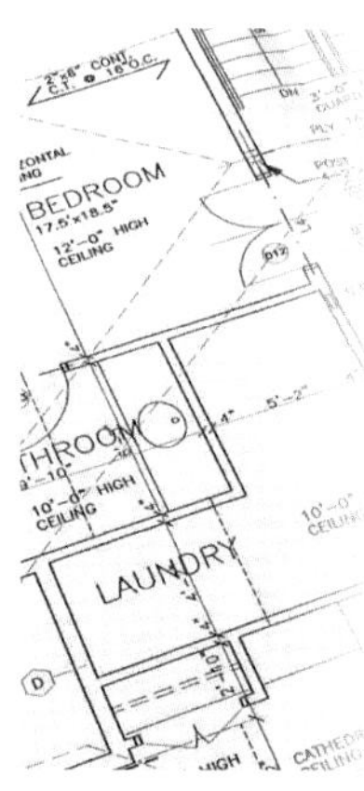

INTERPRETING PLANS

Refer to the floor plan view in Appendix Figure 2 and obtain the following information:

- overall length and width of the building
- internal dimensions of the garage
- smoke alarm points
- type of floor covering installed in the laundry
- the Australian standard to which all pre-fabricated wall framing must comply
- the number of vertical construction joints to be included in the building.

4.3.4 Roof plans

This plan is often superimposed on the floor plan and can include:

1. shape of roof
2. slope of roof (pitch)
3. gutters, downpipes and sometimes levels
4. type of roof cover
5. roof penetrations–chimney, roof lights
6. type of construction (if applicable).

4.3.5 Services plans

Refer to Appendix Figure 6.

These are normally superimposed on floor plans in domestic building plans and can show, for example:

1. electrical layout
2. plumbing and drainage layout
3. air conditioning.

4.3.6 Sections

Refer to Figure 4.11 and Appendix Figure 5.

Section details can include:

1. new and old ground levels
2. heights above ground of rooms and roof
3. construction and thickness of floors, walls, roofs
4. internal and external walls
5. openings in walls
6. location of fittings and fixtures
7. names of rooms
8. slope of roof.

Fig. 4.11 Section

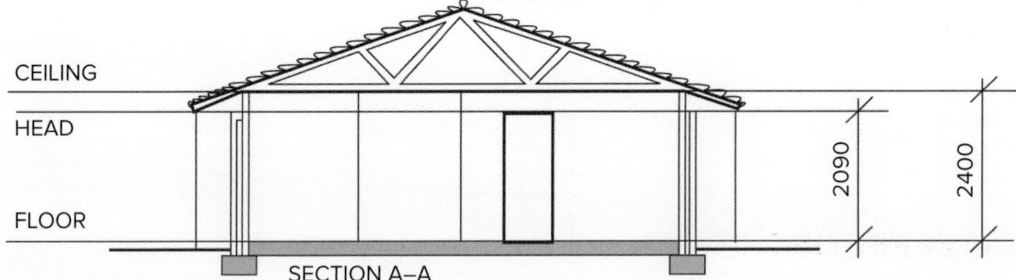

INTERPRETING PLANS

Refer to the elevation views in Appendix Figure 3 or section views in Appendix Figure 4, and identify the roof cladding material and roof pitch.

4.3.7 Elevations

Refer to Figure 4.12, and Appendix Figures 3 and 4.

Elevation drawings will show:

1. size and shape of external walls
2. size and shape of openings
3. external finishes
4. new and old ground levels
5. floor and ceiling levels
6. roof shape and slope.
7. window (joinery) height.

Fig. 4.12 Elevation

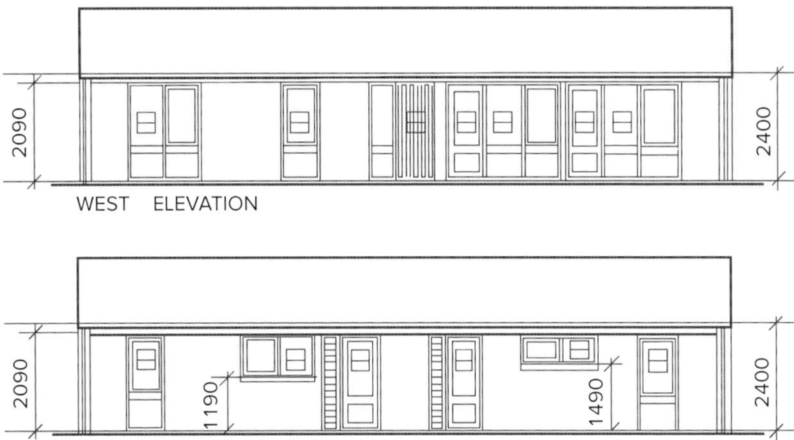

INTERPRETING PLANS

Refer to elevation views in Appendix Figures 3 and 4 and obtain the following information:

- type of roof
- type of windows
- floor to ceiling height.

CHECK YOUR UNDERSTANDING

1. Which plan view will give you the position of a building on an allotment?
2. How are elevations identified on a working drawing to avoid confusion?
3. List five items of information that can be found on the elevation view.
4. List the most common scales used in the following drawings:
   a. site plans ______________________
   b. plan views ______________________
   c. elevations ______________________
   d. sections ______________________ .

4.4 Using drawing programs and tools

The tools used to produce working drawings and plans have come a long way over recent years, with the development of many computer aided design (CAD) programs. You can now produce drawings of the highest quality very quickly on a computer at the click of a mouse.

However, there will still be times when a builder may need to produce their own working drawings and plans for the purposes of quoting or presenting ideas to the client as specified in preliminary discussions on the building project.

Plans and drawings are not just for the construction of new buildings. The builder may also produce their own plans for the manufacturing of smaller items in the workshop such as cupboards and cabinetry, furniture, staircases, windows and doors.

The builder can do this quickly and accurately using traditional drawing methods. Some items may be drawn to scale, while others may be drawn to full size.

Drawing tools include:

- drawing board
- T-square
- set squares
- scale rule
- ruler
- compass
- protractor
- eraser
- pencils
- retractable pencils
- drawing pens
- drawing paper
- graph paper.

Fig. 4.13 Drawing tools on a plan—compass, scale rule, set square and drawing pen

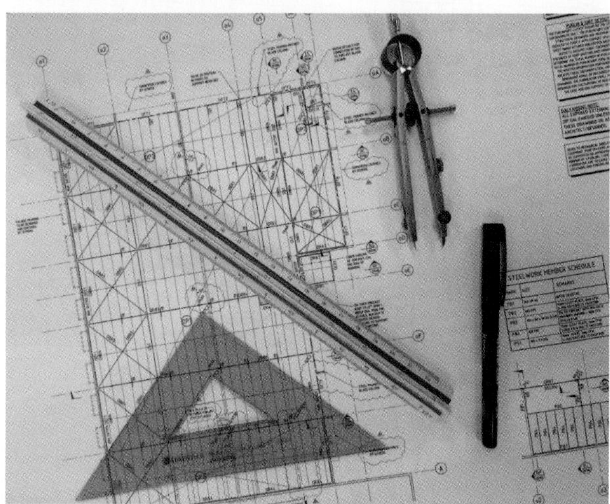

Courtesy of Alister Ford

TIP

- An important point to remember: always treat your drawing equipment the same as you would your tools of trade.
- Keep everything clean and in good order so that when you need to use them, they will not let you down.

Student research

A handy reference for the tradesperson when on-site is www.blocklayers.com. Explore this website and how it could help you.

End of chapter activity

Activity 1: Plan drawing

Using the drawing tools discussed in section 4.4, neatly draw the following three types of plans to the specifications provided.

1. A site plan of your house and the land that it occupies.
   - Include your house, out-buildings, driveway, entrance and any trees.
   - The plan does not need to be drawn to scale or have any measurements included.
   - Try to visualise the size and shape of the block of land and position the house and out-buildings correctly so you have a realistic interpretation.
2. A simple floor plan of your own house.
   - Include all of the rooms in the house as well as all windows and doors.
   - The plans do not need to be drawn to scale or have any measurements included.
   - Try to visualise the size of the rooms in relation to each other and draw them accordingly so that this preliminary plan of your house is realistic.
3. A set of plans for a cubby house that could be constructed in a workshop or on-site.
   - The style of, and dimensions for, the cubby will be provided by your instructor.
   - The drawings will include one plan view and four elevations all drawn at a scale of 1:20.
   - Draw in all dimension lines and measurements.
   - All other information can be recorded in the title box.

AUSTRALIAN BUILDING CODES AND STANDARDS

The following standards are relevant to this chapter:

AS 1100:
Technical Drawing

1100.101
Part 101:
General Principles
(1992)

1100.301
Part 301:
Architectural Drawing
(2008)

Tradie talk—Knock up a sketch showing what is required for this part of the job

Chapter 5

Working effectively and sustainably within the construction industry

Learning Objectives

LO 5.1 Know about the Australian construction industry

LO 5.2 Know about occupations, job roles and working conditions within the construction industry

LO 5.3 Accept responsibility for your work duties and workload

LO 5.4 Work within a team

LO 5.5 Identify personal development needs

LO 5.6 Identify current resource use and implement resource improvements

LO 5.7 Know about sustainability and environmental regulation compliance in construction

Introduction

This chapter will give you the knowledge and understanding of the Australian construction industry you need to prepare you to become an effective contributor, while carrying out sustainable work practices, individually or as a member of a team, within the industry.

You will learn about:

- the structure of the Australian construction industry and the opportunities that lie within it for employment, as well as the specific occupations and job roles, and the associated conditions
- your responsibilities to yourself and your fellow workmates while working within a team, and how to be an effective contributor to the everyday functioning of the job site
- identifying and carrying out safe work methods and practices that meet the standards of national and state legislation
- identifying your own professional development needs, and the steps required to gain the necessary skills and knowledge for future advancement within the construction industry

- dealing with disharmony or disputes within the team effectively before they escalate
- identifying the current resources that are used within your own work role and the opportunities to improve the efficiency of these resources
- environmental regulation compliance and the procedures that must be followed with the correct reporting of breaches or potential breaches.

5.1 The Australian construction industry

The construction industry is one of the largest industries in Australia and it continues to grow as every year passes. It encompasses all areas of construction, from the average family home to the tallest skyscrapers in major cities.

In fact, the construction industry is Australia's largest non-service industry.

5.1.1 Population growth and the construction industry

Australia continues to experience a steady population growth, which is why the construction industry also continues to grow–more people means there is more need for housing and the infrastructure that goes with it, such as shopping centres, hospitals, schools, transport, roads, railway stations, airports and recreational facilities. The list is endless. Everything you see about you that provides shelter, protection and services can in some way be attributed to the construction industry.

As our major cities continue to grow, both in terms of population and area, so too does domestic housing continue to sprawl further and further.

New suburbs are being created and existing suburbs are filling quickly as vacant land is purchased so that new dwellings can be constructed.

In inner-city suburbs, new apartment blocks are being constructed and older manufacturing buildings are being transformed into apartments for more condensed living.

Rural cities and towns are also growing larger as the construction of infrastructure makes the daily commute to work in the city so much easier. This means that everyday living services in these towns and cities need to be catered for, and so new schools, medical centres, hospitals, shopping precincts and recreational facilities are also being constructed.

In June 2019, Australia's 12 major cities contained 75% of the country's entire population of approximately 25.2 million people. This means that nearly 19 million people live in our eight capital cities and four next largest cities combined (Fig. 5.2).

With this type of population growth, it is easy to see why major cities and the surrounding areas are the hotspots for employment in the construction industry.

Currently there are approximately 1.1 million people employed in the Australian construction industry. Taking into account the population growth figures in Figure 5.1, there will continue to be sustained growth in the construction industry as new technologies, new construction materials and modernised construction methods are introduced on a regular basis.

Fig. 5.1 Australian population growth since 1981

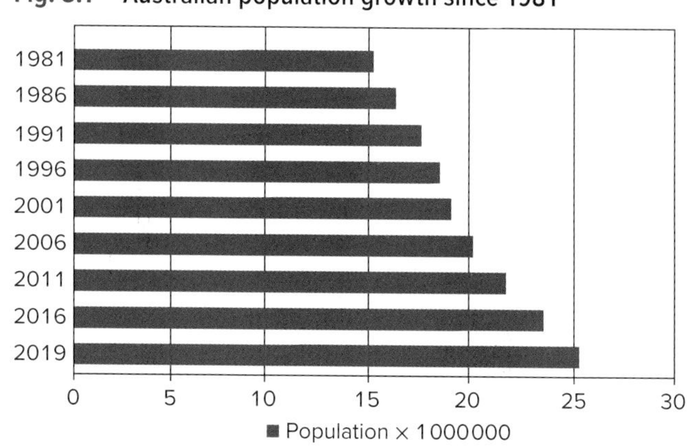

5.1.2 Construction industry sectors

The Australian construction industry can be divided into different sectors depending on the type of work being carried out.

From the perspective of carpentry work, the industry can simply be divided into three sectors:

- residential/domestic sector
- commercial sector
- industrial sector.

These three sectors are generally where someone would enter the construction industry as an apprentice, gain a trade qualification over the designated time of the apprenticeship and then look to pursue employment in any chosen vocation within the industry.

Often, further training will then be undertaken to gain the required qualifications for advancement within the industry.

This could mean being self-employed and taking out registration as a builder or working for a larger company and undertaking training to be able to fill specialist roles within that company's framework.

Residential/domestic construction may consist of buildings such as:

- houses
- townhouses
- flats
- bungalows
- out-buildings associated with the above.

Commercial construction may consist of buildings for the following uses:

- shopping centres
- retail stores
- hospitals
- medical centres
- serviced offices
- nursing homes
- restaurants
- cafes
- hotels
- motels
- sporting/recreational facilities
- police/fire/ambulance stations.

Fig. 5.2 Population in major cities

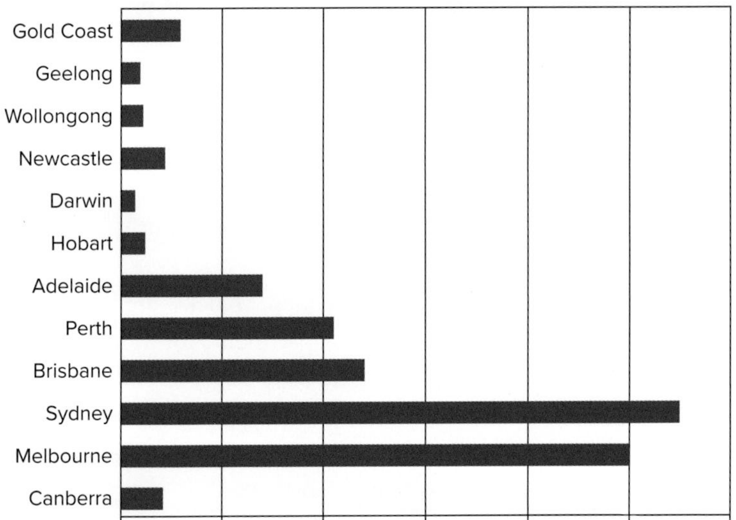

Industrial construction may consist of buildings or structures for the following uses:

- refineries
- bridges
- power stations
- warehouses
- factories
- airports
- railway stations
- desalination plants
- skyscrapers
- damming and hydro plants.

It doesn't matter which construction sector you look at, planning, demolition, construction, renovation, repairs and maintenance work will always be needed; this also extends to other sectors, which may work alongside all three of the major sectors previously listed. For instance:

- site and building surveying sector: surveying, subdividing parcels of land, determining site boundaries, setting required levels, map distances, heights, slopes and land contours
- site development and preparation sector: subdividing land, constructing roads and services, preparing individual building blocks
- general construction and demolition sector: residential and non-residential construction, including carrying out alterations, renovations and additions to existing structures
- building structure sector: bricklaying/blocklaying, concreting, roofing (concrete or terracotta tiles, metal, slate, shingles), steelwork
- installation services sector: plumbing (hot/cold-water supply and systems, gas supply and systems, sewer lines and septic tanks, drainage), electrical (power supply, lighting), telecommunications (phone lines, media, television), fire services, air conditioning, heating, refrigeration, security systems
- building completion sector: glazing, carpentry, plastering, tiling, painting and decorating
- specialist construction sector: waterproofing, scaffolding, rigging, swimming pools, spas, landscaping, fencing, crane operation, dogging, formwork, post-tensioning, pre-tensioning.

5.2 Occupations, job roles and working conditions within the construction industry

5.2.1 Trade occupations and job roles

There are many different career opportunities within the Australian building and construction industry–many begin with someone undertaking and completing an apprenticeship.

It is important to research the industry and learn about each trade area and the tasks that each of these trades perform. This can be done through career study classes at school, trying out specific trade-related subjects at school, attending trade expos, coordinated work experience through school, seminars, workshops, after-hours work or holiday work.

It is important to talk to people who work in the industry. They may be immediate family members, other relatives, family friends or a friend's family members. Seek out as much information as possible. This will help you to decide on a specific trade career path.

The Australian Apprenticeships Pathways website, www.aapathways.com.au, provides information and guidance to assist those who wish to pursue a career in the construction industry.

Qualifications gained by completing an apprenticeship or completing specific training in the building and construction industry can include:

- carpentry
- joinery
- carpentry and joinery
- shop-fitting
- stair-building
- plumbing
- bricklaying/blocklaying
- wall and ceiling lining
- painting and decorating
- solid plastering
- floor and wall tiling

- concreting
- demolition
- glazing
- stonemasonry
- dogging
- formwork and falsework
- rigging
- roof tiling
- scaffolding
- steel fixing
- waterproofing.

Further advancement within the industry can be even more far-reaching if you complete apprenticeship training then continue with specific job role training to gain more qualifications. This training can be done through registered training organisations, which may include industry groups and educational institutes.

Future roles can include:

- management
- supervision
- estimating
- drafting
- surveying
- company representation
- machine and plant operation
- sales and service
- business ownership
- combining trades.

From directly working in a specific job role for a construction company–which means being paid every week or fortnight, accruing sick leave, annual leave, long service leave and having superannuation contributions made on your behalf by your employer–to being self-employed and indirectly working as a sub-contractor for different companies–meaning that all the leave entitlements and contributions would be made by yourself–the range of work is vast, but all of these roles need to be filled if the building and construction industry is to remain vibrant and self-sustaining.

5.2.2 Technology trends

Technology in the construction industry is forever evolving, which in turn means that the employment opportunities are forever evolving as well. Not only are the methods used to carry out many construction tasks being made easier, with tasks now able to be completed by qualified tradespeople in a much quicker time frame than previously, but the employment opportunities within the design and manufacturing processes are also adapting as the technology and materials change.

Examples of such change are the use of high-level electronic technology utilising computer design programs such as BIM (building information modelling) or CAD (computer aided design) to engineer, design and plan modern structures, coupled with modern computer-aided machinery to build pre-fabricated construction items such as wall frames and roof trusses from either timber or steel. These structures are then delivered to a building site to be erected on-site in a fraction of the time it used to take using hand and power tools.

This is just one example of the many pre-fabricated job options available in the construction industry. Another area offering employment opportunities within the industry relates to the packaging and transport requirements of pre-fabricated construction items.

Building materials are always changing with new technology and trends. For instance, composite materials are being produced more often and being used for more applications.

Tools and equipment are also constantly being improved by power tool manufacturers. Battery-powered tools are now being used on construction sites for most tasks. From battery drills to table saws, every conceivable power tool can now be battery powered and fast charged, which avoids the need to run power leads around construction sites.

Another important area is the development of more environmentally friendly construction machinery used with batteries rather than relying on liquid fuel as the power source. Items such as earth-moving equipment, cranes and larger mobile safe working platforms are now being seen more often on construction sites.

5.2.3 Employment conditions

Employment within the construction industry, whether full time or part time, means being employed under a contract of some description. A contract of employment will outline both the legal and the organisation's requirements in relation to such things as the duties, responsibilities and expectations of both the employer and the employee. It will also include company policies and procedures, as well as employment conditions and entitlements, including but not limited to hours of work, rates of pay, penalty rates, leave entitlements and expected employee behaviour.

If at any time you need to clarify anything regarding your contract of employment, speak to your employer. If you do not receive the required information, you can seek it out from other sources. Industry bodies such as building associations and unions can give you advice through their specific newsletters, bulletins, publications and websites. Some can even assist with access to legal advice or representation if required.

Industry awards and wage scales can be accessed through www.fairwork.gov.au.

The website also includes information on subjects such as:

- workplace agreements
- industrial agreements
- enterprise agreements
- access and equity principles and practice
- anti-discrimination and related policies
- dismissal laws.

5.2.4 Safe work methods and practices

Every worker has the right to a safe workplace. It should be the highest priority for all employers and employees to ensure that everyone can carry out all of their daily work tasks safely and efficiently, free from any physical or mental stresses, then return home to their family after the day's work.

All companies within all sectors of the construction industry are subject to federal and state and territory work health and safety laws, regulations and legislation and the conditions within them. This means that everyone employed by the company is also subject to these conditions and must always follow and uphold them.

The *Occupational Health and Safety Act* is a very important legal document, which outlines the safe work practices that need to be adhered to.

Employees must make every effort and take reasonable care to ensure the wellbeing and safety of themselves as well as any coworkers.

They must cooperate with their employer to assist in complying with all requirements of the *Occupational Health and Safety Act* and while at work they must not interfere with or misuse either recklessly or intentionally anything provided for health, welfare or safety.

Employers must provide for the safety and wellbeing of all employees in a way that is free of risk. Duties of employers to employees to achieve this include:

- provision and maintenance of plant and work systems that are safe, practical and pose no risk to health
- making arrangements for the safe use, handling, storage or transport of plant or substances
- making safe, monitoring, maintaining and keeping risk free any places of work under their control
- providing facilities that are adequate for the welfare of any employees at any workplace under their control
- providing information, instruction, supervision or any required training to any employees to ensure that they can perform their roles in a safe manner without risk to their health.

By putting these points from the *Occupational Health and Safety Act* into an easier-to-understand form, the following points are some important construction site safe work practices and methods that the employer would be expected to ensure:

- access to clean, well-maintained site amenities such as toilets, eating areas, first-aid kit/area and drinking water
- OHS policies and procedures observed and maintained on a day-to-day basis
- training in the correct selection, fitting and use of personal protective equipment (PPE) and clothing
- training given for the safe operation of plant and equipment
- training for emergency procedures, including the use of firefighting equipment
- good housekeeping practices to ensure the work area is kept clean, tidy and safe
- training in risk assessment to ensure all workers can contribute to safe work plans
- waste and debris storage procedures and disposal plans, established in accordance with environmental and safe work procedures
- prevention of bullying and harassment of any employee
- establishment of designated smoking areas
- zero tolerance to drugs and alcohol in the workplace.

For further information on the Workplace induction for construction workplaces information sheet go to www.safeworkaustralia.gov.au/doc/workplace-induction-construction-workplaces-information-sheet.

Fig. 5.3 Sample checklist for site induction

WORK INDUCTION FOR CONSTRUCTION WORKPLACES INFORMATION SHEET

3

APPENDIX A - WORKPLACE SPECIFIC INDUCTION CHECKLIST

This checklist provides examples of what could be covered in workplace specific induction training for workers.

No	Items covered	Yes	No	N/A
1.	Have you checked the competencies and qualifications of inductees?			
2.	Have you discussed and clearly stated the procedures for reporting incidents, injuries and hazards?			
3.	Have you discussed the workplace safety rules?			
4.	Has the person been taken through relevant safe work method statements for the tasks to be performed?			
5.	Is specialised equipment required and have they been trained to use the equipment?			
6.	Do people have the correct PPE available e.g. ■ hard hat ■ safety glasses ■ safety boots ■ long sleeve shirt, and ■ high visibility vest?			
7.	Have you shown the person what to do in an emergency and identified the location of the: ■ assembly point and evacuation route ■ closest medical facility ■ contact details of emergency services, and ■ provisions for emergency communications?			
8.	Have you shown the person where all relevant firefighting equipment is located, for example fire extinguishers and hose reels?			
9.	Have you shown the person: ■ the location of the first aid facilities and kits, and ■ who the first aiders are and how to obtain treatment?			
10.	Have you shown the person where all the facilities are located including: ■ crib sheds ■ toilets, and ■ drinking water?			
11.	Have you explained the workplace security procedures?			
12.	Have you introduced the person to the Health and Safety Representative?			
13.	Do they have any further questions or need clarification on any points or topics?			

NOVEMBER 2014

Workplace induction for construction workplaces—information sheet, Safe Work Australia CC BY-NC 4.0 https://creativecommons.org/licenses/by-nc/4.0/

5.2.5 Drugs and alcohol in the workplace

Drugs and alcohol should not be used or consumed in the workplace under any circumstances. Full reference to this topic can be found in your already completed OHS chapter.

5.2.6 Smoking in the workplace

Although not illegal, many construction sites are now run as smoke-free. Full reference to this topic can be found in your already completed OHS chapter (Chapter 1).

5.2.7 Bullying, harassment and discrimination in the workplace

These actions, whether intentional or not, are now illegal and punishable by law. Full reference to this topic can be found in your already completed OHS chapter.

5.2.7.1 Discrimination

Employers are now subject to all federal anti-discrimination laws, meaning that everyone has an equal opportunity to apply for and be given due consideration for any job that may be available to be filled. In other words, everyone gets a 'fair go'. Full reference to this topic can be found in your already completed OHS chapter.

CHECK YOUR UNDERSTANDING

1. List the three main sectors of the Australian construction industry.
2. Give three examples where technology advancements have enhanced productivity in the construction industry.
3. Identify and list five situations or actions that may be construed as workplace bullying.

5.3 Accepting responsibility for your work duties and workload

Your duties and the accompanying workload in your chosen career pathway will be established in consultation with your employer, taking into consideration your qualifications, experience and employment status.

An overview of your duties and workload will be documented in your individual job description.

Specific tasks will be allocated to you as they are required to be carried out, and careful planning and scheduling should go into any task to ensure safe and efficient work practices are employed to complete the given task to the required quality and standards by the agreed deadline.

Sometimes, the tasks and completion timelines will be affected by different factors that may be uncontrollable. These include the weather, delays in the delivery of materials, substandard materials needing replacement, equipment breakdown, servicing or unavailability, or the non-attendance at work of team members through illness or personal matters. No matter what the reason for the delay, re-planning should be done and the delays should be communicated to your supervisor as soon as possible.

If at any time you are unsure about what to do, stop and ask. Don't try to take any short cuts. Speak to your supervisor. It may be the case that more training in specific areas or work processes is required, such as new building materials, machinery operation, tools, plant, equipment or personal protective equipment. Or it may even be necessary to have workers with expertise in specific areas join the team to ensure the tasks are carried out safely and correctly to the required quality and standards.

An excellent format for planning construction tasks is the Gantt chart (Fig. 5.4).

This system of planning was invented in the early 1900s by a man named Henry Gantt.

Fig. 5.4 Gantt chart

PRO-BUILD CONSTRUCTIONS

18–28 South Road
City West, Victoria 3006

Phone (03)987 654 3211
email: probuild@construction.com.au

Construction Schedule: Steel portal framed garage/workshop

Activity	WEEK 1					WEEK 2				
	1	2	3	4	5	6	7	8	9	10
Site establishment and scrape	■									
Building set out	■									
Excavate and prepare for footings and slab		■								
Plumbing and electrical services		■								
Set up formwork, reinforcement			■							
Set up cast ins for portal frame			■							
Pour and finish concrete				■						
Construct portal frame				■						
Erect portal frame					■					
Fit roofing and flashings						■				
Fit window and door frames and flashings						■				
Fit wall cladding and flashings						■	■			
Fit roller door							■			
Rain headers, guttering and downpipes								■		
Plumbing fit out								■		
Electrical switchboard and fit out									■	
Carpentry fit out									■	
Building clean										■
Site clean										■
Hand over										■

Sub-contractors register

City West Excavations
Contact: Jim. 4344 472 837

No Leaks Plumbing
Contact: Rick. 5425 488 663

Shocking Electrical
Contact: Robert. 5444 843 787

Concreting
Contact: Val. 5442 579 633

Steel supplies
Contact: Rod. 9353 464 486

Roller door
Contact: David. 7526 377 488

Courtesy of Alister Ford

Effectively, a Gantt chart is a bar chart that lists the activities and stages of any construction project vertically in the correct order of them happening, with the time frames for each activity listed horizontally, usually in weekly blocks divided into the working days of the week.

These charts can be filled out by hand and displayed on any construction site noticeboard for any of the workers to see.

Gantt charts can also now be developed electronically on spreadsheets and printed out for display or forwarded on to workers via electronic devices. There are also excellent templates available for download on the internet. These can then be automatically updated at any time if the schedule of work changes for any reason.

This is an excellent way for sub-contractors to be kept up to date with the work scheduled for them as it nears commencement time and for any delays to be communicated to all workers.

Such charts are generally formulated by the project supervisors as they know the specific time frames for jobs and the organisation of other trades and materials, etc. However, there is no reason why you cannot, individually, jot down your own schedule for your designated tasks using this system. It doesn't matter whether it's an hourly plan for a single day or a daily plan for a week. If you can be organised in terms of your workload time frames, you will work more effectively, be more productive and produce a higher-quality product for your employer or client.

CHECK YOUR UNDERSTANDING

1. How would you ensure that safe and efficient work practices are employed to complete the given task by the agreed deadline?
2. Which work-team members need to be actively involved in the planning of any tasks to be carried out?

5.4 Working within a team

Work teams, work crews, work gangs–these are all common construction industry terms for a group that works together to get the job done safely and efficiently. A work team generally consists of the workers employed by the construction company, but others can come and go from the team as different tasks need to be performed. These can include sub-contractors, delivery drivers, crane operators, dog-men and specific labour hires. As a team member, you should be actively involved in the planning of any tasks to be completed, in consultation with supervisors, OHS representatives and other team members, including any temporary team members, to ensure that all safety aspects are addressed. Safety matters include the formulation of safe work method statements (SWMS), the use of all tools and equipment, personal protective equipment, hazard identification, building and construction materials handling and use, agreed timelines for stage and final completion, reporting systems for any delays and the quality control requirements of the finished product.

All members need to be clear about their individual role and allocated tasks within the team as designated specifically by supervisors or team leaders. These should be confirmed with fellow team members so that each individual can assist others when required. Looking out for each other on a construction site can be hugely beneficial to the progression of the overall job and the personal safety of all team members. Knowing that someone has 'got your back' can be a great motivator for aiming to achieve the best quality of work in the safest manner. Being appreciated by other team members because you look out for them and assist them when required–and knowing in turn that when you need assistance, all you need to do is ask–will bring harmony to the work group and create a high-quality work environment.

Any confusion can lead to disharmony within the team, accidents or altercations occurring, building materials being damaged and rendered unusable, or tools, plant or equipment being badly damaged.

Such incidents will be costly–whether in terms of lost time through personal injury, the damaging effects of witnessing a colleague being badly injured, time lost waiting for new construction materials, tools, plant or equipment to be delivered, plus any specific training or instruction required to safely use the new products, as well as time lost while management, supervisors and team members work to resolve disputes.

5.5 Identifying personal development needs

The construction industry is forever evolving. Changes are being made rapidly to accommodate new construction materials, work systems and technologies. With these changes come changing job roles. It is important to be able to identify such changes and keep up to date with any new systems so that your employer sees you as an asset to the advancement and future success of the company.

Identifying training and professional development opportunities in these areas can lead to you gaining new qualifications that keep your knowledge base of new industry trends up to date, as well as equipping you with the practical skills required to implement such changes.

Talk to your supervisors and employers about any opportunities that you identify. It may be something that they have not considered but, when bought to their attention, they may see it as an opportunity for one of their employees to learn new skills that will help them advance in the industry as well as becoming an extra asset to the company.

Many construction material, work system and technological changes that occur in the construction industry workplace come with training provided in that specific area. Employers will select those they feel would be best suited to participate in any required training. It is important to always be enthusiastic about the chance to further your knowledge and skills so that when such opportunities arise you will be considered for inclusion.

Not everyone can do every training, so there will be times when others get an opportunity, but you do not. This is a chance for you to keep on working enthusiastically to company requirements and standards, but also to communicate with those supervisors and workers who completed the training to see what they have learned. You can learn from everyone.

As well as learning new skills associated with new construction systems and materials, this is also a chance for you to develop and enhance other skills, such as the communication skills needed to explain about and instruct others in what you have learned and the organisational skills required to plan and carry out the work, whether individually or within a work team.

With regard to the technology required to correctly explain construction procedures, this could include the use of computers and associated programs or tablet and phone apps.

Not all training and professional development opportunities have to be associated with new technologies and systems in the industry, however.

There are many opportunities to train and gain qualifications in traditional construction practices. Ticketed courses such as traffic management control, scaffolding, rigging, dogging, forklift, elevated work platforms or skid steer can enhance your employment opportunities.

Having a test and tag licence can be advantageous as an employer can then have someone within the company carry out this work as required rather than bring in an outside contractor.

There can also be training opportunities to become involved in the OHS aspects of the building industry, such as becoming an OHS representative or officer.

Additionally, there may be opportunities to further your current qualifications. If you are an apprentice, you will be completing your Certificate III studies as required as a component of your apprenticeship. When you complete your allocated trade school studies you can then consider undertaking the certificate IV course. This can lead to you gaining your registered building licence or to progress to another vocation within the construction industry such as staff and site management roles, costing and quoting or even branching into areas such as drafting or surveying.

The training opportunities for furthering your career in the construction industry are many and varied.

CHECK YOUR UNDERSTANDING

1. What could be the advantage of bringing training opportunities that you have identified for your personal development to the attention of your employer?

5.6 Identifying current resource use

The resources that you use in the construction industry will differ from one construction site to the next. These differences may be very minimal, or they could be major.

The term *resources* includes the building materials and all of the associated sundries that are used in the construction that is taking place. Correctly calculating the construction materials, ordering them and organising delivery and storage are all very important factors to be considered as any mistakes can be very costly, requiring new materials and leading to associated job delays.

Other resources may include the tools and equipment that you require, including the means of maintaining and transporting all of your gear.

Additional required resources such as electricity, water, gas, fuel, chemicals, adhesives or paints need to be organised well in advance of the required time of usage. In some cases, specific storage requirements will need to be met.

Having a good knowledge of all of the required resources for the job is vital.

It is also very important to ensure that anything related to construction resources is well documented.

Companies will have specific reporting systems. For example:

- building material stock lists
- building sundries stock lists
- chemical registers
- plant and equipment booking systems
- vehicle booking systems
- employee training and qualification registers.

Such documents must be used correctly and kept up to date. These are some points to bear in mind:

- If building materials or building sundries are running low, they need to be ordered to replenish supplies.
- Failure to update chemical registers can have catastrophic consequences in the event of an emergency.
- Emergency services personnel need to know immediately upon arrival of any potential hazards they may encounter.
- Plant and equipment should be clearly booked for use and logbooks maintained for servicing and maintenance purposes.
- Company vehicles need to be clearly booked for use and used at the required time.
- Employee qualification and training registers need to be kept up to date for all personnel on-site.

If a specific task arises, it is then easy to establish if anyone on-site has the correct qualification or training to carry out the task. For example, traffic management may be required so that delivery trucks or cranes may enter or leave the construction site. The properly trained employee can then be designated to carry out this task.

A great tip for anyone starting a career in the construction industry is to keep your own work journal.

This does not need to be a day-to-day diary documenting every task that you do. Use it instead as a journal, in which you document all of the important things you learn from talking to others and watching how others go about their everyday tasks on the building site. Include details of how to correctly carry out specific tasks, whether related to your own specific trade skills, occupational health and safety, hazard identification and reporting, communication, job site organisation, planning procedures, site meetings–including details of how you actively participated in such meetings to give your opinion, advise others or make suggestions for resource or work procedure improvements–other trade workers or sub-contractors, emergency procedures, first-aid procedures and environmental practices. There are so many new skills to learn and knowledge to be gained. One thing is guaranteed: you will never stop learning while employed in the construction industry.

WORKPLACE SCENARIO

One of the things I dislike most nowadays is that we live in a throwaway society, and this mentality is slowly creeping into the construction industry.

Throwaway building materials.

Throwaway material and equipment packaging.

(continued)

Throwaway containers.

Throwaway tools.

The convenience of it all adds up to a quicker completion time and handover of the finished job.

I recall when I commenced my apprenticeship in 1979. One of the first things that my new boss did was sit down in the carpentry/joinery workshop with me and formulate a list of tools that I would need to purchase over the four years of my apprenticeship so that I would be work-ready as a qualified tradesman when the time came.

He told me to buy quality, explaining that it may cost a bit more, but that these tools will serve you well and will last for your entire working life and beyond.

From my very first pay cheque, I started to put this kit of tools together. I still have most of these tools in my toolbox in the shed, all sharp and well maintained, ready to be used when needed.

Part of this kit comprises three high-quality steel hand saws with solid timber handles—not the type you buy today that are chemically sharpened and with a plastic handle, so that the first nail you hit renders it useless. So, you throw it in the bin and go and buy another one.

As soon as I purchased these new hand saws my boss said, 'Well, now you need to learn how to breast, set and sharpen them, but a lot of practice will be required.' I really had no idea about any of this, but I said I was willing to learn a new skill.

The first thing I had to do was build a set of saw chops (a timber construction in the form of a clamp to hold the saw blade while you sharpen it). Now I was ready to sharpen my saws as required.

After setting the saw into the chops I was shown how to breast it. This meant running a flat file lengthways along all of the saw teeth to bring them all back to the same height.

Then I needed to set the saw. This was done with a small setting tool, which, when squeezed shut, pushed a single saw tooth across to the desired position so that the cut would be wider than the thickness of the steel blade and the saw wouldn't jam up during cutting. Every tooth on the saw had to be individually set—every second tooth down one side and then every second tooth back up the opposite side.

Then the saw was ready to be sharpened, using a small triangular-shaped file held at the correct angle to sharpen each tooth of the saw individually.

When finished, these saws would always be as sharp as they were on the day I purchased them, and they never let me down in terms of the quality of work, whether I was required to produce everything from framing construction to fine joinery work.

After I had completed two years of my apprenticeship an opportunity arose that meant transferring my apprenticeship from the country town where I lived to Melbourne.

I went from a workshop environment with one carpenter and two apprentices to another workshop in the city with three carpenters and I was to be the fourth apprentice—a daunting experience for a second-year apprentice who was still not old enough to get a driver's licence.

Anyway, I hadn't been there long and was learning a lot more about trade, communication and teamwork skills, along with being kept very busy with all of the tasks assigned to me on each day's workshop plan.

I was surprised to walk into the workshop not long after commencing my job and there were probably five or six hand saws, a few portable power saw blades and some of the blades from the large rip saw and planer laid out on one of the work benches. My boss asked me if my hand saws needed sharpening and to put them with the other saws so they could go to the saw-sharpening service to be sharpened. 'Only $5 per saw, great value,' he told me. I did some quick maths in my head: three saws, $5 each, total $15—nearly 20% of my week's wage gone on saw-sharpening costs. I told my boss, 'Thanks for the offer but it's OK, I'll sharpen them myself.' Everyone stared in disbelief. None of them, tradesmen or apprentices, knew how to sharpen a saw. They said, 'Well, we have to see this.'

So, I built a new set of saw chops. And then in half an hour I had my rip saw and panel saw breasted, set and sharpened. Luckily, they were still fairly sharp, and I didn't need to hit them too hard with the file.

By the end of the day I had one saw from each of my workmates on my workbench. Everyone wanted to learn how to sharpen their own saws. I didn't mind, I was happy to show them, so I taught all of my new workmates how to sharpen their own hand saws.

Over the years I've saved a lot of money by spending 10 or 15 minutes of a lunchbreak once or twice a week maintaining and sharpening my saws. And I still do. I don't need to buy throwaway saws.

5.7 Sustainability and environmental regulation compliance in construction

5.7.1 Sustainability in construction

Sustainable building is the term commonly used to describe the activity of building design and construction carried out according to sound environmental principles. This includes the effect of construction and building on the environment during all building phases–from the amount of energy used in the production of building materials and the waste produced during construction to the use of ecologically non-toxic substances and the eventual demolition of the building and the recycling of its materials. Sustainability in the built environment must also be taken into account during the planning of new developments, the extent of land exploited in development areas and the number of roads and amount of transportation needed to service the development.

Not every new building needs to be designed by an architect or be high-tech to be classified as sustainable. It is by being guided by the simple ideas and basic principles of building design and construction that we can ensure that environmentally friendly buildings become the norm. A building 'consumes' energy (see Fig. 5.5).

Fig. 5.5 Sustainable buildings—built with the environment in mind

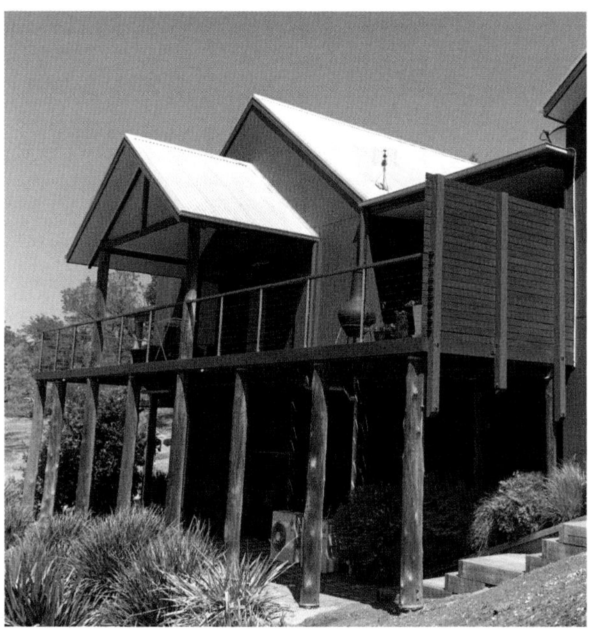

Courtesy of Alister Ford

5.7.1.1 Strategies for sustainability

All types of construction have positive and negative aspects. For example, a timber building is easier to insulate but a masonry building is better suited to utilising passive solar gain; a timber building can be built from locally grown materials, but one built from bricks could be recycled and used again. The choice of design, construction methods and systems and materials must by necessity be related and based on environmentally sound principles for it to be classified as sustainable.

Strategies for energy efficiency include:

- designing for low-energy use by utilising superinsulation, airtightness and ventilation control; active and passive solar energy; day-lighting and solar control; efficient cooling and heating systems; and efficient lighting and appliances
- using low-embodied energy materials, preferably those sourced locally, relatively unprocessed and originating from a renewable resource, and limiting the use of high-energy materials.

Strategies for material use include:

- using unprocessed materials in the first instance such as solid timbers and natural stone, earth, clay and fibres and using local materials, if available
- using materials that can be re-used such as soft mortars for bricks and blocks, screw fixings, and limiting glues and composite materials
- minimising the use of new materials by revamping buildings for new uses, refurbishing whole buildings or building components and using recycled materials, if feasible
- using materials that will form a healthy interior environment, such as organic surface coatings, natural fibres, and not using formaldehyde glues in sheet products.

The energy consumption of buildings is our most significant issue if we are to reduce their environmental impact. Close to 20% of all energy used in Australia is done so in the construction and operation of buildings. As much as 60–70% of this energy could be saved by energy conservation strategies such as bulk insulation, energy-efficient glazing, airtightness in specific climatic zones, use of passive design approaches, efficient heating and cooling systems (active systems), efficient appliances and landscaping to provide shade or wind protection appropriate to our different climatic zones.

Conservation strategies adopted by governments have been introduced through the following:

- building and design standards (e.g. BCA, Building Code of Australia; BASIX, Building Sustainability Index)
- certification of trades processes and labelling programs (e.g. the Green Star program; WERS, the Windows Energy Rating Scheme)
- taxation incentives (e.g. insulation subsidies).

Energy conservation is as much about choosing the appropriate energy source as it is about saving that energy. In modern buildings, energy usually takes the form of electricity, predominantly supplied by the national grid and generated from fossil fuel combustion (coal or gas), hydropower or gas supplied from a natural reserve or as a product of fossil fuel combustion (coal gas).

A building 'consumes' energy (Fig. 5.6):

- through the manufacture of building materials and components (embodied energy)
- in the distribution and transportation of building materials and components to the construction site (grey energy)
- in the construction of the building itself (construction energy)
- in the operation of the building, its systems and appliances each year (operating energy).

These types of energy are defined below:

- *Embodied energy* is the amount of energy used to extract and process a building material. It is expressed by the amount of carbon released into the atmosphere by the manufacturing process. The objective of a sustainable approach to building is to select materials and components that minimise this type of energy. The most straightforward way of doing this is to select materials that are used in their raw state (timber, stone, compacted earth), recycled (steel, crushed brick and concrete) or processed close to the building site.
- *Grey energy* is that energy consumed in transporting building materials from the manufacturer or quarry to the construction site. By supporting local industry and using local materials, this energy can be minimised. Where the building site is at a distance from the supply sources, the energy cost will be higher. This type of energy is termed as 'grey' because it is usually not seen or accounted for.

- *Construction energy* is the energy used in the construction of the building itself and is the smallest energy component used in the overall building process. Like grey energy, it is often overlooked; however, this energy is important for running a construction site, avoiding waste generated by poor building procedures and the uneconomic use of water and irresponsible disposal of demolished or unused materials, along with general health and safety measures.
- *Operating energy* is the energy used in the operation of the completed building, along with its systems and appliances. This is the greatest amount of energy consumed and occurs after the building is handed over to the owner. This energy use receives the most attention by energy conservation measures such as building design strategies, the selection of materials and efficient heating and cooling systems, lighting and appliances. Operating energy is closely tied to the climatic zone in which the building is built, and the response to design and construction taken to minimise energy consumption will vary from climatic zone to climatic zone; however, generally every building will require some seasonal heating or cooling and day-to-night ventilation.

Fig. 5.6 Average embodied and operating energy usage for a house

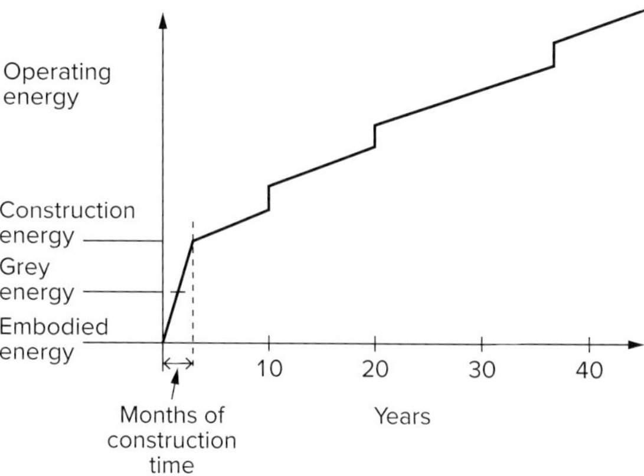

5.7.1.2 Making buildings low energy

Low-energy buildings usually refers to a building's operating costs, that is, the amount of energy and the cost of it for heating or cooling a building to a comfortable temperature for human habitation. A low-energy building needs to be matched to the climatic zone in which it is built and requires sufficient thermal insulation to prevent heat gain or loss through the building envelope. Table 5.1 shows a comparison, by building element, between the approximate heat loss/gain for a traditional and a low-energy house in Sydney. As you can see, low-energy design practices save a large amount of energy.

Every construction material has some ability to conduct heat. This thermal conduction value is called its *U-value* (sometimes expressed as an R-value, which is a reciprocal of the U-value, or U = 1/R). The U-value describes the flow of heat (usually air) passing through a one-square-metre part of the external envelope of the building, for example, a wall constructed in the brick veneer (Fig. 5.7).

Table 5.1 Approximate heat loss/gain for house (Sydney)

External envelope elements	Traditional house	Low-energy house
Air movement/leakage	25%	15%
Floor	10%	2%
Roof	15%	6%
Walls	25%	15%
Windows	25%	13%

Fig. 5.7 U-value—the flow of heat through the external envelope

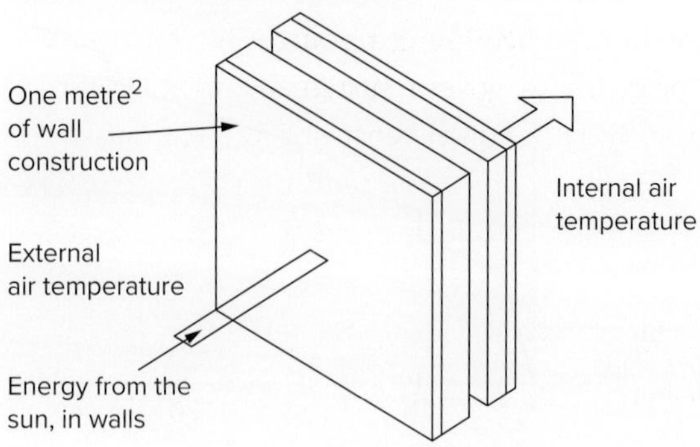

Fig. 5.8 Heat gained or lost through the building envelope

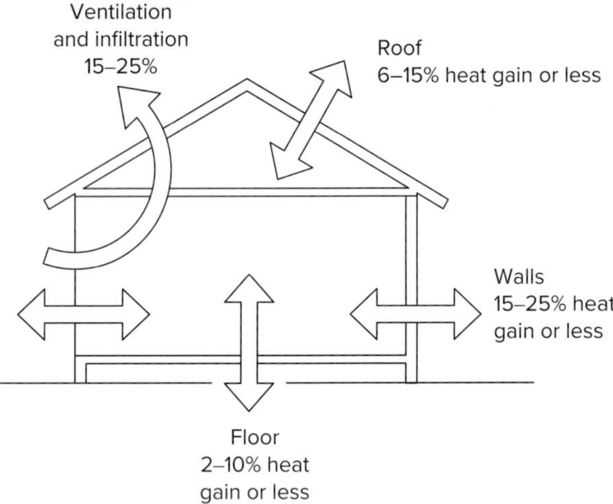

The lower the U-value (or higher the R-value), the less the heat gain or loss moving through the wall from outside to inside. In order to achieve a particular interior thermal comfort level, the thickness of the wall would be increased as needed. In practice, this does not happen, but a combination of different materials and bulk insulation is used in the construction of the wall to achieve the comfort level.

The greatest heat gains or losses are usually caused through air leakage of the external envelope. In traditional buildings this was usually through poor-fitting windows and doors and uninsulated walls, roofs and floors. In a low-energy building, greater attention is paid to the airtightness of the external envelope, which can reduce heating and cooling bills by increasing the efficiency of the insulation. An insulation layer is only as effective as its airtight sealing.

Airtightness is a 'barrier' (a vapour barrier) made up of the combination of interconnected materials, sealed joints and components in the building envelope to provide a barrier against air leakage from outside to inside and vice versa. Air leakage through the building envelope accounts for up to one-third of heat gain and loss in modern buildings (Fig. 5.8). In humid environments, air leakage of warm air causes condensation to occur within the building envelope, leading to mould growth in structural members, particularly in lightweight construction.

Energy conservation at all stages gives the greatest reduction of the impact of a building on the environment. The most important energy measures for achieving low operating energy are:

- using efficient thermal insulation on the external envelope of the building–the roof, walls, windows and floor; use either reflective foil or bulk insulation depending on the climatic zone
- ensuring that thermal or heat bridges are eliminated
- depending on the climatic zone (see the National Construction Code), to construct either an open external envelope or an airtight external envelope, using a design that minimises heat gain or loss through its orientation and through natural or mechanical ventilation
- using energy-efficient windows and glazing
- installing efficient heating and cooling systems that have the lowest energy consumption
- using thermal mass in the colder climatic zones.

CHECK YOUR UNDERSTANDING

1. List and describe the four types of energy that a building consumes.

5.7.2 Legislation and sustainability

State and territory governments in Australia, along with building industry agencies, have taken steps to address the major impacts of waste on the environment and the overuse of resources in construction through enacting legal frameworks, regulations and practices to govern sustainability on building sites. The commonwealth legal framework is that of the overall country, ensuring protection of the natural (land protection, water resources, biodiversity) and cultural (Indigenous culture, heritage works) environments. In addition, the Commonwealth provides advisory information and assessment tools for all those in the construction industry. These include:

- the National Construction Code incorporating the Building Code of Australia
- standards for the use and composition of materials and practices (e.g. AS 1684 Residential Timber-Framed Construction)
- energy efficiency guides
- waste management handling guides
- voluntary accreditation schemes, such as NABERS (the National Australian Built Environment Rating System) and Green Star.

State governments provide practical legal regulations for the building industry in each state, such as:

- applicable building standards
- health, sanitary and workplace safety standards
- environmental protection, for example, environmental protection Acts such as the *Environmental Protection Act 2017* (Vic)
- specific state and territory regulations, for example, ACTHERS (the ACT House Energy Rating Scheme), BASIX (the Building Sustainability Index, NSW) and 5 Star Plus (WA).

The local government focus for buildings is:

- planning and building permits to undertake works
- waste disposal and recycling of materials
- the provision of utility services such as water and sewerage.

Other bodies that operate within the building industry offer training, accreditation and information guides on procedures for sustainable practices for tradespersons and builders. These include:

- state and territory Master Builders Associations (e.g. the Master Builders Association of Victoria: www.mbav.com.au)
- the Master Plumbers' and Mechanical Services Association of Australia: www.plumber.com.au
- the National Electrical and Communications Association: www.neca.asn.au
- Housing Industry Association Ltd: www.hia.com.au
- Safe Work Australia: www.safeworkaustralia.gov.au
- Keep Australia Beautiful: www.kab.org.au
- Ecospecifier Global: www.ecospecifier.com.au
- Good Environmental Choice Australia (GECA): www.geca.eco
- Sustainable Choice: www.lgp.org.au/sustainable-choice.

5.7.3 Sustainability practices on the building site

5.7.3.1 Storage of materials

Many materials are delivered to a building site during the construction phase. On a large building or multi-housing development, the amount of materials to be handled can be extremely large. Thus, the sensible placement and secure storage of materials until they are needed is critical. The construction sequence of the project is aided by the logical placement of materials, which eliminates the double handling of materials and improves the efficiency of the workflow (Fig. 5.9).

Fig. 5.9 Logical placement of building materials

Courtesy of Alister Ford

Good storage decisions are crucial to reduce material wastage due to loss or damage. These include the following considerations:

- placing a waterproof cover over materials to protect them from the weather
- separating materials into different groups relevant for each trade
- establishing enclosed storage areas (e.g. wire cages) to reduce accidental physical damage to materials; protect delicate and fragile materials; and contain loose materials, for example, sawdust, sand and gravels
- providing a support structure for relevant materials to prevent distortions such as the warping, bowing and twisting of timber
- establishing a safety enclosure for hazardous materials and substances and ensuring PPE and related safety data sheets are available
- erecting barriers during the construction phase to prevent heavy downpours washing away loose material and waste and stopping polluted run-off entering the stormwater system.

5.7.3.2 Conservation of energy

The amount of energy used on a building site during construction is called *construction energy*, as mentioned earlier. While small in comparison to the operating energy used over the building's life, it still counts for the builder during construction. The main energy source for construction sites is electricity, which in remote locations is generated via diesel or petrol from a generator.

Sources of energy waste are commonly:

- the use of old or inefficient equipment
- leakage from compressed air and LPG equipment that has been left on
- electricity from equipment being left on when not in use, including work lights, computers, air conditioning and vacuum equipment
- blunt equipment such as power saws, routers, drills, cutters, etc.

5.7.3.3 Dealing with waste materials

A distinction needs to be highlighted between waste materials and site rubbish. Waste on a building site usually comprises timber offcuts, sawdust, plastic packaging materials, metal trimmings and broken glass. Site rubbish comprises domestic rubbish, paper wrapping, plastic cups, empty bottles and the like.

Table 5.2 lists the most common waste on building sites.

The builder or site supervisor is responsible for waste materials and site rubbish. The separation of these rubbish types is an important part of good site management (Fig. 5.10).

Table 5.2 Common waste on building sites

Liquid waste	Solvents; contaminated water from the wash-down of equipment; rainwater run-off; any other liquid
Metal	Offcuts, nails, straps, paint canisters, cans
Paper	Cardboard, packaging, boxes, documents
Plastics	Packaging and wrap, straps, Powernail strips, bottles, solid foams, Corflute sheets
Solid waste	Concrete, bricks, tiles, ceramics, plaster, render, sand
Timber	Off-cuts, sawdust, MDF, plywood protection sheets

Fig. 5.10 Building waste and site rubbish should be separated

Courtesy of Alister Ford

CHECK YOUR UNDERSTANDING

1. Explain the difference between waste and rubbish on a building site.

5.7.3.4 Site management

Across Australia, state and local laws require that a building site is clean and ordered to both protect the environment and the workers on-site. On-the-spot fines for site litter and larger fines for waste material and pollution from a site into the immediate environment can be issued by local governments and state agencies. Any breaches or potential breaches must be reported to the relevant authorities immediately. Good site management can prevent this occurring and produce a professional image for the building industry. Other positive outcomes of good site management include the following:

- a well-organised and clean site is a safer workplace
- clients will equate an efficient site with good workmanship
- a clean and efficient site will save money for the client, the builder and each individual tradesperson by minimising material damage and site accidents.

5.7.4 Sustainability terminology

It is important to identify, source and use environmentally and socially responsible materials and products. Table 5.3 lists common sustainability terms to assist with this goal.

Table 5.3 Common sustainability terms

Access	Inclusive design principles should be used to cater for all people with disabilities and all those with specific needs
Biodiversity	The variety of all living things; the different plants, animals and micro-organisms, the genetic information they contain and the ecosystems they form
Carbon dioxide (CO_2)	A greenhouse gas that is colourless and non-toxic; a product of burning fossil fuels and the main contributor to climate change
Climate change adaptation and mitigation	Commonly thought of as extreme weather events such as drought, storms, cyclones, heavy rains and floods caused by the increase in greenhouse gases in the atmosphere brought about by the burning of fossil fuels
Conservation	Preserving the resources we have and using them in a cautious manner that does not deplete them for future generations
Ecological footprint	A measure of human demand on the earth's ecosystems
Ecosystem	An area of natural habitat that contains living organisms. One ecosystem interacts with another, so a disturbance in one will affect another
Embodied energy	The amount of energy used to manufacture and process a building material; measured by the amount of CO_2 released by the process to extract it and manufacture the material
Embodied water	The volume of fresh water used to produce a particular product, measured at the place where it was produced
Energy	Energy comes in many forms and can be converted from one form to another. Energy is always conserved, as governed by thermodynamic laws
Environment	All that surrounds us: air, earth, water, vegetation
Environmental management	The strategic arrangements made to reduce the environmental impacts of a development or an organisation's operations
Environmental site analysis	An assessment of a building site for its vulnerability to flooding, rainfall run-off, bushfire, habitat contamination and endangered species
Fossil fuel	A naturally occurring substance created over millions of years that is used as a fuel
Greenhouse gases	Natural and human-made gases that absorb and re-emit infrared radiation. The main gases are water vapour, carbon dioxide, methane, nitrous oxide and hydrofluorocarbons
Health and wellbeing	Buildings should promote healthy environments and lifestyles for users. This includes workplace health and safety for all site workers and the promotion of healthy lifestyles through eliminating hazardous substances, health checks and extended education programs
Job safety analysis	An assessment of a building site to identify hazards and risks to workers' health
Life cycle	A series of changes in the life of a material or building component or system. Included is the cost of acquiring the product, installing the product, operating it, repairing it over its useful life and disposing of it
Natural materials	Materials of zero-embodied energy found in nature requiring no processing. These include tree trunks and branches, tree bark, tree wattles, and mud, clays, stones and grasses
Natural resources	Raw materials available for human activities, e.g. forests are a resource for timber
Pollution	Contamination of the environment or an ecosystem by substances that are harmful to living organisms
Processed materials	Materials subjected to high-energy or chemical processes, such as reconstituted timber products including hardboards and chipboards; manufactured bricks, including clay, concrete and silica lime products; roofing tiles including terracotta clay-based tiles; and concrete/cement products
Supporting communities	Consideration of the effect of a building development on the local community. The building industry needs to become involved and establish how development will improve an area
Sustainability	The ability to continue activities without exhausting the available resources or degrading the environment. This includes economic, social and natural aspects
Sustainable development	A development or building project that meets the needs of the present without adversely affecting the ability of future generations to meet their needs
Synthetic materials	Materials (usually of high-embodied energy) that have little or no natural content. These include the category loosely called plastics

Trade and labour standards	Consider if clients, contractors and supply chains are meeting accepted labour standards. Ensure there is no exploitation of cheap labour and that acceptable working conditions are provided
Transport and mobility	Consider opportunities for sustainable transport of workers and materials throughout the building phase. Priority of walking, cycling and public transport should be emphasised
Waste	Materials that are surplus to those needed for a project. These need to be minimised by re-using, recycling and recovering throughout the construction phase and across the supply chain
Worked natural materials	Materials (of low embodied energy) that occur naturally but are mechanically worked to make them more useful in building, such as split and sawn timber, dressed timber, slates, dressed stone, lime and gypsum
VOC	Volatile organic compounds. These compounds become a colourless and pungent gas at room temperature, which causes watery eyes and nausea, skin irritations and difficulty in breathing

5.7.5 Sustainability and suppliers

It is important that the suppliers you use act in a sustainable manner and meet the basic standards of environmental performance. Just as you are applying sustainability principles to your trade, you should expect that others in the supply chain do the same. Most of the highest environmental impact occurs in the supply chain beyond your control.

Recommended sustainability expectations for suppliers:

- They are proactive in improving the sustainability aspects of their products.
- They comply with relevant Australian Government policies and standards.
- They look to reduce the environmental and social impacts associated with the use of their products.
- They are innovative and constantly seek out areas of improvement for the sustainability of their products.
- They are interested in improving sustainability in the supply chain of their products.
- They are responsive to your questions on sustainability.

Student research

1. Visit the Australian Apprenticeships Pathways website, www.aapathways.com.au. Download a copy of the *National Code of Good Practice for Australian Apprenticeships*. Read the document and discuss the following points in written form.
   (a) Why has this code of practice been developed?
   (b) Who should keep a copy of the code of practice in their records at all times?
   (c) What are the five major points in the code of practice that the employer and apprentice will work together to achieve and maintain?
   (d) List the five items regarding the responsibilities and obligations of the employer. Discuss in detailed written form the requirements mentioned in these five items.
   (e) List the two items regarding the responsibilities and obligations of the apprentice. Discuss in detailed written form the requirements mentioned in these two items.

End of chapter activity

Activity 1:

Using Microsoft Excel or Word, create a Gantt chart for a construction project that you have recently worked on with your employer or while participating in any construction training.

- Print the document as if it were to be displayed so that all workers on-site could refer to it.

AUSTRALIAN BUILDING CODE AND STANDARDS

National Construction Code 2016 (Building code of Australia)

AS 1684 Residential Timber-Framed Construction

Tradie talk—The three Rs: reduce, re-use, recycle

Chapter 6

Carpentry hand tools

Learning Objectives

LO 6.1 Understand personal protective equipment uses

LO 6.2 Identify and use measuring and marking equipment

LO 6.3 Identify and use hand saws

LO 6.4 Identify and use chisels

LO 6.5 Construct basic timber joints

LO 6.6 Identify and use hammers

LO 6.7 Identify and use bench planes

LO 6.8 Understand workbench equipment and their uses

LO 6.9 Identify and use cramps

Introduction

Hand tools are a basic necessity for any good tradesperson. Learning how to identify and use hand tools properly and confidently will aid any tradesperson in their day-to-day work. The ability to use hand tools properly will provide the skills and knowledge for the proficient use of hand-held power tools.

In this chapter you will learn how to identify common timber joints and construct these joints using the tools outlined in this chapter. You will be able to tell when your hand-tool techniques are improving because the quality of the joints will improve. You will also learn how to sharpen chisels, hand-plane blades and hand saws.

6.1 Personal protective equipment (PPE)

Personal protective equipment (PPE) is the most important 'tool' a tradesperson will use. It will help keep you safe and protect you from being injured while you work by minimising risks to your health and safety. This section lists the essential forms of PPE you will need when working. Always look for the Australian Standard logo so you know the item of PPE meets the relevant standard.

6.1.1 Clothing

Protective clothing includes overalls, work pants, work shorts, high-visibility shirts, jackets and hats specifically designed to be used while working in the building industry. These are made of materials that won't burn or melt, can't be easily ripped and provide protection from UV radiation. Essentially, they are much more durable than normal clothing and are designed to protect or shield your body from workplace hazards and injury.

6.1.2 Work boots/shoes

Safety footwear is designed to protect your feet from getting injured and support your body so you can work effectively. They come in many shapes and sizes with different levels of protection offered for the building industry. This includes various grades of toe cap, reinforcement that provides built-in protection at the front of the boot/shoe. There is a huge range of designs available, some looking like regular runners or traditional leather boots.

6.1.3 Eye protection

Eye protection for the building industry is designed to protect you from hazards such as flying particles, dust and harmful gasses. A range of eye protection is available, including goggles and safety spectacles, some with the appearance of designer sunglasses. Made of materials that don't shatter or break easily, all eye protection must be manufactured to the relevant Australian Standard. Eye protection is designed to be close fitting so there are no large gaps between the eyewear and the face to prevent objects from getting in.

6.1.4 Ear protection

In the building industry, ear protection is necessary when working in or near a noisy environment. The most common protection devices are earmuffs that cover the entire ear or plugs that are inserted into the ear canal. Both are designed to dampen down loud noises so your hearing isn't damaged. The better the quality of device, the better the protection.

6.2 Measuring and marking equipment

6.2.1 Four-fold rule

The four-fold rule (or carpenter's rule) is made of boxwood or plastic and is for general use, especially benchwork. It is one metre in length and marked in millimetres. For accurate measuring, hold the rule on edge so that the markings will be in contact with the timber (Fig. 6.1).

> **TIP** The four-fold rule can be used for more than just measuring. It can also be used to gauge lines on timber and to determine the size of a gap. For example, the thickness of the rule when folded up is generally 10 mm, which is the same thickness as the plasterboard found on most house walls.

6.2.2 Retractable measuring tapes

Retractable tapes are available in lengths from 2 m to 8 m and are particularly useful for construction work.

6.2.3 Long fibreglass/steel tapes

Long fibreglass/steel tapes are available in lengths of 20 m and 30 m and are used for measuring on-site, setting out and large construction work (Fig. 6.2 and 6.3).

Fig. 6.1 Four-fold rule: (a) general view; (b) holding and measuring

(a)

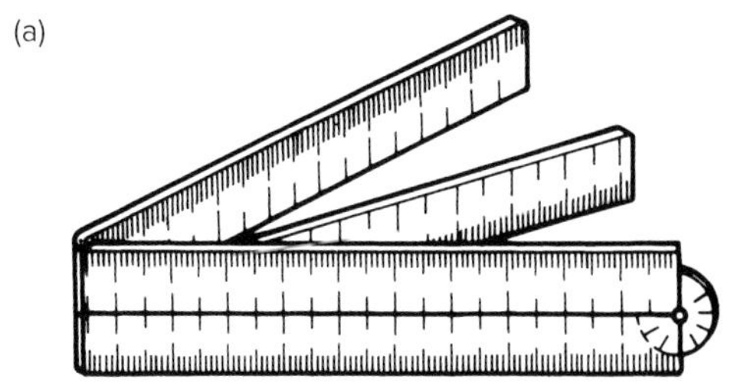

(b)

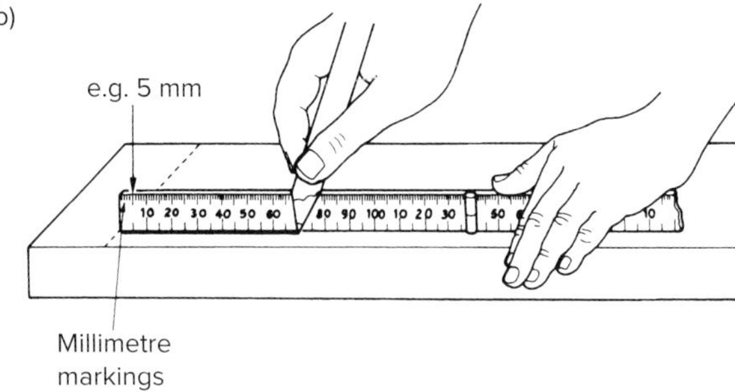

Fig. 6.2 Retractable measuring tape

Fig. 6.3 30 m tape

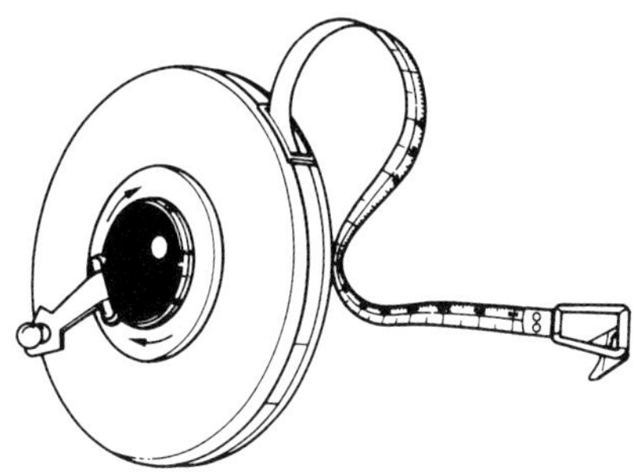

Fig. 6.4 Trammel points

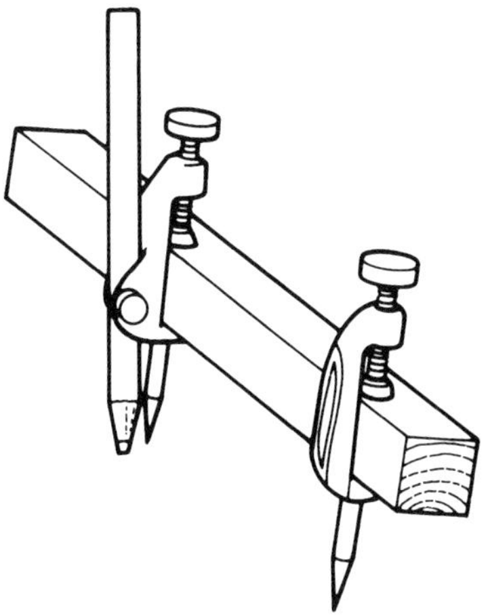

6.2.4 Trammel points

Trammel points can be attached to a wooden bar to form a pair of dividers and are used for stepping off large distances or marking out circles (Fig. 6.4).

6.2.5 Carpenter's pencil and crayons

A carpenter's pencil has a large flat lead, which stays sharp for longer, especially when working with rough-sawn timber (Fig. 6.5). The pencil is available in soft, medium or hard grades. Medium is suitable for most purposes.

TIP Generally, carpenter's pencils are red in colour (medium grade) and bricklayer's pencils are green (hard grade). Bricklayer's pencils would not really mark timber but would rather dent it because they are so hard.

Lumber crayons are variously coloured markers used for conspicuous marking on rough-sawn timber surfaces.

6.2.6 Try square

A try square is used for marking lines at right angles to a face or edge and for checking that timber is square (Fig. 6.6). It consists of a stock and a blade, which is permanently fixed at 90° to the stock. The size of the square is determined by the length of the blade, which can vary from 100 mm to 300 mm.

Fig. 6.5 Carpenter's pencil

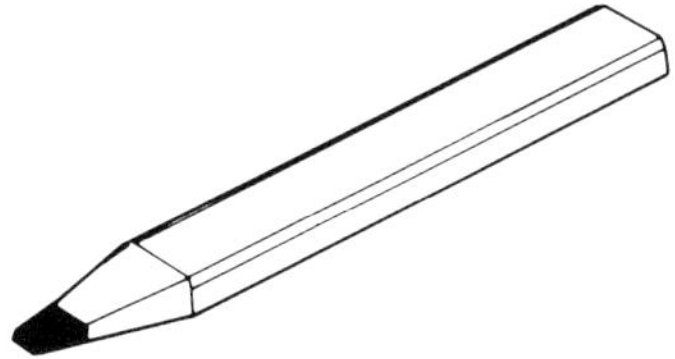

For accuracy, try squares must be checked from time to time (Fig. 6.7). This is done by selecting a board with a straight edge and using the try square to mark off a line at right angles. Reverse the square, and if the blade of the square coincides with the line it is accurate.

Fig. 6.6 Try square

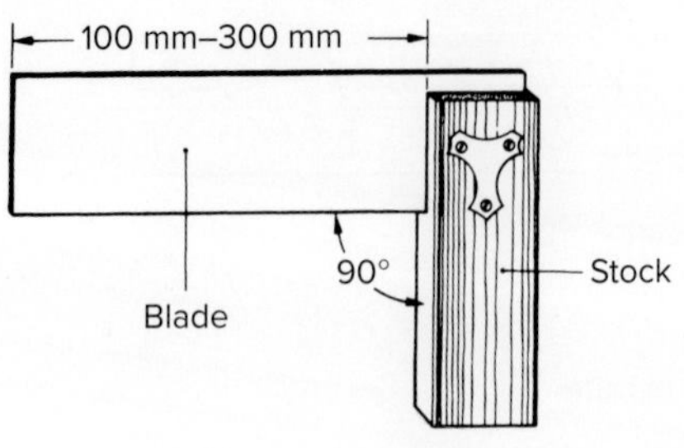

Fig. 6.7 Checking a try square

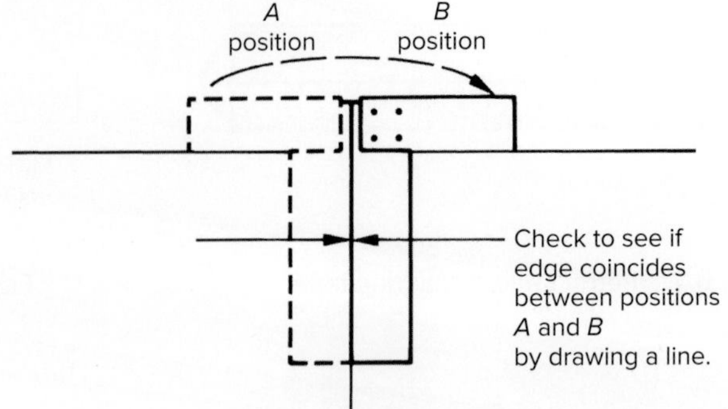

Fig. 6.8 Combination square

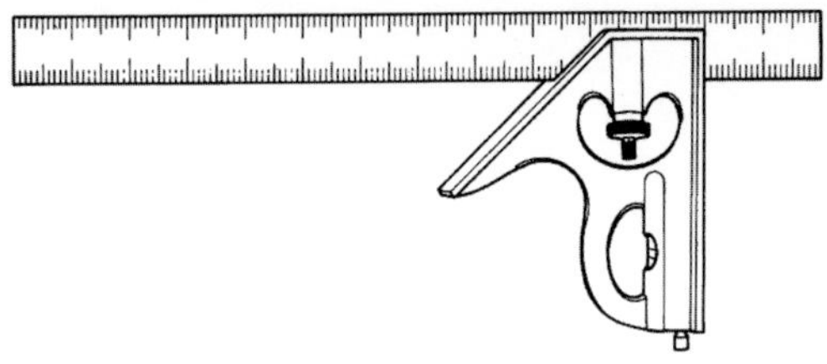

6.2.7 Combination square

A combination square can be used to set lines at 90° and 45° (Fig. 6.8). The blade is 300 mm long and is adjustable on the stock. Sometimes a level bubble is built into the stock.

6.2.8 Steel carpenter's square

The carpenter's square is used for squaring large sheets of material and for setting out in roofing and stair construction (Fig. 6.9). It has a tongue and a blade, usually 400 mm × 600 mm, marked in millimetres.

Fig. 6.9 Steel carpenter's square

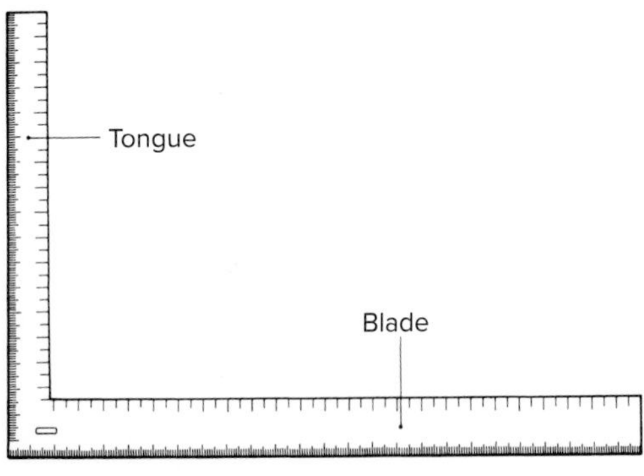

6.2.9 Sliding bevel

A sliding bevel is used to set various angles (Fig. 6.10). The blade is adjustable to the stock.

Fig. 6.10 Sliding bevel

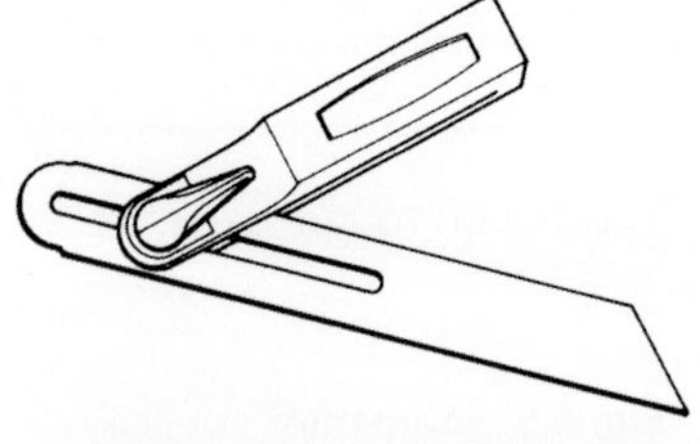

6.2.10 Marking gauges

Marking gauges are a group of tools used for marking lines parallel to an edge or a face (Fig. 6.11). The gauge is usually made of beech wood and the stock is adjustable on the stem.

6.2.11 Mortise gauge

A mortise gauge has two spurs, one of which is movable (Fig. 6.12). It gauges two lines at once and is used mainly in marking out mortise-and-tenon joints, slip tongues and in similar situations.

CHECK YOUR UNDERSTANDING

1. How long is a carpenter's four-fold rule? In what units is it marked?
2. What tool in the carpenter's kit could be used to mark off angles at 35°?
3. What is the purpose of a marking gauge?

Fig. 6.11 Marking gauge

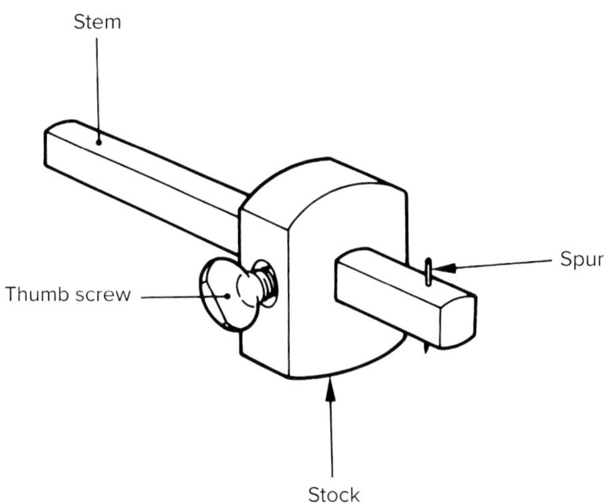

Fig. 6.12 Mortise gauge

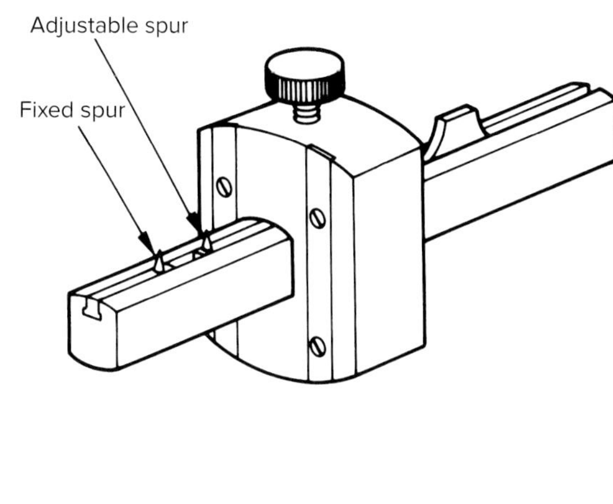

6.2.12 Chalk line reels

Chalk line reels are a convenient way to mark straight lines, mainly on construction work (Fig. 6.13). The string is dusted with chalk, and stretched between two given points. It is then flicked to produce a visual straight line.

Fig. 6.13 Chalk line reel

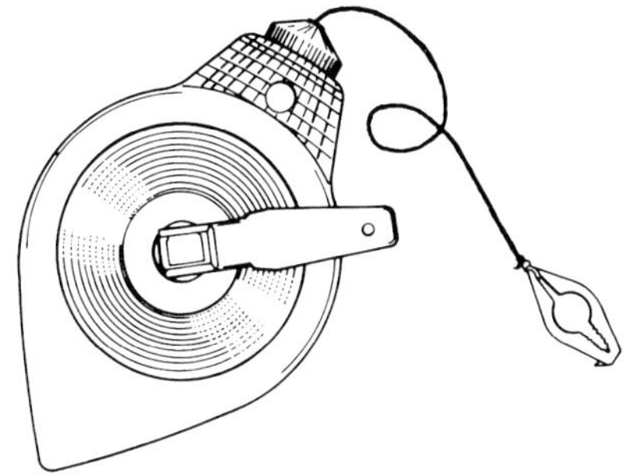

6.2.13 Marking knife

This is used as an alternative to a pencil. It has a sharp blade used to mark accurate lines, where a saw or chisel is to be used. It is ideal for marking shoulders of tenons (Fig. 6.14).

SAFETY TIP Always make sure the hand tools you are using are sharp—a sharp tool is far safer than a blunt tool.

6.3 Hand saws

Woodworking hand saws are manufactured from high-quality steel, specially formulated and tempered to provide the right degree of hardness and flexibility so that the teeth will maintain their sharp edge but will not break off while in use or when being set and sharpened (Fig. 6.15). A good-quality saw, when held by the handle and tip, should bend into a uniform curve and return to a straight line when released. The handle is made of wood or plastic moulded to fit the hand comfortably. The best guide to quality is undoubtedly the reputation of the manufacturer and the price. Saws suitable for professional use may not always be the cheapest initially, but in skilled hands they will give many years of faithful service.

Fig. 6.14 Marking knife

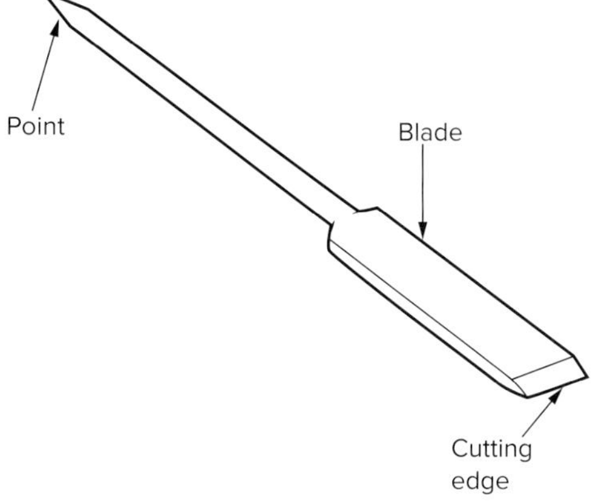

Fig. 6.15 Typical hand saw

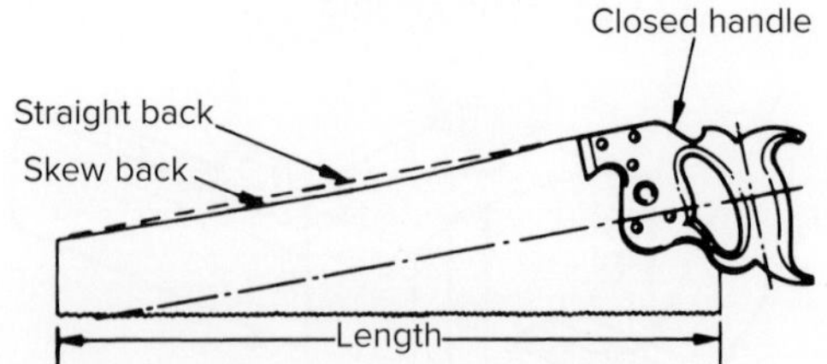

In recent years a number of new types of saws have come onto the market to meet the changing needs of the trade; for instance, saws not only for use in timber but also for plastic, metal and other sheet materials.

Teeth are formed on the cutting edge of the saw and are shaped according to the purpose of the saw.

6.3.1 Shape of saw teeth

Timber is composed of wood fibres overlapping and lying in the direction of the grain. The saw teeth must sever the fibres cleanly and remove the waste from the saw cut. *Rip saws* are designed to cut in the direction of the grain; the teeth form a series of chisel points that sever the fibres and the wood crumbles away in front of the teeth in the form of sawdust (Fig. 6.16).

Fig. 6.16 (a) Rip saw teeth; (b) cutting action

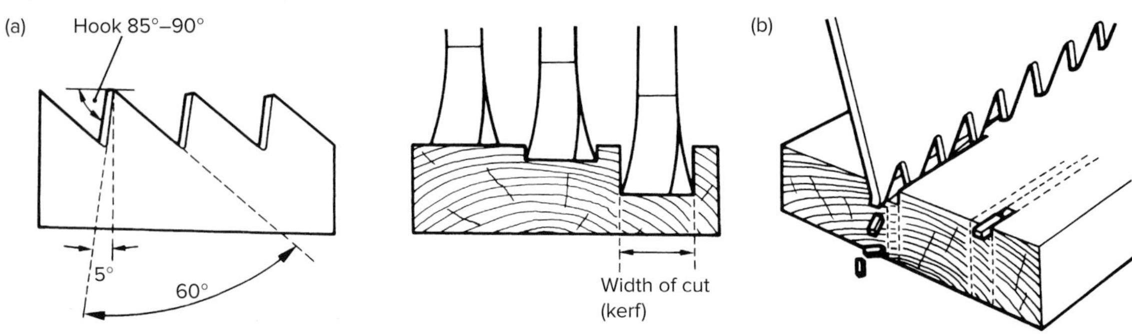

Saws designed to cut across the grain of the timber are shaped like a series of knife edges that sever the fibres in two places; again, the wood crumbles away between the cuts. These are called cross-cut saws (Fig. 6.17).

The angle formed between the face of the teeth and the length of the blade is referred to as the *hook* and is important to the performance of the saw. With a large angle of hook–up to 90° for rip saws–the

Fig. 6.17 (a) Cross-cut teeth; (b) cutting action

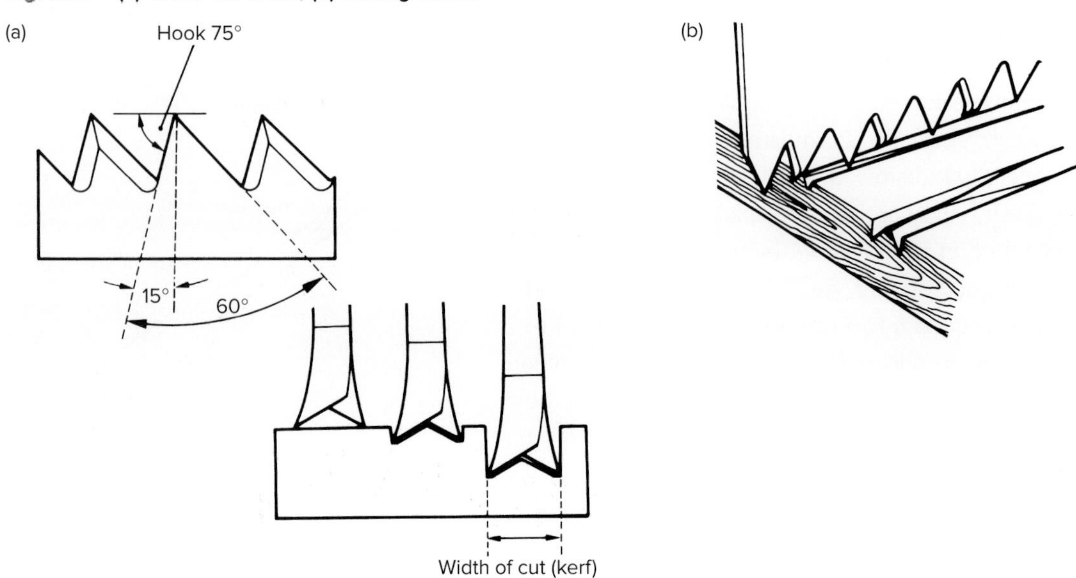

Fig. 6.18 Size of saw teeth

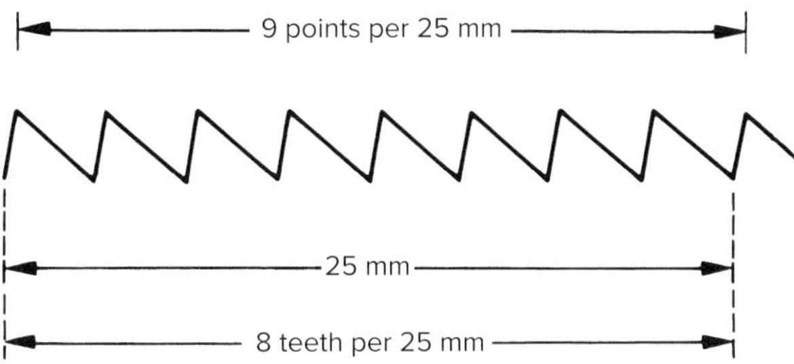

saw will cut more quickly, but the cutting will be harsh and absorb more power. If the face of the tooth is laid back by only a few degrees, the cut will be smoother and require less effort.

Fig. 6.19 (a) Saw grip; (b) stance; (c) supporting the end

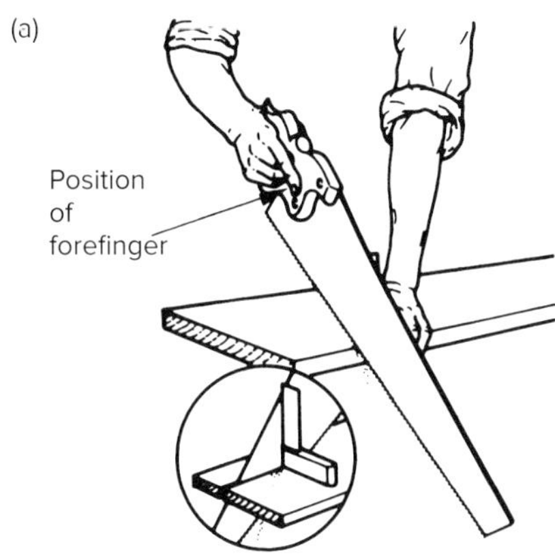

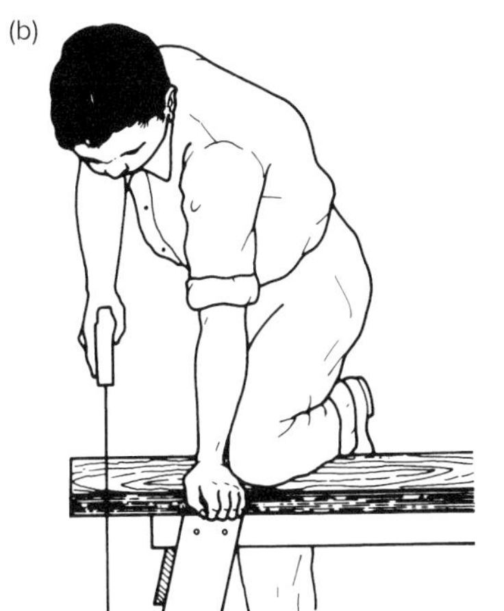

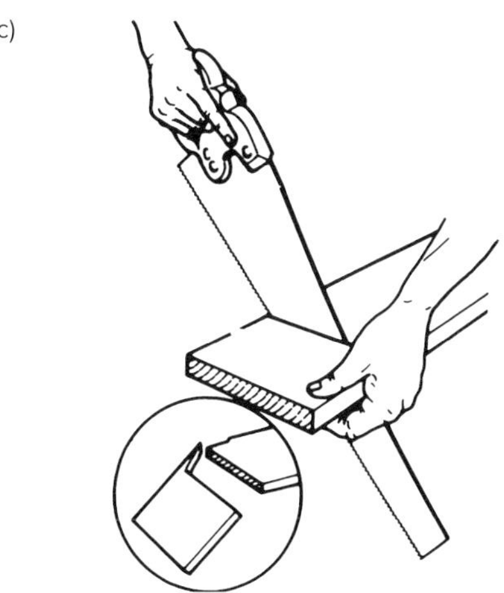

6.3.2 Clearance to the saw blade

So that the blade will not bind in the saw cut, or **kerf**, the teeth of the saw are *set*. This is the practice of bending the top half of each tooth in opposite directions, so that the saw kerf will be approximately 1.5 times the thickness of the blade. Soft green timber will require more set than hard dry timber.

Good-quality saws are also *taper ground*, which means that the back of the saw blade is thinner than the cutting edge, thus also providing clearance in the cut.

6.3.3 Size of saw teeth

The size of the teeth is expressed as the number of teeth to 25 mm (Fig. 6.18). Sometimes the term 'points per 25 mm' is used, which includes the point at the start and finish of the 25 mm so that eight teeth per 25 mm (8 TP 25 mm) equals nine points per 25 mm (9 PP 25 mm).

6.3.4 Using hand saws

The correct grip for all saws is shown in Figure 6.19(a). This helps to relax the hand muscles and gives direction to the saw blade. Timber should be supported on saw stools and the correct stance adopted (Fig. 6.19b). Notice how the saw, the forearm and the shoulder are in a straight line. Commence with the saw at a low angle, in line with the direction of the cut, and use short, light strokes. Guide the saw with the thumb of the other hand. Gradually increase the angle to approximately 45° when **cross-cutting**, and 60° when **ripping** (Fig. 6.20).

Fig. 6.20 Sawing angles: (a) cross-cutting and (b) ripping

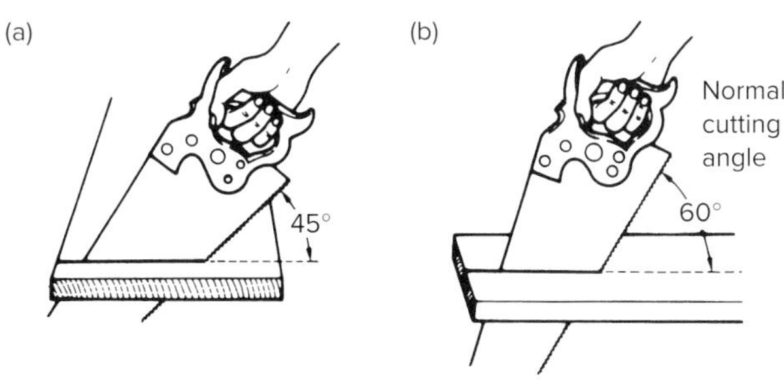

6.3.5 Hand saws in general use

Without putting any downward pressure on the saw blade, maintain the stance and continue the cut using the full length of the blade, with the arm swinging freely in line from the shoulder. Support the end of the timber so that it will not collapse and splinter at the end of the cut (Fig. 6.19c).

Most saws manufactured in the West have adopted a tooth shape that is 'leaning forward' (Fig. 6.21a). This type of tooth arrangement will cut more aggressively on the forward stroke and will more or less slide across the fibres on the backward stroke. By altering the angle on the front of the tooth, it will cut well both across and along the fibres of the timber. The angle between the teeth is maintained at 60° so that the teeth can be sharpened by using a triangular saw file.

Another tooth pattern now receiving wide recognition is the *straight tooth* (Fig. 6.21b). This will cut in the same manner both on the forward and backward stroke. It is used on some hand saws and bow saw blades and for cutting across the grain. (Some 'old timers' in the trade may recall seeing this tooth shape on rare occasions; they would refer to them as 'shark tooth'.)

The 'straight' toothing on carpentry saws was originally a Scandinavian type; in modern production this tooth shape is obtained by a grinding wheel (rather than a file), which gives a superior **bevelled** surface and an extremely sharp cutting edge. The precision required in saw toothing is illustrated by the fact that only 0.1 mm to 0.3 mm of the outermost tooth points actually cut into the workpiece. It is the shape of these tiny tooth tips that determines the efficiency of the saw.

Fig. 6.21 (a) Conventional saw teeth and (b) straight saw teeth

(a)

(b)

Table 6.1 Common hand saws

Name	Length (mm)	Teeth (25 mm)	Purpose
Rip saw	700	3–5	Ripping with grain
Half rip saw	650	5.5 & 6	General purpose, small amounts of ripping and cross-cutting
Hand saw	650	7, 8, 9	General purpose, cross-cutting
Panel saw	500, 550, 600	10	Light interior fitting and cross-cutting

Fig. 6.22 Area of hardened point

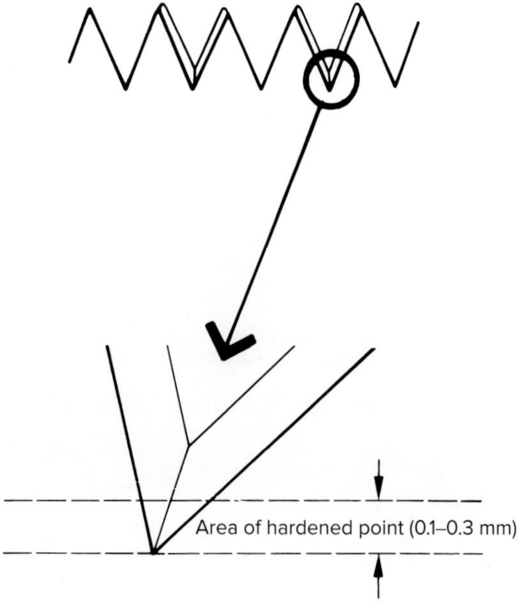

6.3.5.1 Hard-point saws

Manufacturers, having produced saws with tooth points shaped for maximum efficiency and to very fine tolerances, and also realising the difficulties of resharpening them, have turned their attention to making teeth last so long that they will need resharpening only after considerable time and use.

Hard-point high-frequency hardening was the solution to increasing the lifetime of the tooth points (Fig. 6.22). An extra hardening may seem an easy and natural solution, but it is a rather intricate manufacturing process; the steel has to be heated to the exact temperature and then quickly cooled.

Sometimes hard-point saws are criticised on the grounds that they cannot be resharpened. This is only partly correct as they can be resharpened after they have become dull with use (Fig. 6.23), as follows:

1. Put the saw flat on the work bench and take a fine-grained oilstone with a flat surface.
2. Place the stone so that it rests on the blade and just covers the tooth points.

3. Slide the stone with light pressure from the tip of the saw to the handle twice.
4. Turn the saw and repeat the process on the other side.

This process will grind a little off the tip of any teeth that were bent outwards during the setting and will put a neat and sharp edge on the critical 0.1 mm to 0.3 mm of the point. Press only lightly with the stone as it will remove some of the set, and if the saw develops a tendency to jam in the cut, it will have to be reset to give a wider kerf.

Fig. 6.23 'Sharpening' hard-point teeth

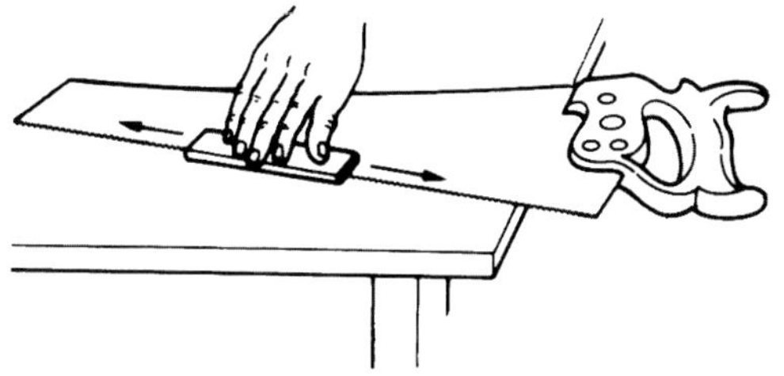

Fig. 6.24 Tenon saw

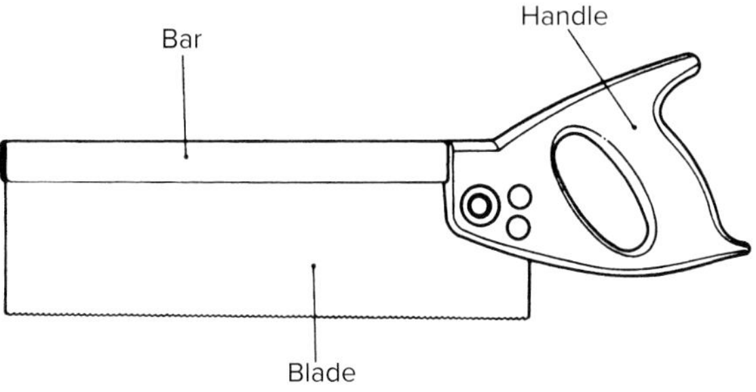

6.3.6 Tenon saw

The tenon saw is used for general benchwork (Fig. 6.24). The blade is straight and parallel, reinforced on the back edge by a bar of steel or brass. Consequently, the saw may sometimes be referred to as a *back saw*.

The length of the tenon saw is 250 mm, 300 mm or 350 mm, with 11 to 15 teeth per 25 mm. A popular choice for the joiner is a 300 mm by 12 teeth per 25 mm tenon saw. As the saw is used mainly for cross-cutting with some ripping operations, the teeth are sharpened with a bevel as for cross-cutting and will provide satisfactory overall performance.

To use the tenon saw, the timber must be held firmly in a bench hook or vice. Grip the saw in the same manner as you would grip other hand saws (Fig. 6.25). Adopt a stance so that the saw cut, the saw and the operator's arm and shoulder are in a straight line. Lift the handle and start the cut with a series of short strokes. Gradually lower the hand and follow the line across the width of the timber until the saw is cutting horizontally. Follow the line down the edge, cutting slightly into the bench hook.

At times, due to misuse, the blade may become buckled and slightly displaced in the back. This can sometimes be corrected by holding the saw around the handle and giving the saw one sharp tap on the back, opposite the buckle, using a block of wood or a mallet. If this does not correct the fault, further attention from a saw doctor must be sought.

Fig. 6.25 Stance for using the tenon saw

CHECK YOUR UNDERSTANDING

1. List at least three differences between saw teeth intended for ripping and those used for cross-cutting.

6.3.7 Mitre box

The mitre box is a jig in which timber can be held and cut accurately at an angle of 45°. It is most commonly used for making mitred or scribed joints to mouldings. The simplest form of the mitre box is made up of timber in the form of a three-sided box, with slots cut to guide the saw.

6.3.7.1 Constructing a mitre box

Material required: 90 × 35 mm pine 1/450 mm, 110 × 19 mm pine 2/450 mm, PVA glue, 4/30 mm bullet head nails, 6/45 mm × 8 gauge countersunk p2 screws.

1. Mark the face and face edge of each piece of timber.
2. Gauge a line using the marking gauge 17.5 mm from the edge of the 19 mm pine along its entire length.
3. Mark in 50 mm along the gauged line from both ends of 19 mm thick timber. This mark is where the nails will go.
4. Make a mark 100 mm from each end and in the centre of the 19 mm thick timber along the gauged line and drill a 3 mm pilot hole at these marks for the screws.
5. Tap the nails into the timber at the corresponding marks. Be sure only to start the nails and not hammer them home.
6. Run a bead of glue along both edges of the 35 mm piece of timber.
7. Position the sides of the mitre box along the base and drive the nails home, pinning the sides and the base together.
8. Drive the screw into the corresponding marks.

6.3.7.2 Setting out a 45° angle

The 45° angle is accurately set out as follows:

1. Measure square across the overall width (*w*) of the box.
2. Set off the distance *w* along the length, and form a square.
3. Mark the diagonals of the square across the top edges.
4. Square down the face and cut the slots accurately with the tenon saw.

Fig. 6.26 Timber mitre box

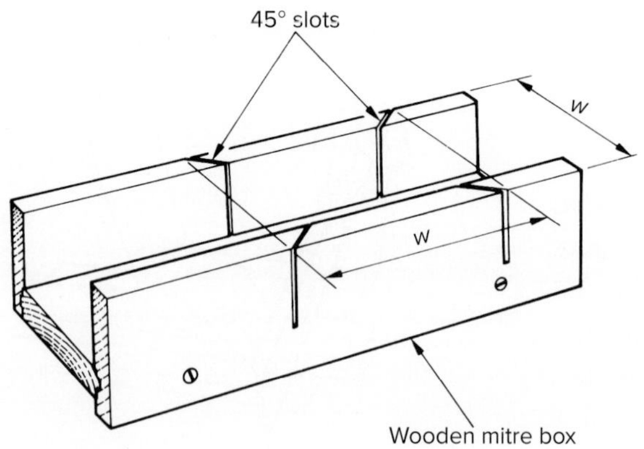

To use the mitre box, hold the moulding firmly against the back of the box with the length mark aligned with one side of the slot, and the saw on the waste side of the line (Fig. 6.27). Make the cut with a fine saw, preferably a tenon saw, cutting slightly into the bottom of the box.

6.3.8 Mitre block

To mitre small mouldings, it is often more convenient to use a mitre block (Fig. 6.28).

This is used in a manner similar to the mitre box. Glue and screw the block together. Mark a 45° angle and cut the slots with a fine tenon saw.

Often timber can be held in the mitre block or box more securely and with less effort if a strip of abrasive paper (say 120 grit) is glued along the back face. A square cut can be included to quickly square the ends of small sections.

6.3.9 Dovetail saw

A dovetail saw is a smaller version of the back saw (Fig. 6.29). It is used for fine, accurate cutting; largely, as the name implies, for ripping dovetail pins. The length of this saw is 200 mm or 250 mm, with 14 to 16 teeth per 25 mm, sharpened for ripping.

The dovetail saw can have an open wooden handle or a straight handle.

Fig. 6.27 Using the mitre box

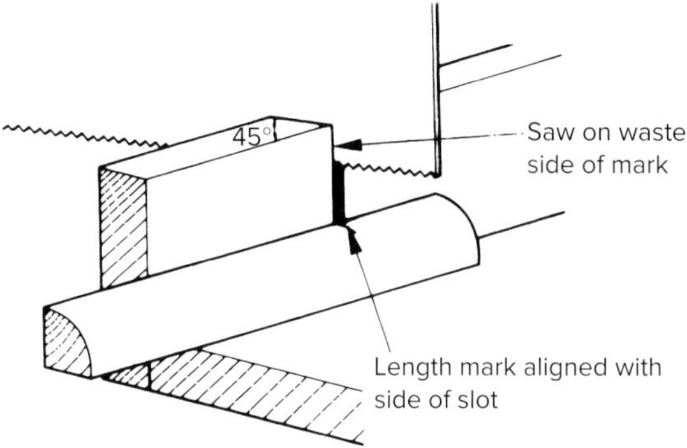

6.3.10 Saws for cutting curves

Cutting to curved lines is now a task more likely to be performed with power tools. A hand saw used for cutting around a curve must have a narrow blade, which may have to be held under tension in a spring-loaded frame.

Fig. 6.28 Mitre block

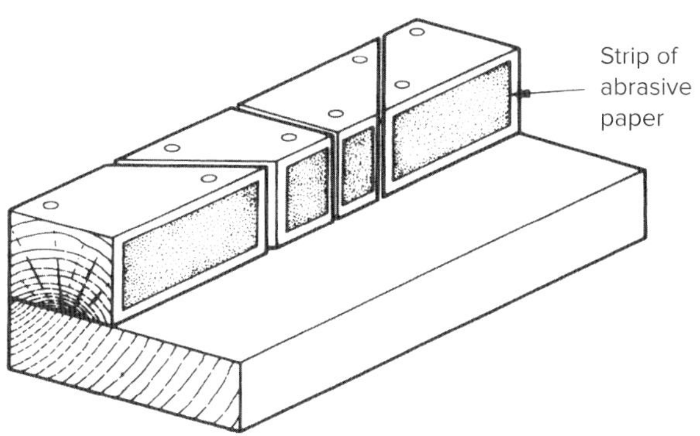

6.3.10.1 Coping saw

The coping saw is used for cutting around tight curves such as scribing mouldings or when removing waste from dovetail pins (Fig. 6.30). The blades are replaceable. They are approximately 150 mm long and 3 mm wide, and are tensioned in the frame by turning the handle. They can be angled to cut in any direction by twisting the pins holding the blade.

Fig. 6.29 Dovetail saw

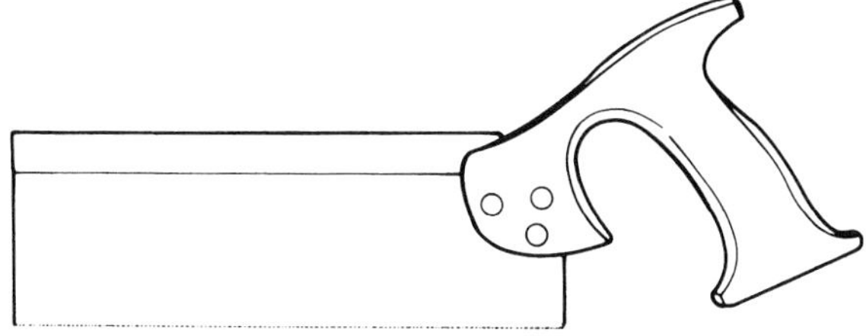

6.3.10.2 Keyhole saw

The keyhole saw is used for cutting around curves, starting closed cuts (see also the power jig saw) or, as the name implies, opening up the straight lower part of a keyhole after the top has been drilled (Fig. 6.31). Traditionally, the keyhole saw was included in a set, consisting of a wooden handle with three blades of different sizes, which was sold as a 'nest of saws'. The three blades are, from the smallest, the *keyhole blade*, the *compass blade* and the *pruning blade*.

Fig. 6.30 Coping saw

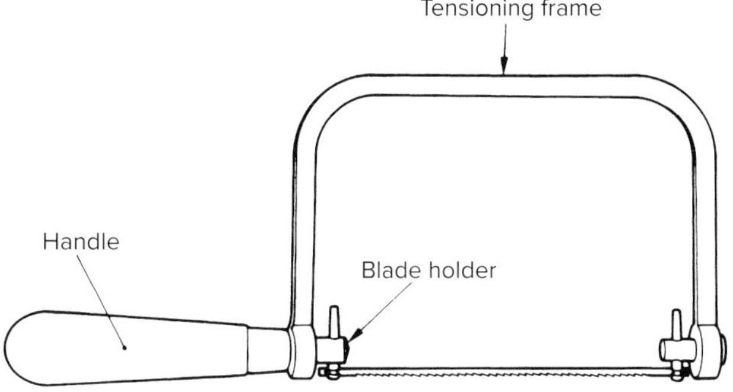

The modern keyhole saw has a metal handle with interchangeable blades of different sizes, suitable for cutting timber, plasterboard and light-gauge sheet metal.

CHECK YOUR UNDERSTANDING

1. What is the main purpose or common use of the following saws?
   a. Tenon saw
   b. Coping saw

6.4 Chisels

Chisels are designed to meet the demands of different operations: some are intended purely for hand use, while others used for heavy work may require the generous use of a mallet. A suitable chisel should be selected for use in each work situation. The size of the chisel is based on the width of the blade, ranging from 3 mm to 51 mm.

Fig. 6.31 Keyhole saw

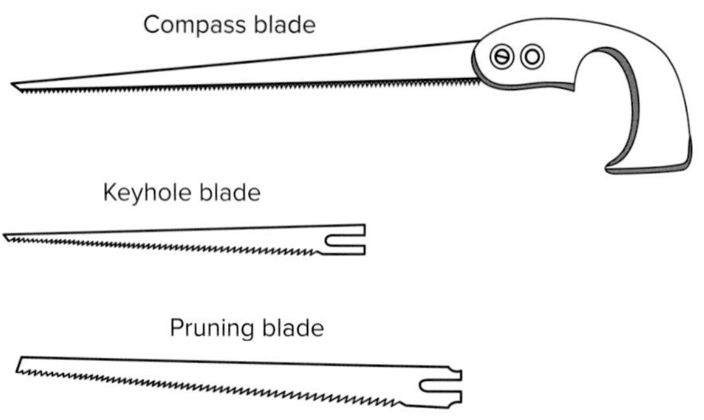

Fig. 6.32 Firmer chisel

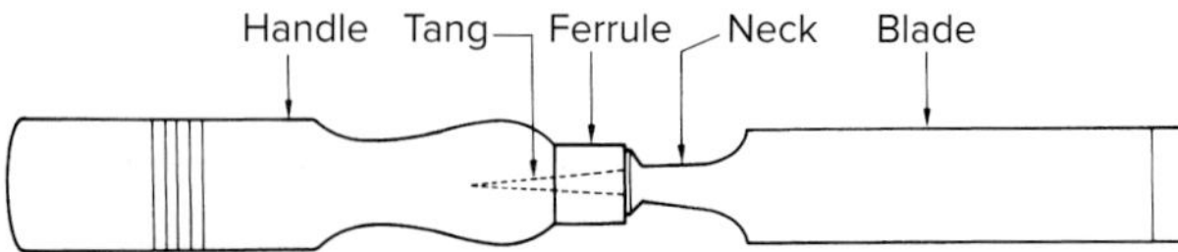

The wood chisel is one of the most basic tools in the carpenter's kit, and consists of a straight blade of specially tempered tool steel, attached to a handle of wood or tough plastic. Other chisels in general use are as follows:

- *Firmer chisel.* This is a general purpose chisel for benchwork or light construction work, where the mallet could be used sparingly. The handle is attached to the blade by the tang and a brass ferrule helps prevent the handle from splitting. Firmer chisels with handles of tough-impact plastic material give excellent service under the strain of heavy work (Fig. 6.32).
- *Bevelled-edge firmer chisel.* The edges of the blade are bevelled off to reduce resistance to the blade, particularly when paring by hand and working into a corner (Fig. 6.33).
- *Registered pattern chisel.* This is designed for heavy work in hard timber. The main features are a heavy blade, a short thick neck and a leather washer between the shoulder and handle. There is also another ferrule at the top of the handle. The registered pattern chisel (Fig. 6.34) is designed to be used with a mallet.

Fig. 6.33 Bevelled-edge firmer chisel

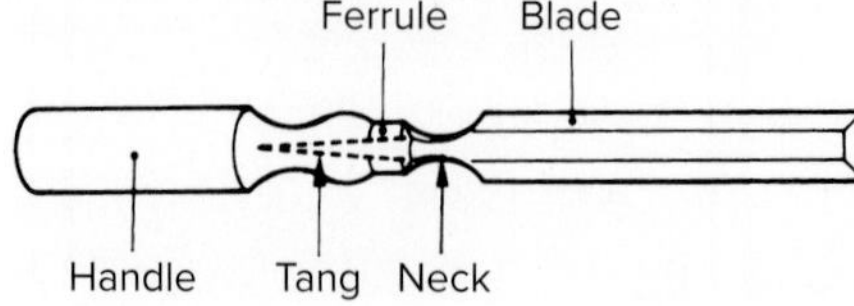

Fig. 6.34 Registered pattern chisel

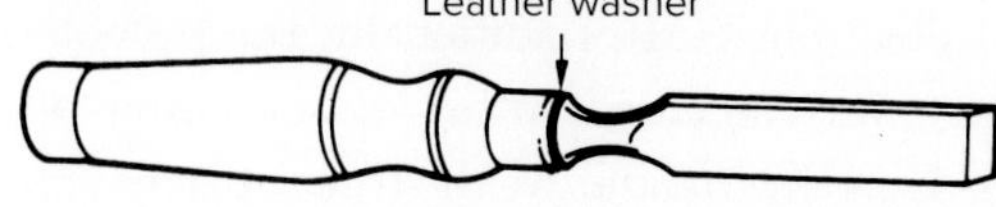

- *Mortise chisels.* These chisels are intended for chopping mortises, using the mallet. Usually they are a socket-type chisel with a thick blade to withstand the heavy use and leverage to which they are subjected.
- *Butt chisel.* This is a short-bladed bevelled-edge chisel for accurately fitting hinges and locks.

The mallet is used for driving the chisel (Fig. 6.35). The weight of approximately 1 kg provides the necessary force without causing damage to the chisel handle.

CHECK YOUR UNDERSTANDING

1. What type of chisel would you select for chopping out housings in hard timber? Give three reasons for your selection.

6.4.1 Oilstones

Fig. 6.35 Mallet

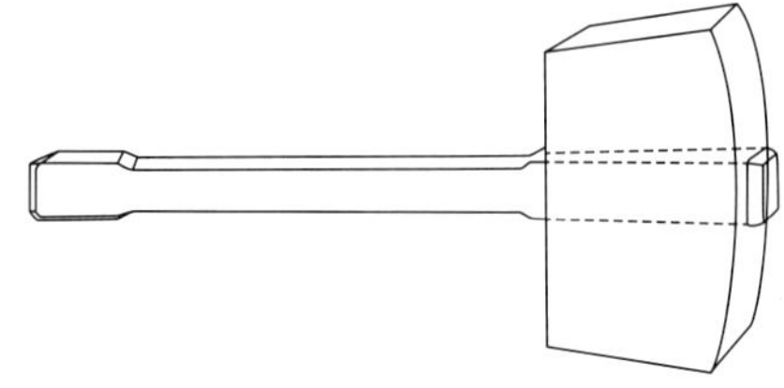

The sharp cutting edge on chisels and other carpenter's edge tools is produced by honing on an oilstone, which is composed of abrasive particles bonded together to form a solid stone (Fig. 6.36a). The length of the stone may vary from 150 mm to 200 mm, and they are usually 50 mm wide by 25 mm thick. **Aluminium oxide** is the most suitable abrasive for obtaining a keen edge on good-quality tools.

As the name 'oilstone' implies, some lubrication is necessary to reduce friction and float away the waste particles of steel, and to prevent the pores of the stone from clogging up. The simplest way to provide a suitable lubricant is to mix motor oil and kerosene in about equal parts, and reserve it for use on the oilstone only. Special lubricants such as neatsfoot oil can also be used.

Stones are available in grades of coarse, medium and fine. The fine stone will produce a sharper edge, but the cutting will be very slow. A combination stone with coarse on one side for quick removal of waste, and fine to produce a sharp edge on the other, can be a very useful stone for all occasions.

Natural stones such as the 'Washita' and the 'Soft Arkansas' will give the keenest of cutting edges on quality tools but they are expensive; the cost is hardly warranted except for those who engage in fine hand woodworking. Oilstones should always be housed in a strong wooden case to protect them from dirt and damage (Fig. 6.36b).

Also available are diamond sharpening stones. These are in the shape of an oil stone, but are impregnated with diamond fragments. They are used in the same way as oil stones, although only water is used as the lubricant. Oil will damage these stones and render them useless.

Fig. 6.36 (a) Oilstone and (b) oilstone case

(a)

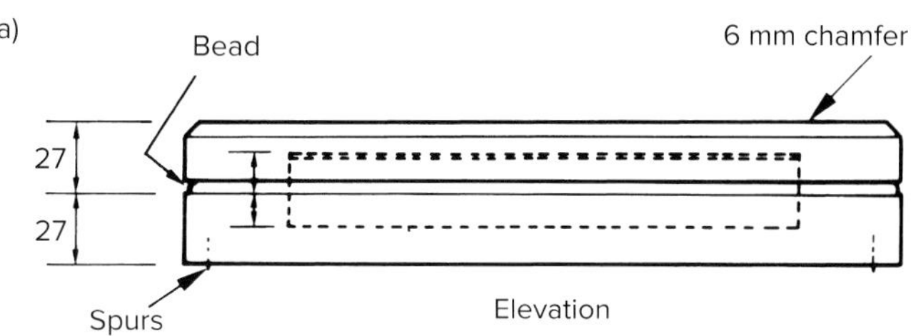

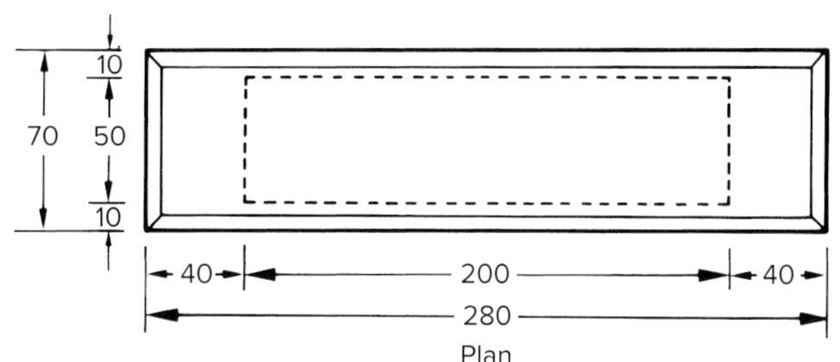

(b)

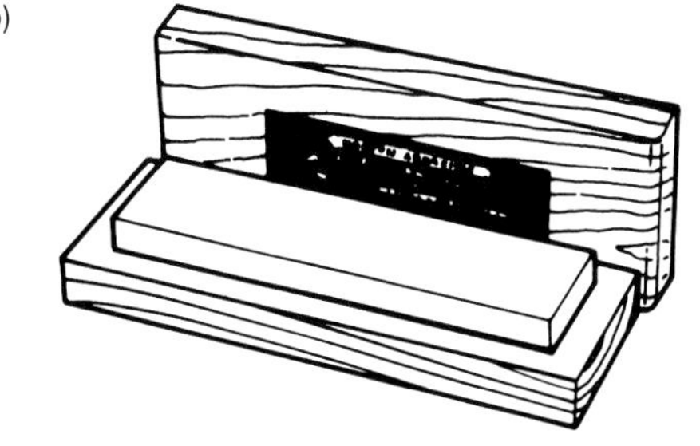

6.4.2 Sharpening chisels

The sharpening and honing of chisels and most carpenter's edge tools involve two angles–a grinding angle and a sharpening angle.

Fig. 6.37 Grinding angle

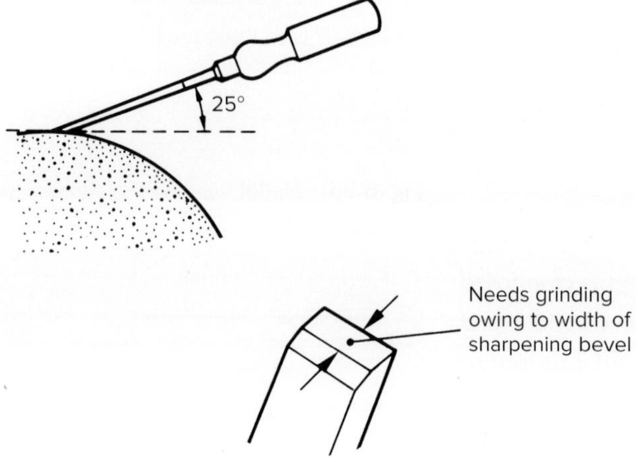

Fig. 6.38 Sharpening angle

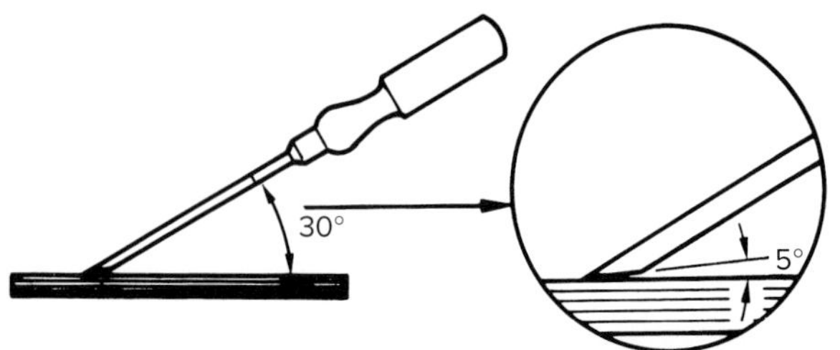

6.4.2.1 The grinding angle

The grinding angle, produced on a grinding wheel, can vary from 25° to 30° and must be square to the blade (Fig. 6.37). When using a grinding wheel, care must be taken that the chisel does not become overheated, so it should be cooled frequently with water.

6.4.2.2 The sharpening angle

As just the point of the blade will do the cutting, only the point needs to be sharpened to a keen edge on the oilstone. The sharpening angle will vary from 30° to 35° (Fig. 6.38). At the lower angle, the chisel would give satisfactory service on light paring work but in hard timber the edge may tend to break away, and sharpening to a greater angle will ensure more reliable service.

To sharpen the chisel, follow these steps:

1. Spread sufficient lubricant on the oilstone and place the grinding angle flat on the face of the stone. Lift the blade a little, up to 5°, to obtain the correct sharpening angle.
2. With a light pressure, rub the chisel backwards and forwards over the stone, or use a figure-eight motion until a burr, sometimes called a *wire edge*, appears on the back of the blade (Fig. 6.39). A figure-eight motion is used to prevent the stone from wearing unevenly.
3. Turn the chisel over and, holding the back perfectly flat on the stone, rub to and fro until the wire edge is removed (Fig. 6.40). It may be necessary to reverse the blade a couple of times, for a few strokes only, until the wire edge finally comes away. This is often seen left lying on the stone. Wipe the stone down before further use.

Fig. 6.39 Chisel on oilstone—feeling for wire edge

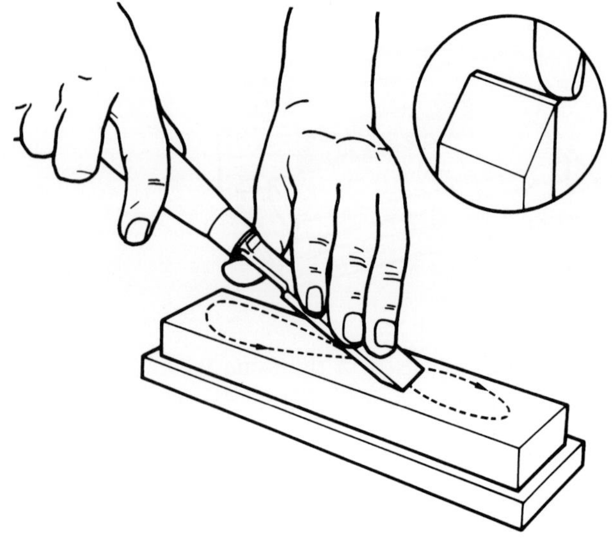

Fig. 6.40 Removing wire edges—turning chisel flat on oilstone

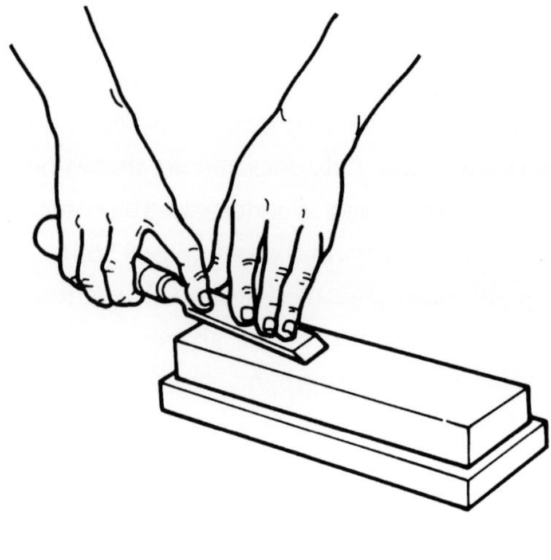

WORKPLACE SCENARIO

It's inevitable that at some point every apprentice will end up hurting themselves with a hand tool. One of the most common injuries I have seen occurs while using a chisel. It's very easy to forget that the leading edge of a chisel is sharp, and apprentices hold the timber they are working on with their non-dominant hand in front of the chisel.

While they are paring the timber, they often push too hard, the cutting-edge slips and pushes into the hand holding the timber and blood starts gushing. The number of stitches they need depends on the sharpness of the chisel.

The following joints can all be produced using the tools that have already been discussed.

6.5 Basic timber joints

6.5.1 Half-lapped/halving joint

The **half-lapped joint** is one of the most commonly used joints, both in detail joinery and construction work and can be referred to as a *scarf joint* when joining timber in length (Fig. 6.41). The setting out of most joints in woodwork requires working from a given face; this is usually clearly indicated by marking the face with a distinguishing mark called a 'face mark', as shown in Figure 6.42.

Fig. 6.41 Half-lapped joint: (a) corner halving; (b) tee-halving; (c) tee-halving with cut plate; (d) extending timber in length (scarf joint)

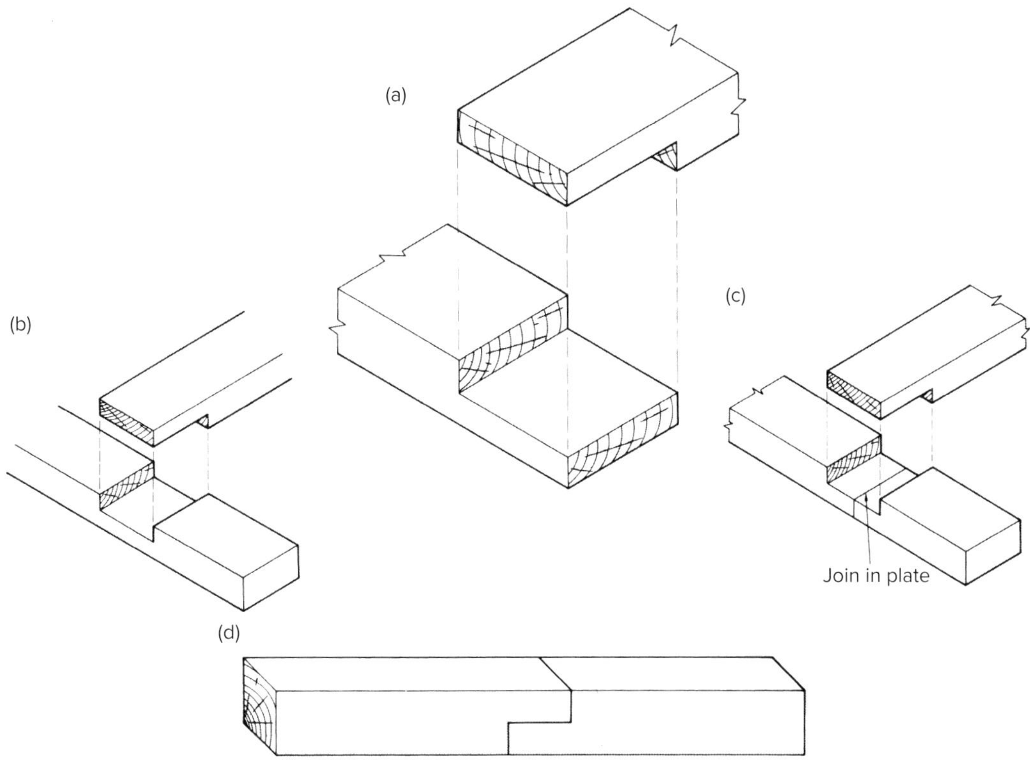

Fig. 6.42 Scarf joint/halving joint set-out

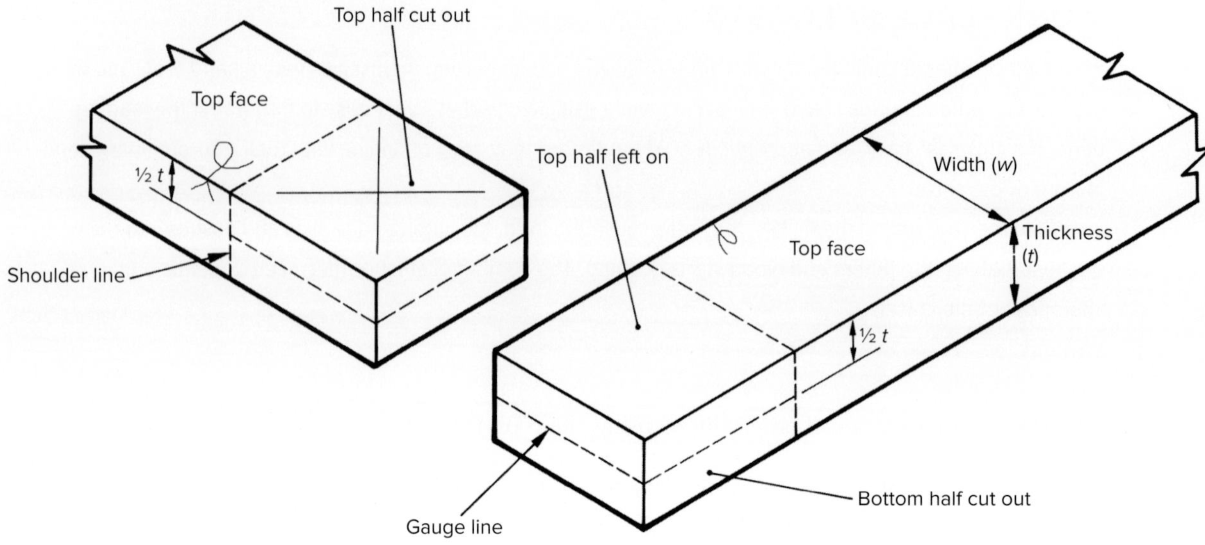

Fig. 6.43 Setting marking gauge

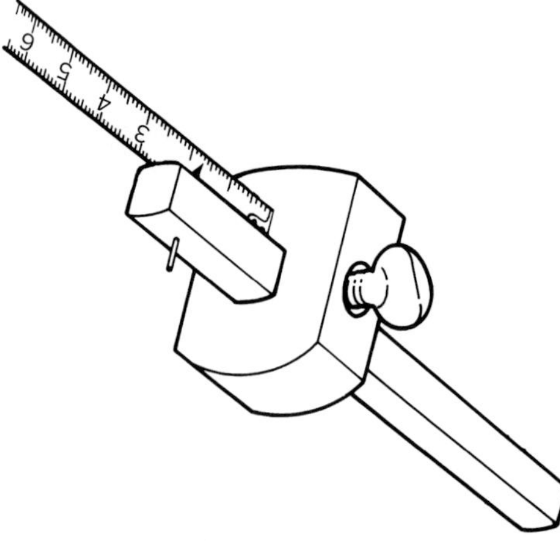

6.5.1.1 Setting out the half-lapped joint

The procedure is as follows:

1. Square the ends of the two pieces of timber to be jointed and mark the working faces.
2. Measure back from the ends the distance equal to the width of the material *(w)*, and square the shoulder line around the timber.
3. Set the marking gauge to half the thickness of the material *(t)*, and with the stock of the gauge against the face, gauge around the three sides of the joint on both pieces of timber (Fig. 6.43).
4. Mark the waste half to be cut out with a distinct cross.

> **TIP** Ensure that both pieces of timber are gauged the same distance from the face so that when the joint is cut out, the amount left on one piece should be the same as the amount cut out of the other, and the face will be flush.

Fig. 6.44 Holding marking gauge

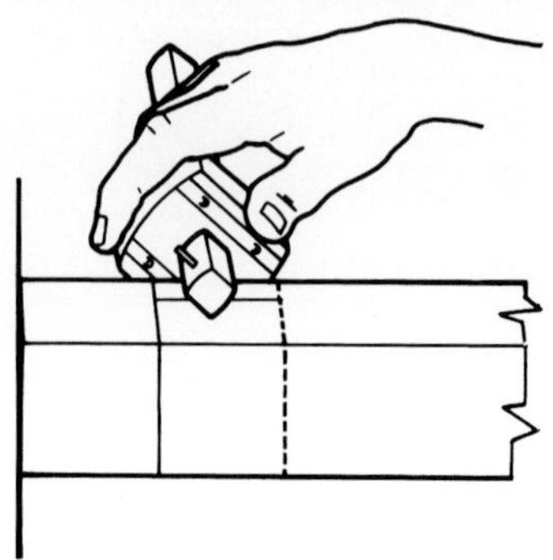

6.5.1.2 Cutting out the half-lapped joint

Select a suitable saw. This job involves small amounts of ripping and cross-cutting without the need for a high-quality finish, so a panel saw would be an appropriate choice.

When cutting away a 'cheek' from the end of the timber, as in this case as well as a number of other joints, the golden rule is, *rip first and cross-cut second.* The reason for this is that we can rip slightly beyond the shoulder line and the strength of the joint will not be affected. Then cross-cut just up to the first cut. To cross-cut beyond the gauge line would seriously reduce the strength of the joint.

6.5.1.3 Ripping the half-lapped joint

1. With the timber supported on stools, rip diagonally on the waste side of the line for half the joint.
2. Reverse the timber and continue the cut from the opposite side, taking care to keep on the waste side of the line. Lift the handle of the saw to finish the cut slightly beyond the shoulder line.
3. Lay the timber flat on the stools and make the cross-cut on the waste side of the shoulder line, just up to the first cut. Check for accuracy–the shoulder should be a close fit, and the joint should be square and the face flush.

Fig. 6.45 Process of sawing—ripping

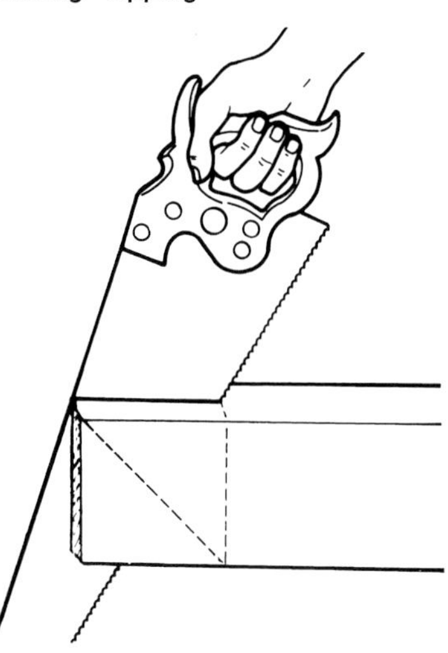

6.5.2 Housed joint

The housed joint is also a commonly used joint in all kinds of carpentry work. Figure 6.47 illustrates an example.

6.5.2.1 To set out the housed joint

1. Locate the position of the housing on the face of the timber and measure the distance equal to the width of the corresponding piece of timber. Square two lines across the depth and partly down the sides. Figure 6.47 shows the designation of the terms *w* (width) and *d* (depth).
2. Set the marking gauge to the amount left on the timber and gauge the housing from the top or bottom face as the case may be.

Fig. 6.46 Process of sawing—cross-cut up to first cut

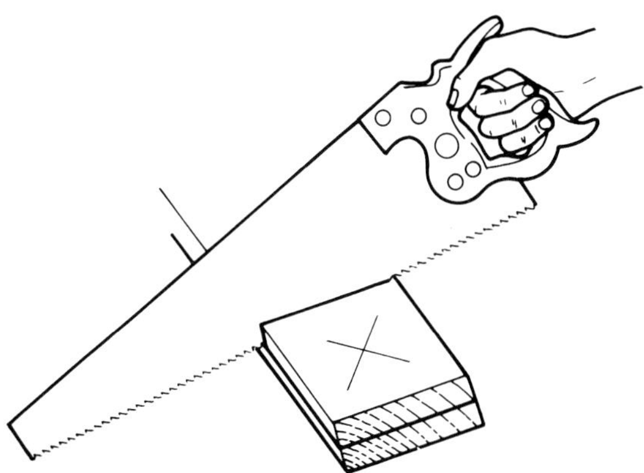

Fig. 6.47 Housed joints (a) horizontal and (b) vertical

(a)

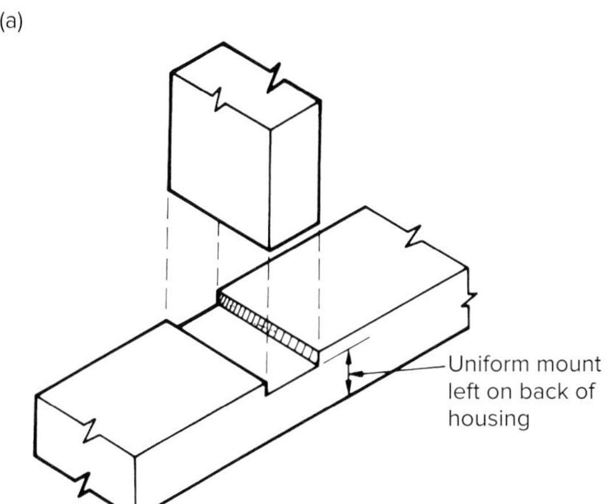

(b)

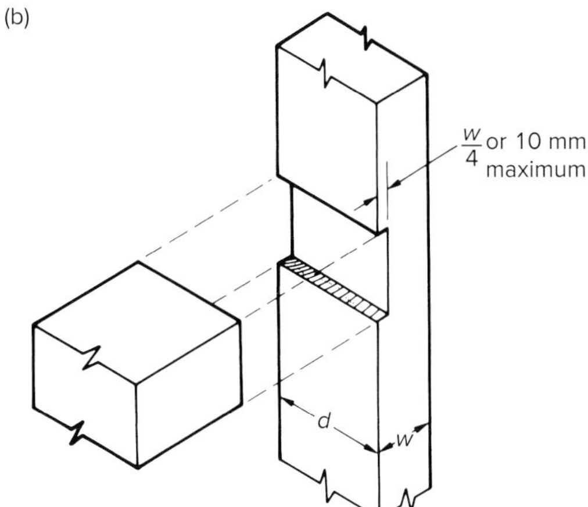

6.5.2.2 Cutting the housed joint

1. Using a hand saw, cut down the sides of the housing, keeping on the waste side of the line and taking care not to cut beyond the gauge line.

Fig. 6.48 Housed joint set-out

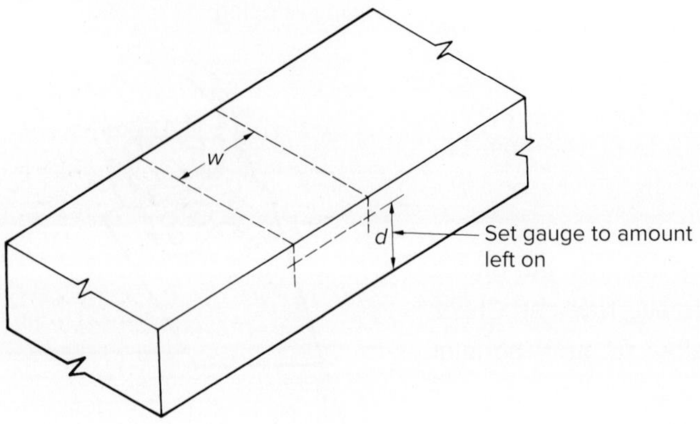

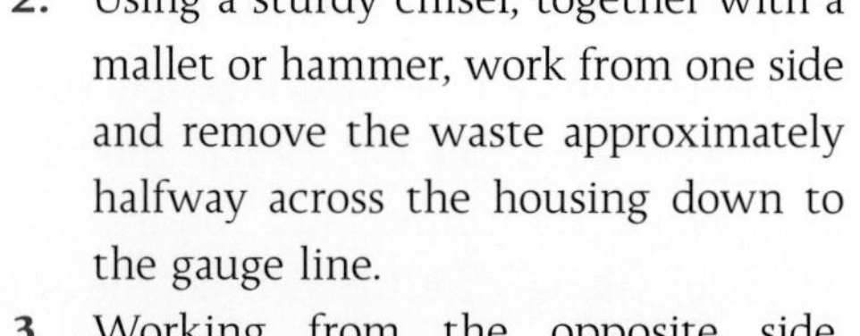

2. Using a sturdy chisel, together with a mallet or hammer, work from one side and remove the waste approximately halfway across the housing down to the gauge line.
3. Working from the opposite side, remove the remainder of the waste.
4. Paring by hand, smooth the bottom of the housing. Note how the chisel is gripped between the thumb and forefinger, which keeps it under control at all times (Fig. 6.49).

Fig. 6.49 Cutting the housed joint

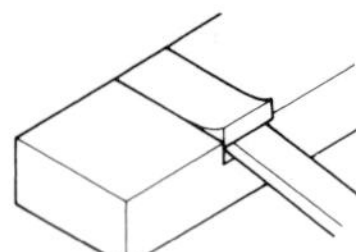

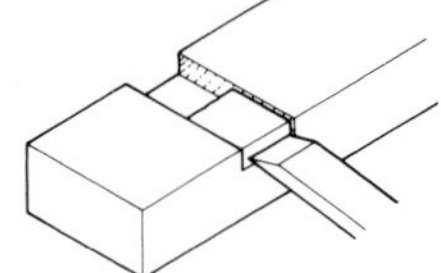

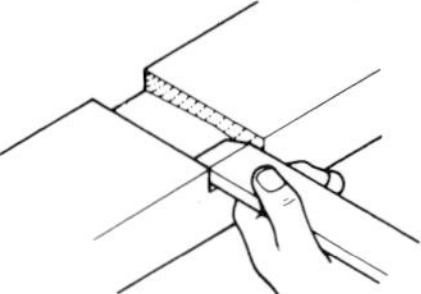

6.5.3 Butt joint

The **butt joint** shown in Figure 6.50 is extensively used in timber framing.

6.5.4 Stopped housed joint

The housing in a **stopped housed joint** does not continue right across the face of the timber; Figure 6.51 shows an example.

Fig. 6.50 Butt joint

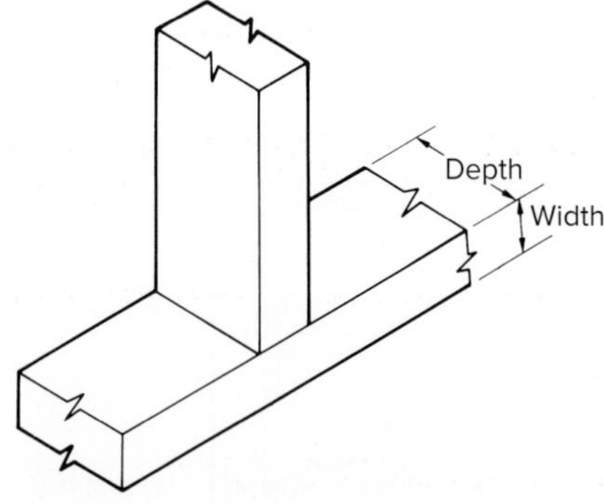

Fig. 6.51 Stopped housed joint

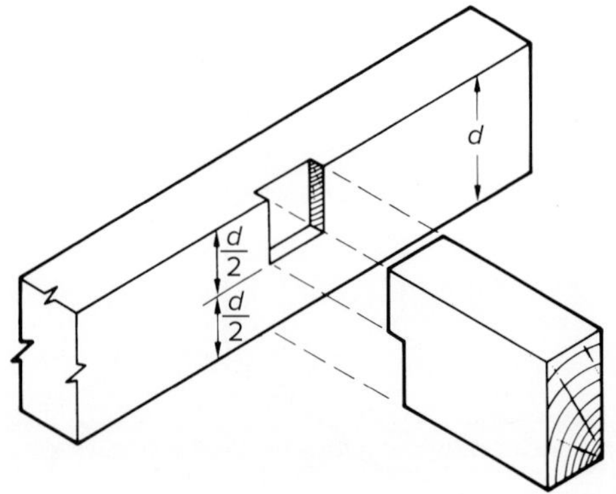

6.5.5 Splayed housed joint

The splayed joint shown in Figure 6.52 will serve the same function as the stopped house joint.

6.5.6 Notched joint

The example shown in Figure 6.53 is another application for floor framing. The joist has been notched to leave a uniform depth where it will be supported on a timber bearer, or in some cases a steel beam. A modification of the notched joint is the *birdsmouth*.

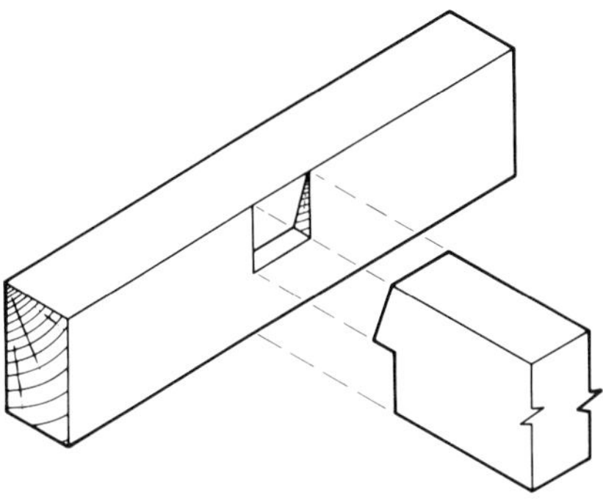
Fig. 6.52 Splayed housed joint

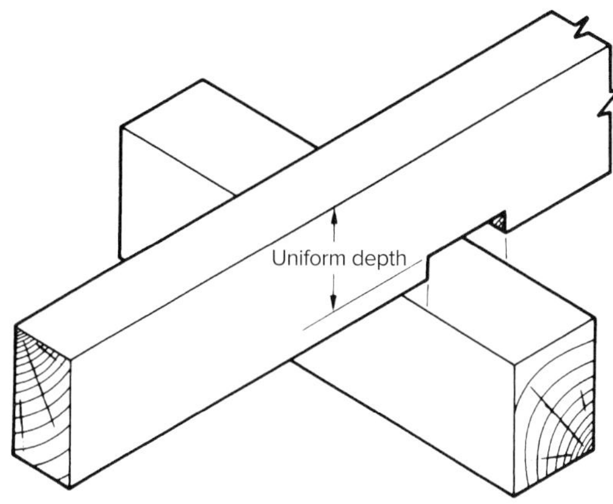

Fig. 6.53 Notched joint

6.6 Hammers

6.6.1 Claw hammers

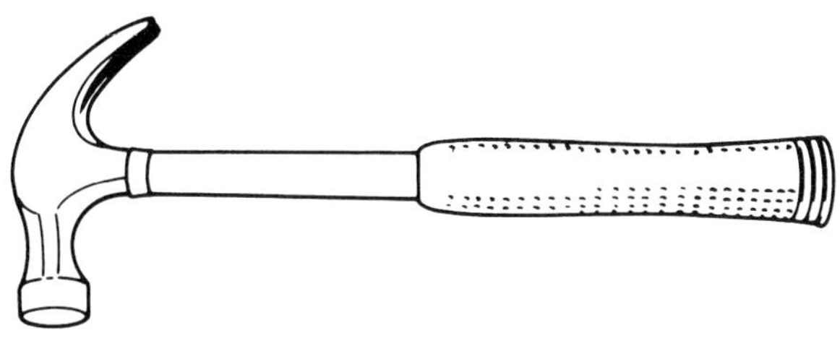
Fig. 6.54 Claw hammer

Claw hammers are essential equipment for the construction carpenter (Fig. 6.54) and are used, of course, for the driving and extracting of nails (Fig. 6.55). Hammers have a drop forged head with a handle or shaft that may be made of wood, fibreglass or steel, with a shockproof hand grip, which is very popular. The size of the hammer is expressed as the mass of the head and common sizes are 450 g, 570 g and 680 g, or, in imperial units, 16 oz, 20 oz and 24 oz. The face of the claw hammer is case hardened and should never be used to strike another hammer or other hard metals, as the hardened face may chip away. For working on construction carpentry, a heavy hammer is the most suitable.

TIP There are many varieties of claw hammers; some have straight claws, some curved claws and some larger heads. Make sure you purchase a hammer that is suitable for the type of work you will be doing and that is not too heavy. Many carpenters end up with tennis elbow because the hammer they use is too heavy.

6.6.2 Warrington hammer

This pattern of hammer is often favoured for light benchwork (Fig. 6.56). It has a tapered cross pein, which is used to start panel pins held between finger and thumb. The polished head varies in mass from 100 g to 450 g and is fitted with an ash handle.

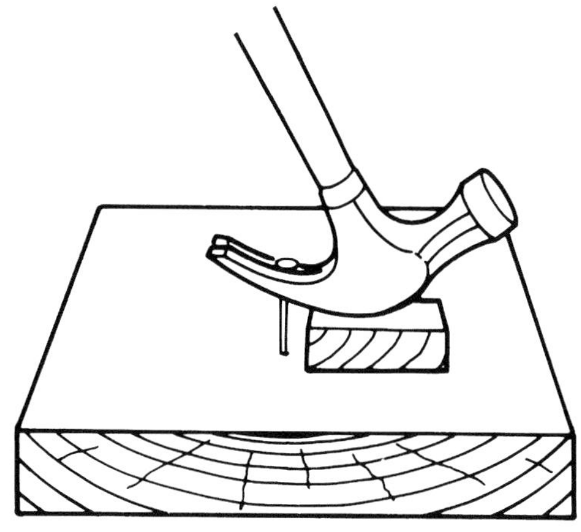
Fig. 6.55 Using the claw to extract a nail

6.6.3 Nail punches

Bullet head nails, used in most finish work, are punched slightly below the surface with a nail punch, which has a concave tip to prevent it slipping off the nail (Fig. 6.57). The diameter of the tip can be 0.8, 1.5, 2.3, 3.3 or 4 mm. Select a tip size approximately equal to the dimension of the nail head.

Fig. 6.56 Warrington hammer

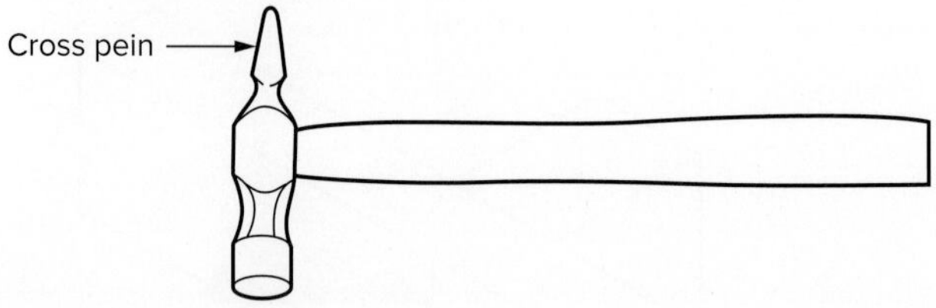

6.6.4 Pinch bar

Pinch bars are also called *wrecking bars* and are used for demolition work, pulling large nails or for levering building units, such as wall frames, into position. The pinch bar has a claw at one end, and an offset chisel point at the other (Fig. 6.58).

Fig. 6.57 Nail punch

Fig. 6.58 Pinch bar

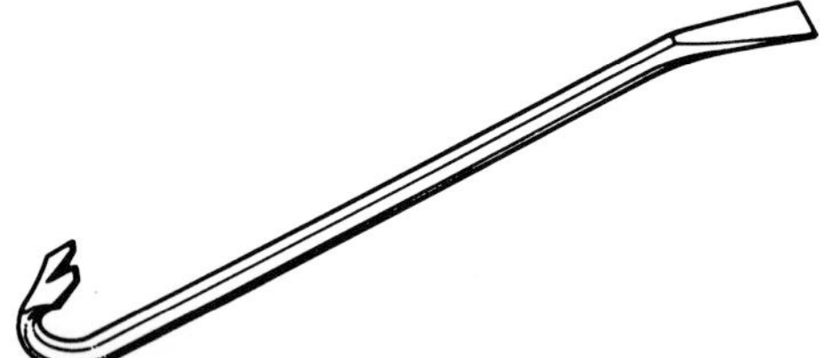

6.7 Bench planes

Bench planes generally refer to a group of planes that are similar to each other in construction but vary in size. Each has a particular function to perform. Figure 6.59 shows a smoothing plane. Planes in the group are listed in Table 6.2.

The term 'fore plane' has now fallen into disuse. It may be regarded as a short trying plane since it is sharpened in a similar manner.

The construction of a bench plane is illustrated in Figure 6.60.

Table 6.2 Types of bench planes

Name	Length (mm)	Blade width (mm)
Smoothing plane	240 & 260	45, 50 & 60
Jack plane	355 & 380	50 & 60
Fore plane	455	60
Trying plane or jointer	560 & 610	60 & 65

Fig. 6.59 The smoothing plane

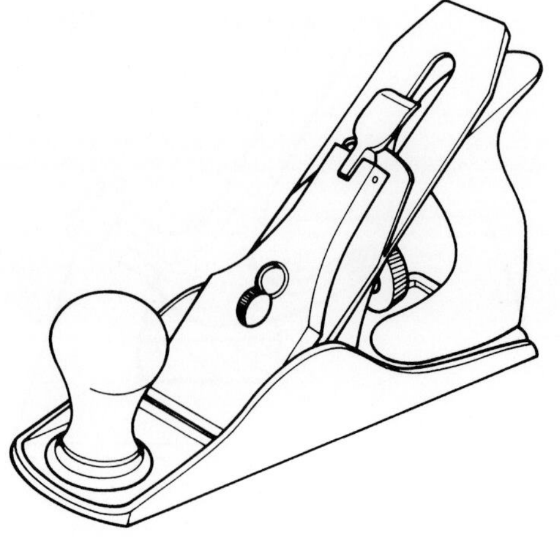

Fig. 6.60 Exploded view of the bench plane

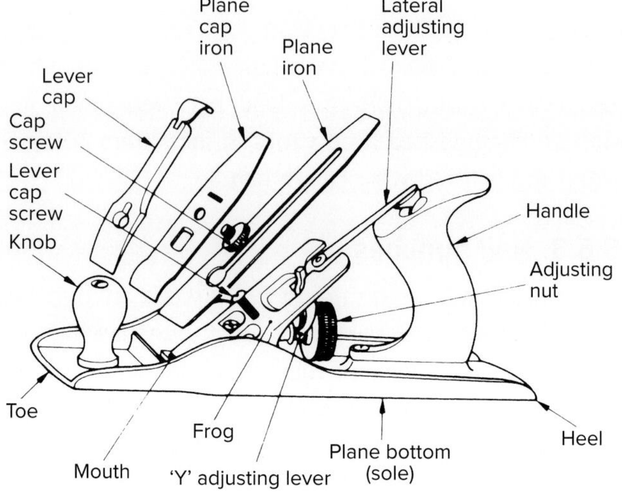

6.7.1 Plane iron

The plane iron consists of two main parts:

1. the cutting iron
2. the cap iron or back iron.

These are held together by the *cap iron screw*. The cutting iron does the actual cutting and the cap iron is provided to stiffen the cutting edge and prevent vibration and 'chattering' (Fig. 6.61). The cap iron also performs the important function of putting the curl into the shavings so that they will roll out and clear the waste from the mouth of the plane.

The cap or back iron must fit tightly down on the cutting iron; otherwise the shavings will become wedged in the gap and clog the mouth of the plane.

The distance the cap iron is set back from the cutting edge (Fig. 6.62) depends mainly on the nature of the timber. Difficult curly-grained timber requires the cap iron to be set as close as possible, say 0.5 mm, to the cutting edge and a very fine shaving taken. On milder timber, the cap iron can be set back as far as 2 mm and a heavier shaving taken.

The cutting iron, in common with most other edge tools, is sharpened with two distinct bevels (Fig. 6.63):

1. the grinding bevel (25°)
2. the sharpening bevel (30°).

The shape of the blade across the face varies (Fig. 6.64); the explanation for this will become apparent when considering the purposes of the bench planes.

Fig. 6.61 Action of the plane iron

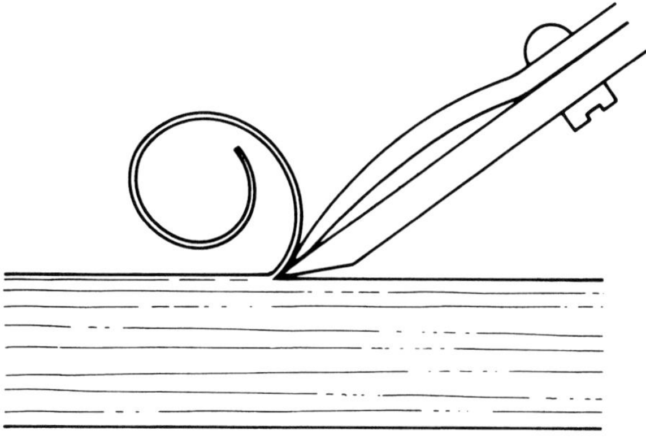

Fig. 6.62 Adjustment of the back iron

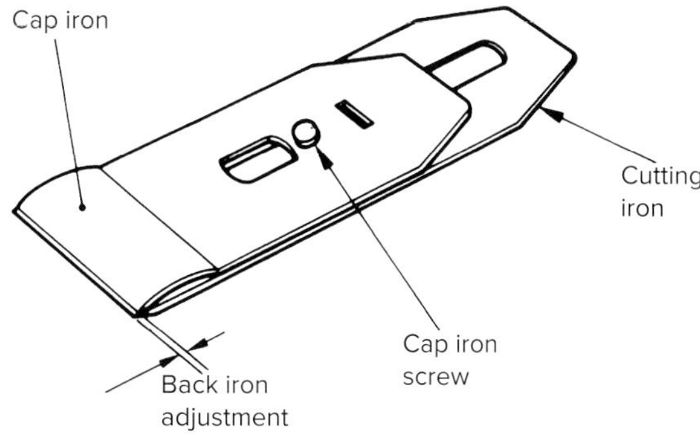

Fig. 6.63 Sharpening angles

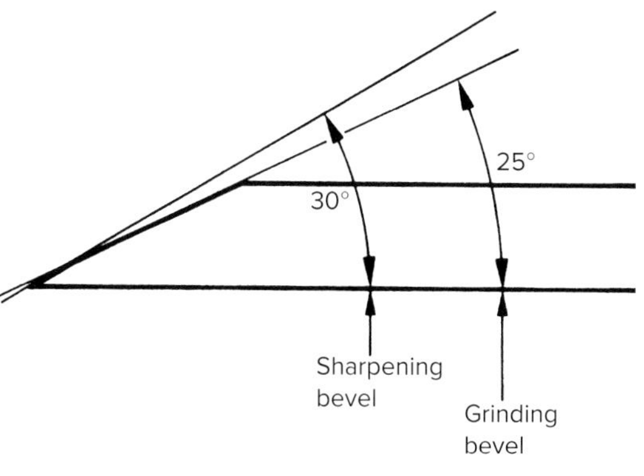

6.7.1.1 Grinding the cutting iron

To remove the blade from the plane, lift the lever and remove the lever cap (Fig. 6.65). The plane iron can then be lifted from the body of the plane. Separate the cap iron from the cutting iron by laying the blade flat on the bench and using a large screwdriver to loosen the cap iron screw. Turn the blade over, slide the cap iron back and turn it through 90°. Now slide the cap iron forward and remove the screw through the hole in the blade.

To grind the blade:

1. Set the tool rest square to the grinding wheel, lay the blade on the rest and, moving it from side to side, grind the edge just sufficiently to remove any gaps. The grind should finish straight and square, or in the case of the jack plane, with a slight camber (Fig. 6.66). Check with the try square.
2. Adjust the tool rest to grind at the correct angle, which is 25°. Lay the blade on the rest and, moving it from side to side, grind it at this angle until the thick edge is almost removed.

Fig. 6.64 Blade shapes

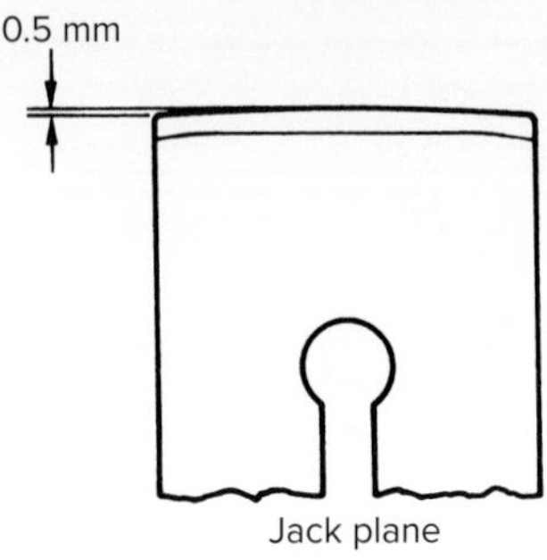

Fig. 6.65 Removing the cap iron

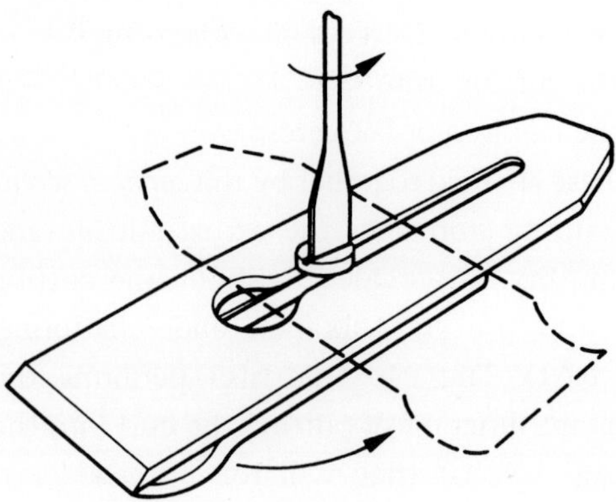

Fig. 6.66 Squaring the plane iron

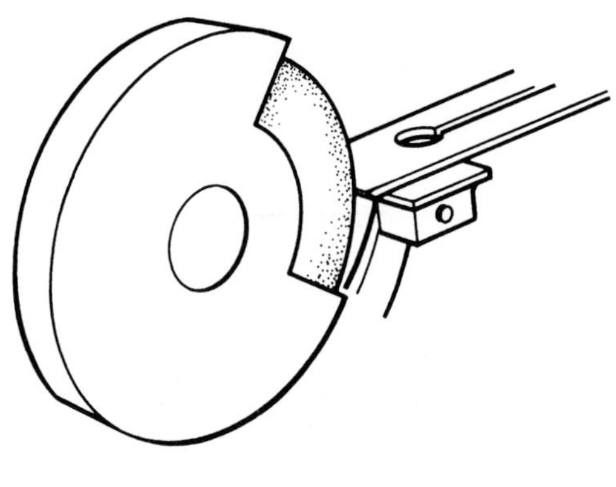

Fig. 6.67 Grinding the bevel

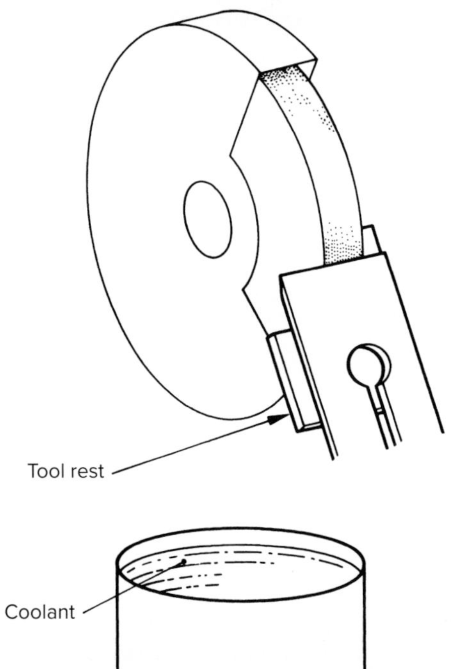

3. Hold the edge between the fingers and up to the light; any remaining thick edge will be visible, showing where further grinding is required. However, do not attempt to produce the final cutting edge on the grinder.
4. Tilt the blade sideways and, with just a slight touch, remove the sharp corners if necessary.

SAFETY TIP During grinding, it is most important that the blade is kept cool. Keep a container of water nearby and cool the blade frequently. The first sign of overheating is when the surface of the blade turns a light straw colour. Stop immediately and cool the blade. Overheating can ruin a good plane iron as the cutting edge becomes soft due to the loss of temper, and it will not hold a keen edge.

6.7.1.2 Sharpening the cutting iron

Producing a sharp cutting edge requires a fine oilstone or, even better, a good natural abrasive stone. Figure 6.68 shows the way to grip the plane iron to hold it at a constant angle.

1. Rest the grinding angle of the blade on the stone and raise the blade through 5°. Move the blade back and forth using the full length of the stone. Alternatively, move it with a figure-eight motion until a burr appears on the back of the blade.
2. Turn the blade over and, holding the blade perfectly flat on the stone, remove the burr. It may be necessary to reverse the blade for a few strokes to finally remove the burr, which will probably be seen left lying on the stone.

Assembly of the cap iron is the reverse of the removal procedure, except that the cap iron is slid forward to the required setting from the cutting edge. The blade is then turned over and the screw tightened.

Fig. 6.68 A & B Gripping the plane iron

(a)

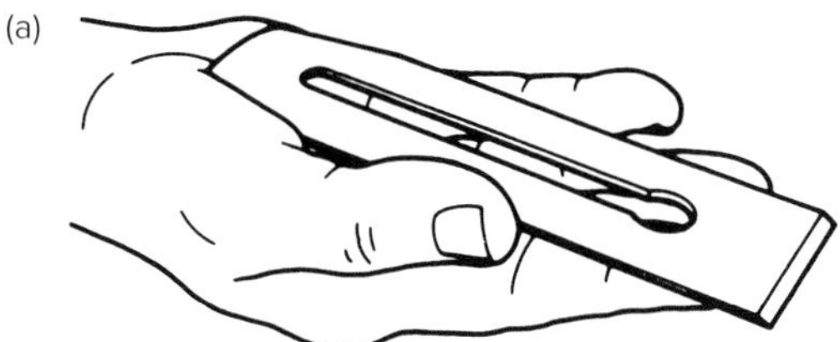

(b)

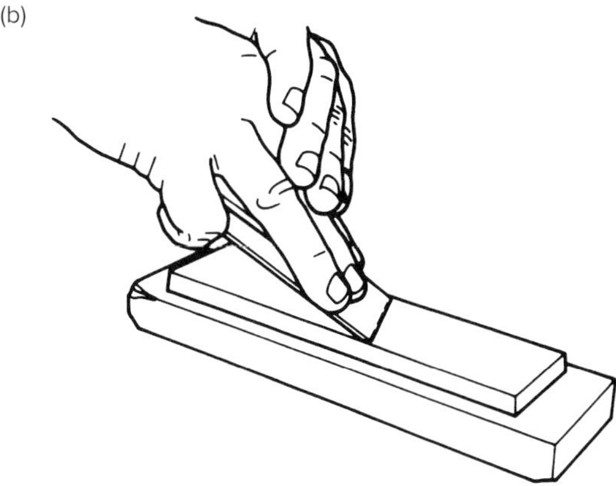

6.7.2 Adjusting the plane for use

Replace the plane iron by placing the cutting edge into the mouth of the plane and resting it on the frog, with the 'Y' adjusting lever and the lateral adjusting lever in the slots provided for them. (Refer to Figure 6.60 to identify the parts of the plane.) Replace the lever cap and press down the lever to lock the blade in place.

Turn the plane upside down and sight down the sole (Fig. 6.69). If the blade is not visible, turn the cutter adjusting nut (usually clockwise) until the blade can be seen projecting from the mouth. If one corner is projecting more than the other, use the lateral adjusting lever to move the blade sideways until it is projecting uniformly across the full width of the blade. Readjust the projection of the blade to take just a fine shaving. Always commence planing with a fine shaving and, if desired, increase it as conditions permit.

Fig. 6.69 Sighting down the sole

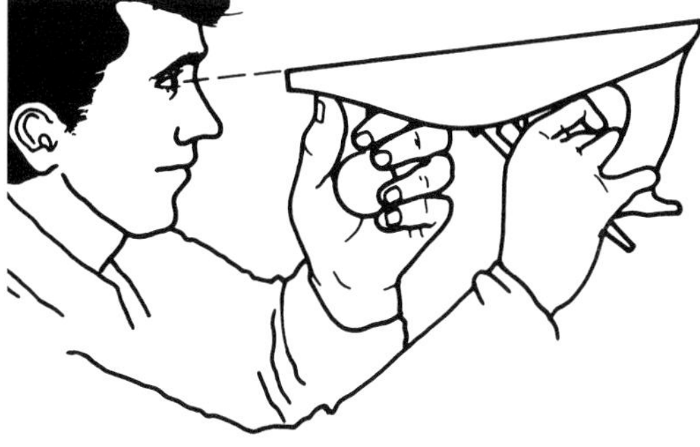

6.7.3 Smoothing plane

The smoothing plane is very often the first plane to be added to the carpenter's and joiner's tool kit. Its function is to smooth off timber, leaving the surface flat and free of planing defects. For this purpose, the blade is sharpened perfectly straight across, with just the corners rounded off to prevent them digging in and leaving small ridges on the surface.

The smoothing plane will be constantly at hand for the joiner, to flush off joints and clean off the face of framing, as in doors, windows, etc. It is also used to remove cutter marks from the surface of machine-dressed timber. Although not primarily intended for planing timber to size, it may sometimes be used for this purpose on short lengths.

The work on which a smoothing plane is to be used should always be held securely against the bench stops or in a vice. In order to smooth a wide surface, plane it in strips commencing from the nearest edge and work across the job until the whole surface is covered. Always plane in the direction of the grain. If necessary, reverse the direction of planing to determine which way produces the best result.

To achieve a satisfactory result with the smoothing plane under all conditions, it may be necessary to adjust the mouth (Fig. 6.70). This needs to be carried out on a smoothing plane only rarely. The amount of adjustment, although very small, can be significant when attempting to produce a smooth surface on difficult curly-grained timber.

Figure 6.70(a) shows that when the mouth is adjusted for normal planing, the distance from the cutting edge to the sole of the plane is approximately 1 mm to 1.5 mm. When planing difficult timber, the tendency is for the shaving to split and tear away in front of the cutting iron. Closing the mouth of the plane further (Fig. 6.70b) means that the sole is holding down the timber closer to the front of the cutting iron, preventing the shaving from lifting and breaking away. It naturally follows that if the mouth of the plane is closed up, only a very fine shaving can be taken–a thicker shaving would only become wedged in the narrow mouth. So, if a coarser shaving is to be taken, open the mouth.

Fig. 6.70 A & B Different mouth adjustments

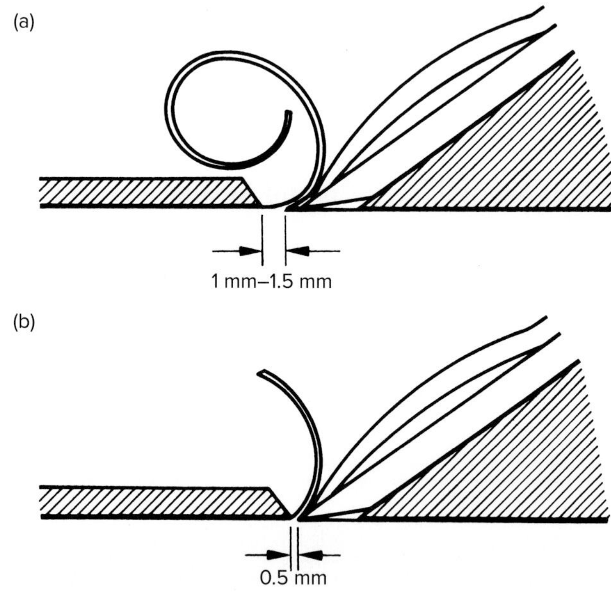

To adjust the mouth opening, first remove the plane iron. Then loosen the two screws securing the frog and move the frog backwards or forwards by turning the frog adjusting screw. Tighten the screws when the correct setting is obtained.

In summary, to produce a smooth flat surface with the smoothing plane under difficult conditions, it is necessary to:

1. have the cutting iron correctly shaped and razor sharp
2. set the cap iron close to the cutting edge
3. close the mouth of the plane
4. take a very fine shaving.

Experience and experimentation will establish the best settings for any particular job.

6.7.4 Jack plane

The jack plane is similar to the smoothing plane and is used to dress timber to its approximate size and shape. The greater length of the jack plane makes it easier to produce a straight flat surface. The blade is sharpened with a slight camber, which makes it easier and quicker to remove the waste; however, the surface will be left with a series of corrugations, which must be removed later using a smoothing plane (Fig. 6.71).

If the timber is of a long length, the final truing up may be done with a trying plane.

Fig. 6.71 Corrugations to be removed by a smoothing or a trying plane

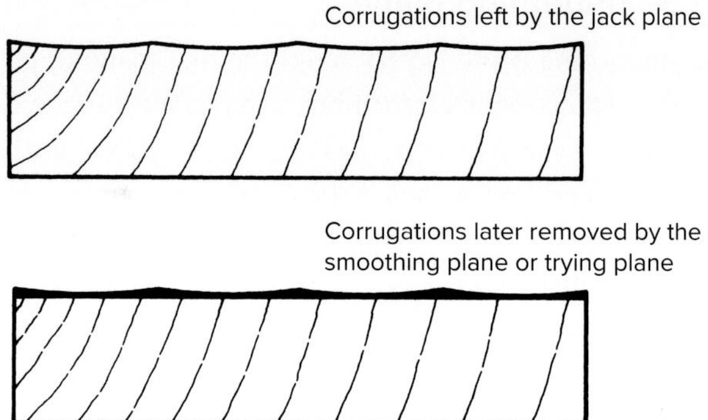

6.7.5 Trying plane

The longer trying plane is most suitable for straightening long lengths of timber, particularly where the carpenter is edge jointing boards, as may be necessary for table or counter tops. The object is to produce a true flat surface. The blade is therefore sharpened straight across with the corners rounded off.

Many experienced tradespeople like to use the heavier trying plane wherever possible, but it must be conceded that much of the work of the trying plane is now being taken over by power tools. However, when accurate joining is required, nothing can equal the trying plane guided by skilled hands.

CHECK YOUR UNDERSTANDING

1. What is the function of the back iron?
2. At what angle is a plane iron ground?
3. Why might it be necessary to adjust the mouth of the plane?

6.7.6 Planing rough-sawn stock to section size

For all those seeking to become skilled in the use of hand planes, this is a basic exercise that should be practised. Note not only the way in which the plane is handled, but the procedure that is followed to reduce the rough rectangular stock to a given sectional size. (Trade terms are often used rather loosely, and differ in some areas. In the workshop, the raw unworked timber may sometimes be referred to as stock.)

This is one example of a basic procedure that is followed in the same order whether hand tools or machines are used. The jack plane is the best one to use for this purpose, but practise with the smoothing plane if it is the only one available. Assume for this exercise that a piece of 100 mm × 38 mm rough-sawn Pacific maple, 600 mm in length, is to be dressed to 90 mm × 32 mm finished size. The procedure will be described in four steps; each step will be confined to one of the four sides of the stock to be dressed.

1. Select the best face of the timber to become the face side. Lay the timber flat on the bench with the face side up, and the end against the bench stop. Set the jack or trying plane to take a fine to medium shaving, and grip the handle with the finger lying parallel with the edge of the blade (Fig. 6.72).

   To commence planing, rest the toe of the plane on the face of the timber and exert a downward pressure. Use the full length of the plane by keeping it parallel with the timber; move it forward maintaining pressure on the front.

   Fig. 6.72 Gripping the plane

   As the plane commences to cut, the high spots will be removed first, in short shavings. When the plane is fully supported over the timber, exert forward and downward pressure equally with both hands; towards the end of the stroke, transfer downward pressure to the back to prevent the toe dipping down as it moves over the end of the timber and becomes unsupported. With practise, the distribution of pressure becomes an automatic reaction–pressure on front hand, distribute equally, transfer pressure to back hand (Fig. 6.73).

   If the planing is difficult and rough, reverse the timber to see if the grain is running in the opposite direction.

Continue planing in strips across the face of the timber until full shavings are rolling out of the mouth of the plane. Test the face for accuracy in the following three ways:

(a) Use a straight edge, or the sole of the plane tilted over, to check that the face is straight from end to end (Fig. 6.74).

(b) Use a short straight edge or try square to check that it is flat across the width (Fig. 6.75). Hold the timber up to the light; any irregularity can be seen more readily under the straight edge.

(c) The face could also be twisted or, to use the trade term, be in 'wind'. *Wind sticks* are two short lengths of timber, say 300 mm × 40 mm × 15 mm. Their section size can vary, but they must be parallel. Lay a wind stick square across the timber near each end and sight the two top edges; any wind will be exaggerated, indicating the high corners of the stock (Fig. 6.76).

To remove the wind, take shavings diagonally across the high corners. Check again, and when all is correct, mark the face with the traditional *face mark*–a large 'e' with the tail continued to one edge (Fig. 6.77).

Fig. 6.73 Three pressure steps: (a) pressure at commencement of stroke; (b) pressure distributed equally; and (c) pressure at end of stroke

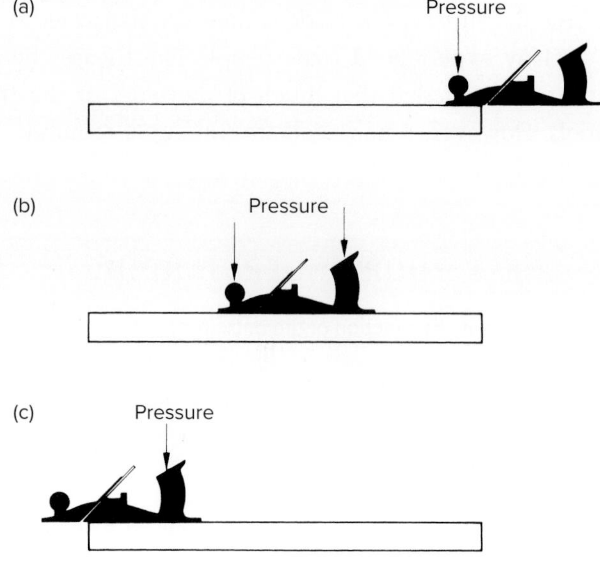

Fig. 6.74 Checking the length for straightness

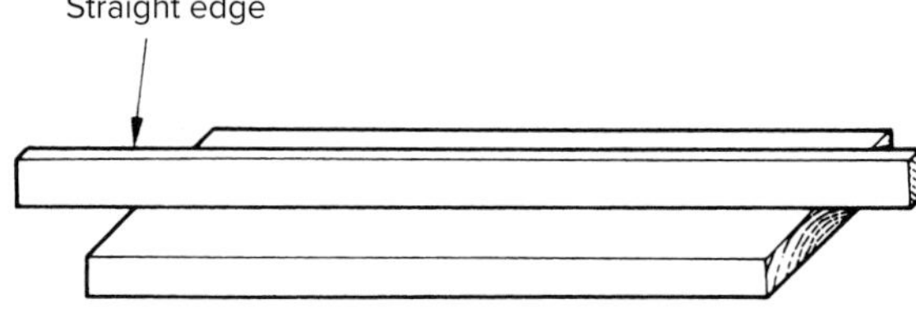

2. On the extended tail of the face, mark points to the face edge, which must now be planed straight and square to the face side; this is called 'shooting the edge' (Fig. 6.78). To do this, first secure the timber in a vice. Grip the plane as shown in Fig. 6.78. Holding the plane square to the face side, plane the edge, distributing the pressure in a similar manner as for the face side. Test the face edge in the following two ways:

   (a) Use a straight edge to check that it is straight from end to end

   (b) Use a try square to check that it is square to the face side (Fig. 6.79).

Fig. 6.75 Checking the width for flatness

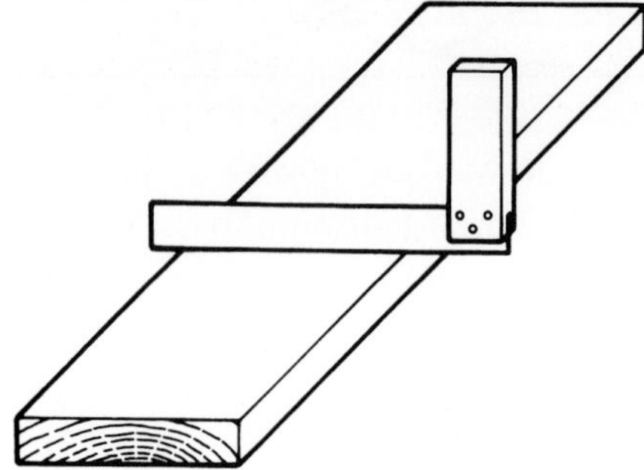

Fig. 6.76 Checking the face for wind

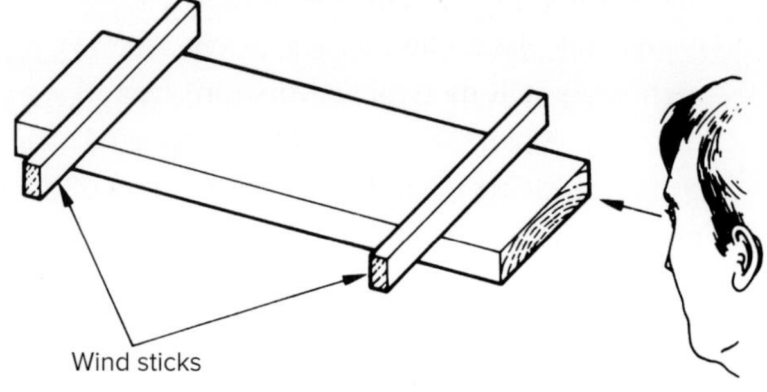

Fig. 6.77 Face mark

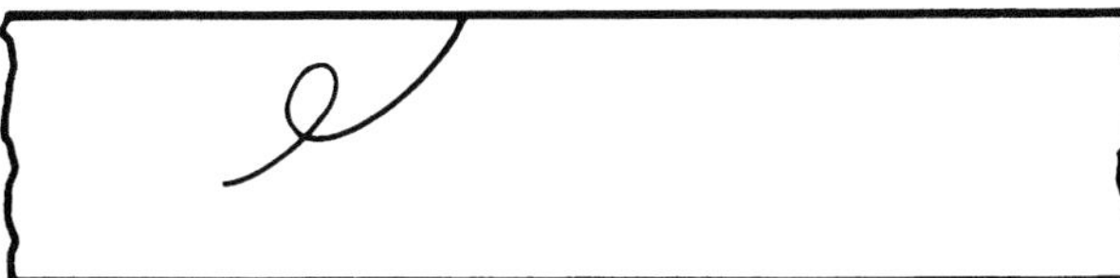

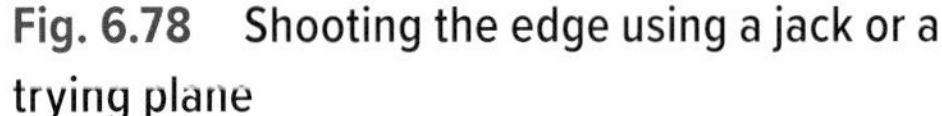

Fig. 6.78 Shooting the edge using a jack or a trying plane

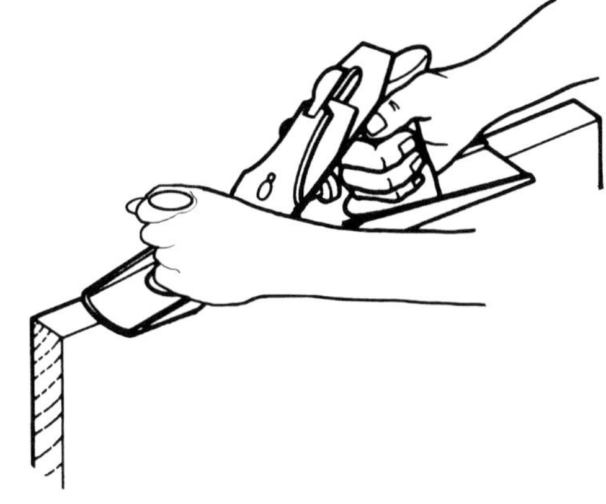

Fig. 6.79 Checking the edge for squareness

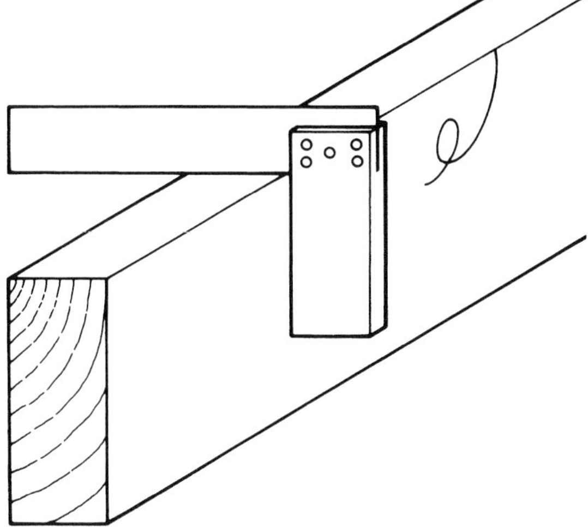

Fig. 6.80 Correcting the edge

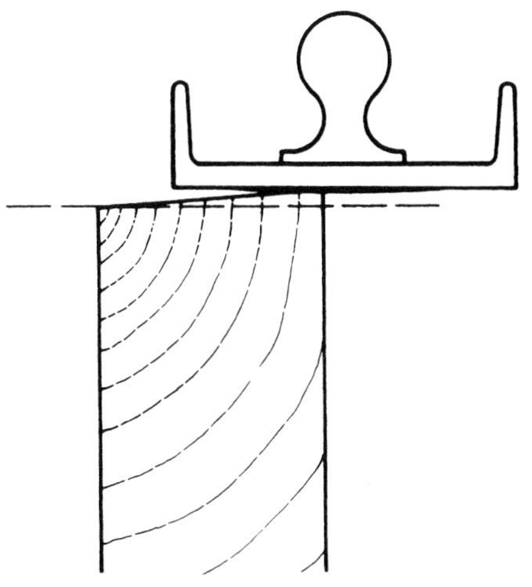

Fig. 6.81 Face edge mark

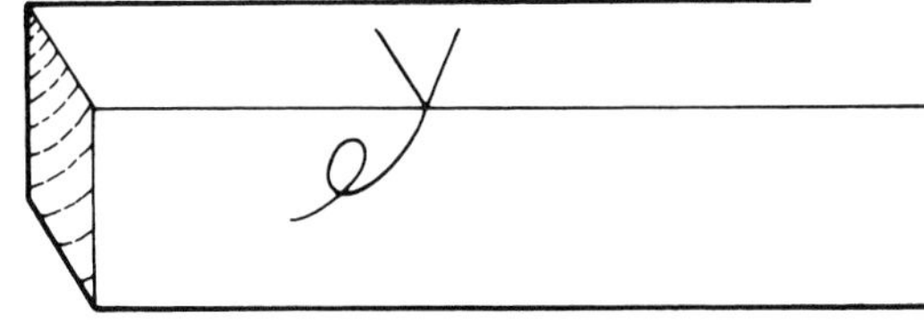

If it is not square, move the plane over to take shavings off the high edge only (Fig. 6.80), finishing with a shaving of full thickness. Test again, and when all is correct, mark with the *face edge mark*–a 'V' with the point to the long tail of the face mark (Fig. 6.81).

When preparing the face side and edge, do not lose sight of the fact that the timber must be finished to a given section size; so do not plane unnecessarily with the result that the timber is finished undersize.

3. The timber must now be reduced to the specified width. Set the marking gauge to width, 90 mm in this case, and with the stock of the gauge to the face edge, gauge the width along the two faces. If there is an excessive amount of waste to be removed in the width, use the rip saw and cut to about a millimetre on the waste side of the gauge line, which is sufficient to smooth the rough edge and finish accurately to size with the jack plane.

   When planing away waste, keep a careful watch on the gauge lines and stop planing when you just start to split the lines.

4. The timber must next be reduced to its specified thickness. Set the gauge to the correct thickness, 32 mm in this case, and with the stock to the face side, gauge both edges and ends to thickness. Lay the stock on the bench against the stop and, watching the gauge lines on both edges, plane away the waste, stopping at the gauge lines.

Fig. 6.82 Block plane

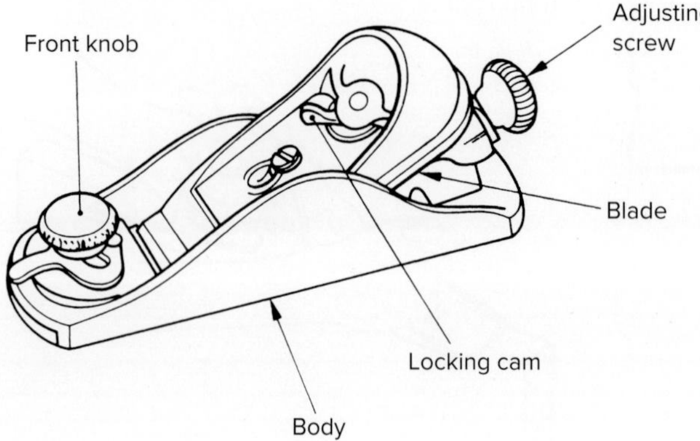

Fig. 6.83 Planing end grain

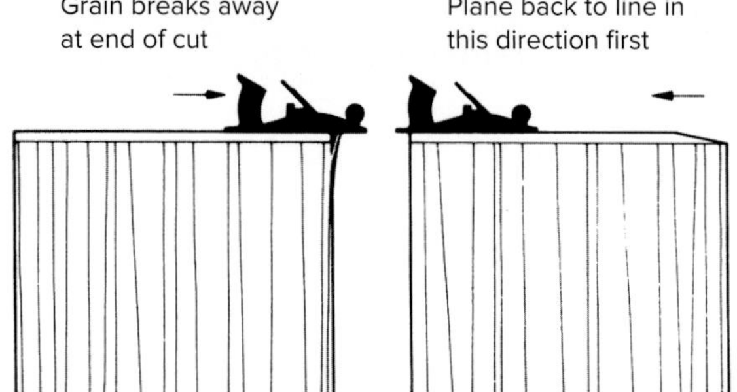

6.7.7 Block plane

The block plane is primarily intended for planing **end grain** but can also be used for trimming other small items (Fig. 6.82).

This plane is approximately 180 mm in length, with a blade 41 mm in width, set at a low angle of 20°. There is no back iron and the blade is sharpened at the same angle as the bench planes–a grinding angle of 25° and a sharpening angle of 30°. It is mounted in the plane with the bevel side up. Different makes will vary in the way the blade is fitted and adjusted; however, the better-quality planes have full screw adjustment for depth of cut and lateral movement. To use the block plane for planing end grain, the blade must be razor sharp and should be set for a very fine shaving (Fig. 6.83). To prevent the timber splitting away at the end of the cut, plane from both directions.

6.7.8 Lubricating the sole of the plane

Planing will be made much easier if the sole of the plane is lightly lubricated. However, this must be done sparingly, and care must be taken to ensure that the surface of the timber is not contaminated with oil or wax. Candle grease or other dry lubricants are suitable. Alternatively, an oil pad can be made up. Roll a strip of felt into a small tin can and apply neatsfoot oil. Allow the oil to soak into the felt until it seems almost dry. To use the oil pad, wipe just the toe of the plane across the pad so that any residue will be removed from the surface of the timber by the shaving following.

6.7.9 Spokeshave

The spokeshave has a cutting action similar to the planes and its purpose is to clean up curved edges of timber (Fig. 6.84).

The sole is narrow to enable it to follow curves. The general practice is to use a spokeshave with a flat sole to follow convex curves and one with a curved sole on concave surfaces (Fig. 6.85).

The blade is 43 mm to 53 mm wide. There is no cap iron; the single cutter is clamped in place by a lever cap.

Fig. 6.84 Spokeshave

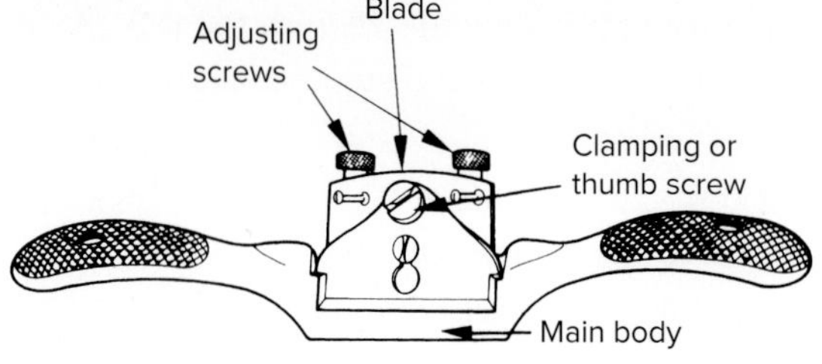

Fig. 6.85 Flat and curved spokeshaves

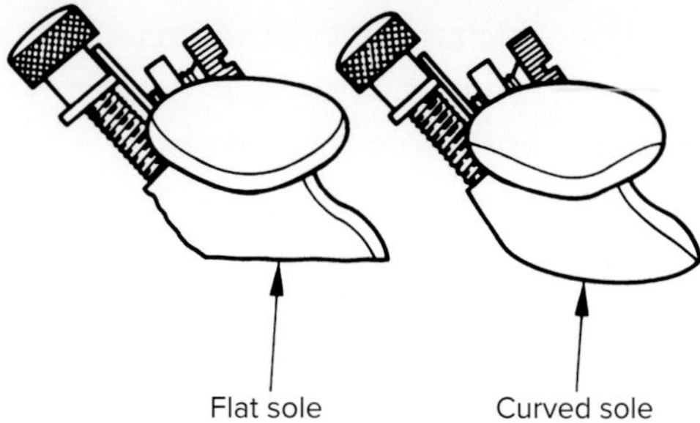

To set the depth of cut, loosen the thumb screw, use the two adjusting screws to make the adjustment and then tighten the thumb screw.

In use, the spokeshave is set to take a fine shaving and must always follow the direction of the grain. Ideally, the spokeshave should be held square across the timber; however, it is often found that it cuts more cleanly if held at a slight angle to make more of a slicing cut.

To clean up a concave curve, grip the spokeshave by the handles, with the thumbs resting on the back edge in the small depressions provided for the purpose.

Hold the job securely in the vice and commence at the top of the curve; push the spokeshave away from you, following the direction of the grain to the bottom of the curve. Reverse the timber and work the other half of the curve.

Fig. 6.86 Dressing a concave curve

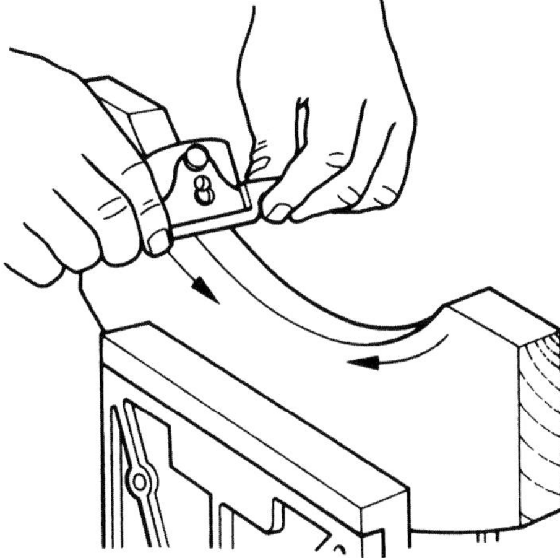

A convex curve can be worked in a similar manner: commence at the centre of the curve and dress each way in the direction of the grain (Fig. 6.87).

The blade of the metal spokeshave is sharpened in a similar manner to the blade of a plane (Fig. 6.88). If it is difficult to hold the small blade, secure it in a stock fashioned from a piece of timber.

The stock consists of a block, approximately 100 mm × 45 mm × 19 mm, with a slot in which the blade is held. The block is rounded off to fit the hand comfortably.

TIP The planes detailed previously are the ones that are now commonly available or used in Australia. Many more hand planes exist and most can still be purchased. They are all designed for specific purposes. It will be useful to do some research and see how many planes you can find.

CHECK YOUR UNDERSTANDING

1. What mark is used to indicate the face of the piece of timber being worked on?
2. Why are the two pieces of timber forming a half-lapped joint both gauged from the same face?

6.8 Workbench

The workbench is an essential item of workshop equipment. The basic requirements of a bench suitable for trade use are illustrated in Figure 6.89.

The bench must be solid and rigid and mortised or bolted together. A double-sided bench where the joiner can work around all sides is preferable, and it must be large enough to lay out items of joinery, doors or large panels of sheet material. The overall dimensions indicated can be taken as a guide–the height can be adjusted to suit the individual.

The top must be made from a fine-grained timber (e.g. Klinki pine) that will not bruise other material worked on the bench. Bolts for fixing the

Fig. 6.87 Dressing a convex curve

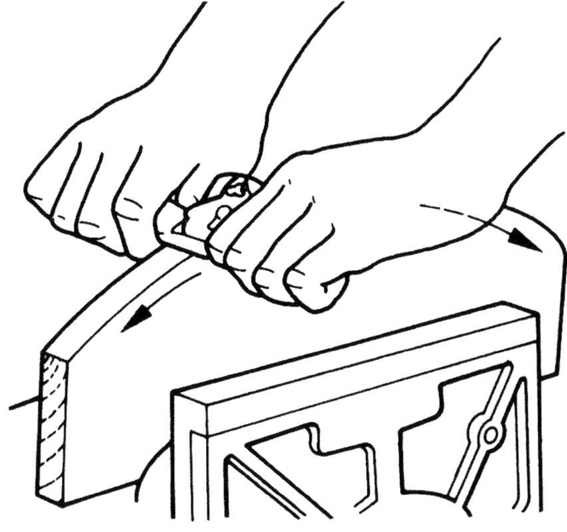

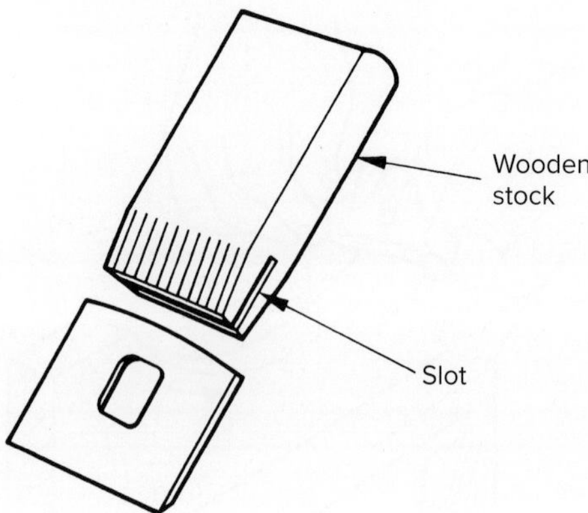

Fig. 6.88 Sharpening a spokeshave blade

top to the frame should be counterbored, and a wooden plug should be set in to cover the head of the bolt.

Attachments to the bench will include a woodworker's vice, a bench stop and a bench holdfast.

6.8.1 Woodworker's vice

The size of the vice is indicated by the width of the jaws and varies from 150 mm to 250 mm. The movable jaw is operated by a handle that revolves a screw, which in turn engages in a nut at the back of the vice. A useful feature on some vices is a *speed screw*. This consists of a quick-release lever that, when pressed, releases the nut holding the screw, allowing the movable jaws to slide freely so they can be quickly adjusted to any thickness of material.

Fig. 6.89 Workbench

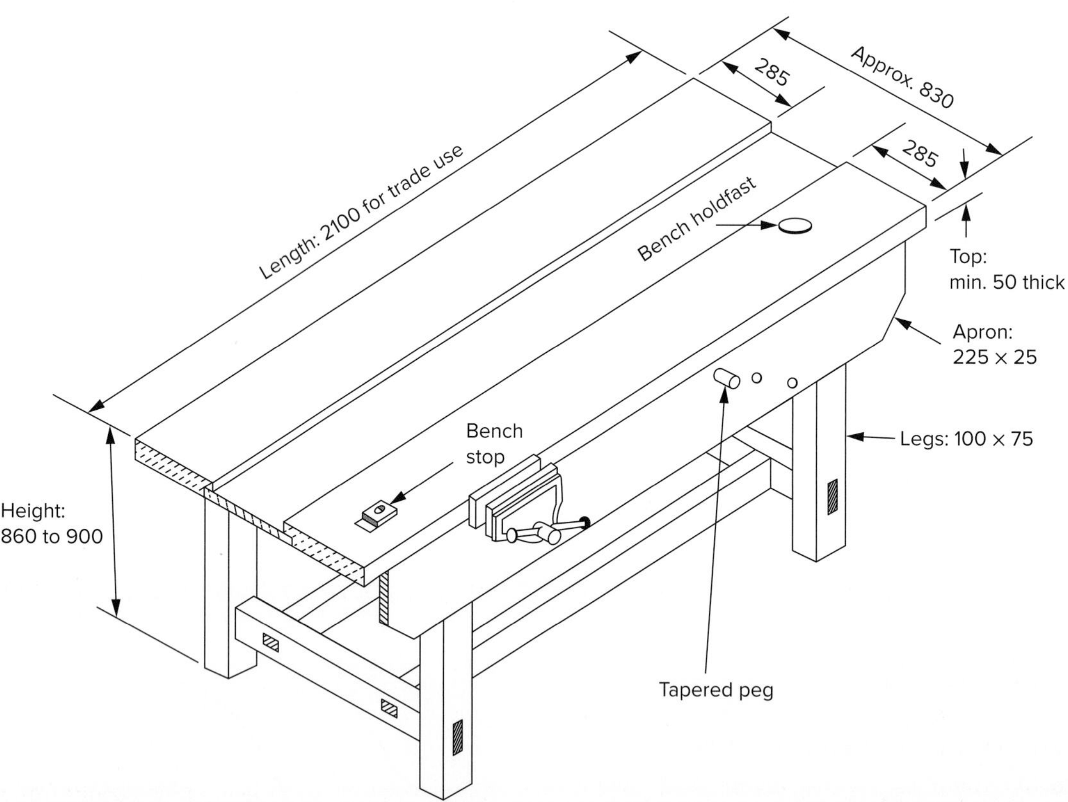

When fixing the vice to the bench, the top of the metal jaws must be kept down at least 10 mm from the top of the bench; wooden liners are fixed inside the jaws and are level with the benchtop (Fig. 6.90).

6.8.2 Bench stop

The bench stop is usually located at the left-hand end of the bench. It can be the metal type, set flush into the benchtop and adjustable in height, or a wooden peg projecting from underneath, also adjustable in height (Fig. 6.91a).

Another useful type of stop is a *'V' block* (Fig. 6.91b). It serves as a stop and also holds timber upright.

6.8.3 Bench holdfast

One or more collars are set into the benchtop, which engage the shaft of the holdfast (Fig. 6.92). As the screw bears down on the top of the shaft, the arm pivots, holding the work piece firmly down on the benchtop.

6.8.4 Other useful features of the workbench

Holes approximately 22 mm in diameter, drilled in the apron of the bench, are used to fit a tapered wooden peg that can support the ends of long pieces of timber when held in the vice.

Fig. 6.90 Bench vice

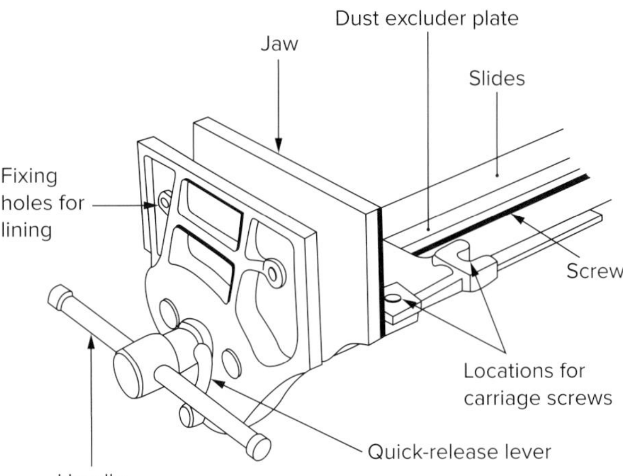

Fig. 6.91 (a) Bench stop; (b) 'V' block bench stop

(a)

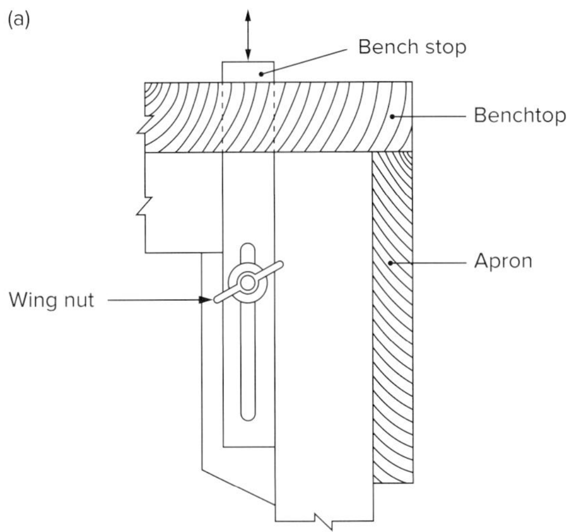

(b)

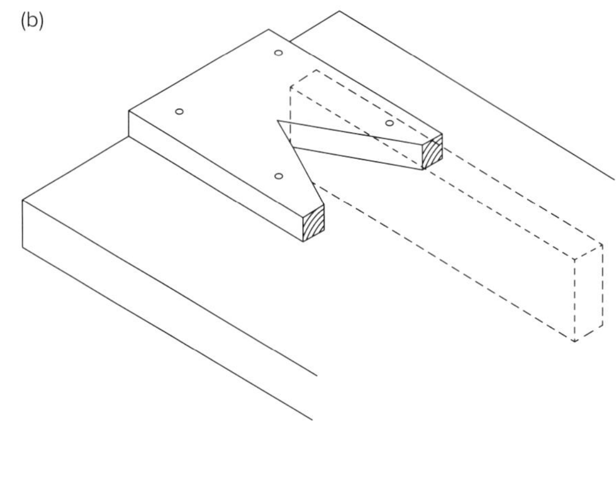

Individual tradespeople will make their own additions to a workbench–for example, a shelf or drawer underneath, or racks across the end, to store tools and small items when not in use.

6.8.5 Workbench accessories

6.8.5.1 Bench hook

The bench hook is used to hold timber firmly and to protect the bench during sawing operations (refer to tenon saws) (Fig. 6.93). The **cleats** can be glued and pegged to the base with a 10 mm dowel. If the cleats are screwed, the screws should be well countersunk to prevent tools being damaged.

6.8.5.2 Mortise board

Timber being mortised, drilled or chopped can be held securely against the face of the block of the mortise board (Fig. 6.94) with a G-cramp. The mortise board also protects the benchtop.

6.8.5.3 Shooting board

A shooting board is another item made up by the joiner (Fig. 6.95). It assists when planing the edges of thin boards straight and square, or when trimming the square ends of timber. Its length is approximately

Fig. 6.92 Bench holdfast

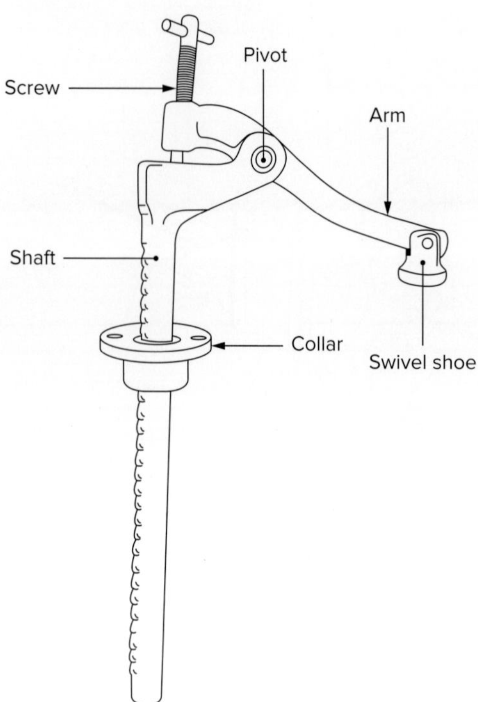

Fig. 6.93 Bench hook

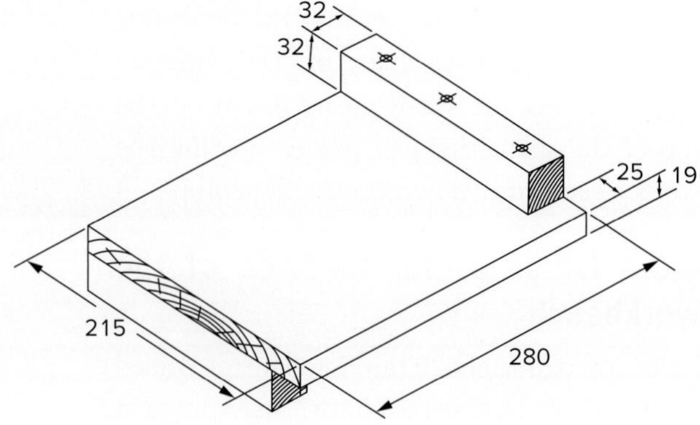

Fig. 6.94 Mortise board

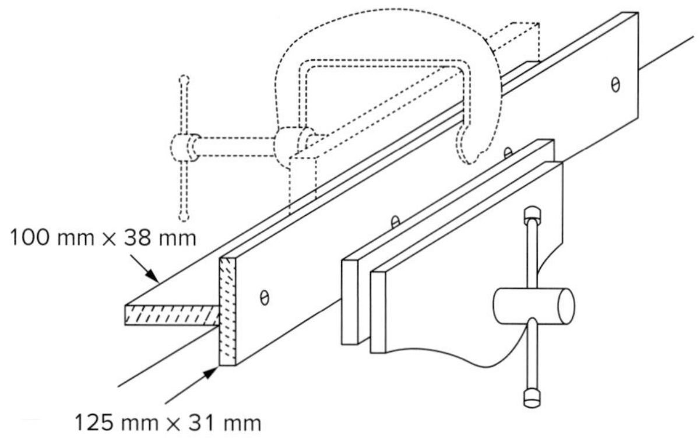

Fig. 6.95 Shooting board

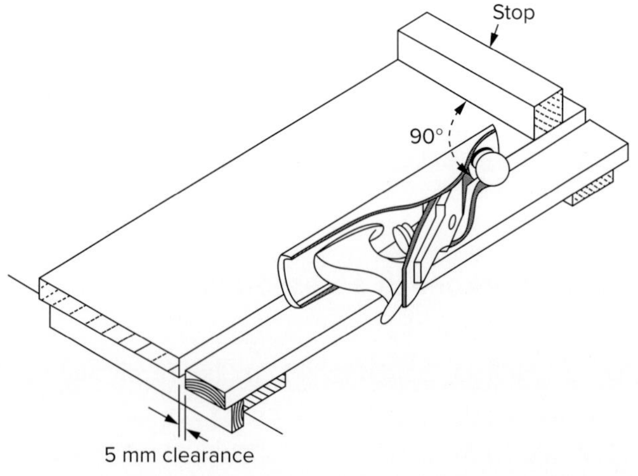

600 mm. The shooting board consists of two boards forming a rebate in which the plane will slide. A clearance of say 5 mm is allowed for in the corner of the rebate so that the waste can clear and not accumulate, which would affect the accurate running of the plane. The stop must be rigid, and set at right angles to the direction of the plane.

When planing end grain, work from both directions (see the 6.7.7 Block plane section) to prevent the corners splitting away. Turn the work piece over, hold at an oblique angle, and trim the corner back to the line. Reverse the timber, and holding it firmly against the stop, trim the end square.

6.8.5.4 Saw stools

Saw stools are essential equipment; they vary in dimensions and construction. Figure 6.96 shows the details of a saw stool suitable for average conditions. The height can vary from approximately 560 mm to 600 mm, and can be adjusted to enable the carpenter to assume a comfortable posture. Note that the legs must be set out in pairs by reversing the direction of the bevels.

Approximate dimensions:

Top 100 × 50

Legs 75 × 38

Cleats 150 × 25

Fig. 6.96 Saw stool

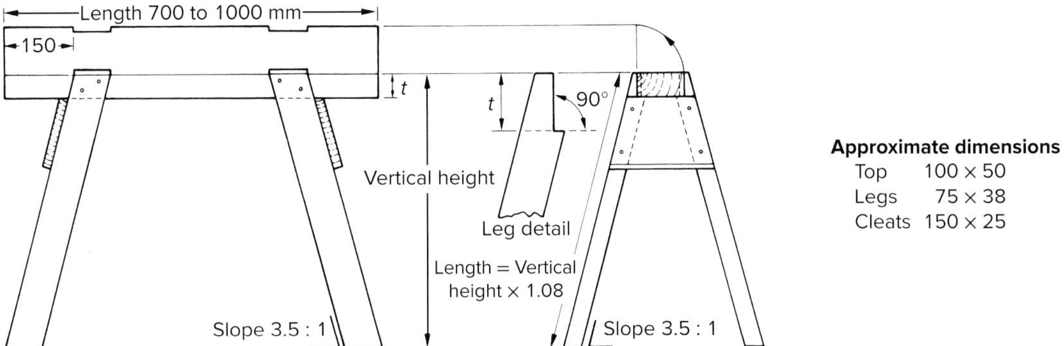

6.9 Cramps

Cramps come in various forms and, as will be shown later, an important consideration when making up any item of joinery is how it can be cramped or pulled together tightly.

6.9.1 G-cramp

The G-cramp is the most versatile and popular cramp (Fig. 6.97). It is used for holding material securely on the bench and cramping glued joints together; a heavy G-cramp is a valuable aid for holding members together during fixing in construction work.

G-cramps are commonly made in a wide range of sizes from 50 mm to 300 mm, the size being the maximum distance between the jaws of the cramp. For trade use, cramps should be made from forged steel with a steel screw; some smaller sizes may be aluminium or pressed metal.

Fig. 6.97 G-cramp

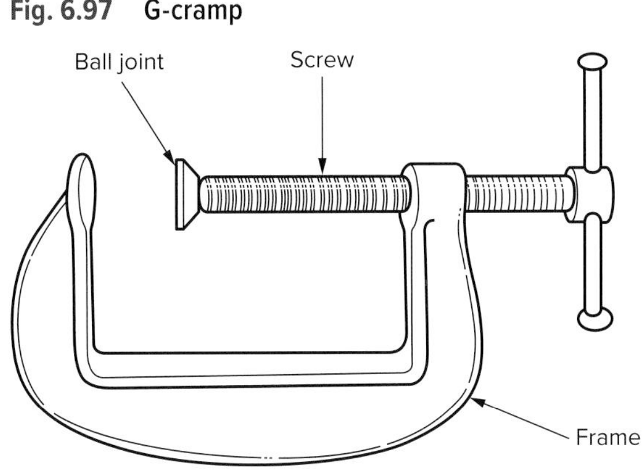

Always place a block of scrap timber under the shoe of the cramp when using it on finished work to prevent the surface being marked, and only hand-tighten the cramps.

6.9.2 Edge cramps

Edge cramps are a comparatively recent development of the G-cramp, and are very useful for cramping edge strips to prefinished boards (Fig. 6.98).

Fig. 6.98 Edge cramp

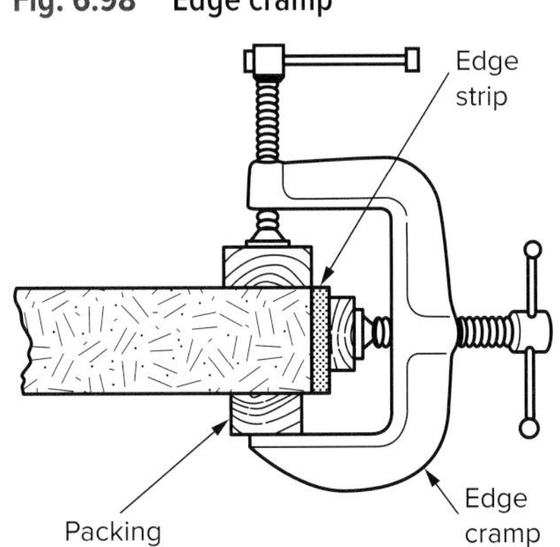

6.9.3 Quick-release or fast-action cramp

In a quick-release or fast-action cramp, a sliding jaw moves on a steel bar (Fig. 6.99a). It can be quickly adjusted to any opening and tightened by the steel screw. The size can vary from 100 mm to 1000 mm. They are used in similar circumstances as the G-cramp, and the carpenter should adopt the same precautions.

Fig. 6.99 A & B Quick-release or fast-action cramps

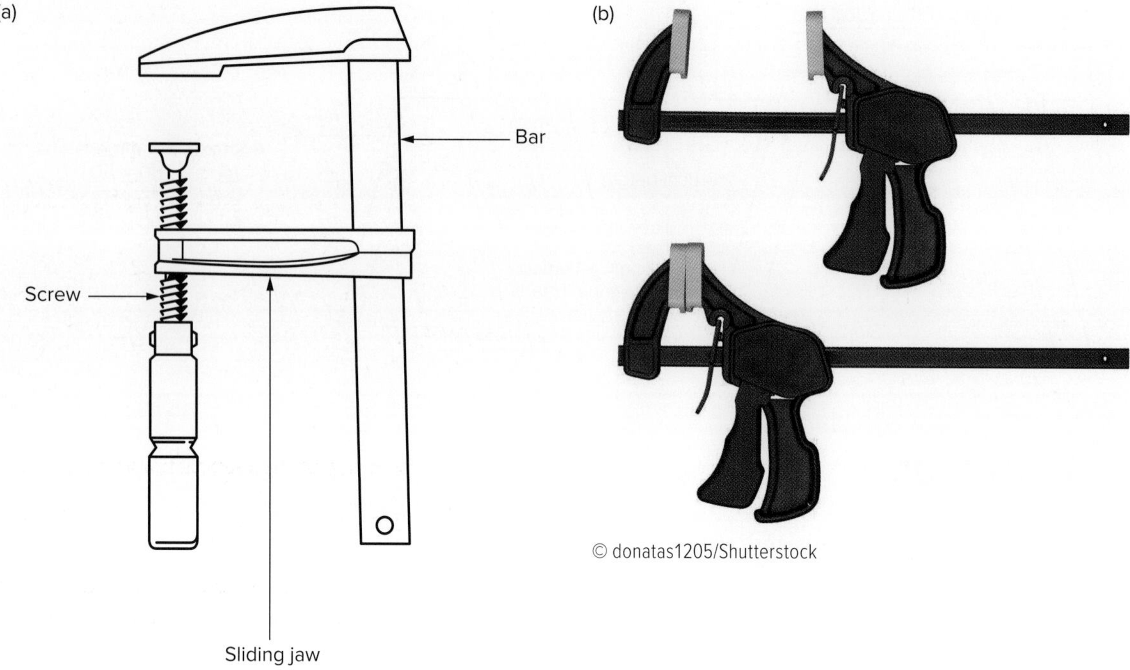

© donatas1205/Shutterstock

Fig. 6.100 Sash cramp

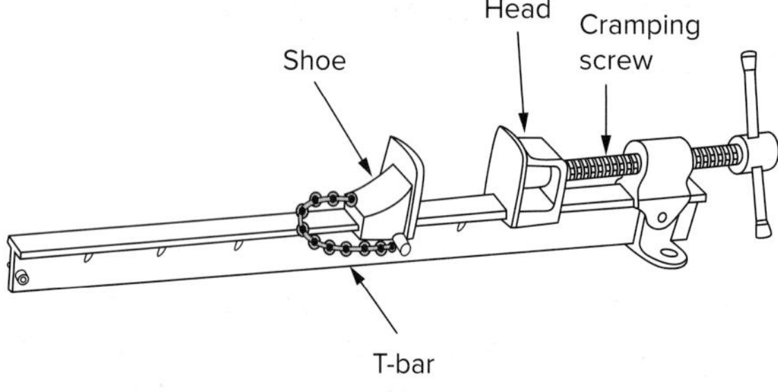

Fig. 6.101 Cramp heads

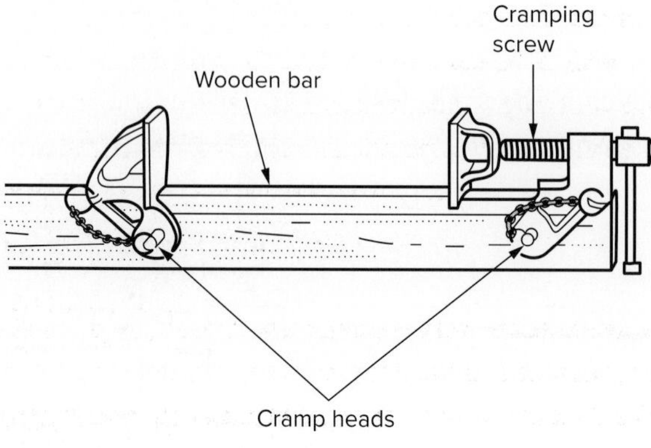

6.9.4 Sash cramps

Sash cramps may vary from a light, flat bar cramp to one with a heavier T-bar section, but all are now generally referred to as sash cramps (Fig. 6.100). The size of the cramp is the length of the steel bar, and may range from 600 mm to 1800 mm.

Sash cramps are used for tightly cramping together joinery items, window sashes, doors, wide boards or cabinet framing during gluing-up operations.

The opening of the jaws can be adjusted by moving the sliding shoe along the bar and fixing with a pin through the nearest hole. The final adjustment and tightening is then done using the screw and adjustable head. Blocks of scrap timber should be used under the shoe and head to prevent damage to finished work.

6.9.5 Cramp heads

Cramp heads are a convenient way to make up a long sash cramp of any length (Fig. 6.101).

The cramp heads are fixed by a pin to a bar of timber 25 mm thick. Another type of cramp head can be fixed to a length of galvanised water pipe, which will serve the same purpose.

6.9.6 Corner or mitre cramp

The corner cramp is used to cramp mitre joints together and hold them securely in place while gluing, nailing or otherwise fixing (Fig. 6.102). It has two sets of screw-adjusted 'feet', which will hold the two halves of the joint at right angles to each other against a fence.

Fig. 6.102 Corner or mitre cramp

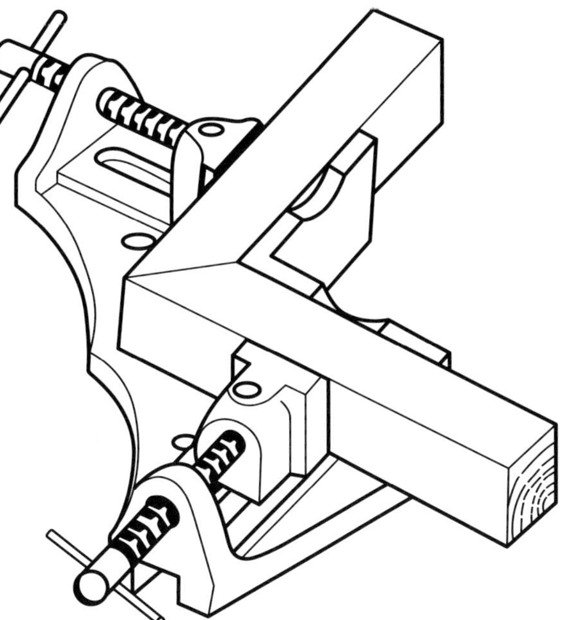

Student research task

Search online for the PPE requirements of your state or territory workplace safety authority (e.g. WorkSafe Victoria, NT WorkSafe). Read the information about PPE and where and when it should be used.

End of chapter activity

Please refer to the joints chapter (Chapter 7) for this activity.

AUSTRALIAN BUILDING CODES AND STANDARDS

Standards Australia, **AS/NZS 62841.4.1:2018**

Electric Motor-Operated Hand-Held Tools, Transportable Tools and Lawn and Garden Machinery—Safety—Part 4.1

Particular Requirements for Chain Saws

Tradie talk—**Don't strangle your hammer**

Chapter 7

Timber joints

It is important for those students who want to become carpenters within the building industry to understand the content in this chapter, even though it is not explicitly covered in this qualification. This content will enhance the training being undertaken and also provide essential knowledge of the function and application of timber joints.

Learning Objectives

LO 7.1 Construct framing joints

LO 7.2 Identify carcase joints and their uses

LO 7.3 Identify joining methods used for widening timber

Introduction

In this chapter you will learn about the construction techniques employed and how best to set out and construct the more advanced joints used in the production of furniture, timber doors, timber windows and so on. The techniques you are shown in this chapter can be practised using either hand or power tools. In preparing these joints, attention to detail is a must; it is essential to ensure clean working practices while preparing and constructing the joints–and you must have a thorough knowledge of the tools covered in previous chapters, plus an ability to use them proficiently, if you are to produce high-quality work.

7.1 Framing joints

Framing joints are those joints that usually join members at right angles, such as those found in doors, windows and other framed units. However, before proceeding with the details of the various joints, note the names given to the framing members in Figure 7.1. The *stiles* are the vertical side members. The *rails* are the horizontal members connecting the stiles and are named according to their location–top, middle or intermediate, or bottom.

WORKPLACE SCENARIO

For centuries, the only effective way to join two pieces of timber was to construct tight-fitting joints. Such joints were generally based on a mortise and tenon, and were held together using a wooden peg. Many specific joints were developed for particular purposes, but they generally relied on tight-fitting construction and wooden pegs to hold the joint together. In large framing joints, the timber used was green. So once the joint had been constructed, put together and pegged, the drying timber would shrink, causing the joint to tighten and increase in strength.

With the development of modern fixings, such as bolts, nails, screws, brackets and glues, timber can be joined using one of these methods without constructing a joint. This has decreased the time it takes to construct a building, piece of furniture, etc.

Fig. 7.1 Framing members

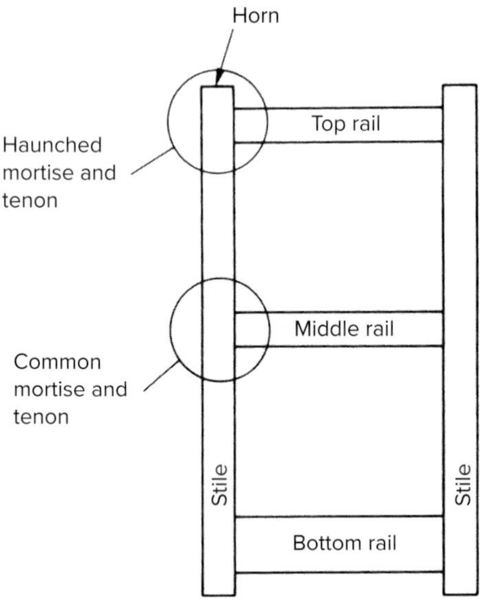

7.1.1 Mortise-and-tenon joint

The mortise-and-tenon joint is probably the strongest and most universally used joint in framing and can be adapted to many situations.

7.1.1.1 Common mortise-and-tenon joint

The common mortise-and-tenon joint (Fig. 7.2) is one in which the tenon is the full width of the rail and goes fully through the stile. A typical application would be the joint between a middle rail and stile.

The tenon is made to be one-third of the thickness of the stile or piece to be mortised, or it is made to the nearest chisel size. This ensures equal strength between the tenon and the sides of the mortise. 'Through' tenons are held securely in the mortise by wedges that should penetrate the stile at about two-thirds of the stile's width. To be effective, wedges should be only slightly tapered and not made to a sharp point.

Figure 7.3 shows the effects of both a correctly shaped wedge and one with too much taper. A tapered wedge allowance (WA) is cut back on each end of the mortise. A suitable WA is one-twelfth the width of the stile.

It is not suggested that you should start splitting millimetres or delving into the mortise to measure the wedge, but with experience and by bearing these few rules in mind, you will be able to estimate and proportion wedges that will serve their purpose effectively.

7.1.1.2 Haunched mortise-and-tenon joint

The haunched mortise-and-tenon joint is used where the joint is made at the ends of the stile, i.e. between the top or bottom rail and the stile (Fig. 7.1).

During manufacture, waste of 10 mm to 20 mm called *horn* is left on the ends of the stiles and is not removed until the item is finally being fitted. The horn serves a number of purposes: it strengthens the end of the stile while it is being worked (mortising and wedging); it can often be used to secure the framing during cramping-up and cleaning-off operations; and it will protect the ends during stacking

Fig. 7.2 A & B Common mortise-and-tenon joint

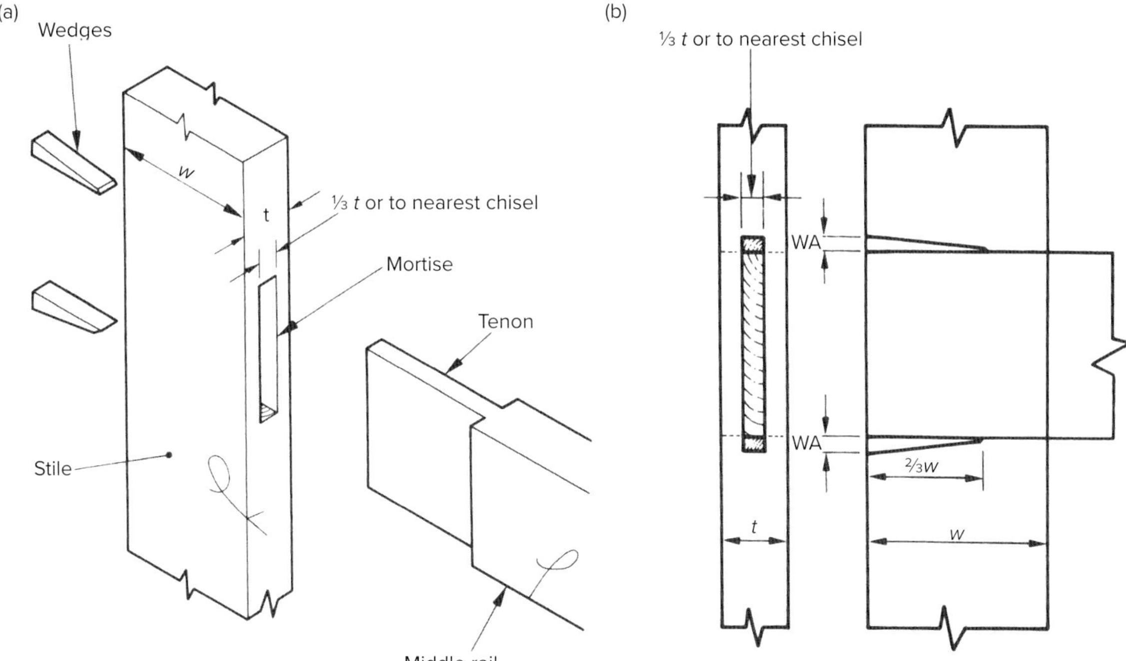

and transporting. When the joint is finished, it can eventually be flushed off accurately.

If the tenon were made the full width of the rail, there would be nothing left to hold it in once the horn was removed. Therefore, it is reduced in width and sufficient material is left on the end of the stile for the wedge to hold. A short portion of tenon, the actual *haunch,* is left near the shoulders to prevent the rail from twisting. The length of haunch is made the same as the thickness of the tenon, to a maximum of 12 mm. A rule for the width of the tenon is that it should be made two-thirds the width of the rail minus one WA (Fig. 7.4).

Other parts of the joint are proportioned as for the common mortise-and-tenon joint.

Fig. 7.3 Wedge shape

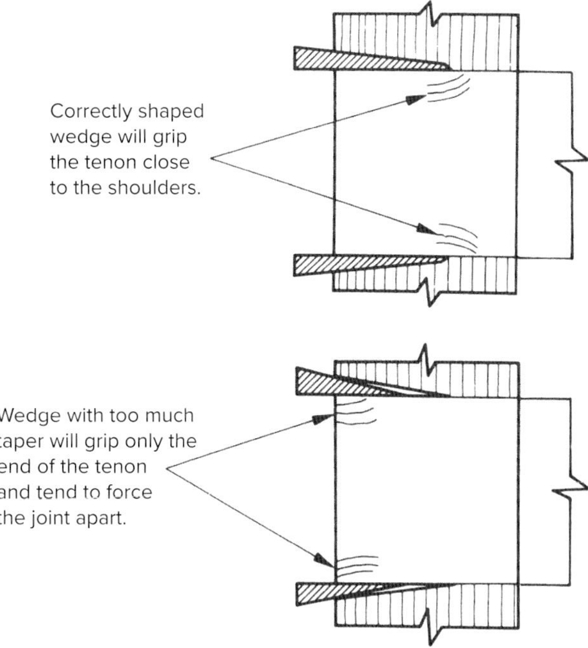

7.1.1.3 Faults

Some possible faults in mortise-and-tenon joints are shown in Figure 7.5.

7.1.1.4 Modifications to the mortise-and-tenon joint

In joinery work, the mortise-and-tenon joint is seldom as simple as has so far been explained. Where the joint is used for framing up doors or windows, the edges of the members may be plough grooved or rebated to receive panels, and the joint must be modified accordingly.

Fig. 7.4 A & B Haunched mortise-and-tenon joint

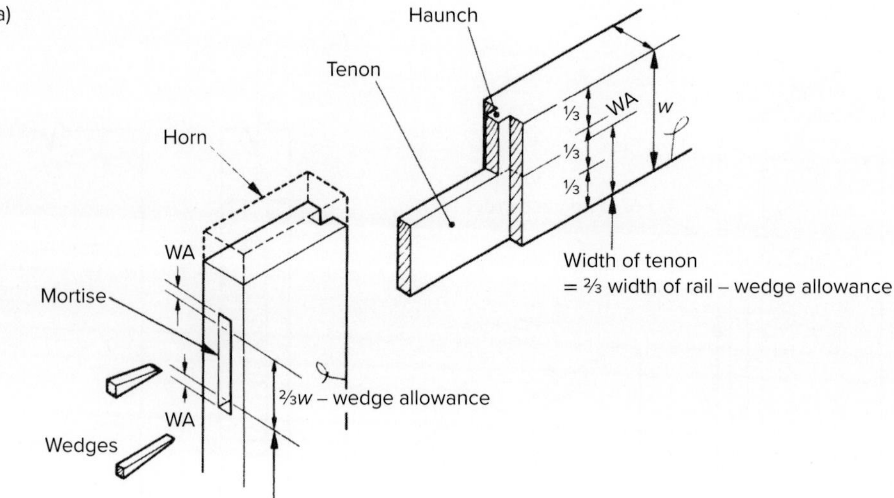

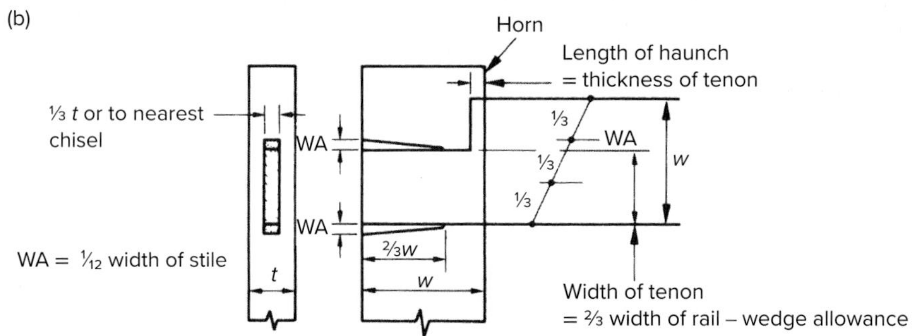

Fig. 7.5 A–E Faults in mortise-and-tenon joints

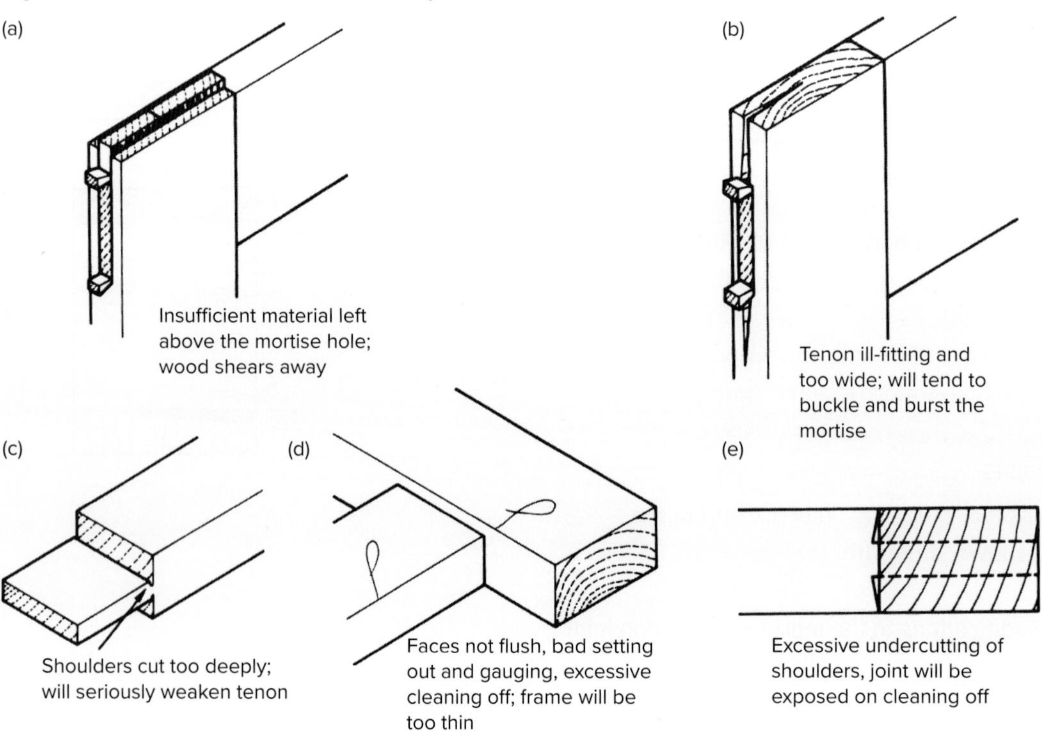

Figure 7.6 shows a haunched mortise-and-tenon joint between a rail and stile where a plough groove has been run on the inside edges. To proportion the width of the tenon, the word 'effective' must be added to the rule: the tenon is made two-thirds the effective width of the rail minus one WA.

The effective width of a rail can be defined as the width of the middle one-third, this being the one-third on which the mortise and tenon are located. In this joint, the width of the plough groove is made the same as the thickness of the tenon; the haunch will fill the groove.

7.1.1.5 Double mortise-and-tenon joint

The double mortise-and-tenon joint is used where the rail is considerably wider than the stile, for example, the middle or bottom rail to a framed and panelled door. The stile would be badly weakened if a mortise were cut out long enough to receive a tenon the full width of the rail.

To correct this, a double mortise-and-tenon joint is used. The stile is not weakened, and the wedging power is doubled. The width of a tenon should not exceed five times its thickness. The setting out is shown in Figure 7.7.

Note that the effective width is divided into three parts and each tenon is minus one WA, taken from the middle of the rail.

Figure 7.8 shows the setting out for a wide bottom rail to a stile; besides a double mortise-and-tenon, an allowance has been made for a haunch. The effective width is divided into 11 equal parts and proportioned as shown. One WA is taken from the bottom of each tenon.

7.1.1.6 Long and short shoulder mortise-and-tenon joint

The inside edges of stiles and rails may be rebated to receive panels or glass; the shoulders on the rails are different in length. The difference in length is equal to the depth of the rebate.

When a neat appearance across the edge is more important than maximum strength, as for instance on showcase doors, the splayed haunch can be used. Note that the effective width is again divided into three equal parts when setting out the haunched tenon (Fig. 7.9).

Fig. 7.6 Joint between top rail and stile

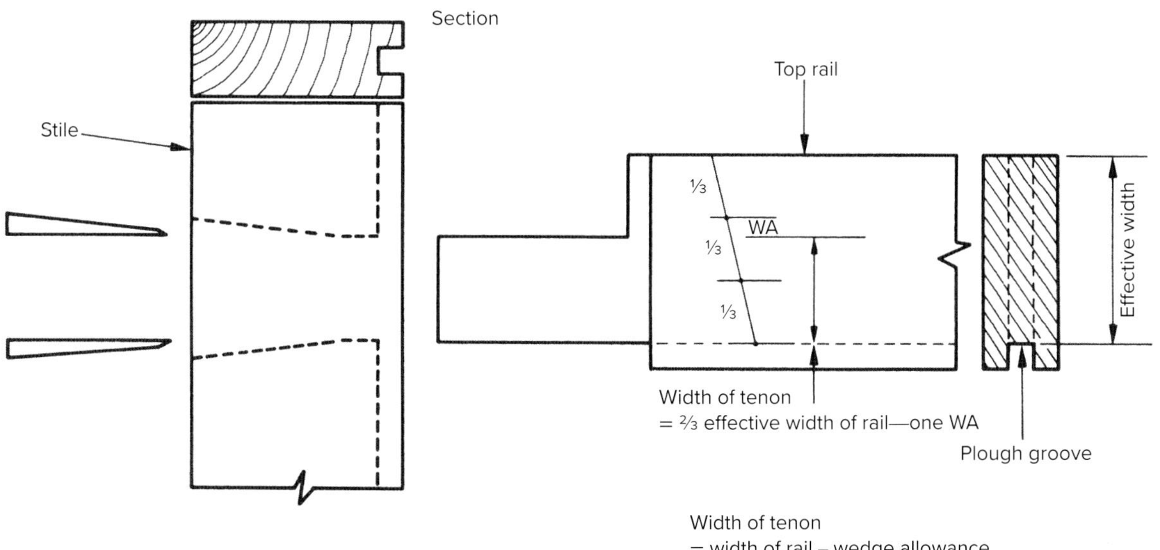

Fig. 7.7 Double mortise-and-tenon joint at the middle rail

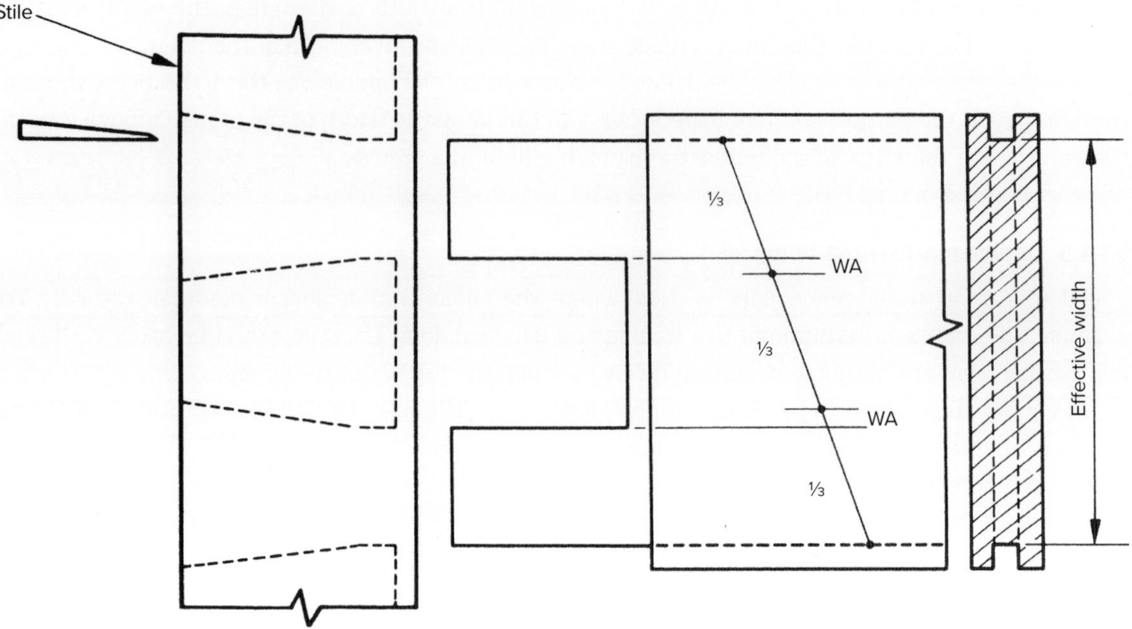

Fig. 7.8 Joint between bottom rail and stile

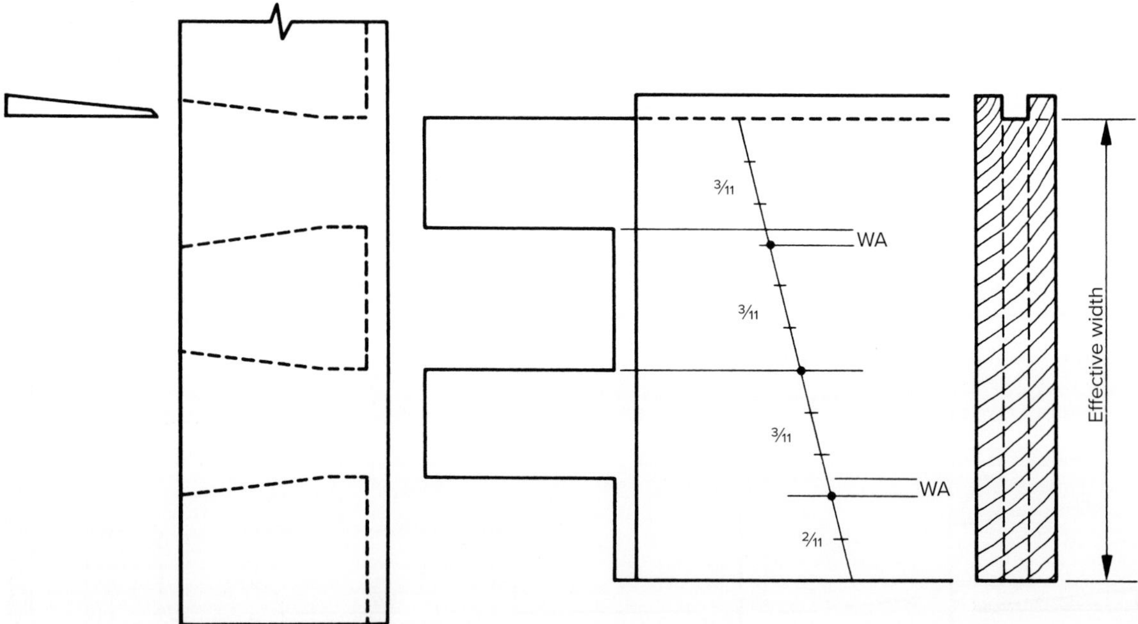

7.1.1.7 Barefaced mortise-and-tenon joint

The barefaced mortise-and-tenon joint is used where the stile is thicker than the rail; this is the case in some doors where the sheeting is continuous over the face of the rail. In this type of mortise-and-tenon joint there is only one shoulder on the rail (Fig. 7.10).

The *double* barefaced mortise-and-tenon joint is suitable for a wide rail (Fig. 7.11).

Fig. 7.9 (a) Long and short shoulder mortise-and-tenon; and (b) splayed haunch

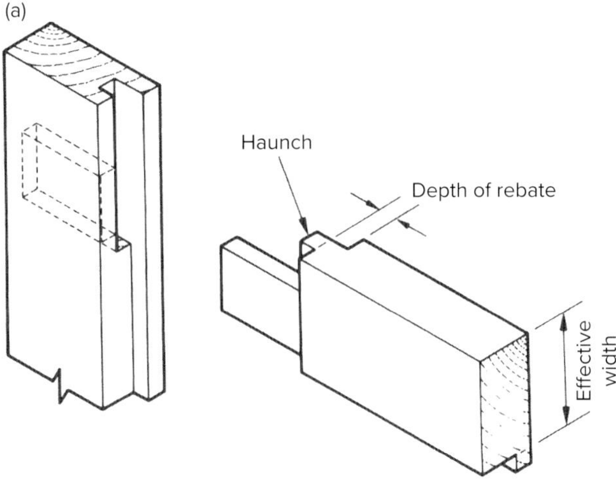

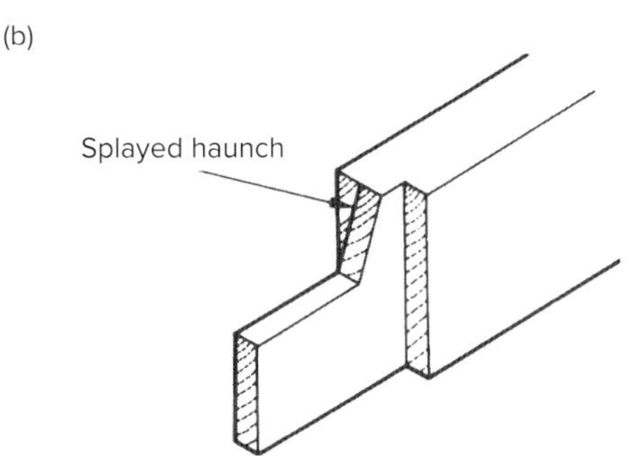

7.1.1.8 Twin mortise-and-tenon joint

When the rail is very thick, and a tenon one-third the thickness would mean the joint is weakened by an unusually large mortise, the twin mortise-and-tenon joint is used (Fig. 7.12). The rail is prevented from twisting and there is twice the gluing area.

7.1.1.9 Stump mortise-and-tenon joint

The tenon need not always pass fully through the stile. In cases where the end grain and wedges showing on the edges would be objectionable, for example, on cabinet doors, the stump (or stub) mortise-and-tenon joint could be used (Fig. 7.13).

The depth of the mortise is made one-half to two-thirds the width of the stile. The tenon cannot be wedged from the outside, so it can be fitted neatly to the mortise and held with an adhesive, or alternatively it can be wedged using fox wedges. Fox wedging consists of inserting two small wedges into saw cuts in the tenon, which, on striking the bottom of the mortise, wedge the tenon as it is cramped into the mortise.

Fig. 7.10 Barefaced mortise-and-tenon joint

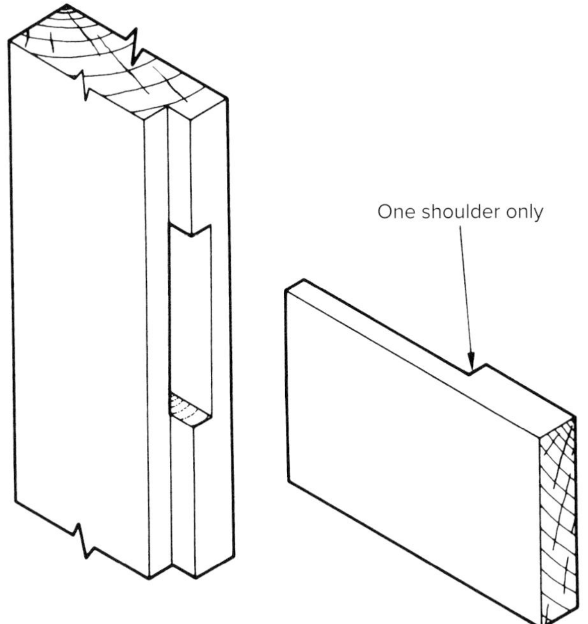

Fig. 7.11 Double barefaced mortise-and-tenon joint

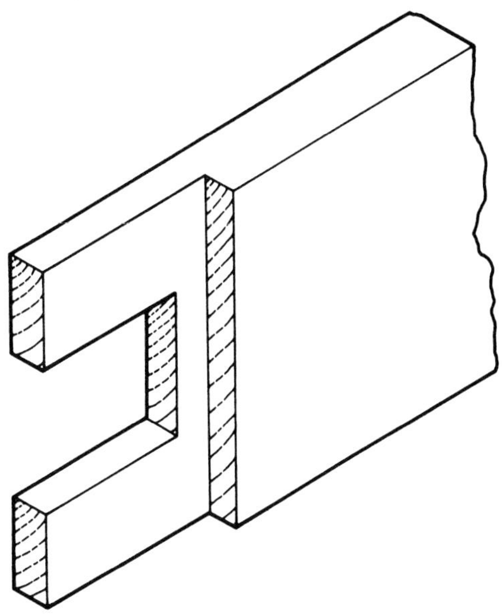

Fig. 7.12 Twin mortise-and-tenon joint

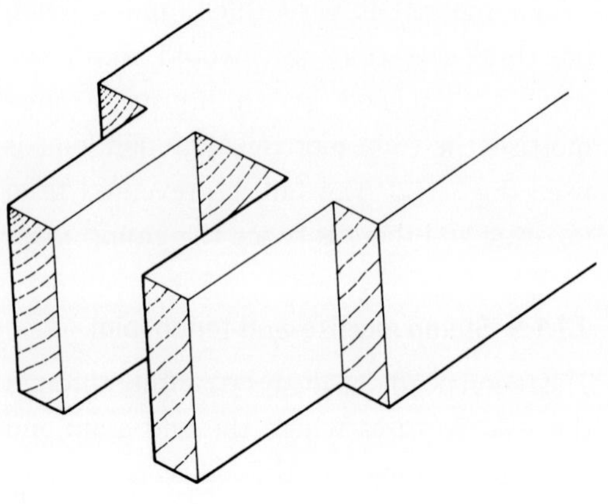

Fig. 7.13 Stump mortise-and-tenon joint

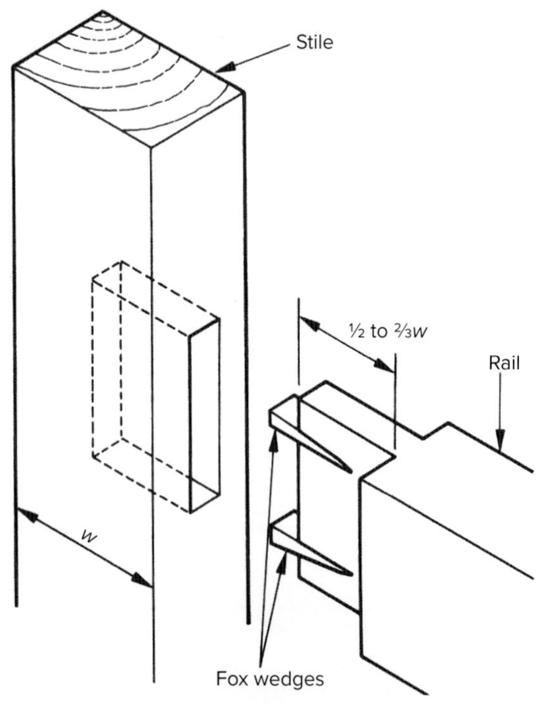

7.1.2 Making the mortise-and-tenon joint

The setting out of joinery items and the marking off of components in preparation for machining a mortise-and-tenon joint will now be dealt with more fully. Some pointers as to the use of hand tools in the marking out and cutting of mortise-and-tenon joints are included.

1. **Gauging.** Set the distance between the spurs of the mortise gauge to the width of the actual chisel used for mortising (Fig. 7.14).

   Gauge the mortise and the tenon, one immediately after the other, using the same gauge and keeping the stock firmly against the face side (Fig. 7.15).

2. **Chopping the mortise.** To chop the mortise, hold the stile in the mortise block, face edge down.

   Stand as near as possible to, and in line with, the mortise to get a good view to ensure that the chisel is held straight (Fig. 7.16a). Commence near one end of the mortise and chop a 'V' with the chisel. Turn the chisel around and continue cutting in sections, allowing the waste to move into the space made by the 'V' (Fig. 7.16b).

   Cut halfway through and finish by squaring the ends down accurately. It will be easier to remove if you make a number of cuts to chop up the waste more finely. The mortise should finish up exactly the same width as the chisel, and it should never be necessary to turn the chisel and pare away at the sides.

   Turn the timber over and repeat the process from the face edge. Chop up the waste more finely and then remove it with a core driver, which is a tapered piece of hardwood slightly smaller than the mortise hole. Finish by cutting back the WAs tapered to about two-thirds the width of the stile.

3. **Ripping the tenon.** Small tenons can be cut with the tenon saw; however, an experienced tradesperson would do the job much more quickly and easily with a hand saw. Hold the timber in the vice, leaning away at about 45°. Keep on the waste side and, following the two visible lines, rip one half of the tenon diagonally. Turn the timber around and follow the line along the third

side, finishing with the saw square to the shoulder.

The tenon should fit straight off the saw with only very minor smoothing of the cheeks. Always rip first and then cut the shoulder line.

4. **Cutting the shoulder line.** The initial setting out is usually done with a sharp pencil. However, a much cleaner and more accurate shoulder may result if the shoulder line is cut using a marking knife.

   Place the blade of the knife on the line, slide the try square up to it and cut the line. If a small recess is chiselled on the waste side of the line, it can then be continued with the tenon saw (Fig. 7.19), only slightly undercutting the shoulder.

5. **Cutting the wedges from the waste.** Remember that wedges will be required when a haunch is being cut. Wedges can be cut from the waste, which will have the exact thickness required.

Fig. 7.14 Setting the mortise gauge

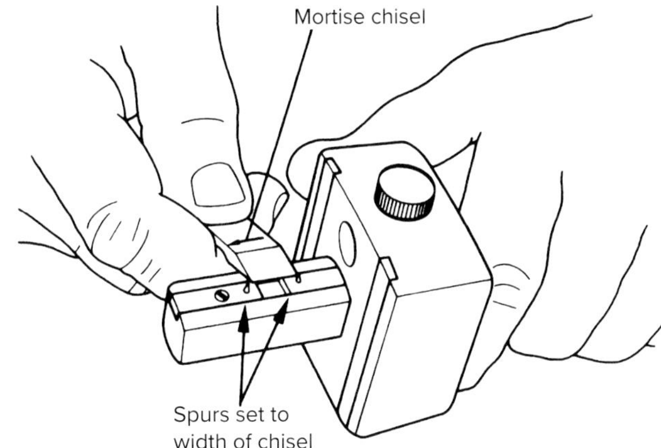

Fig. 7.15 Gauging the mortise-and-tenon joint

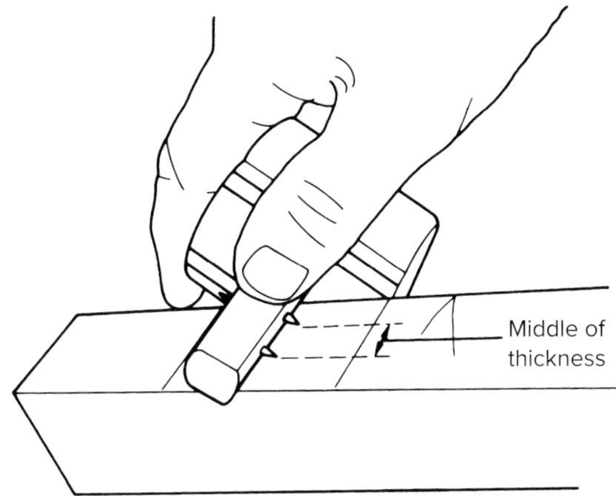

Fig. 7.16 Chopping the mortise: (a) hold the chisel straight; and (b) the waste from each cut moving into the previously cut space

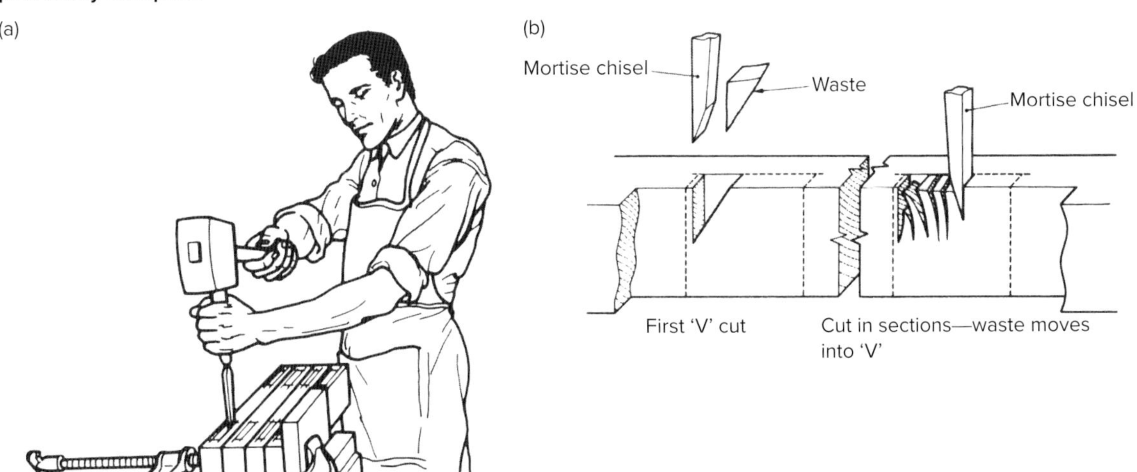

Fig. 7.17 Ripping the tenon

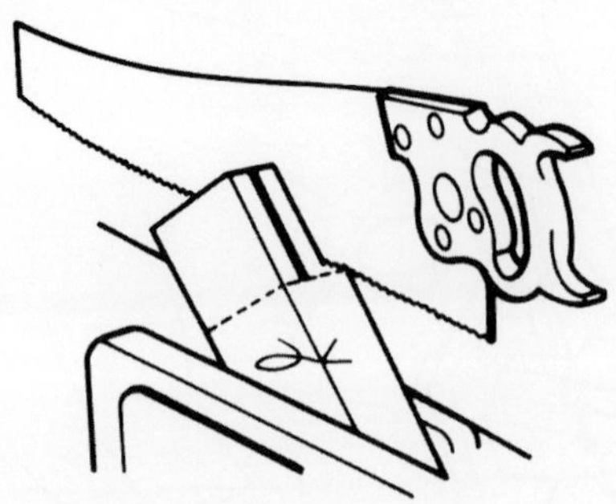

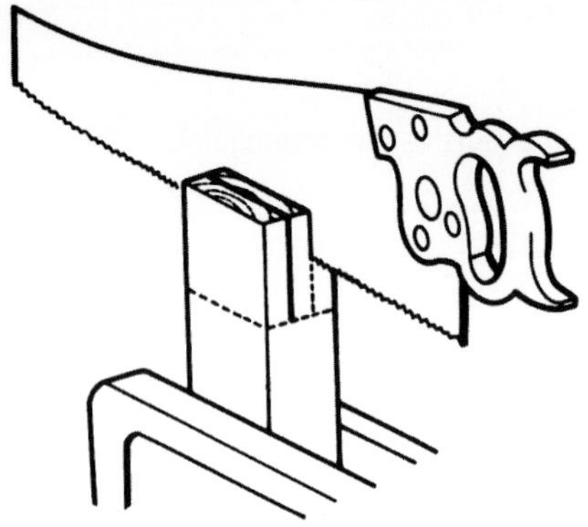

Fig. 7.18 Marking knife

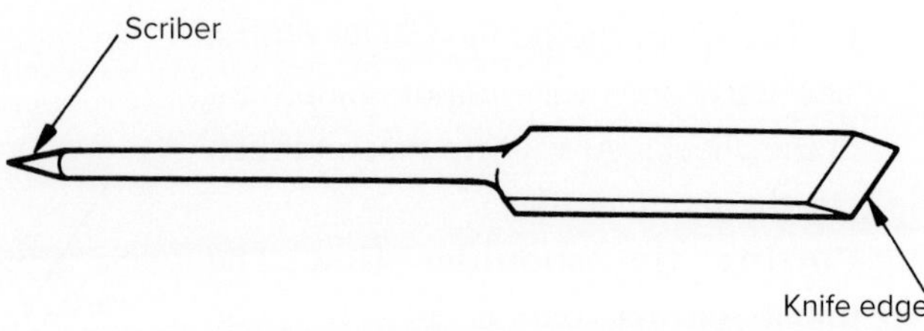

Fig. 7.19 Cutting the shoulder line: (a) knife cut to shoulder line; (b) form recess; and (c) complete cutting shoulder line

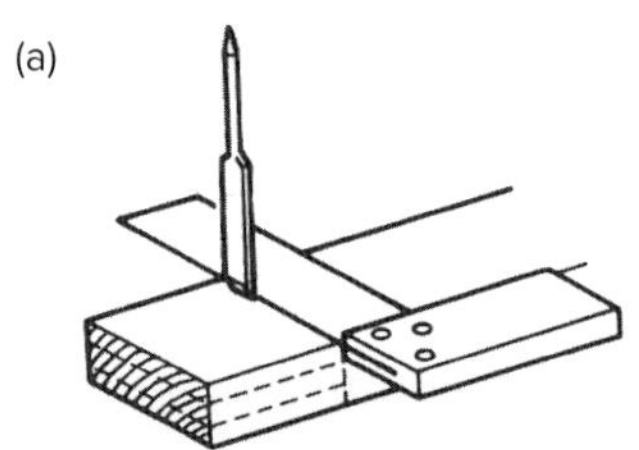

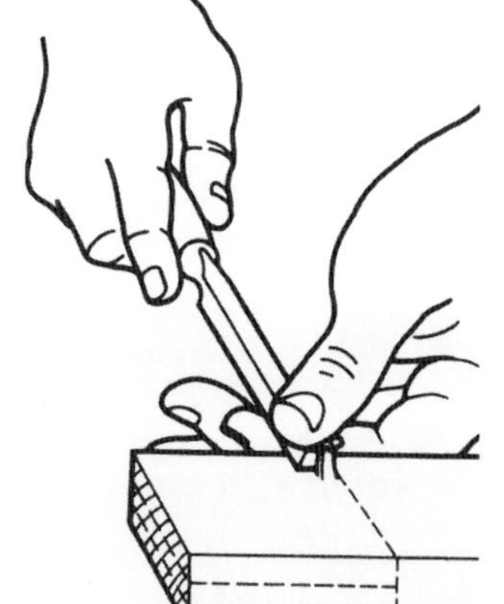

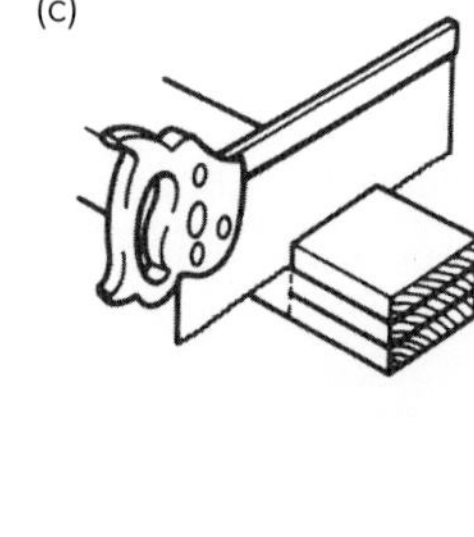

Fig. 7.20 Cutting wedges from waste

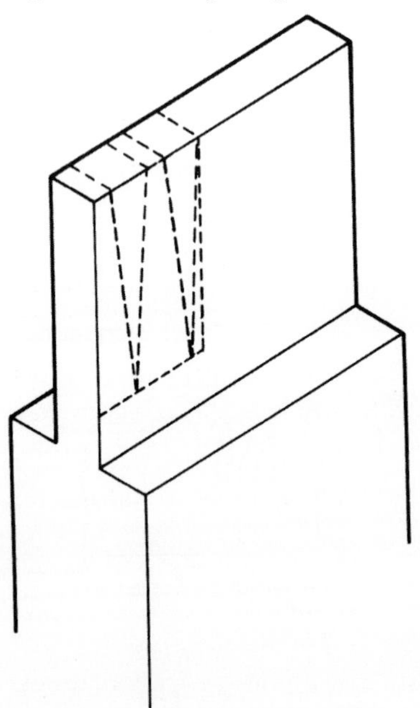

CHECK YOUR UNDERSTANDING

1. Where would a haunched mortise-and-tenon joint be used, and why is it suitable in this situation?
2. How would a mortise gauge be set when gauging a mortise-and-tenon joint?

Fig. 7.21 (a) Corner halved joint; (b) cross halved joint; (c) tee halved joint

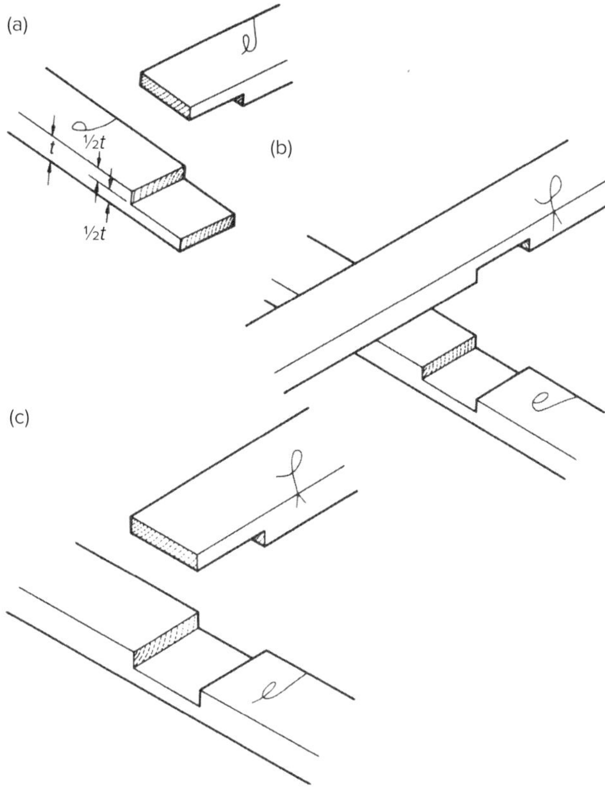

Fig. 7.23 Marking the dovetail

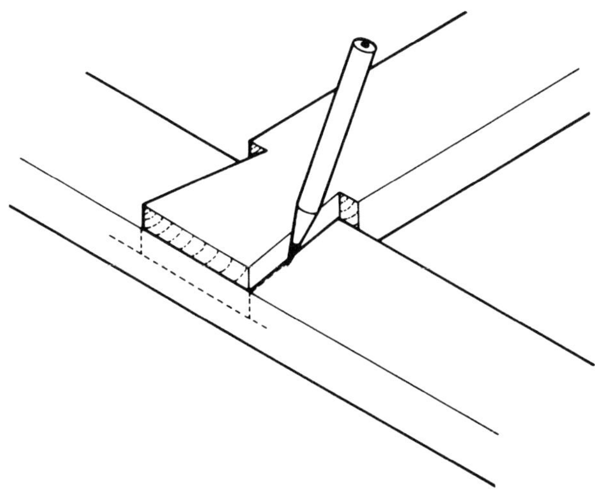

Fig. 7.22 Dovetailed tee halved joint

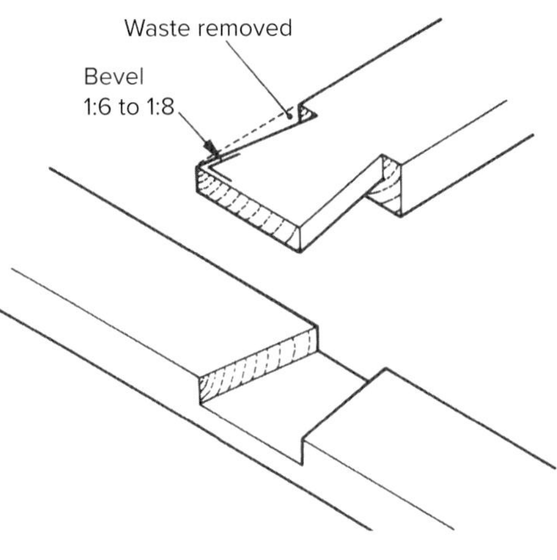

7.1.3 Half-lapped joint

The half-lapped joint is also called the *halved joint*. The half-lapped joint in timber-frame construction is described in Chapter 6; this joint also finds applications in joinery and cabinet work.

Set the gauge to one-half the thickness of the timber and gauge both parts from the face side. Cut the halvings, following the same procedure as for the cheeks of the tenon–that is, rip first, cross-cut second.

A modification of the half-lapped joint is the dovetailed tee halved joint (Fig. 7.22). From the dovetail, lay it in position on the second piece and with a sharp pencil mark the bevel (Fig. 7.23). The sides of the tapered trench are then cut on the waste side, leaving the pencil line clearly visible. The dovetail should be a close fit, acting as a wedge to draw the shoulder together.

7.1.4 Cutting a trench

The four steps in cutting a trench are illustrated in Figure 7.24. This is applicable to housed joints as well as to the dovetailed tee halved joint and the bridle joint.

Fig. 7.24 Four steps in cutting a trench

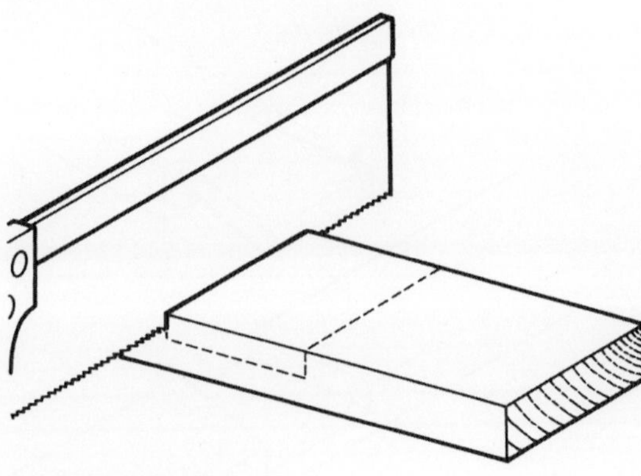

1. Cut sides of trench on waste side of line.

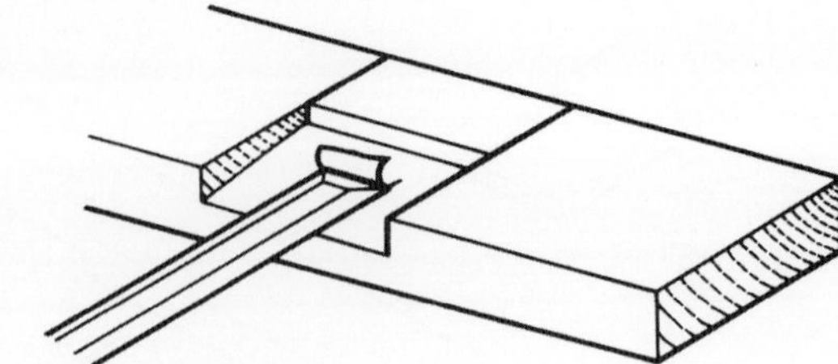

2. Remove waste half-way through and down to gauge line. Use chisel and mallet.

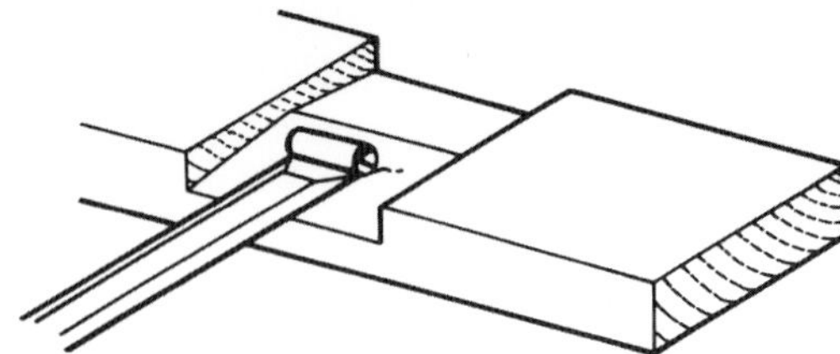

3. Turn timber, and chisel from other side down to gauge line, leaving trench high in centre.

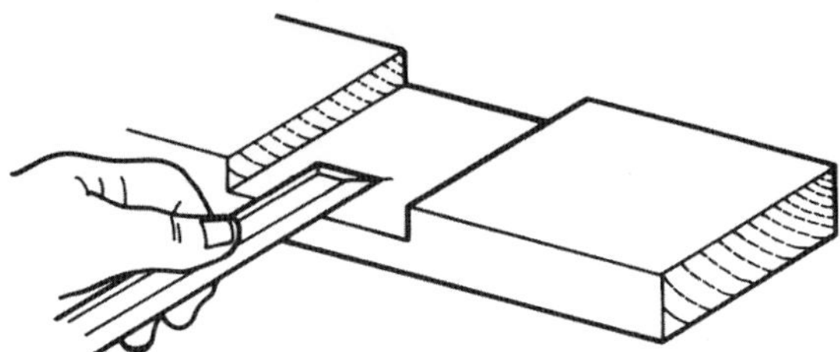

4. By hand, pare bottom of trench flat. Note how chisel is held between thumb and fingers and kept under control at all times.

7.1.5 Dowelled joint

The dowelled joint consists of mating holes, drilled into the two parts, into which round pegs called *dowels* are inserted, joining the parts together. When correctly proportioned and assembled with a suitable adhesive, the joint is strong enough for most purposes; there is a saving in timber as the rails are cut off at the shoulder line.

Dowels are made from hardwood or softwoods, in the standard diameters of 6, 8, 10 and 12 mm. Dowels have a fluted circumference, which ensures a firm grip on the dowel hole. The fluting also allows the surplus glue and air to escape from the bottom of the hole as the dowel is driven home. Dowel rod is available in long lengths from 1 m to 2 m, or dowels may be ready-cut to length and pointed.

TIP Dowels most commonly have a diameter of 10 mm. The dowel should not be an overtight fit in the hole; if the dowel is too tight it will probably split the timber.

The setting out of dowels on the ends of the rails is shown in Figure 7.27. The ideal proportions are shown.

A little compromise may be necessary at times to fit the dowels onto a narrow rail. Remember, however, that a stronger joint may result if a small size of dowel is used rather than a dowel that is too large for the job, which may seriously weaken the timber around the dowel.

Table 7.1 is a guide to the diameter of dowels to use relative to the thickness of the timber to

Table 7.1 Dowel selection chart

Diameter of dowel (mm)	Thickness of timber (mm)
6	10–14
8	15–18
10	19–26
12	27–32

Fig. 7.25 Dowelled framing joint

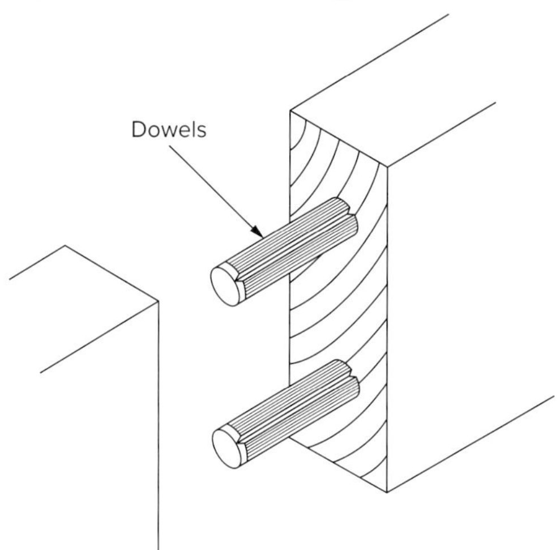

Fig. 7.26 Dowel profiles

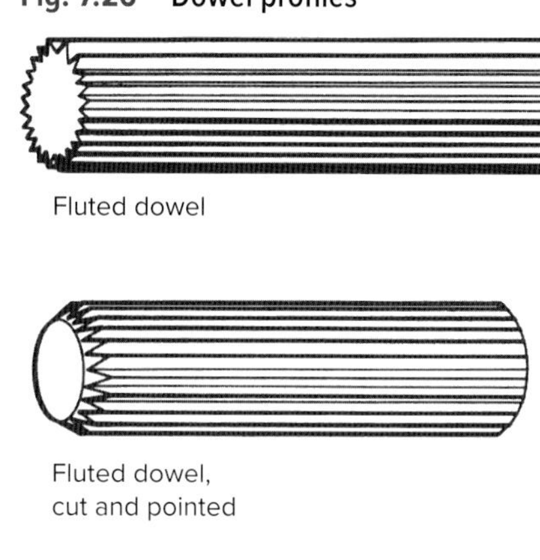

be dowelled. Because 10 mm is the most common diameter of dowel, it is used over a slightly wider range of timber thicknesses than are the other sizes. If the two parts forming the dowelled joint are of unequal thickness, the dowels must be chosen to suit the thinner of the two.

The dowels used must be long enough to penetrate each part of the joint sufficiently so that maximum strength is attained. Penetration can vary from two to a maximum of three times the diameter.

The dowel holes must be drilled deep enough to allow at least 2 mm clearance beyond the end of the dowel. The use of a large countersink bit to form a small glue clearance around the hole will also help when cramping the shoulder of the joint tightly together.

On wide rails, more than two dowels can be used provided there is sufficient material left between the dowels–say a minimum of twice the diameter.

Fig. 7.27 Setting out dowels

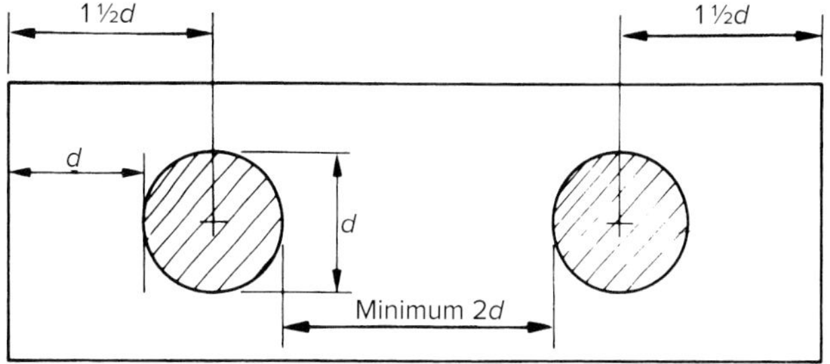

7.1.5.1 Making the dowelled joint

The procedure for making the dowelled joint, as found on the corner of a piece of dowelled framing, is outlined below. Note the direction of the face and face edge marks in Figure 7.29.

1. **Cutting shoulder lines to rails.** To cut the shoulder lines to the rails using the tenon saw, first square completely around the shoulder line, cutting with the marking knife, and chisel a small groove beside the cut line in a similar manner as for the mortise-and-tenon joint.

Fig. 7.28 Dowel penetration

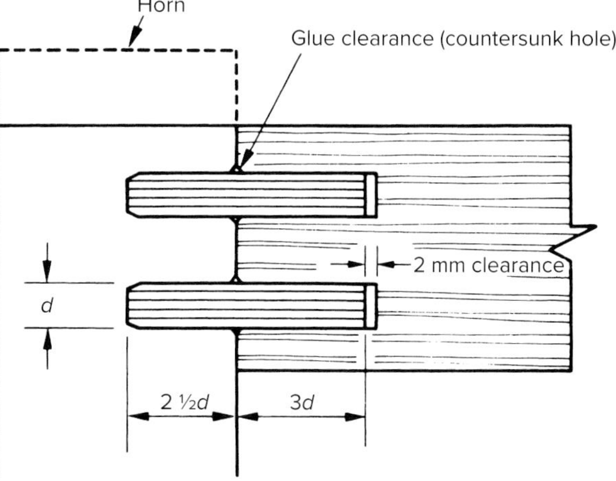

Fig. 7.29 Corner of a dowelled frame

Fig. 7.30 Cutting the shoulder

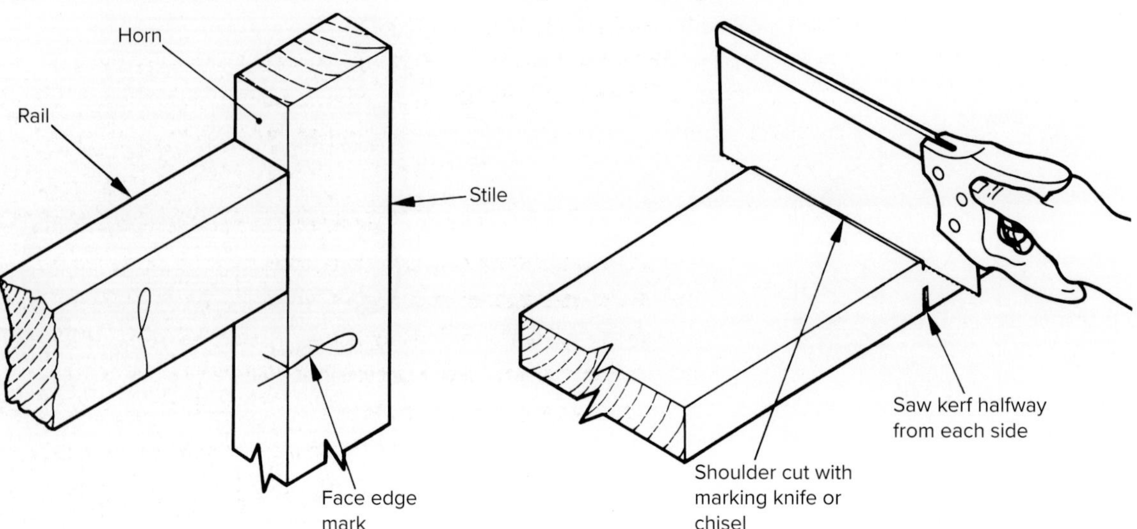

Continue to cut the shoulder with the tenon saw, only very slightly undercutting. Cut until halfway through, turn the rail over and repeat the process, meeting up with the first cut.

2. **Marking out.** The dowel holes in the stile and rail must mate together. First, mark the position of the rail on the face edge of the stile and mark the dowel centres.

   Transfer the dowel centres onto the end of the rail and square them across.

   Set the marking gauge to half the thickness of the material and, keeping the stock to the face side, gauge the centres for the holes. Depending on the equipment available to bore the holes, it may be necessary to gauge the position of one hole only, which can then be used as a guide to setting up a dowel jig or machine to the required setting.

3. **Boring the dowel holes.** The *dowelling jig*, shown in Figure 7.32, used in conjunction with a dowel bit in the electric drill, is a fast and accurate way to reproduce holes.

   The jig is supplied with a number of bushes to suit the size of dowel drills–6, 8, 10 and 12 mm. The bush provides the guide for the drill and can be adjusted to drill holes a fixed distance from the face side. A mark on the front of the jig enables it to be lined up accurately with the line of the hole.

Fig. 7.31 Setting out dowel centres

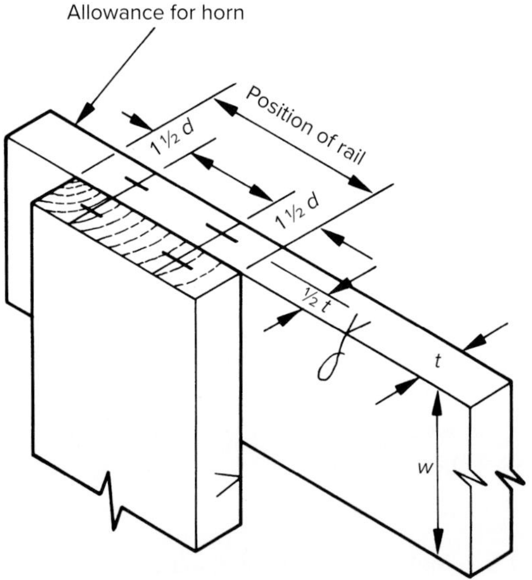

Fig. 7.32 Dowelling jig

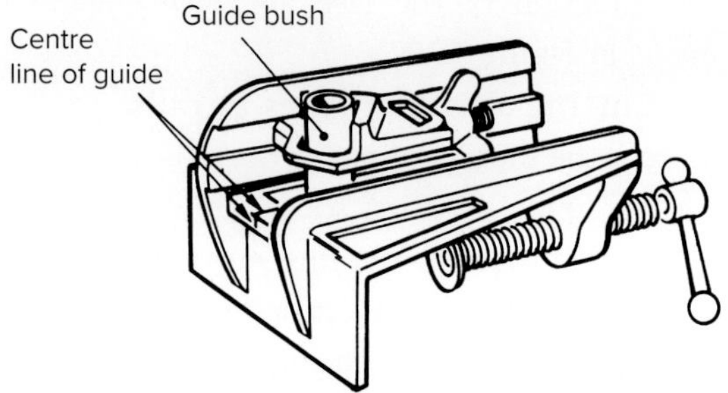

To use the dowelling jig, select and fit the required guide bush to the jig and adjust the centre of the bush to the gauge line for the centre of the dowel holes. Clamp the jig to the timber, aligning the mark on the jig with the line of the hole, and drill the hole to the required depth. A depth stop is necessary on the drill and there are a number of types available that clamp onto the drill. A most effective stop can also be made up as follows: drill a hole lengthwise through a piece of wood and cut off to length; round the corners and fit over the drill (Fig. 7.34).

The *dowel borer* is basically a horizontal drill mounted over a table that is adjustable in height (Fig. 7.35). To set up a dowel borer, fit a drill of the required size and adjust the height of the table so that the distance from the table to the centre of the drill is equal to the distance from the face of the material to the centre of the dowel hole. Set the depth stop to drill the holes to the required depth.

To use the dowel borer, lay the timber on the table face side down, grip firmly and slide the table forward, feeding the timber onto the drill until the depth stop is reached. In a modern workshop, horizontal dowel-boring machines may be much more refined than described here; however, the basic principles remain the same.

4. **Assembling.** After the dowels have been cut to length and pointed, the joint is ready to assemble. Glue the dowel by dipping the ends into the adhesive. Adhesive can also be applied to the dowel holes by taking a short length of smaller-diameter dowel or stick, dipping the end into the adhesive and wiping it around the hole.

Drive the dowels into the stile first. Apply a little adhesive to the holes and shoulder line of the rail, then tap the joint together. Pull together firmly with a sash cramp to ensure a tight fit. The well-made dowel joint should be square, with close-fitting shoulders on both sides and with the face side flush, and of course the timber must not be split. If splitting occurs this usually indicates that the dowel is too long or the hole is too tight.

Fig. 7.33 Setting up the dowelling jig

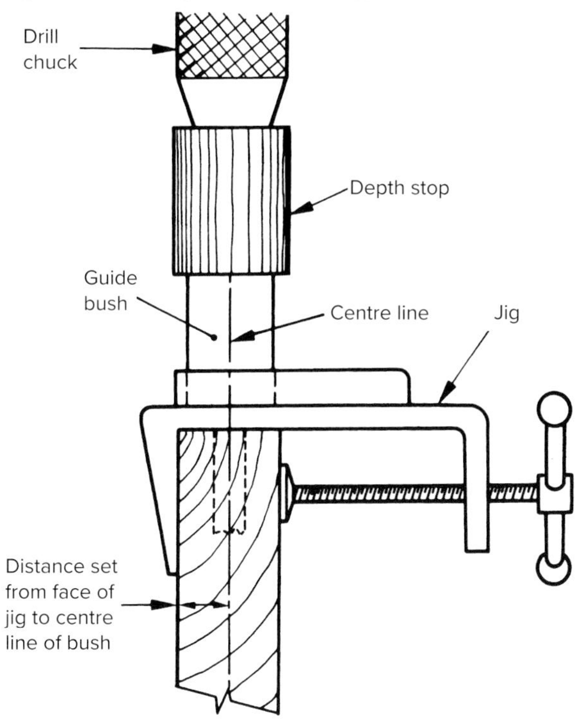

Fig. 7.34 A & B Depth stops

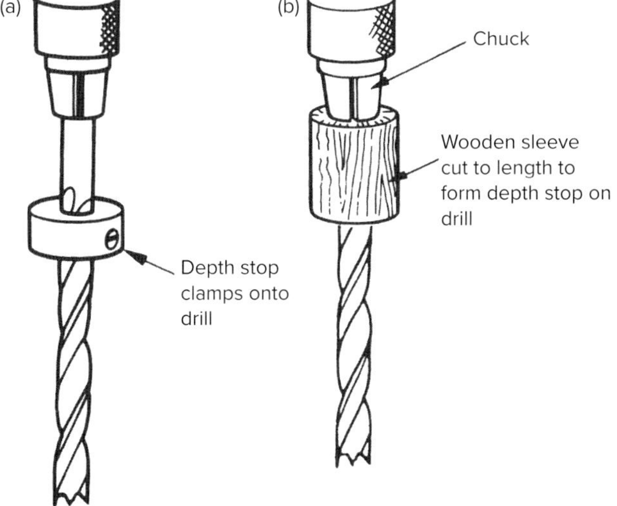

Fig. 7.35 Dowel borer

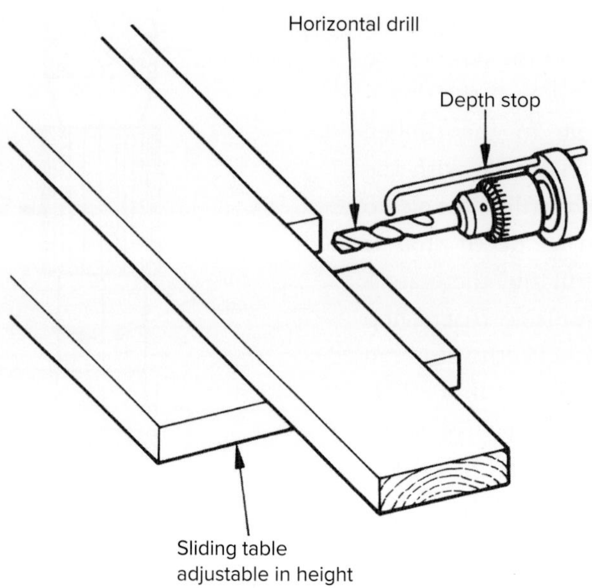

CHECK YOUR UNDERSTANDING

1. When setting out dowels on the end of a rail, what is the minimum distance the dowels should be kept from the edge?
2. Describe briefly how the dowel jig would be set up for boring dowel holes.

7.1.6 Biscuit joints

Biscuit joints are an alternative to dowel joints. Figure 7.36 shows two examples of corner joints using biscuit joints for butt and mitre joints. The advantage with this method is that the machine for cutting the joint is portable and highly accurate. This method has many applications, including widening boards (Fig. 7.37) and corner joints. This method of joining can also be used without glue, if a means of locating components in large pieces is required.

TIP Setting out biscuit joints follows a similar procedure to dowel joints.

Fig. 7.36 Two examples of corner joints using biscuits for: (a) butt joints; and (b) mitre joints

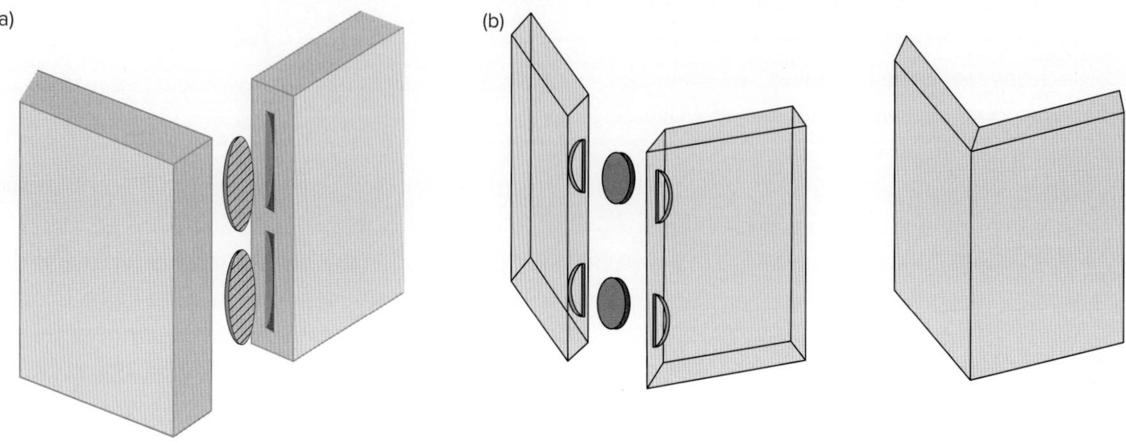

7.2 Carcase joints

Carcase joints are sometimes referred to as *angle box joints*. Joints in this group are distinguished by the fact that the angle formed is between the face sides. They are used to join up the carcase to cabinets, corners of drawers, shelving and so on.

Again, for satisfactory results, joints must be accurately set out and cut to detail. Fixing is often done using nails and screws in conjunction with an adhesive or, if concealed fixing is required, dowels or 'feathers' can be used. If the joint is cramped, some adhesives on their own may be adequate.

Fig. 7.37 Widening boards using biscuits

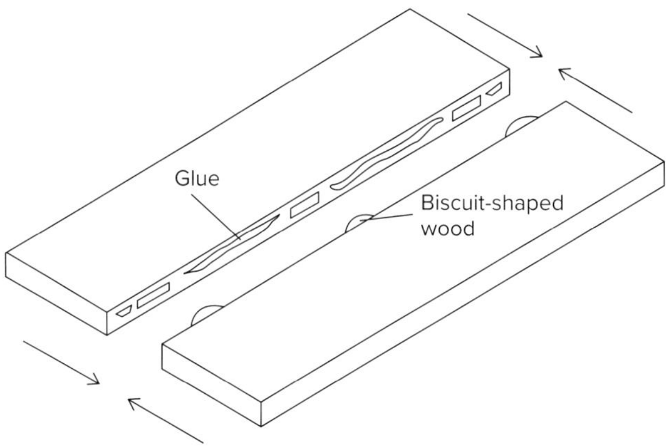

Fig. 7.38 Power tool used to cut biscuit slots

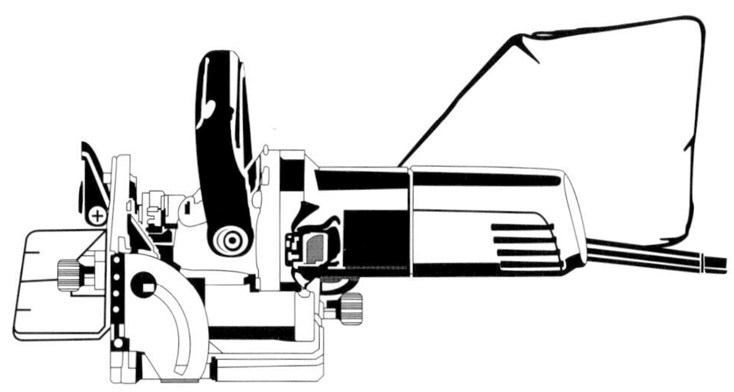

7.2.1 Butt joint

The butt joint is the simplest carcase joint and it can be nailed or screwed (Fig. 7.39). The ends must be cut square across both the face and edge. To nail the joint, start the nails in the top piece, 'dovetailing' them at an angle to secure a better grip, apply adhesive, bring the joint together and complete the nailing. Finish by punching the nails slightly beneath the surface before the glue sets. Otherwise, if the nails are punched after the glue sets, the glue joint could fracture. The butt joint can be strengthened through the addition of glue blocks to the inside corner.

Fig. 7.39 Butt joint

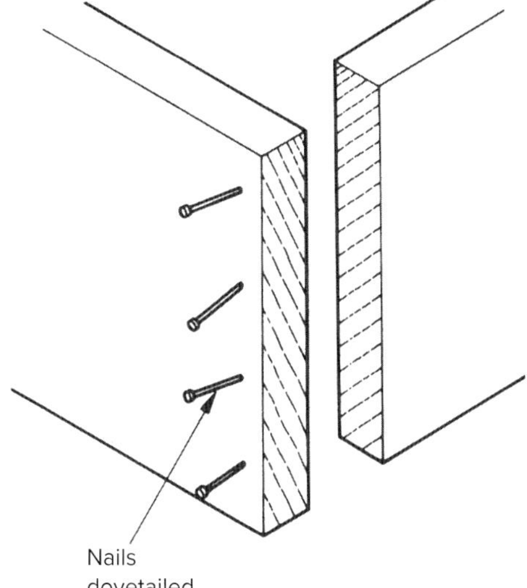

7.2.2 Rebated joint

The rebated joint is commonly used and can be adapted to a variety of materials and applications, including drawers (Fig. 7.40). The depth of the rebate can vary from one-third to two-thirds the thickness of the material, and therefore shows little end grain. Fixing is by adhesive and nailing or screwing; the nailing can be done on either side of the joint.

7.2.3 Tongued and trenched joint

The tongued and trenched joint makes a strong neat joint that can be fixed adequately using only an adhesive (Fig. 7.41). The short grain beside the trench is a weak point that makes it unsuitable for use with manufactured boards. The tongue should be no more than one-third the thickness of the timber.

Fig. 7.40 Rebated joint

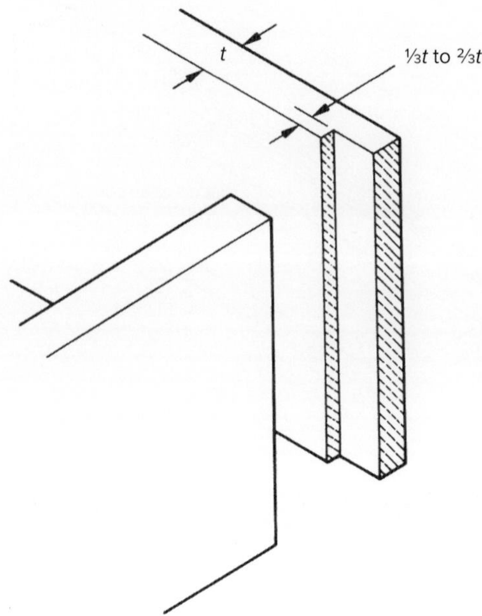

Fig. 7.41 Tongued and trenched joint

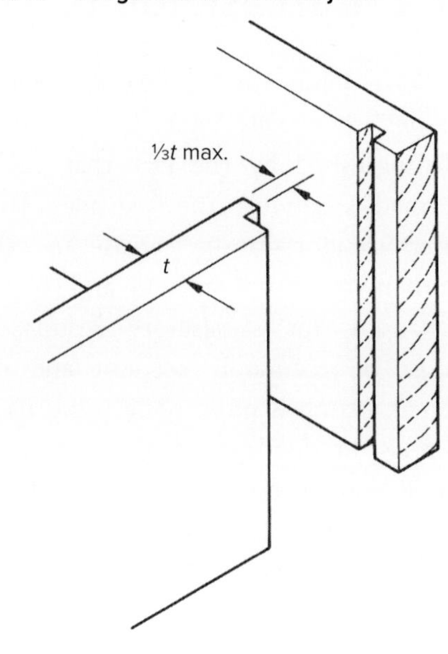

7.2.4 Mitred joint

The mitred joint shows no end grain (Fig. 7.42). The parts of the joint must be cut at 45° where the joint is to be at right angles and must be dressed accurately. It can be fixed by an adhesive, nailing or screwing, or some combination of the three.

In this case, the nails should be started in the top piece (Fig. 7.43). Apply a little adhesive and bring the joint together.

Keep the top piece a little uphill, because when the joint is nailed, it will slide down into place, pulling the parts together and expelling any surplus adhesive. Nailing can be carried out from both sides; the nails are then punched below the surface. The recess is filled with a suitable stopping compound.

Concealed fixing can be achieved by use of a 'feather' or loose tongue (Fig. 7.44).

Fig. 7.42 Mitred joint

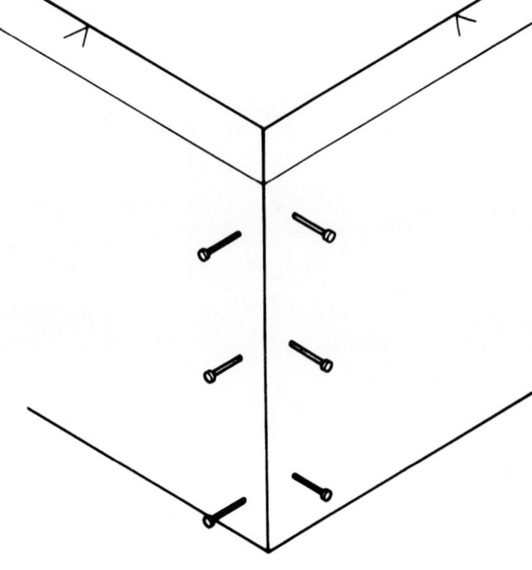

Fig. 7.43 Nailing the mitred joint

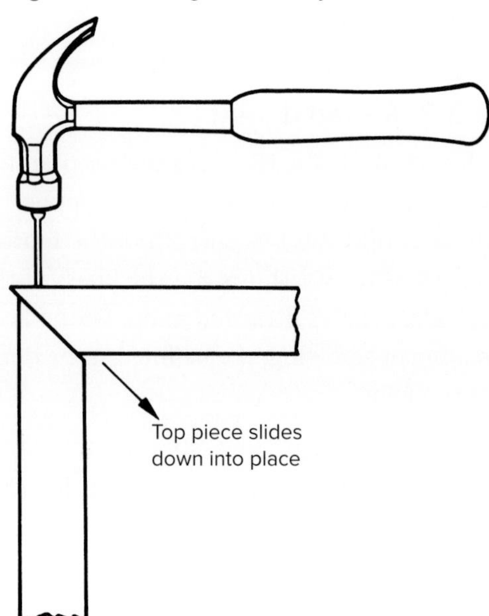

Make a saw cut, or use the router to run a small groove, across the joint. Into this can be fitted a tongue of thin plywood or veneer.

Due to the thin feathered edge of the mitre, this joint is sometimes unsuitable for use in veneered or manufactured boards as the face veneer can chip away easily. Figure 7.45 shows a modification of the joint to make it more suitable to this material.

7.2.5 Housed joints

Housed joints are used in solid timber furniture construction, supporting shelving and so on. A *through housing* will show the joint on the exposed edge (Fig. 7.46).

If this is undesirable, a neater joint is the *stopped housed joint* (Fig. 7.47).

In this joint, the housing is stopped short of the edge and the shelf is cut to butt against the side.

To cut the stopped housing using hand tools in solid timber, first mark the position of the housing. Use the thickness of the board to be fitted as a template for marking the exact width, and cut the sides with the marking knife or pencil.

At the stopped end, use the chisel to start the housing sufficiently to allow the

Fig. 7.44 Loose tongue in the mitred joint

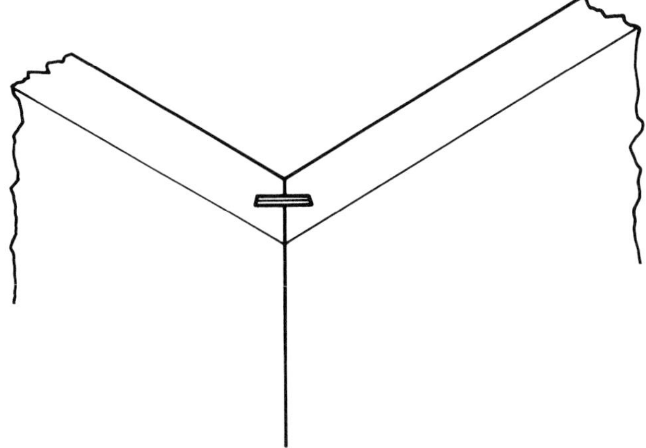

Fig. 7.45 Modification of the mitred joint for veneered boards

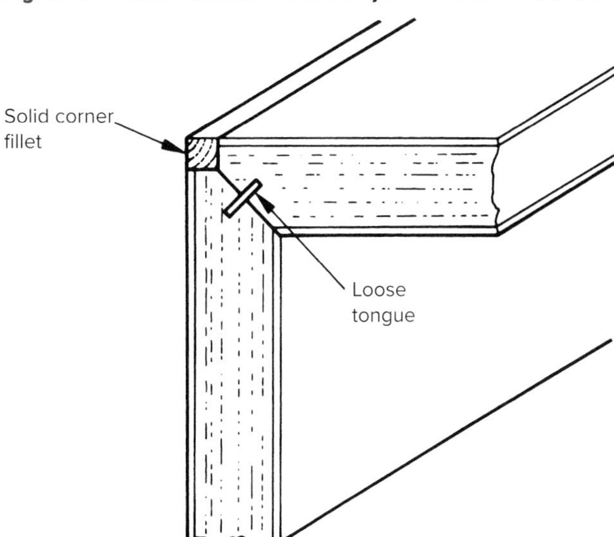

Fig. 7.46 Through housed joint

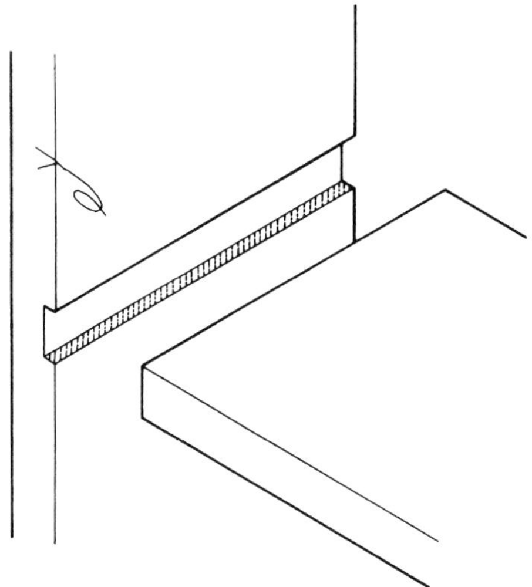

Fig. 7.47 Stopped housed joint

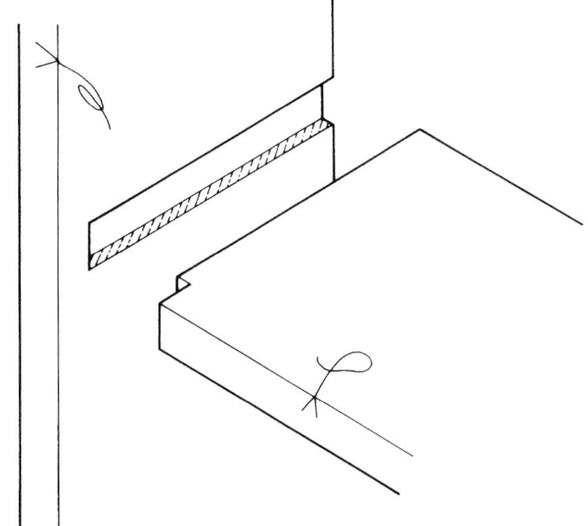

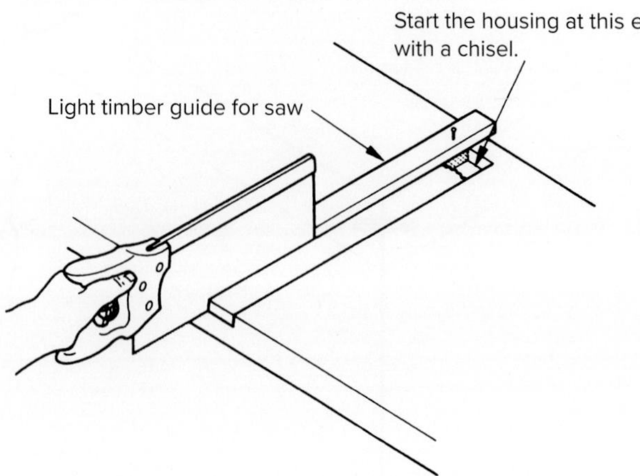
Start the housing at this end with a chisel.
Light timber guide for saw

Fig. 7.48 Cutting the stopped housed joint

tenon saw to cut down the sides. A light timber strip, pinned or clamped along the line, is a useful saw guide over a long cut. After sawing both sides, chisel away the waste and smooth the bottom of the housing to a uniform depth, preferably using the hand router plane.

Housed joints can be made readily using an electric router with an appropriate cutter the thickness of the board to cut the housing.

CHECK YOUR UNDERSTANDING

1. List three types of carcase joints.

7.2.6 Shelf supports

Another popular method for supporting shelves in carcases is to use adjustable shelf supports.

7.2.7 Dowelled joint

The dowelled joint can also be used as a carcase joint and is particularly adaptable for use with particleboard where the range of other suitable joints is limited (Fig. 7.49).

Details on the setting out and construction of the dowelled joint have already been described. Precut 8 mm dowels are convenient for use with 16 mm or 18 mm particleboard. Apply a suitable adhesive to the joint and cramp together until set.

Fig. 7.49 Dowelled carcase joint

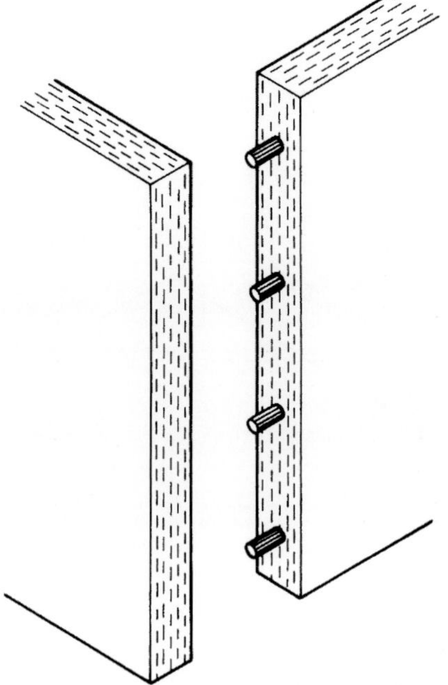

7.2.8 Dovetail joints

The dovetail joint is the strongest of the carcase joints. The presence of handmade dovetail joints in manufactured articles has long been regarded as a sign of quality.

Each joint must be individually marked and accurately fitted. Face sides and face edges must be clearly marked and followed so that the joints will come together in the finished article the correct way round. The face sides can be either *all inside* or *all outside*. Face edges must *all* point in the same direction. In the examples shown, the face side has been marked on the inside. The most common use for dovetails in joinery and furniture is to join the fronts and sides of drawers.

All parts must have face and edge marks to avoid confusion. Dovetail joints are set out from the end of the timber, and parts must be cut to their finished length and the ends dressed square as a first step to making the joints.

Types of dovetail joints are now described.

> **TIP** This section covers handmade dovetail joints; however, proprietary dovetail jigs are readily available for use with an electric router fitted with a dedicated 'dovetail' router bit to suit a wide variety of applications.

7.2.8.1 Common dovetail joint

The common dovetail joint is one in which the pins and sockets go fully through the timber and, when assembled, show end grain on both of the outside faces. The angle or ratio of the pins illustrated in the following examples can be between 1 in 4 and 1 in 7.

To make the common dovetail joint, proceed as follows:

1. **Squaring.** After the ends have been dressed square, measure back from the end the thickness of the material and square around both pieces to be joined. Use a sharp pencil and a try square.
2. **Setting out the pins.** Set out the pins on the end of the timber. The details of setting out are shown in Figure 7.51. Take three-quarters the thickness (¾ *t*), which is the maximum width of the pins, and round to the nearest millimetre; set a sliding bevel to 1 in 4 (Fig. 7.52).

   Mark a half pin (so named because it has a slope on one side only) on each edge of the timber. If there are to be two intermediate pins, there will be three spaces. Adopting the 'in and over' method, divide the width into the required number of equal spaces. Mark out one pin to each space (Fig. 7.51). Then mark the bevel of the pins across the end of the timber. The maximum distance between the pins is three times the width of the pin (3 × ¾ *t*). To ensure a close-fitting shoulder line, use the marking knife to cut along the line of the waste between the pins.
3. **Ripping the pins.** Using a fine dovetail saw, rip the pins, keeping on the waste side of the line (Fig. 7.54).

Fig. 7.50 Common dovetail joint

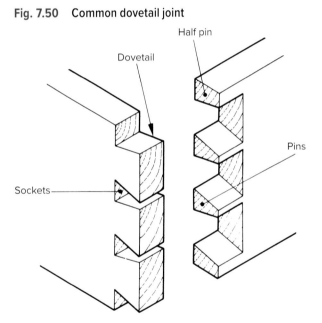

Fig. 7.51 Setting out pins for a dovetail joint

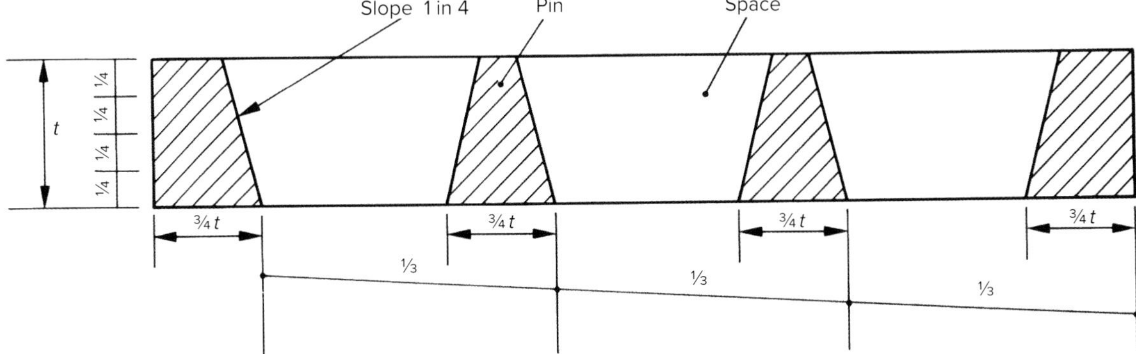

Fig. 7.52 Setting the bevel

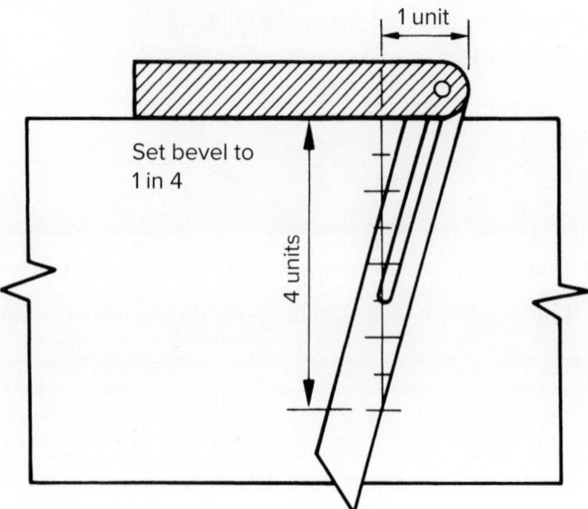

Fig. 7.53 Completed set-out of dovetail pins

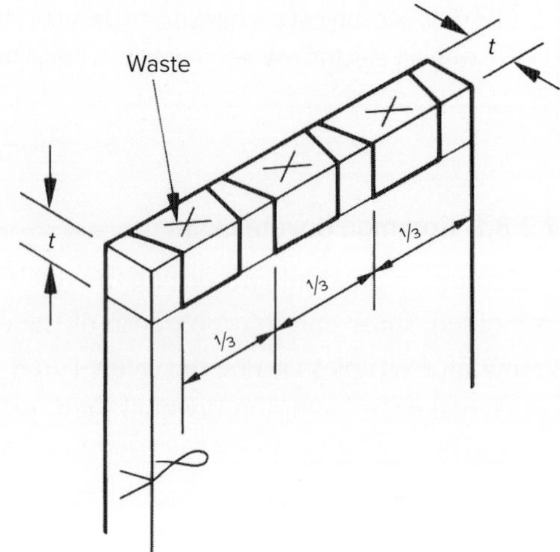

Fig. 7.54 Ripping the pins

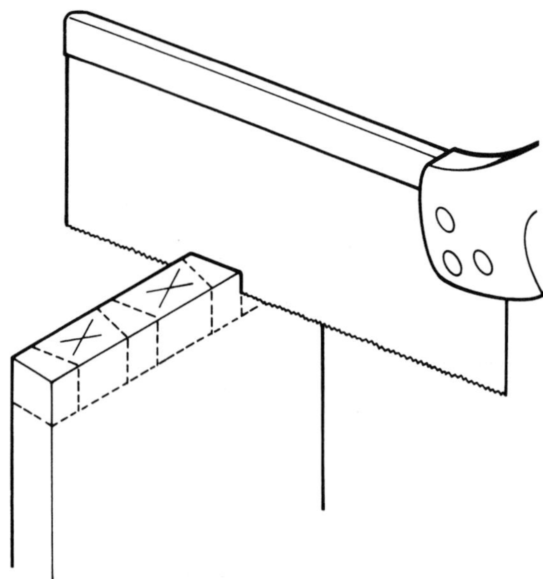

Fig. 7.55 Cutting the waste with a coping saw

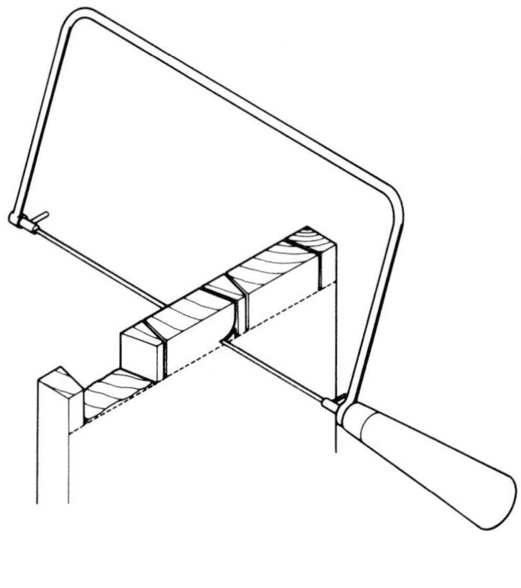

4. **Removing the waste.** There are two recognised methods of removing the waste from between the pins. The first uses the coping saw to cut away the waste (Fig. 7.55). Keep just on the waste side of the line and use a paring chisel to true up the shoulder line. Very slightly undercut, working from both sides of the timber.

   The second method is to chisel down the shoulder line and remove the waste in the form of a 'V' for half the thickness (Fig. 7.56). Turn the timber over and remove the other half of the waste in a similar manner; cutting in this way, and leaving the front end of the

Fig. 7.56 Chiselling out the waste

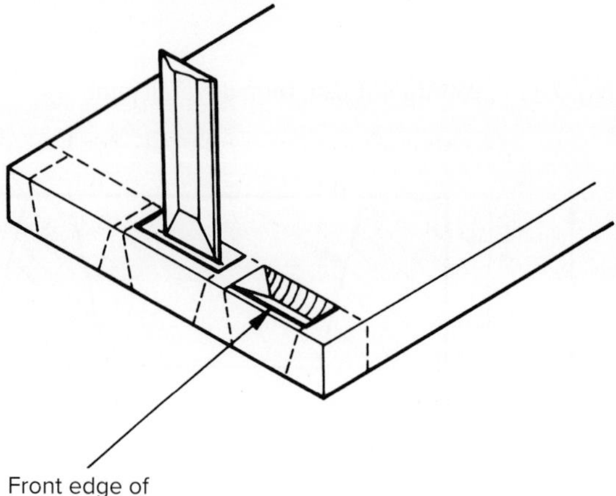

waste intact, provides support for the end when the timber is turned over. There is always a strong chance that the timber will break out between the pins, so use a very sharp bevelled-edge chisel and proceed cautiously in the manner described.

5. **Marking the sockets.** Use the pins as a template to mark out the sockets on the other piece (Fig. 7.57). Use a sharp pencil and square the line across the edge (Fig. 7.58).
6. **Ripping the sockets.** To rip the sockets, keep on the waste side of the line. The pencil line should remain visible and completely intact (Fig. 7.59).
7. **Removing the waste.** Using a suitable narrow paring chisel, remove the waste, again working from both sides for the intermediate pins and saw away the waste to the half pins. If necessary, use a medium- to fine-grade abrasive paper to sand the inside faces. The joint is ready to assemble.
8. **Assembling.** If the joint has been cut out accurately, it should fit off the saw with only minor adjustments to the fit. The tapered pins should fit neatly, tending to pull the shoulder lines together without the need for any other form of cramping. Apply a small amount of adhesive to the mating surfaces, enter the pins into the sockets and tap the joint together. A diagonal brace may be required to hold the article square while the adhesive sets, after which the outside faces can be flushed off with the smoothing plane and sanded.

Fig. 7.57 Marking the sockets

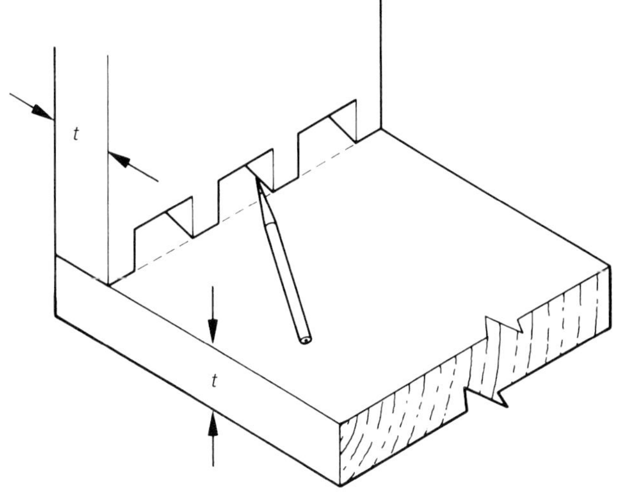

Fig. 7.58 Squaring the sockets

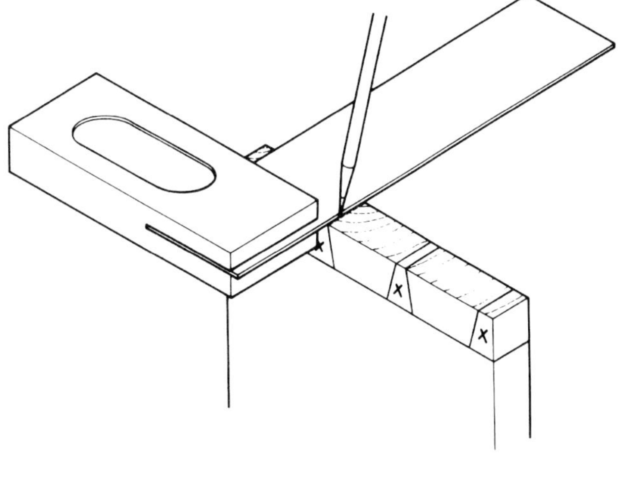

7.2.8.2 Lapped dovetail joint

The lapped dovetail joint is so named because of the lap formed on one piece, so that when the joint is assembled, end grain will be seen on one side only (Fig. 7.60). The most common application of the lapped dovetail is the joint between the front and side of a drawer. The thicker piece containing the lap is the *front*, and the other piece is the *side*. They will be referred to as such in the description of the procedure.

The setting out of the pins on the drawer front is shown in Figure 7.61. Divide the thickness into three equal parts and make one-third the lap. Dovetail pins are then set out on the remaining two-thirds, their width referred to as p. It can be seen from the set-out that if the maximum width of pins is ¾ p, this is the same thing as ½ t (t being the thickness of the front), rounded off to the nearest millimetre.

Fig. 7.59 Ripping the sockets

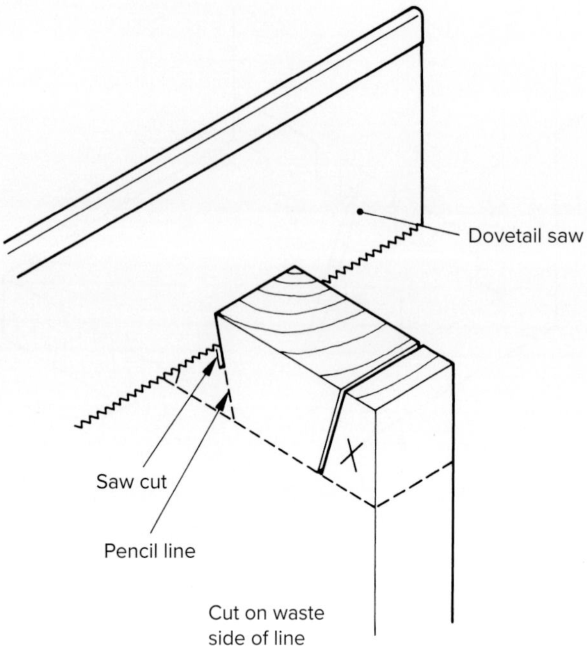

Fig. 7.60 Lapped dovetail joint

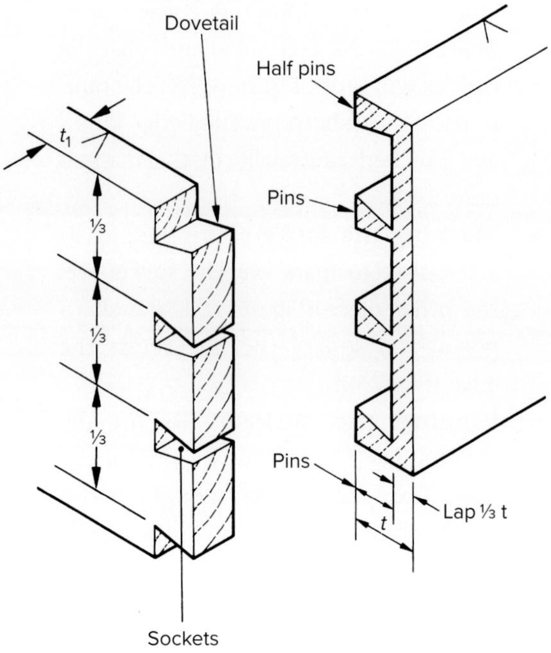

Fig. 7.61 Setting out of pins for a lapped dovetail joint

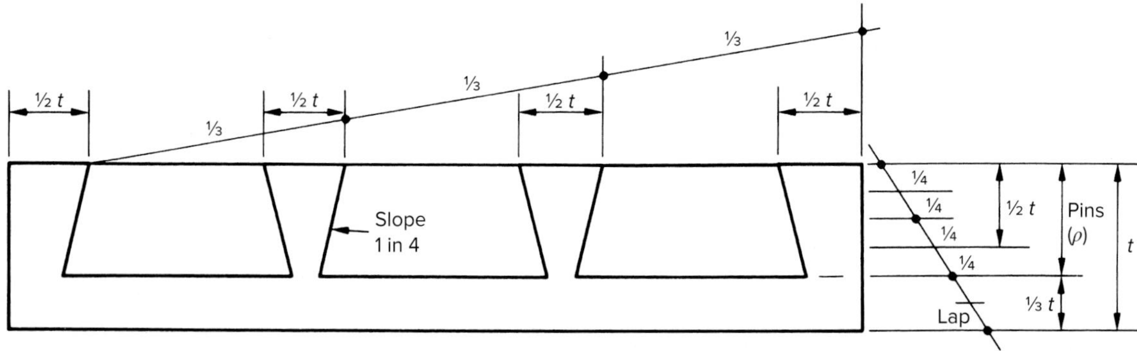

To make the lapped dovetail joint, proceed as follows:

1. After squaring the ends, gauge the distance *p* on the end of the front piece and also from the end of the side piece across both faces and edges. Gauge the distance t_1 (the thickness of the side from the end of the front) (Fig. 7.62).

> **TIP** A marking gauge is adequate if a cutting gauge is not available.

Gauging these distances can be done with a cutting gauge, which is similar to a marking gauge except that a small knife replaces the spur (Fig. 7.63). Its purpose is to gauge across the end grain or to cut parallel strips of thin timber. First set the gauge to *p* and gauge the front and around the end of the side–gauge lightly across the outside face where the mark may be difficult to remove. Reset the gauge to t_1 and gauge across the front.

2. Set out the dovetail pins on the end of the front, adopting the 'in and over' method, and mark the slope on the pins with a sliding bevel set to 1 in 4 (Fig. 7.61).

Fig. 7.62 Gauging for a lapped dovetail joint

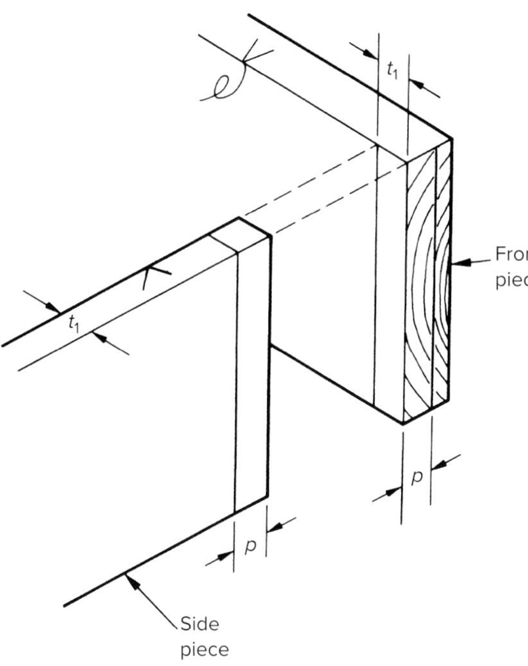

Fig. 7.63 Cutting gauge

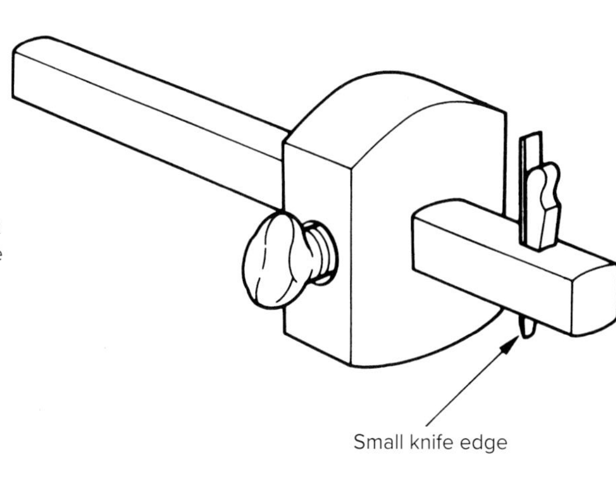

Fig. 7.64 Ripping the dovetail pins halfway

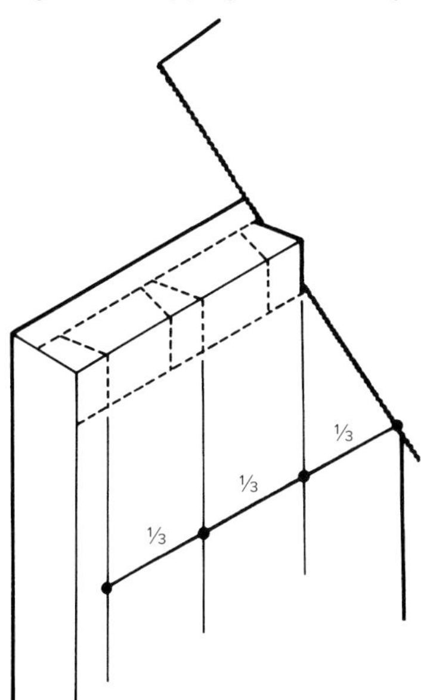

Fig. 7.65 Removing the waste

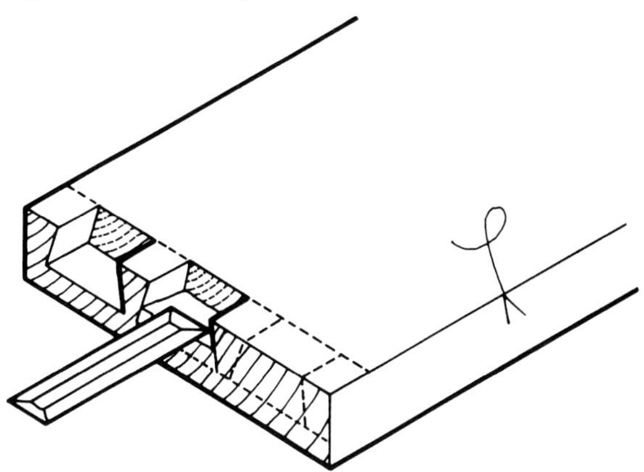

3. Rip the dovetail pins halfway diagonally (Fig. 7.64).
4. Remove the waste from between the pins using a sharp bevelled-edge chisel: cut off the waste, slightly undercutting, and pare it away (Fig. 7.65).
5. Use the dovetail pins as a template to mark the sockets on the side (Fig. 7.66). Stand the front piece on the side and with a sharp pencil mark the pins.

Fig. 7.66 Marking the sockets

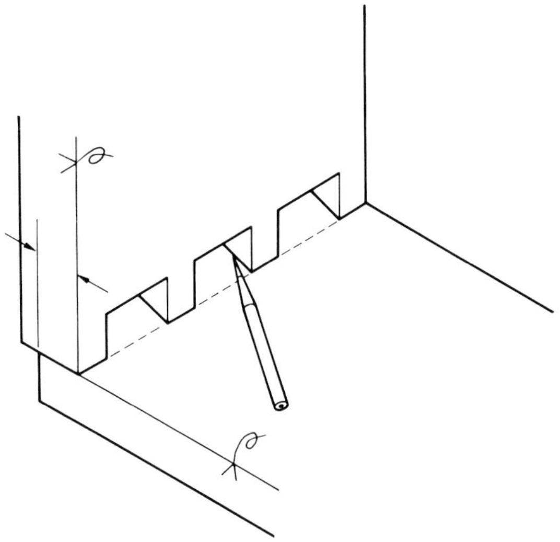

6. Remove the waste from the sockets in a manner similar to the common dovetail.
7. Check the joint for fit; if necessary, sand the inside faces. Apply adhesive lightly and tap the joint together. As with the common dovetail, the tapered pins should be a close fit, tending to draw the joint together.
8. After the adhesive has set, clean off and sand the outside face.

CHECK YOUR UNDERSTANDING

1. What ratio/angle can be used when constructing dovetail joints?

7.2.8.3 Alternative method for making the dovetail joint

The alternative method for making a dovetail joint involves marking the pins from the sockets. Proceed as follows:

1. Set out the sockets on the side, proportioning in the same manner as for the pins.
2. Use a fine dovetail saw and rip the sockets but do not remove the waste. A number of pieces, say three or four, can be cramped together and the sockets ripped from the one set out.
3. Lay the side on the end of the front, moving it forward sufficiently to compensate for the thickness of the saw kerf.

   This distance is illustrated in Figure 7.67. Measure square across the sloping saw kerf and note the distance z. If using a fine dovetail saw, this distance may be approximately 2 mm but must be checked with the actual saw being used on the job.

   Hold the side in position firmly, place the point of the saw into the saw kerf and mark onto the end of the front for each pin.
4. Remove the side and square down the face of the front of the line of the saw cut marked on the end; rip the pins on this line.
5. Remove the waste from the sockets and between the pins. Fit and assemble the joint as previously described. Note any correction considered necessary to the distance z for future use.

Fig. 7.67 A & B Setting out from the sockets

(a)

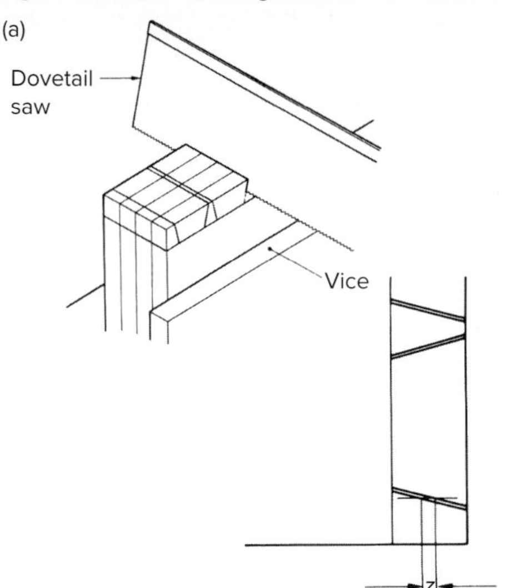

(b)

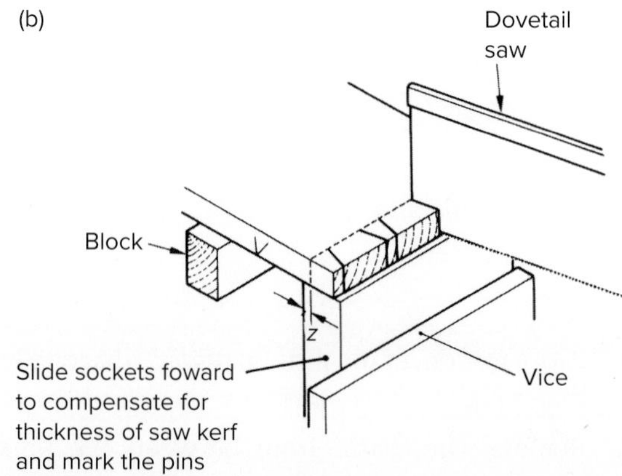

CHECK YOUR UNDERSTANDING

1. Show how the pins for a common dovetail joint would be set out on the end of a piece of timber 100 mm × 12 mm.
2. What are two forms of dovetail joints?

7.3 Joints used for widening timber

With the widespread use of plywoods, medium-density fibreboard (MDF) and other manufactured boards in large sheet sizes, the need to edge joint solid timber to make up wide boards has greatly diminished. However, for items such as tabletops, countertops and so on, there is often a preference for the individual warmth of solid timber.

Fig. 7.68 Butt joint for widening

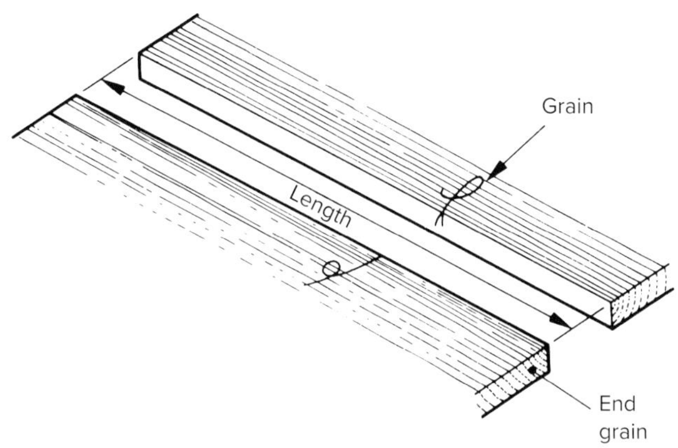

7.3.1 Butt joint

The butt joint is suitable for use where the joint is subject to only minor stresses and has a stable moisture content. The jointed edges must be planed or machined straight and square, and be close fitting before being glued and cramped together with a suitable adhesive (Fig. 7.68).

The butt joint can be strengthened through the addition of dowels or biscuits, positioned so that they align with the face of the boards (Fig. 7.69). This is probably the most commonly used joint for this purpose. The dowels/biscuits are set out at approximately 75 mm from the ends and 200 mm apart. See the section 'Dowelled joint' (p. 158) for the proportioning of dowels.

Fig. 7.69 Dowelled joint for widening

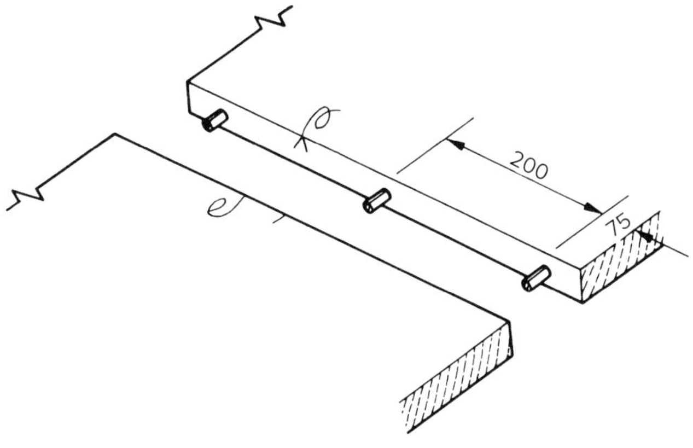

7.3.2 Loose-tongue or 'feather' joint

The butt joint, after preparation, can also be strengthened through the addition of a loose tongue (Fig. 7.70). The tongue can be of plywood, cut with the greater number of veneer layers lying across the line of the joint. Alternatively, a tongue can be cut from solid timber, with the grain running diagonally across, and a thickness of no more than one-third the thickness of the boards. Cut the width of the tongue about 1 mm less than the combined depth of the plough grooves. The groove that is to receive the tongue is gauged from the face side, which should be flush when the joint is assembled.

7.3.3 Matching boards for edge jointing

Boards should be matched for grain and colour to the best advantage for both ease of working and final appearance. First check and mark the direction of the grain in each board, taking a trial shaving if necessary. If the boards are cut from one long length of timber, take care that their direction is not reversed. Joint the boards with the grain running in the same direction. Besides the difference in colour caused by the light falling on the grain in opposite directions, cleaning off is also made more difficult if the grain in one board is running in the opposite direction to the one adjacent to it.

Fig. 7.70 Loose-tongue joint

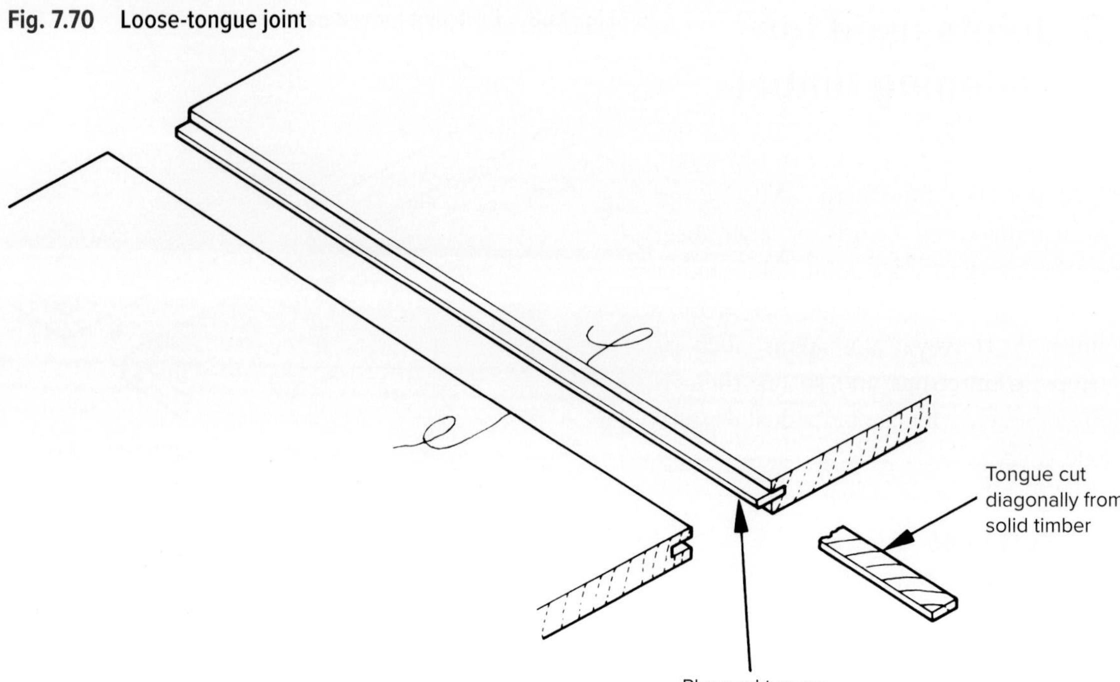

Lay out the boards, and when it is thought that they are in the best combination, mark the face sides pointing to the edges to be jointed; if there is more than one joint, indicate by marks or numbers the pairs to be joined (Fig. 7.71).

7.3.4 Jointing the boards by hand

Hold the two boards to be jointed back-to-back and secure them in a vice. Straightening a long length of timber is a job for the trying plane. The technique is shown in Figure 7.72.

Shoot the edges of the two boards together. One theory is that if the boards are together and are planed slightly off-square, and the edges are then brought together, the face will lie on a straight line as the bevel on one board will complement the bevel on the other. Experience shows, however, that it is better to keep the edges square by testing with the try square from the face side.

After the boards have been planed straight and square, they can be tested for accuracy as shown in Figure 7.73.

Fig. 7.71 Matching boards

Fig. 7.72 Planing the edges

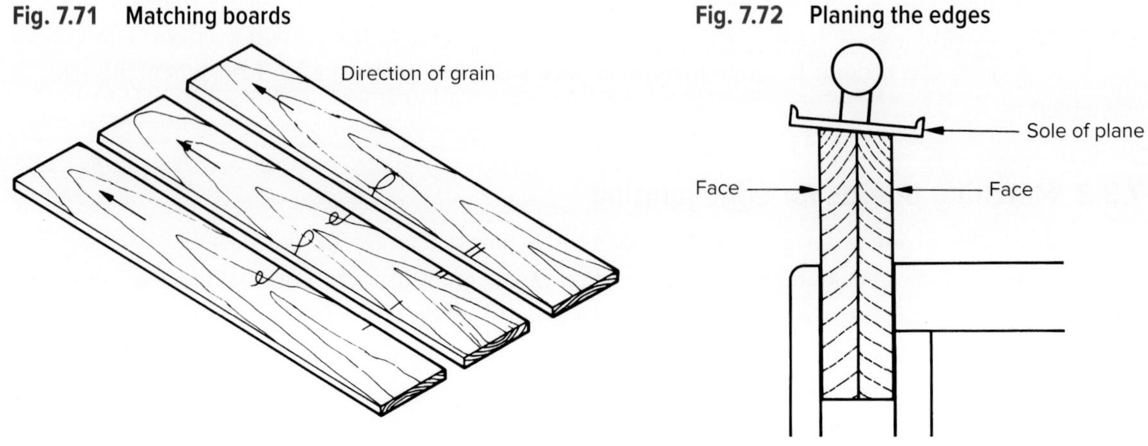

Hold one board in the vice and stand the second board on top. First, see that the faces are in a straight line by sighting down or using a short straight edge. Then check that the boards are a close fit along their length. To test, grip one end of the top board and move it from side to side. If there is a high spot, the top board will pivot on this point, which can be easily detected and removed with a fine shaving. If the board pivots on an end point, it is probably too hollow in the centre and the high point at the end is removed. Continue testing until the boards are a close fit along their length and no high spots can be detected.

Fig. 7.73 Testing for accuracy

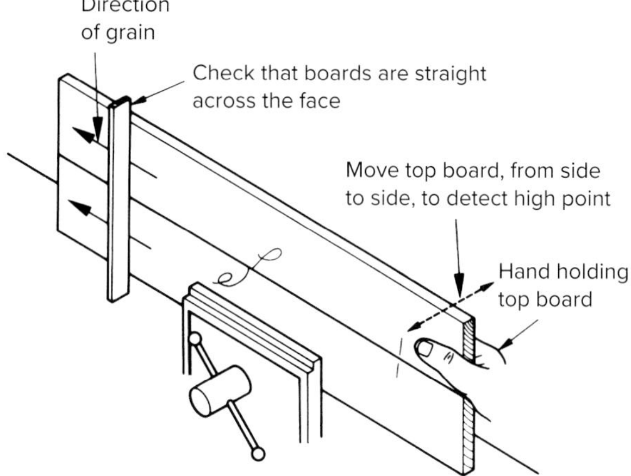
Direction of grain
Check that boards are straight across the face
Move top board, from side to side, to detect high point
Hand holding top board

Another theory is that when boards are planed, they should be finished slightly hollow along their length towards the centre of the board so that a narrow strip of daylight can just be seen when held to the light. This is because when the boards are jointed together, and are adjusting to their environment over a period of time, the ends will tend to shrink to a greater degree than will the centre of the boards. The hollow planing will compensate for this uneven timber movement.

Fig. 7.74 Marking for the dowels or biscuits

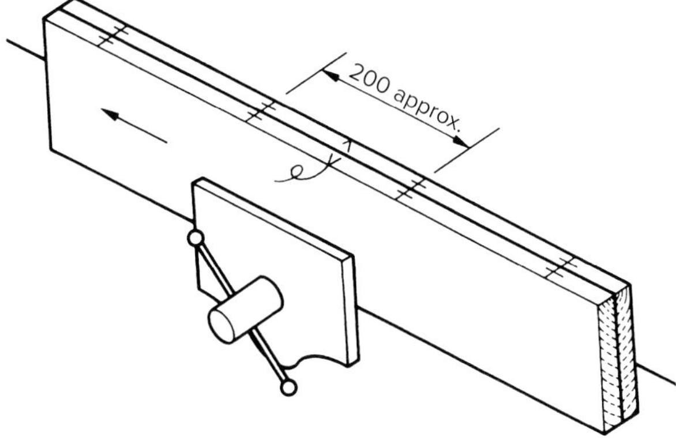
200 approx.

If the boards are to be dowel or biscuit jointed, again hold them back-to-back in the vice, mark the position of the dowels or biscuits (Fig. 7.74), and square them across both boards; then gauge the holes from the face side on both pieces.

TIP See the procedure for the boring of dowel holes under the heading 'Making the dowelled joint' (p. 159) earlier in this chapter.

7.3.5 Gluing up jointed boards

Lay out sash cramps and have available blocks of scrap timber that can be used under the shoe of the cramps to prevent the edges of boards being damaged.

Hold the two edges to be jointed together, apply adhesive to the dowelled joints or the tongue and groove joints as the case may be, and enter the dowels or the tongue. Align the remainder of the joint, bring it together and tap it lightly. Lay the boards in the sash cramps and, holding them hard down on the bar of the cramps, cramp them together lightly. Lay another cramp across the top and, keeping the bar hard down on the boards, tighten the cramps, as shown in Figure 7.75. This will hold the boards flat and prevent them from buckling as the cramps are tightened uniformly to bring the joint together and expel any surplus adhesive. Do not overtighten the cramps to the point where the boards are distorted.

Finally, wipe away any surplus adhesive with a dry cloth (Fig. 7.76).

Fig. 7.75 Cramping the boards

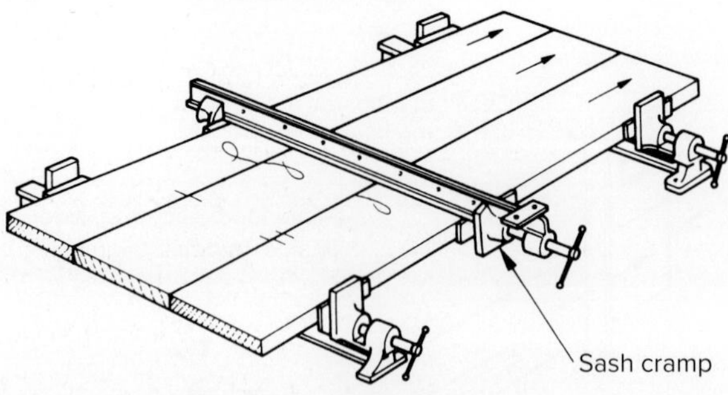
Sash cramp

Fig. 7.76 Cleaning off wide boards

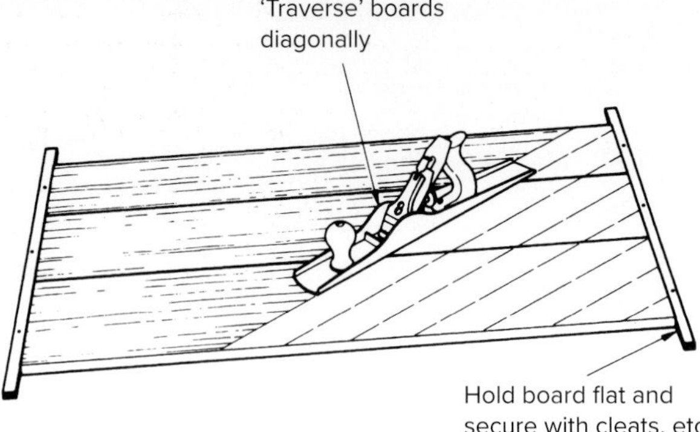
'Traverse' boards diagonally
Hold board flat and secure with cleats, etc.

Use a trying plane with a sharp blade that is set to take a fine shaving, and commence by planing diagonally across the boards, a practice termed *traversing*. Work in strips the width of the plane iron, overlapping each step slightly. Continue planing in a similar manner across the opposite diagonal until any small irregularities are removed. Complete the cleaning off by planing in strips in the direction of the grain until any surface blemishes are removed.

Sanding the surface of wide boards is carried out in the same sequence as for planing, whether by hand or by electric sander. Start with a fine grade of paper, say 120 grit, so as not to make any deep scratches that will be difficult to remove later, and sand diagonally across the boards in strips. Next, sand in the direction of the grain with a medium grade of paper, say 100 grit, until any scratches are removed, and then finish with a fine paper, 120 or 150 grit.

TIP If you have access to a static machine called a 'thicknesser', the planing of the jointed boards by hand can be skipped—provided the machine is wide enough to accept the piece. If it is not, the accepted method is to produce the piece in narrower sections, allowing it to fit the thicknesser and later to be glued together and finished as described in the previous section.

7.3.6 Fixing wide boards of solid timber

Wide boards made of solid timber will shrink and expand with changes in moisture content. When they are fixed in place, provision must be made for them to move freely to accommodate these changes.

A good example of what can happen is to be seen in many old tables that have a split right down the middle. Close examination would probably show that the top was simply nailed to the frame around its outer edges and was therefore restrained from shrinking. If this were the case, the stresses set up during one or more long dry summers would be so great that the boards simply split down the centre. The damage would probably not have occurred if the top had been attached to the frame by means of 'buttons', which hold the top down but allow it to shrink or expand with atmospheric changes (Fig. 7.77).

Another example is where the wide board is held flat by cleats screwed across the back, but the screw holes are slotted to allow the board to move as its moisture content changes (Fig. 7.78).

Another procedure that minimises the shrinkage in wide boards is to run a number of plough grooves in the back surface. This means that the shrinkage across the surface will be localised into a number of smaller widths and will be less overall.

CHECK YOUR UNDERSTANDING

1. What are three methods of joining boards to make a wide piece of timber?
2. How is a tabletop held down to the frame?

Fig. 7.77 Buttons for a tabletop: (a) close up; and (b) full view

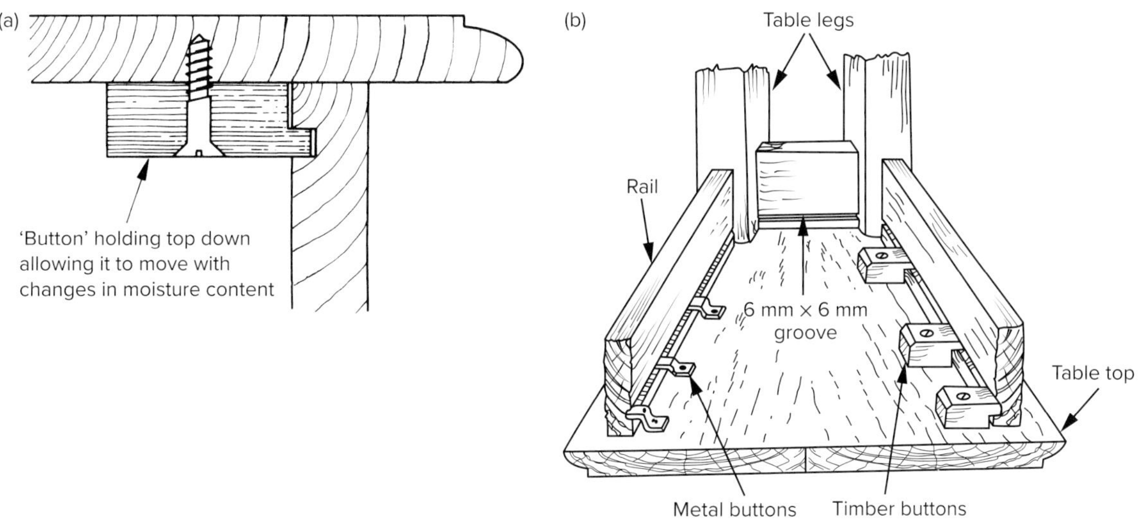

Fig. 7.78 A–C Slotted hole in cleat method

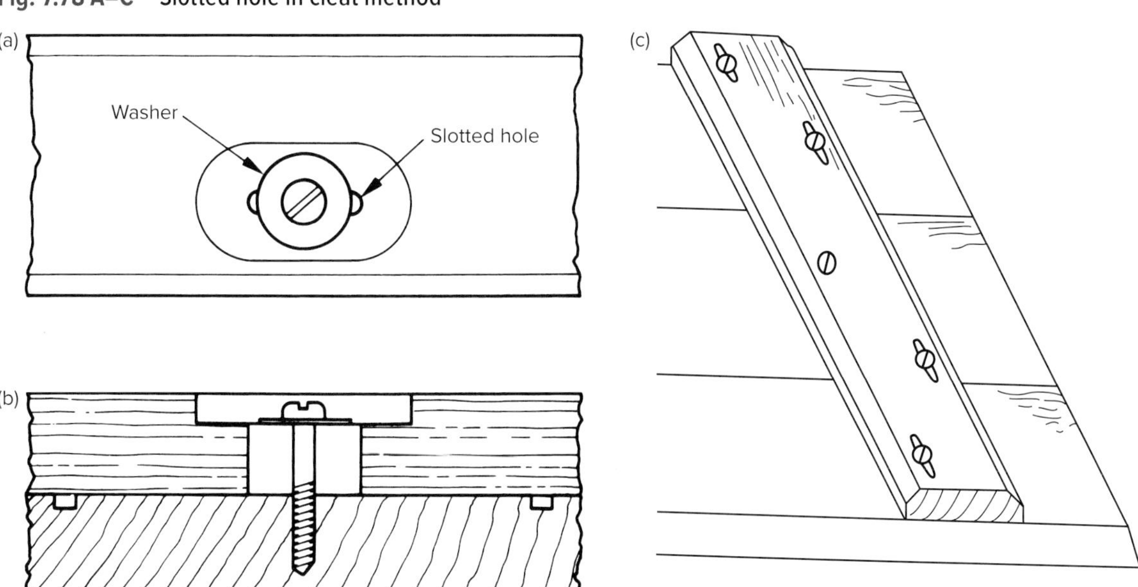

Student research

Search the internet for 'heavy framing joints' and look at the other types of joints that can be used to join timber together in large timber-framed buildings.

End of chapter activity

Activity 1:

Construct the object below. The task combines a number of joints covered in this chapter so far. Read the plans (Fig. 7.79) carefully as some of the measurements need to be calculated.

The timber size required to construct this project is 85 mm × 30 mm.

Fig. 7.79 A Object to be constructed

(a)

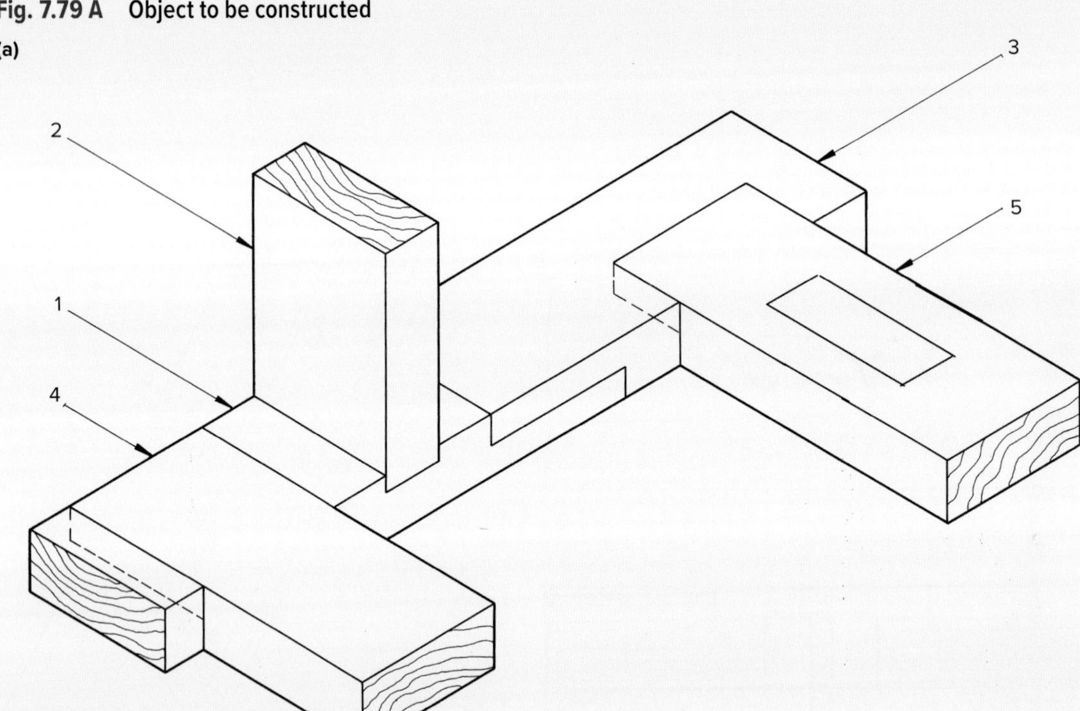

Adapted from NSW Department of Technical and Further Education, School of Building, Carpentry and Joinery, Stage 1

Fig. 7.79 B Plans

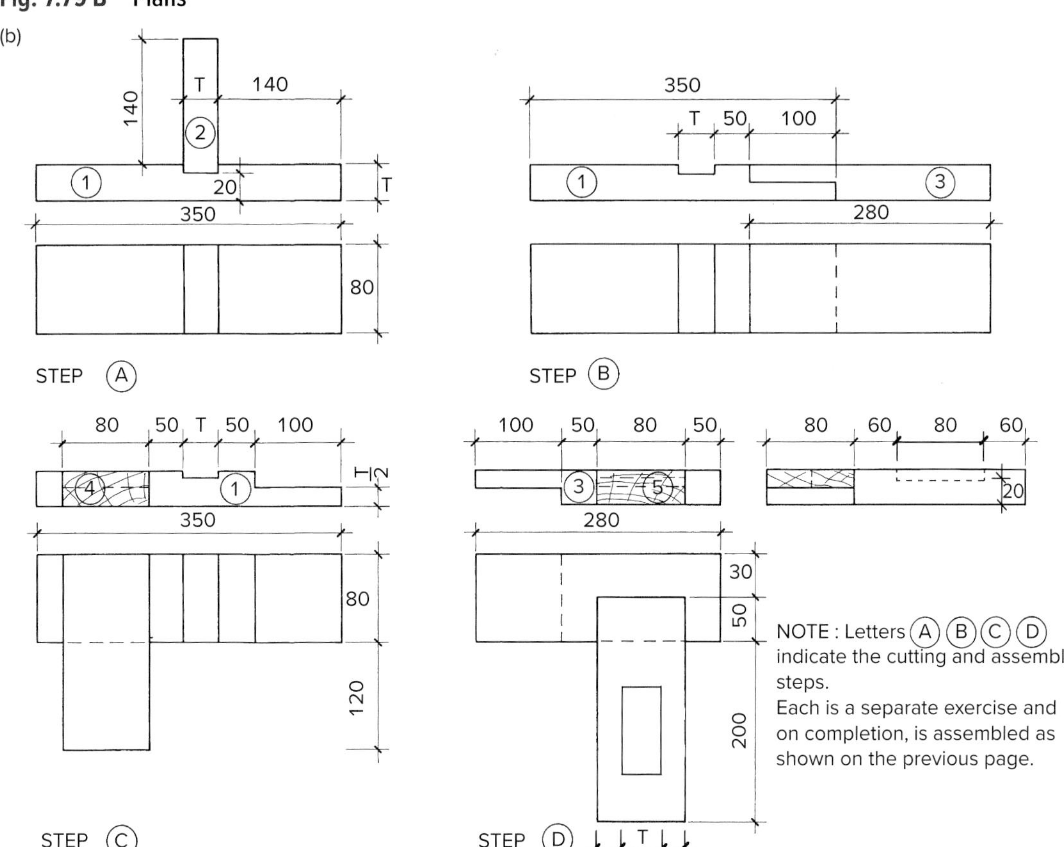

Adapted from NSW Department of Technical and Further Education, School of Building, Carpentry and Joinery, Stage 1

AUSTRALIAN BUILDING CODE AND STANDARDS

AS 1684

Tradie talk—Chop a mortise

Chapter 8

Construction fastenings and adhesives

It is important for those students who want to become carpenters within the building industry to understand the content in this chapter, even though it is not explicitly covered in this qualification. This content will enhance the training being undertaken and also provide essential knowledge of the function and application of construction fastenings and adhesives.

Learning Objectives

LO 8.1 Identify nail types and their uses

LO 8.2 Identify the varieties of woodscrews and their application

LO 8.3 Identify bolts, nuts and their uses

LO 8.4 Apply associated tools used with fasteners

LO 8.5 Identify masonry anchors

LO 8.6 Understand adhesives and their application

Introduction

There are many ways to connect two pieces of timber together and to connect timber to another material. In this chapter, you will learn about the various types of fixings, fasteners and adhesives available in the construction industry. You will learn about the associated tools used with some of these fixings and the various applications and scenarios to which the fixings and fasteners are suited.

8.1 Nails

Nails developed from the wooden peg. They have a long and varied history, from the first hand-forged nails to cast nails and nails cut from strips of flat steel sheet, to the mass-produced nails manufactured from steel wire that we use today.

Throughout the building industry, nails are one of the most common methods of fastening materials together and they do so quickly and efficiently using the friction created between the nail and the material into which it is driven. Nails are used predominantly for fixing timber to timber, but many types have been developed for use in fixing a range of other materials.

Nails are described by their length and diameter in millimetres and by the shape of the head. One of the most common types is the bullet head nail (Fig. 8.1). It is used in framing and fixing work where the small head is punched below the surface and is relatively inconspicuous.

The flat head nail is used mainly for nailing in low-density timbers and materials where the bullet head would tend to pull through the soft timber (Fig. 8.2). For exterior use, steel nails are **galvanised** to provide protection from corrosion, and this detail is added to their description. A typical description would be 75 × 3.75 bullet head, or 65 × 3.15 galvanised flat head.

The load-carrying capacity of a nail in timber depends primarily on its diameter (*d*). For structural work where maximum strength is required, the thickest possible nail that can be driven without splitting the timber can be taken as the correct size. Where the required number of nails for a joint cannot be driven without splitting the timber, the nail holes should be predrilled using a drill having four-fifths (80%) of the diameter of the nail.

Fig. 8.1 Bullet head nail

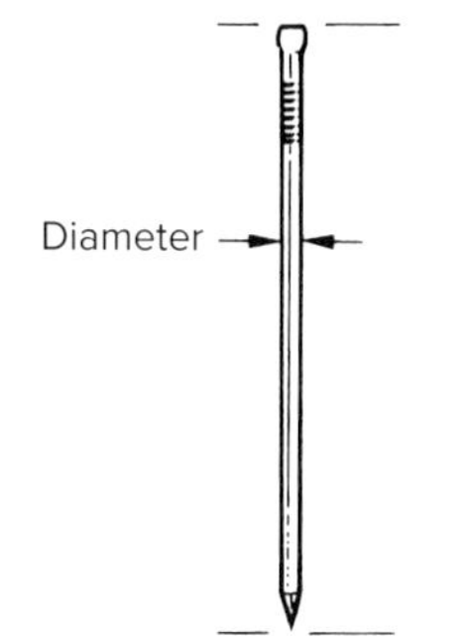
Diameter

Fig. 8.2 Flat head nail

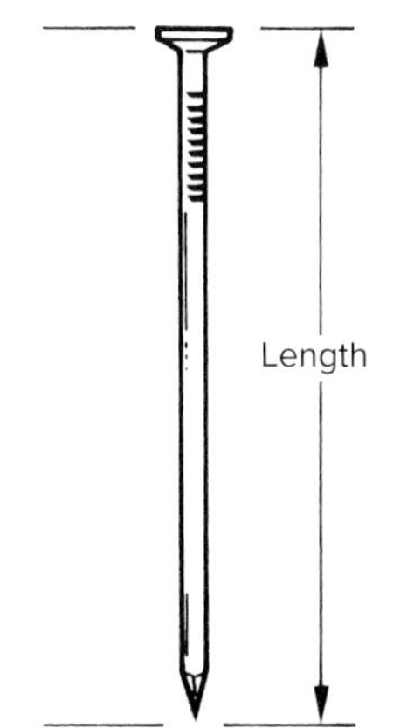
Length

TIP There are rules governing the use of nails—how many, what length, what type and how they should be used—on a timber-framed house. Refer to AS 1684 Residential Timber-Framed Construction for more information.

8.1.1 Nail types

8.1.1.1 Lattice head nails

Lattice head nails are used for fixing plasterboard linings to timber frames (Fig. 8.3). The head is driven just below the surface but not far enough to break the heavy paper surface of the plasterboard around the head.

Fig. 8.3 Lattice head nail

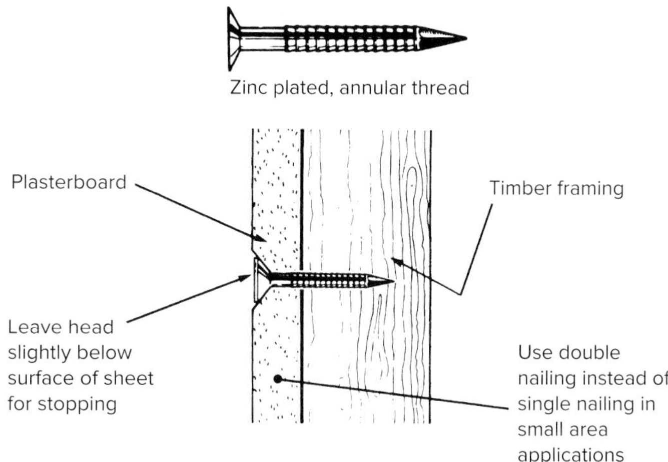
Zinc plated, annular thread
Plasterboard
Timber framing
Leave head slightly below surface of sheet for stopping
Use double nailing instead of single nailing in small area applications

8.1.1.2 Soft sheet nails

Soft sheet nails are zinc plated and designed for fixing thin materials such as fibre cement sheet used as eave lining (Fig. 8.4). The head is left

Fig. 8.4 Soft sheet nails

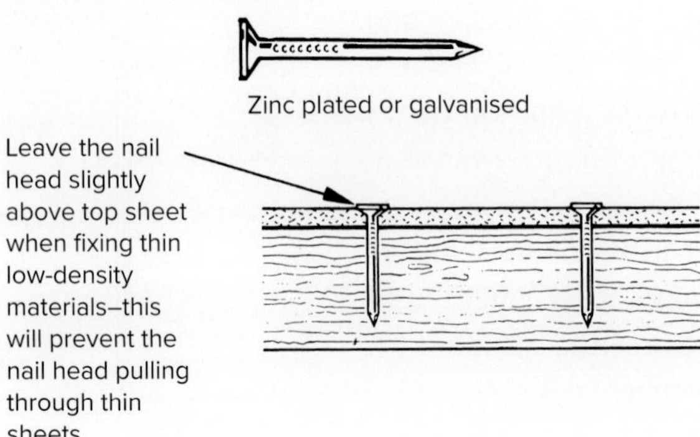

protruding slightly above the surface of the sheet to prevent the head pulling through the sheets.

8.1.1.3 Bracket nails

Bracket nails are galvanised or zinc plated and used in the fixing of metal brackets in timber framing (Fig. 8.5). These must be nailed into solid timber.

Fig. 8.5 Bracket nails

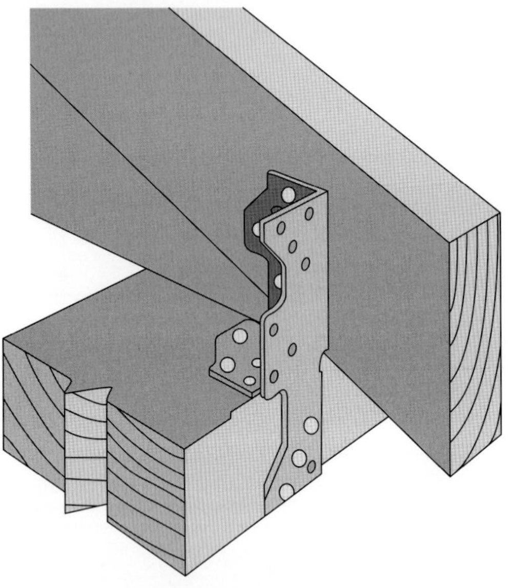

8.1.1.4 Clout nails (clouts)

Clouts are galvanised with a large flat head and are designed for fixing thin sheet metal to timber. It is important that the nail is driven at 90° to the surface, so that the head remains flush at the finish of work (Fig. 8.6a). The head should not be driven through the material being fixed (Fig. 8.6b).

Fig. 8.6 Clouts: (a) correct; (b) incorrect

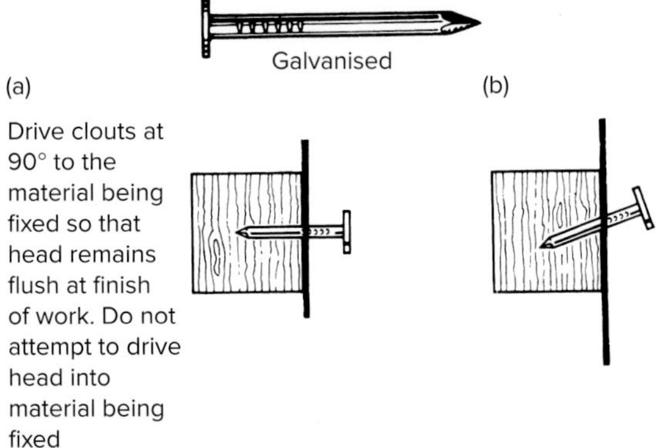

Galvanised
(a)
(b)
Drive clouts at 90° to the material being fixed so that head remains flush at finish of work. Do not attempt to drive head into material being fixed

Fig. 8.7 Timberlok nail

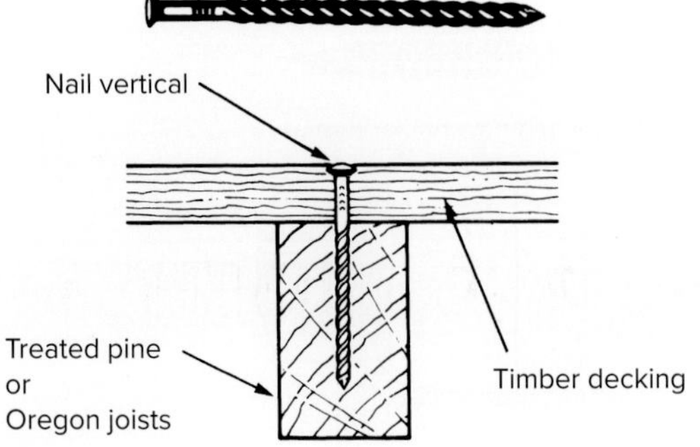

8.1.1.5 Deformed nails

Deformed nails such as 'Timberlok nails' (Fig. 8.7) are designed to provide extra holding power when fixed into soft timber, such as treated pine or Oregon. The nails have a helical thread, so there is no need to skew nail. When nailing through hardwood decking it is necessary to drill the top piece to avoid splitting.

8.1.1.6 Staples

Staples are used for fixing wire strands to timber–for instance, in fencing (Fig. 8.8). To minimise the chance of the timber splitting, the staple is driven at an angle to the grain. Always choose the smallest staple to suit the wire.

8.1.2 Nailing practice

To drive a nail, select a hammer of sufficient weight, grip it by the hand grip and swing the hammer freely (Fig. 8.9). Aim to strike the head of the nail squarely and with sufficient force to drive it home.

Fig. 8.8 Staple

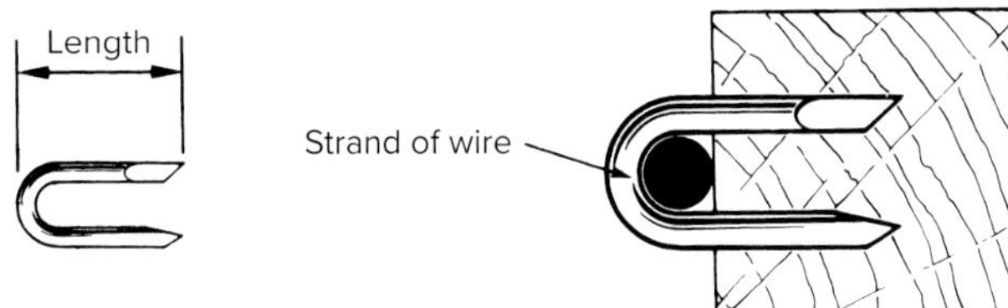

TIP Tapping at the nail with short light strokes and only sufficient force to bend the nail, but not enough to drive it home, can have only one obvious result.

Nailing into end grain requires a nail to be longer than when nailing into cross grain, if the same holding power is required.

Figure 8.10 is a guide to selecting a nail of satisfactory length. Where the minimum depth of penetration for a nail cannot be achieved, the equivalent holding power may be obtained by using nails with a deformed shank, for example, Timberlok nails (see Fig. 8.7). When nailing into thin timber, extra holding power can also be obtained by **clenching** the nail (Fig. 8.11), whereby a nail long enough to protrude through the back of the materials being nailed is hammered in and bent over in the direction of the grain. The point is then punched back into the timber. Extra grip can also be obtained by dovetailing nails (Fig. 8.12), and increasing the strength of the nailed joint against a direct pulling force.

Skew nailing (Fig. 8.13) is widely adopted as a means of fixing structural members in timber framing. First, pull the two members firmly together and skew the nails at an angle somewhere between 45° and 60°. For finished work, complete the process of driving the nails with a nail punch to avoid damaging the timber.

Splitting the timber is one of the most common failures when nailing. Select a nail of a diameter that is no greater than required for the job in hand and avoid nailing too close to the edges. Stagger nails so that they are not in the same line of grain (Fig. 8.14).

Fig. 8.9 Hammering action

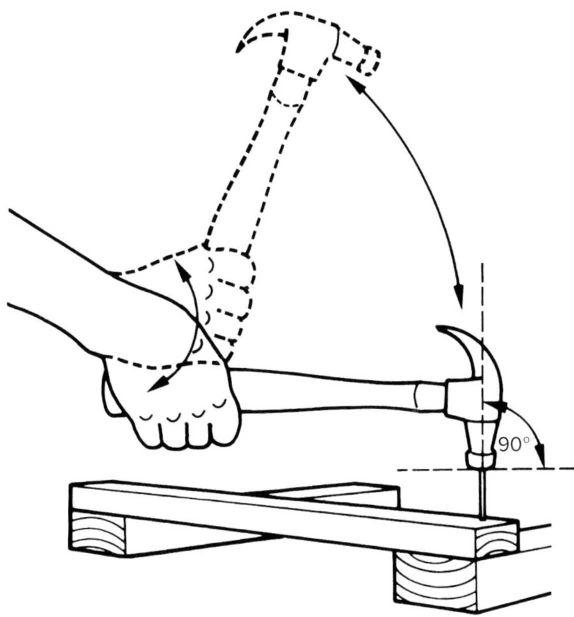

Fig. 8.10 Holding ability of nail: (a) into the side grain; (b) into the end grain

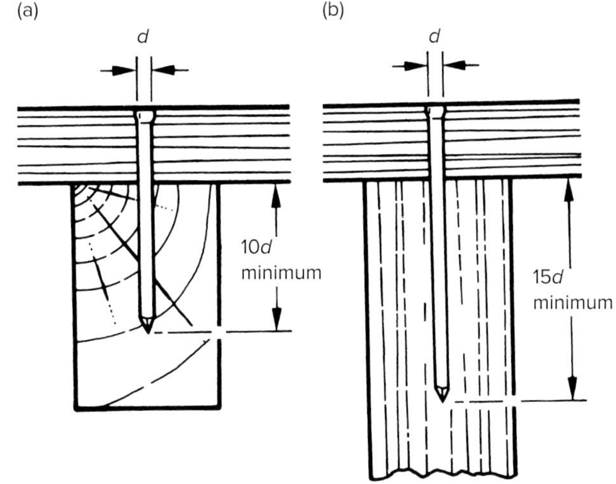

A handy tip for nailing without splitting in a difficult situation is illustrated in Figure 8.15. Place the head of the nail on the timber and strike the point with the hammer. This makes an indentation that will later accommodate the nail head and which slightly flattens the point of the nail so that it will tend to punch its way through the fibres without separating them and splitting the timber. Reverse the nail and drive on the spot marked by the nail head.

If all else fails, it then becomes necessary to drill **pilot holes** for the nails in order to avoid splitting.

Fig. 8.11 Clenching of nail

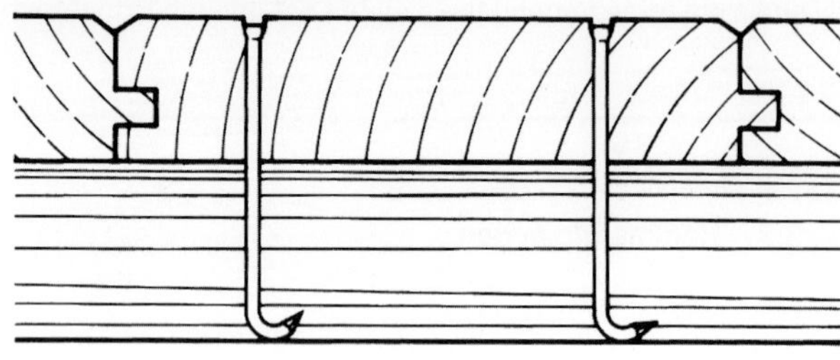

Fig. 8.12 Dovetailing with nails

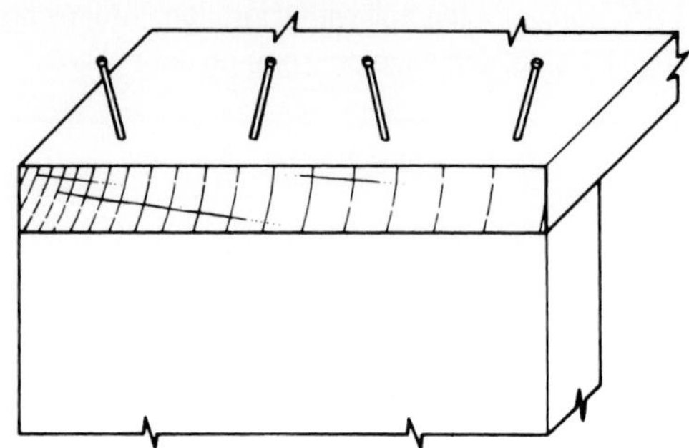

Fig. 8.13 Skew nailing

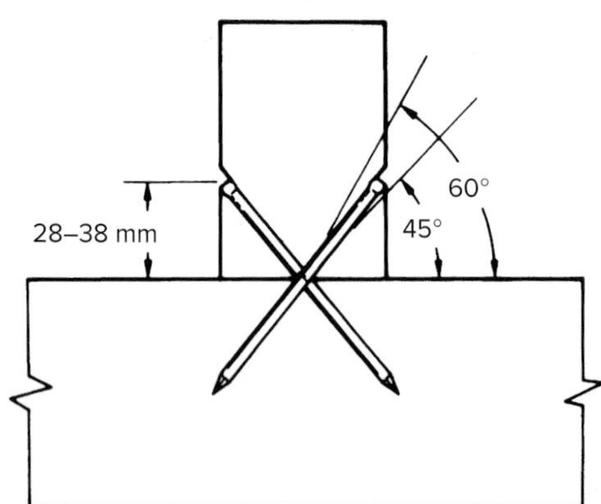

Fig. 8.14 Staggering of nails

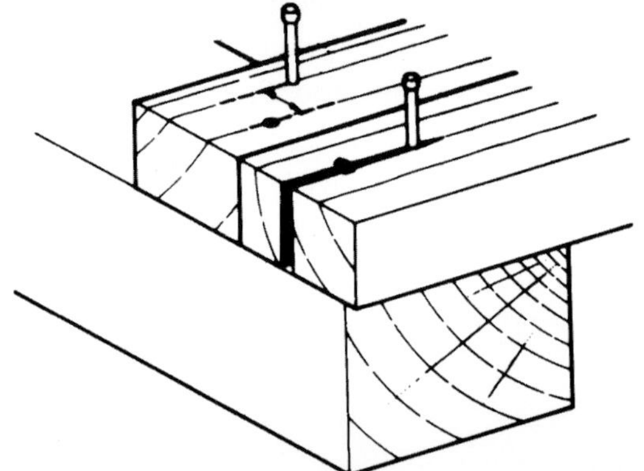

TIP Driving nails into hard timber can often be helped by some form of dry lubrication on the nail. Often the most convenient method is to stroke the nail through your hair a few times. Even the natural oils in your hair will provide just that little bit of lubrication needed to drive the nail successfully.

CHECK YOUR UNDERSTANDING

1. By what means do nails achieve their holding power when driven into timber?
2. When nailing into cross grain, how far should a 75 × 3.75 bullet head nail penetrate the final receiving piece of timber to obtain a satisfactory grip?
3. What are three ways in which the holding power of nails can be increased when the minimum amount of penetration cannot be achieved?
4. For what purpose would a lattice head nail be used?

8.1.2.1 Nail punch

The nail punch is an essential accessory when nailing and consists of a tapered point that is slightly hollow (Fig. 8.16). This prevents the punch from slipping off the nail head. In finished work, nail heads are punched just below the surface and the hole is filled during the painting or other finishing process.

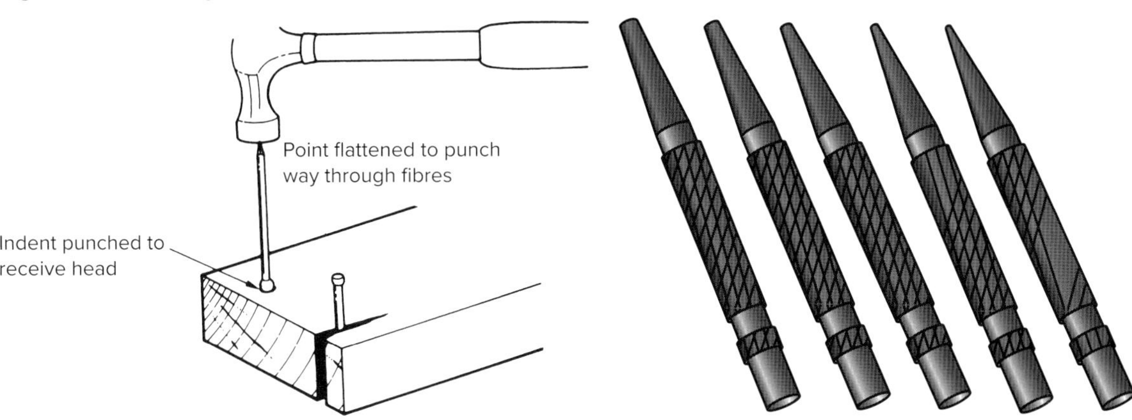

Fig. 8.15 Flattening of nail effect (punching through)

Fig. 8.16 Nail punches

> **TIP** Nail punches come in a range of diameters. Choose the punch that will suit the nail being used.

CHECK YOUR UNDERSTANDING

1. What tool is used to push the head of a bullet head nail below the surface of the timber?
2. List two methods to prevent timber from splitting when nails are being driven close to the end of a piece of timber.

8.2 Woodscrews

Woodscrews provide great holding power, when correctly inserted, and are used in the assembly of work requiring firm, strong joints (Fig. 8.17). Unlike nails, screws can also be removed readily, without damage to the surrounding material, and are particularly useful for the assembly of temporary work or for fixing metal fittings such as hinges, locks and similar items to timber that may require some adjustment. Screws can also be inserted without vibration, which could cause damage to the adjacent structure.

Woodscrews develop their holding power from the thread cutting its way into the fibres of the timber. They are unsatisfactory for screwing into end grain, as the thread will sever the fibres and the wood around the screw will crumble away. The size of a woodscrew is expressed as the gauge by the length. In general, use woodscrews in the range from 3G to 16G. The higher the gauge number, the thicker the screw.

The length of the screw is expressed in millimetres and is measured from the part of the head normally finishing flush with the surface of the timber to the point.

Fig. 8.17 Screw types

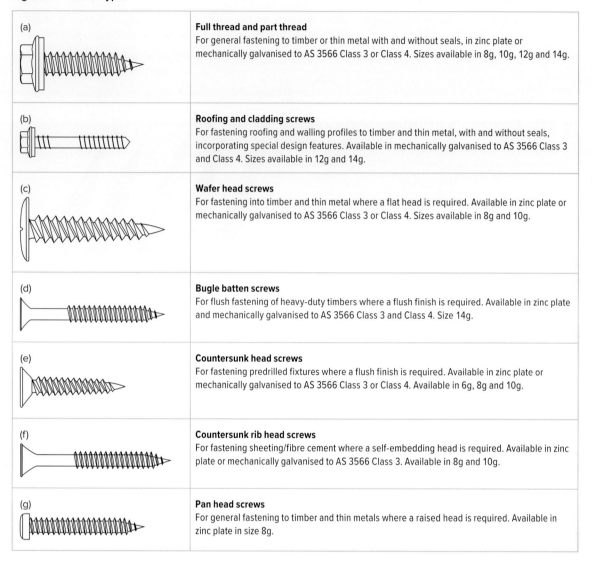

(a)	**Full thread and part thread** For general fastening to timber or thin metal with and without seals, in zinc plate or mechanically galvanised to AS 3566 Class 3 or Class 4. Sizes available in 8g, 10g, 12g and 14g.
(b)	**Roofing and cladding screws** For fastening roofing and walling profiles to timber and thin metal, with and without seals, incorporating special design features. Available in mechanically galvanised to AS 3566 Class 3 and Class 4. Sizes available in 12g and 14g.
(c)	**Wafer head screws** For fastening into timber and thin metal where a flat head is required. Available in zinc plate or mechanically galvanised to AS 3566 Class 3 or Class 4. Sizes available in 8g and 10g.
(d)	**Bugle batten screws** For flush fastening of heavy-duty timbers where a flush finish is required. Available in zinc plate and mechanically galvanised to AS 3566 Class 3 and Class 4. Size 14g.
(e)	**Countersunk head screws** For fastening predrilled fixtures where a flush finish is required. Available in zinc plate or mechanically galvanised to AS 3566 Class 3 or Class 4. Available in 6g, 8g and 10g.
(f)	**Countersunk rib head screws** For fastening sheeting/fibre cement where a self-embedding head is required. Available in zinc plate or mechanically galvanised to AS 3566 Class 3. Available in 8g and 10g.
(g)	**Pan head screws** For general fastening to timber and thin metals where a raised head is required. Available in zinc plate in size 8g.

8.2.1 Shape of head

The head of the screw can be a countersunk head, round head or raised head, and is slotted or recessed so that the screw can be turned. The most common is the Phillips or **Pozidrive** head, but other types are available (Fig. 8.18). The screw heads can be aligned more easily and the finished appearance is more decorative. Plastic caps in different colours are available to fit the Pozidrive slot and can be used to cover the heads of countersunk screws.

Screws are manufactured from steel or brass for most purposes, but for special uses there are also screws made in silicon bronze, monel metal, stainless steel and aluminium. The surface can be plain or galvanised, nickel plated, cadmium plated, zinc plated, Florentine bronze or brass plated.

Fig. 8.18 Types of screw heads

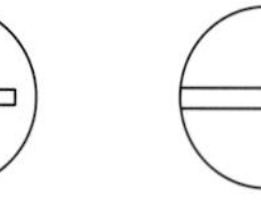
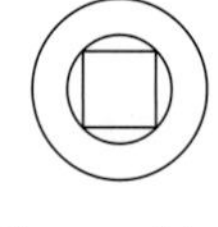
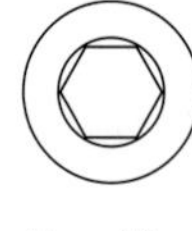

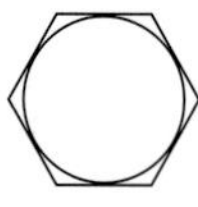

8.2.2 Drilling for woodscrews

When inserting woodscrews, it is essential to drill suitable pilot holes. The first hole for the shank of the screw must provide a clearance for the shank so that the joint can be pulled together (Fig. 8.19).

Table 8.1 gives a list of the appropriate drill sizes for pilot holes.

The diameter given for the shank of the screw has been rounded off to enable a quick comparison with drill sizes, but lies within the tolerances permitted and is sufficiently accurate to determine pilot holes in timber. The drill size is adjusted to enable the use of drills most likely to be carried by the carpenter in a metric or imperial drill set.

The second pilot hole for the thread of the screw is approximately equal to the core diameter of the thread. This permits the thread to cut its way into the timber in order to develop its holding power. As a general rule, the pilot hole for the thread in timber of average density is about half the diameter of the shank. Hard timbers may require a slightly larger pilot hole, soft timbers a little less; and in very soft timber, screws up to about 6G may be inserted quite satisfactorily without a second pilot hole.

When fixing metal fittings such as hinges and locks, it is always necessary to drill a small pilot hole so that the screw will go in straight and the head will lie flat in the face of the fitting (Fig. 8.20a). In some timbers, Oregon for example, when a pilot hole is not bored, screws will simply follow the soft grain and go in at an angle, so that the head will project above the face of the fitting (Fig. 8.20b). Screws inserted at an angle are a common cause of trouble with 'binding' in hinges.

Fig. 8.19 Hole drilling for screws

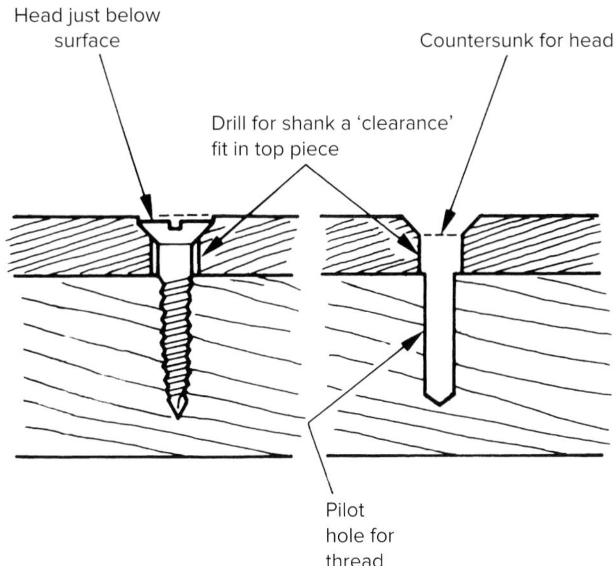

Table 8.1 Woodscrew pilot holes

Gauge	Nominal diameter shank	Pilot hole for shank	
		mm	inches
3	2.40	2.5	5/64
4	2.75	3.0	7/64
5	3.10	3.5	1/8
6	3.45	3.5	9/64
7	3.80	4.0	5/32
8	4.20	4.5	11/64
9	4.50	4.5	3/16
10	4.90	5.0	13/64
12	5.55	6.0	15/64
14	6.30	6.5	1/4
16	7.00	7.0	19/64

When inserting countersunk screws into timber, the pilot hole must be countersunk using a **countersinking** bit so that the screw head can pull into the timber and finish just slightly below flush.

Screws will always drive more easily if the thread is lightly lubricated. Tallow or beeswax is sometimes used, but probably the most readily available dry lubricant is a piece of common soap rubbed lightly on the thread.

TIP If using flat head screws, it is considered neat practice to leave the slots in the heads of screws all running in the same direction, usually in the direction of the grain.

Fig. 8.20 Correct seating of screw: (a) correct; (b) incorrect

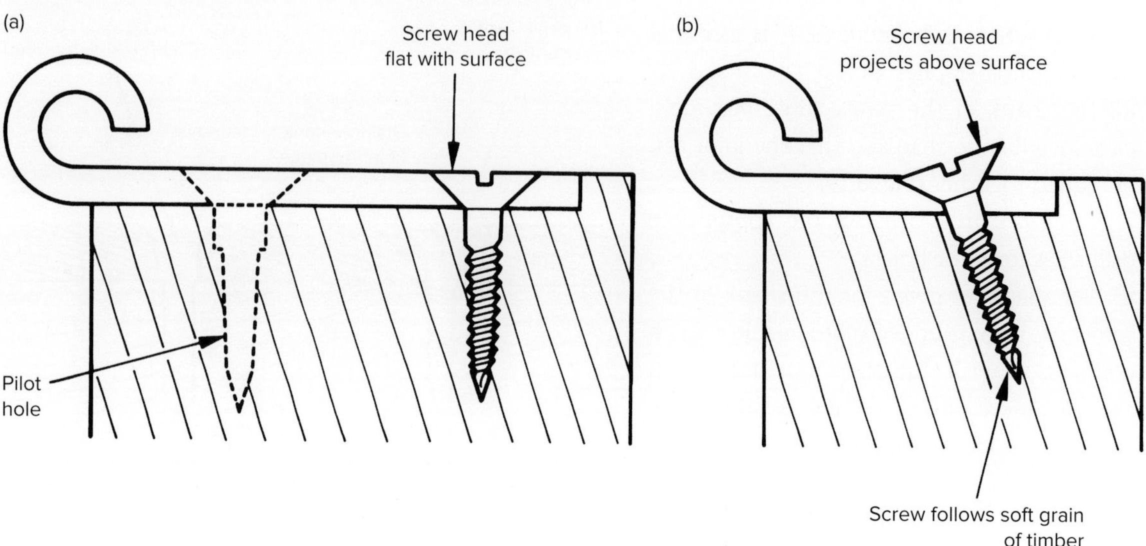

CHECK YOUR UNDERSTANDING

1. What size of drill would be used to drill a pilot hole for the shank of a 10-gauge woodscrew?
2. List at least three reasons why woodscrews would be used for fixing in preference to nails.

8.2.3 Coach screws

Coach screws are a range of large screws similar in principle to woodscrews, but with a hexagonal head that can be turned using a spanner or socket (Fig. 8.21). They are used to attach heavy metal fittings to timber, or timber to timber, where nailing or woodscrews would be inadequate.

The length and the diameter of coach screws are expressed in millimetres. The diameter is often described as 'M6', 'M8' and up to 'M20', which is the diameter of the shank in millimetres.

Drilling for coach screws first requires a hole for the shank that is a clearance fit and then a second hole for the thread. Drilling for an M10 coach screw in securing a heavy metal bracket to timber is shown in Figure 8.22. A washer is used under the head of the coach screw when it is pulling into the timber.

Lubrication will greatly assist the entry of the coach screw and also its removal, if necessary.

Fig. 8.21 Coach screw

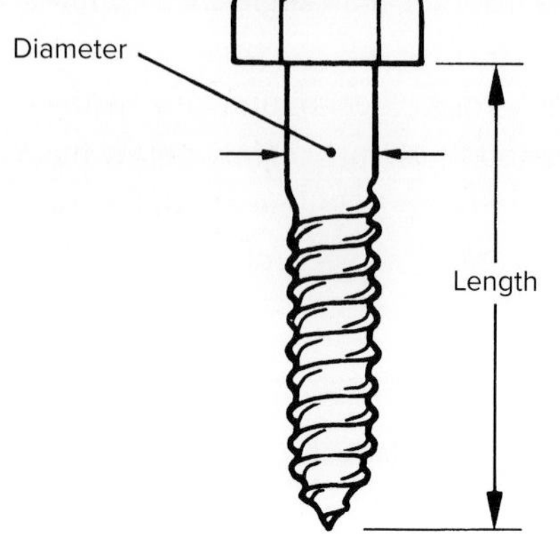

Fig. 8.22 Use of coach screws

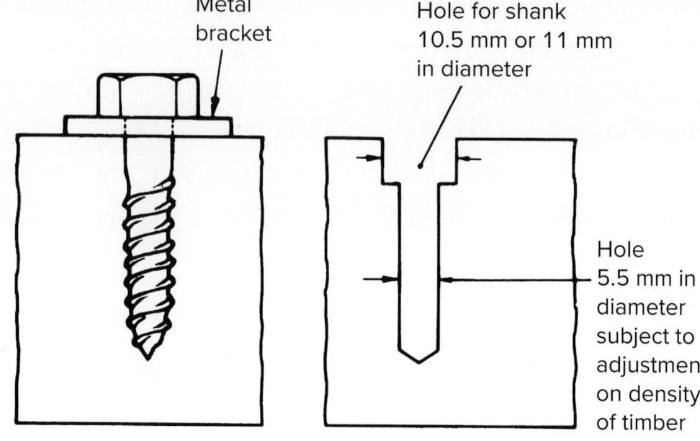

WORKPLACE SCENARIO

With advances in battery technology, and the portability and power that a modern impact driver/battery drill now has, it is preferable today to use screwing connections and joints in many cases as the fixing method of choice. Until recently nails were generally preferred, due to the ease and speed with which they can be driven, either by hand or using a nail gun. However, the development of the impact driver and the greater holding power of screws have meant that screws are now preferred in many instances over nails.

8.3 Bolts and nuts

Bolts and nuts are sometimes referred to as carriage or coach bolts. They provide a fixing of maximum strength, used largely in construction work that is subject to movement or vibration.

In carpentry work, for use in timber, the most commonly used bolt is the cup head (Fig. 8.23). The bolt has a domed head with a square shank underneath, which prevents the bolt from turning as the nut is being tightened with a spanner. For some applications, the hexagon head bolt may be more suitable. For protection against corrosion, bolts can be zinc plated or, for maximum protection in exposed locations, they can be heavily galvanised.

Fig. 8.23 (a) Cup head bolt; (b) hexagon head bolt

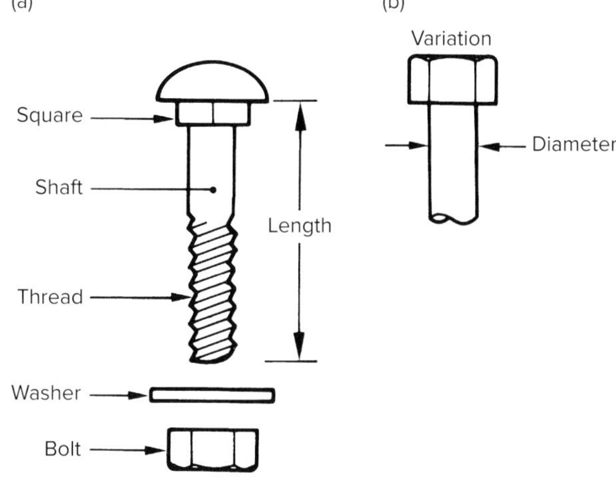

Bolts are described by the length × diameter × shape of the head, and surface finish. Length and diameter are measured in millimetres, listed as M6, M8, M10 up to M20, with a metric thread. Bolts are still available in imperial units (length and diameter given in inches), with a Whitworth thread.

TIP Be careful not to mix up imperial and metric nuts and bolts as they are not interchangeable.

A clear hole must be bored through the pieces of timber being joined with bolts, providing clearance for the shaft of the bolt so that the joint can be pulled tightly together. Washers are used under the nuts when pulling into timber; in soft timber a large washer may also be included under the head.

CHECK YOUR UNDERSTANDING

1. List at least four details that must be supplied when ordering bolts and nuts.

8.4 Associated tools

8.4.1 Screwdrivers

Screwdrivers have evolved through various designs; today most screwdrivers will have a blade 75 mm to 300 mm in length with a fluted plastic handle (Fig. 8.24). They are available in a range of head patterns such as flat, Pozidrive or Phillips head. Alternatively, screws can be driven using a battery-powered drill/driver or electric screw gun for which a range of screw tips are available (Fig. 8.25).

Fig. 8.24 Screwdriver

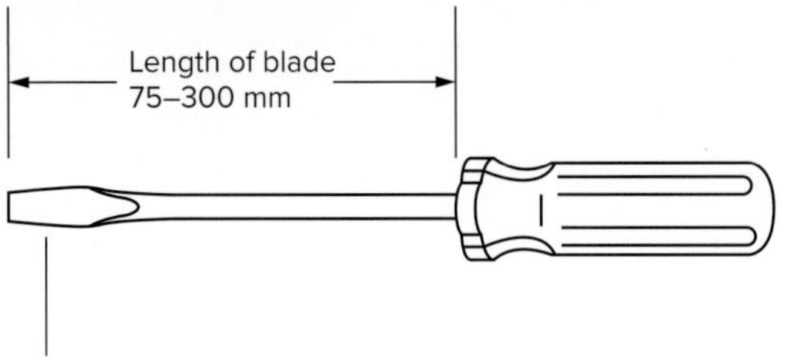

Length of blade 75–300 mm

Point-cross ground

Fig. 8.25 Screw tips

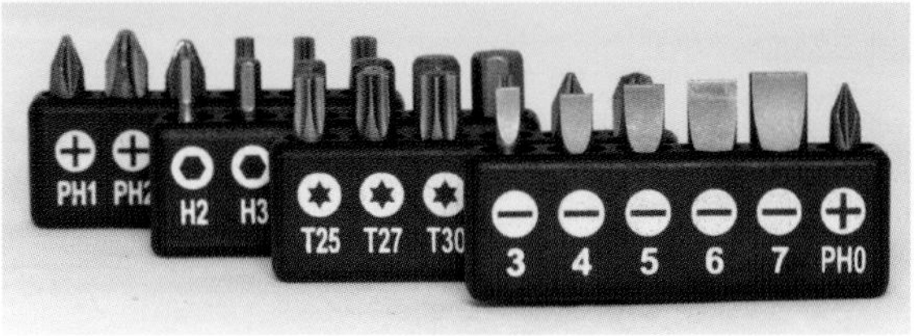

CHECK YOUR UNDERSTANDING

1. List three different types of screwdriver.

Fig 8.26 Countersink bit

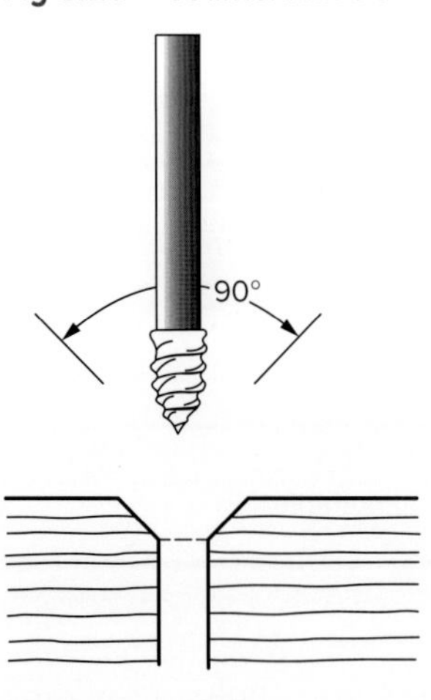

8.4.2 Countersink bit

The countersink bit is used to recess a hole to accept a countersunk screw (Fig. 8.26). The rose countersink is the most common design and its diameter varies from 9 mm to 15 mm.

8.4.3 Bradawl

When inserting very small screws (up to about 4G) in soft timbers, a pilot hole can be made quickly using a bradawl (Fig. 8.27). To prevent the timber from splitting, the bradawl has a small chisel point or square shank sharpened to a point and is twisted up to 45° in each direction to drill a starter hole. It may be found that small screws can be positioned more accurately using a bradawl sharpened to a round point, provided care is taken not to split the timber.

Fig. 8.27 Bradawl

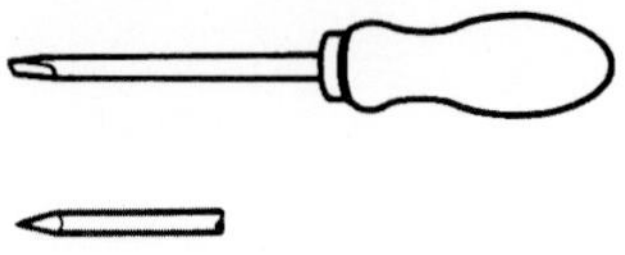

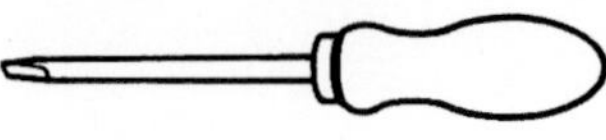

8.5 Masonry anchors

This type of hardware is very important in the building industry. The hardware allows other materials such as timber and steel to be attached to masonry products/components. There are numerous types of anchors and each one has a different application. The Hilti chart on the next page, shows examples of some of the anchors available.

Fig. 8.28 Various types of anchors available for different applications

	Size		Head Type				Base Material				Corrosion Resistance			Features		Approvals
	Minimum Size Diameter	Maximum Size Diameter	Stud (Externally Threaded)	Flush or Internally Threaded	Hex Bolt	Round/Mushroom	Concrete	Lightweight Concrete	Grout-filled Block	Hollow Block/Brick[12]	Hot-dip Galvanized Steel	304 Stainless Steel	316 Stainless Steel	Seismic	Cracked Concrete (per ICC-ES)[13]	1: FM 2: UL 3a: ICC Concrete 3b: ICC Grouted Masonry 3c: ICC Hollow Masonry 3d: ICC Unreinf. Masonry 4: COLA 5: Miami-Dade 6: NSF 61 7: LEED 8: 2010 FBC HVHZ 9: 2009 IBC 10: 2012 IBC
HDA Undercut Anchor	M10	M20	■				■	□			■[15]		■	■	■	3a, 4, 7, 9, 10
HSL-3 Heavy-duty Expansion Anchor	M8	M24	■		■		■	□						■	■	3a, 4, 7, 9, 10
KWIK Bolt-TZ Expansion Anchor	3/8"	3/4"	■				■	■	□			■	■	■	■	1, 2, 3a, 4, 7, 9, 10
KWIK Bolt 3 Expansion Anchor	1/4"	1"	■	■ 1/4"-3/8"			■	■	■		■ 3/8"-3/4"	■	■	■ A-B		1, 2, 3a, 3b, 4, 7, 8, 9, 10
KWIK Bolt-V Expansion Anchor	1/4"	3/4"	■				■	■	■							2, 7
KWIK HUS-EZ Screw Anchor	1/4"	3/4"			■		■	■	■					■	■	3a, 3b, 4, 7, 8, 9, 10
KWIK HUS Screw Anchor	3/8"	3/4"			■		■	■	□							7
HCA Reusable Coil Anchor	1/4"	3/4"			■ 1/4"		■	□								-
HLC Sleeve Anchor	1/4"	3/4"	■	■ 1/4"-3/8"	■	■	■	□	■	■		■				7
KWIK HUS-EZ-I Screw Anchor	1/4"	3/8"		■ 1/4"-3/8"			■	■						■	■	1, 3a, 4*, 7, 8, 9, 10
HDI/HDI+/HDI-L+ Drop-in Anchor HDI+ Stop Drill Bit with Setting Tool available	1/4"	3/4"		■			■	■				■ 303 SS				1, 2, 7
HDI-P Flush Anchor	3/8"	3/8"		■			■	□								7
HCI Cast-in Anchor Color Coded:	1/4"	3/4"		■			■	■								1, 2, 7
KWIK-CON II + Concrete and Masonry Screw	3/16"	1/4"		■	■		■	□	□	■		■ 410 SS				5, 7

Hilti (Aust.) Pty Ltd

CHECK YOUR UNDERSTANDING

1. Using Figure 8.28, which anchors can be used in hollow block/brickwork?

8.6 Adhesives

Glues play an important role in joining timber and other construction materials. Adhesives can help provide additional strength to a joint that has been constructed using mechanical fasteners such as screws or nails. Alternatively, glues and adhesives can be used without the assistance of mechanical fasteners; if this method is used, it is imperative that the manufacturer's specifications are followed in the application of the adhesive to achieve maximum strength.

Consult the manufacturers' instructions prior to application, taking note of drying times, required ambient temperatures, clamping requirements and pressures, factors affecting adhesion such as moisture or dust and any potential occupational health and safety (OHS) hazards that could be encountered when using the particular adhesive.

Many products on the market have been developed for specific situations. For example, interior use, exterior use, marine use, etc.

Student research

Research four different types of adhesives commonly used in the construction industry. What application are they designed for, how do they cure and how should they be applied?

End of chapter activity

Please refer to Chapter 9, Carpentry Power Tools for this activity.

AUSTRALIAN BUILDING CODES AND STANDARDS

AS 1684.2-2010
Residential Timber-framed Construction

AS 5216:2018
Design of Post-Installed and Cast-In fastenings in Concrete National Construction Code (NCC)

AS/NZS 2699.2:2000
Built-In Components for Masonry Construction - Connectors and Accessories

Tradie talk—**Smack a nail into it to hold it**

Chapter 9

Carpentry power tools

Learning Objectives

LO 9.1 **Establish a power supply**

LO 9.2 **Safely handle a portable power saw**

LO 9.3 **Safely use a mitre saw**

LO 9.4 **Safely use an electric drill**

LO 9.5 **Safely use a portable jig saw**

LO 9.6 **Safely handle a portable power planer**

LO 9.7 **Safely use electric sanders**

LO 9.8 **Safely use a portable electric router**

LO 9.9 **Identify battery-powered tools**

LO 9.10 **Know about pneumatic tools and their application**

LO 9.11 **Safely use an angle grinder**

LO 9.12 **Safely use a drill press**

LO 9.13 **Identify load-handling equipment**

Introduction

Power tools are an important part of the building industry. Without them it would take longer to perform basic tasks, which would therefore extend the time it takes to construct a building.

In this chapter, you will learn how to identify various power tools and plant. You will learn how to prepare tools for use, including basic maintenance plus changing blades and cutters, as well as identifying the dangers associated with the electricity supply and with tools and equipment, so you will become confident and proficient in using them. It is always a good idea to establish a safe operating procedure (SOP) before using any power tool. (A generic SOP for power tools can be seen in Figure 1.3.) Remember always to refer to the manufacturer's specifications prior to using a tool for the first time.

Fig. 9.1 Domestic temporary power pole

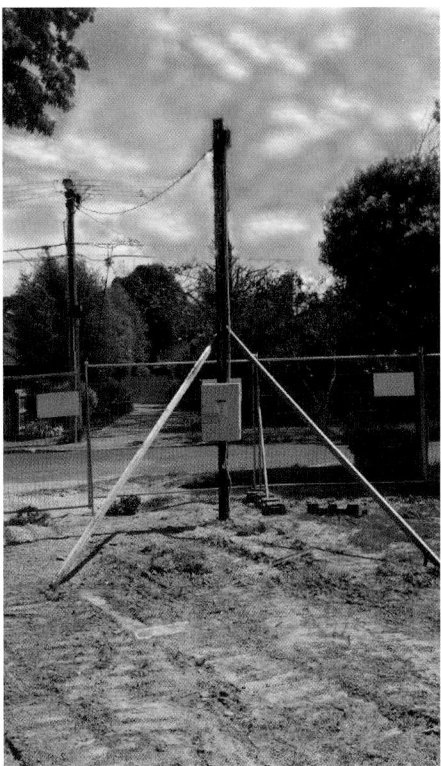

Courtesy of Super Safe Hire

9.1 Power supply

All construction worksites require an electricity supply for lighting, operating the power tools and the plant, and heating and cooling. The power can usually be found in a temporary power board cabinet for larger construction sites or on a temporary power pole, which is usually found on domestic building sites (Fig. 9.1 and 9.2). It is important that power leads do not extend more than 30 m. Multiple supplies are therefore required on large construction sites.

The mains power supplied is generally 240 volts (V), 50 hertz (Hz); portable power tools suitable for use with this supply will have these figures indicated on a plate attached to the machine.

These temporary power supplies must be installed by a qualified electrician and tested according to state or federal requirements. In addition, the power supply must be fitted with a safety switch.

Fig. 9.2 A & B Commercial temporary power boards

(a)

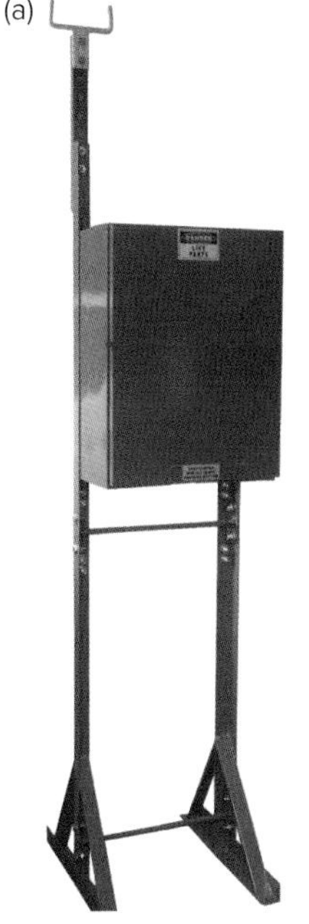

(b)

Courtesy of DNN Australia Pty Ltd

Fig. 9.3 Portable generator

© canoness/123RF.com

9.1.1 Portable generators

Portable generators (Fig. 9.3) can provide a source of electric power in situations where mains power is unavailable, but some precautions must be observed to ensure their safe and efficient operation. Always check that the quality of electricity being generated by a portable generator is suitable for the efficient operation of the tools in use. It must have a power output sufficient to drive all of the power tools in use at any one time. Refer to the plate attached to the machine for the power required to operate it efficiently, which will be expressed in watts (W) or amperes (A), but most probably both.

Some generators may have a built-in circuit breaker, which will protect the operator by cutting off the power supply should a fault develop in any of the power tools or appliances in use. For example, take the tools shown in Table 9.1. As the total load of these tools is 3100 watts or 13.6 amps, a suitable generator must have this output in watts or amps.

TIP Portable generators generally have a petrol motor as their power source. With this in mind, check to see if the motor is a 2-stroke motor (oil and petrol mixed together) or 4-stroke motor (straight petrol and oil located in a separate sump). Ensure the correct fuel is used and that the oil is appropriate for the machine. Refer to the manufacturer's instruction before operating any machine you are unfamiliar with.

Table 9.1 Requirements of common power tools

Tool	Watts		Amps
One power saw	1750	Allow	7.5
One planer	750	Allow	3.5
One hammer drill	600	Allow	2.6
Total	3100	Allow	13.6

9.1.2 Residual current device (RCD)

Residual current devices (RCD), sometimes referred to as *safety switches,* are mandatory on all building sites. They protect the workers from being electrocuted by monitoring the current flowing in the active and neutral wires supplying the item of equipment.

Under normal circumstances, the current flowing in the two wires is equal. When a fault in the circuit or an accident with the equipment occurs, an imbalance in the current flowing in the two wires is detected by the RCD, which automatically cuts off the power before injury or damage can happen.

RCDs are essential if the work is carried out in an older building as not all buildings have had safety switches installed. So, as a rule, plug the RCD into the wall socket and then connect all of your tools to the RCD so you will be protected from being electrocuted if something happens.

9.1.3 Power leads

Fig. 9.4 A power lead

Courtesy of Daniel Bonnici

Power leads are an important tool for any tradesperson as they allow power to be brought to the area where it is needed (Fig. 9.4). They come in various forms, lengths, thicknesses and ratings. When purchasing power leads, ensure that they are suitable for the tools to be connected to them and will allow enough current through to enable the tool to work properly.

Power leads should:

- be completely unwound when in use–if not they can heat up and melt
- be wound in big loops when in storage (avoid tying knots in them as this will put stress points on the lead and can cause cracking in the lead casing)
- always be checked before use to make sure the internal wires are not exposed, especially at the ends where the plug is connected (if the internal wires are exposed, do not use the lead and get an electrician to check it)
- never be pulled out from the plug by the lead as this will stretch the lead and will pull the casing out of the plug connection
- never be rolled up when plugged in–always unplug the lead first and then roll it up.

9.1.4 Double-insulated and all-insulated appliances

All-insulated appliances are totally enclosed in a casing made of insulating material and have no external metal whatsoever. This form of construction eliminates the need for further earthing or for a second internal barrier of protective insulation.

Portable electric tools of the double-insulated or all-insulated types must not be earthed and so are fitted with a two-core flexible cord. If a three-pin plug is fitted to a two-core lead, no connection is made to the earth terminal of the plug.

Portable electric tools of the double-insulated type are easily identified as they are required to be clearly marked with the words 'double insulated' or with the symbol shown in Figure 9.5. Some older double-insulated tools may be marked 'do not earth double insulated'.

Fig. 9.5 The double-insulated symbol

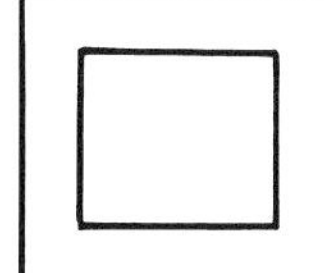

All-insulated tools are also marked 'double insulated' and can be identified by the casing of non-metallic material.

9.1.5 Electrical hazards checklist

1. **Are construction switchboards properly constructed and set up?**
   Ensure electricity switchboards are of robust design and build and are mounted securely. Where electricity metre and fuse assembly is fitted, make sure the switchboard has an endorsed service fuse locking device.
2. **Are there enough switchboards for the job?**
   Switchboards should be spaced according to the maximum allowable lengths of the extension leads being used. The maximum lengths of the most common types of flexible leads are listed below (generally the shortest length of lead should be used when setting up switchboards):
   - 25 metres for 10 amp extension lead with 1.0 mm^2 flexible cables cores
   - 35 metres for 10 amp extension lead with 1.5 mm^2 flexible cables cores

- 5 metres for 15 amp extension lead with 1.5 mm^2 flexible cables cores
- 40 metres for 15 amp extension lead with 2.5 mm^2 flexible cables cores.

3. **Are the correct socket outlets being used?**
   Check that only 240 volt socket outlets with double pole switches are used on relocatable structures and portable equipment.
4. **Are portable outlet devices suitable?**
   Double adapters and other types of domestic multi-plug power boards are not suitable for use on construction sites.
   Only use multi-plug portable devices that comply with AS/NZS 3105:2007 (Approval and Test Specification–Electrical Portable Outlet Devices) and are the industrial type fitted with over current and RCD protection.
5. **Is construction cabling identified and protected from mechanical damage?**
   Make sure construction cabling is clearly identified and has protection where there is a risk of mechanical damage.
6. **Are electric tools and flexible leads in a safe condition?**
   Ensure that electric power leads and tools brought on-site are suitable and in good condition.
   - Use only heavy-duty type extension leads and Australian-approved power tools.
   - Have leads and tools inspected, tested and tagged by a suitable qualified person at intervals of three months; keep a register of inspections.
   - Visually inspect leads and tools for signs of damage before use.
   - Make sure unserviceable tools or leads are immediately withdrawn from service.
7. **Have Certificates of Electrical Safety been issued?**
   Ensure the electrical contractor provides you with Certificates of Electrical Safety for all construction wiring, including switchboards.
8. **Has the electrical installation been tested?**
   The electrical installation, including all construction wiring and switchboards, needs to be inspected and tested by a licensed electrician or electrical inspector before use and re-tested every six months for the duration of the project.
9. **Are RCDs (safety switches) fitted and have they been tested?**
   Ensure that all fixed and portable RCDs are tested for tripping current and time by a licensed electrician every calendar month while used on-site. The tripping function of portable RCDs should be checked before use by operating the RCD's test button.
10. **Are portable generators suitable?**
    If portable generators are used on-site, ensure the generator:
    - is fully serviceable and has been properly maintained where supplying a fixed installation
    - is installed by a licensed electrician.
11. **Are extension leads being used safely?**
    Check that leads are not lying in mud or water, or in areas where they can be damaged or become tripping hazards. Use stable, insulated lead stands to keep them above head height. Do not allow leads to be wrapped around scaffolds or falsework; use s-shaped off-cuts of steel reinforcing bar sheathed in cut-off lengths of garden hose instead.
12. **Are plant and temporary structures a safe distance from powerlines?**
    Make sure there is always a safe distance between live powerlines and cranes, concrete booms, earth-moving equipment, elevating work platforms, hoists, scaffolds, falsework and portable ladders by strictly observing 'no go zone' safe clearances.
13. **Is electrical installation and repair work being done safely?**
    Make sure that all electrical installation work and any repairs to the electrical installation are undertaken only by licensed electricians working for registered electrical contractors.

This checklist is a summary of one found on the WorkSafe Victoria website. To view the complete list, visit: www.worksafe.vic.gov.au/forms-and-publications/forms-and-publications/electrical-hazards-checklist.

CHECK YOUR UNDERSTANDING

1. What device should be fitted to the electricity supply to ensure the safety of workers? What does it do?
2. How often should leads and power tools be tested and by whom?
3. Are power leads allowed to lie on the ground?

9.2 Portable power saw

The portable power saw is one of the most used tools by carpenters. The size of a saw is expressed as the diameter of the largest saw blade the machine will accommodate; with only minor variations, the commonly available sizes are 160, 185, 210 and 235 mm. When selecting a saw for a particular purpose, it is necessary to consider its maximum cutting capacity, when the blade is at both 90° and 45° to the base. For example, a 235 mm saw at 90° may have a maximum depth of cut of 84 mm, or 58 mm at 45°.

The parts of a portable saw are shown in Figure 9.6. The saw blade is coupled to the motor by a series of gears, and cuts as it rotates towards the saw base. A chute ejects the sawdust to the side. The base can be tilted at an angle of between 45° and 90° to the saw blade and is locked in place. The depth of cut can also be varied by an adjustment knob, usually located at the back of the saw, which allows the base to pivot on a point at the front, and to be raised up or down and locked into position.

The retractable guard is an important safety feature. This will retract automatically as the cut is started and spring back into position at the conclusion of the cut. The guard must always be kept in good working order, and never be held back, leaving the blade exposed. Power saws incorporating a riving knife behind the saw blade, which will prevent the saw cut from closing on the blade and the saw 'kicking back' dangerously, are also available.

A moulded handle also contains the starting trigger. The larger models have a second handle to provide greater control of the saw.

WORKPLACE SCENARIO

Back when I was an apprentice, I was doing a fit-out for a new fast food restaurant. The site foreman asked me to cut some large holes in pieces of plywood. When I went to my van, I realised I did not have my jig saw with me. The holes had to be cut urgently as some other tradies were waiting to finish this particular job. So, with youthful exuberance, I came up with the bright idea of using my power saw to do the job. I managed to cut some rough, round-shaped holes, but the way in which I had to do the job was extremely dangerous. I was lucky no one saw me and that I didn't injure myself. With hindsight, I should have tried to borrow a jig saw or bought a cheap one from the local hardware store. The moral is, always use the tool as it was designed to be used. Lots of things seem like a good idea at the time but the consequences could be horrific.

Fig. 9.6 Circular saw

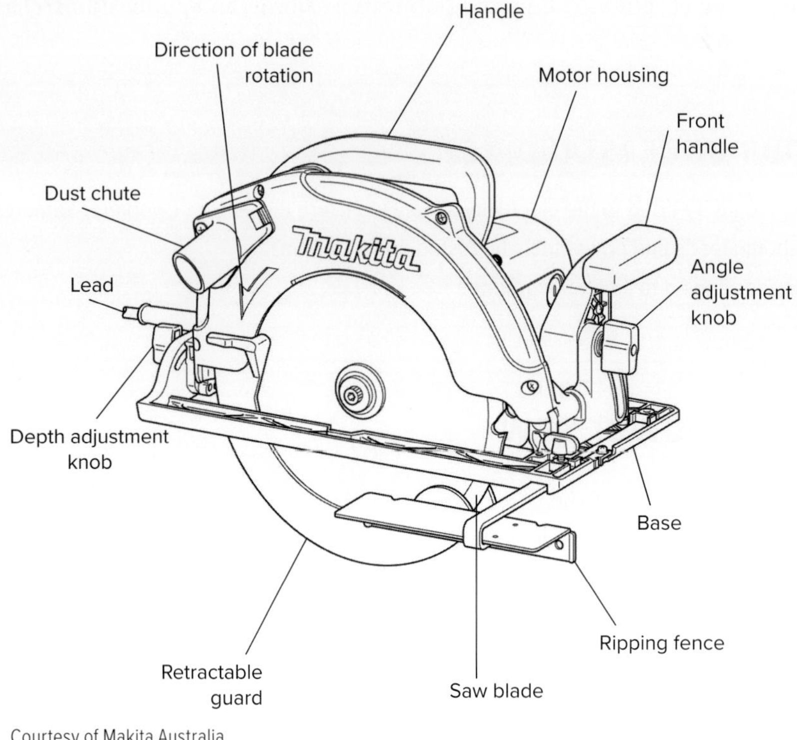

Courtesy of Makita Australia

9.2.1 Circular saw blades

There are a number of saw blades available for use with the powered circular saw and these are designed for different purposes. Teeth are shaped to cut with maximum efficiency under a variety of conditions. To describe a particular blade, it is necessary to provide the following information:

1. the outside diameter of the blade
2. the diameter of the centre hole (also called the **bore**), which, of course, must fit the spindle of the machine; sometimes blades are sold with a large centre hole and adapter rings to reduce the size of the hole to suit various machines
3. the number of teeth on the blade.

For example, a blade for a particular machine may be described as 'carbide tipped, 235 × 25 × 40 tooth', or as 'a cross-cut blade, 160 × 20 × 120 tooth'. One of the types of blades in general use is discussed in section 9.2.1.1. Not every type of blade may be available to fit any particular machine, but manufacturers usually provide details in the literature supplied with the machine.

9.2.1.1 Tungsten-tipped blade

The tungsten-tipped blade consists of a saw plate made from tough, high-quality steel with inserts of tungsten carbon alloy welded in to provide the cutting points (Fig. 9.7). The blades' teeth are ground to close tolerances and will give a very smooth cut. It is claimed that these teeth will last approximately 40 times longer than a traditional steel-toothed blade before they need to be resharpened, but they must then be returned to the saw doctor for reconditioning. Take particular care not to hit hard objects with the teeth as the tungsten carbon alloy is extremely hard and can chip easily or an insert can be knocked out completely.

Fig. 9.7 Tungsten-tipped saw blade

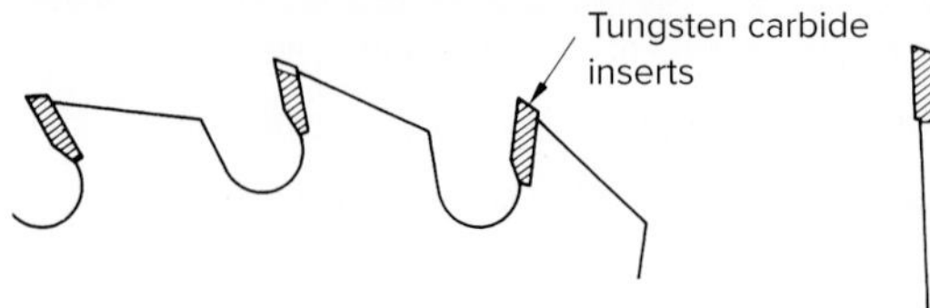

The number of teeth on a tungsten-tipped blade varies considerably. Blades with greater numbers of teeth can be used for the clean cutting of plywoods, etc. Blades with fewer teeth are suitable for ripping and more general building work where a clean cut is not required.

9.2.1.2 Changing the saw blade

To change the saw blade, first disconnect the saw from the power supply. Check whether the particular saw is fitted with a shaft lock. If it is, rest the saw on the bench and press the lever to engage the lock. Using a correctly fitting spanner (usually supplied with the saw), undo the bolt and remove the **collar** holding the saw blade.

If there is no **shaft lock**, lay the base of the saw on a flat board, retract the guard and press the blade against the edge of the board to stop the blade turning. Using the appropriate spanner, remove the bolt and collar holding the blade. The blade is then lifted and removed through the slot in the base plate.

Replacing the blade is the reverse operation, taking care that it will rotate in the right direction. Looking at the machine from the blade side, the saw blade revolves in an anti-clockwise direction and the teeth on the front edge will be moving up towards the base. Saw blades commonly have an arrow indicating the direction of rotation and this should be facing the outside and be visible when the blade is fitted. Replace the collar and bolt, and tighten with the spanner.

9.2.2 Range of work

9.2.2.1 Straight cross-cutting

1. Support the timber to be cut on stools in such a way that it will not collapse and jam against the saw blade as it is cut. Beware of long unsupported ends, which may cause overbalancing or splitting of the timber. Restrain the timber from moving: a cramp or cleats nailed to the stools will provide a stop and prevent the timber from moving away or rotating as the cut is made.
2. Adjust the depth of cut of the saw, so that it will project approximately 6 mm through the bottom of the work (Fig. 9.8). Plug the saw into a power point and switch the power point on.
3. Grip the saw firmly by the grips provided and lay the base on the timber at the start of the cut. On the front of the saw base there are two notches (refer to the instruction manual). One, usually the most conspicuous, is the guide for following the line with the saw cut when the blade is set at 90° to the base.

   The second notch is the guide for following the line when the blade is set at 45° to the base. When the blade is set at angles of between 45° and 90°, sight over the saw and follow the line visually. In the example of straight square cutting shown in Figure 9.9, select the relevant notch and align it with the line marked for the saw cut.
4. With the blade clear of the timber, start the motor using the trigger switch and allow it to gain full speed before feeding it through the timber, following the marked line. The rate

Fig. 9.8 Depth of cut

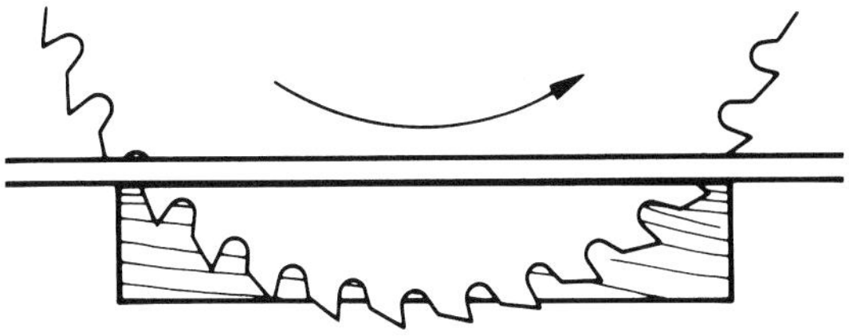

Fig. 9.9 Showing the saw notch and cutting line

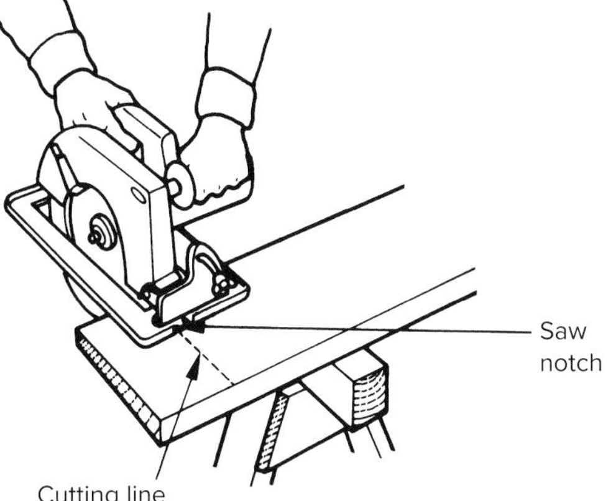

of feed can vary depending on the size and nature of the material being cut. The best guide is to listen to the motor. If it tends to labour and slow down, reduce the rate of feed, but if the motor is running freely, the rate of feed can be increased provided the saw is kept under control at all times. It will soon be seen that the notch on the front of the saw base passes over the edge of the timber, and the line is lost from view before the cut is completed. With practice, there is usually no great difficulty in keeping the saw moving in a straight line to complete the cut. Once the cut is complete, release the trigger and let the blade come to a stop. **Never use your fingers to slow down the blade**.

When it is necessary to cut a number of pieces to the same length–wall studs, for example–it is faster and more accurate to lay a number of pieces side by side and clamp them together (Fig. 9.10). Square a line across and follow this line with the saw.

Fig. 9.10 Multiple cut

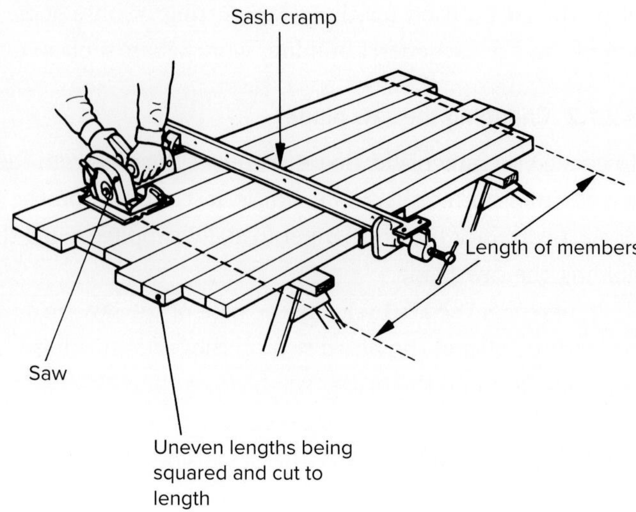

9.2.2.2 Single bevel cutting

Single bevel cutting, such as the plumb bevel to roof common rafters, can be carried out without any further adjustment to the saw provided the timber is of the same thickness (Fig. 9.11).

Fig. 9.11 Single bevel cut

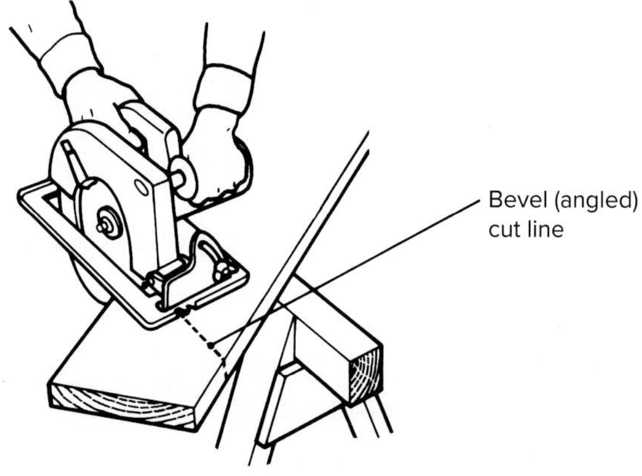

9.2.2.3 Compound bevel cutting

Compound bevel cutting is commonly encountered with the roof creeper rafters (see Chapter 14), which are bevelled in two directions–a plumb bevel across the face and an edge bevel. For most roofs, tilt the blade at 45° to the base for the edge bevel.

To perform the cut, the following steps should be followed:

1. Disconnect the saw from the power supply and tilt and lock the base to 45°, as indicated on the protractor scale on the saw (Fig. 9.12). The depth of cut must now be adjusted so that the blade will again project about 6 mm through the base of the work.

Fig. 9.12 Tilting and locking base

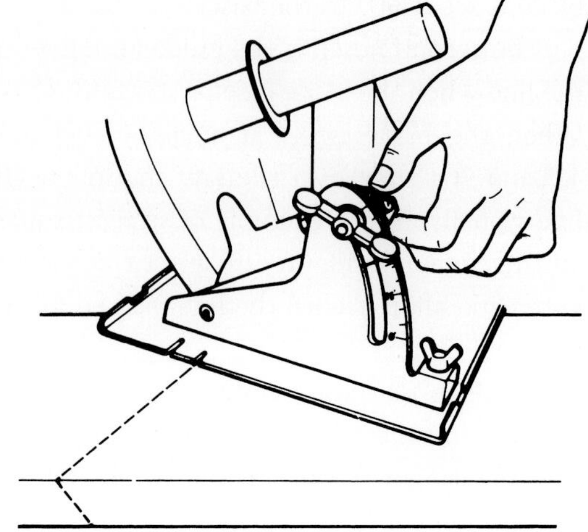

2. Connect the saw to the power and switch on the power.
3. Place the saw base on the face of the timber at the start of the cut and align the second notch on the saw (for 45° cutting) with the line to be cut (Fig. 9.13). Switch on the motor and, when full speed has been reached, feed the saw onto the timber with the guide notch following the line.

Sometimes, when the saw is tilted at an angle, the retractable guard will tend to bind at the start of the cut. In this case, before starting the cut, take the hand from the front handle and lift the guard by the lever provided (Fig. 9.14), for just sufficient time to start the saw feeding freely. Release the lever and again grip the front handle.

Fig. 9.13 Compound bevel cut

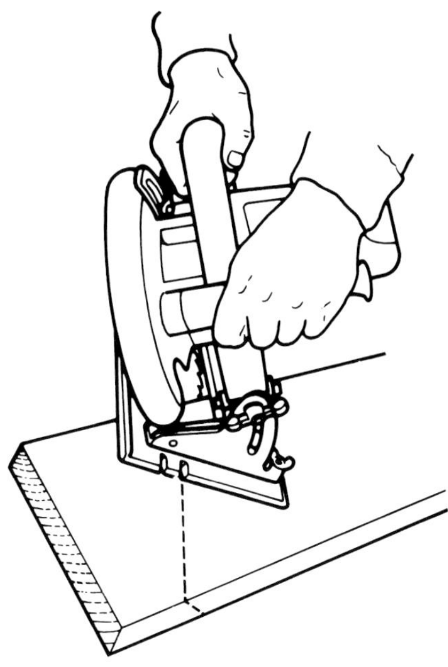

9.2.2.4 Ripping cuts

Ripping cuts can be made with the saw using the ripping **fence** as a guide–this can be used on either side of the saw. To set the fence up:

1. Disconnect the saw from the power and adjust the angle of the blade and the depth of cut–again about 6 mm through the bottom of the material.
2. Fix the ripping fence in position and adjust to the width of material to be cut (Fig. 9.15). Using a rule, measure from the fence to the side of the saw teeth.
3. Lay the board to be ripped across the saw stools. So that the saw blade projecting through the bottom of the board will not cut into the stools, pack it up with a piece of scrap material, held in place with small skew nails (Fig. 9.16). The board itself must be held securely with a clamp or pinned with a nail, whichever is suitable, without fouling the saw.
4. Reconnect the saw to the power and place the base of the saw at the start of the cut, with the ripping fence against the edge of the board. Start the motor and feed into the timber, keeping the fence against the edge. When making a long ripping cut, make sure there is sufficient power lead freely available so that the cut can be completed without having to stop to untangle the lead (Fig. 9.17).

This method can also be used to groove or rebate a length of timber, as shown in Figure 9.18. In this case, the blade is set to the depth required for the groove. A wider groove can be cut by gradually altering the setting of the fence, and performing successive cuts to build up the required width of groove.

Fig. 9.14 Start of compound cut

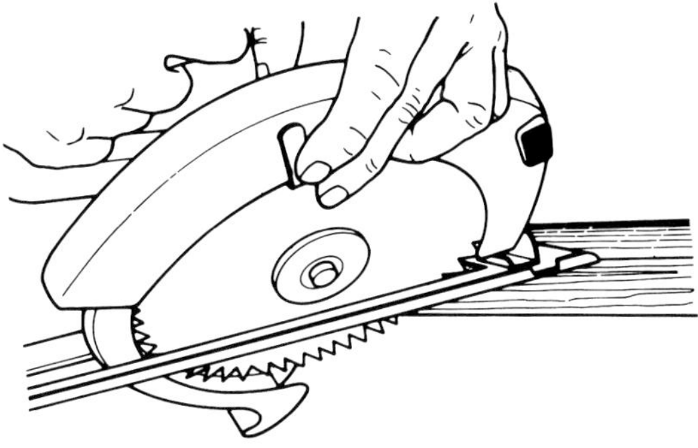

Fig. 9.15 Setting width

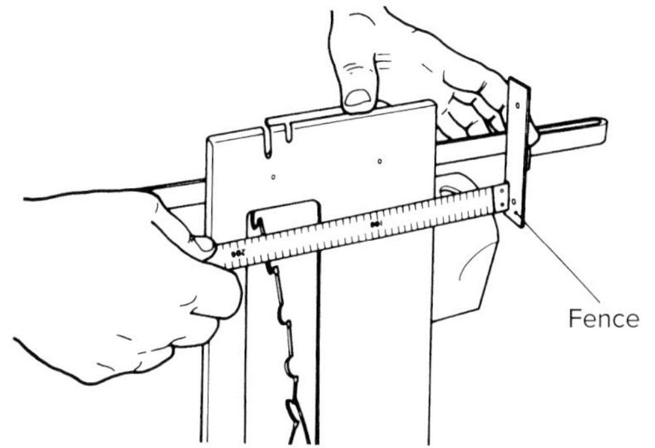

Fig. 9.16 Packing and support for base

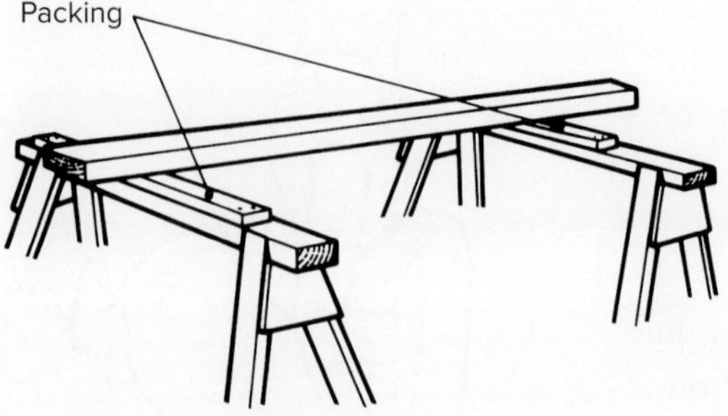

Fig. 9.17 Ripping cut

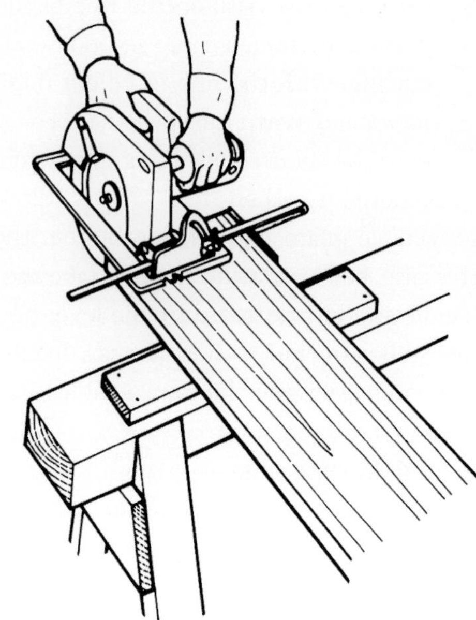

Fig. 9.18 Grooving

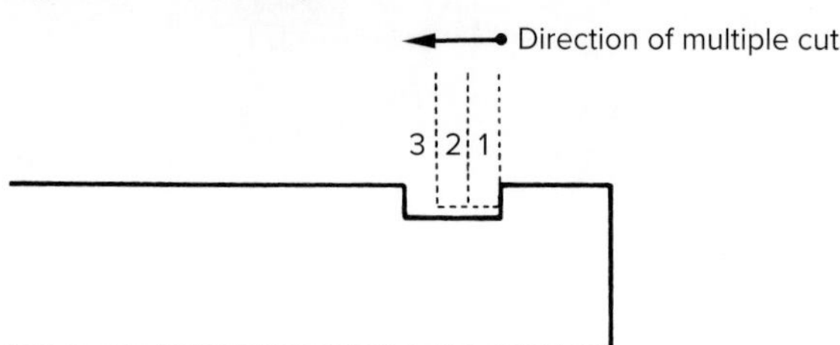

CHECK YOUR UNDERSTANDING

1. List the two basic adjustments that can be made to the power saw. What is the effect of each?

9.2.2.5 Cutting sheet material

Sheet materials such as plywood, particleboard and the like are easily cut with a power saw. Select and fit a suitable saw blade for the purpose–a small-toothed cross-cutting blade. Fit the blade and adjust the depth of cut to no more than 6 mm below the sheet to be cut.

1. Large sheets of material must be supported in such a way that they cannot collapse when sawn through (Fig. 9.19). The bearers directly under the sheet will be marked by the saw blade projecting below the sheet but can be reserved for this purpose in the future and regarded as expendable over time.
2. For rough work, sheets can be cut freehand, marking a line and following it with the sawing guide located on the front of the base of the saw.
3. For more accurate work, where the cut is too far from an edge to use the ripping fence, a batten can be clamped across the sheet and used as a guide for the edge of the base plate in order to accurately cut a straight line.
4. For speed and convenience, the guide batten can be made up in the form of a T-square (Fig. 9.20). Screw the blade to the stock at 90°. Leave the stock a little long and when the first cut is made, it will be cut off to the correct distance from the saw blade to the edge of the base. The square can be held in place with small

Fig. 9.19 Cutting sheet material—layout

cramps or fine nails. Another method, particularly suitable for particleboard with a prefinished face, is to drive two nails into the edge of the stock, cut them off at about 5 mm long and sharpen the points. Located in position and lightly driven into the edge of the board, they will hold that end securely while a small G-cramp at the other end will secure a long length (Fig. 9.21).

5. When cutting prefinished sheets such as plywood panelling and veneered particleboard, keep the face side of the sheet down, and mark out and cut from the back of the sheet (Fig. 9.22). The saw teeth cut from the bottom as they rotate up towards the base and tend to leave splinters on the top of the cut. This often does not matter, but it is undesirable on finished surfaces. Setting the saw blade only 6 mm below the base plate helps to minimise the tendency to chip the top surface of the material. Furthermore, the prefinished surface will not be scratched by the saw base passing over it.

Fig. 9.20 Pinned square

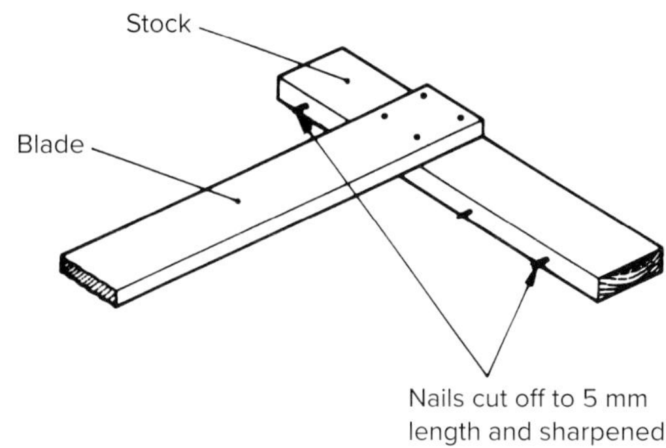

Fig. 9.21 Sawing to the batten

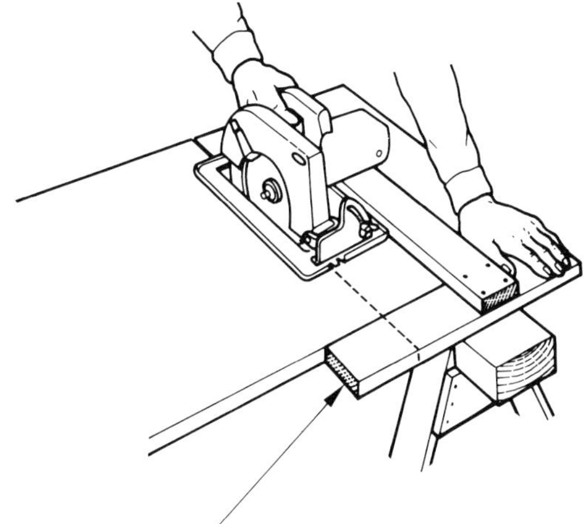

9.2.2.6 Protractor and saw guide

The protractor and saw guide is a valuable guide for the accurate cutting of straight and bevel cuts, as in Figure 9.23. Place the saw base on the start of the cut and align the relevant sawing guide with the line to be cut. With the protractor set to the required bevel, slide it up to the side of the saw base. Grip firmly and proceed with the cut, keeping the edge of the saw base against the arm of the saw guide.

Fig. 9.22 Cutting prefinished sheets (face down)

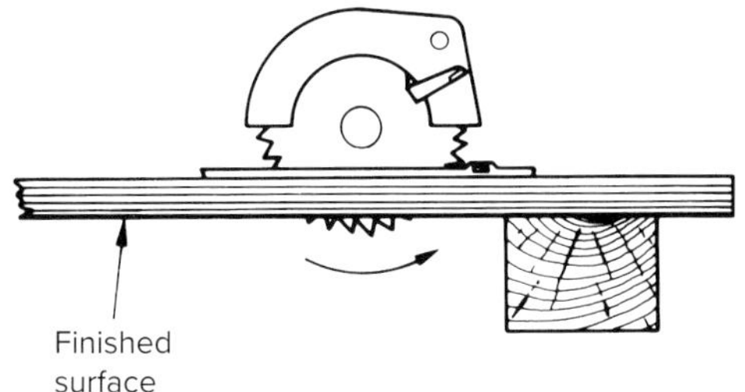

Fig. 9.23 Use of protractor

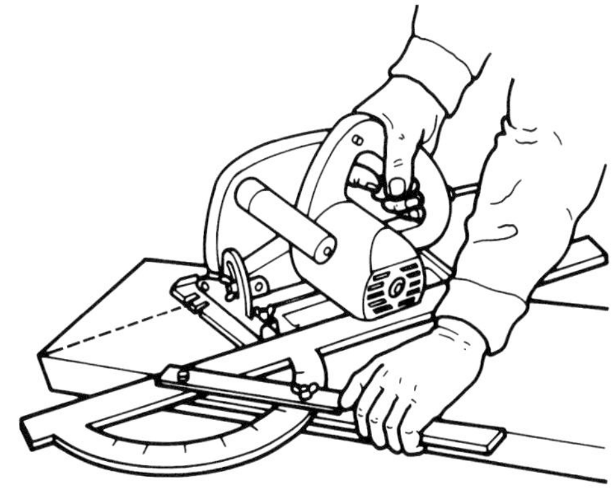

Safety measures for the power saw

For safe working practice with the portable power saw, follow these steps:

1. Always disconnect the saw from the power source before making any adjustments.
2. Carry out all adjustments using correctly fitting spanners and ensure that all bolts or thumb screws are properly tightened before putting the saw back into service.
3. Ensure that all guards–particularly the retractable lower guard–are in place and working correctly at all times.
4. Check lead to ensure there are no nicks or exposed wires.
5. The circular saw is intended only for cutting in a straight line; never attempt to cut around a curved line as the saw may jam and 'kickback'.
6. The most dangerous place near any circular saw is the area directly behind the blade. Neither the operator nor any assistant should ever be allowed to reach behind the blade to remove waste or support the timber, lest the saw kicksback with serious results. Sawing machines incorporating a riving knife afford greater protection in this sensitive area.
7. Do not stand directly behind the saw blade when operating the saw: again, kickback could lead to a badly gashed leg.
8. Always use two hands and grip the saw by the handles provided and learn to control the saw from this position.
9. Keep the sole of the base plate in contact with the face of the timber being cut.
10. When using the saw above ground level, work only from a stable platform that has been erected to state or federal regulations and adopt a comfortable, well-balanced position.
11. Wear safety goggles/glasses and ear protection at all times. A dust mask or respirator may be needed depending on the material being cut. Keep the electric lead away from the saw blade at all times–some operators prefer to sling the lead over their shoulder for this purpose–and at the conclusion of the cut allow the saw blade to stop rotating before laying the saw down.

TIP A good explanation of kickback can be viewed at www.youtube.com/watch?v=R4kxQqBu3ME.

9.3 Mitre saw

The compound **mitre saw** is designed to dock timber off to length and can be adjusted to cut at right angles or any angle down to 45° (Fig. 9.24a). It has two adjustments: the saw can lay over to 45° and the saw's table is adjustable to 45°. The slide compound saw has the added extra that it can cut timber generally up to 300 mm wide as well as being adjustable to cut compound angles (Fig. 9.24b). Some saws are designed so that the depth of the cut can be adjusted and slots or trenches can be made.

There are many makes and sizes available, and each can operate in a slightly different manner. Fitted with a suitable blade, the saw can quickly make smooth, accurate cuts and is widely used for interior fixing, fitting architraves, skirtings and so on. The larger models have the capacity to handle most structural timbers and when treated with care will give reliable service.

Ensure you are familiar with the saw and its features by reading the manufacturer's instructions before you use it. The saw is safely guarded with a retracting lower guard, which is sometimes made of transparent plastic, leaving the saw blade visible. Ensure the saw is set up on a stable surface, for example, a bench with extension tables and adjustable stops.

Fig. 9.24 (a) A compound mitre and (b) a slide compound saw

(a) © Dmitry Solmashenko/Shutterstock

(b) © pryzmat/Shutterstock

9.3.1 Operation of a mitre saw

1. To operate the mitre saw, unlock the rotating table and set the angle of the saw to the desired cutting angle. This can be square to the fence or up to 45° either side of square.
2. Feed the timber onto the saw table and hold firmly in position against the fence by hand. Use the clamping device for large sections of timber, aluminium or plastic.
3. Grip the handle and start the saw by pressing the on/off switch.
4. Lower the saw slowly but with a uniform pressure down on to the material and complete the cutting.
5. Release the downward pressure on the saw and it will return, by spring loading, to its raised position.

Safety with the mitre saw

- Ensure that you never cross your hands when using a mitre saw; that is, never hold a piece of timber with your left hand on the right-hand side of the saw and then operate the saw with your right hand. This action increases the risk of being cut by the blade of the saw if the saw jams.
- Never rip using a mitre saw.

9.4 Electric drill

Electric drills, such as the one shown in Figure 9.25, comes in a range of sizes. The size is expressed as the maximum size of drill bit the chuck will accommodate. Common sizes are 8, 10 and 13. Most drills on the market today will take at least a 10 mm drill and this is a good size for a general-purpose machine. The maximum diameters for drilling in steel and wood are often included in the drill specifications, for example 'steel 10 mm and wood 21 mm'.

The speed at which the drill **chuck** rotates with no load is called the *no load speed* and is important. For general use in woodworking and building construction, the following speeds are given as a guide in the selection of a suitable machine. A number of drills on the market will, no doubt, meet these requirements.

On a two-speed drill, the speed is varied by a series of gears and a selector switch. A suitable drill would have a low speed of approximately 950 rpm (revolutions per minute) and a high speed of about

Fig. 9.25 Electric drill

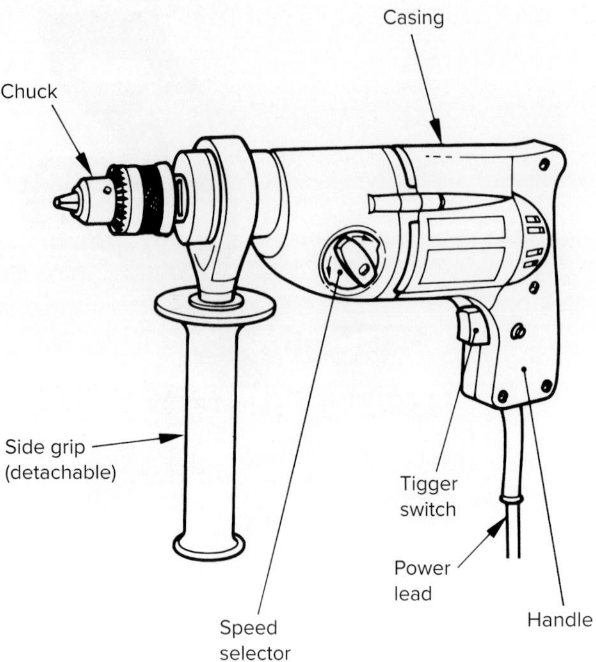

Fig. 9.26 (a) A hammer drill and (b) SDS bit

Hilti (Aust.) Pty Ltd

2300 rpm. This provides a speed suitable for most drilling operations in a variety of materials and a range of diameters.

Another option is the variable-speed drill, which has an electronic control: by exerting pressure on the trigger the speed can be varied. Most 10 mm drills will have a no-load speed of between zero to 1800 rpm and zero to 3500 rpm, and the larger-capacity drills will have a lower speed range. In some instances, this type of drill will have a reversing switch.

9.4.1 Hammer drills

Hammer drills can be used for normal drilling operations but also have a hammer feature, which is selected when drilling into any form of masonry (Fig. 9.26a). Special masonry drill **bits** are necessary. When this feature is selected, not only does the drill turn, but hammer blows are also delivered to the head of the drill bit. This gives it a superior performance when drilling into masonry such as rock, brick and concrete.

A number of hammer drill models are available and the features and capacity vary considerably. Some will have variable speed and reversing capabilities. A typical drill in this class may have a 10 mm chuck and a capacity to drill up to 13 mm in concrete. More powerful drills may have a 13 mm chuck and a capacity for drilling concrete of up to 20 mm in diameter.

Two-speed hammer drills also provide reliable service and are a popular choice. A typical two-speed drill will have a 13 mm chuck and a capacity to drill up to 19 mm in concrete.

Some hammer drills use a system called SDS (Special Direct System). The chucks and bits are of a special design, as shown in Figure 9.26(b). This system is designed for masonry drilling, and provides much better results than traditional straight shanked drill bits that are held rigidly in the chuck. No key is required to change the bits; simply insert the bit, twist and the bit is secured. The bit is loose once secured in the chuck. Some of these drills can also be used for chiselling into masonry as the rotation can be turned off and special chiselling bits can be installed.

Safety rules for the electric drill

The electric drill is not always regarded as a dangerous machine; however, observance of a few simple rules will avoid any possible mishap and many broken drill bits:

1. Hold all work firmly when drilling. Small work should not be held in the hand but clamped in a suitable vice.
2. Beware of the drill 'grabbing', particularly when drilling thin metal, causing it to be wrenched from the operator's hands, possibly causing injury. Always grip the drill firmly with both hands, using the detachable side handle where necessary for maximum grip.
3. Beware of any loose clothing or hair getting caught and wrapped around the drill, particularly when using large-capacity and hence more powerful drills.
4. Use safety goggles when drilling metal, and also use ear protection when drilling masonry or other hard materials.
5. Ensure the chuck is done up tightly and never leave the chuck key in the drill chuck; do not attempt to stop a coasting drill by hand.
6. Allow any coasting of a drill to stop before laying it down.
7. Never force a drill beyond its capacity, and do not run a large capacity drill at low speeds for extended periods or cover up ventilation holes with the hands. Lack of ventilation will cause the drill to overheat.

9.4.2 Accessories for drills

9.4.2.1 Twist drills

Twist drills are probably the most commonly used accessory to the drill, and are used for boring holes into wood, metal and various plastic materials (Fig. 9.27). Drilling machines are usually equipped with either a three-jaw keyless chuck or a chuck that is tightened using a chuck key.

Sizes vary from 1 mm to 25.4 mm in diameter with over 300 sizes in between. The woodworker has no need for this vast range of sizes, and drill sets are made up in a range of sizes to meet most requirements. Standard sets may consist of:

1. 12 drills from 1 mm to 6.5 mm in rises of 0.5 mm
2. 19 drills from 1 mm to 10 mm in rises of 0.5 mm
3. 25 drills from 1 mm to 13 mm in rises of 0.5 mm.

Drills are contained in a metal or plastic case and it is advisable to purchase a set this way to keep the sizes in order. Many sets may be purchased in imperial units and a set may range from 1⁄16" to 1⁄4", 3⁄8" or 1⁄2" in rises of 1⁄64".

Reduced shank drills permit the larger-sized drills to be used in a machine with a smaller chuck capacity, and can work satisfactorily in timber without overloading the machine.

Drilling into metal requires the hole centre to be marked with a centre punch (Fig. 9.28), sometimes included with the drill set. This engages the point of the drill and stops it wandering off-centre.

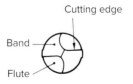

Fig. 9.27 Twist drills

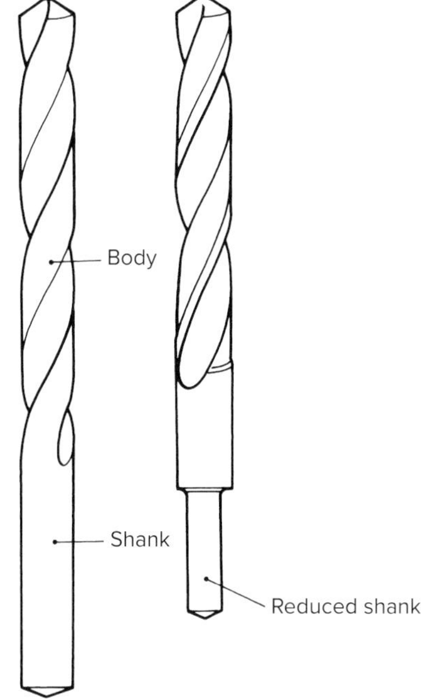

Fig. 9.28 Centre punch

A rough rule, to help select the correct speed on the drilling machine when using twist drill bits, is to use the higher speeds for smaller-diameter drills and the lower speeds for the larger drills.

When drilling, particularly into metal, exert sufficient steady pressure on the drill to keep it cutting, otherwise it will simply rub on the metal, generating heat and blunting the drill. Whenever possible, add a lubricant such as light or soluble oil or metal-cutting fluid, which will help to keep the drill cool and performing efficiently.

Always keep drills sharp and remove any waste that becomes clogged in the flutes.

Fig. 9.29 Dowel bit

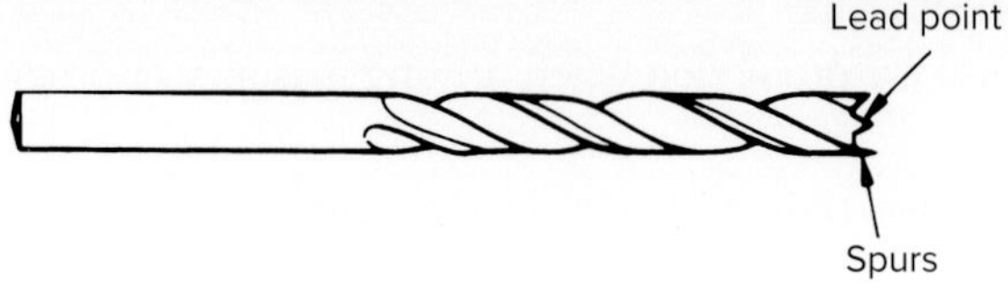

9.4.2.2 Dowel bit

Dowel bits are available in sizes of 3 mm to 13 mm for boring into cross or end grain (Fig. 9.29). They are similar in construction to twist drills but have two spurs and a centre point that prevents them wandering off-centre and following the grain.

9.4.2.3 Spade bit

Machine bits have either a *brad point* (Fig. 9.30) or a *screwed point.* Spade bits are available in sizes of 6 mm to 38 mm. The spade bit has the long brad point, which holds it positively on centre even when boring diagonally to the direction of the grain. There are two cutting edges, and flat surfaces are ground on the shank to be gripped securely in the three-jaw chuck without turning.

Spade bits are suitable for boring clearance holes for bolts and other fittings. They will not cut particularly clean holes and to work efficiently they must be able to rotate at a high speed, up to approximately 2000 rpm. Long spade bits are available up to 400 mm long in sizes from 6 mm to 25 mm.

Never bore a hole right through wood or particleboard with a flat bit as the wood will splinter away from the edge of the hole. When the tip of the bit emerges, turn the work around and bore from the other side.

9.4.2.4 Wood bit extensions

Wood bit extensions are available in 300 mm and 450 mm lengths and can be used to extend the reach of power wood bits over 16 mm diameter (Fig. 9.31). The shank of the standard bit fits into the socket of the extension and is locked with a grub screw. The other end of the extension fits the drill chuck.

9.4.2.5 Machine auger

The machine auger comes in sizes from 6 mm to 32 mm and is used for boring clean holes in timber; it is particularly useful in joinery and cabinet work (Fig. 9.32). The spur cuts the fibres cleanly, and the

Fig. 9.30 Spade bit—point tip (high speed) (a) rough cutter and (b) smooth cutter

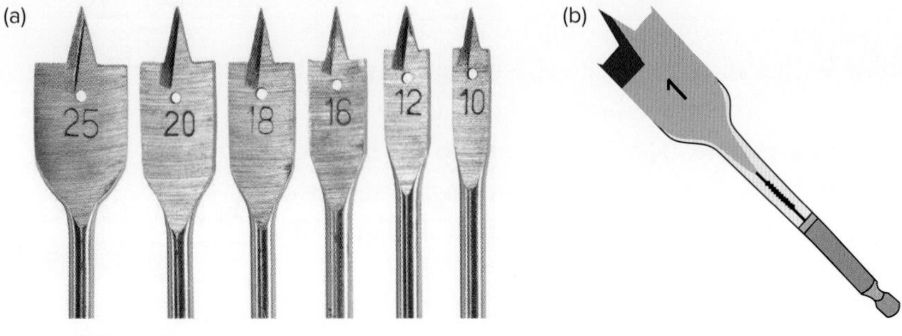

Fig. 9.31 Wood bit extension

Fig. 9.32 Machine auger—screwed point (slow speed)

cutting edge removes the waste. It is operated at low speed, allowing the screwed point to draw the bit into the timber at its own pace. The shank is fluted to be gripped securely in the drill chuck. Normally machine augers are about 150 mm long, but they also can be obtained in a longer series, 400 mm in length, with diameters of 6 mm to 25 mm.

9.4.2.6 Hole saws

Hole saws are used for cutting round holes in wood, particleboard, sheet metal and fibreglass (Fig. 9.33). They consist of a centre drill to which a mandrill is attached. The mandrill contains a number of slots into which various diameters of curved saw blades can be fitted. Blades vary from 19 mm to 63 mm in diameter.

To use the hole saw, set the drilling machine to its slowest speed, mark the centre of the hole, and commence with the centre of the drill, feeding the saw into the timber with a light but firm pressure. Bore the hole from both sides; when boring in thin material, back up the sheet with a timber block and bore right through.

Another pattern of hole saw has its blade made up in the form of a cup that is fitted to an arbor held in the drill chuck. Blades range from 14 mm to 152 mm in diameter. They have a cutting edge of high-speed steel, and when used with a suitable drilling machine can cut a range of metal and wood.

Fig. 9.33 Hole saw

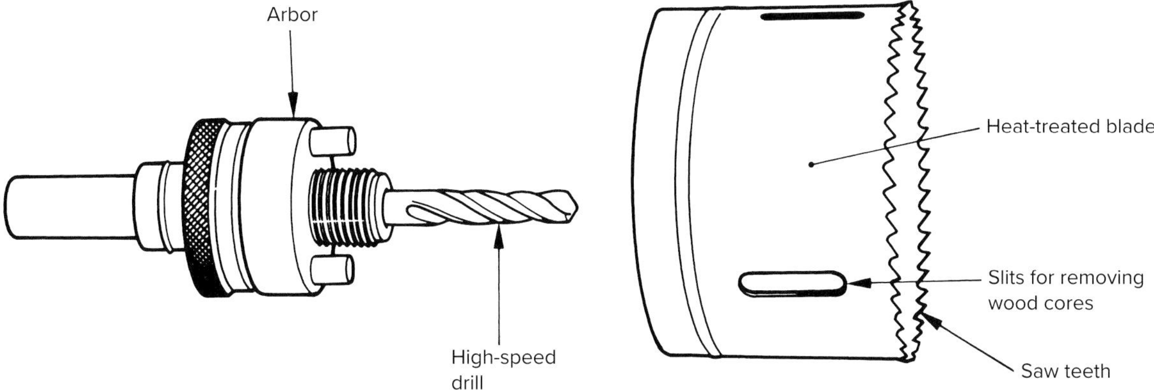

9.4.3 Masonry drills

Masonry drills are used when hammer or rotary drilling in masonry such as brick, concrete, stone, ceramic tile and similar materials (Fig. 9.34). The sizes available range from 3 mm to 25 mm. The larger sizes have a reduced shank that will fit a 10 mm or 13 mm chuck.

The tungsten carbide tip is brazed to a toughened steel shank, and is sharpened to grind rather than cut into the masonry. Operate the drill at a slow speed and exert enough force to keep it cutting, otherwise it will simply rub against the masonry, generating extreme heat and blunting the drill. When drilling deep holes, withdraw the drill occasionally and clear away the spoil.

Fig. 9.34 Masonry drill

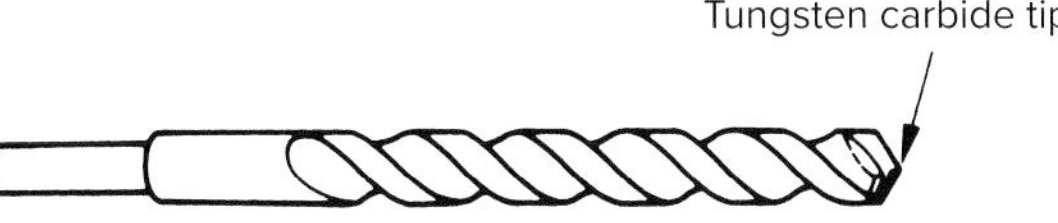

9.5 Portable jig saw

The portable jig saw is used for making closed cuts and circular cuts, and with the aid of a rip fence can be adapted to perform light ripping operations (Fig. 9.35).

The saw blade is approximately 60 mm to 80 mm in length, and moves up and down in a reciprocating action at up to 3700 strokes per minute (Fig. 9.36). Two-speed or variable-speed models permit the selection of a speed more suitable for the job in hand. Some models allow you to change the cutting action from straight (up and down) to an orbital straight line cutting stroke. This option can increase the speed of cut on some materials. Some models come with an inbuilt blower to blow the sawdust away and some models can be attached to a vacuum or extraction system to suck the dust up.

The high speed is selected for cutting timber, while the lower speeds, together with a suitable blade, are selected for cutting metals. Most models are capable of cutting timber up to 50 mm thick and steel up to 6 mm thick.

The cutting is done on the up-stroke and the face of the material being cut should be laid downwards to avoid any chipping out on that surface.

A number of blades are available for the cutting of wood, metals and plastics (Fig. 9.37). They vary in width and in the size of teeth. Consult a selection chart to find out which blade is the most suitable for any particular job. Different makes vary in the way the blade is fitted to the plunger, so it is important to check that the blade is the correct one to fit your machine. Read the manufacturer's instructions to find out.

Fig. 9.35 Jig saw

Fig. 9.36 Cutting action of the jig saw

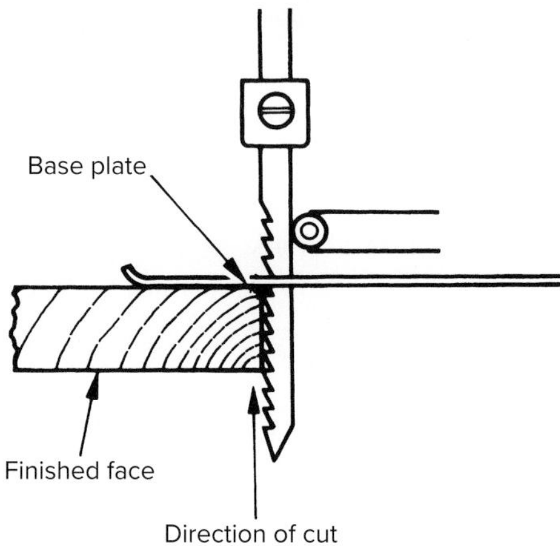

Fig. 9.37 Typical jig saw blades: (a) coarse wood blade; (b) fine wood blade; (c) metal-cutting blade; (d) carbide chip blade

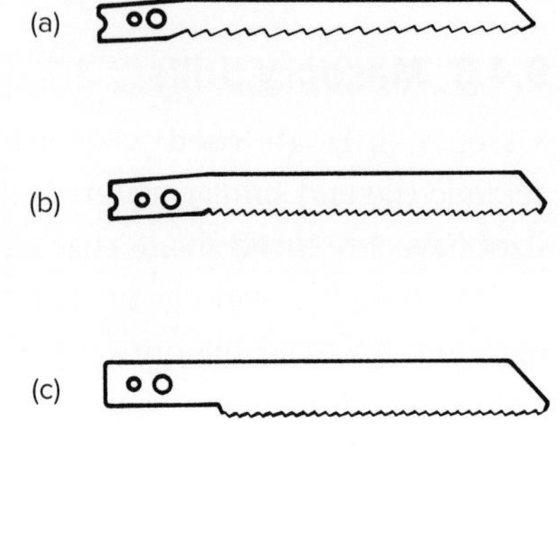

9.5.1 Using the jig saw

The material to be cut must be supported clear of the bench or on stools to allow clearance underneath for the projecting saw blade. Most work should be clamped securely, but light work may be held with one hand and the saw operated with the other.

To make the cut:

1. Grip the saw by the handle and rest the front of the base plate on the material, so that the blade is aligned with the start of the cut (Fig. 9.38).
2. Switch on the motor and, with the base plate held firmly on the material, feed the saw forward, adjusting the rate of feed so that it will cut freely without labouring. Some machines will have a blower behind the blade that will blow away the sawdust, leaving the line to be followed readily visible.

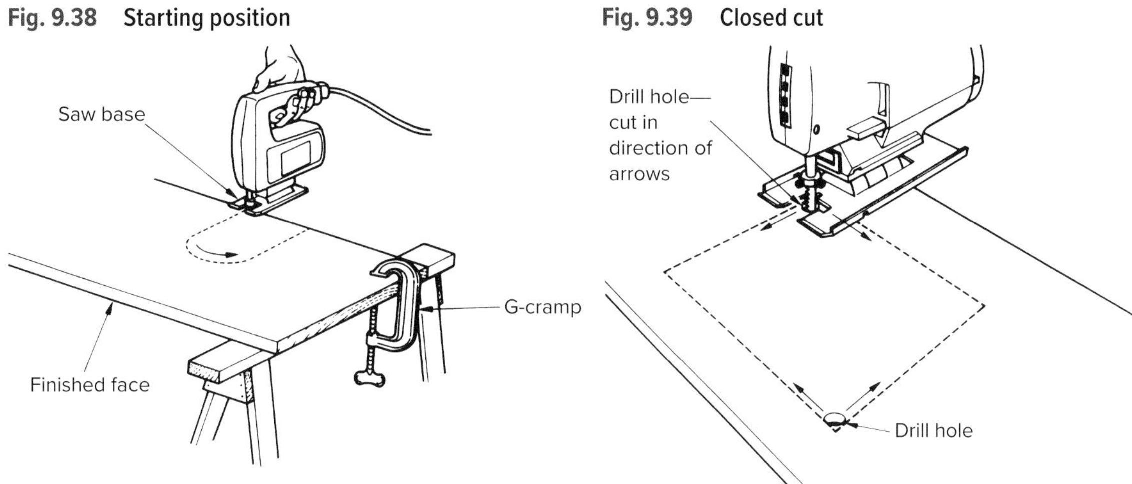

Fig. 9.38 Starting position

Fig. 9.39 Closed cut

3. At the conclusion of the cut, switch off the motor, lift the saw from the surface and lay it on its side with the blade pointed away from the operator.

To make a closed cut, such as a cut-out for a basin or an opening in floorboards (Fig. 9.39), drill a hole large enough for the blade to fit through to start the cut, and proceed as previously described.

Safety with the jig saw

The following safety precautions should always be taken when using the jig saw:

1. Wear safety eye protection to protect the eyes from chips of brittle material that will flick upwards, and ear protection to protect your hearing.
2. Do not place a hand or power cord underneath to support material–the blade is projecting below.

9.6 Portable power planer

The portable power planer is a valuable on-site or workshop machine capable of removing surplus timber at a fast rate, while at the same time producing a smooth planed surface (Fig. 9.40).

The portable power planer has a cutter block to which the cutters are fitted and which rotates at a high speed (Fig. 9.41). The rear base is set flush with the rotating cutters and the front base is adjustable to the depth of cut. Adjustment is made by rotating the front knob; a graduated scale indicates the depth of cut.

Fig. 9.40 Portable power planer

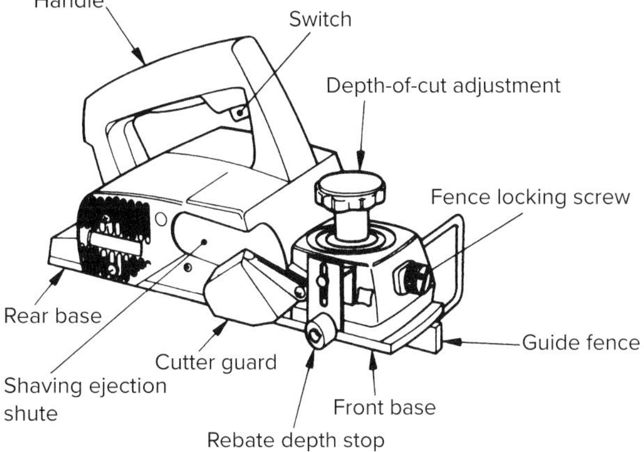

The size of the planer is indicated by the width of cut and is commonly 82 mm. However, planers that can cut up to 155 mm in width are available.

The cutter head contains two blades that are made of high-speed steel or are tungsten carbide tipped. These revolve at 15 000 to 16 000 rpm. A planer revolving at say 15 000 rpm, may be described as making 30 000 cuts per minute (15 000 × 2) (Fig. 9.42).

Fig. 9.41 Cutting action of the power planer

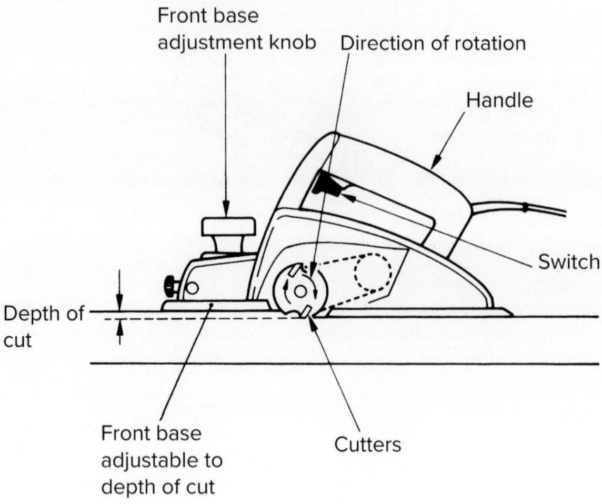

The power planer is frequently used for normal dressing of timber surfaces, fitting doors, and cutting rebates and chamfers.

9.6.1 Using the power planer

To plane straight, flat surfaces, the timber must be held firmly. The action of the planer tends to push the timber forwards, in a manner similar to the hand plane, so the material should be secured in the vice or against a bench stop.

Adjust the planer to the depth of cut. For fast rough planing, the cut can be up to 3 mm deep, but where a fine smooth finish is required on difficult timber with interlocking or curly grain, reduce the depth of cut to no more than 1 mm and advance the planer more slowly. Grip the planer by the two handles provided and lay the front base on the timber at the start of the cut, keeping the cutters clear. Start the motor and allow the machine to reach full speed.

Apply pressure to the front of the planer and advance it forward at a steady rate. Listen to the sound of the motor–if it tends to labour, lessen the rate of feed. When the planer is fully supported by the timber, transfer the pressure equally to the front and rear. Then, as the front base passes over the forward end, apply firm pressure to the rear base to prevent the front dipping down.

Wide boards are planed with a series of parallel passes. At the completion of planing, lift the tool from the surface and switch it off.

9.6.1.1 Planing narrow edges

The guide fence shown in Figure 9.43 will help to hold the planer square to the face of the timber. The guide fence is attached by rods to the side of the planer and can be adjusted and locked in any position across the base. To make the cut, hold the guide fence against the face of the timber and operate the planer as previously explained.

Fig. 9.42 Planing motion

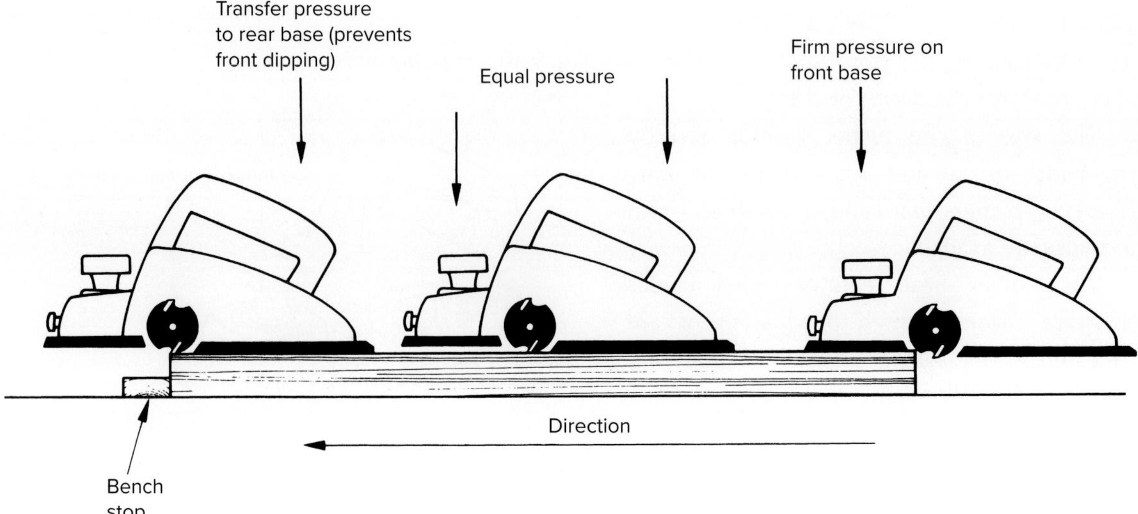

9.6.1.2 Cutting rebates

Rebates are cut by making a series of passes to the required depth. To set up the planer for rebating, three adjustments must be made:

1. Set the *depth of cut* to the amount required to be removed at each pass, 2 mm to 3 mm.
2. Set the *depth of rebate* required, measured from the face of the rear base, by adjusting the depth gauge found on the side of the planer (Fig. 9.44). (The maximum depth of rebate may be up to approximately 20 mm but will vary on different machines.)
3. Set the *width of rebate* required, i.e. the distance required from the guide fence to the edge of the base (Fig. 9.45).

To cut the rebate, grip the planer by the two handles, hold the guide fence against the face of the timber, and operate the planer as previously described, making a series of passes until the depth stop limits the rebate to the pre-set distance.

It is advisable to cut a rebate on scrap timber to check the adjustments before working on the finished timber.

Fig. 9.43 Planing narrow edges

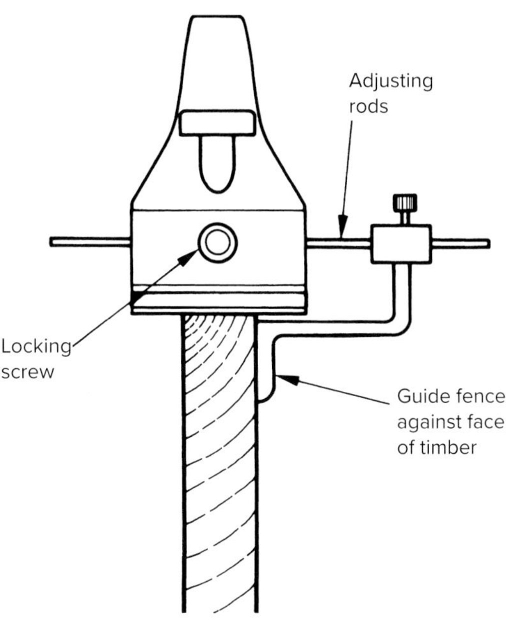

9.6.2 The cutter head

The cutter head is the most important part of the planer. The cutter blades must be sharp and correctly adjusted at all times for maximum efficiency. The design of the cutter head will vary with different manufacturers.

Some form of adjustment will be provided to set the cutter blades correctly or, alternatively, a cutter gauge may be included, as an accessory, to set the blades before assembly to the cutter head. (Refer to the manufacturer's instruction manual for details.) If the adjustment is correct, the cutting circle should align with the rear base (Fig. 9.46).

Fig. 9.45 Setting up for width of rebate

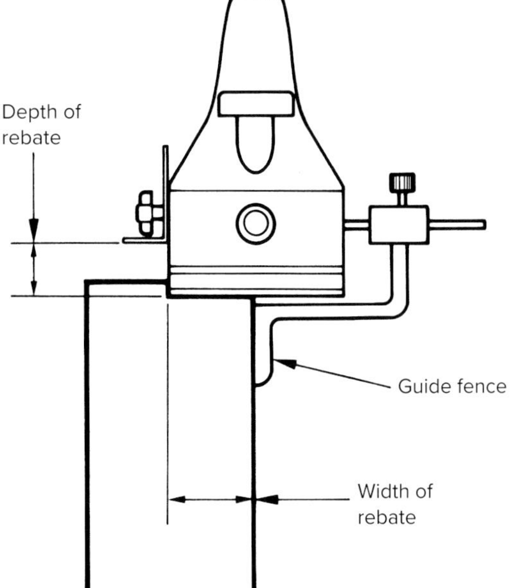

Fig. 9.44 Adjusting the depth for rebate

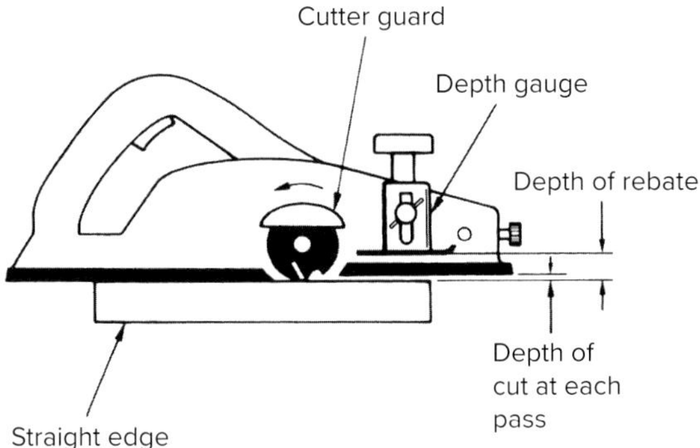

Fig. 9.46 Correct adjustment of cutter blade

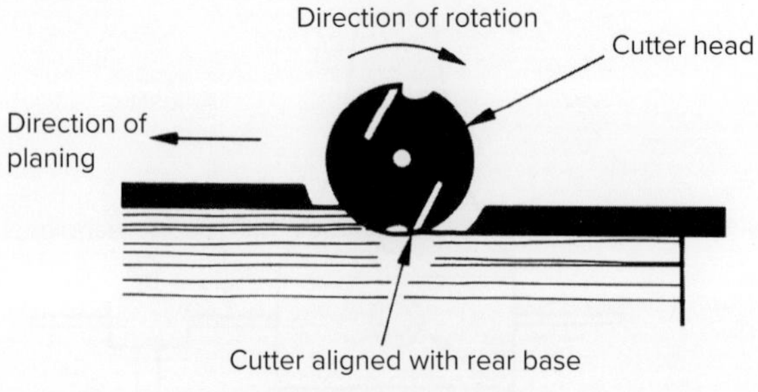

Fig. 9.47 Blade not set deeply enough

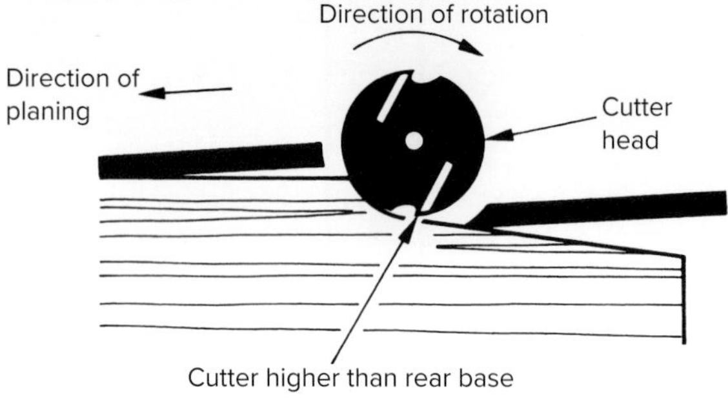

Fig. 9.48 Blades projecting too far

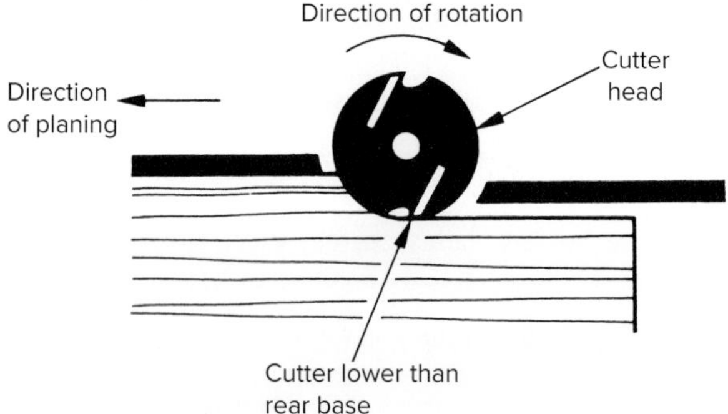

If the blades do not project far enough, they will gouge out at the beginning of the cut, creep uphill and probably not cut at all (Fig. 9.47). Blades that project too far will leave gouge marks along the cut and at the ends (Fig. 9.48).

Safety with the power planer

Always take the following precautions when using the power planer:

1. Disconnect the planer from the power source before attempting to remove the cutter blades or make adjustments.
2. The blades, of course, are exposed across the mouth of the planer. Always grip the tool by the handles provided, and never place the fingers underneath the base.
3. Wait for the machine to stop before placing it on the bench.
4. To protect the blades and the operator when the tool is not in use, lay it across the tool well of the bench or elevate the front on a block of wood and switch off the power.
5. Keep the power lead away from the blades.
6. Always keep the blades sharp and correctly adjusted.
7. Blades will retain their sharp edge for longer if special care is taken to remove any hard obstructions, such as nails and any dirt or grit from the surface, before planing.

CHECK YOUR UNDERSTANDING

1. If the blades fitted to the cutter head of the power planer do not project far enough, what is the result?

9.7 Electric sanders

9.7.1 Portable belt sander

One of the fundamental components of the **belt sander** is a belt of coated abrasive made up into a continuous loop that runs around two rollers–one a drive roller with a rubberised surface and the other a tensioning and tracking roller (Fig. 9.49). On the sole of the sander, the belt passes over a steel plate or platen, backed by a cork-rubber compound plate that provides a flat working surface where the belt comes in contact with the timber.

The size of the machine is indicated by the width of the sanding belt, for example, 75 mm or 100 mm.

With a suitable belt, the sander can be used on wood, paint, plastics or metal. However, only the flat sanding of timber surfaces will be discussed here. The 100 mm belt sander is recommended for this purpose.

Fig. 9.49 Portable belt sander

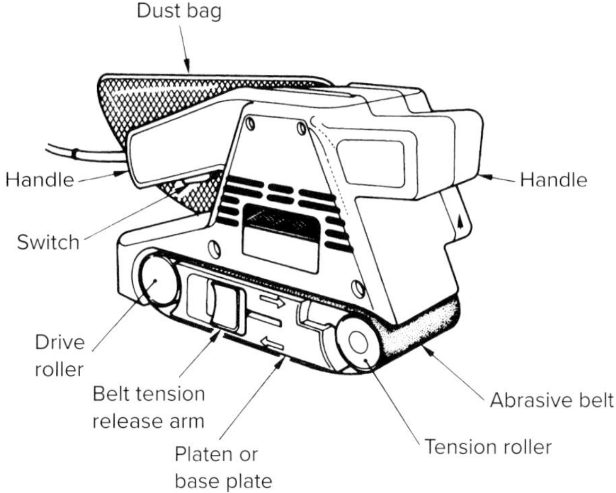

9.7.1.1 Fitting the abrasive belt to the sander

To fit the abrasive belt to the sander, disconnect the tool from the electrical power source and lay the machine on its side (Fig. 9.50). Find the tensioning release arm. Operate this lever to release the spring tension on the front roller. The belt can now be fitted over the two rollers. Return the tension lever to its former position. It is important that the belt travels in the correct direction so that the lapped joint of the belt will not catch and come apart, therefore reducing its working life.

There is an arrow on the belt that indicates its direction of travel, and as the belt passes over the base plate on the sole of the machine, this arrow must be pointing towards the back. To assist in fitting the belt, the machine may have an arrow on one side indicating the direction of travel; the arrow on the belt must point in this direction.

Fig. 9.50 Fitting the belt

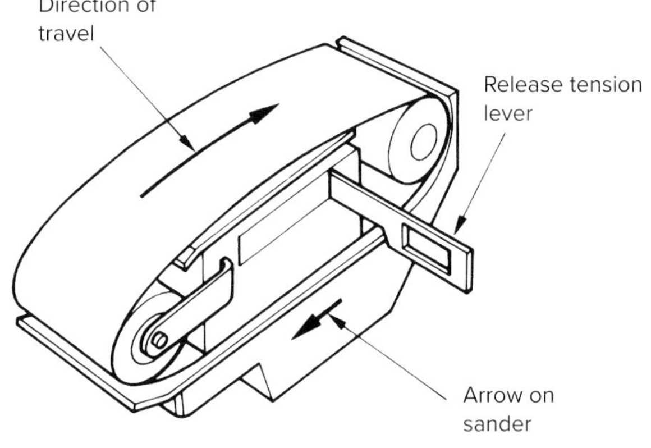

9.7.1.2 Adjusting the belt

To adjust the belt, turn the sander over on its back, connect the power and switch it on in short bursts to observe the way in which the belt is tracking (Fig. 9.51). If it tends to move sideways, use the adjusting knob to correct its position until it runs freely, with the edge flush with the outside edge of the base plate.

Fig. 9.51 Adjusting the belt

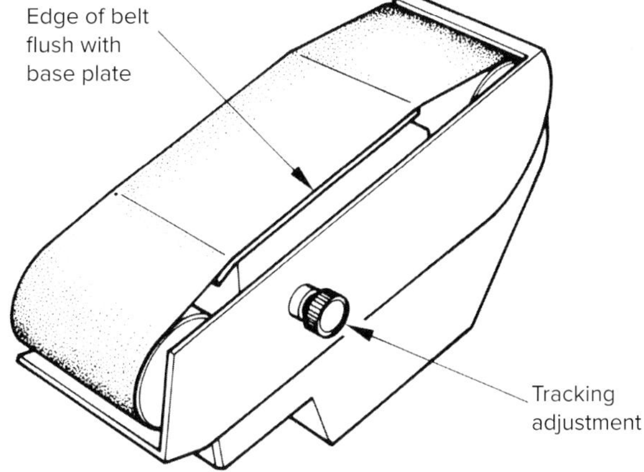

9.7.1.3 Using the belt sander

The material to be sanded must be secured firmly, as when in use the sander will tend to crawl forwards, throwing the timber backwards.

1. Clamp the timber in place or fix a cleat behind the timber to restrict movement (Fig. 9.52).
2. Grip the sander by both hand grips, hold it clear of the material, switch on and allow the motor to reach full speed.

Fig. 9.52 Action of the sander

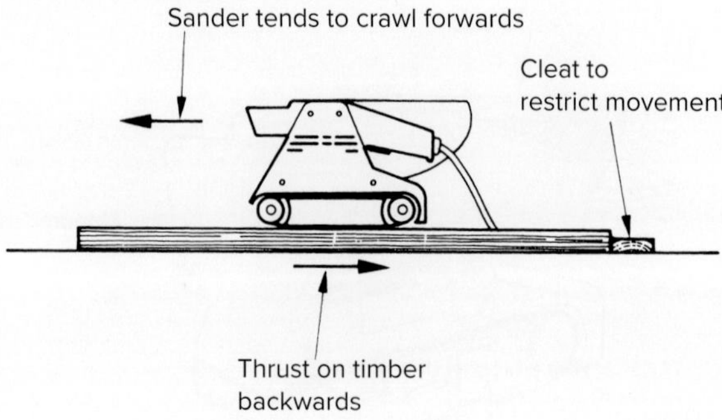

Fig. 9.53 Sanding motion

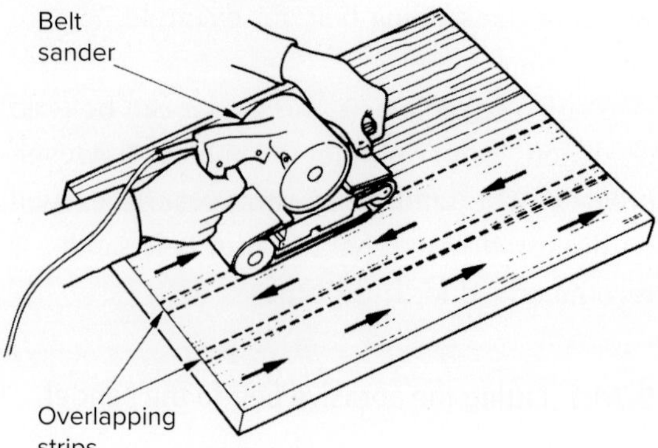

3. Apply the base plate to the surface of the timber with a forward motion and sand to and fro in strips in the direction of the grain, allowing the sander to move along at a steady uniform pace from one end of the surface to the other. Do not apply downward pressure as the mass of the machine is sufficient for efficient cutting.
4. Overlap each strip by approximately one-quarter the width of the belt (Fig. 9.53). Allow the sander to rest evenly on the surface and sand each area for the same amount of time. Do not allow it to dwell in one spot. This will result in a depression that will show up on the finished surface.
5. Lift the sander from the surface before switching off.

The belt sander used with a coarse belt, 60 to 80 grit, is capable of the fast removal of surplus material but it will leave scratch marks. To produce a fine finish to dressed timber, sanding may have to be done in at least two steps. For the first step, use a medium-grade belt, say 100 grit, followed by a fine-grade belt, 120 to 150 grit. These suggestions can be taken as a guide only; each job must be considered separately as it will depend on the species of timber and the standard of finish required.

Safety with the belt sander

Always follow these procedures and safety precautions when using the belt sander:

1. Keep the power lead clear of the abrasive belt at all times. Experience shows that the belt can pick up the lead and jam it against the tool housing.
2. Empty the dustbag regularly during use.
3. Sweep away surplus dust from the path of the sander as it will tend to clog the abrasive belt.
4. Remove any oil or grease from the surface, which will also clog the belt.
5. Always wear ear and eye protection. A dust mask may be appropriate as well.

CHECK YOUR UNDERSTANDING

1. Why is it important for the belt of a portable belt sander to rotate in the correct direction?

9.7.2 Orbital and random orbital sanders

The **orbital and random orbital sanders** consist of a motor housed in a plastic casing (Fig. 9.54). An orbiting base plate, faced with a rubber or Velcro pad to which can be fitted the abrasive paper, is coupled to the motor. Some models may be fitted with a blower and dustbag.

Fig. 9.54 (a) Random orbital sander and (b) orbital sander

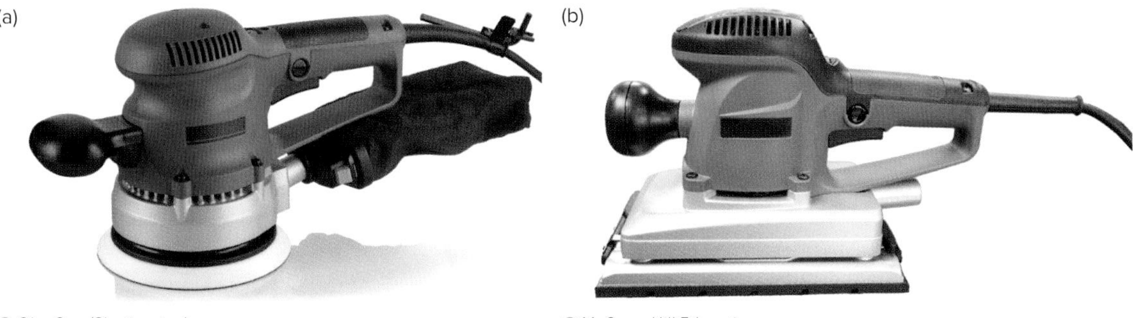

(a) © OlegSam/Shutterstock

(b) © McGraw-Hill Education

To understand the action of the orbital sander, hold the sander upside down, make a pencil dot on the orbiting base, switch on the sander and observe the dot, which will revolve in a small orbit in the same manner as the abrasive particles.

There is a wide variation in the specifications of the considerable number of orbital sanders available.

The size of the orbital sander is commonly expressed as that portion of a standard 280 mm × 230 mm sheet of abrasive paper that will fit the machine. For example, a ½-sheet sander takes abrasive paper of 280 mm × 115 mm and a ⅓-sheet sander takes paper of 230 mm × 93 mm.

The size of the random orbital sander is expressed as the diameter of the disc.

For the orbital sander, the paper must be fitted tightly over the rubber pad and be held by the clips. Refer to the instructions with each individual model for fitting details.

The random orbital is fitted with a Velcro pad, so ensure the disc is centred accurately before turning the machine on.

The orbital and random orbital sander will not remove a great amount of timber, but if used with suitable grades of paper, will produce a fine flat finish.

9.7.2.1 Using orbital-type sanders

Timber to be sanded must be held firmly; otherwise it will tend to oscillate with the orbits of the sander. The timber is sanded in overlapping strips in the direction of the grain (Fig. 9.55). To use the sander, grip with both hands, switch on and place the orbiting pad flat on the face of the timber. Apply a light pressure and slowly move the sander along each strip, allowing the orbiting action of the base plate to do the work.

Fig. 9.55 Sanding procedure

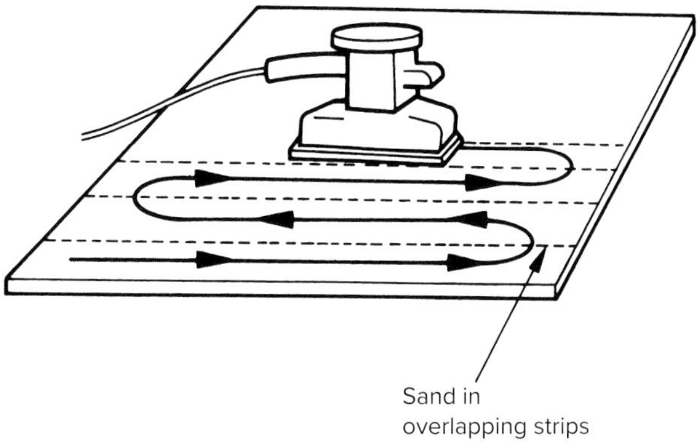

Sometimes, after orbital sanding, fine circular scratch marks can be detected on the surface, particularly if it is stained or lacquered. This is most probably caused by the use of an abrasive paper that is too coarse, and by moving the sander along too quickly. When sanding dressed timber for a stained finish, commence with a medium-grade paper, say 100 grit, and finish with a fine-grade paper, 150 or 180 grit. Commencing with a paper that is too coarse will leave scratches that are difficult to remove, and sometimes a satisfactory finish may be obtained only after a final sanding by hand. Lift the sander clear of the timber before switching off.

9.7.3 Sander safety

Obviously, sanding operations produce a quantity of fine dust and it is no doubt a wise move to use the dustbag to reduce this hazard. Most models can be attached to a shop vacuum to reduce the dust produced. The use of a dust mask is also recommended, and many operators, sensitive to the dust produced by some timbers, will find it essential.

CHECK YOUR UNDERSTANDING

1. What precautions are taken to prevent circular swirl marks showing on a finished surface when using the orbital sander?

9.8 Portable electric router

An electric router can be adapted, with suitable cutters and accessories, to laminate trimming, shape straight or curved edges, and reproduce intricate shapes using a template (or templet) guide.

The **router** consists of a wide base that can be controlled by two conveniently placed handles (Fig. 9.56). Fitted to the base is a high-speed electric motor with an on/off switch. Directly coupled to the motor is a collet-type chuck into which the various cutters or *bits* can be fitted. The height of the motor body can be adjusted within the base, and this is how the depth of the cutter, relative to the base plate, is adjusted. Accessories such as a guide fence and template guides are fitted to the base to perform various operations. The speed of the motor may vary from 24 000 to 30 000 rpm, and the power range is from approximately 750 to 2000 watts.

Fig. 9.56 Electric router

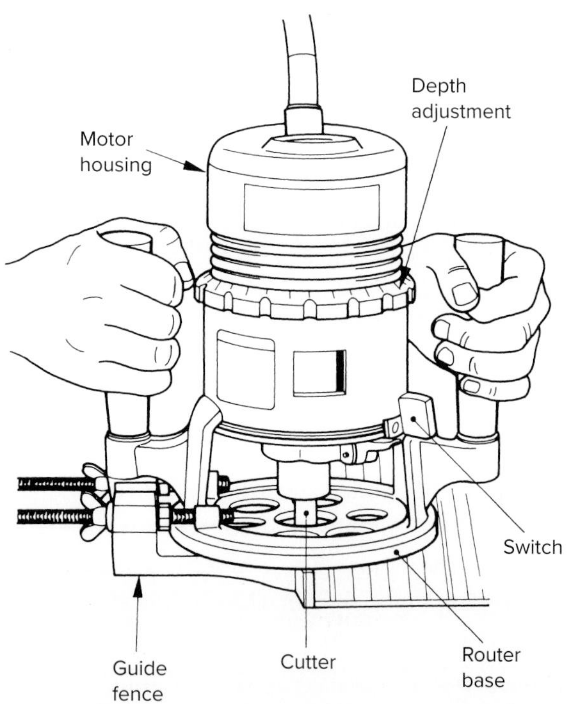

The following description of the router is fundamental to the great number of routers available on the market. However, details of their operation may vary considerably, so always study and follow carefully the manufacturer's instructions when setting up any particular machine for use.

9.8.1 Router cutters or bits

The best way to demonstrate the versatility of the router is to consider the large number of cutters–over 300 types and sizes–that can be adapted to a wide variety of operations. A few examples are illustrated below.

9.8.1.1 Straight bits

Straight bits are most commonly used for cutting rebates, plough grooves, housings, etc. (Fig. 9.57). They are manufactured from high-speed steel with tungsten carbide cutting edges brazed onto the shank. This provides a good working life for the bit.

The shank diameter varies from 6 mm to 12 mm, and many cutters are still available where the shank diameter is given in imperial sizes: ¼" (6.35 mm), ⅜" (9.5 mm) and ½" (12.7 mm).

With suitable adapters, bits can be fitted to most machines. The diameter of cut varies from 2 mm to 38 mm. A very useful set of tungsten-tipped cutters consist of bits that give a diameter of cut of 10, 13, 16 and 18 mm, these being the same as the thicknesses of standard particleboard. Straight cutters can have either a double flute, which will cut more smoothly, or a single flute for the smaller sizes up to 6 mm.

Some of the straight bits are shaped on the point to cut a specific shape, e.g. the core box bit and the V-grooving bit (Fig. 9.58).

When using straight cutters, the router must be provided with some form of guide such as the guide fence, the template guide or the guide batten.

Fig. 9.57 Straight cutters

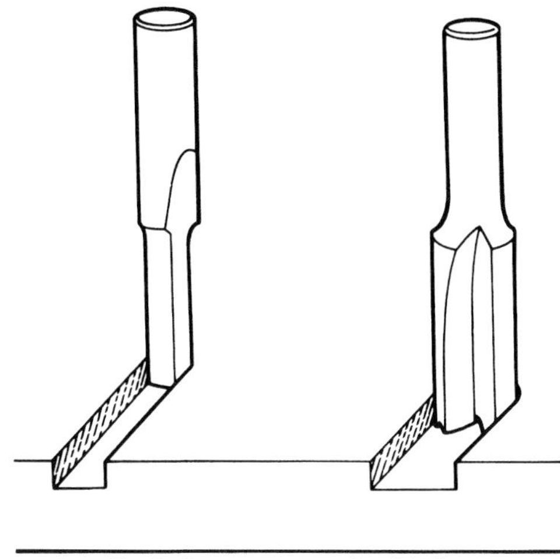

Fig. 9.58 End shapes

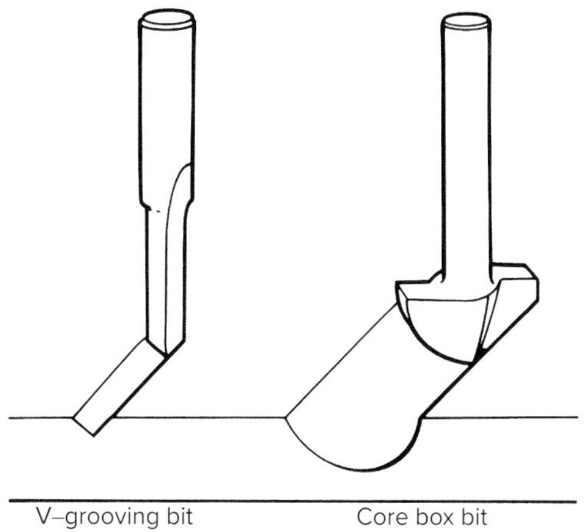

9.8.1.2 Edge-forming bits

Edge-forming bits are a group of bits used to give an edge a specific shape (Fig. 9.59). They include a pilot or ball-bearing that guides the bit along the edge of the work.

The edge must be clean and true, as the pilot will follow any irregularities. The bit must be kept moving along the edge or the solid pilot tip will overheat and burn the timber. A much more satisfactory arrangement is with the ball-bearing guide, which rolls along the edge, guiding the cutter. Examples of ball-bearing edge-forming bits are shown in Fig. 9.60.

9.8.1.3 Trimming bits

The trimming bit consists of an arbor, which fits the router chuck, and on which is mounted

Fig. 9.59 Edge-forming bits with solid pilot

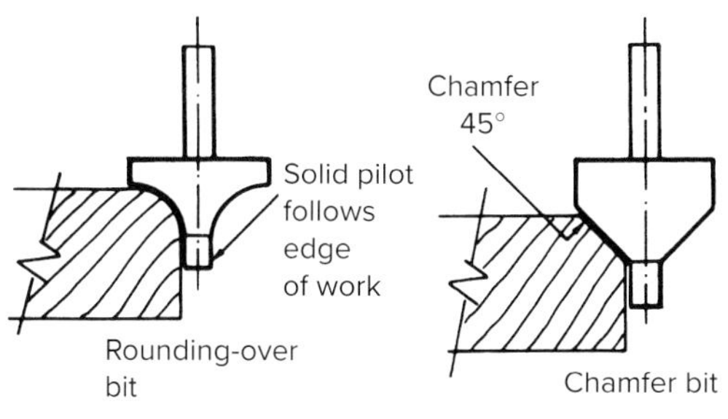

Fig. 9.60 Ball-bearing edge-forming bits

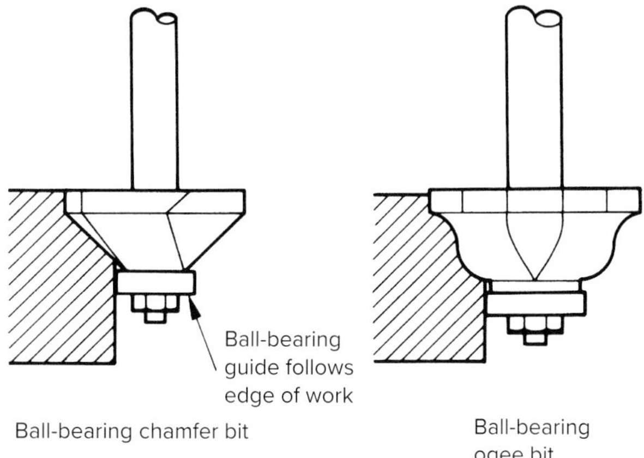

the cutter and a ball-bearing guide (Fig. 9.61). Flush or bevelled edges can be produced with the appropriate cutter.

Fig. 9.61 Trimming bits

Cutter

Arbor

Ball-bearing guide

Ball-bearing guide

Flush trimming bit

Bevel trimming bit

9.8.1.4 Slotting cutters

Slotting **cutters** are manufactured in a limited range of sizes for the repetitive cutting of a groove or slot, usually to the edge of boards. Their design varies–some are fitted with a ball-bearing guide, others are used with a guide fence.

9.8.1.5 Fitting the cutter

Before attempting to fit any cutter to the router, disconnect the machine from the power source.

To unlock or tighten the chuck, it is necessary to lock the shaft to prevent it turning. Some machines may be fitted with a shaft lock; otherwise it is quite common for two spanners to be supplied as accessories, one to hold the shaft and one to turn the chuck.

So that cutters of different shank diameters can be fitted, the router may be supplied with a number of reduction sleeves and/or collet cones (Fig. 9.62).

Fig. 9.62 Reduction sleeves and collet cones

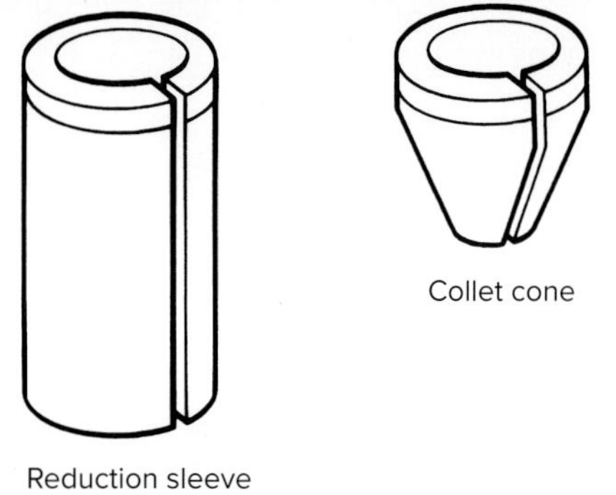

These are available in a range of metric and imperial sizes. If difficulty is experienced in fitting cutters, consult the instruction manual of the machine and check that the correct accessories are being used. When all is correct, insert the shank of the cutter a safe distance, say at least 15 mm, into the chuck, then tighten the chuck.

Fig. 9.63 Setting the depth of the cutter

9.8.2 Adjusting the router

The following adjustments are made when setting up the router for use:

1. To set the depth of cut, unlock the motor body clamp and move the motor up or down using the means provided for adjusting (Fig. 9.63). The calibrated dial helps to make fine adjustments; however, it is often more reliable to check the depth of cut from the sole plate. When correct, tighten the motor clamp.
2. To set the guide fence, fit the guide rod together with the fence to the side of the router (Fig. 9.64). Slide the fence along the rods to the required distance from the cutter, using the adjusting screws to make the final setting, and then lock by tightening the wing nuts.

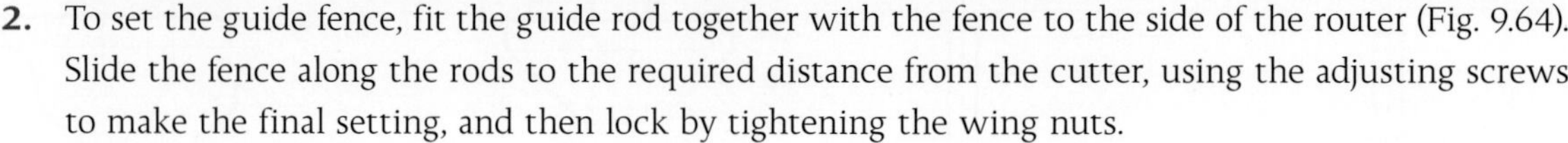

Make a trial cut on a piece of waste timber to check the settings before proceeding to the finished work.

9.8.3 Operating the router

Grip the router firmly using the two hand grips, switch on the motor and allow it to reach full speed. Lay the front portion of the base plate on the timber and slowly feed the cutter into the timber. Proceed with the cut, moving from left to right against the rotation of the cutter (Fig. 9.65).

The rate of feed is best controlled by listening to the sound of the motor–if it tends to labour, slow the rate of feed. It should never slow down to the point where it is almost stationary as this will generate heat and burn the cutter, the motor and the timber. At the conclusion of the cut, withdraw the router from the timber and switch it off.

To prevent the timber splitting away at the end of the cut when cutting completely across end grain, make a short cut, against the direction of feed, where the main cut will finish. Return to the start and make the main cut moving from left to right (Fig. 9.66).

9.8.4 Curved guide fence

The router can be made to follow around a convex edge (e.g. on a round table) by using the curved guide fence (Fig. 9.67).

Fig. 9.64 Setting the fence

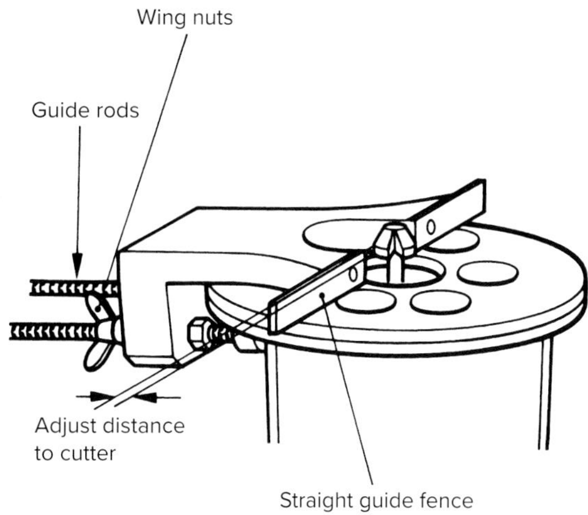

Fig. 9.65 Motion of cutting

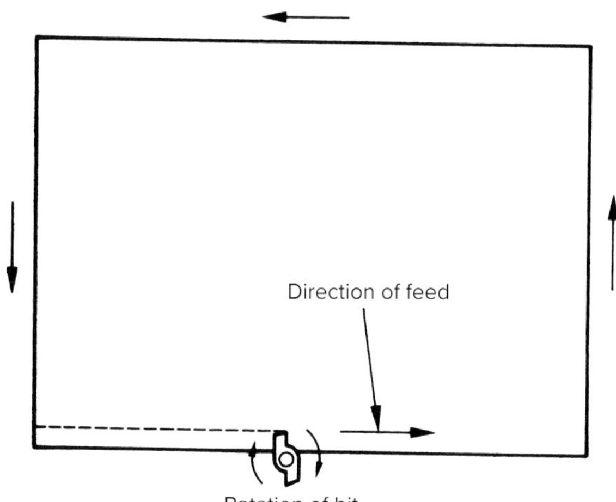

Fig. 9.66 Cutting end grain

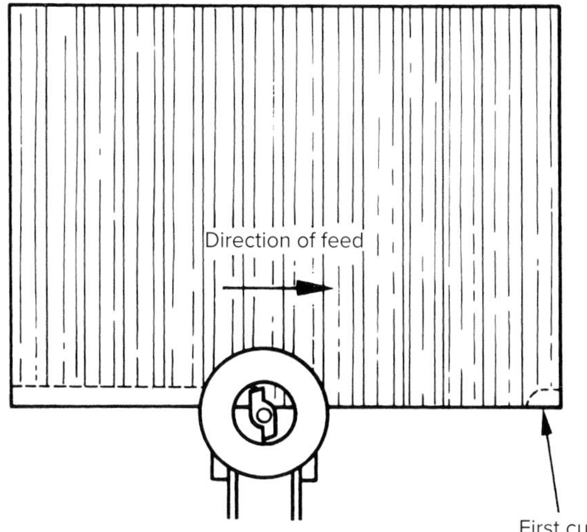

Fig. 9.67 Curved guide fence

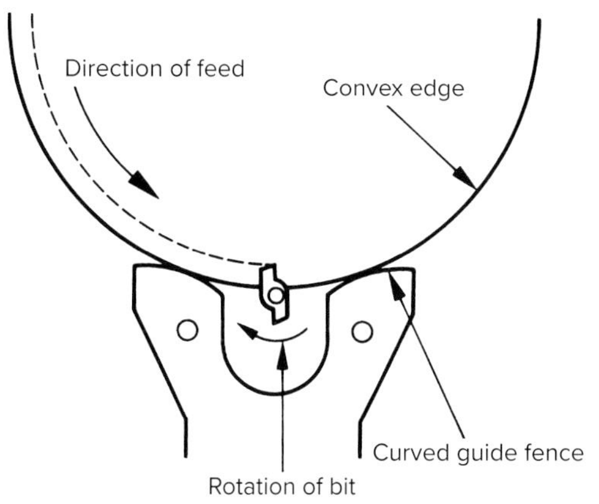

Setting the position of the curved fence to obtain the required width of cut is largely a matter of trial and error. Proceed slowly, making small adjustments, and constantly check until the required width is obtained. Make the cut as previously described, again working from left to right.

Fig. 9.68 Template guide

9.8.5 Template guide

Where a number of similar shapes are to be duplicated, a template of the shape can be prepared, and this shape can be copied by the router using the template guide (Fig. 9.68). The guide consists of a metal disc with a tubular collar that projects below the base plate.

In use, the collar follows the template, and the bit passing through the collar makes the cut.

Template guides can be obtained in a number of sizes for various machines. When marking out the template, compensate for 'A'–the distance from the edge of the cutter to the outside face of the collar (Fig. 9.69).

Fig. 9.69 Action of template guide

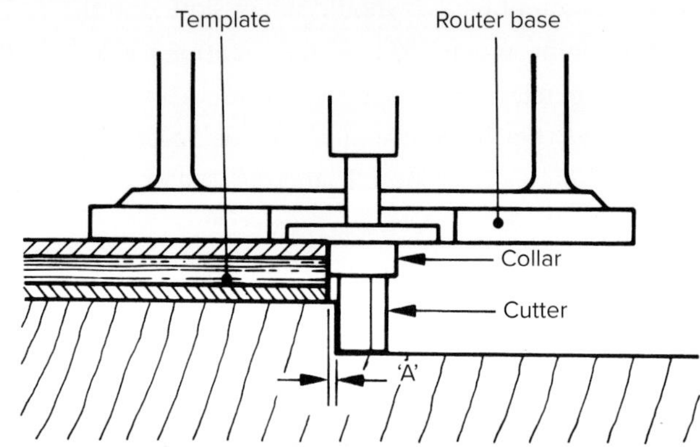

CHECK YOUR UNDERSTANDING

1. In which direction should the router be moved when it is in operation? Why?

9.8.6 Cutting housings

Cutting housings, often used in the assembly of cupboard shelving and so on, can be made using a batten clamped across the material to act as a guide for the router base plate. If the guide is made up in the form of a T-square, the first cut made will set the distance from the cutter to the edge of the router base; the guide can then be quickly set on the mark to produce further housings (Fig. 9.70).

To cut a housing wider than the width of the single cutter, clamp two battens across the material (Fig. 9.71). Follow these to cut the width of the housing, and guide the router freehand to remove any superfluous material in between.

Fig. 9.70 Using a T-square batten guide

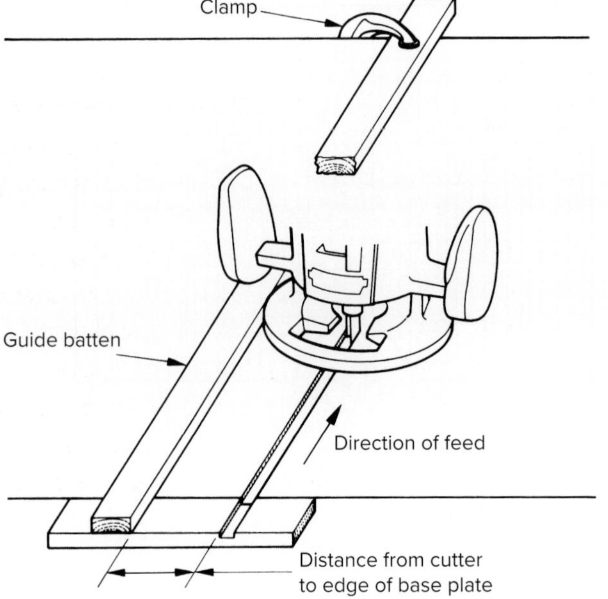

9.8.7 Cutting parallel to an edge

Use the guide fence to follow the edge and guide the cutter (Fig. 9.72). If the edge is narrow, provide added support for the router by clamping a support block to the material.

9.8.8 Cutting to an internal corner

Cutting to an internal corner can be carried out using a corner guide, which consists of a right-angled block of wood screwed to the face of the guide fence. Adjust the corner guide up to the cutter so that there is equal projection on both sides.

9.8.9 Plunge router

On the plunge router, the motor is supported above the base on two spring-loaded rods; a lever is set close to the hand of the operator to lock the motor in position (Fig. 9.73). Located on the base are usually three depth stops, which can be pre-set to control the depth of cut or 'plunge' of the router.

The plunge design is most convenient for use on repetitive work, particularly if areas to be cut are of a different depth, or if deep router cuts are to be made in light progressive stages to avoid overloading the motor.

To use the plunge router, unlock the motor clamp; the motor will lift by spring action to its full height. Rest the base plate on the material and set the first stop to the required depth of cut. Switch on the motor and press down. The cutter will enter the timber and stop at the pre-set depth. Lock the motor clamp and proceed to cut the material over the required area to this depth. Unlock the motor clamp; the motor will rise. Flick the depth stop around to the next pre-set depth, press the motor down again and continue the cutting process. Repeat the operation with the further stop as required and, at the completion of the work, unlock the motor to allow it to rise clear of the material, then switch it off.

The plunge router can be used with the standard accessories, i.e. guide fence, template guide, etc.

Fig. 9.71 Cutting wide housings

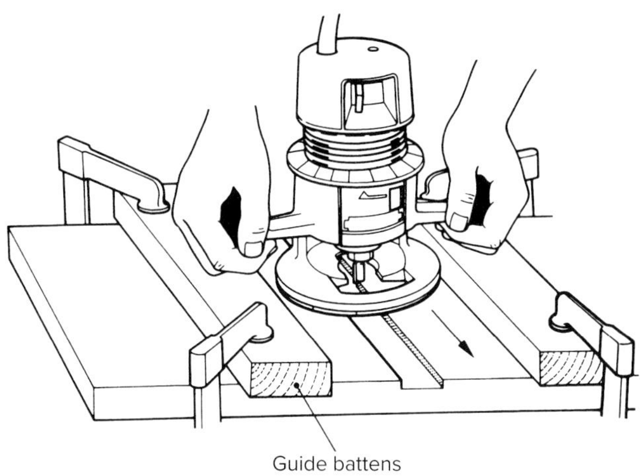

Fig. 9.72 Cutting parallel to an edge

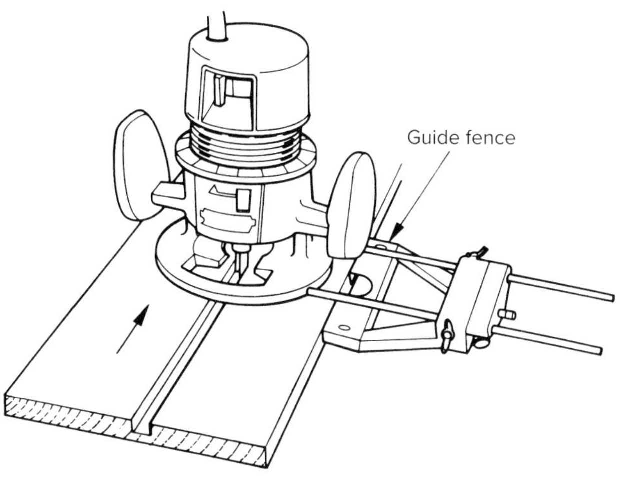

Fig. 9.73 Plunge router

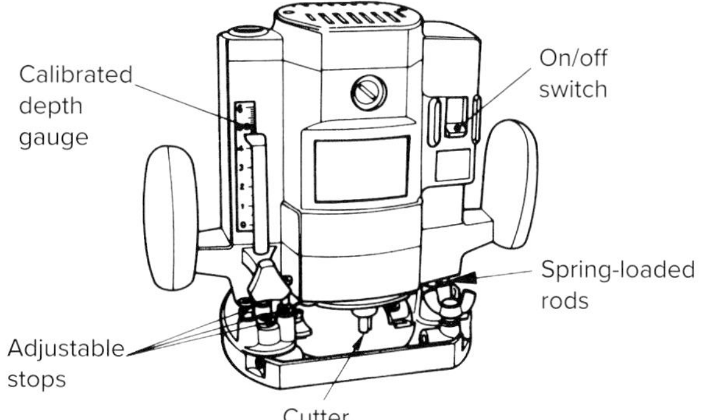

Safety with the router

1. Disconnect the machine from the power source before fitting cutters.
2. Use only correctly designed and manufactured router bits. Never attempt to fit any other type of drill bit or cutter to the machine.
3. Ensure that the chuck and adjusting screws are securely tightened before use.
4. Wear safety goggles to protect eyes from splinters or chips and ear protection to protect your hearing.
5. Hold the router firmly when switching on. The initial torque may twist it out of the hands.
6. Keep cutter clear of the timber when switching on and off.
7. Hold the router by the two handles provided during use.

9.9 Battery-powered tools

All of the tools detailed in this chapter are available in battery-powered versions (Fig. 9.74). Using the most up-to-date battery technology, these tools can perform as well as their corded equivalent in most applications. The advantage battery-powered tools have over corded tools is their portability and the

Fig. 9.74 **Battery-powered: (a) drill; (b) circular saw; (c) jig saw; (d) sander; (e) impact driver; (f) demolition/reciprocating saw; (g) multi-tool; (h) angle grinder**

(a)

Courtesy of Makita Australia

(b)

© McGraw-Hill Education

(c)

© Ansis Klucis/Shutterstock

(d)

Courtesy of Makita Australia

(e)

Courtesy of Makita Australia

(f)

© Derek Hatfield/Shutterstock

(g)

Courtesy of Makita Australia

(h)

Courtesy of Makita Australia

tool itself not requiring testing and tagging every three months. Only the battery charger needs to be tested and tagged.

Additional battery-powered tools include:

- impact driver (Fig. 9.74e)
- demolition/reciprocating saw (Fig. 9.74f)
- multi-tool (Fig. 9.74g)
- angle grinder (Fig. 9.74h).

9.10 Pneumatic tools and nail guns

9.10.1 Nail guns

'Nail guns' or nailers, are available in a wide variety of models to suit particular applications.

The models illustrated in Figure 9.75 are most adaptable to the assembly of house framing as they have the capacity to drive nails from 50 mm to 90 mm in length.

The power source for the pneumatic nailer is compressed air supplied via a flexible hose connected to a portable air compressor. The 'Impulse' nailer power source is a disposable gas canister and battery inserted into the tool.

Fig. 9.75 Pneumatic nailers: (a) Senco; (b) Paslode

(a)

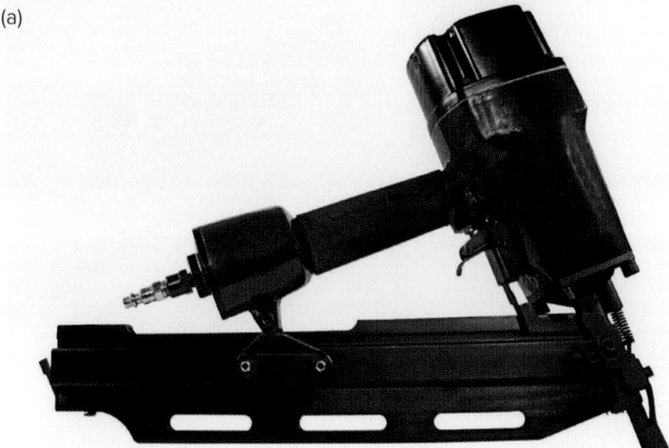

© Dusty Cline/Shutterstock

(b)

Courtesy of Paslode (a division of ITW Australia Pty Ltd)

The volume of air for the pneumatic nailer must be sufficient to maintain a continuous flow. It is measured in cubic metres per second (m^3/s) or litres per second (L/s).

Ensure the compressor being used will supply the sufficient volume of air needed for correct operation. The compressor must be fitted with a pressure regulator that can be set to supply the air at a constant and correct operating pressure.

To meet safety standards, some nailers are fitted with a safety catch that will not allow tool operation unless the safety catch attached on the nailer is pressed firmly against the surface of the timber. (This is also known as a 'trigger release', 'work contact' or 'safety attachment'.)

To use the nailer, first set the correct force (or depth of drive) for that nail to be driven, and then simply press the point against the timber in the direction in which the nail is to be driven and squeeze or pull the trigger. The compressed air will operate the nailer, driving the nail and feeding the next nail into position ready to repeat the cycle.

SAFETY TIPS Nailers

1. Never put your hand behind the timber you are nailing, in case the nail protrudes through.
2. Always wear ear and eye protection.
3. Always keep your hands clear of the tool nose piece.
4. Always know what you are nailing into.

CHECK YOUR UNDERSTANDING

1. What safety precautions are built into the pneumatic nailer?

9.10.2 Air compressors

An air compressor is a mechanical device either powered by electricity or an internal combustion motor that draws air from the atmosphere, compressing and pressurising the air, storing the air in a tank, which can be released in quick bursts to power various tools (Fig. 9.76).

Tips for using an air compressor:

- Always check there is sufficient oil in the sump of the compressor.
- Always ensure a water filter is installed in the outgoing air supply, so any moisture in the compressed air is removed before it gets to the tool.
- Drain the tank of water periodically.
- Never run the tool over its recommended operating pressure.

Fig. 9.76 Air compressor

© yevgeniy11/Shutterstock

CHECK YOUR UNDERSTANDING

1. What liquid should be checked in an air compressor to ensure it runs smoothly? What does it do?

9.11 Angle grinders

Angle grinders are versatile tools that can cut or grind numerous materials using dedicated blades. Each type of blade is generally suitable for cutting only one type of material and can become a hazard if used to cut or grind an incompatible material.

Materials that can be cut or ground include:

- metal
- masonry
- plastic.

The tool is not limited to cutting materials only–it can also be used to sand timber, buff metal or paintwork and remove paint or rust using a wire brush (Fig. 9.77 and 9.78). Special accessories are produced for the angle grinder for these applications.

The two sizes commonly found on a building site are 100 mm or 225 mm; this is the size of the blade the machine will take. However, the 225 mm model is becoming less common as it is extremely dangerous and has caused many serious injuries.

The tool has guarding around the disk that directs the sparks away from the operator when cutting steel. It has a trigger or switch operated by one hand and a handle closer to the blade for the other hand.

SAFETY TIPS

Always remember that not all of the blade is guarded, so the operator must always be aware of the position of the blade/disc in relation to their body.

Wear appropriate PPE at all times when using an angle grinder—some of the hazards encountered when using this tool include:

- kickback
- sparks
- dust
- discs shattering
- loud noise
- excessive vibration.

The necessity to change the discs/blades of an angle grinder can vary depending on the model and brand, so be sure to read the manufacturer's instructions prior to using the tool.

Angle grinder safety tips

1. Inspect discs for cracks or chips and replace if needed.
2. Wear appropriate PPE.
3. Grip the machine firmly and ensure a comfortable stance.
4. Start the machine and allow it to reach full speed.
5. Start cutting, maintain consistent pressure and make sure the motor is not slowing down when cutting.
6. Complete the cut, lift the machine and allow the blade to stop rotating before putting the machine down.

Fig. 9.77 Angle grinder with sanding disc

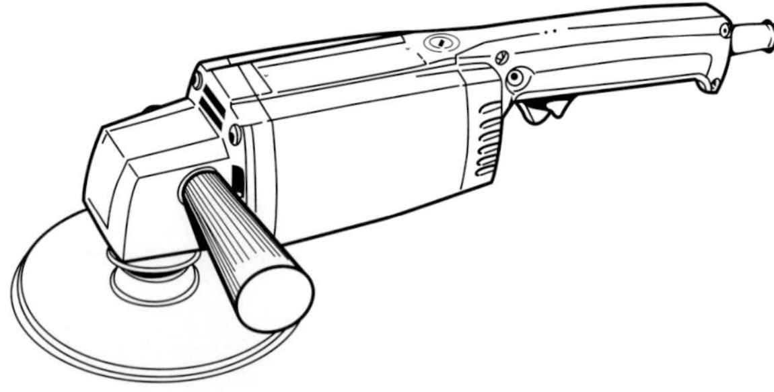

Fig. 9.78 Angle grinder, grinding metal

CHECK YOUR UNDERSTANDING

1. What materials can an angle grinder cut?

9.12 Drill press

The drill press is a very useful workshop machine (Fig. 9.79). Its main function is to drill accurate holes into various types of materials, including:

- wood
- metal
- plastic.

General features of a drill press include:

- adjustable depth stop
- adjustable table height
- multiple speed settings.

When drilling small or metal items, a drill press vice can be used to secure the item. Before using this tool, ensure you don't have any loose clothing, long hair or anything dangling as this can get caught when the chuck is spinning. Eye protection is required to stop chips, swarf and dust from getting into the eyes. When using the drill press, ensure the material being drilled is secured adequately and that the appropriate speed has been selected for the size of drill bit being used and the type of material.

Fig. 9.79 A & B Drill press

(a)

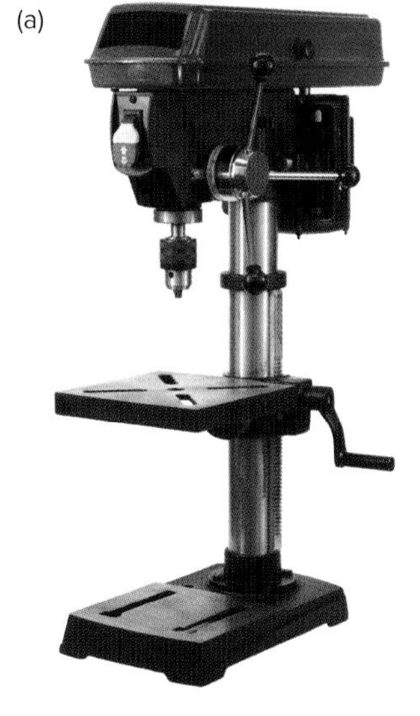

© Photoexpert/Shutterstock

(b)

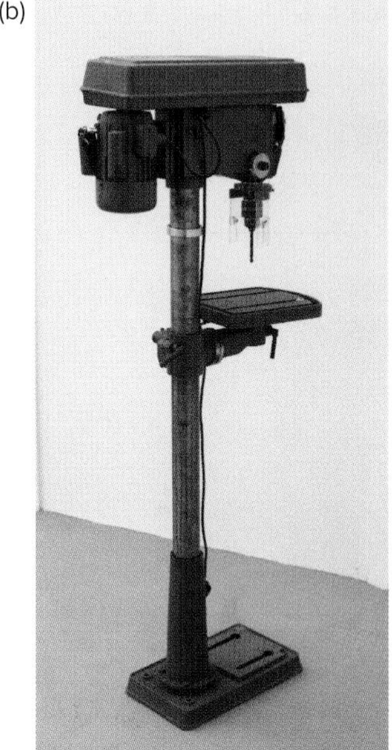

© alan payne/Shutterstock

CHECK YOUR UNDERSTANDING

1. List three features of a drill press.

9.13 Load-handling equipment

This equipment makes it easier to move heavy objects around. There are various types, including:

- hand trolley
- pallet jack
- pallet stacker
- panel trolley.

Load-handling equipment protects workers' bodies from injury as lifting and moving heavy objects manually can cause muscular and skeletal injuries.

Fig. 9.80 Hand trolley

© gualtiero boffi/Shutterstock

Fig. 9.81 Pallet jack

© Baloncici/Shutterstock

Student research

There are other tools available in the building industry that have not been mentioned in this chapter. Investigate what these are. Compare your findings with those of your classmates.

End of chapter activity

Activity 1:

Try building a saw stool. These are an essential item for on-site work. A well-constructed saw stool will last for years.

Fig. 9.82 Saw stool

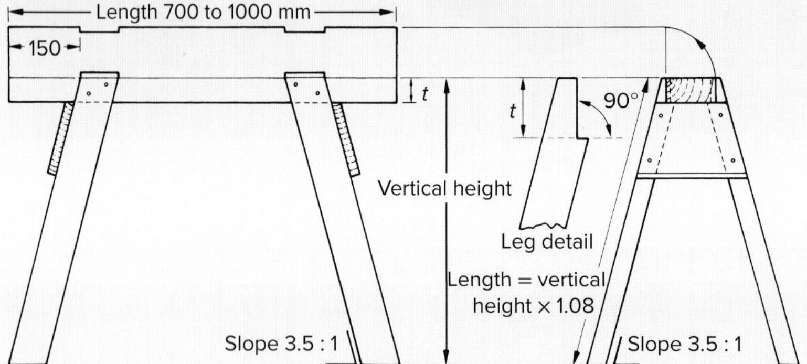

Approximate dimensions

Top	100 × 50
Legs	75 × 38
Cleats	150 × 25

AUSTRALIAN BUILDING CODES AND STANDARDS

AS/NZS 62841.4.1: 2018
Electric Motor-Operated Hand-Held Tools, Transportable Tools and Lawn and Garden Machinery—Safety—Part 4.1: Particular Requirements for Chain Saws

Tradie talk—Don't strangle your hammer

Chapter 10

Site setting out and basic levelling

Learning Objectives

LO 10.1 Identify setting out equipment

LO 10.2 Identify and use levelling equipment

LO 10.3 Set out on-site

LO 10.4 Set out on sloping sites

LO 10.5 Set out screeds for concrete slab

Introduction

Before starting to set out any building it is essential to define the exact boundaries of the building allotment. This can be done only by a registered surveyor who will make a 'peg out' survey, which consists of locating or relocating the boundaries and marking them with a peg or mark on a fence, for example. The surveyor will issue a survey certificate showing a plan of the allotment, where the boundary marks are to be found and how they can be identified. If requested, he or she can also provide further information, such as the levels of the land relative to adjacent streets and drainage services.

The setting out and levelling of a building is the most important part of the building process. The fact that it is the first task done in the construction process means any mistakes made at this point will be carried throughout the entire construction. It is essential that the building is set out accurately in line with the drawings and that the base structure is square and level. If these tasks are achieved, the construction process will be much smoother.

It is crucial to have a clear site when setting out, as any obstruction could pose a tripping hazard or impede the setting out of the building. Maintaining a tidy site throughout the construction process is also important as it will help with site safety as well as ensuring good access when building. Finally, once the setting out has been completed, it is important to ensure the equipment is cleaned and stored appropriately.

In this chapter, you will learn about the various types of equipment used for setting out and levelling a structure on flat and sloping sites. You will also learn about the techniques used when levelling and setting out structures and how to set up concrete screeds.

10.1 Setting out equipment

The position of the building on the site is indicated by setting up **profiles** and string lines. Profiles may be saddles, hurdles or a continuous profile.

10.1.1 Saddles

Saddles consist of hardwood pegs about 400 mm long, with a cross member or ledger approximately 600 mm long (Fig. 10.1). They are economical in the use of material and are favoured on sites that are level or very nearly so.

Fig. 10.1 Saddle

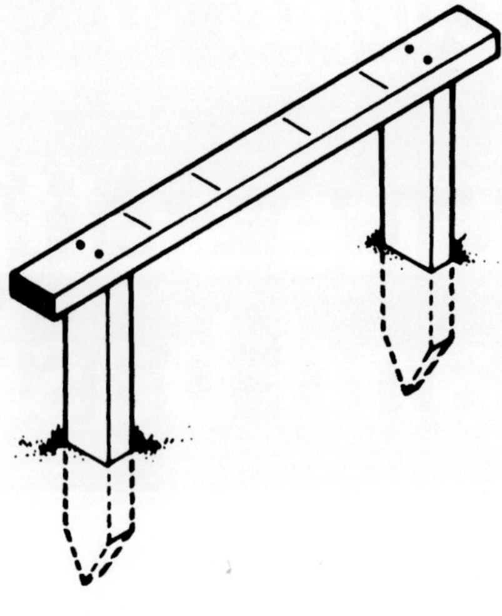

10.1.2 Hurdles

The cross member of a hurdle can be adjusted in height (Fig. 10.2). In general, hurdles are more suitable on a sloping site where all of the cross members can be set level. In some cases, very tall hurdles may require some diagonal bracing to keep them rigid.

Fig. 10.2 Hurdle

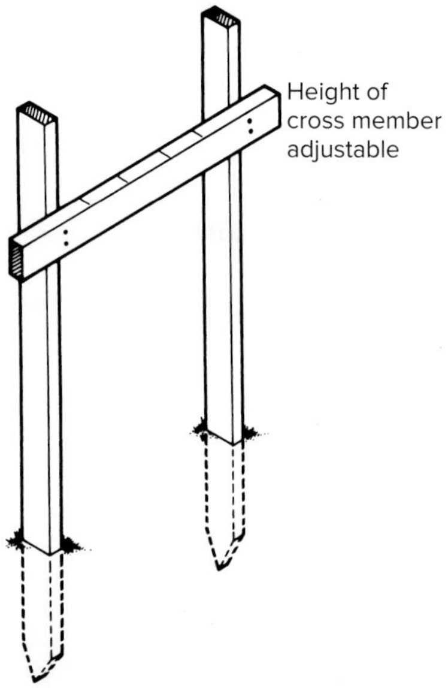

10.1.3 Continuous profile

In some areas, builders prefer to set up a continuous profile where a level cross member is set up completely around the perimeter of the building at a convenient working distance (Fig. 10.3). String lines can be readily set up on the level at any point and plumbed down to the sloping ground line.

Other requirements for setting out include a heavy hammer for driving pegs, string lines, a long steel tape, plumb bob and a spirit level. A large builder's square can be made of timber, where a triangle is set up in the ratio of 3,4,5 to form a right angle (Fig. 10.4).

Fig. 10.3 Continuous profile

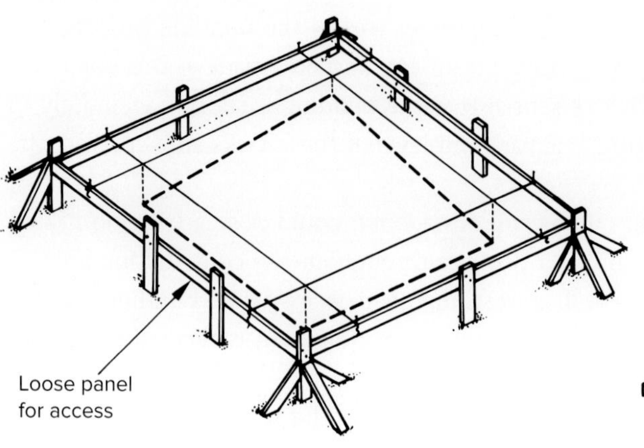

Fig. 10.4 Builder's square

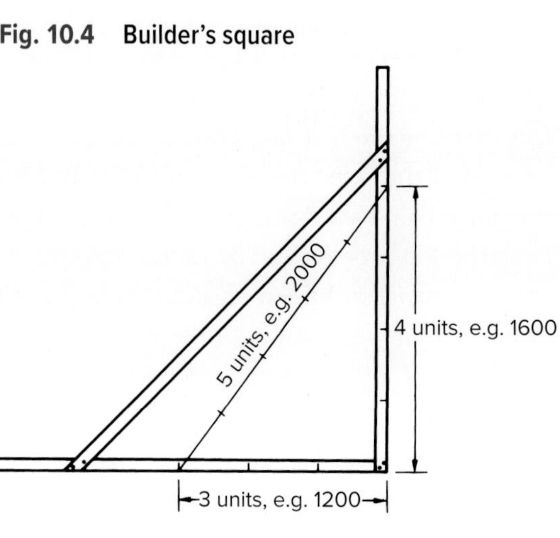

CHECK YOUR UNDERSTANDING

1. Why should a 'peg out' survey be carried out before any building work commences?
2. What are three types of profiles used to indicate building outlines?

WORKPLACE SCENARIO

When setting out any building, it is essential to ensure that it is set out in the correct position and on the correct block. There have been instances of houses being built on the wrong block in a new area, which is usually down to the carpenter not checking to make sure they are on the right site from the plans, not checking lot numbers or street names, etc. It doesn't take long to double check, but it can be costly if you build on someone else's property.

10.2 Levelling

Accurate levelling is one of the most basic requirements of building construction. Floor levels must be set out so that they lie on a true horizontal plane at the correct height, as specified by the building designer.

A horizontal line, commonly referred to as a *level line,* is one that is at a tangent to the curvature of the earth's surface at any given point (Fig. 10.5). Perhaps the simplest practical example of a horizontal line is to imagine a line lying on the surface of still water.

A vertical line is at right angles to a horizontal line and is usually referred to by the trade term *plumb line*. Other terms used in levelling procedures are as follows.

Fig. 10.5 Horizontal and vertical levelling lines

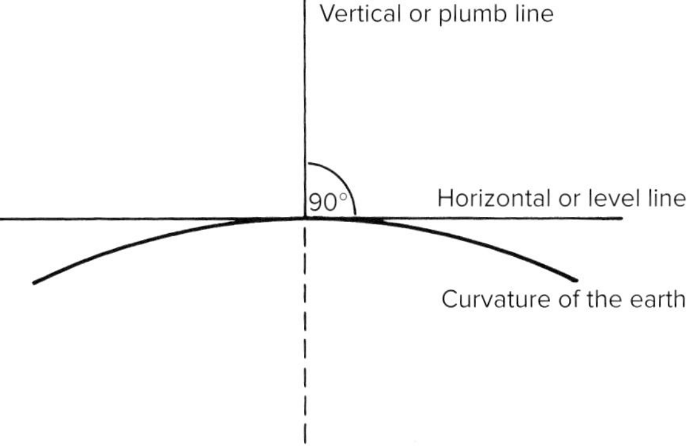

10.2.1 Bench mark

The bench mark is a fixed point of reference, the elevation of which is known. It may take the form of a metal plug set in concrete or a mark chiselled in a wall. Bench marks can sometimes be observed set in the kerb near street corners, the location and elevation of which have been recorded on surveyors' drawings. On building sites, a temporary bench mark is sometimes established and given an assumed elevation.

10.2.2 Datum

The datum is a fixed point of known elevation; it is the reference point from which the elevations of other points are calculated. The temporary bench mark of assumed elevation, say 30.000 m, becomes the datum for that particular building site and may be referred to as the *job datum.*

10.2.3 Reduced level (RL)

The term *reduced level* (RL) is given to the elevation of a point relative to a given datum. For example, on Figure 10.6, if point *A* is equal to an RL of 31.000, it is 1.000 m above datum, and if point *B* is equal to an RL of 29.000, it is 1.000 m below datum. Thus, the difference in level between points *A* and *B* equals 2.000 m.

The assumed elevation given to the datum is sufficiently high to ensure there will be no minus quantities where RLs are below the datum.

Fig. 10.6 Reduced level of an elevation point

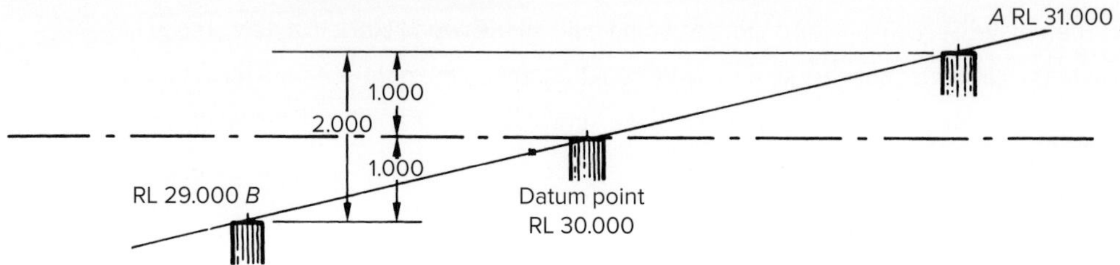

There are two common methods used to calculate reduced levels: height of collimation, and rise and fall. When taking levels, it is important that the readings are recorded correctly and accurately using a table. Table 10.1 is an example of a table used when taking levels using the rise and fall method. This type of table is also useful to record levels even if the reduced level is not going to be calculated.

Table 10.1 Table used to record readings using the rise and fall method

Reading	Back sight (BS)	Intermediate sight (IS)	Fore sight (FS)	Rise	Fall	Reduced level (RL)	Distance	Remarks
1								
2								
3								
4								
5								
6								
7								
8								
9								
10								

CHECK YOUR UNDERSTANDING

1. What is the purpose of establishing a datum?

10.2.4 Levelling equipment

Levelling equipment is essential for measuring accurately on-site. The instruments and methods used include the following.

10.2.4.1 Spirit level and straight edge

A spirit level and straight edge can be used together over short distances with great accuracy. The straight edge is normally constructed from aluminium, or sometimes timber, and must be constantly checked for straightness or errors in setting out may occur. When taking levels with a spirit level, reverse the level to ensure that it is reading accurately. Temporary intermediate pegs, as shown in Figure 10.7, can be driven between the points to increase the range of the straight edge.

Fig. 10.7 Levelling with a straight edge

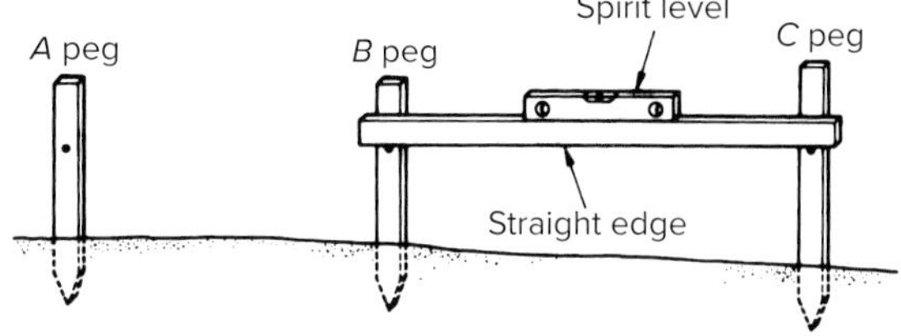

10.2.4.2 Water level

A water level consists of a length of clear plastic hose or tube about 10 mm in diameter, which is filled with clean potable water (Fig. 10.8). As the measurement is carried out, each end of the tube is held on the vertical. Care must be taken to remove all air bubbles when the tube is filled.

Fig. 10.8 Water level

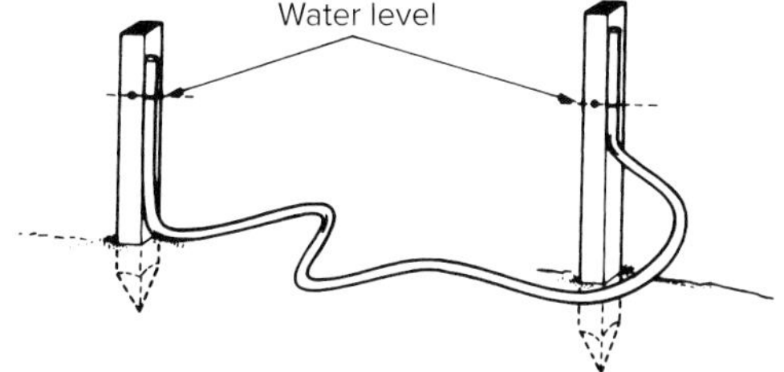

TIP Filling a water level without getting air bubbles in the water can be achieved by siphoning water from a clean bucket.

The level works on the principle that water will always find its own level and therefore each end of the tube will find the water at the same level. This method can be very accurate, and the range is limited only by the length of the tube. A particular advantage is that it does not require a direct viewing from point to point, and can be taken around obstructions. It has been found to be very useful for internal work, levelling from room to room. Plug the ends with a length of dowel when not in use; however, the ends must be open when in use.

10.2.4.3 Dumpy/automatic level

A **dumpy level** is an optical instrument that consists of a telescope fixed to a vertical axis, with a three-screw base used to level the instrument (Fig. 10.9). A set of crosshairs is located in front of the eyepiece of the telescope, and attached to the telescope is an accurate levelling bubble. When the telescope is levelled it will project a horizontal 'line of sight' as it is rotated around its vertical axis. It can also be used to set off horizontal angles.

Fig. 10.9 Dumpy level

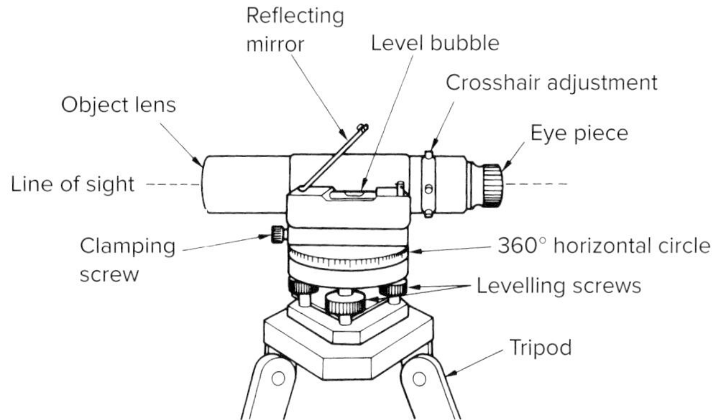

The dumpy level is accurate over long distances when used in conjunction with a surveyor's staff. Levelling for most domestic buildings, however, can be carried out by measuring between points, marked on a timber rod with a metric rule.

Fig. 10.10 A & B The two-peg test

(a)

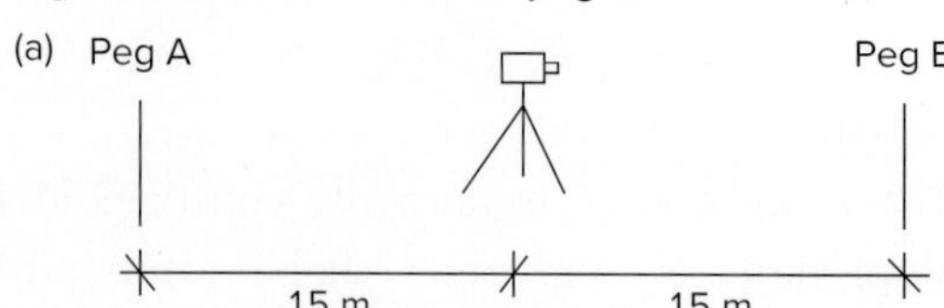

(b)

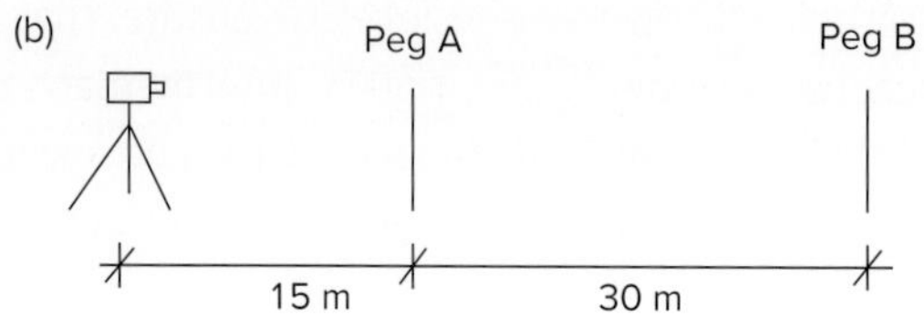

10.2.4.3.1 Checking the dumpy

To check the dumpy level for accuracy, a two-peg test needs to be carried out (Fig. 10.10). Set up a dumpy level in the centre of two pegs. Ensure that the dumpy is exactly 15 m from each peg. To find this measurement you will need to use a plumb bob suspended from the screw on the tripod (Fig. 10.11). Using either a surveyor's staff or a piece of timber, place the staff on top of Peg A and take a reading. Transfer this reading to Peg B (both pegs need to be the same height).

Move the dumpy past one of the pegs, ensuring it is 15 m away from the peg and in line with both of them.

Take the reading on Peg A; this will be a different reading, but don't worry, just note it down. Go to peg B and take a reading. If both of the new readings are the same, the dumpy is working accurately. If there is a difference in the two new readings, then the dumpy needs to be recalibrated by a specialist company that services surveying equipment.

10.2.4.3.2 Looking through the dumpy

To look through the dumpy, first adjust the instrument to your eye strength. This is done by rotating the eyepiece until the crosshairs are clear and sharp (refer to Fig. 10.12). You will see large horizontal and vertical crosshairs. The horizontal crosshair is the one that is used for taking height readings

Fig. 10.11 A & B A plumb bob

(a)

Courtesy of Daniel Bonnici

(b)

from a surveyor's staff or a piece of timber. The vertical staff can be used to ensure the staff is plumb. There are also two smaller horizontal lines. These are called stadia lines and can be used to calculate distance.

To determine the difference in level between two points, *A* and *B*, set up the level in a convenient position where the two points will be visible, and adjust for level. Hold a wooden rod/staff vertically at point *A* and mark the level of the crosshairs on the rod/staff using a pencil, chalk or felt marker. Hold the rod on point *B* and mark the level of the crosshairs on the rod. Measure between the two marks; this measurement is the difference in level between points *A* and *B*. The distance from point *B* to the line of sight is the lesser distance; therefore, there is a rise from point *A* to point *B* (Fig. 10.13).

Fig. 10.12 Crosshairs in dumpy eye piece

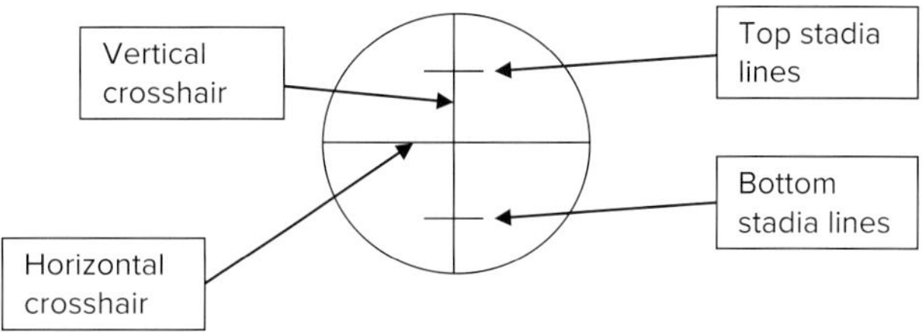

Fig. 10.13 Using the dumpy level

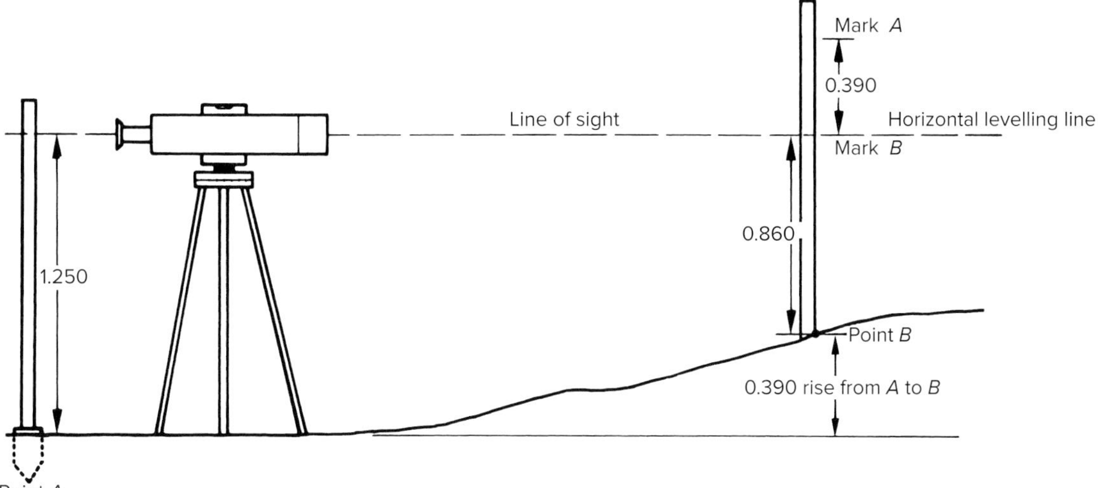

10.2.4.4 Laser level

A laser level is an electronic instrument that has a rotating head that emits a laser beam (Fig. 10.14). It is set up on a tripod like a dumpy. Most modern lasers are self-levelling. Once the instrument is level, the head will start to rotate and the laser beam will start emitting. A receiver is required to detect the laser beam. The receiver normally has an LED screen, which will show when the beam has been detected. Most receivers also have an audible beep. There are normally three different beeps: one pattern to indicate to move the receiver up, one to move the receiver down and a continuous beep that indicates when the beam and receiver are aligned.

Fig. 10.14 Laser level

The laser level can be used in the same manner as a dumpy level. The main difference is that it is a one-person operation. Instead of having a second person looking through a lens as on the dumpy level, a receiver is used to locate the beam being emitted by the laser level. The receiver is attached to the staff and the required heights can be transferred as with a dumpy level.

Height differences can also be calculated by moving the receiver up or down along the staff when determining two heights from two points.

It is always a good idea to read the manufacturer's instructions/specifications for the specific laser level being used, to ensure the level is being used appropriately and that you are familiar with all of the laser level's features.

10.2.4.5 The builder's square

The builder's square is a large square fabricated from timber and is used to set out right angles. It is commonly made on the site from available material. Two straight pieces approximately 90 mm × 35 mm and 3 m to 4 m long are required for the arms, together with a suitable brace. The two arms are joined at the corner and then set square by the 3,4,5 principle.

Measure along the arms three units and four units respectively, making the units as large as possible for greatest accuracy. Fix the brace at one end and adjust the arms to measure five units across the hypotenuse of the triangle and finish fixing the brace. To set out horizontal right angles accurately with the builder's square, it is laid on the ground and at least one arm of the square must be level.

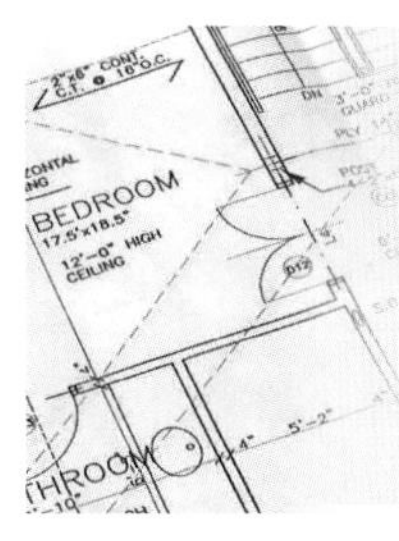

INTERPRETING PLANS

Refer to the floor plan in the appendix.

- Set out the double garage. Set out the entire area, showing all walls.
- Set the height of the hurdles to 200 mm above a datum point set by your teacher.

CHECK YOUR UNDERSTANDING

1. List three levelling devices that can be used over large distances.
2. What is a builder's square and how would you check that it is 'square'?

10.3 Setting out on-site

The outline of a domestic building and its position on a near-level allotment are shown in Figure 10.15. The building is of timber-framed construction, and there is a strip footing around the perimeter with a single brick dwarf wall up to under-floor level. The floor level is specified as RL 31.200.

The procedure begins by first clearing the site of long grass, tree stumps and roots. Assemble all of the necessary tools and materials on-site and proceed as follows. (The order of setting the string lines is shown in Figure 10.16.)

10.3.1 Step 1

Set out the necessary markings on the saddle/hurdle cross members. Figure 10.16 indicates the width of the excavation and the position of the brick dwarf wall. For a concrete footing, the width of the excavation is the same as the width of the footing. If a brick footing is specified, approximately 150 mm is added to the footing width to provide working clearance. Partly drive a nail at the points where the string lines are to be fixed. To set out the simple outline, twelve profiles are required.

10.3.2 Step 2

Working from the survey pegs, accurately locate the side boundaries. If this is done by setting up a long string line, take care that it is not affected by breezes, thus causing inaccuracies. Placing some temporary intermediate pegs will help to overcome this problem.

10.3.3 Step 3

The general principle adopted in setting out is to take the basic rectangle of the building and set it out first. Then check for accuracy, and add or subtract any minor projections. In this case the rectangle *ABCD* will be set out first.

It has been decided to set up line *AB* first, because it is close to the minimum clearance from the side boundary and is the longest side. It seems logical to set it out accurately, parallel to the boundary, this being the most critical dimension. Some may prefer to reverse the order of strings 1 and 2, particularly if it were a very wide building lying across the middle of the block. Theoretically, it should not matter which order is adopted but in practice small discrepancies can arise, and it is comforting to know that at least they will not be allowed to happen where they could lead to any serious consequences.

Measure in from the boundaries and roughly locate points *A* and *B*. Beyond point *B*, set up a profile as shown in Figure 10.17. To erect the saddles/hurdles, refer to Figure 10.18. Lay the cross member on the ground and measure from the boundary to ensure that it is in its correct position. Drive in two pegs of equal height adjacent to the cross member. Lift the cross members up to the top of the pegs, check their position and secure with nails to the pegs.

Fig. 10.15 Site layout

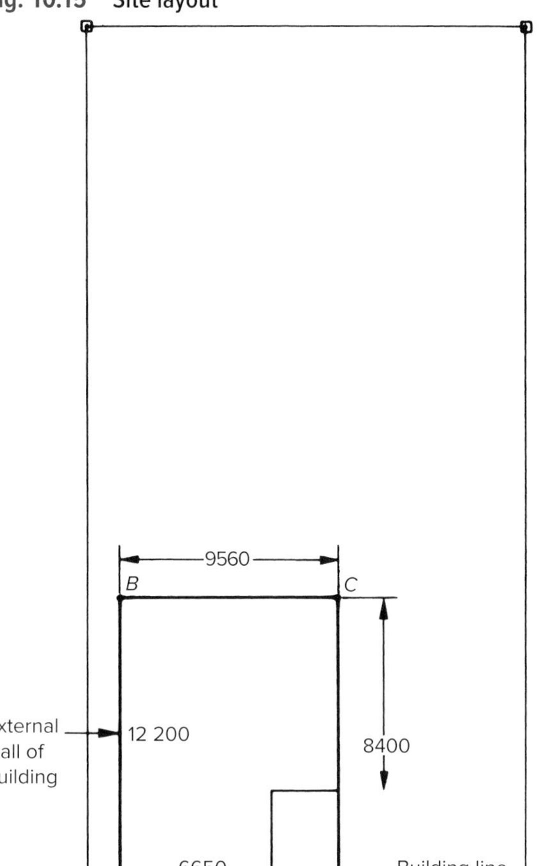

Fig. 10.16 Basic building rectangle—sequence of operations

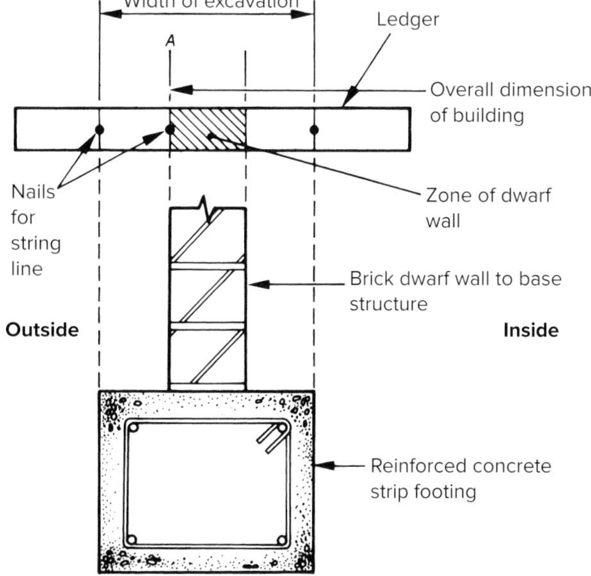

Fig. 10.17 A & B Ledger set-out for timber-framed construction

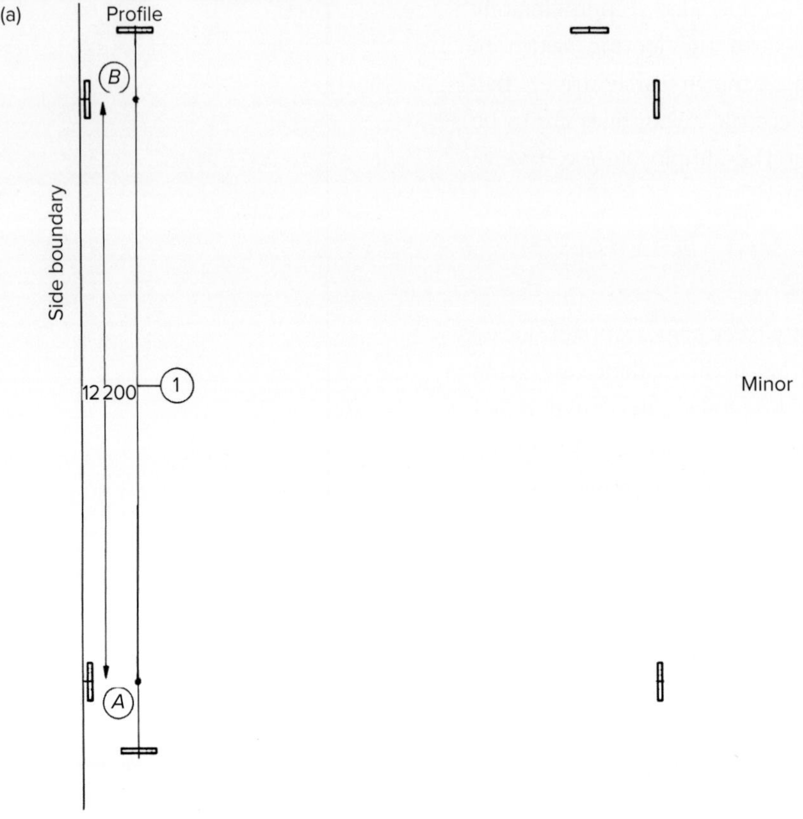

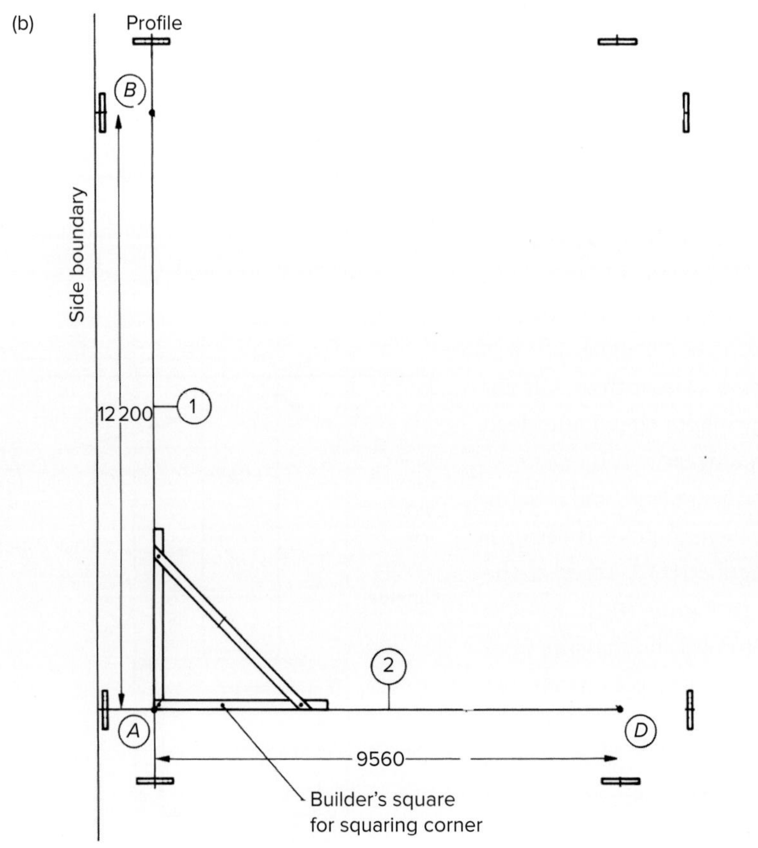

Fig. 10.17 C & D Ledger set-out for timber-framed construction

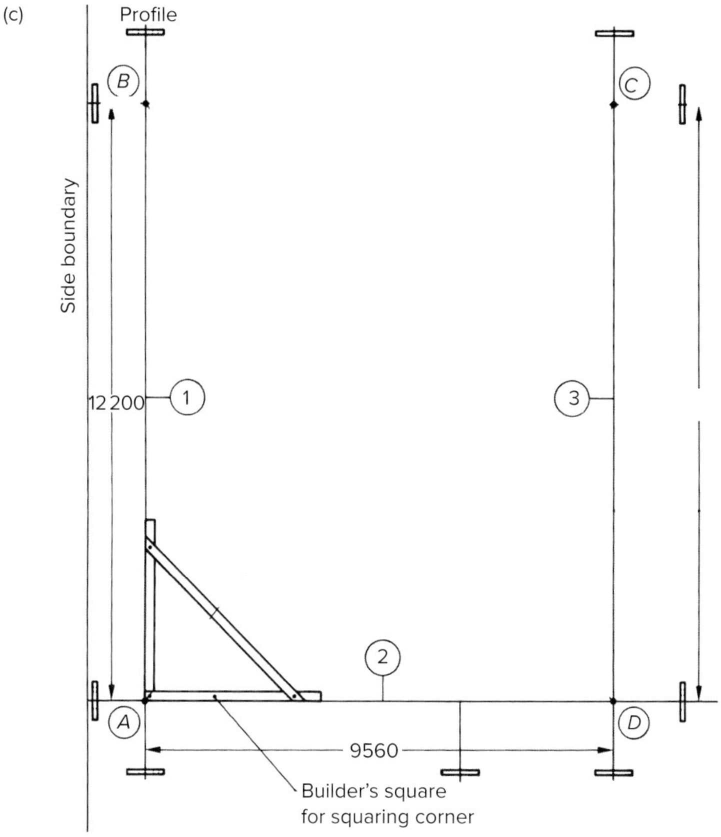

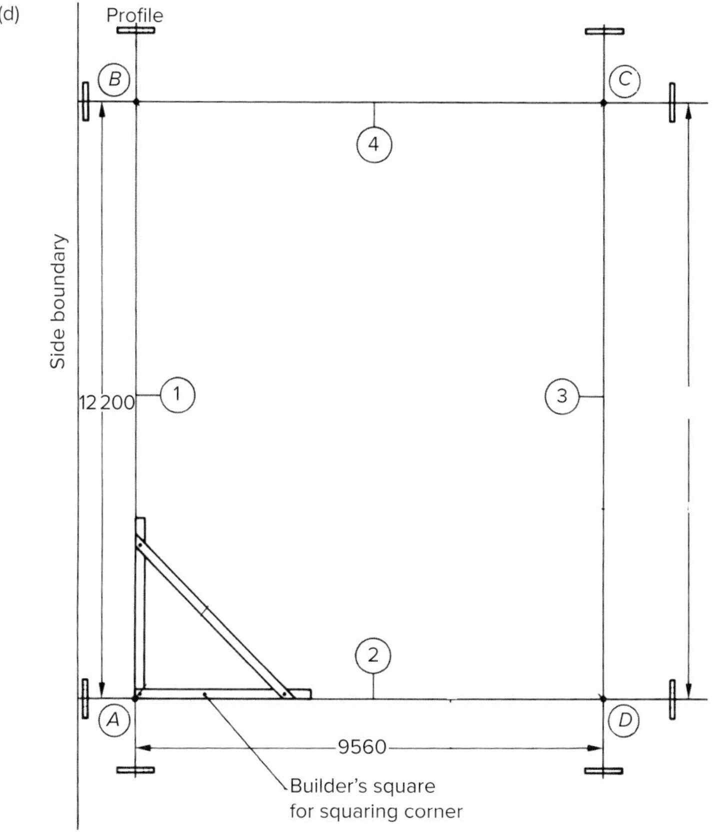

TIP Ensure that the height of the saddle/hurdle profile is set to a predetermined height (e.g. the height of the top of the dwarf wall, piers or stumps).

The depth of the flooring and timber floor members must be deducted from the RL for the top of the brickwork (Fig. 10.19). If the depth of floor timbers is 220 mm then the RL to the top of brickwork is equal to 30.980 m (i.e. 31.200 less 0.220 m). Therefore, the top of the brickwork is 0.980 m above datum. (Figure 10.20 shows the hurdle being levelled using the dumpy level.)

Set the level in a convenient position where the hurdles and datum can be sighted, and adjust the line of sight for level. Stand a wooden rod on the datum and mark the line of the crosshairs, point *E*. Measure down 0.980 and make a second mark *F*. Stand the rod beside the pegs and raise it until point *F* is in line with the crosshairs. Mark the bottom of the rod on the pegs and attach the saddle/hurdle profile to the pegs. The top of the saddle/hurdle should be level with the marks made. The other saddles/hurdles should be levelled in a similar manner.

Fig. 10.17 E Ledger set-out for timber-framed construction

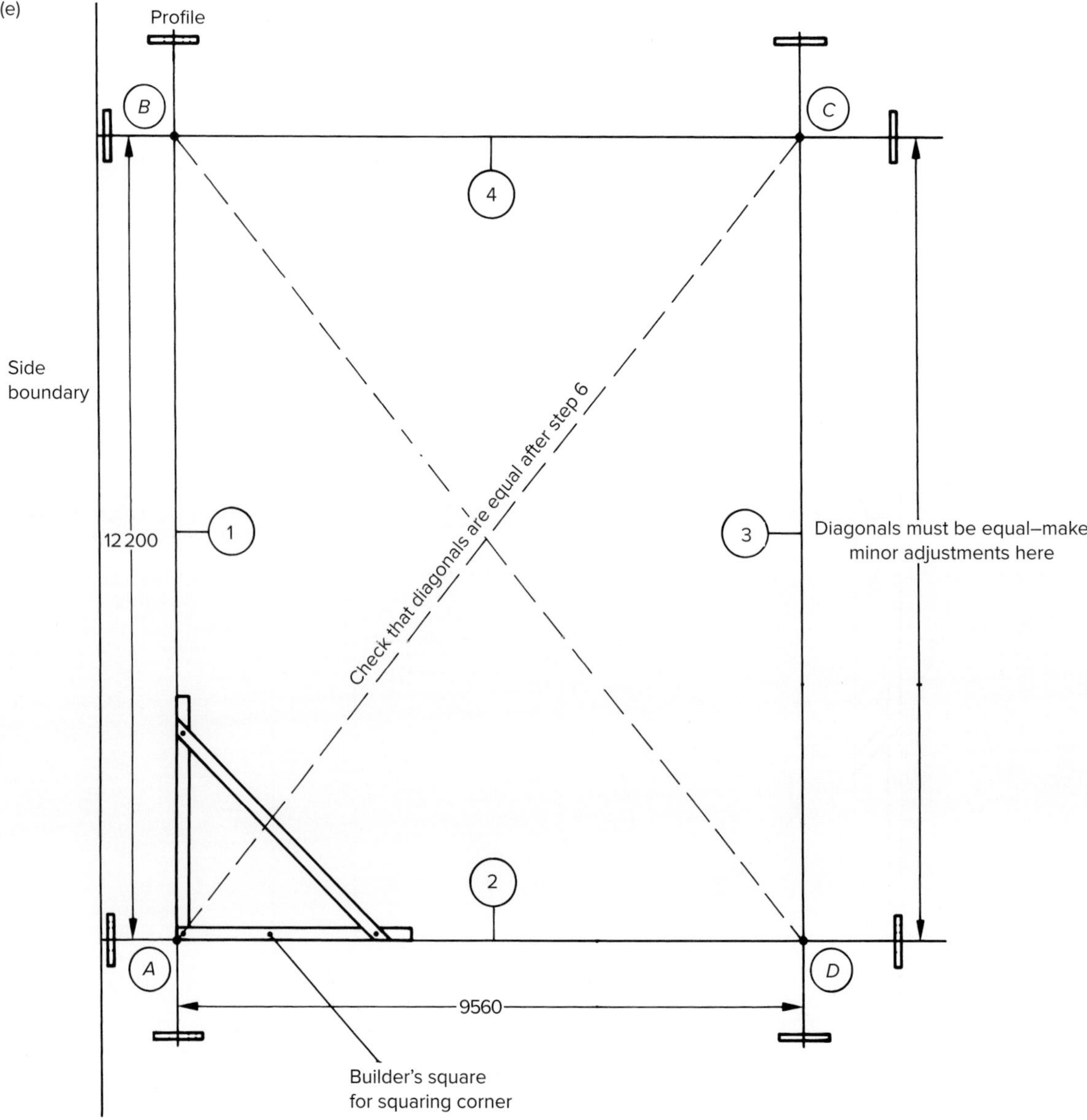

Building regulations require that a minimum clearance be maintained from the ground to the underside of the floor framing timbers and this minimum must always be observed.

Beyond point *A* set up another profile in a similar manner, ensuring it is levelled to the same height as profile A, and stretch a string line indicating the outside face of the brickwork. Profiles must be set back far enough to allow working space when excavating the trenches. If excavating by hand, a 1 metre clearance should be sufficient, but if excavating by machine, 2 or 3 metres may be necessary depending on the particular backhoe being used.

10.3.4 Step 4

To set up line *AD*, measure from the front boundary to the building line and erect and level a profile beyond point *A*. Attach a string line and secure the other end to a temporary peg beyond point *D*. The builder's square is commonly used to set this line at right angles to line 1 in Figure 10.21. Plumb down and lay one arm of the square directly under line 1. Adjust line 2 so that it also lines up with the other arm of

Fig. 10.17 F Ledger set-out for timber-framed construction

(f)

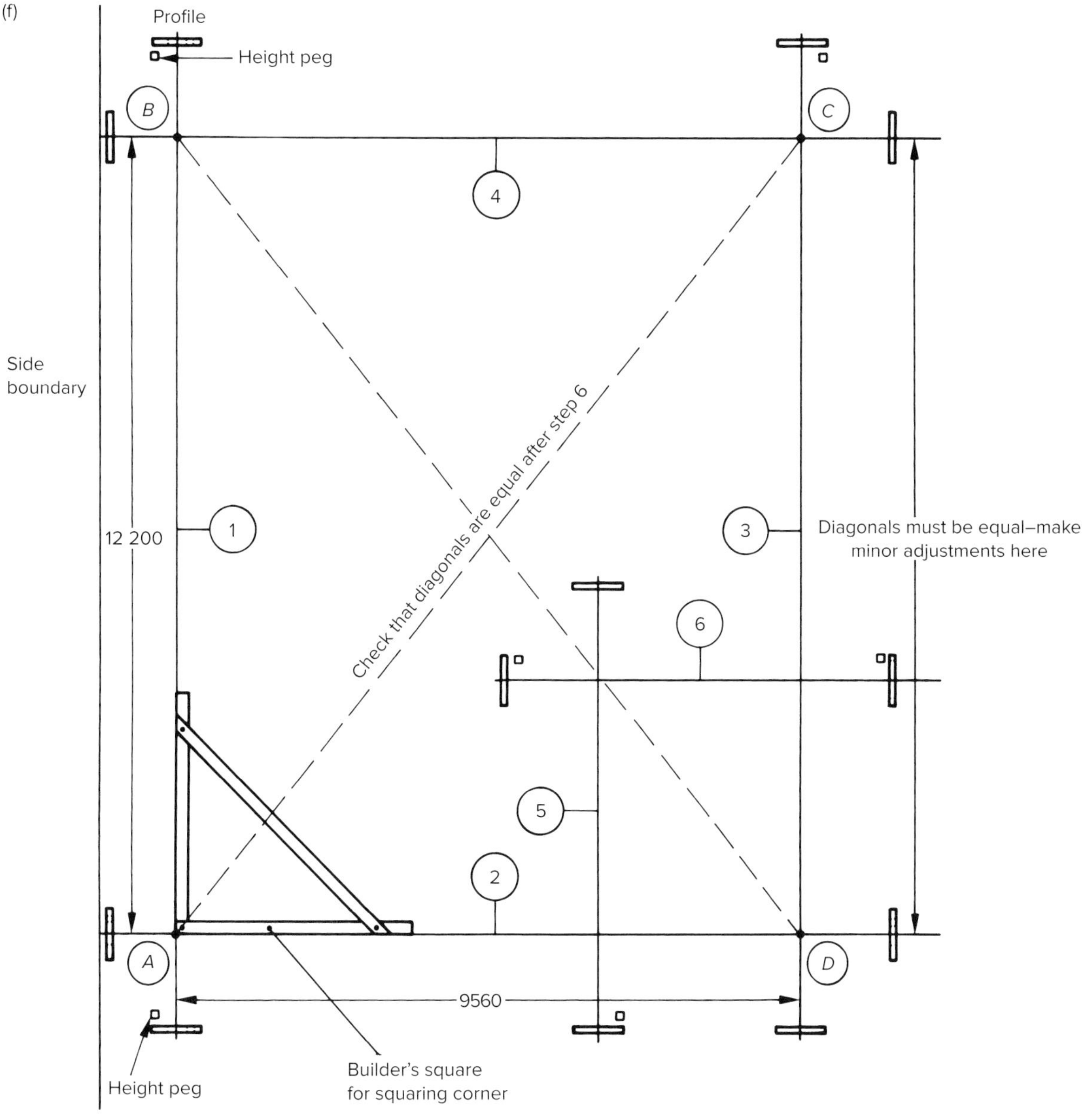

the square. Erect and level a profile beyond point *D* and attach the string line, indicating the face of the brickwork. Care in setting out an accurate right angle at this stage avoids the need to make adjustments at a later stage.

10.3.5 Step 5

Measure the width of the building for the profile beyond points *C* and *D*. Continue to follow the procedure for erecting and levelling the saddles/hurdles (Fig. 10.18). Attach a string line, which is now parallel to *AB*.

10.3.6 Step 6

Measure the depth of the building from the profiles at *A* and *D* and erect and level the profiles beyond *B* and *C*. Attach a string line, now parallel to *AD*.

10.3.7 Step 7

This completes the basic rectangle, but before proceeding further it must be checked for squareness by measuring the diagonals, which must be equal. Hold the tape as taut as possible. It helps if someone can take the sag out of the tape by supporting it near the centre. If minor adjustments have to be made, the two profiles beyond *C* and *D* will both have to be adjusted either way by the same amount (Fig. 10.17).

Fig. 10.18 Set-up of saddle profile

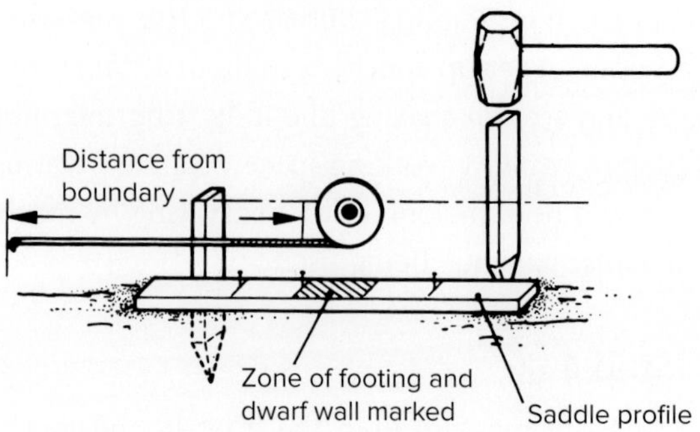

Fig. 10.19 RL from top of ground to floor structure—timber-framed construction

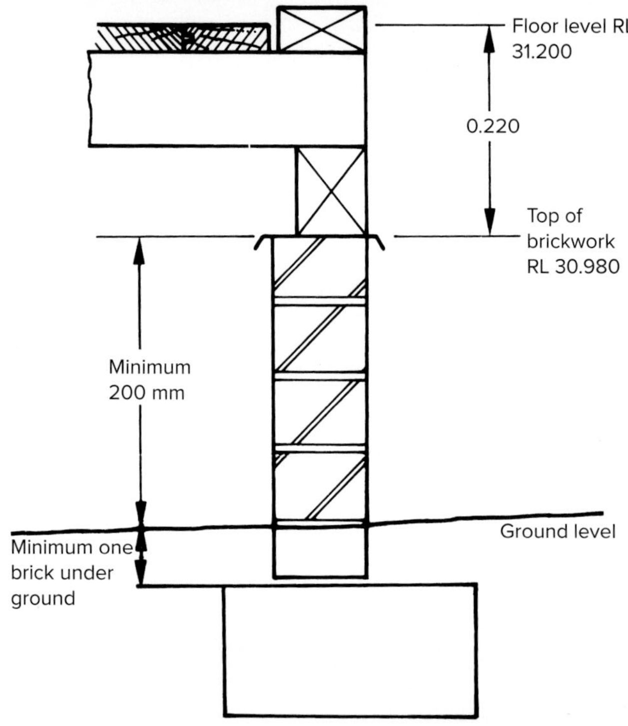

10.3.8 Step 8

Set up lines 5 and 6 parallel to the sides of the main rectangle to indicate the face of the brickwork.

The 3,4,5 method

If a builder's square is not available, the 3,4,5 method can be used.

- Measure along line *AD* 3 metres and drive a small peg into the ground; mark the 3-metre measurement with a nail on the top of the peg.
- Measure along line *AB* 4 metres and drive a small peg into the ground; mark the 4-metre measurement with a nail on the top of the peg.
- Measure the distance between the two pegs diagonally. This measurement should equal 5 metres. If it isn't, adjust the position of the nails, ensuring the 3-metre and 4-metre distances are maintained.
- Once a 3,4,5 triangle has been created, this will give us a square (90°) corner. The lines *AD* and *AB* need to pass directly over the two small pegs we have just used. You can plumb up from the nails to the string line using a spirit level, to check the lines are in the right position.

10.3.9 Step 9

Excavation of the trenches can now proceed. For each wall, set string lines to indicate the position of the trenches (Fig. 10.22). Plumb down onto the ground and scratch a line or mark the line with a few small pegs. If excavating by machine, a sprinkling of lime to mark the line of the trench can be a good guide for the operator to follow.

The depth of the trench must be adjusted so that the brickwork can be finished in full courses to the height of the hurdles. The standard metric brick measures 230 mm × 110 mm × 76 mm. Joints are 10 mm, so one course of bricks is 86 mm. Prepare a brick gauge rod marked off in brick course increments of 86 mm, and if the footing is of concrete, tack a loose batten on the bottom, equal to the depth of the footing (Fig. 10.23).

Level across from the saddles/hurdles and use the rod to adjust the depth of the trench to work to brick courses (Fig. 10.24). Establish the depth of the trench at each corner and level between them to complete the excavation.

Pegs, commonly short lengths of 10 mm steel, are placed at suitable intervals along the trenches to indicate the height to the top of the concrete footing.

Fig. 10.20 Levelling hurdle

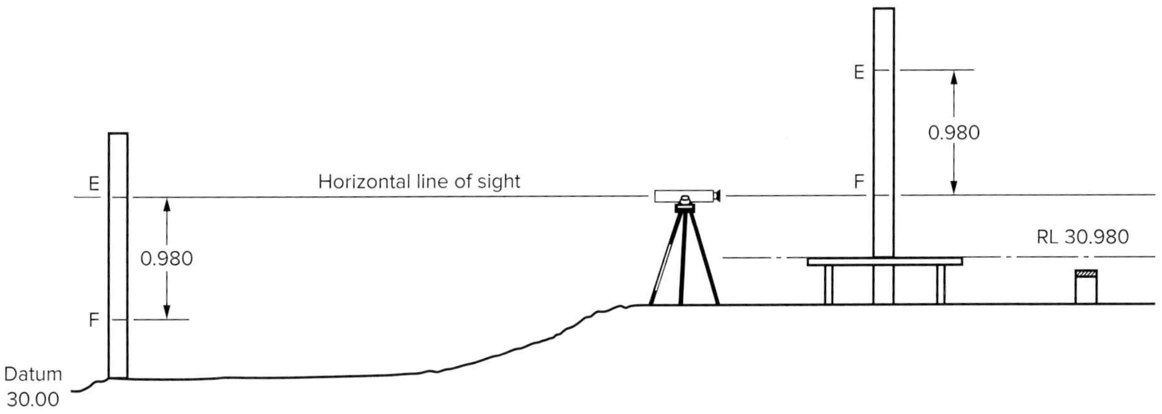

Fig. 10.21 Corner set-out

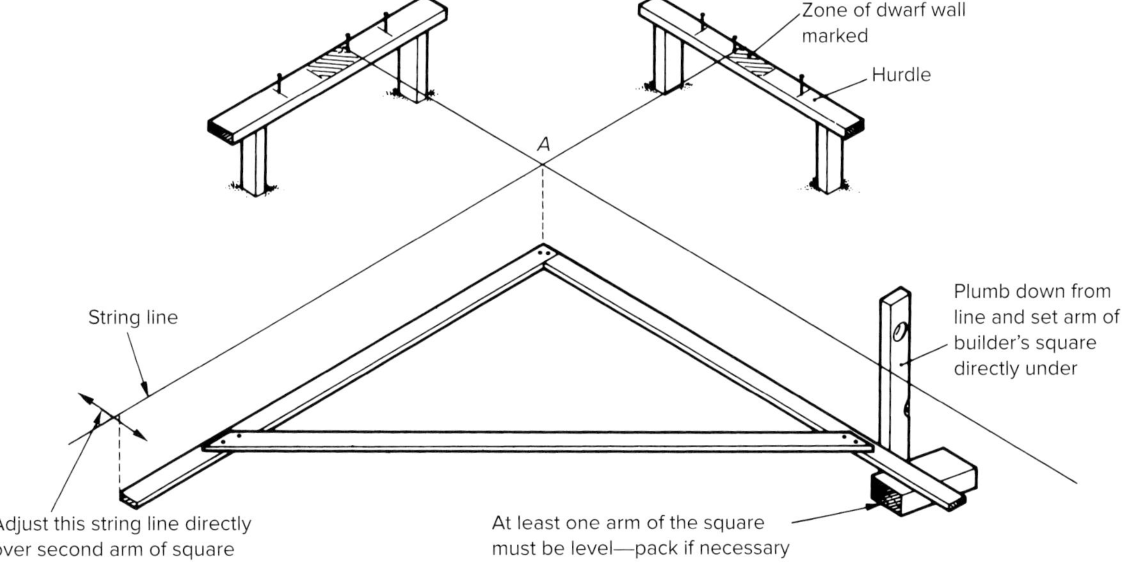

Fig. 10.22 Excavation of trenches

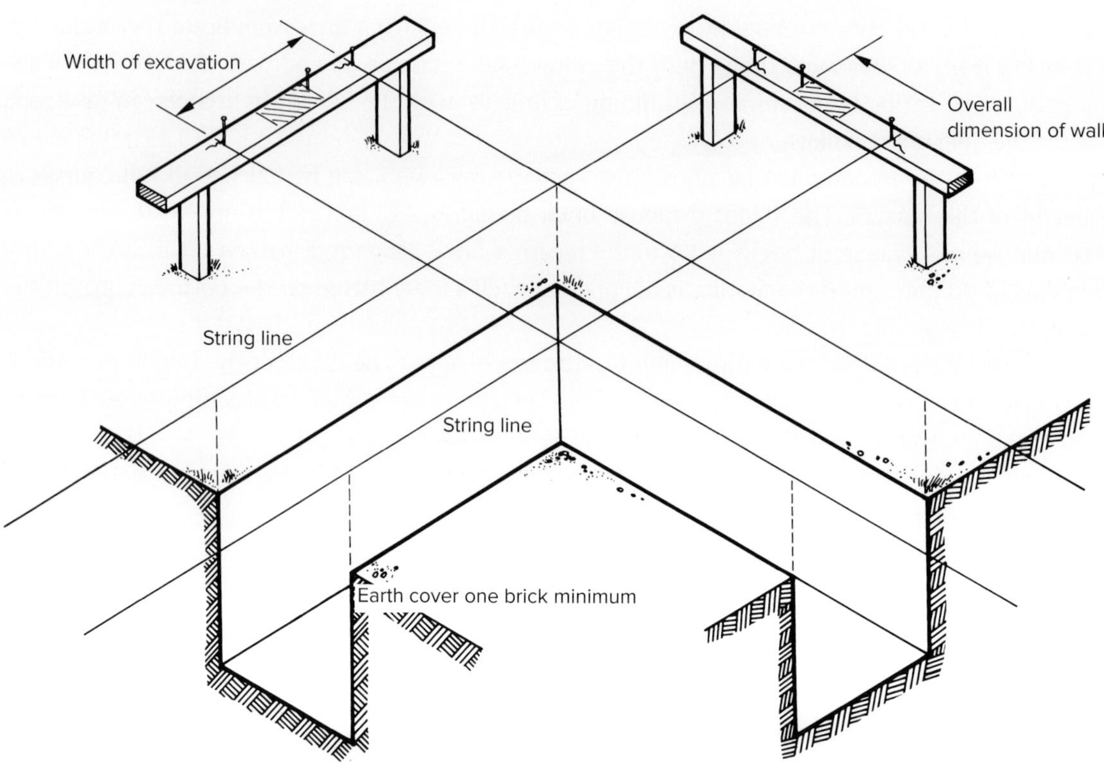

Fig. 10.23 Set-out of a brick or gauge rod

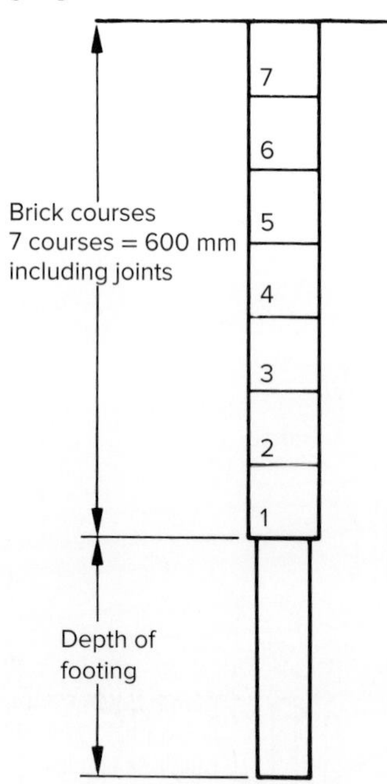

Fig. 10.24 Establishing the depth of trenches

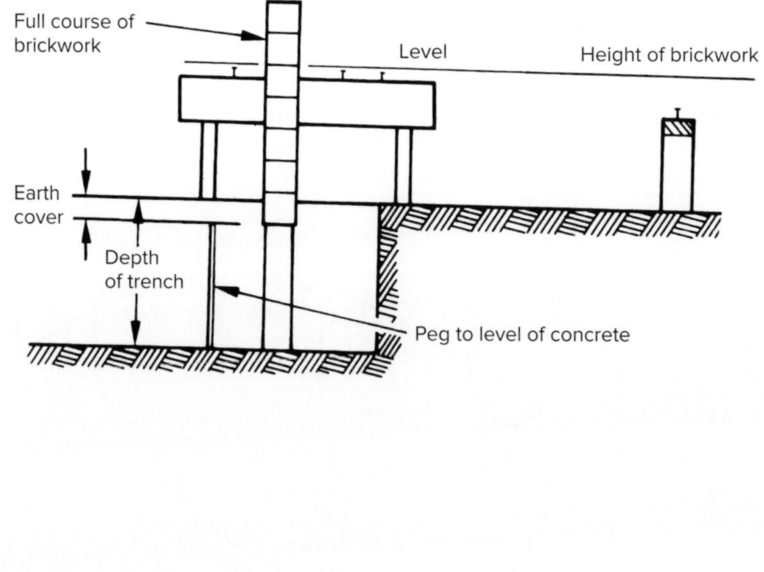

CHECK YOUR UNDERSTANDING

1. What marks are made on a profile cross member when setting out for a brick wall on a concrete footing?
2. How would you check your basic set-out for square?

10.4 Setting out on sloping sites

When setting out on a sloping site, hurdle profiles are often used and are thought to be more convenient and accurate. All of the cross members on the profiles are set level at the height of the brickwork (Fig. 10.25).

Setting out of the string lines can be done in the same order as is shown in Figure 10.17, but it is usually more accurate to set out the marking on the cross member after the profile has been erected. Placing a cross piece on the ground in its approximate position is a good guide to where the profile should be placed.

The bottom of the trenches must always be level, and on a sloping site they must be stepped to form level benches. The height of the steps is a multiple of brick courses or 86 mm (Fig. 10.26).

Start at the lowest corner and excavate to the finished level. Continue along at that level until the depth of earth cover over the footing indicates a step could be made, then continue for another 600 mm to allow for an overlap in the footing, and step up a distance equal to brick course multiples. Benches should preferably be no less than approximately 1200 mm long and not made any closer to a corner. The maximum height of a single step is 600 mm.

Fig. 10.25 Set-out on sloping site

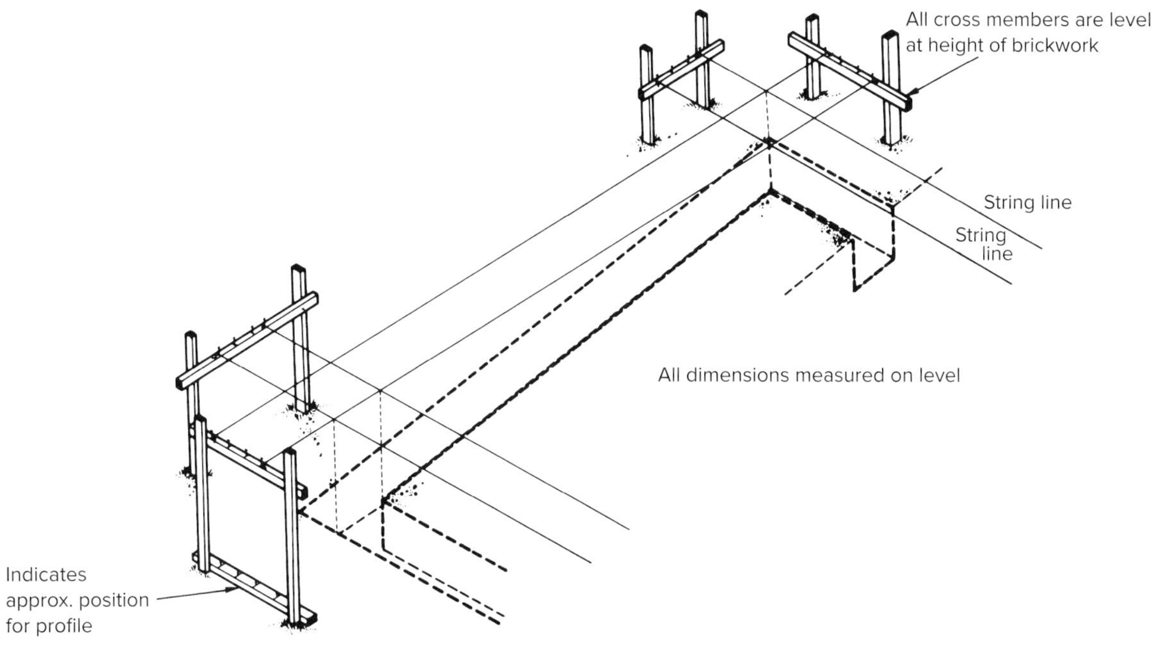

Fig. 10.26 Excavation for footings on sloping site

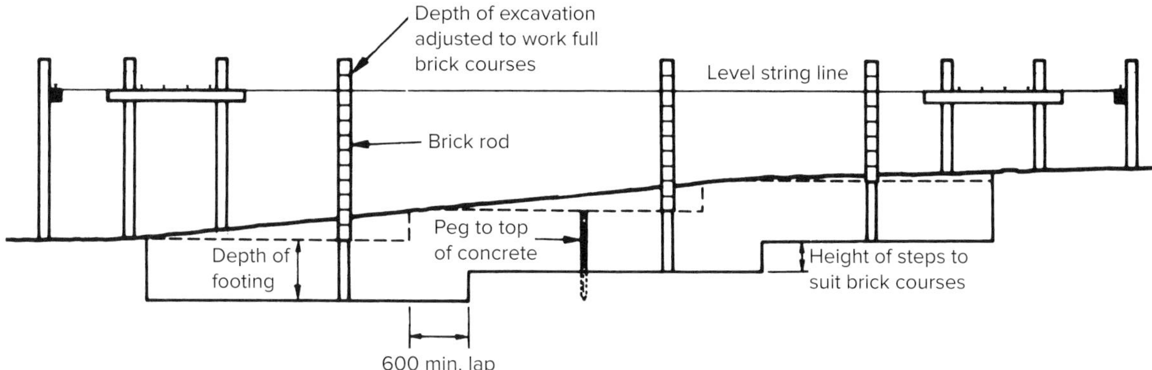

10.4.1 Setting out for sleeper piers

Sleeper piers are also called intermediate piers, isolated piers or solitary piers. To support the ends of bearers, engaged piers are bonded to the dwarf perimeter walls, and sleeper piers provide intermediate support when the length is more than a single span (Fig. 10.27). To set out for brick sleeper piers, stretch a string line between the engaged piers, indicating the height of the piers, and plumb down to the ground. Mark the position on the ground for the excavations, not exceeding the spacing specified. Excavate for the footings using the brick gauge rod to establish the depth of the excavations, so that two courses will be below ground level and will finish with full courses of brickwork to the string line.

Fig. 10.27 Excavation for isolated or sleeper piers

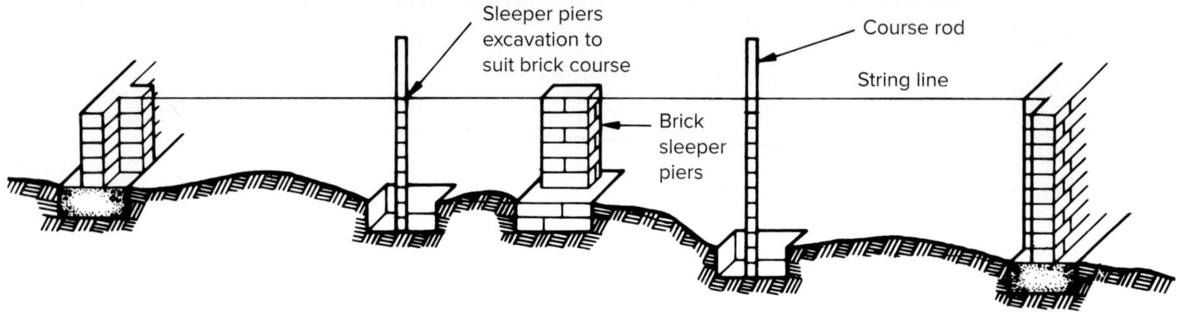

CHECK YOUR UNDERSTANDING

1. How would the depth of footings be determined for a brick dwarf wall to a timber-framed cottage?

10.5 Setting out screeds for concrete slab

Screeds are boards set up to establish the height at which the concrete is to be levelled off. A common example is the boards that are used to form up the sides of a concrete path in order to retain the concrete, and that indicate the height at which the path will be finished. A concrete garage slab is another typical situation that requires the setting out of screeds (Fig. 10.28).

For example, a concrete garage slab 6300 mm × 3600 mm × 100 mm thick is to be set out on a near-level site. Besides the setting out equipment previously used, also required will be screed boards sized at least 100 mm × 25 mm and long enough to form the perimeter of the slab, as well as enough pointed pegs–approximately 50 mm × 25 mm–to support the screeds.

First, level and clear the site of any grass or roots. Soil used for filling should be free of all vegetable matter and well consolidated by ramming to a depth not exceeding 600 mm. Greater depths of fill using road base can be achieved with engineering specifications.

Fig. 10.28 Set-out for screeding boards

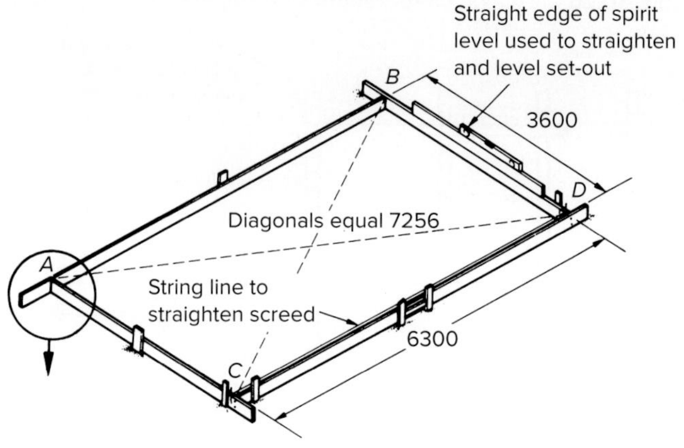

10.5.1 Step 1

Lay a long screed on the ground beside its final location, square one end and mark off the length of the slab *AB* to 6300 mm (Fig. 10.29). Aim to arrange the screeds so that their length need not be cut any more than necessary. If close to a boundary line, measure from the boundary a distance equal to the clearance, less the thickness of the screed. Drive a peg at each end, which will support the back of the screed.

Establish the height to the top of the slab and mark on one peg, and level this height through to the second peg.

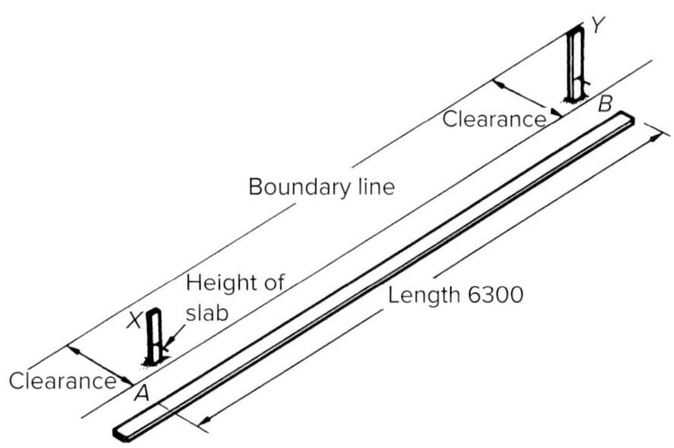

Fig. 10.29 Setting up screeding board length

10.5.2 Step 2

Nail or screw the screed to the pegs with the top edge on the level marks. If nailing, dolly up behind the pegs with a heavy hammer to avoid loosening the pegs.

To straighten the screed, attach a string line (Fig. 10.30), packing it out at each end with wooden blocks approximately 10 mm thick. Using a loose block of the same thickness, adjust the screed to the string for both line and height. Drive pegs behind the screed approximately 1 metre apart and secure with a nail or screw.

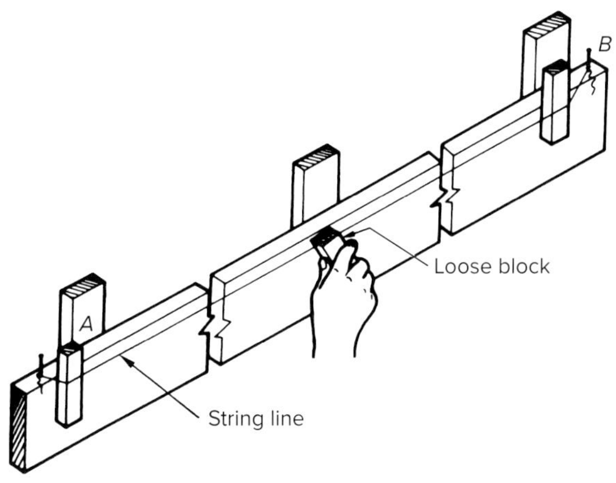

Fig. 10.30 Straightening screed board

10.5.3 Step 3

Select a screed for side *AC*, square one end, mark off the width (3600 mm) and nail in place at *A* with the inside face of *C* on the length mark. If one is available, a builder's square could be used to square the corner *AC*, or use the 3,4,5 method mentioned earlier (Fig. 10.31).

Drive a peg behind the screed at *C*, level from *A* to *C*, and secure with a nail or screw. Straightening screeds over shorter distances can be carried out with a straight edge.

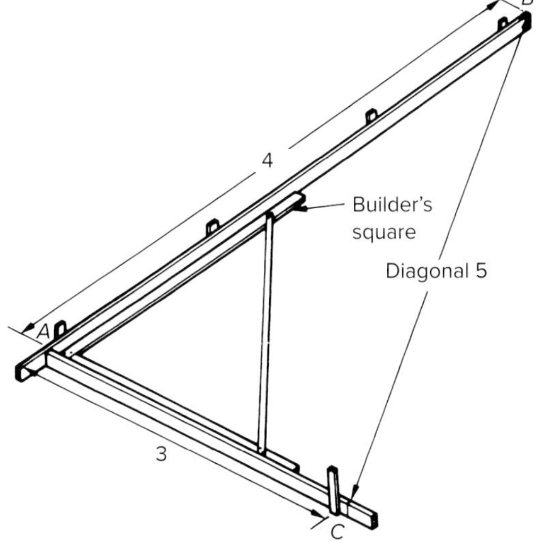

Fig. 10.31 Squaring up corner of screed

10.5.4 Step 4

Select screeds for *CD* and *BD*, square one end and measure off the length and width respectively. If screeds must be joined, a short cleat across the back will keep them in alignment. Assemble the screeds by nailing them together at *B*, *D* and *C*. Level *BD* and secure to peg at *D*. Before proceeding and driving in all of the pegs, it is no doubt a wise move to check the accuracy of the setting out at this time. For squareness, measure both of the diagonals *AD* and *BC*. Both measurements should be the same. If they are not the same, some adjustments will need to be made. Refer to Figure 10.17 and step 7; a check of the levelling can be made by levelling across the diagonals *A* to *D* and *B* to *C*.

10.5.5 Step 5

Set up a string line from *C* to *D* at the thickness of a loose block from the corner. Adjust the screed to the line for both line and height, and secure to pegs. A straight edge can be used to straighten the shorter screeds and secure with the remaining pegs about 1 metre apart.

To calculate the quantity of concrete required:

$$\begin{aligned} \text{Volume} &= \text{length} \times \text{width} \times \text{height} \\ &= 6300 \times 3600 \times 100 \text{ mm} \\ &= 6.3 \times 3.6 \times 0.1 \text{ m} \\ &= 2.268 \text{ m}^3 \end{aligned}$$

CHECK YOUR UNDERSTANDING

1. What is the function of a screed?
2. Define the following terms:
   a. plumb
   b. level.

Student research

Research what is meant by 'reduced level' and why it is used. Then investigate the different methods for obtaining reduced levels.

End of chapter activity

Activity 1:

With a partner, set out a rectangular building 4 m × 3 m on some grass.

- Use hurdles and string lines.
- Ensure all of the hurdles are the same level (ask your teacher to establish a datum and to allocate a height above the datum for your hurdles).
- Remember to set up the first corner using the 3,4,5 method and check that the final set-out is square by measuring the diagonals.

AUSTRALIAN BUILDING CODES AND STANDARDS

AS 1684

NCC

Guide to Standards and Tolerances 2015 —VBA

Tradie talk—Is that string line tight enough?

Chapter 11

Subfloor and flooring

Learning Objectives

LO 11.1 Identify types of flooring

LO 11.2 Know about brick base structures

LO 11.3 Install posts and stumps

LO 11.4 Install bearers and joists

LO 11.5 Lay strip flooring

LO 11.6 Lay fitted floors

LO 11.7 Know how to install platform floors (using strip flooring)

LO 11.8 Lay sheet flooring

LO 11.9 Lay wet area flooring

LO 11.10 Know about alternative subfloor materials and their uses

Introduction

In this chapter you will learn about how the **subfloor** is the first part of any timber-framed house to be constructed. The subfloor is what the rest of the structure is built upon. All of the loads are transferred to the foundations through the subfloor framing and it can be a conduit for insect attack or damp infiltration to the rest of the building. It is important that the subfloor is constructed level and square and made of suitable materials to deal with the prevailing ground conditions. Mistakes made in the construction of the subfloor can cause problems later in the build that can be hard to overcome.

The subfloor framing to timber-framed domestic buildings, as shown in Figure 11.1, consists of bearers supported clear of the ground on piers, stumps or dwarf walls. Running at right angles to the bearers, and supported by them, are joists that in turn support the actual flooring.

Fig. 11.1 Subfloor framing

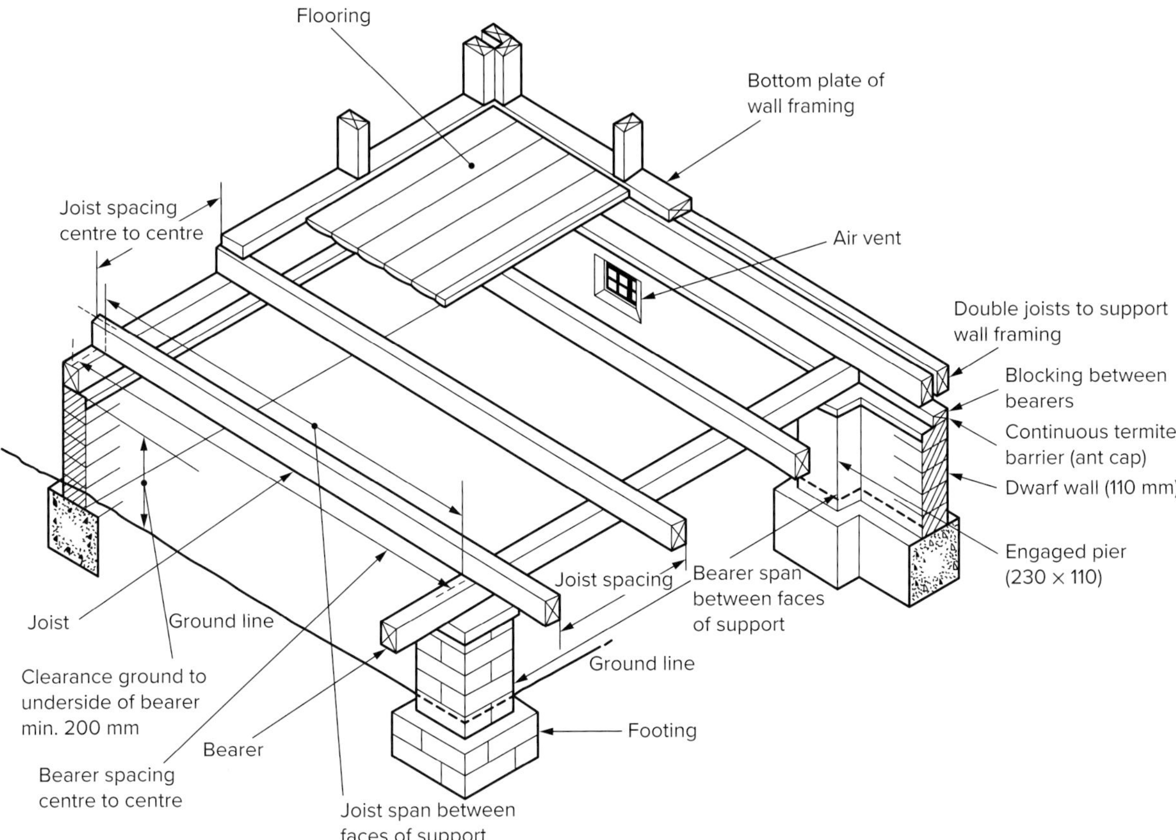

The spacing of bearers and joists is taken as the centre-to-centre distance apart; however, when setting out a number of members of equal thickness in practice, it is more convenient to adopt the 'in and over' method of spacing, which is the same distance as centre to centre. 'In and over' is the practice of measuring from the face of one member, over its thickness, and to the face of the next similar member. The span of the bearers and joists is the face-to-face distance between points of support.

11.1 Types of flooring

Flooring can be strip flooring made up of individual tongued and grooved boards varying from approximately 75 mm to 150 mm in width, or sheet flooring manufactured from special grades of particleboard or structural plywood, in sheets of up to 3600 mm × 900 mm.

There are two procedures that can be followed for laying flooring to the joists.

11.1.1 Fitted floor

The **floor is fitted** to each room after the wall framing has been erected (Fig. 11.2). In this case it is necessary to provide double joists under all external and internal walls that run parallel to the floor joists. These are spaced to provide support for the wall framing and for the ends of the floorboards. The

Fig. 11.2 Fitted floor

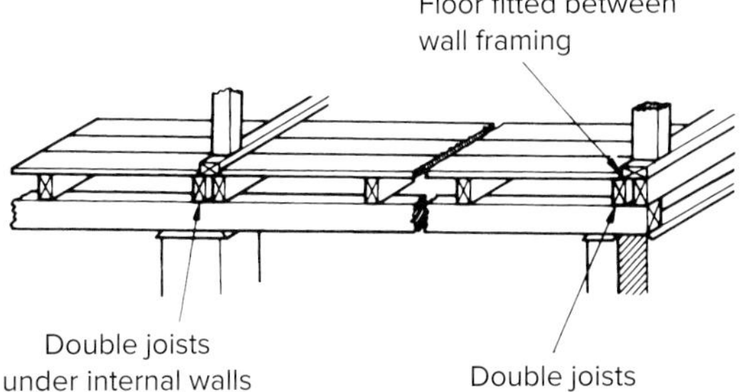

Fig. 11.3 Platform floor

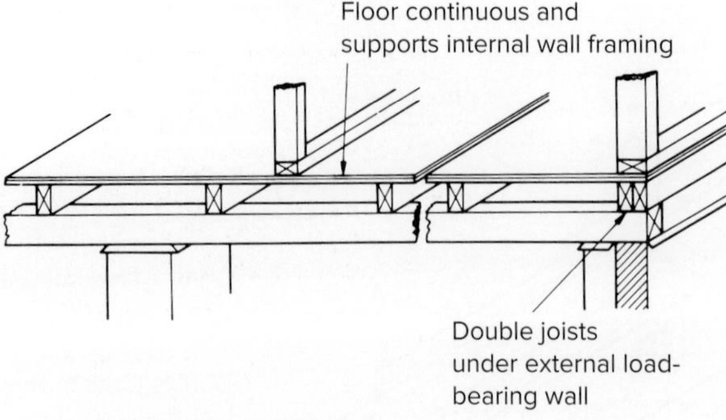

flooring is not finally laid until the building has been made watertight.

11.1.2 Platform floor

This type of floor is laid over the whole area of the floor joists before the wall framing is erected (Fig. 11.3). Double joists are provided under external load-bearing walls that run parallel to the floor joists, but for the remainder of the building, joists may be spaced without reference to the positioning of the walls. An exception is in the case of particleboard flooring where a single joist is required under all load-bearing walls that run parallel to the floor joists.

Platform floors can be laid quickly; they provide a convenient and safe working platform during the construction process. Generally, manufactured sheet flooring is used that has a water-resistant treatment; however, construction should proceed promptly as the flooring cannot be exposed to the weather for long periods.

Before discussing the timber flooring members in greater detail, we should consider the footings and supporting structure relevant to timber-framed buildings, sheeted externally with weatherboard, fibre cement products or other lightweight cladding materials. (Refer to Chapter 10 for information on setting out the site.)

CHECK YOUR UNDERSTANDING

1. Describe, with a sketch if necessary, the difference between a fitted floor and a platform floor.

Fig. 11.4 Base structure

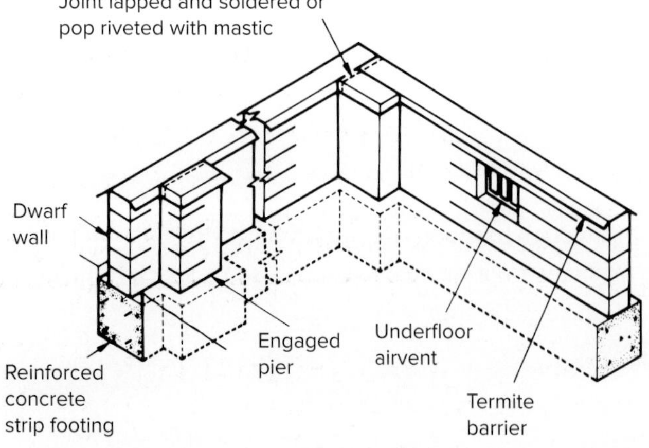

11.2 Brick base structure

Foundation walls constructed of brick are 110 mm thick and are strengthened by having engaged piers, 230 mm × 110 mm, bonded or tied to them at 1800 mm centres to support bearers (Fig. 11.4). Footings for a base brick structure are constructed from reinforced concrete. Between external foundation walls, bearers can be supported by isolated piers or **stumps**.

Foundation walls and piers must be built to sufficient height to allow for adequate clearance between the ground and the underside of the bearer.

TIP For dimensions for strip footing and reinforcement requirements, please refer to Australian Standard AS 2870: 2011 Residential Slabs and Footings.

11.2.1 Underfloor ventilation

The space between the ground and the underside of the floor must be thoroughly ventilated; failure to provide adequate measures can lead to serious problems. Damp, ill-ventilated places can become a haven for termites or fungal growth, and flooring boards can become distorted due to the absorption of a large amount of available moisture under the house.

Any tendency for excessive underfloor dampness must be corrected to the satisfaction of the local government authority; this can usually be provided by adequate drainage. Regulations that specify the amount of ventilation may vary in different areas.

Figure 11.5 shows three methods of providing ventilation in a base brick structure.

11.2.2 Termite barriers

Termite barriers, or ant caps as they are sometimes known, are made of galvanised steel or other approved metal, and are laid over all foundation walls, piers and stumps at the level of the underside of the floor timbers (Fig. 11.4).

Ant capping must project 40 mm beyond the internal face of the wall and be turned down at an angle of 45° (Fig. 11.6). All joints must be lapped and soldered, or otherwise sealed in an approved manner.

Fig. 11.5 Air vents: (a) terracotta; (b) Acme-type vent; (c) open brick

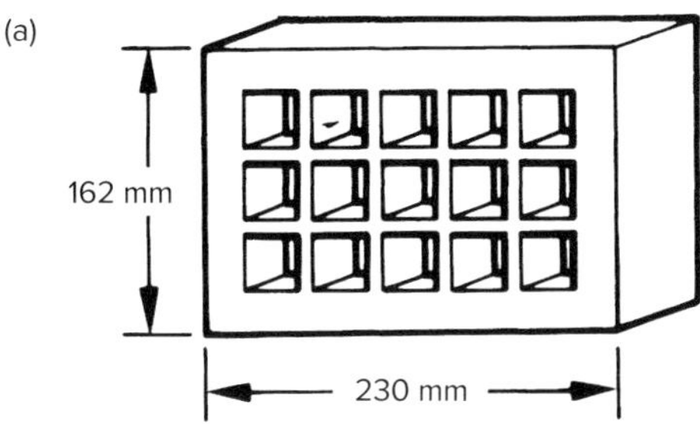

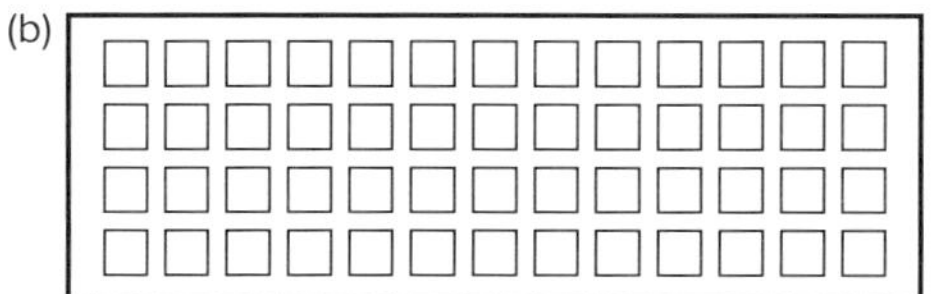

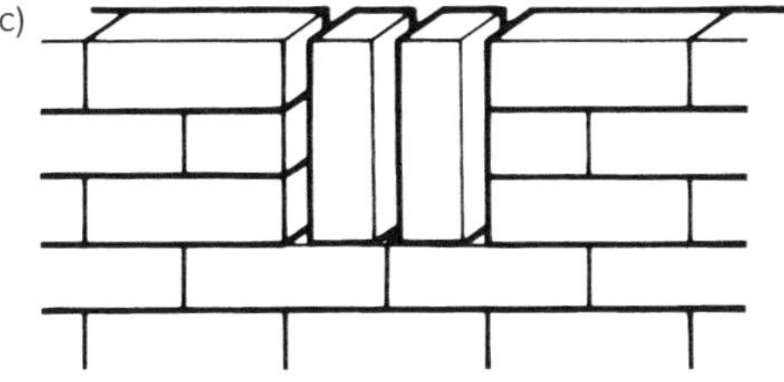

11.3 Posts and stumps

Sawn timber posts placed on timber sole plates is one method used to support bearers in areas where termite attack is not a serious threat (Fig. 11.7). Posts and sole plates must be of a durable timber species as they are in contact with the ground. Suitable species are ironbark, white cypress pine, red gum or treated radiata pine that has been pressure preservative impregnated.

Fig. 11.6 Ant capping—continuous

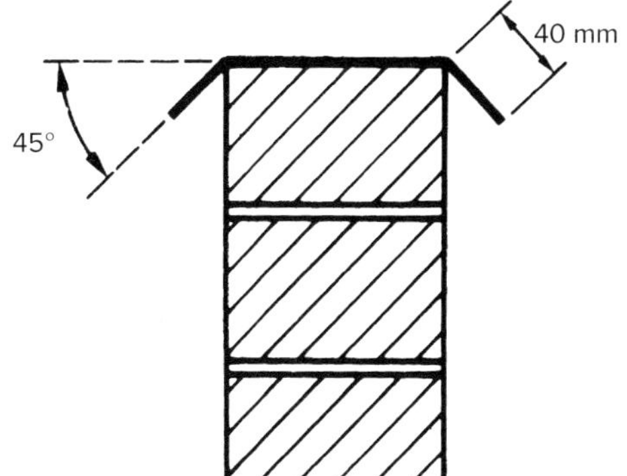

Fig. 11.7 Timber stump and bearer

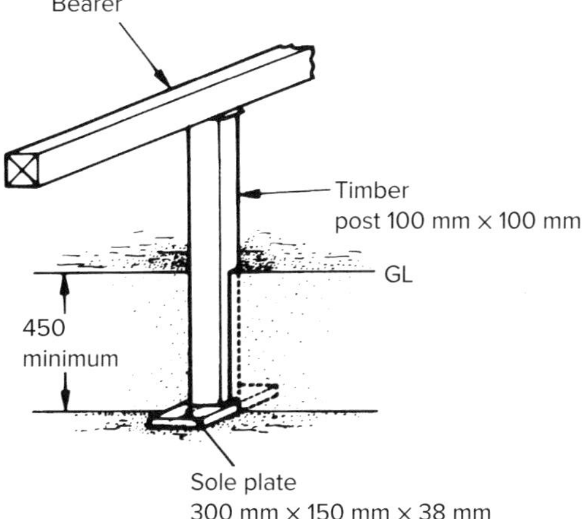

Fig. 11.8 Corner bracing

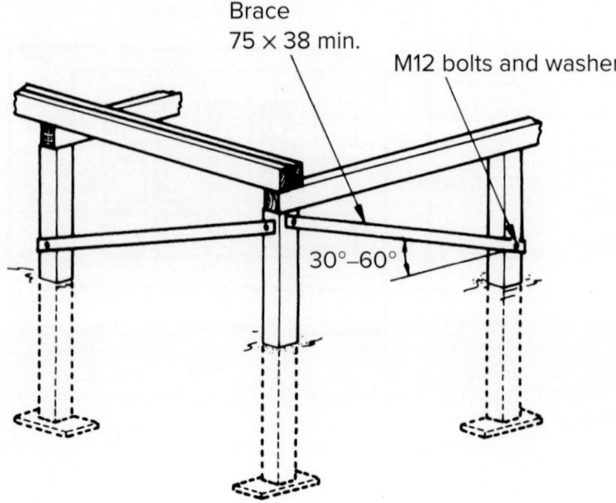

Traditionally, timber stumps are 100 mm × 100 mm, set a minimum of 450 mm into the ground and at a maximum height of 1200 mm above the ground. Sole plates can be 300 mm × 150 mm × 38 mm or as specified.

For information regarding timber stumps, posts and sole plates, please refer to AS 1684 Residential Timber-Framed Construction. The soil type will need to be established before designing the footing structure.

Where the height of posts and stumps exceeds 1200 mm, bracing is required (Fig. 11.8). Refer to AS 1684 for the requirements. It is usual trade practice to space stumps 1200 mm apart along the length of the bearers, with bearers spaced 1800 mm apart, centre to centre.

11.3.1 On-site setting out of timber stump/posts

The procedure outlined for measuring and levelling in Chapter 10 remains constant and it is again suggested that it is sound practice to set out the main rectangle of the building by measuring first from the side boundary, often the most critical dimension to be observed.

11.3.1.1 Step 1

Establish the side boundary and, if necessary, set up a string line.

11.3.1.2 Step 2

Measure from the front boundary and locate the approximate position of the building. Set up hurdles parallel to the side boundary and at the required distance for the side of the building nearest to the boundary.

11.3.1.3 Step 3

Measure back from the front boundary to locate the position of the front of the building set-up, two more hurdles parallel to the front boundary. At this point a square corner will need to be established. Add the depth of the building in order to locate the back of the building accurately. Set up all required hurdles to establish the perimeter of the building (eight in total for Figure 11.9). Set up two extra hurdles to locate the centre row of stumps (refer to Figure 11.9); the string line for the centre row of stumps should be offset 50 mm from the centre line of the stumps. This will mean the string line will be running along the side of the stumps and not on the top. Ensure all hurdles or profiles are set up to the height of the top of the stump so the string line can be used to level the stumps when being installed.

The posts can be cut slightly long and tapped down with the head of the crowbar to their correct height (Fig. 11.10). Backfill the soil in layers, no more than 150 mm thick, and ram firmly.

11.3.1.4 Step 4

Excavate holes to the required depth, ensuring the floor of excavation is reasonably flat and straight with no loose soil. Place the correct-sized sole plate on the floor and 'ram' it with the head of a crowbar until it won't sink any further into the ground. Measure from the sole plate to the string line, which has been set to the height of the underside of the bearer; cut the stump to that length then place the stump, plumb and to the string line. Backfill and compact the soil as you go to ensure the stump

Fig. 11.9 Site set-out

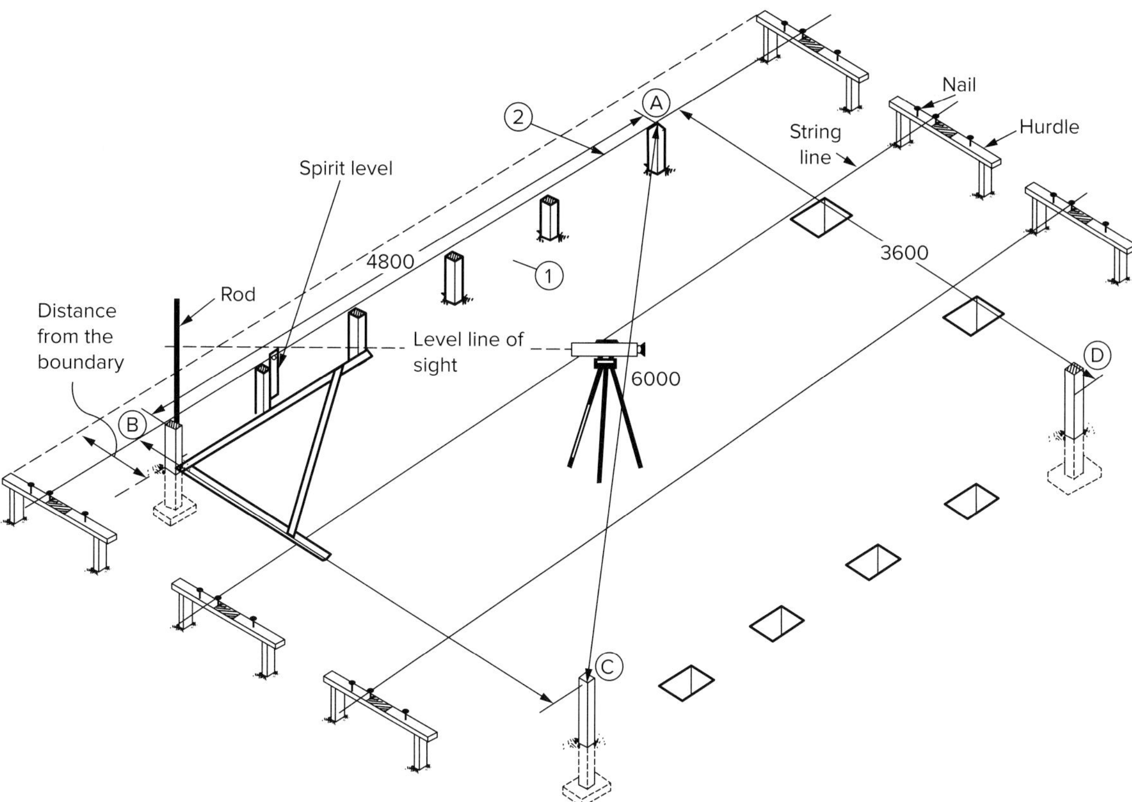

remains in the correct position. To check the height of the stumps, use a dumpy level set up near the centre of the area and use a timber rod to transfer the height from the hurdles to the top of the stumps.

11.3.1.5 Step 5

Set out the position of the intermediate stumps/posts between *A* and *B* and excavate to the required depth. Set the stump/posts in position, aligning them with the string lines, and plumbing up in the other directions. Tap each post down to the height of the string line.

11.3.1.6 Step 6

Repeat the process with the next two rows of stumps.

Fig. 11.10 Setting the timber stump in position

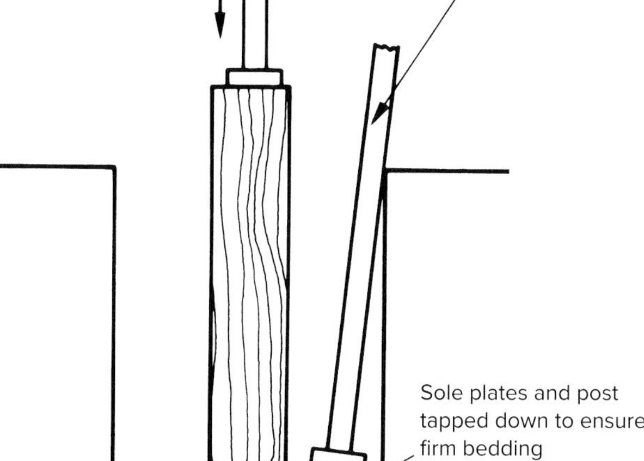

11.3.2 Pre-cast concrete stumps

Pre-cast concrete stumps have superseded timber posts in many places. This type of stump is 100 mm × 100 mm, sunk no more than 25 mm into a concrete base/pad at least 150 mm thick (Fig. 11.11).

TIP For information of pad thickness, diameter and depth, the soil type will need to be established and then AS 1684 will need to be consulted.

Fig. 11.11 Pre-cast concrete stump

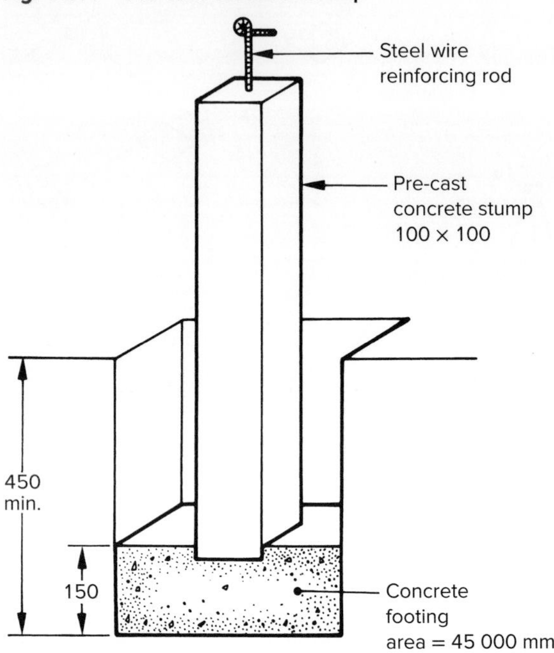

Fig. 11.12 (a) Determining the stump height; (b) positioning the stump

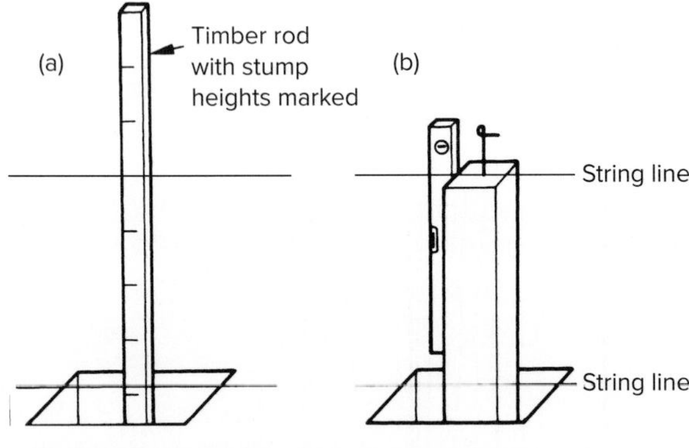

Pre-cast concrete stumps are reinforced with steel wire rods, which are left projecting from the top and bent over in order to secure the bearers; a threaded rod is sometimes used instead of wire reinforcement and left projecting 150 mm for the same purpose.

When ant caps placed over the stumps are pierced, the hole must be sealed with a bitumen compound or other suitable means to prevent the entry of termites.

Concrete stumps are manufactured in lengths in multiples of 100 mm so that the depths of excavation must vary to make minor adjustments in the height of the stumps. Use a light timber rod marked off in the length of stumps to determine the depth of excavation (Fig. 11.12). See the procedure steps listed below.

Calculating and setting pre-cast concrete stumps in position

Calculate the amount of concrete required to form the stump pads. Generally, the stump holes will be round, so use the formula $\pi r^2 \times h$, where:

$\pi = 3.142$ r = radius of the stump hole h = thickness of the concrete pad

Once you have calculated the amount of concrete required for one pad, multiply that answer by the number of holes that have been dug. Do not forget that the final answer should be in cubic metres, so you can tell the company how many cubic metres of concrete are required. You can order concrete in increments of 0.2 of a cubic metre–it is always a good idea to order an additional 0.2 of a metre of concrete just to make sure you don't run short.

To set the stump in position:

1. Using a wheelbarrow, place the concrete into the stump holes to form the base, Ensure the concrete mix is sufficiently dry to prevent the heavy concrete stump from sinking too fast or too far into the soft liquid concrete.
2. Stand the stump in position and align in one direction with string lines and plumb in the other direction.
3. Tap the stump down to the correct height–if necessary, using a wooden block to protect the head of the stump from damage.

4. Backfill the soil in layers not more than 150 mm thick; pack the dirt gently as too much force will cause the stump to move as the concrete is still wet.
5. Thoroughly clean all of the tools and equipment that were used to place the concrete and store them appropriately.
6. Let the concrete set overnight. The following day, to further compact the soil around the stump, pour half to a full bucket of water around the stump.

TIP To determine the stump hole depth, diameter and the thickness of the concrete pad, a soil test must first be carried out by a geotechnical engineer to determine what soil type is found on the site.

WORKPLACE SCENARIO

The use of concrete stumps is commonplace today and, as a product, they are far superior to red gum stumps as they do not rot. However, the installation of concrete stumps can be quite a frantic process to carry out. Generally, the concrete is ordered and comes to the site already mixed. The first job is to distribute the correct amount of concrete into each hole using a wheelbarrow, then, working quickly, place each stump in the correct position, ensuring it is plumb and in the right line. The speed at which the stumps have to be placed into the already-hardening concrete can vary depending on the weather: the hotter the day, the quicker you have to work. Cooler overcast days are ideal as the concrete sets slower, therefore providing additional time to get the job done.

11.3.3 Steel stump supports

Steel stump supports are widely available as an alternative to timber and concrete stumps (Fig. 11.13). Setting out is carried out in the same manner as concrete stumps. Steel stumps can be installed in a similar manner to concrete stumps and cast into concrete. Alternatively, a concrete pad is poured to the top of the ground level and no backfilling with dirt and tamping is required. The posts are bolted down to a concrete pad and are screwed or bolted to the bearer. Many of the types available are adjustable in height, so the use of packers is not required. They are generally galvanised, so they are suitable in damp environments. The steel posts are also termite resistant.

Fig. 11.13 Steel stump supports

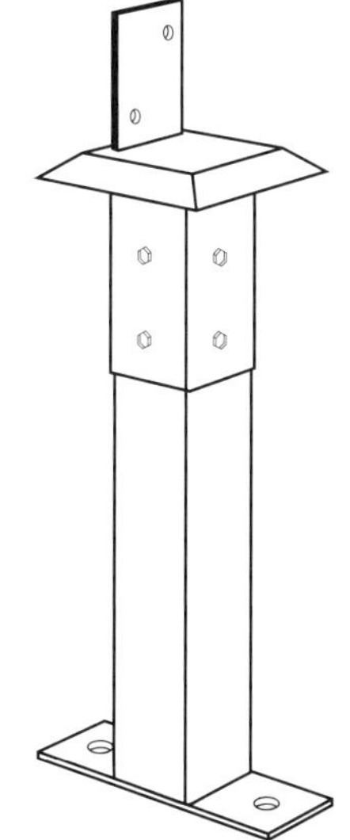

CHECK YOUR UNDERSTANDING

1. What precautions must be taken to prevent termite attack on ground-floor timbers?

11.4 Bearers and joists

The complete details of the possible combinations of spans, spacings, dimensions and stress grades of timbers that could be used in floor framing can be found in AS 1684. These can be quite extensive. The figures given below can be taken as an example of what is traditionally common trade practice.

> **TIP** Steel can also be used as an alternative to timber. These sections would need to be engineered for a specific purpose or bought off the shelf from a manufacturer of steel subfloor members.

Traditionally, bearers would be 100 mm × 75 mm hardwood, F8 or F11, laid on edge with a maximum span of 1200 mm for **single span** bearer, and 1500 mm for **continuous span** bearer supporting, for example, load-bearing external walls carrying roof sheet. AS 1684.4 Part 4: Non-cyclonic Construction details alternative spans and spacings for all timber framing members achievable by using different timber grades and/or sizes. Care should be taken to ensure compliance.

Traditionally, joists are commonly 100 mm × 38 mm or 50 mm hardwood, F8 or F11, with a maximum span of 1800 mm, spaced at 450 mm or 600 mm centre to centre.

The common practice today is to use timber that has been kiln dried to construct a subfloor. Kiln-dried timber is stable and consistent in size, unlike green timber. The commonly available species and stress grades of kiln-dried timber are: radiata pine (treated and untreated), F5, MGP10, MGP12; native hardwoods (mountain ash), F17, F27; Baltic pine, F5, MGP10.

11.4.1 Laying bearers and joists

The laying of bearers and joists is an important operation and must be carried out in an orderly manner. The objective is to finish up with the tops of the joists level and in a straight line (Fig. 11.14). This is essential if they are to provide an accurate base on which the rest of the framework can be erected: they must provide firm support for floorboards so that there are no uneven and creaking floors.

Fig. 11.14 Laying bearers and joists

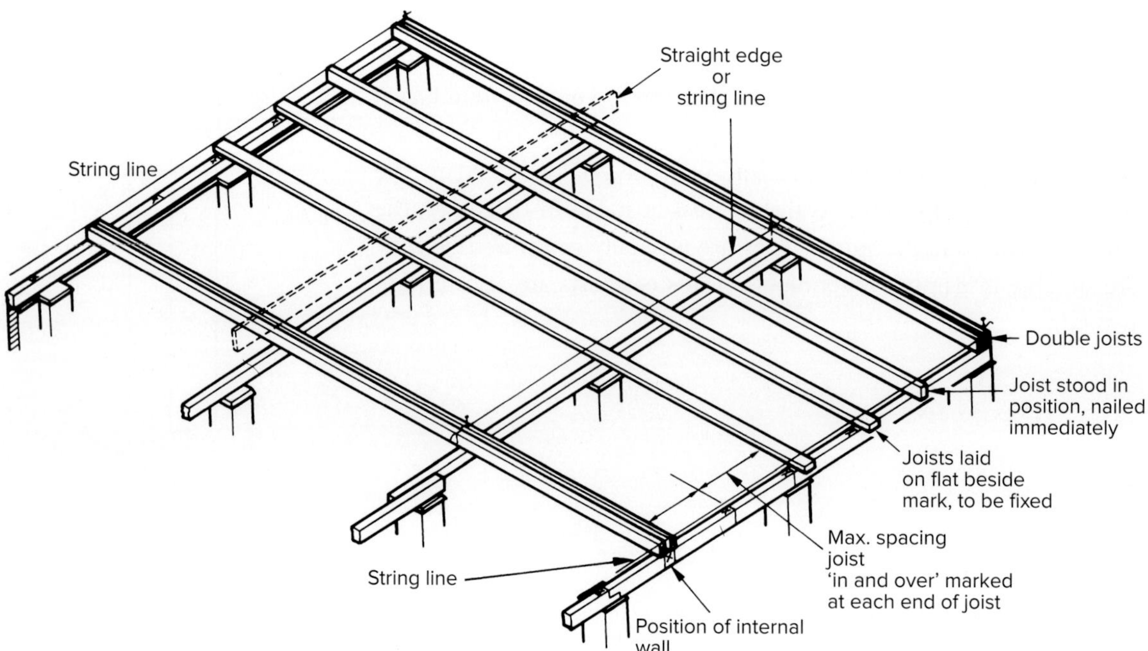

After the initial work of clearing the underfloor area and setting the termite capping in place, the laying of the bearers and joists can commence.

The procedure is as follows:

Fig. 11.15 Joining bearers

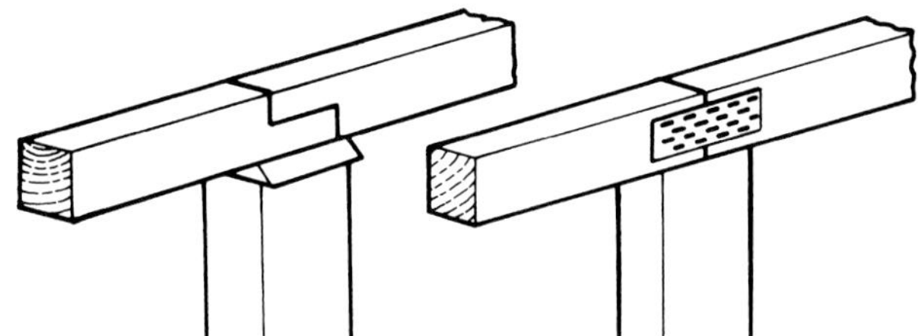

11.4.1.1 Step 1

Lay out the bearers over piers, stumps or foundation walls and join in length (Fig. 11.15). Intermediate bearers can be lapped and skew nailed together, but outside bearers must be kept in a straight line and joined with a scarfed joint or butted with a nail plate.

Fig. 11.16 Seating bearers on piers

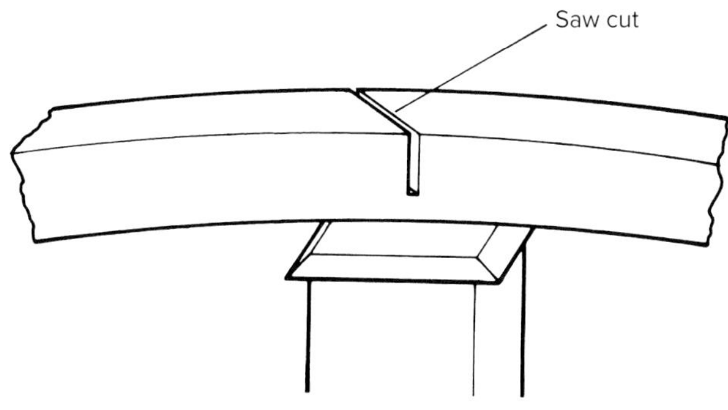

Bearers are laid with the round edge up and must bed down firmly on each pier or stump (Fig. 11.16). Where a bearer has an excessive round and will not bed down securely, it is sometimes cut partially through over the pier. The cut is made diagonally across the edge and over the centre of the pier, only deep enough to weaken the bearer so that it will rest down firmly on the pier. If it is supported by wooden stumps, the bearer is skew nailed to the stump.

TIP When a bearer or joist is cut over a point of support to remove an excessive bow, it can no longer be deemed 'continuous span' and may not comply with AS 1684.4.

11.4.1.2 Step 2

From the stack of joists available, select the straightest. Use these for double joists.

11.4.1.3 Step 3

Plumb up from the foundation walls, stumps or piers. Measure and mark the overall dimension of the wall along the bearers. Measure and mark the length of the double joists and, by skew nailing, fix the pairs of double joists to the external walls in position. Ensure they are straight by using a string line or by sighting. More secure nailing will be obtained if the surplus ends are left on and then cut off after nailing.

11.4.1.4 Step 4

If the building is to have a fitted floor, set out and mark the location of the interior walls on the bearers. Fix double joists to provide support for the wall and the ends of the floorboards and to maintain the correct overall dimension of the building.

11.4.1.5 Step 5

Between the pairs of double joists, set out the spacing for the intermediate joists using the 'in and over' method of measurement.

Fig. 11.17 Straightening bearers

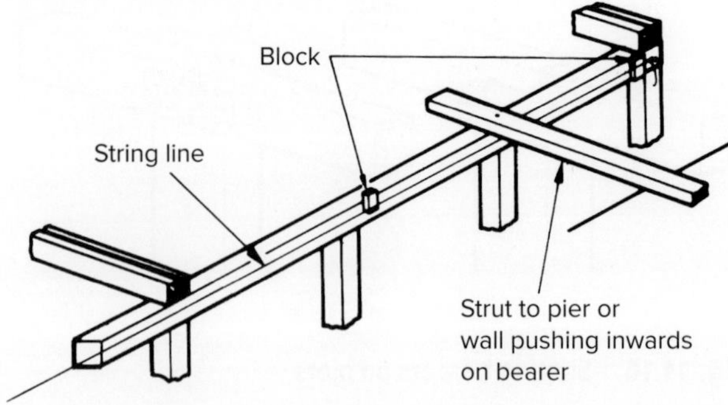

11.4.1.6 Step 6

Before proceeding to fix the remainder of the joists, straighten the outside bearers by setting up a string line packed out from the side of the bearers (Fig. 11.17). Some light raking struts from the ground may be necessary to hold them straight.

11.4.1.7 Step 7

Lay the joists on flat across the bearers beside each mark.

TIP Never, under any circumstances, leave a joist standing on edge until immediately before it is to be nailed permanently into position. Should anyone step on a loose joist on edge it can easily roll, leading to severe injury to the carpenter or an unsuspecting workmate.

11.4.1.8 Step 8

If using **green hardwood** or recycled timber of inconsistent sizes, set up the external joists, then string a line above the edge of the outside bearers and at the level of the top of the joists (Fig. 11.18). To align the tops of the joists over the intermediate bearers, a long straight edge can be a suitable alternative, although it may be cumbersome to use.

If using dressed seasoned timber or a manufactured product, the joists can be positioned and nailed into position. Step 9 is not required.

11.4.1.9 Step 9

If using green hardwood or recycled timber with the joists laid on flat beside each mark, measure the distances from the top of the bearer to the string line or straight edge (distance *x*). In many cases the joist may be the correct width and can be stood on edge and skew nailed to the bearers. If it is too wide, however, measure down from the top the distance *x* and notch out the bottom edge. After checking each point of contact, stand it on edge and nail it into position.

If the narrowest joists are selected and used as external double joists to establish the height of the joists, theoretically none of the joists should need packing to bring them up to the correct height. In practice, however, the odd joist may need packing up to the correct height at some point. The material used for packing should be non-compressible and should cover the whole bearing area. Heavy fibre cement board has been found to be satisfactory. It can be nailed through with no chance of cracking or 'shivering' out. Small chips of wood are not permissible.

Fig. 11.18 Notching joist to obtain level floor (green hardwood or recycled timber)

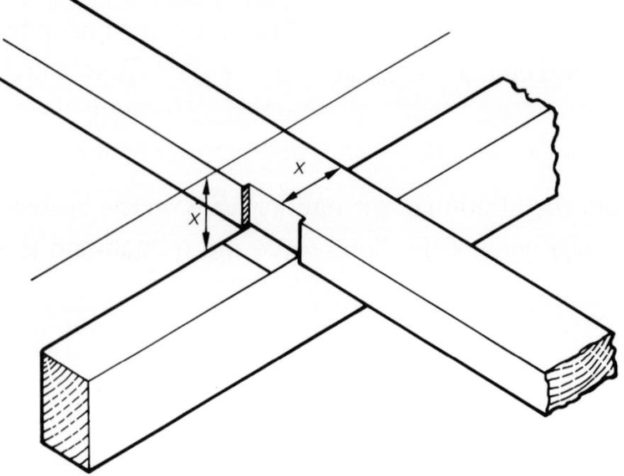

Joists are laid with the round edge up; if they are badly rounded, they are sometimes partly cut through diagonally over the bearer to secure a firm fixing (Fig. 11.19).

Fig. 11.19 Straightening rounded joists (spring)

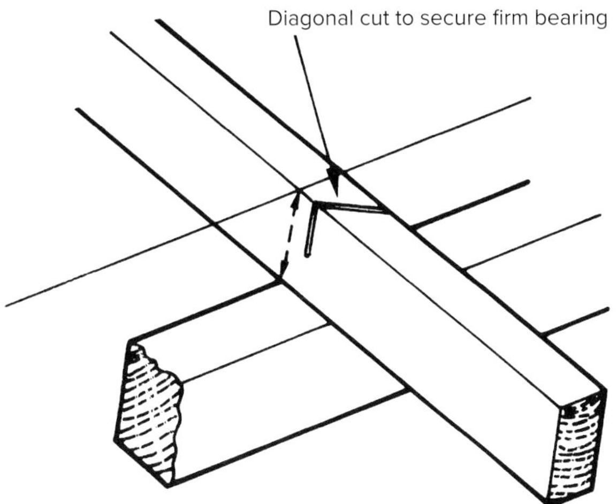

> **TIP** When a bearer or joist is treated in this way, it can no longer be deemed 'continuous span' and may not comply with AS 1684.4.

11.4.1.10 Step 10

Mark joists to length from the outside string lines and cut them to length.

Laying joists for a platform floor can be carried out with just a few variations. Set up the pairs of outside double joists and space out the intermediate joists over the full area of the floor. Using string lines, permanently fix some of the intermediate joists in position at their correct height. Use a straight edge or string line to fix the remaining joists in between, as previously described.

CHECK YOUR UNDERSTANDING

1. Define the following terms:
   a. sole plate
   b. pad
   c. stump/post
   d. bearer
   e. joist.
2. Which species and section size of timber is traditionally used for the following?
   a. Bearers
   b. Joists
3. Describe two methods of joining bearers over a stump/pier.
4. Briefly describe how the tops can be kept in line when laying green hardwood or recycled timber floor joists.
5. List three materials from which stumps can be made.
6. Where should double joists be located for the following?
   a. Platform floor
   b. Fitted floor

Fig. 11.20 Trimming around openings

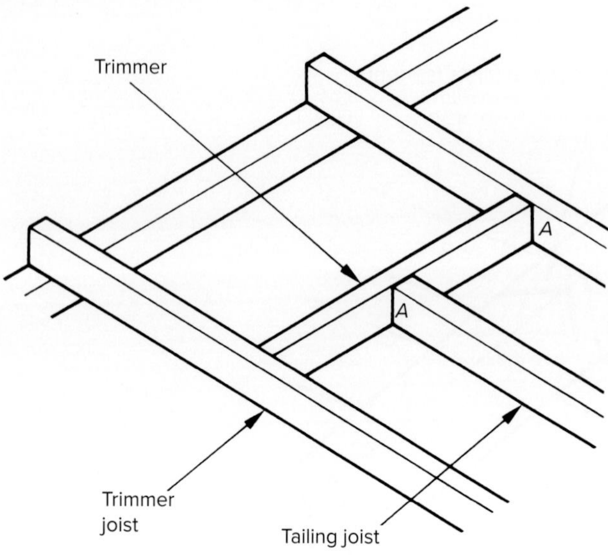

11.4.2 Trimming and openings in floors

Where there are openings in floors, it is necessary to trim around the opening. This procedure consists of shortening one or more of the joists. They are then called *curtailed* or *tailing joists*. The ends of the tailing joists must be supported by a trimmer, which is then supported by the adjacent joists, known as *trimming joists* (Fig. 11.20).

Joints between these joists must be made in such a way as to transfer the load without seriously weakening the joists. They also must not rely on nails in the end grain to carry the load. Suitable metal fittings are used to strengthen the joints. Where the trimmer exceeds 1 metre in length, it is increased in width to carry the increased load. The trimming joists must be of the same section as the trimmer they support.

Trimming out under wall plates (blocking) is used when there is a join in the bottom plate that does not land on a joist. Joist material is used to block under the wall plate, in line with the plate; therefore, these trimmers run at 90 degrees to the joists. The entire plate must be supported, so a double trimmer would be used.

11.5 Strip flooring

Strip flooring consists of individually milled tongued and grooved (T & G) boards. Throughout Australia and New Zealand, many species of timber are available for flooring, including mixed hardwoods or selected hardwoods such as tallowwood, brush box, blue gum, merbau, blackbutt, turpentine, mountain and alpine ash, as well as the softwoods, cypress pine and radiata pine.

Flooring is described by its species and *nominal sectional size* (Fig. 11.21), that is, the width and thickness of the sawn board from which it is milled. Widths vary from 75 mm to 150 mm; thicknesses are usually 19 mm or 25 mm. For example:

- 85 × 25 tallowwood T & G flooring
- 110 × 25 cypress pine T & G flooring.

To calculate the number of linear metres of flooring required to cover 1 square metre, divide the 'cover' into the 1 metre, expressed as 1000 mm. For example, the linear metres needed to cover 1 square metre is equal to 1000 divided by 85, which equals 11.77 m. Cutting waste must be added to

Fig. 11.21 Strip flooring profile (cross-section)

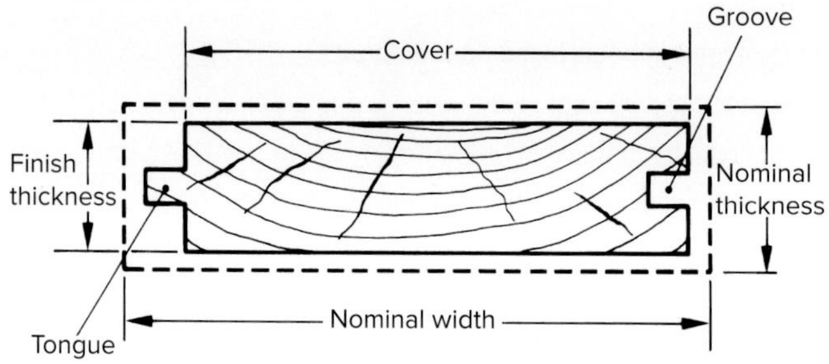

Groove
Cover
Finish thickness
Nominal thickness
Tongue
Nominal width

this total in order to allow for cutting out defects, cutting boards back over joists, or cutting to room lengths. The amount of waste will vary; however, a figure of, say, 6% is a fair average for good-quality flooring.

End joints in floorboards are usually made over a joist and the ends are only slightly undercut so that they will butt tightly at the face (Fig. 11.22). End-matched flooring has the tongue and groove continued across the ends, and the joints are not necessarily placed over a joist, thus there will be less waste in cutting down.

To calculate the linear metres of flooring required to cover a given area, apply the following formula:

Fig. 11.22 (a) Butt joint and (b) end-matched joint

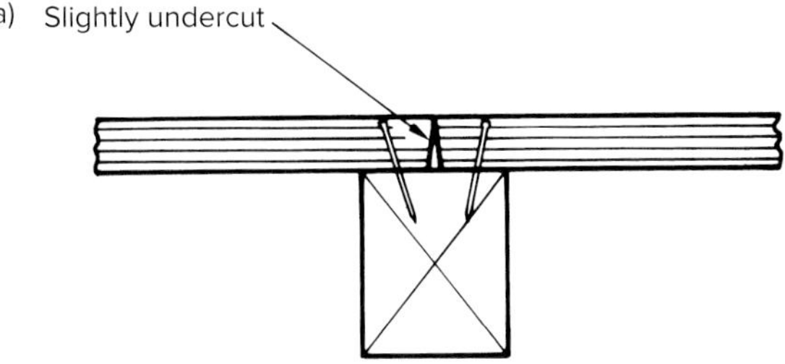

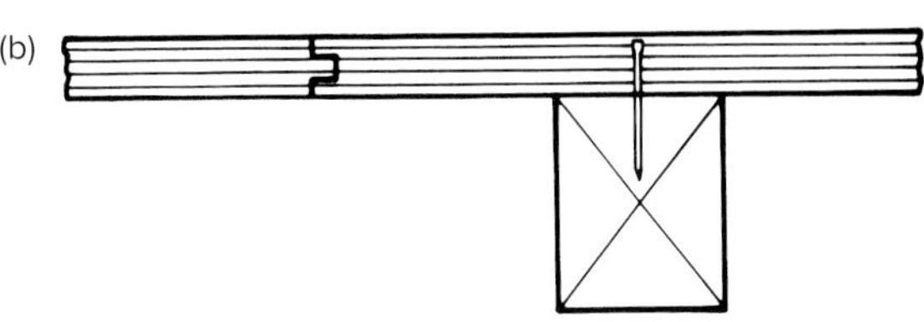

$$\text{linear m} = \text{area} \times \text{metres of flooring per m}^2 + \text{cutting waste}$$

For example, an area 12.60 m by 5.65 m is to be covered in 85 mm × 25 mm tallowwood flooring. Allow 6% for wastage.

$$\begin{aligned} \text{linear metres} &= 12.6 \times 5.65 \times 11.77 \\ &= 838 \text{ m} \\ \textit{plus} \text{ cutting waste} &= 838 \times 6/100 \\ &= 50 \\ \text{Total} &= 888 \text{ m} \end{aligned}$$

Alternatively,

$$\begin{aligned} \text{linear metres} &= (12.6 \times 5.65 \times 11.77) \times 1.06 \\ &= 888 \text{ m} \end{aligned}$$

11.5.1 Secret nailed flooring

Secret nailing is used where a top-quality floor is required, such as a dance floor (Fig. 11.23). The boards are no more than 75 mm wide so that there is a minimum of shrinkage, and they are specially milled so that they can be nailed through the tongue and end-matched. Each board must be cramped and nailed separately. These floors are obviously comparatively slow to lay and thus expensive.

11.5.2 Moisture content

The moisture content of the flooring boards at the time of laying is important. Boards that are unseasoned or have a high moisture content will shrink excessively, leaving unsightly gaps or, in extreme cases, will allow cold air to seep up past the tongue and groove.

Boards with too low a moisture content will absorb moisture and expand, lifting the joists and bearers up from the piers, thus causing a very springy and unstable floor–a condition for which there is no simple cure.

The moisture content of boards should be tested at the time of laying. Alternatively, the boards should be brought in directly from

Fig. 11.23 Secret nailing

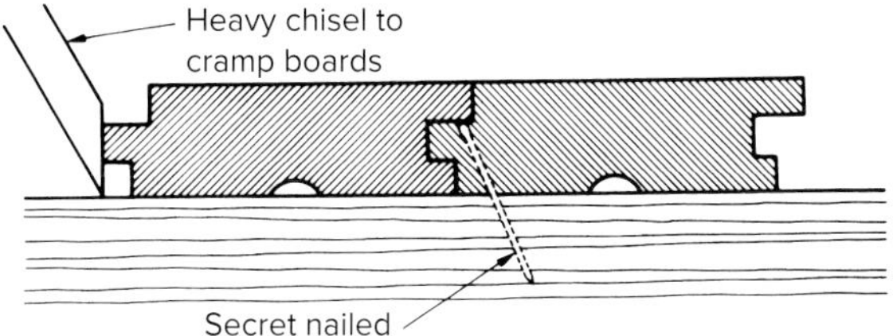

Fig. 11.24 Starting point for fitted floor

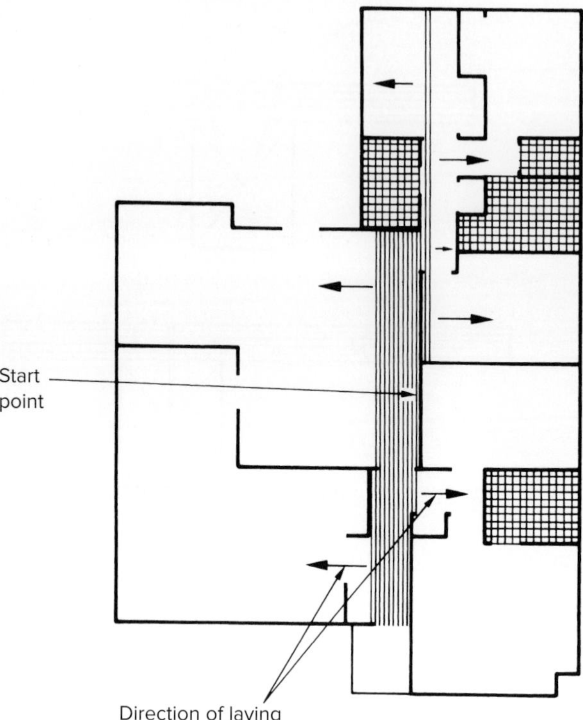

a mill at a specified moisture content. Even where the regulation amount of underfloor ventilation is provided, there is a tendency for the area to remain damp. Generally, fewer flooring problems will arise if the boards are laid with a moisture content of about 2% above average for the area.

In some areas, particularly where cypress pine flooring is used, it is common practice to use unseasoned flooring that is cut down, turned upside down and left to lie loose for a few weeks. It provides a safe working platform while construction proceeds and also allows the moisture content to stabilise. When the building is finally enclosed and the roof covered, it is then turned over and fixed before the interior finishing commences. Strip flooring is more suited to a fitted floor.

Fig. 11.25 Floor cramp

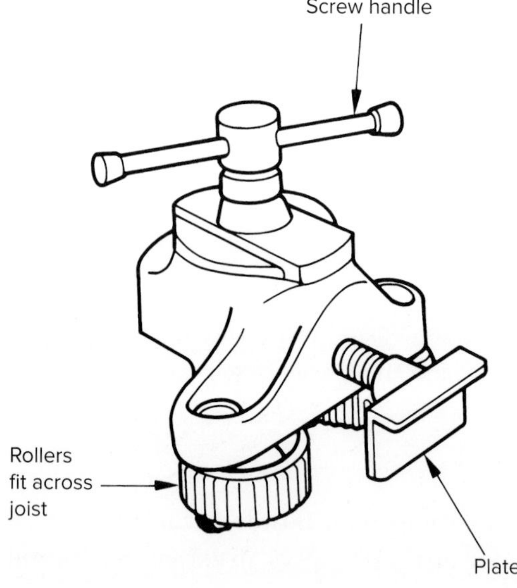

11.6 Fitted floors

The starting point for laying a fitted floor will vary with each individual plan, but as a general rule the longest run of boards can be found somewhere near the centre of the building, probably along a passageway and into some of the rooms (Fig. 11.24). From this starting point, continue laying the flooring in both directions through doorways and into individual rooms.

AS 1684 states that the boards should be kept 10 mm clear of wall plates, parallel to the direction in which they are laid, and that the ends of the flooring boards should not bear against the bottom plate at right angles to the direction of laying.

Refer to AS 1684 or the manufacturer's recommendations for the fixing of floorboards. Boards are generally nailed and the nails punched well below the surface.

The flooring cramp has two cams that separate and grip the joists, plus a screw or ratchet action that cramps the floorboards (Fig. 11.25). Two are usually needed for a satisfactory job.

11.6.1 Step 1

Start by fixing the first board 10 mm clear of walls and keeping it straight, particularly where it continues into other rooms. Use a string line if necessary.

11.6.2 Step 2

Cut down enough boards to cover up to about 900 mm, enter the tongues into the grooves and tap up lightly.

11.6.3 Step 3

Lay a straight buffer piece along the edge of the boards and fix the floor cramps behind them (Fig. 11.26). Lay a short piece of board at right angles over the boards, and stand on this to prevent the boards buckling while cramping them up to a pressure relevant to the moisture content. Mark the centre line of the joists across the face of the boards and start nailing from the cramped edge, omitting the first nail to expedite the entry of the next board.

11.6.4 Step 4

It is usually sufficient to cramp boards at every second joist. To keep them straight, cramp only lightly near the ends where the boards will pull in more easily.

11.6.5 Step 5

Continue laying in a similar manner until the last few boards–these will have to be cramped with a lever or heavy chisel (Fig. 11.26). Cut the last board to fit 10 mm clear of the wall.

Fig. 11.26 Nailing and cramping a fitted floor

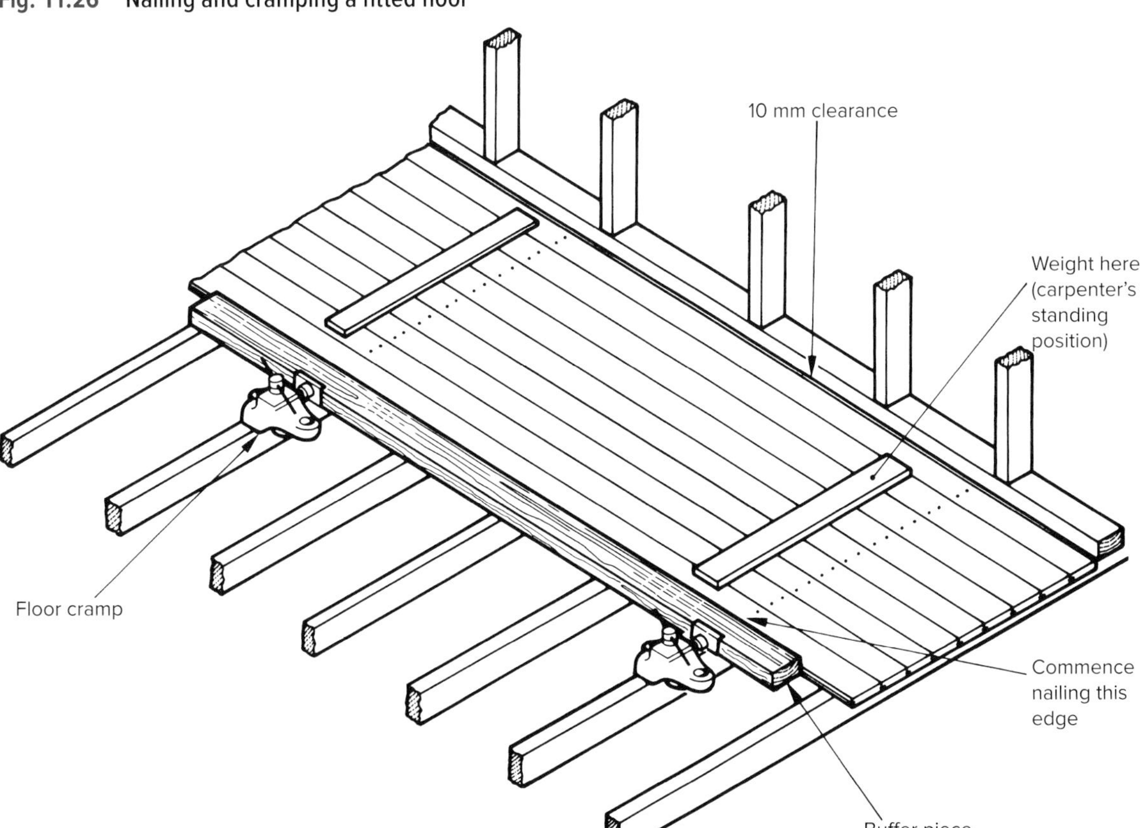

11.7 Platform floors (using strip flooring)

When laying a floor over a large rectangular area–a platform floor or the floor of a public hall–start by laying the first boards down the centre of the area, using a string line to keep them straight. Continue laying each side of the centre towards the outer walls. Take care that the boards are kept straight and parallel at each cramping (Fig. 11.27).

Fig. 11.27 Laying a platform floor (strip)

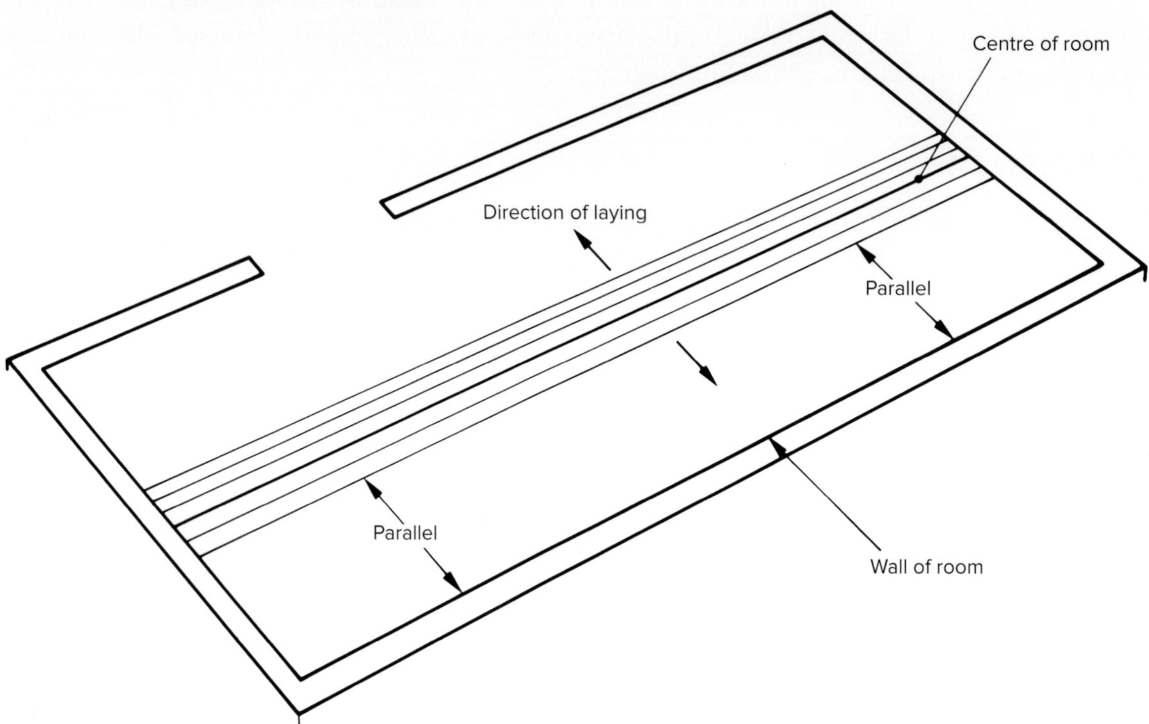

11.8 Sheet flooring

Sheet flooring is widely used and permits floors to be laid very quickly, particularly when applied as a platform floor.

11.8.1 Particleboard

Structural particleboard sheet flooring is manufactured to AS 1859 (Reconstituted wood-based panels) and is available in two thicknesses–19.5 mm for joists spaced at 450 mm centres and 22 mm for joists spaced at 600 mm centres. The common sheet sizes are 3600 mm × 900 mm, 3600 mm × 600 mm, and 1800 mm × 900 mm. The edges of the sheets are grooved and have a factory fitted plastic tongue (Fig. 11.28). The ends of the sheets are sealed and are butted together over joists.

Sheets are fixed to joists with 50 × 2.8 galvanised nails with a flat or bullet head. It is also recommended that a 5 mm bead of structural adhesive be laid on the joists before positioning the sheets and nailing off.

Figure 11.29 illustrates a common pattern for nailing; however, always refer to the manufacturer's recommendations.

Regulations may vary between states and territories throughout Australia; the manufacturer's instructions should always be referred to before using particleboard flooring.

11.8.2 Procedure for fixing platform floor

11.8.2.1 Step 1

Fix joists to suit the thickness of sheet being used: 450 mm or 600 mm centre to centre. Add an extra joist under any load-bearing walls running parallel to the direction of the joists.

11.8.2.2 Step 2

Set a string line to the outside edge of the building and at right angles to the joists.

11.8.2.3 Step 3

Position the first sheet with its tongue edge to the string line and nail to the joists. Do not hard nail the grooved edge: this will make mating of the next sheet easier. It is highly recommended that construction-grade adhesive be used on the joists before positioning the sheets. The adhesive will dry quickly, and the sheet should be placed no more than 10 minutes after laying the adhesive, then nailed to joists within 15 minutes of placing.

11.8.2.4 Step 4

Fix the second sheet again with its tongued edge to the string line and ensure that it is butted firmly end to end with the first sheet.

11.8.2.5 Step 5

Repeat Step 4 for subsequent sheets to complete the first row.

Fig. 11.28 Detail of tongue for sheet flooring

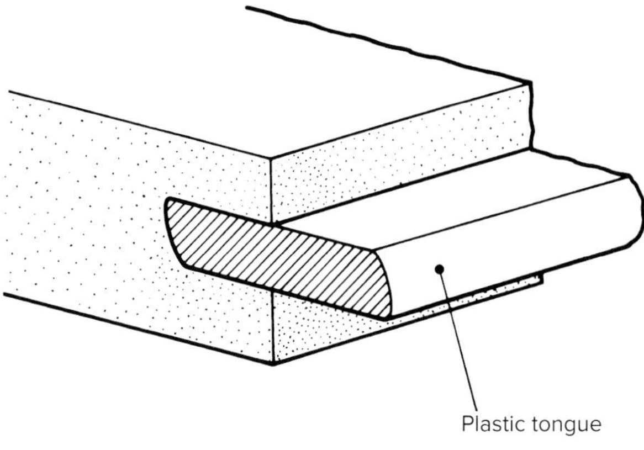

Fig. 11.29 Nailing pattern for sheet flooring

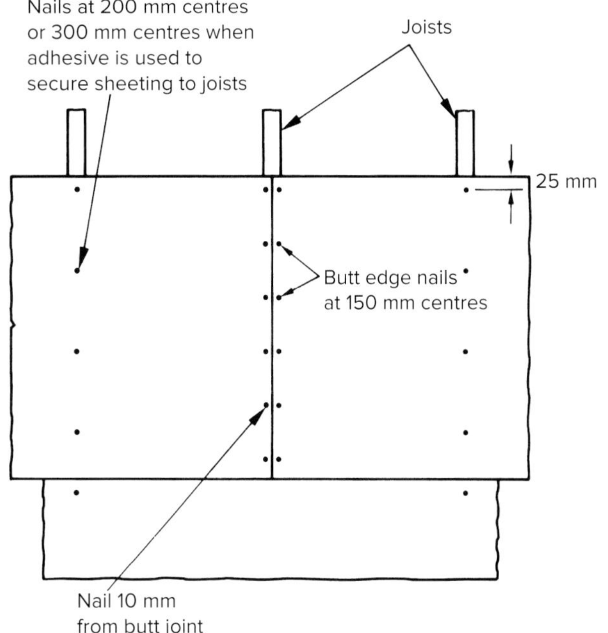

11.8.2.6 Step 6

Fix the second and subsequent rows of sheets, ensuring that the butt joints are staggered and that the plastic tongues are entered into the grooves. After each row has been laid, nail off the previous row.

11.8.3 Structural plywood flooring

The installation of structural plywood flooring is a similar process to that described for particleboard, and it is particularly suitable for platform floor installations (Fig. 11.30). The material may be softwood or hardwood plywood, and is classified as a structural 'A' bond ply. This means that the adhesives used to hold the laminations of the board together are waterproof. Structural plywood is available in two thicknesses for flooring–12 to 25 mm.

Refer to the manufacturer's instructions for fixing of these products.

Installing plywood flooring

Some important factors to consider when installing plywood flooring are as follows:

1. The grain of the face of the boards must run parallel to the span or perpendicular to the joists and the sheets must be continuous over more than one span.

2. The sheets should always be laid with the face side upwards. The back side is normally marked.
3. The joists must be butted over bearers, and cannot be lapped, as is often the case with strip flooring. All butted joints must be central over a joist.
4. Nail spacing is similar to other sheet materials, but it is suggested that a flooring adhesive also be used.

Fig. 11.30 Laying sheet flooring

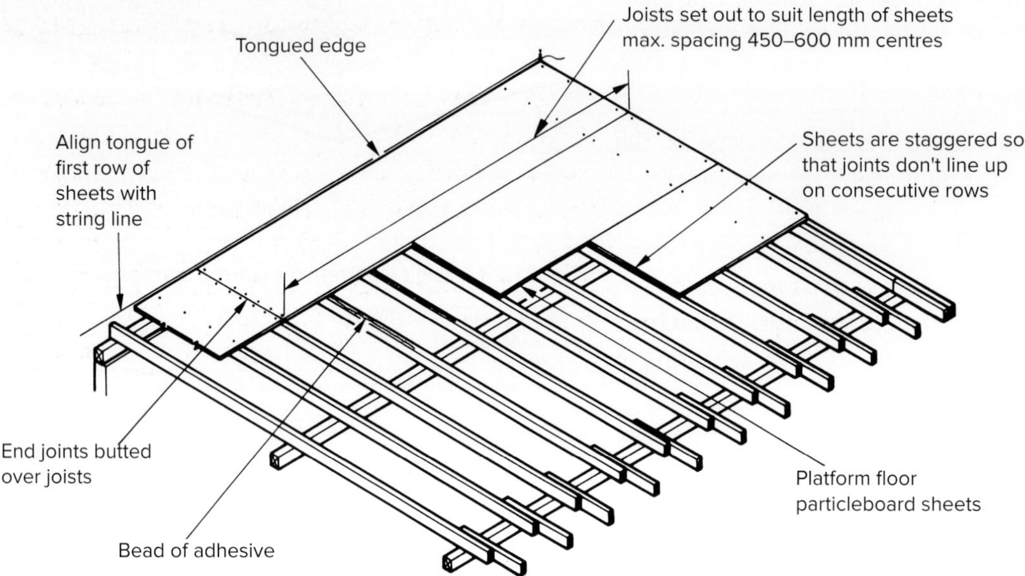

11.9 Wet area flooring

Wet area flooring is required in rooms such as bathrooms and laundries.

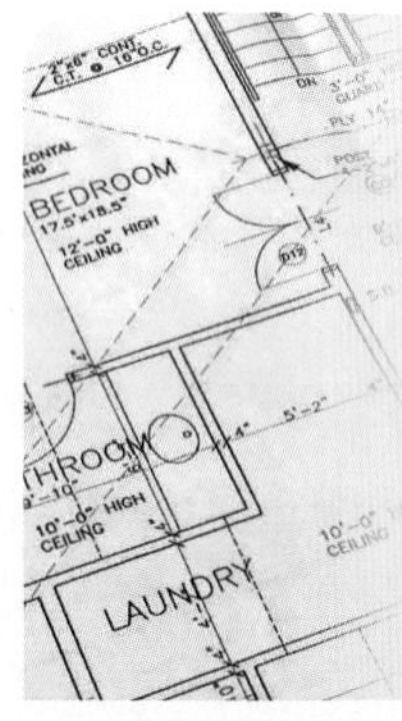

INTERPRETING PLANS

Refer to the set of plans located in the appendix.

1. How many rooms in the building would be considered as wet areas?
2. What is the total floor area of all of the wet areas combined? Please provide your answer in square metres.
3. If the wet area sheet flooring is 3.6 m × 9 m wide, how many sheets would be needed to floor all of the wet areas? Add 10% for waste.

11.9.1 Fibre cement compressed sheets

Fibre cement compressed sheets are high-density water-resistant sheets that can be laid directly over the timber floor joists to provide a highly durable floor for wet areas. The sheets are available in thicknesses of 15 mm to 24 mm. It is preferable to lay sheets before the wall frames have been stood in place.

11.9.1.1 Fixing fibre cement sheets

Sheets are fixed to joists by the use of screws placed no more than is recommended by the manufacturer. The ends of sheets must be fully supported. Butt joints in adjacent sheets are sealed by a liberal application of epoxy adhesive.

11.9.2 Particleboard flooring to wet areas

Manufactured particleboard flooring that contains formulated additives can be used in wet areas (Fig. 11.31). The thickness and sheet sizes are compatible with standard particleboard and are marked in a distinctive manner on both faces to prevent them from being confused or misplaced.

To use wet area particleboard, simply cover the wet area with the wet area sheets. The wet area sheets may extend beyond the designated wet area in order to bond with other sheets.

Fig. 11.31 Wet area particleboard

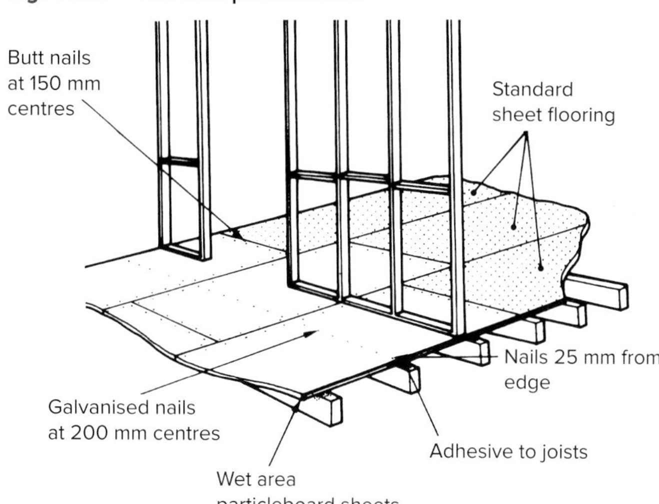

The wet area sheets are fixed in the same manner as standard sheets except that the use of structural adhesive together with galvanised nails is mandatory. The necessary materials and detailed instructions are available from the manufacturer and should be carefully followed.

11.9.3 Structural wet area plywood

Plywood may also be used in wet areas, as long as it is a tongue and grooved structural grade plywood with a type 'A' waterproof glue bonding the laminations. Either softwood structural or hardwood structural plywoods are acceptable.

The thickness of plywoods should be matched to the joist spacing. The installation of plywood wet area flooring is similar to the sheets already described; however, the basic requirements are as follows:

1. The face grain of the sheets should run at right angles to the direction of the joists and should be continuous over two or more spans.
2. Sheets should be butted over joists, and these joints should be staggered.
3. Plywoods in wet areas should be glued and nailed. The adhesive and fixing methods should follow the manufacturer's recommendations.
4. Prior to tiling, the floor should be coated with a waterproof epoxy sealer approved for the material.

CHECK YOUR UNDERSTANDING

1. Using suitable sketches, show the meaning of the following terms applied to strip flooring:
   a. secret nailed flooring
   b. end-matched flooring
   c. standard strip flooring.
2. Briefly describe how you would proceed to lay structural sheet material over a platform floor.

11.10 Alternative subfloor materials

11.10.1 Laminated veneer lumber

Laminated veneer lumber (LVL) is a high-strength engineered wood product used primarily for structural applications. It is comparable in strength to solid timber, concrete and steel and is manufactured by bonding together rotary peeled or sliced thin wood veneers under heat and pressure. LVL was developed in the 1970s and is used today for permanent structural applications, including bearers, joists, beams, lintels, underpurlins, truss chords and formwork. LVL can be used wherever sawn timber is used. However, one of its main advantages is that it can be manufactured to almost any length, restricted only by the availability of transportation to site.

11.10.2 Steel

Steel is another material that can be used to construct a subfloor. There are a number of cold-formed steel profiles available to use as either bearers or joists, as shown in Figure 11.32.

Used as a subfloor material, cold-formed steel is robust and termite resistant; it also resists corrosion as the section is galvanised and is lightweight. The components can be used in either a traditional configuration, as shown in Figure 11.33, or an inline system, as shown in Figure 11.34.

The method used to construct a subfloor using steel is very similar to that for a timber subfloor. However, where nails would be used to connect the components, screws are commonly used instead, as shown in Figure 11.35.

As with any construction, the steel components' size and application, connection specifications and tiedown requirements need to comply with relevant codes and standards. The National Association of Steel-Framed Housing (NASH) produces the technical information for the use of cold-formed steel-framed components–in particular, the standards that are referenced in the National Construction Code (NCC).

NASH also produces fact sheets, PowerPoint presentations, trade instruction sheets and the General Guide to Steel-Framed Building; refer to the NASH website at www.nash.asn.au. NASH standards and handbooks may also be purchased through the website.

Fig. 11.32 Typical joists and bearers

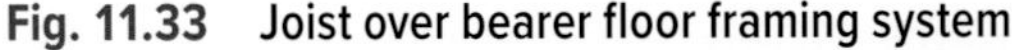

Reproduced with permission of NASH Inc. www.nash.asn.au

The building envelope

This is a concept that describes all of the exterior parts of a building–the foundation, structure, the 'skin', roof, walls, floor, windows and doors, and building services (water, sewerage, electricity, gas).

Fig. 11.33 Joist over bearer floor framing system

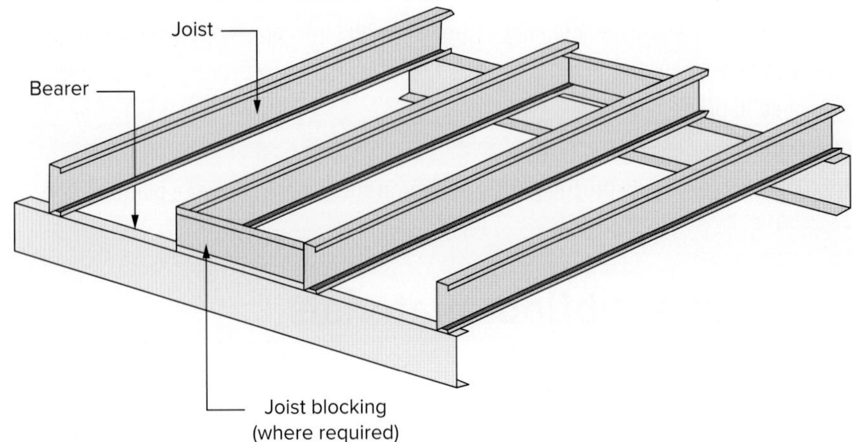

Reproduced with permission of NASH Inc. www.nash.asn.au

Fig. 11.34 Inline floor framing system

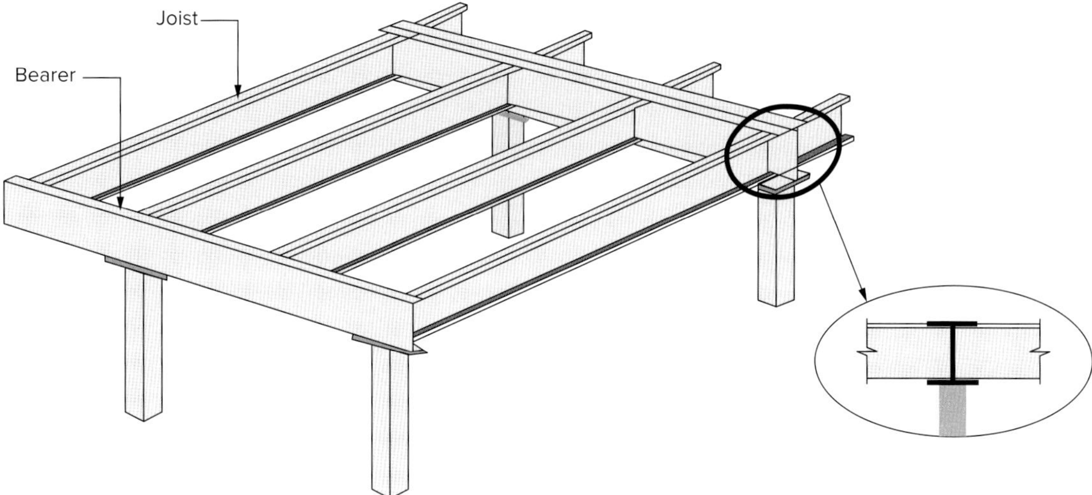

Reproduced with permission of NASH Inc. www.nash.asn.au

Fig. 11.35 Floor joists to bearer connection: lower storey

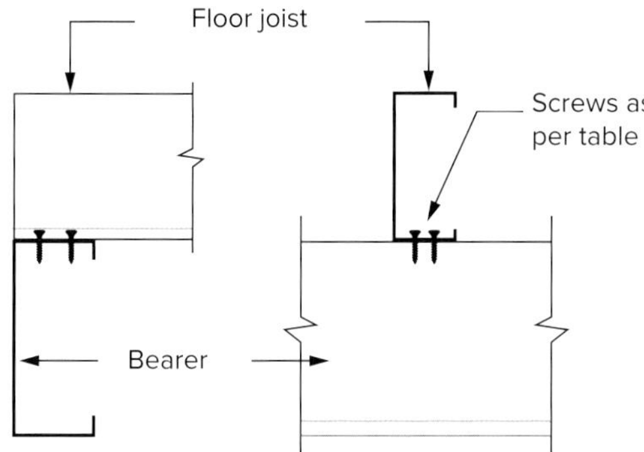

Reproduced with permission of NASH Inc. www.nash.asn.au

Fig. 11.36 The building envelope

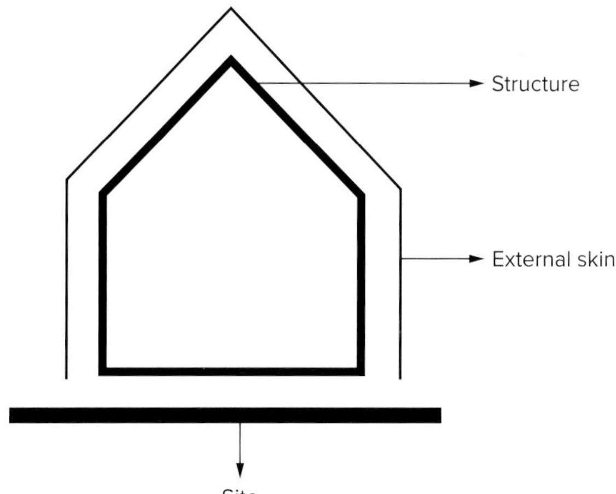

The function of a building's envelope (Fig. 11.36) is to:

- provide structural support to the building parts
- manage or control the flow of air, energy, water and matter from the outside to the inside
- allow the building's materials to enclose the space, connect together sensibly and provide security.

The effectiveness of the envelope centres on the management of weather and climate, indoor air quality, thermal comfort and energy efficiency. Achieving this depends on a sound structure, suitable materials and barriers to water, heat and cold penetration.

In addition, the ability of the envelope to drain water away, and control internal moisture are paramount. For example:

- a brick veneer wall is a dual-stage system–it comprises a primary barrier (the brick skin), through which water penetrates, and a secondary waterproofing barrier or membrane and flashing system to prevent water penetration and divert it back to the exterior.
- a timber-frame wall with timber cladding–any water penetration or condensation (due to humid weather) that is behind the exterior skin becomes trapped and begins the process of decay.

Student research

Obtain a copy of AS 1684. With your teacher's help, go through the section that deals with subfloor construction. Research the following:

- stress grading—what is it and what are its forms?
- green timber definition
- seasoned timber definition.

End of chapter activity

Activity 1:

As part of your training, you will be required to set out and construct a timber subfloor, including digging holes, placing stumps and installing bearers and joists suitable for a platform or cut in the floor.

AUSTRALIAN BUILDING CODE AND STANDARDS

AS 1684 National Timber Framing Code

National Construction Code (NCC)

The National Association of Steel-House Framing (NASH)

Tradie talk—Breaking the timber's back

NEXT-GENERATION CARPENTERS

Structural timber building products have been used for generations. Like no other building product, timber is versatile, sustainable, renewable and greenhouse positive (it absorbs carbon when growing and stores carbon once cut). Carpentry skills have been used to build everything from homes and playground equipment through to bridges. These skills have evolved as new, innovative engineered structural timber products have been manufactured that enable the construction of longer-span, taller and bespoke timber buildings.

Forté Apartment Melbourne

Forté Apartments, Melbourne: Lendlease Design; Photography Boris Iskra

Traditionally, single and two-storey timber-framed houses were manually constructed 'stick-by-stick' using sawn structural softwood and hardwood timber products from the ground up: stumps, subfloor framing, wall framing through to pitching the roof framing. Today, much of this construction can be pre-fabricated, which enables precision-engineered designs, leading to increased speed in construction, less waste, fewer delays due to inclement weather and safer working environments. Carpentry skills have evolved to cater for these new methods of construction.

Changes introduced to the National Construction Code, or the Building Code of Australia, as it is also known, enable timber buildings to be built up to an effective height of 25 metres under the Deemed-to-Satisfy provisions (25 metres is measured from the ground-floor exit to the topmost floor) using *fire-protected timber* and sprinklers throughout the building. This change enables eight-storey timber-frame and massive timber apartment and hotel buildings to be built using traditional sawn structural timber-framing products; and large open-plan office buildings through to hospitals to be built using new, massive engineered timber products such as laminated veneer lumber (LVL), cross-laminated timber (CLT) and glued laminated timber (glulam).

These new engineered timber products can also be used as the exposed structure, and are used to 'frame up' buildings as long-spanning beams and tall columns using LVL or glulam; and continuous floor plates using LVL or CLT connected using thousands of metal screws and bolts, in the order of 300–400 mm in length, metal brackets and other new innovative connectors. No special carpentry tools are typically required except for the humble cordless drill.

International House Sydney

International House Sydney: Tzannes Architects for Lendlease; Photography Ben Guthrie

From a construction point of view, the advantages of using timber building systems in tall buildings are that they are lightweight compared to traditional building materials (up to half the weight of concrete systems), are typically pre-fabricated (door and window openings are precut), precise (panel tolerances within ±2 mm), delivered on a flatbed truck, quick and easy to install, and have been shown to have a positive impact on occupants' health and wellbeing.

Using innovative computer numerical controlled (CNC) cutting machines, intricate shapes, forms and details can be pre-manufactured into massive timber products to the designer's specifications and can be erected on-site by skilled tradespeople. This computer-aided design, detailing and manufacture results in modern buildings that were previously viewed as impossible to construct.

International House Sydney

International House Sydney: Tzannes Architects for Lendlease; Photography Ben Guthrie

From a carpentry point of view, timber is a familiar construction material, is readily available, can be more easily worked and is easier to erect than other building materials. It can now be used in the construction of taller timber buildings where high-strength, safety, reliability, sustainability and aesthetically beautiful construction is desired ... constructed by the next-generation carpenters.

Boris Iskra
National Codes and Standards Manager
Forest and Wood Products Australia

Image acknowledgements:
Forté Apartments, Melbourne: Lendlease Design; Photography Boris Iskra.
International House Sydney: Tzannes Architects for Lendlease; Photography Ben Guthrie.

Chapter 12

Wall framing

Learning Objectives

LO 12.1 Identify wall framing parts

LO 12.2 Arrange studs at wall junctions

LO 12.3 Identify bracing types

LO 12.4 Identify timber sizes

LO 12.5 Set out wall plates

LO 12.6 Assemble timber wall frames

LO 12.7 Use tie-downs and brackets

LO 12.8 Assemble steel wall framing

LO 12.9 Know about two-storey construction

Introduction

In this chapter you will learn how to plan, prepare, set out, construct and erect wall frames for different types of roof loadings. This includes the setting out, cutting and fabrication of both timber and steel wall frames, and the erection, connection and bracing of wall frames to appropriate standards. You will also learn how to plan, prepare and install window frame units to required plans and specifications.

The notes contained in this section on timber wall framing are based on the requirements of the following construction standards:

- Australian Standard AS 1684.2: 2010 Residential Timber-Framed Construction–Part 2: Non-cyclonic Areas
- Australian Standard AS 1684.4: 2010 Residential Timber-Framed Construction–Part 4: Simplified–Non-cyclonic Areas.

12.1 Wall framing parts

The wall framing must provide fixing support for interior and exterior wall sheeting as well as support for the ceiling, roof framing and roof covering. The members comprising the wall framing are named and illustrated in Figure 12.1.

Walls can be *load-bearing* or *non-load-bearing*. The exterior walls of a timber-framed building are load-bearing; in the case of the untrussed roof, where some of the roof loads are transferred to internal walls by way of underpurlins and struts, internal walls also become load-bearing.

A building with a trussed roof has load-bearing external walls, but the interior walls maintain a clearance between the top plate and the truss and carry none of the roof loads; these walls are termed non-load-bearing. When selecting suitable section sizes for members of wall framing, it is necessary to consider how much of the roof loads will be supported by each wall. For this reason, span tables refer to the **roof load width (RLW)**.

Two common examples are shown in Figure 12.2 and include the formula used for determining the RLW factor for member sizing. The trussed roof carries all of the roof loads on the external walls and the RLW is equal to the full span of the top chord as shown; however, in the untrussed roof where some of the loads are diverted to the interior walls, the roof load is reduced as shown.

The RLW is used as an indicator of the loading carried by some roof members as well as the load-bearing wall frame members and supporting substructure.

The range of sectional sizes for wall framing members is considerable and can depend on many factors such as spacing, RLWs, timber species and stress grade. Dimensions quoted in the following notes are examples only of common trade practice. Always refer to the relevant span tables included within AS 1684 Residential Timber-framed Construction for firm details.

Fig. 12.1 Wall frame—general view

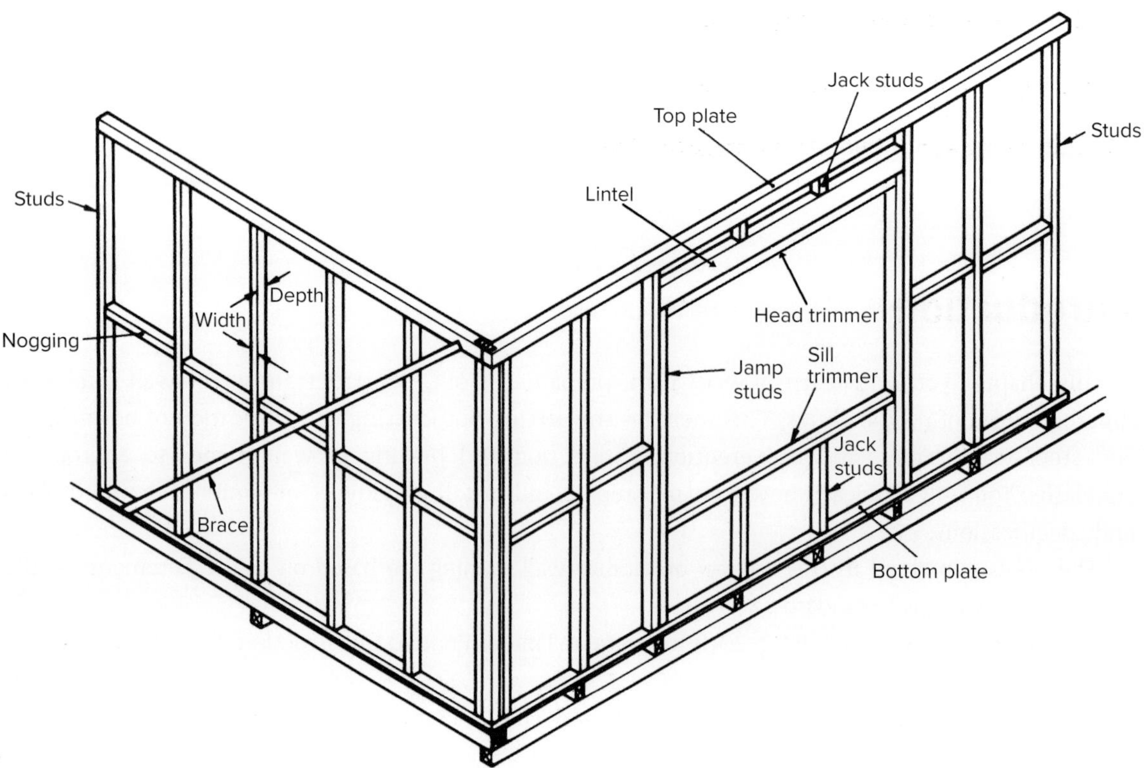

Fig. 12.2 Roof load width (RLW): (a) trussed roof; (b) coupled roof with underpurlins

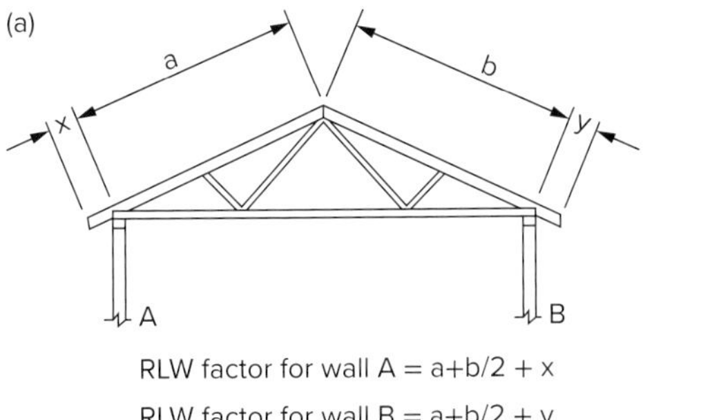

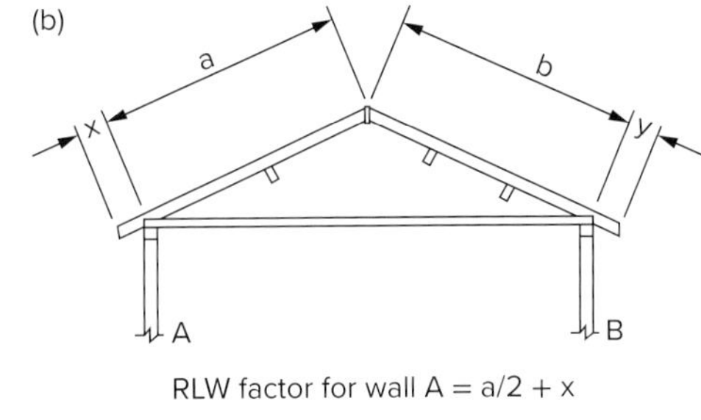

For details of alternative roof designs refer to AS 1684.

12.1.1 Top plate

Fig. 12.3 Laminated top plate

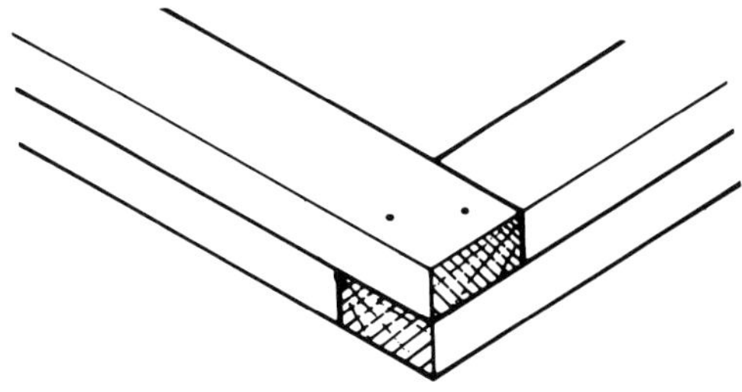

The top plate frames the top of the wall and supports the ceiling joists and rafters (Fig. 12.1). Top plates can be described as having **direct load** where the rafters are located directly over the studs, or are placed within 1.5 times the thickness of the plate from the stud. A **random load** is where the rafters are placed without reference to the location of the studs and exceed the limit. When the top plate is directly loaded, the section size can be reduced to a minimum 35 mm thickness, but if random loaded they must be increased in thickness. Deep top plates can be built up in two laminations, for example, a 75 mm × 75 mm plate can be two lengths of 75 mm × 38 mm (Fig. 12.3).

> **TIP** Where a concentrated load such as roof struts, strutting beams, hanging beams and girder trusses sit on the top plate but are not directly supported by studs designed for that purpose, the top plate shall be stiffened in accordance with the requirements of AS 1684.10 Section 6: Wall Framing (Fig. 12.4).

Fig. 12.4 Top plate stiffening

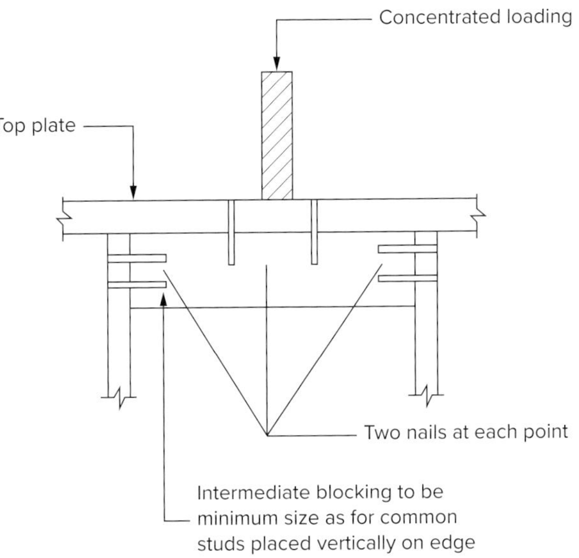

12.1.2 Bottom plate

The **bottom plate** frames the bottom edge of the wall frame and, provided it is supported under the studs by joists, **blocking** or a concrete slab, it can be the same section size as the studs.

Plates can be trenched to a maximum depth of 3 mm to adjust any difference in plate thickness. This is gauged from the outside face of the plate to leave a uniform thickness on each plate. When trenching exceeds this depth, the remaining net depth of the plate shall be used to determine the section size of the plate in accordance with the design limits outlined in the AS 1684 span tables. Trenching is unnecessary when plates are machine gauged to a uniform thickness.

Traditionally, the joints between top plates were done using a butt joint (Fig. 12.5) with a nail plate fixing or half-lapped joints with the joints being nailed together. Bottom plates were permitted to be butt joined at corner intersections as well as the straight joints in continuous wall lengths.

The same joint method applies today for bottom plates, providing both ends are fixed and supported by floor joists, solid blocking or a concrete slab.

Top plates are now joined at corner intersections with the butt joint/nailing plate method, while straight joints in top plates will need to be one of the methods shown in Figure 12.6.

Fig. 12.5 Joining of plates: (a) butt joint; (b) half-lapped joint; (c) tee half-lapped

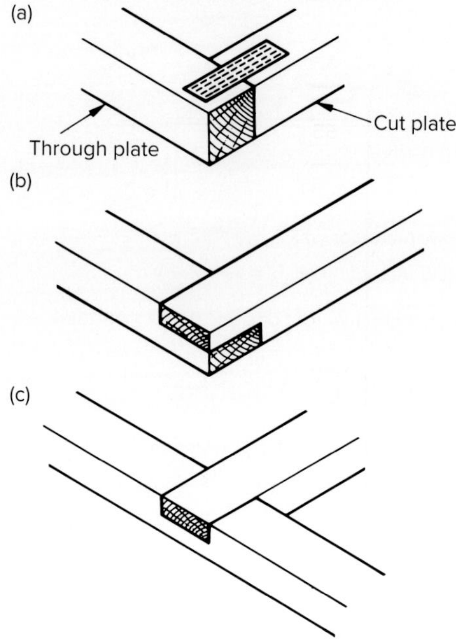

Fig. 12.6 Joining of top plates: (a) top plates joined using plate connection method; (b) ribbon plate construction; (c) plate joints secured with nogging support

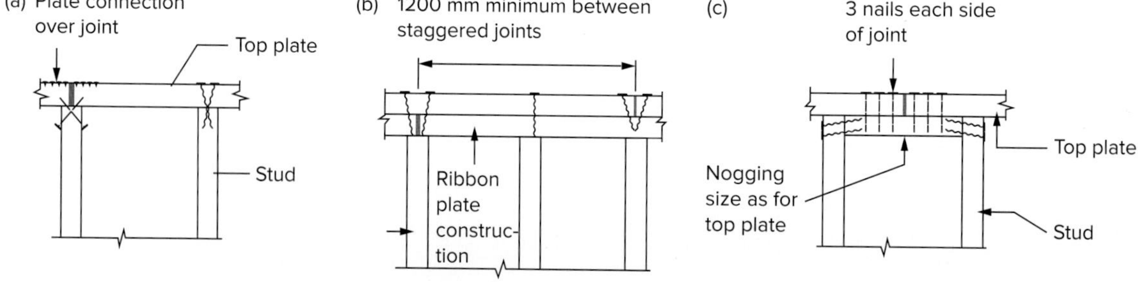

CHECK YOUR UNDERSTANDING

1. When selecting a section size for wall top plates, what is the difference between random load and direct load?

12.1.3 Studs

Studs are vertical members fixed between plates spaced at 450 mm or 600 mm centre to centre. The length of the stud must be calculated to preserve the specified dimension from floor to ceiling, the minimum for habitable rooms being 2400 mm.

Fig. 12.7 Pattern stud

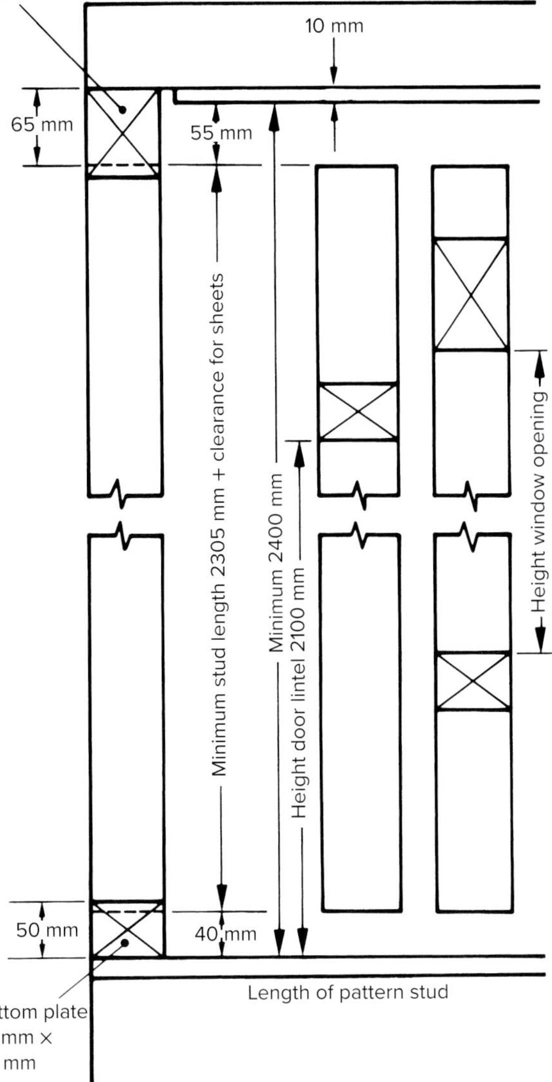

The example in Figure 12.7 shows the calculations for determining the length of the stud for a wall on a platform floor and a ceiling of minimum height. Where plasterboard sheets are fixed to the wall, add a clearance of, say, 20 mm. The bottom sheet of plasterboard is 1200 mm wide and fixed 10 mm from the floor and the top sheet is installed on top with a working clearance of 10 mm at the top. In this example the studs would be cut:

2305 mm + 20 mm clearance = 2325 mm

Studs can be notched up to 20 mm to receive braces and, in some cases where timber is of the lower stress grades (F4, F5), they may have to be increased in depth to compensate for the notching (Fig. 12.8).

12.1.4 Straightening studs

Studs that are badly rounded can be straightened by partly cutting through on the hollow side and inserting a wedge to pull it into line (Fig. 12.9). The cut stud must then be strengthened by placing anther stud next to it and laminating the two studs together.

Fig. 12.8 Stud notching to receive bracing

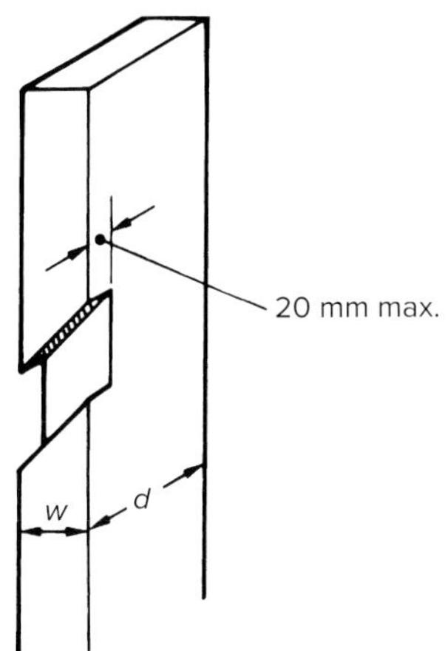

Fig. 12.9 Straightening studs

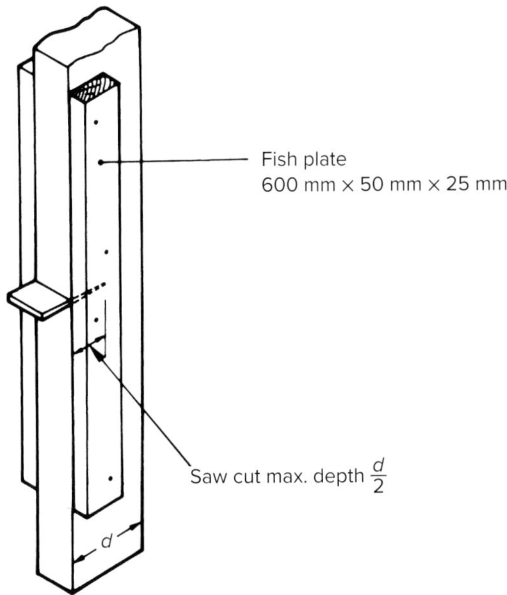

12.2 Arrangements of studs at wall junctions

Studs must be arranged at wall junctions to support the corner and to provide firm ground for fixing the interior linings and the outside sheeting (Fig. 12.10). Corner junctions require three studs.

The placing of the studs also depends on the order in which the walls are erected and the way in which the wall plates are joined. The three studs must be securely nailed together using blocking pieces at least 200 mm in length where necessary. In the case of brick veneer construction, two studs cleated together are sufficient (Fig. 12.11).

At other wall intersections, three studs can also be used (Fig. 12.12), or an alternative arrangement is to use a single stud at the end of the intersecting wall with a backing board. The backing board is 25 mm thick and 50 mm wider than the depth of the end stud, and is let in flush with the face of the intersected wall to provide grounds for fixing internal linings (Fig. 12.13).

The advantage of this method is that the studs in the external wall can be spaced along the full length of the building at their maximum spacing and the interior walls attached in this manner wherever they occur, leading to a saving in the number of studs required at wall junctions. Also, if the roof rafters are spaced to come over the studs, the section size for the top plates can be taken as for direct loading, permitting further economy in timber.

Fig. 12.10 Corner junctions

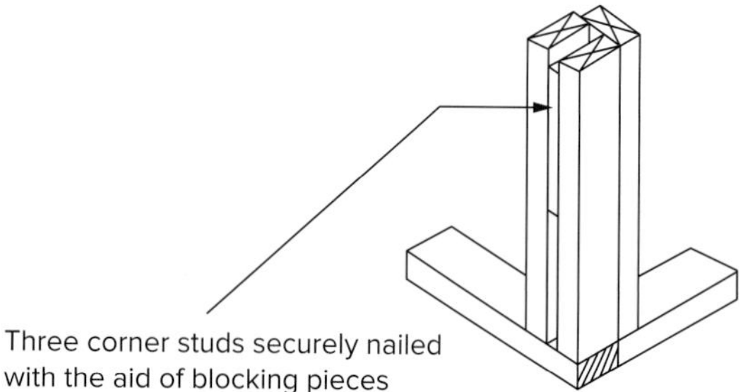

Fig. 12.11 Corner junction—cleating method

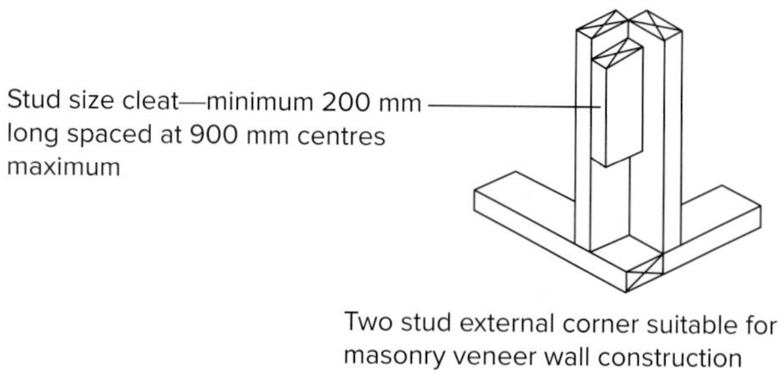

Stud size cleat—minimum 200 mm long spaced at 900 mm centres maximum

Two stud external corner suitable for masonry veneer wall construction

Fig. 12.12 Internal junction

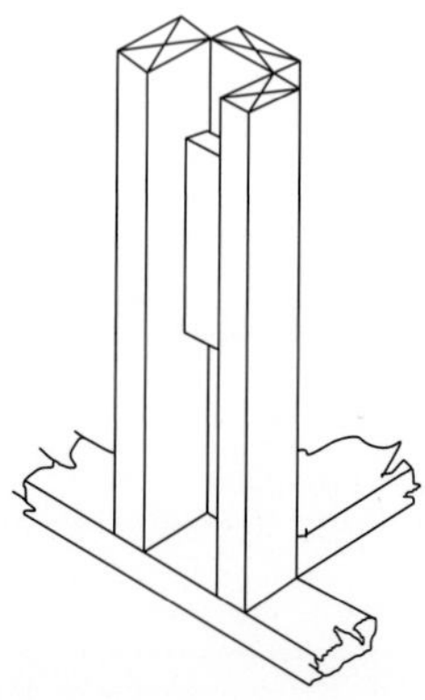

Internal corner securely nailed with the aid of blocking pieces

Fig. 12.13 Internal junction—backing board

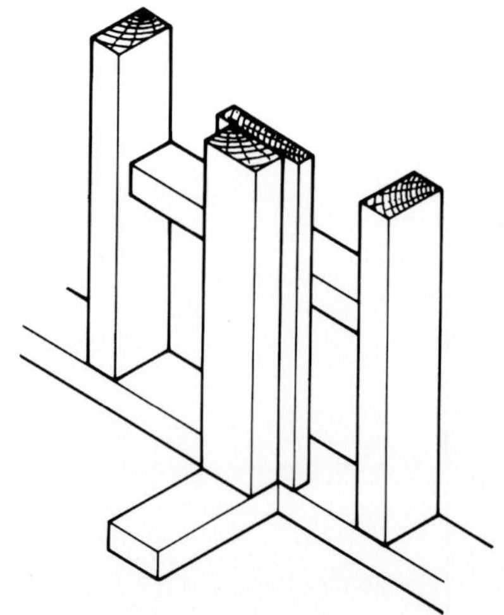

CHECK YOUR UNDERSTANDING

1. Sketch two arrangements of wall studs suitable for an external corner of wall framing.

12.2.1 Lintels

Where openings occur in the wall framing to provide for a door or window, the load over the opening is supported by a lintel. The section size of the lintel will depend on the stress grade of timber selected, the span of the lintel, the rafter/truss spacing and the RLW.

> **TIP** A minimum clearance of 15 mm must be provided from the top of the window frame to the underside of the lintel or lintel trimmer when determining the opening height to allow for weight deflection of the roof members.

12.2.2 Jamb studs

The ends of lintels must be supported by studs at each side of an opening and the studs must be increased in size to carry the additional load. These are referred to as *jamb studs*.

The joints between lintels and jamb studs must provide adequate bearing for the lintel and also secure fixing to the studs. The examples shown in Figures 12.14 to 12.16 can be adopted as appropriate.

This method is designed to be used for spans not exceeding 900 mm (load-bearing walls) and for spans not exceeding 1800 mm (non-load-bearing walls).

Fig. 12.14 **Lintel housing into jamb stud**

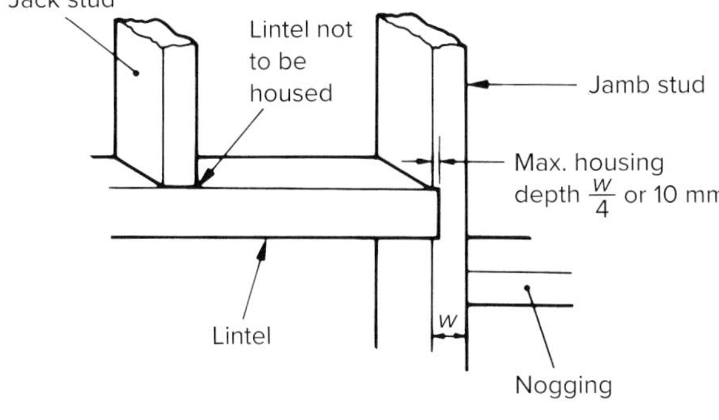

Fig. 12.15 **Lintel thickness less than or equal to half the stud depth and notched through jamb stud**

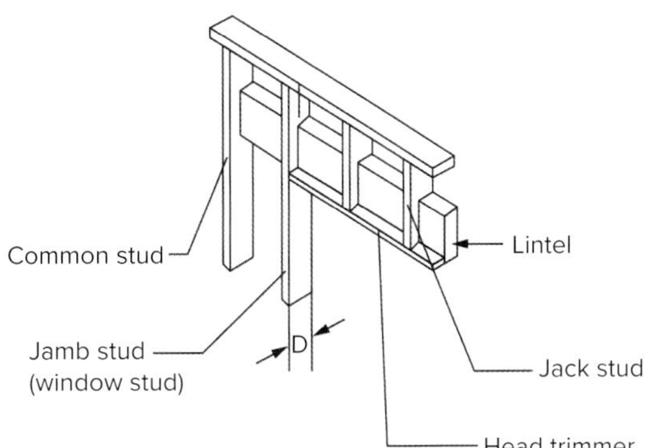

12.2.2.1 Sill trimmers

Sill trimmers are generally of the same section size as the common studs in that wall.

12.2.2.2 Jack studs

Jack studs are short studs above and below window and door openings.

12.2.2.3 Noggings

The studs in each panel of framing are stiffened by the addition of noggings closely fitted between the studs at not more than 1350 mm centres in height. Noggings are usually made in the same section size as the common studs in that wall.

Fig. 12.16 **Lintel supported by jamb stud**

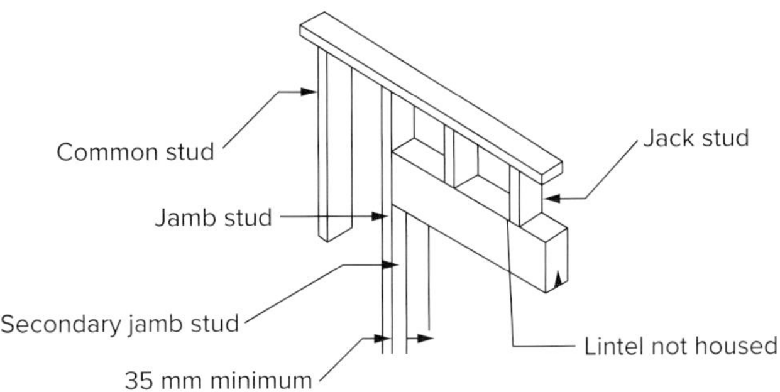

12.3 Bracing

Stud walls must be stiffened against longitudinal distortion by means of permanent bracing; otherwise, a rectangular frame can be readily pushed out of shape (Fig. 12.17).

Fig. 12.17 Need for bracing frame: (a) rectangle readily distorted; (b) brace corrects rectangle to two triangles

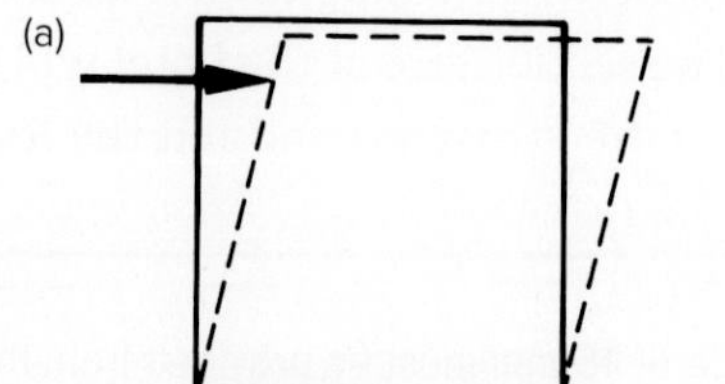

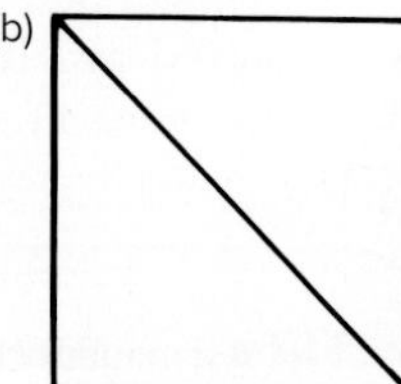

Wall-bracing design, as referred to in AS 1684, describes two types: nominal wall bracing and structural wall bracing.

1. **Nominal wall bracing.** This refers to the lining fitted to the wall frame such as plasterboard, fibre cement, hardboard, plywood or similar materials that are fixed in accordance with the relevant Australian Standards. Nominal bracing can provide up to a maximum 50% of the total bracing required.
2. **Structural wall bracing.** This is purpose-fitted bracing that can be of sheet type, diagonal timber or steel bracing (Fig. 12.18). The introduction of a brace converts the rectangle into two triangular frames and a triangle cannot be distorted without breaking at least one of the sides or joints. The principle of the triangular frame will occur frequently in building when considering the strength of structural methods. The most effective angle for a brace is 45°, with each wall including at least two braces acting in opposite directions.

Fig. 12.18 Structural wall bracing: two diagonally opposed timber or metal angle braces

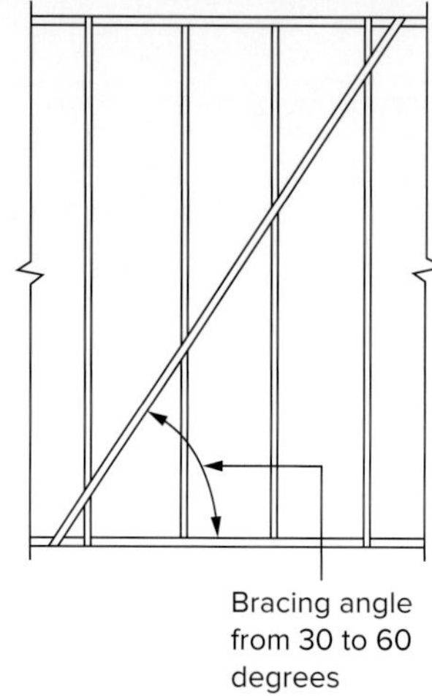

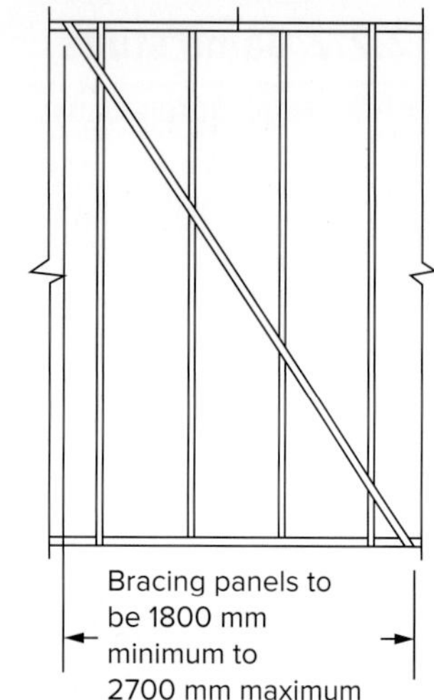

CHECK YOUR UNDERSTANDING

1. What function is served by the following members in timber wall framing?
   a. Top plate
   b. Jamb studs
   c. Head trimmer
   d. Nogging
   e. Diagonal brace
   f. Backboard

12.3.1 Timber bracing

The minimum section size for timber bracing in single-storey construction is 45 mm × 19 mm hardwood. The studs must be notched and the brace let in flush. To calculate the approximate length of material required for a **diagonal brace**, calculate 1.5 times the height of the wall.

For example, for a wall height of 2.4 m, then:

$$\begin{aligned} \text{brace} &= 2.4 \text{ m} \times 1.5 \\ &= 3.6 \text{ m}. \end{aligned}$$

12.3.2 Metal bracing

A galvanised steel angle, 20 mm by 20 mm with a series of preformed holes, can be used as bracing (Fig. 12.19). A single saw cut is necessary to let in one leg of the angle to finish flush with the stud, and the stud must then be regarded as being notched for the purpose of selecting a suitable size.

Fig. 12.19 Steel angle bracing

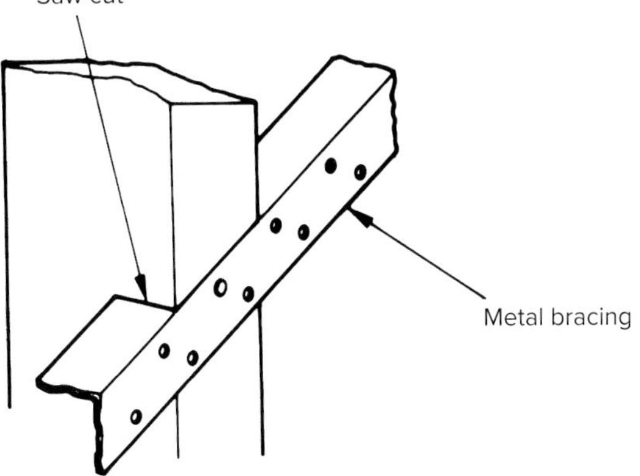

12.3.3 Tension bracing

Tension bracing can be used in walls with un-notched studs (Fig. 12.20). The bracing consists of flat galvanised steel at least 19 mm × 0.8 mm. As it is effective as a brace in tension only, two braces must be used on each panel of wall framing, acting in opposing directions to each other. The braces must be fixed from the top to the bottom plate at an angle of between 30° and 60° to the horizontal and be adequately tensioned.

Fig. 12.20 Tension bracing

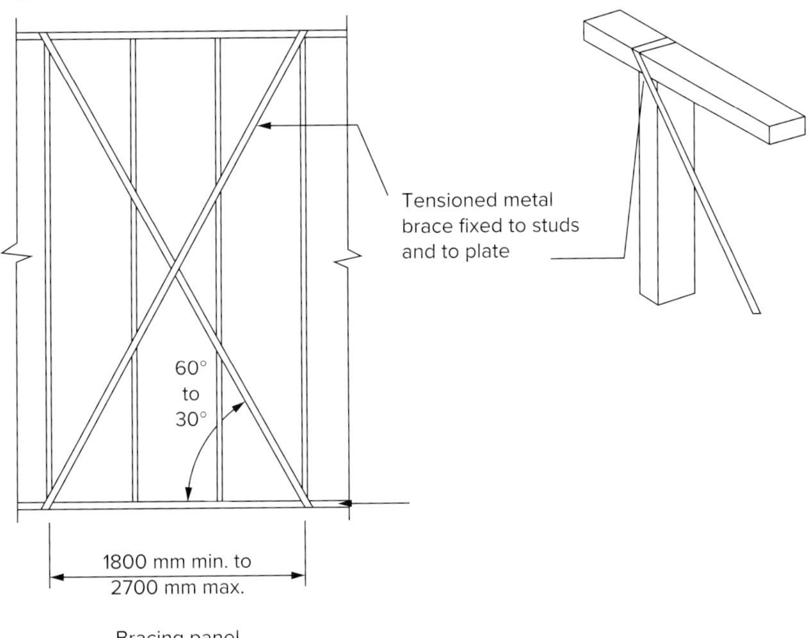

12.3.4 Plywood bracing

Structural plywood from 4 mm to 9 mm in thickness can provide effective bracing when applied to un-notched studs (Fig. 12.21). The minimum length of plywood bracing to each straight wall is 1.8 m and is usually fixed as two 0.9 m panels at the corners.

Fig. 12.21 Surface bracing

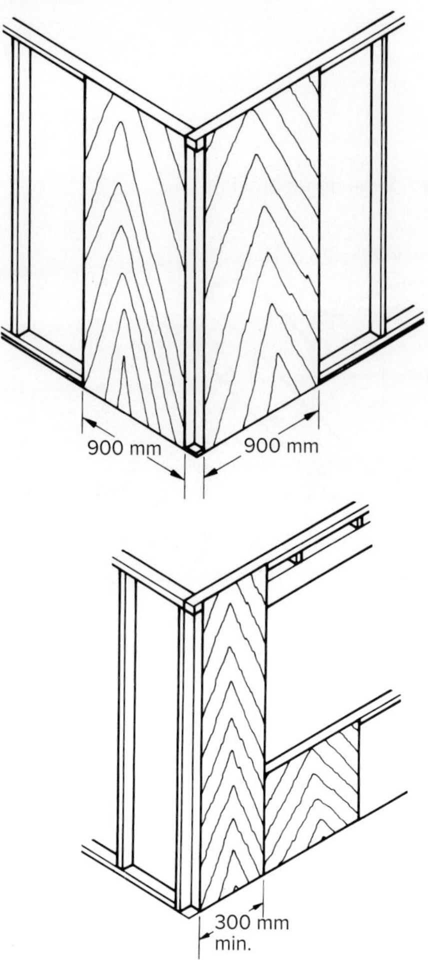

The plywood can be fixed to the internal or external face of the walls, and narrow panels (300 mm minimum) can be applied beside window or door openings to make up the required length of plywood bracing. Structural particleboard, fibro cement and hardboard are acceptable as alternative sheet bracing materials.

12.4 Timber sizes

The selection of suitable sizes for members of wall framing depends on a variety of factors.

A full range of span tables for designing various timber frame components is available in the supplement to AS 1684.2. This supplement contains a selection of design tables for all critical framing members, from floor bearers to roof battens, and covers a range of the common stress grades (strength) of timber available. Table 12.1 shows one example that details how to select the correct **lintel** size for a door or window opening when using F17 seasoned hardwood and meets the requirements for non-cyclonic areas. The example is for lintels that are to support the load of a tile roof and are built into a single- or upper-storey load-bearing wall. Table 12.1 shows only the key heading of the required input for determining the allowable spans and lintel sizes. For firmer details you will need to refer to the current AS 1684.2 supplement Timber Framing Span Tables.

Table 12.1 Span tables for lintels: tile roof, single- or upper-storey load-bearing walls, seasoned hardwood, stress grade F17

Table 18 for LINTELS—Supporting tile roof on single- or upper-storey load-bearing walls										
	Roof load width (mm)									
	1500		3000		4500		6000		7500	
Rafter or truss spacing (mm)	600	1200	600	1200	600	1200	600	1200	600	1200
Lintel beam size D × B (mm)	Maximum lintel span (mm)									

Adapted from AS 1684.2 N1/N2 Supplement 8 2010

> **TIP** You will need to calculate the RLW and also give input for the rafter/truss spacing and the lintel span when using this table.

12.5 Setting out wall plates

Setting out wall plates is a responsible task that is generally entrusted to an experienced tradesperson. An ability to read plans and interpret information from specifications and trade literature is essential. The procedures outlined here have been found to work successfully and emphasise the basic principles.

It is well recognised, however, that some minor modifications may be necessary to adapt these principles to different flooring systems. The following example looks at the procedures of setting out wall plates on a timber platform floor system. However, similar methods would be applied if the wall plates were to be set out on a concrete floor slab.

CHECK YOUR UNDERSTANDING

1. Why does the roof load width (RLW) become significant when designing external wall framing?

12.5.1 Procedure

12.5.1.1 Step 1

Mark out wall frame positions.

Before wall plates can be laid out, the position of the walls will need to be accurately marked out on the floor with the aid of chalk lines. Some wall marks will have been set out previously on the external subfloor bearers to assist with the positioning of floor joists under walls. These marks can now be transferred onto the platform floor. Carefully check that measurements are accurate to the floor plan dimensions. Set out all remaining wall positions and mark off the location of both sides of the plates for their full length. Show all junctions and where walls may end.

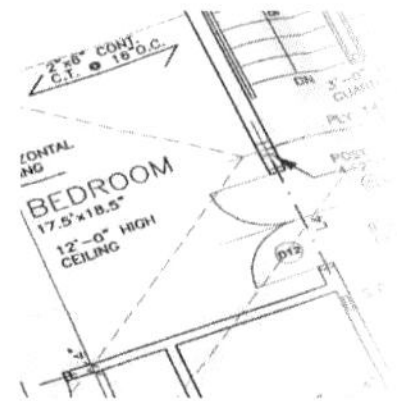

INTERPRETING PLANS

Refer to the floor plan in the Appendix and identify the internal width that the robes in Bedrooms 2 and 3 need to be set out at when marking out the wall plate.

12.5.1.2 Step 2

Select suitable lengths of plate material from the timber ordered.

Always choose the straightest lengths for the **top plates** as this will make it easier for straightening at a later stage. The plates should now be cut to length in pairs and tacked together in place on the floor with the top plate uppermost (Fig. 12.22).

Plates are joined at the wall junctions and can be made to span more than one small room, but they should be constructed in sections that suit ease of handling. Consideration will need to be given to deciding which wall plates will run through and which plate will butt into the other at junctions. A general rule is that the long walls running the length of the building at right angles to the roof span will run through. End walls will run between these, butting into them to form the corner junction.

Where an internal partition wall meets an external wall ('T' junction), the external wall will run through with the internal wall butting into it (Fig. 12.23).

Joints in long walls can be made away from a corner junction. (For information on the joining of top plates, refer back to Figure 12.6.)

Fig. 12.22 Wall plates tacked in position on the floor

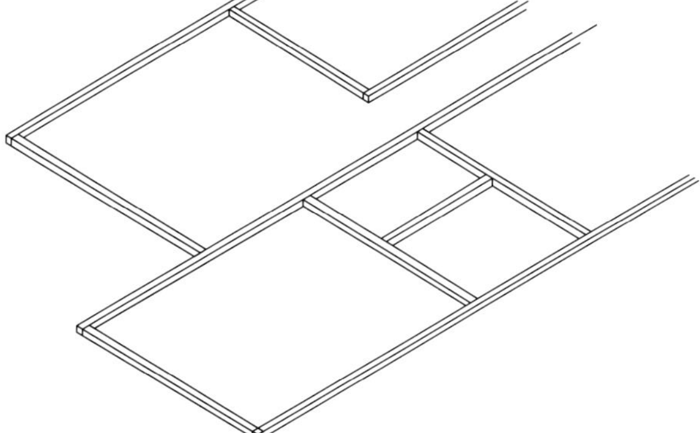

Fig. 12.23 Joining top plate at a 'T' junction

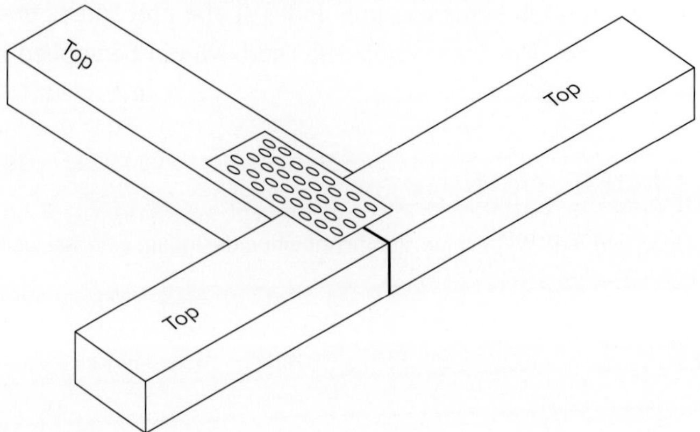

Joins made in top plates should not be made over a door or window opening.

When the plates are all cut to length in pairs and tacked together on the floor as described, the next task will be to accurately mark out the position of the wall studs.

Set out and mark stud positions in the following order:

- wall junctions
- door and window openings
- studs identified to support concentrated loads
- intermediate common studs including jack studs.

12.5.1.3 Step 3

Mark set out for wall junctions.

Where the walls intersect, mark two lines halfway across the top plate and identify with the letter 'W'. This identification is placed on the face side where the two plates meet. This is to make recognition easier when assembling wall frames later. Using these two intersecting lines you can now locate and mark the position of the corner studs. The stud positions are marked on the edge of both the top and bottom plate as shown in Figure 12.24.

Refer to Figures 12.10 to 12.12, which display various corner stud layouts.

12.5.1.4 Step 4

Set out door and window openings (Fig. 12.25).

Locate the positions for window and door openings and mark for jamb studs. Plates are left running through door openings and this portion of the bottom plate is not cut out until the walls are permanently fixed in position. The stud opening size for a window and door must be made larger than the overall size of the frame, so that the frame can be accurately levelled and plumbed up when it is fixed. In the case of timber window frames, add approximately 20 mm to the overall frame size in height and width to obtain the stud opening size.

Fig. 12.24 Wall junction and stud marking

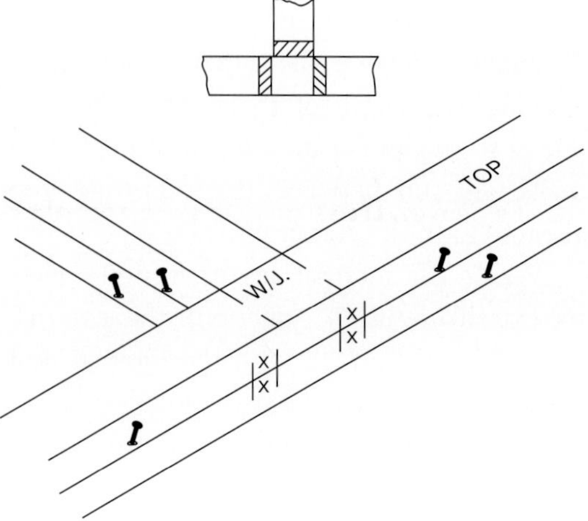

Fig. 12.25 Marking door and window openings

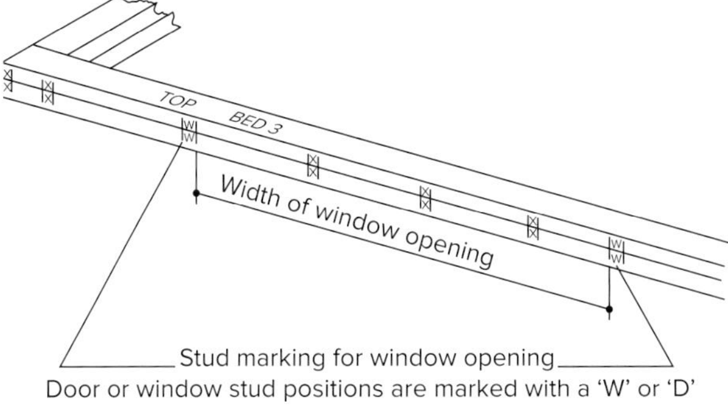

Fig. 12.26 Window manufacturer's information

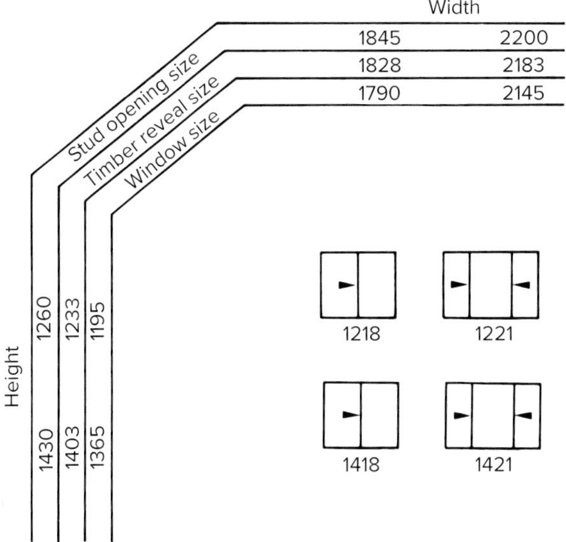

Some manufacturers of stock size frames provide the necessary information in their trade literature.

The stud opening size for aluminium frames can be obtained from the manufacturer's instruction sheets, an extract from which is shown in Figure 12.26.

The nominal size of an aluminium frame is represented in the ordering code as the height times the width, to the nearest 100 mm. For example, a 1218 frame is approximately 1200 mm in height by 1800 mm in width. The exact dimensions, however, are shown at the side and top in line with the frame diagram. For example, the 1218 frame is exactly 1195 mm in height and 1790 mm wide. The dimensions overall, including timber reveals, is also given as 1233 mm × 1828 mm, and the stud opening size is 1260 mm × 1845 mm. A full selection chart will give this information for the manufacturer's full range of windows.

Standard door openings are made 2100 mm in height, and around 700, 800, 900 and 1000 mm in width. These openings will accommodate a standard-sized door plus the jamb and necessary clearances. If the use of narrow architraves is anticipated, it may be possible to close those openings in a little. Use a lumber crayon to mark the extent of the window and door openings.

12.5.1.5 Step 5

Mark studs supporting any concentrated loads that may be required.

These studs are designed to carry major loads such as roof struts, strutting beams, and girder beams, and their design will be in accordance with the requirements of the AS 1684 Span Tables–Studs Supporting Concentrated Loads.

12.5.1.6 Step 6

Set out rafter/truss positions to top plate.

At this stage, it is a good idea to set out the rafter/truss positions now while the plates are still tacked in position near ground level. This will make the task much easier than later when the walls are stood up in position. It may also be necessary to locate common studs under the roof members. This will depend on the thickness of the top plate and spacing of roof members (refer to the details given earlier in this chapter on top plates). If the roof members are directly loaded over the studs, marking the position of the rafters now will assist in locating and marking the common stud positions, which is our next task.

12.5.1.7 Step 7

Set out intermediate common studs.

Mark in the position of all remaining common studs, referring to details given earlier in the chapter (Fig. 12.27). Space out and mark the studs in between using the 'in and over' method where possible, not exceeding the maximum specified spacing.

12.5.1.8 Step 8

Mark plates with an identification mark (Fig. 12.28).

It will be necessary to provide clear identity marks on the plates with a crayon to locate their position in the building. This is done before the plates are removed from the floor in readiness for assembly of the walls. Mark 'TOP' and 'BTM' (bottom) on the face of each plate. Also mark on the plate sufficient information that will locate the position of the plate in the building. For example, 'LOUNGE', 'BED 1', 'LAUNDRY', etc. Some builders identify, close to one end of both plates, the direction in which they are facing–that is, 'NORTH', 'SOUTH', etc. Also near the end of the plate, an identification number or letter (in brackets) can be placed on the edge of the paired plates with a corresponding number marked alongside on the floor. This will help to relocate plates in their correct position should they get mixed up during the wall assembly stage.

Fig. 12.27 Marking common studs

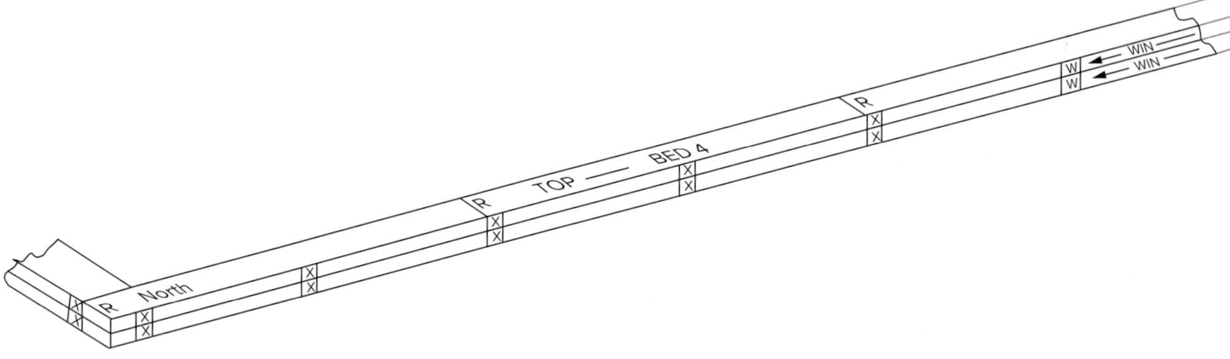

Fig. 12.28 Identity marks on plates

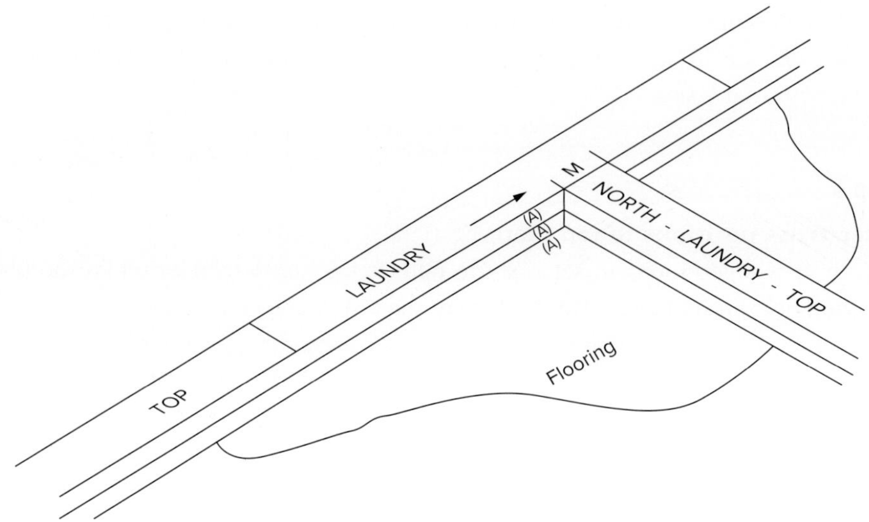

12.5.2 Setting out studs

The length of the studs can be calculated and a pattern stud marked to length (Fig. 12.7). On the face of the pattern stud, mark the height of the window and the door lintels and sill trimmers. It is common practice to carry the window openings up to the **eaves soffit**, in which case the drop of the eaves soffit must be determined. Mark the door heads to provide an opening 2100 mm in height where standard sized doors are specified. Select the straightest of the studs and mark out the jamb studs in pairs, ready for trenching if required.

12.6 Assembly of wall frames

To assemble wall frames on-site, lay out the wall plates and studs on the platform floor as close as possible to the position where the wall frame will finally be erected (Fig. 12.29).

A number of frames can often be made on top of each other, so keep in mind the order in which the walls will be erected and assemble first the frame to be erected last.

As the studs are laid out, quickly sight them in length and lay all of the round edges facing one way, usually round edge up. Some may prefer to lay studs the opposite way round, but the important thing is that they are all laid the *same* way round.

With a carpenter at each end, nail the plates to the studs with two nails at each joint. Cut to length and fix the lintels, trimmers and jack studs. To fix the **noggings**, measure the length along the plate between the studs and cut to length. If the material has been **machine gauged**, much of the nogging can be cut to a standard length to suit the studs spaced at their maximum spacing.

Noggings are spaced at no more than 1350 mm centres in their height. They are fixed in a straight line wherever nail-fixed sheet material is to be butt jointed along that line or, in order to provide more secure nailing, they can be offset by not more than their own width from that line.

12.6.1 Squaring up wall frames

Wall frames are squared up by making the two diagonals equal. Sight that the plates are lying straight and secure one plate against movement by two skew nails.

Fig. 12.29 Layout of wall frame for assembly

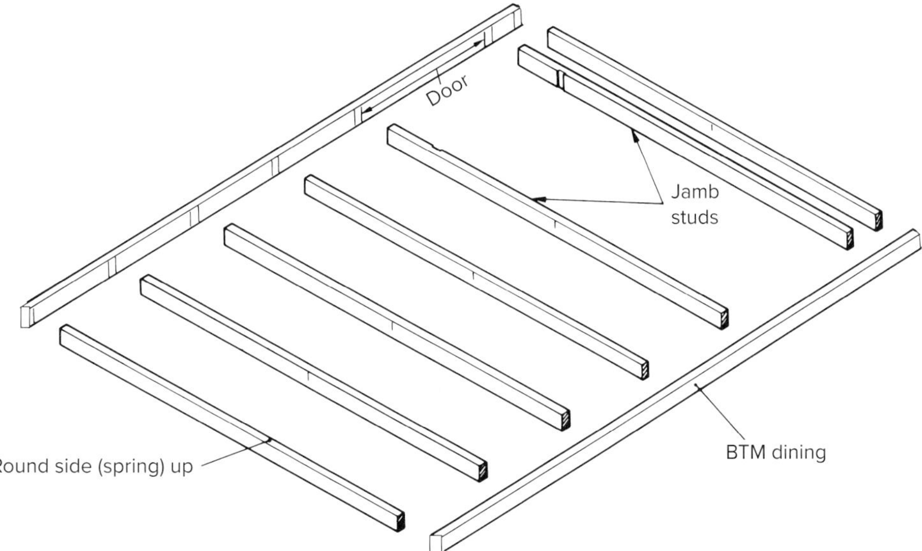

The diagonals can be measured with a long squaring rod or, perhaps more conveniently, with a steel tape. Measure diagonal 1 and assume it measures 4340 mm. Measure diagonal 2, which measures, say, 4310 mm. The difference between diagonals is then 30 mm. Take half this difference, 15 mm, and add it to the shorter diagonal or subtract it from the longer diagonal to give a dimension of 4325 mm. Move the loose plate sufficiently to make the two diagonals equal at 4325 mm and secure the plates with a skew nail ready for fitting the brace.

12.6.2 Bracing

To fit a timber brace, lay the brace across the studs as near as possible to the ideal angle of 45° and tack it lightly in place (Fig. 12.30). Mark each side of the brace, lift it aside and gauge the depth of the notching to allow the brace to finish flush with the framing. Remove the waste and replace the brace, but do not nail it permanently at this stage.

Fig. 12.30 Squaring up wall frames

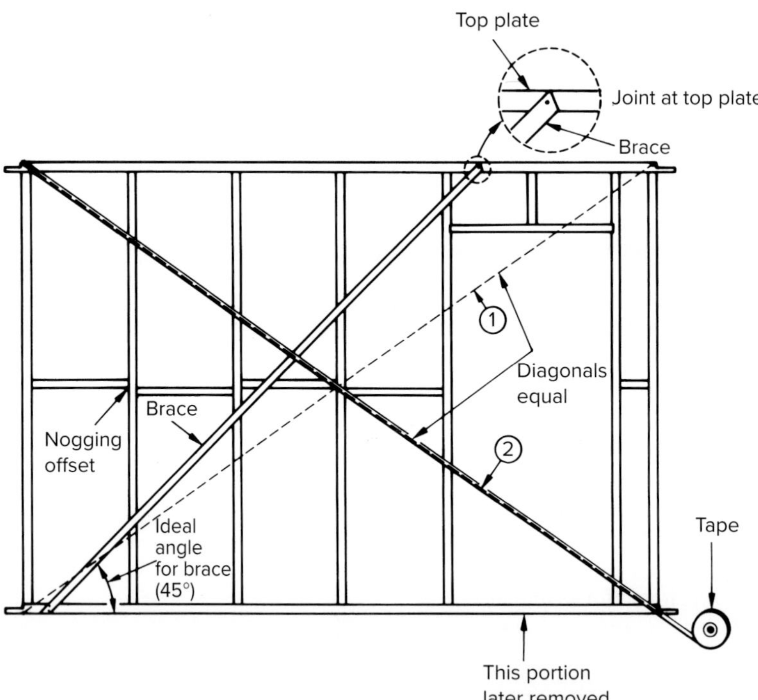

CHECK YOUR UNDERSTANDING

1. How are wall frames squared up prior to being erected?

12.6.3 Erection of wall frames

The erection of wall frames is an operation that warrants special attention to safe working procedures, particularly when working on open floor joists (Fig. 12.31). Ensure there is an ample labour force available to provide a margin of safety when lifting the frames into position. Short cleats nailed to the end of joists will prevent the frames from slipping over the edge. When frames have been stood upright, secure with a temporary brace until a more permanent fixing is possible.

Fig. 12.31 Erection of wall frames

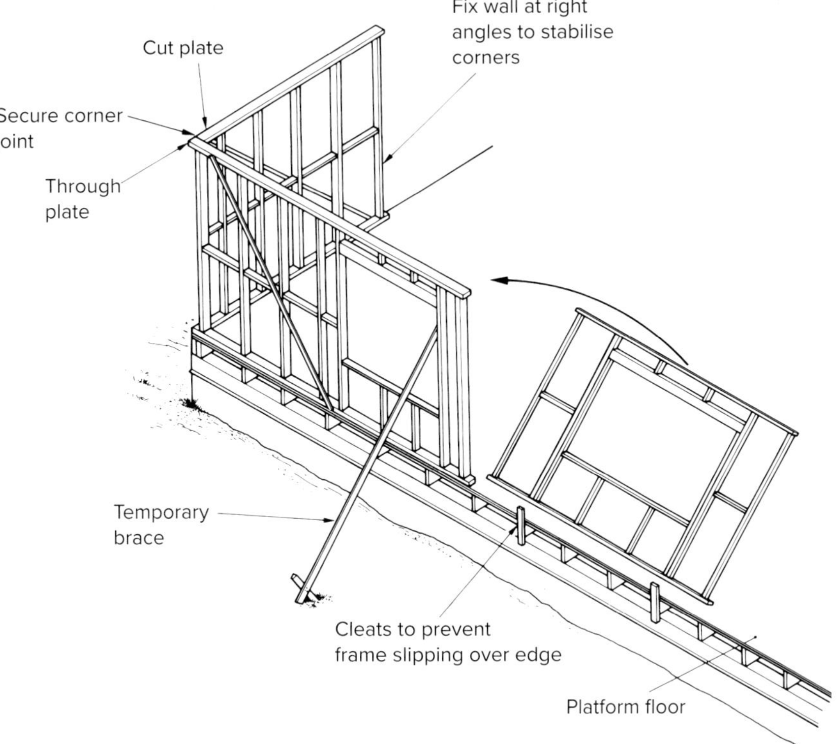

Walls can be most effectively stabilised by erecting frames at right angles to each other and by securing the top joints together with the aid of metal nailing plates. A big plus for the platform floor is that it renders this part of the work easier and much safer.

Proceed to stand all of the wall frames. Do not fix any permanently in position, but proceed with the next operation.

For fixing requirements of plates to joists or slabs once in the final position, refer to AS 1684.

12.6.4 Trueing wall framing

The wall framing must be trued accurately and the job carried out in an orderly manner, as follows:

1. At each external corner, plumb up from the joists and bearers, and fix the bottom plates directly above, nailing through into the joists.
2. Between external corners, straighten the bottom plates by attaching a string line packed out from the plate with blocks of equal thickness. Nail through into the joists. Straighten and fix the bottom plates to the internal walls with a string line or long **straight edge**, or fix to the chalk line marks previously set out on the platform floor (Fig. 12.32). Refer to details given earlier in this chapter.
3. Plumb external corners (Fig. 12.33). Suspend a plumb bob from the corner

Fig. 12.32 Fixing external corners

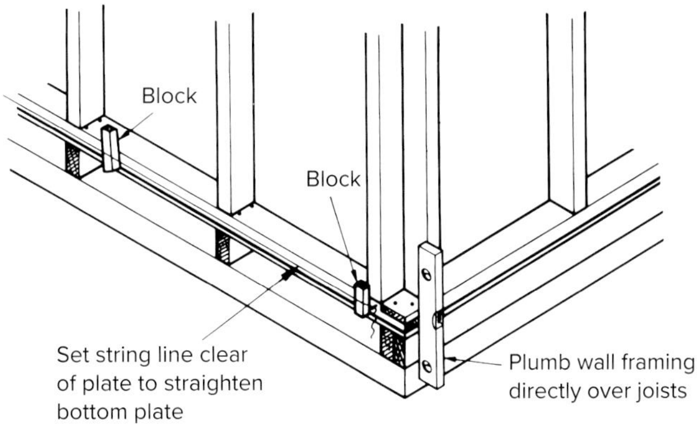

Block
Block
Set string line clear of plate to straighten bottom plate
Plumb wall framing directly over joists

of the top plate and release the braces where necessary to force the frame to plumb up directly over the corner of the bottom plate. Alternatively, use a long straight edge and spirit level in place of the plumb bob. Replace the braces and nail permanently in position. Apply the same procedure if using alternative bracing materials.

4. Straighten the top plates by attaching a string line packed out from the plate (Fig. 12.34). First adjust the plate to the string line at each intermediate wall junction by releasing the brace and adjusting the internal wall over to the string line, where necessary. Replace and permanently fix internal wall braces. Straighten the plates between the wall junctions and secure them with temporary braces. The top plates to internal walls are straightened by sighting or by the use of a straight edge, and are stabilised by temporary braces where necessary until the next operation (fixing of the ceiling framing) is completed.
5. Complete fixing of corner studs and blocking as required.

Fig. 12.33 Plumbing external corner

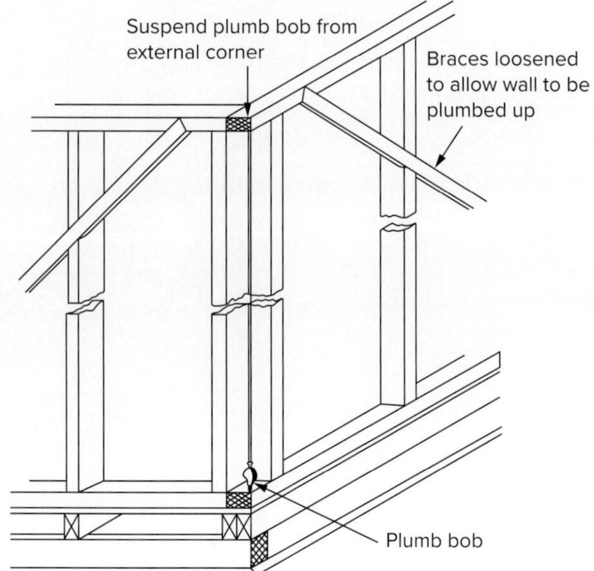

Fig. 12.34 Straightening top plate

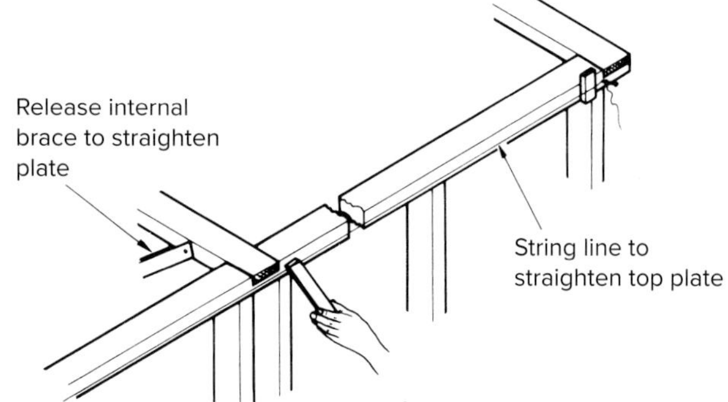

CHECK YOUR UNDERSTANDING

1. What safety precautions should be taken when erecting wall frames?
2. Briefly outline, in sequence, the procedure taken to true up the wall framing after it has been stood in position.

12.7 Tie-downs and brackets

12.7.1 Tie-downs

Tie-downs are extremely important as they assist in anchoring the entire frame down to the foundations in order to resist wind loads, whether on the side of the building or passing over a roof–there is potential for lift to be generated when lower pitch roofs are constructed.

The idea behind tie-downs is that instead of relying solely on nails, screws, etc. to hold one component to another, like the top plate to the stud, additional holding power can be achieved using a length of hoop iron that runs up one side of the stud, across the top plate and then down the other side of the stud. The hoop iron is then nailed in place using the appropriate fixings. This process is repeated at other connections so that the roof is literally tied down to the foundation using hoop iron or specific brackets. Tie-down details and specifications can be found in AS 1684.

12.7.2 Brackets

There are many types of brackets available on the market. Most have a particular function and, generally, they strengthen the intersection of two pieces of timber. Additionally, there are some metal brackets available that replace traditional timber joints.

12.8 Steel wall framing

Steel-framed house construction is used in around one in seven new homes in Australia and provides an alternative method to traditional timber framing. Steel framing systems vary, with manufacturers using different sections, steel thickness and joining methods such as screws, rivets, bolts and welding. However, all systems are designed to the same structural standards and the general principles of wall frame construction on-site are the same for all systems and are quite similar to timber construction.

There are two approaches for constructing steel house frames. The most common method is to have the wall frames pre-fabricated at the factory. These are then transported to the building site and erected in position on the prepared slab or floor system. The other method is to deliver all of the framing component members to the site precut to length, with all fabrication and assembly done on-site. The latter is the less popular approach in urban areas, but it is more commonly used in remote areas.

In either case, the wall component specifications and panel layouts are generated by computer design and detailing software operated by trained frame detailers.

Fig. 12.35 Steel wall framing

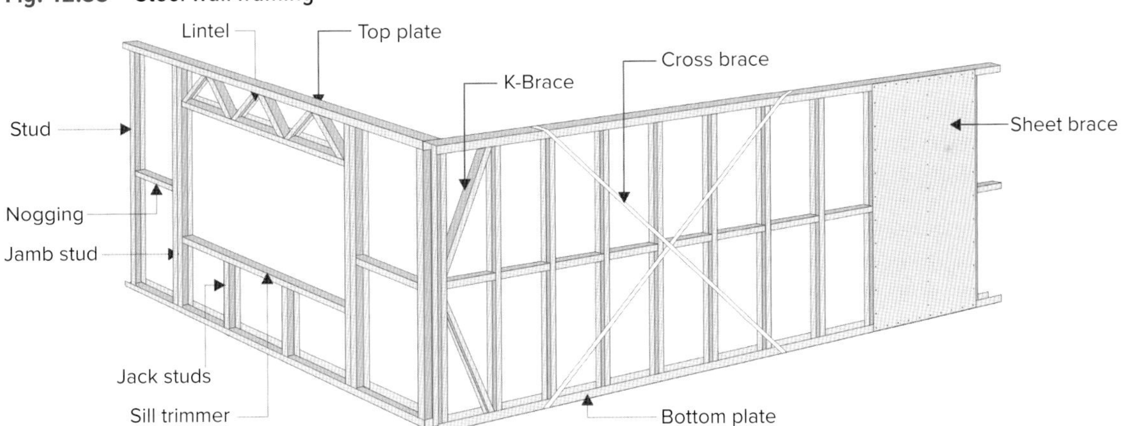

Reproduced with permission of NASH Inc. www.nash.asn.au

CHECK YOUR UNDERSTANDING

1. Name two ways in which steel wall framing can be supplied.

12.8.1 Steel wall framing erection

Pre-fabricated steel wall framing can be erected on all standard flooring systems. The framing is erected using the same set-out techniques and follows the same sequence that you would use for timber framing. The frames are pre-fabricated in lengths up to 7.5 m maximum for ease of transport and site handling. Fixings used on-site will include screws, rivets, masonry anchors, power-driven nails and occasionally bolts and special clips.

Factory prepunched holes are usually provided in the studs and top plates at convenient locations to accommodate the installation of plumbing and electrical wiring. Plastic grommets are fitted to the holes to avoid metal-to-metal contact with copper piping or electrical wiring. Holes may also be swaged or flared to prevent cable damage, but service pipes must be supported when passing through studs and plates.

Wall locations are set out on the slab or platform using a chalk line, starting with the external walls. To compensate for possible differences in slab dimensions, mark the inside of the external wall frame positions. Continue on to the internal walls, checking constantly for squareness.

Where perimeter wall frames are placed on a concrete slab-on-ground, a durable impermeable membrane should be placed between the bottom plate of perimeter wall frames and the slab and extend up the weather side flange of the bottom plate. The membrane is not required beneath internal

Fig. 12.36 Wall frame layout

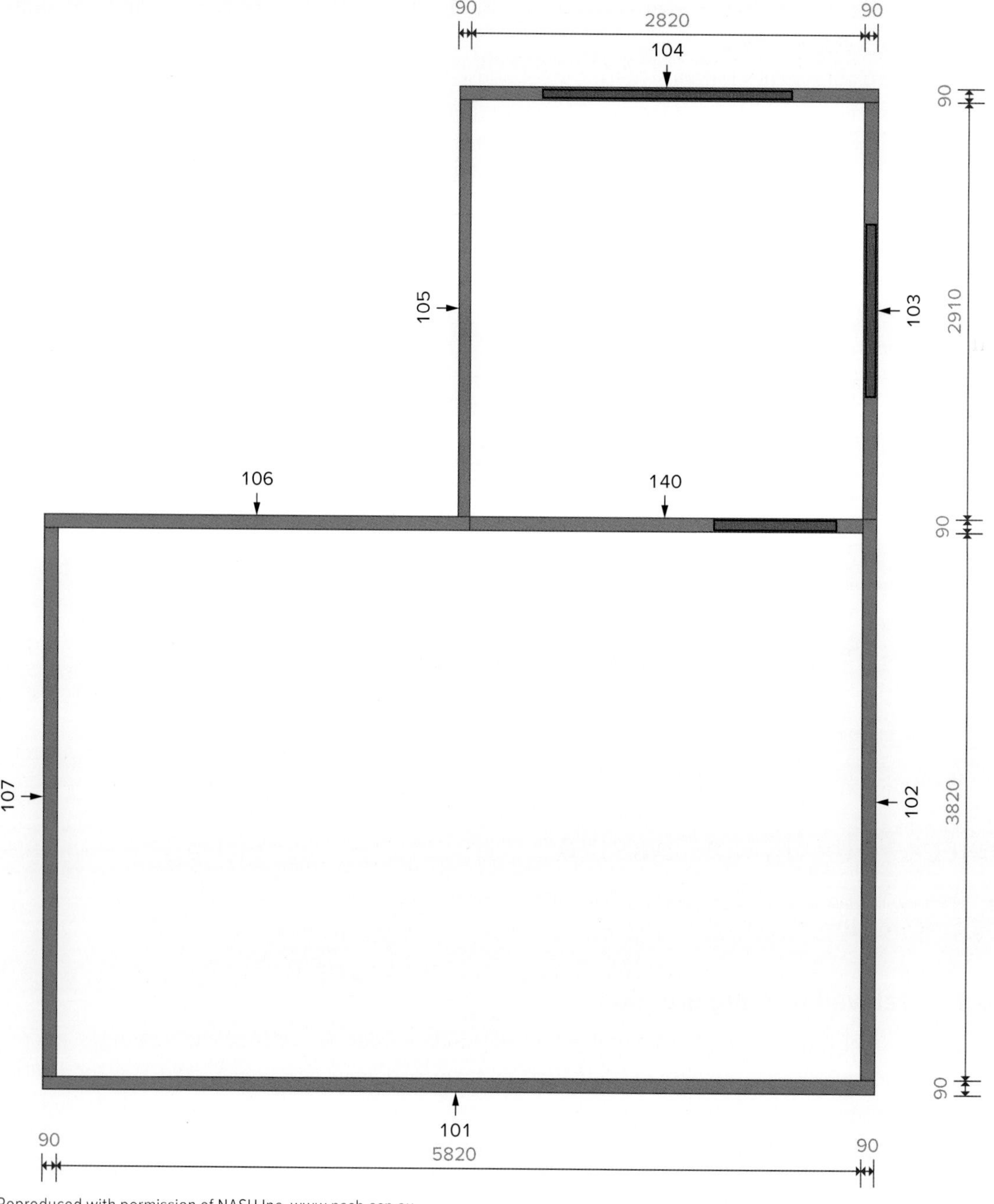

Reproduced with permission of NASH Inc. www.nash.asn.au

Fig. 12.37 Preparation of door opening

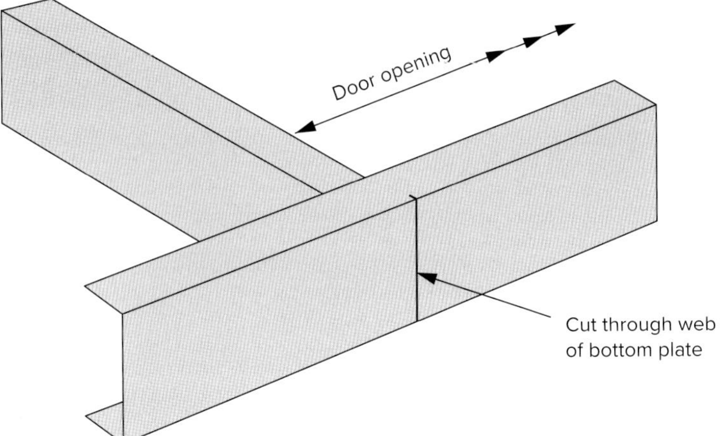

Reproduced with permission of NASH Inc. www.nash.asn.au

wall frames or upper-storey wall frames. The membrane may be the same as that used for damp-proof courses, a paintable bitumen product or a self-adhesive polyethylene.

Note: Drawings and specifications should be carefully checked as the membrane may be required as a condition of product warranty.

Before standing the wall frames:

- mark out the positions of the roof trusses on the top plates
- drill any holes in the bottom plate that may be required for fixing to the slab/floor frame
- cut the web of the bottom plate at either side of the doorways using an angle grinder with a metal cutting disc or preferably using a power saw with a cold-cutting metal blade. This makes the removal of the bottom plates in doorways easier after the frames are erected.

When all wall locations are marked out and the frames prepared, place the frames near their required location as shown on the layout plan. Stand the frames, commencing with an external corner, fixing them together as indicated by the frame supplier on the panel layout drawing. This is typically done with self-drilling screws but may also involve special brackets or clips. Continue to work around the building adding one frame at a time, checking that each frame is aligned with the set-out marks and plumb before proceeding to the next. Use temporary bracing as required to stabilise the frames until the wall and roof framing is complete and fully braced.

CHECK YOUR UNDERSTANDING

1. What methods are used to join steel frames together?

12.8.2 Hold-down

The frames are then fixed down to the slab or floor using methods and spacings specified on the frame supplier's drawings. The specifications will vary depending on the type of floor, type of roof and wind region. Particularly important are the fixings beside the openings, at wall ends and corners, and where bracing straps or sections meet the bottom wall plates. Screw bolts are a common fixing method for concrete slabs as they are fast and reliable. Where a steel floor frame is used, self-drilling screws of the correct size and length are commonly used. In cyclonic areas where hold-down requirements can be very large, the frame supplier will pay particular attention to the continuity of the load path from roof structure to foundations. This may involve specialised fastening arrangements for the wall frames.

Fig. 12.38 Connection details for wall frame to slab and floor joists

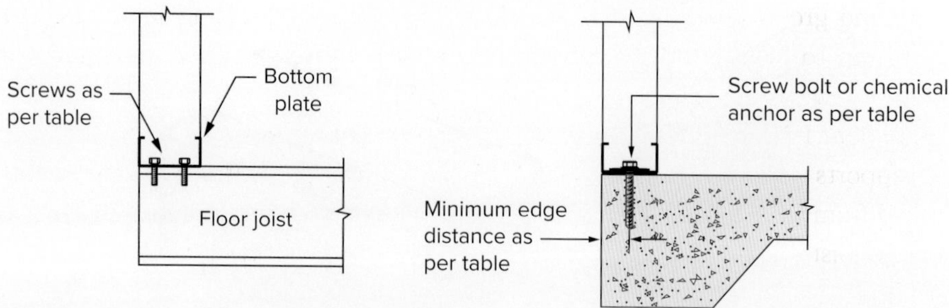

Reproduced with permission of NASH Inc. www.nash.asn.au

12.8.3 Bracing

Bracing strength is critical to the structure both during and after construction. Bracing usually consists of a combination of methods such as metal strapping, K-braces, sheet materials such as steel, fibre cement or plywood, and plasterboard lining. You will need to provide and fix all bracing as specified by the frame supplier.

Fig. 12.39 Bracing of wall frames

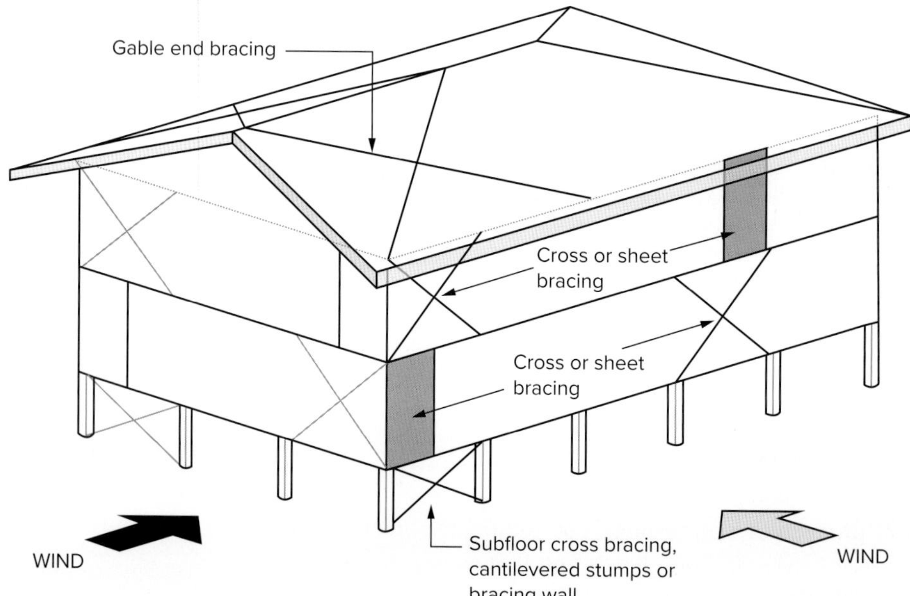

Reproduced with permission of NASH Inc. www.nash.asn.au

The final step in wall framing is to remove the bottom plates in doorways by cutting the flanges at each jamb stud down to the precut plate webs.

For further information, including fact sheets, PowerPoint presentations, trade instruction sheets and the General Guide to Steel-Framed Building, refer to the National Association of Steel-House Framing (NASH) website at www.nash.asn.au. NASH standards and handbooks may also be purchased through the website.

12.9 Two-storey construction

12.9.1 Timber-framed buildings

The principles of timber framing described for single-storey buildings in Chapters 11, 12 and 16 are equally applicable to two-storey buildings, with the addition of details of the construction between the first and second floors.

To support the increased load due to a second storey, some minor modifications are necessary to the **footings** and ground-floor timbers. Specifications usually require an increase in the width of the footings–80 mm to 100 mm or as otherwise approved by the local building authority–with an appropriate increase in the amount of steel used for reinforcing.

In the ground-floor framing of a two-storey construction, a bearer should be located directly under any wall that supports loads from upper floors and is running at right angles to the joists. (Their size must be obtained from AS 1684.2.) Also, a joist should be located directly under all load-bearing walls that are parallel to joists.

WORKPLACE SCENARIO

When stick building timber wall frames, consideration must be given to the quality of the timber being used. Even with kiln-dried timber, not all of it is straight and true, so selecting the straightest lengths and reserving them for the wall plates, corner and jamb studs is best practice. The very worst and most twisted timber should be cut up for blocking and noggins. The remaining timber should be used for the rest of the wall frame components.

12.9.2 Upper-floor joists

The joists to upper floors must be increased in depth to span the various rooms without the need for intermediate support. (Their section size can be obtained from tables in AS 1684.2.) To obtain the size, span and spacing of the joists, the **stress grade** of the timber species being used must be known. As an example of how these factors will affect the maximum allowable joist spans, refer to Table 12.2, which is for joists spaced at 450 mm centres. For other joist spacings, refer to AS 1684.2.

Table 12.2 shows only an example of joist spans for selected timber species. In practice, and for much firmer detail, you should refer to AS 1684.2: 2010, Supplement 10.

Table 12.2 Maximum allowable joist spans (joists spaced at 450 mm centres)

Joist size	Unseasoned				Seasoned softwood	
	F4	F5	F8	F11	F5	F8
150 × 50	2600	2800	3200	3400		
175 × 50	3300	3400	3700	3900		
200 × 50	3700	3800	4100	4300		
250 × 50	4400	4500				
300 × 50	5100	5200				
140 × 35					2300	2600
190 × 35					3500	3700

12.9.3 Platform construction

A common method of two-storey construction, as shown in Figure 12.40, is referred to as **platform construction**. The **studs** are not continuous from the ground floor to the roof, and the joists bear on, and are spiked to, the top plates of the lower-storey load-bearing walls at points immediately above the wall studs.

Where the required depth of joist is equal to, or exceeds, four times the joist width, it is deemed to be a deep joist and is subject to some special provisions. If no continuous trimming joists are provided to the ends of joists above external wall plates, their ends should have solid blocking or herringbone strutting, at least between the outer pairs of joists and between intermediate pairs, at no more than 1800 mm centres. Trimmers or solid blocking should be approximately 25 mm less in depth than the joists and should have a minimum thickness of 25 mm.

Where the unsupported span of joists exceeds 2700 mm for timber less than stress grade F11, then solid or herringbone strutting should be provided between joists and along their span at no more than 1800 mm centres.

The depth of solid strutting is again 25 mm less than the depth of joists; they are fixed centrally to allow for unobstructed through-ventilation of the floor cavity. Herringbone strutting should be 38 mm × 38 mm, and a method of using the steel square to set out a pattern for strutting is shown in Figure 12.41.

An alternative arrangement for supporting upper floors over external load-bearing walls running parallel to the joists is shown in Figure 12.42. Here, the depth of the deep joists is built up by a pair of double joists, 100 mm in depth, plus trimmers fixed at right angles to the wall. This arrangement eliminates a deep end joist and is suitable for either a platform or a fitted floor.

12.9.3.1 Shrinkage of floor joists

One problem that arises when platform construction is adopted for two-storey buildings is the shrinkage of the floor joists. There can be well in excess of 400 mm to 450 mm of cross-grained timber that may shrink considerably, depending on the species and its moisture content.

In single-storey brick veneer construction, a clearance of 10 mm is allowed under all windowsills and at the eaves soffit. In two-storey buildings, where timber with low shrinkage is used in both levels of floor framing, 20 mm clear space should be provided to the underside of first-floor windowsills and

Fig. 12.40 Platform construction

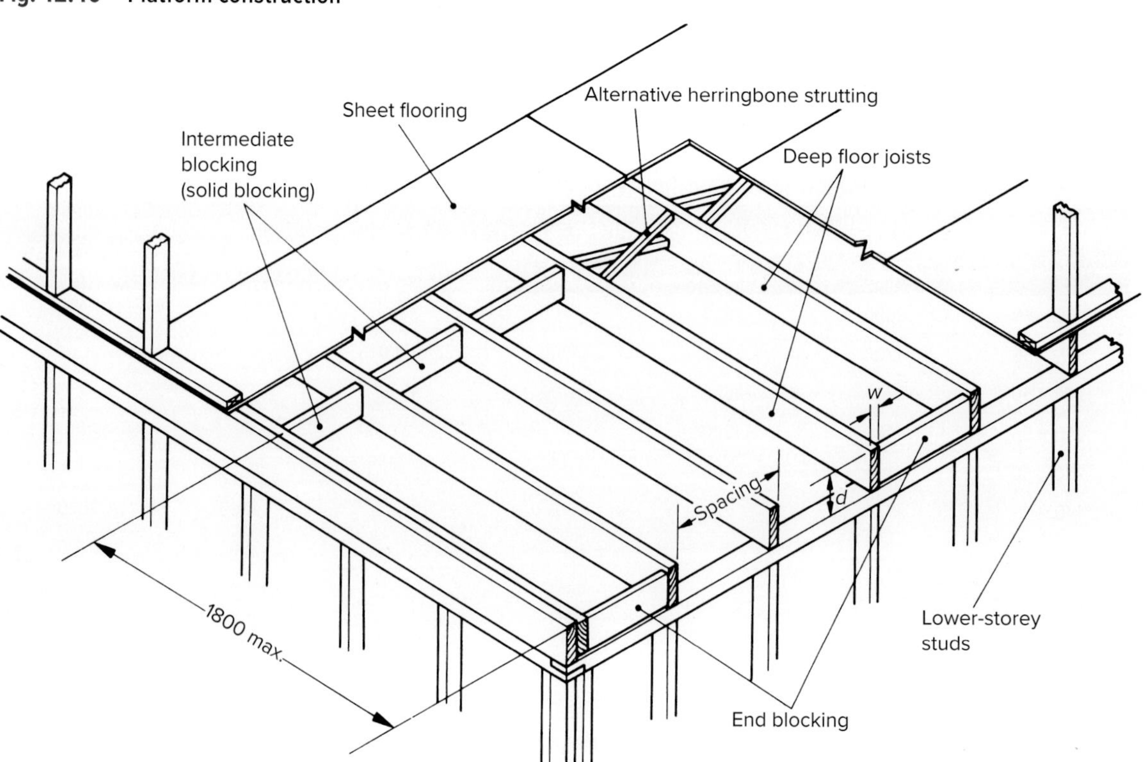

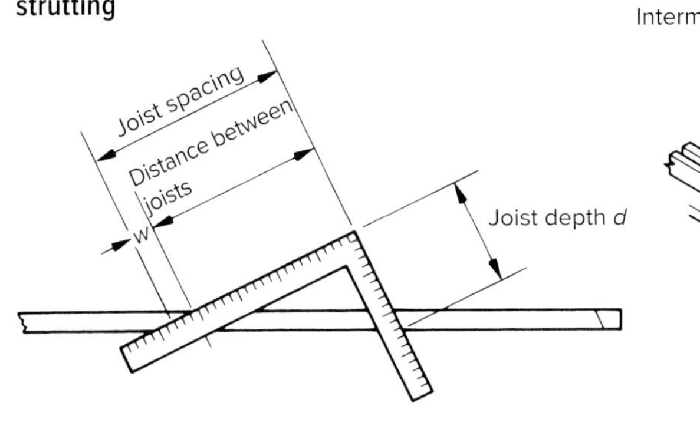

Fig. 12.41 Set-out pattern for herringbone strutting

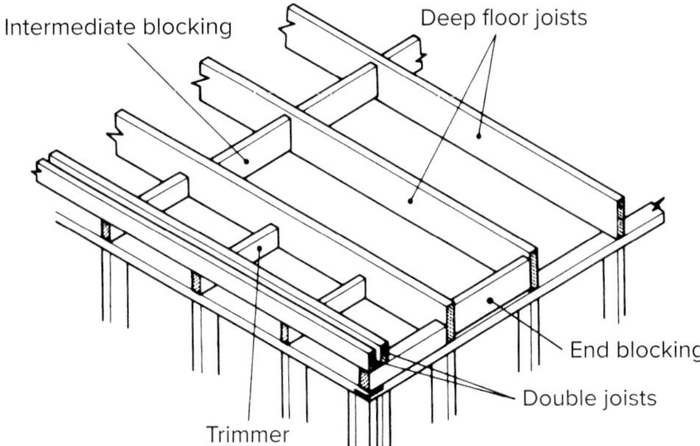

Fig. 12.42 Alternative second-storey floor framing

at the eaves soffit, while 10 mm is still sufficient for the lower windows. Where unseasoned hardwood is used in both levels of floor framing, the above clearance should be increased by 50%.

A method of construction referred to as *balloon framing,* where studs are continuous from the bottom plate of the ground floor to the top plate of the upper storey, is used in the United States to overcome the problem of shrinkage, but it is rarely used in Australia today.

In concrete slab on-ground construction, windowsills on the ground floor do not require a clear space, but those on the first floor require a space of 10 mm.

12.9.3.2 Drilling or notching deep floor joists

It is important that deep floor joists are not weakened by indiscriminate boring or notching to provide access for such things as plumbing and electrical services. Any drilling or notching should meet the requirements as illustrated in Figure 12.43.

1. Holes not exceeding *d*/8, or 25 mm maximum, can be drilled at random provided the holes are not spaced at less than the depth of the member. Also, no more than three drillings should occur in any 1800 mm of span (Fig. 12.43a).
2. For holes not exceeding *d*/4, or 50 mm maximum, where *d* is less than 200 mm, the drilling must be contained entirely within the centre one-third of the face depth, with no more than one drilling in any 1800 mm of span (Fig. 12.43b).

Fig. 12.43 A–D Requirements for notching and drilling deep floor joists

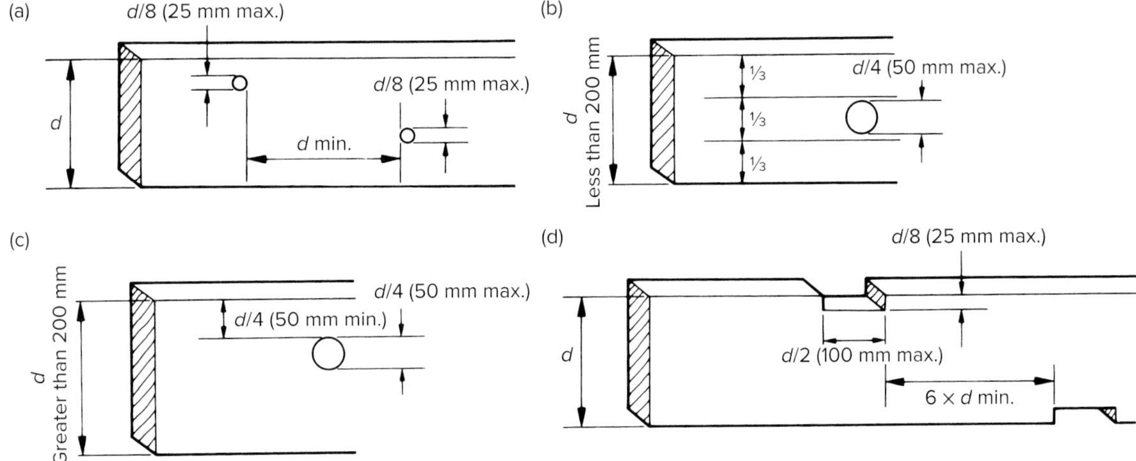

3. Where *d* exceeds 200 mm, holes must be no closer than the diameter of the holes (*d*/4 or 50 mm maximum) to either edge, with no more than one drilling in any 1800 mm of span (Fig. 12.43c).
4. Notching the edges of joists should be avoided, but where unavoidable, must fall within the limits illustrated in the figure (Fig. 12.43d).

12.9.3.3 Engineered timber floor joists

Today, engineered timber beams are becoming more popular as an alternative to using solid timber joists. These products are manufactured in a range of sizes and designs that are suitable to span over small to large open areas. Design factors, span tables and installing information are available from the various manufacturers (Fig. 12.44).

Fig. 12.44 Engineered timber upper floor joist

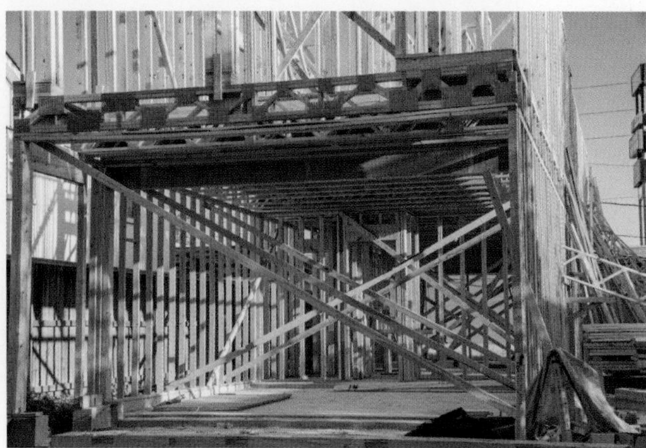

Fig. 12.45 Fire cutting floor joists

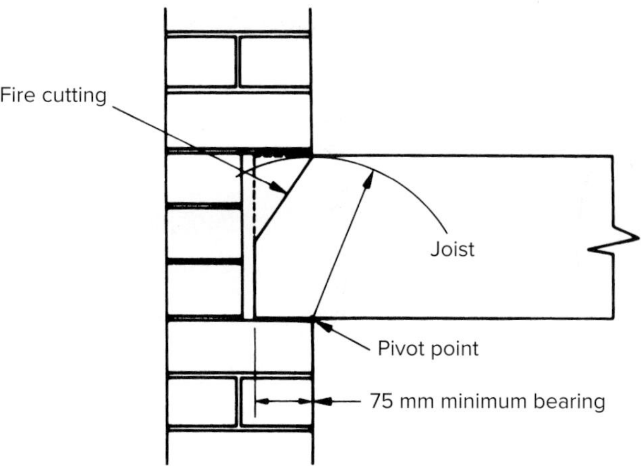

12.9.3.4 Upper floor joists in masonry construction

Where load-bearing walls of a building are of masonry or concrete construction, the ends of upper-floor joists may bear in pockets formed in the wall. This practice was common in the era when load-bearing walls were often of solid brick construction, but would rarely be seen in today's building practice for new work.

Where pockets formed in the wall are adopted to support ends of members, they must allow at least 12 mm of clear air space at the sides and end of members, and provide a solid bearing in the wall of at least 75 mm in depth.

12.9.3.5 Fire cutting

Where deep joists are used in conjunction with masonry or concrete wall structures, the ends of the joists that are bearing in wall pockets should be splayed towards their tops so that they are not restrained by walls in the event of collapse of the floor system caused by fire or overload. This is referred to as *fire cutting* (Fig. 12.45).

Note that if the joist was not splayed off, the top corner would lift as the joist rotated about the pivot point, no doubt collapsing the wall.

CHECK YOUR UNDERSTANDING

1. State what blocking or strutting is required to strengthen deep joists.

Student research

Refer to the following standards, which are relevant to this chapter:

- AS 1684.2–2010 Residential Timber-Framed Construction—Part 2: Non-cyclonic Areas
- AS 1684.4–2010 Residential Timber-Framed Construction—Part 4: Simplified—Non-cyclonic Areas.

  In particular, see the following sections:
- Section 4.2.2.3 Deep joists
- Section 6 Wall framing
- Section 8.3.6 Wall bracing
- Section 9.2.8 Joining of top plates.

  Refer also to the National Association of Steel-House Framing (NASH) website at www.nash.asn.au.

End of chapter activity

Activity 1:

As part of your training, you are required to construct a timber wall frame, containing a door and window opening, noggins, bracing, lintels, jack studs, sill trimmers and head trimmers.

AUSTRALIAN BUILDING CODES AND STANDARDS

AS 1684 Residential Timber-Framed Construction

National Construction Code

Wood Solutions

The National Association of Steel House-Framing (NASH)

Tradie talk—Sight that stud for straight

Chapter 13

Basic roofing

Learning Objectives

LO 13.1 Identify roof structures

LO 13.2 Express pitch

LO 13.3 Calculate roof pitch

LO 13.4 Install ceiling framing

LO 13.5 Identify gable roof components

LO 13.6 Set out rafters

LO 13.7 Prepare the ridge and underpurlins

LO 13.8 Erect the gable roof

LO 13.9 Calculate roof member sizes

LO 13.10 Know how to construct a skillion roof

Introduction

The roof is one of the most important components in a building and generally involves some of the most skilled work in the carpentry trade. The purpose of the roof is to protect the interior of the building and its occupants from the weather. Thus all roofs are built with some slope to their surface so that water can run off.

The structure must be sufficiently strong to withstand the load of the roof covering, rain (snow in some areas) and collected water, and wind loads.

In this chapter, you will learn basic roofing calculations and how they are applied to the construction of a basic gable-end roof, with a complete ceiling frame, as well as how to set out and cut roofing members, the sequence of pitching the roof and how to construct a skillion roof. You will also learn about the importance of AS 1684 Residential Timber-Framed Construction, and how it must be referred to so that the correct roof framing member sizes and stress grade can be determined.

13.1 Roof structures

Roof forms in common use are illustrated in Figure 13.1(a)–(f). The elements that make up a roof are as follows.

Fig. 13.1 Common roof types: (a) lean-to (skillion); (b) gable; (c) boxed gable; (d) hip; (e) hip and valley; and (f) gambrel

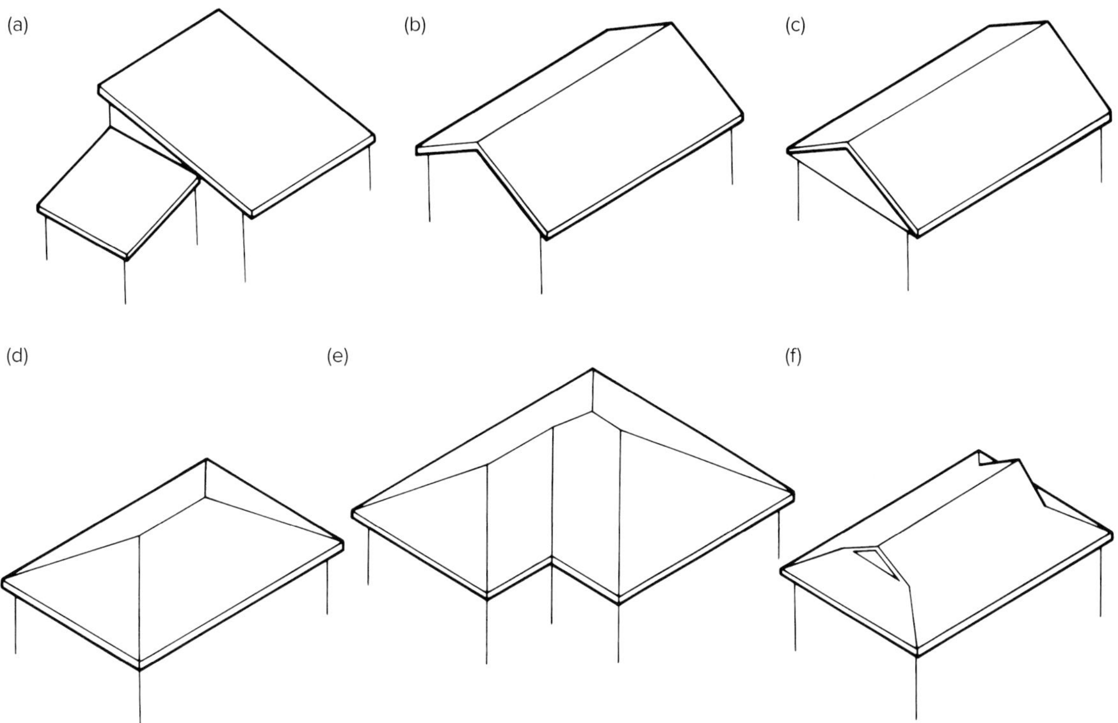

13.1.1 Pitch

As previously mentioned, all roof surfaces are built with some slope so that water can run off. The angle formed between the sloping roof surface and a horizontal surface is termed the **pitch** of the roof. In Figure 13.2, the line *XY* represents a horizontal surface, and *AB* and *BD* are sloping roof surfaces. The angle *BAC* is the pitch of the roof. If the angle *BDC* is the same angle, then the roof is referred to as being *equally pitched*. This is the case with the vast majority of roofs.

13.1.1.1 Span

AD is the **span** of the roof and is measured over all of the wall plates that support the roof.

Fig. 13.2 Basic roof geometry/pitch triangle

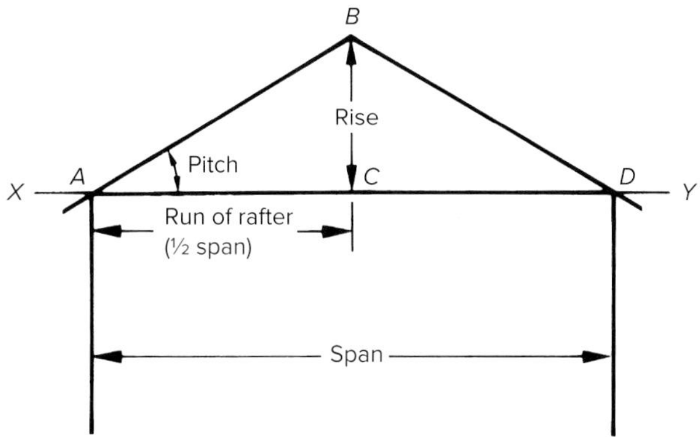

13.1.1.2 Rise

BC is the **rise** of the roof measured from the *XY* line to the apex of the roof.

13.1.1.3 Run

The **run** (half span) is the horizontal distance covered by a sloping roof member. It can also be referred to as the *plan length of the member* and can be found by reading directly from the plan drawing. For example, if *AB* is a sloping roof rafter, then *AC* is the run (half span) of that rafter, if the roof is equally pitched, as in the roof types shown in Figure 13.1.

13.2 Roofing materials and pitch

The pitch of the roof is the angle at which the rafters sit in relation to the **horizontal**. The roof pitch can be expressed in a number of different ways (Fig. 13.3). The pitch can be expressed as an angle (e.g. 22.5°) or a ratio (e.g. 1:2), that is, the **vertical** component being given first. For slopes up to 45°, the vertical component is given as one, and for slopes over 45°, the horizontal component is given as one.

13.3 Calculate roof pitch

Fig. 13.3 How roof pitch is expressed

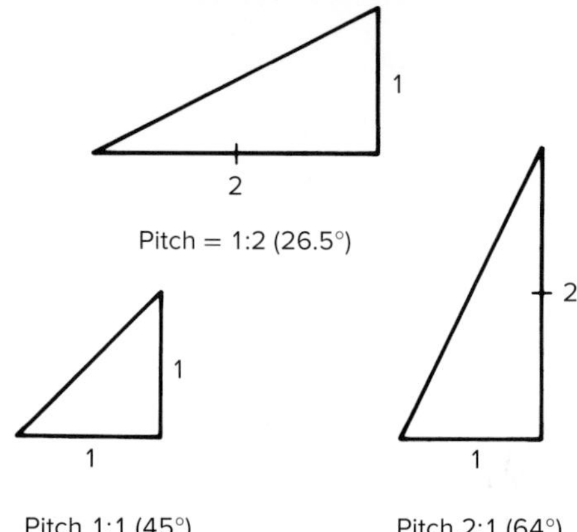

13.3.1 Pitch as a ratio

Note that the ratio is given a non-dimensional ratio. Putting it into trade terms, it is referred to as the **rise/run ratio**, where the rise is the vertical component and the run is the horizontal component. Therefore, a rise/run ratio of 1:2 can be 1 mm:2 mm, 100 mm:200 mm or 500 mm:1000 mm.

13.3.2 Pitch in degrees

If the pitch is given in degrees, then the use of tables or manuals is possible. There are a number of these publications available, and they will give the carpenter the required information to construct the roof. If the carpenter needs the pitch in degrees to be converted to a rise/run ratio, then the following calculation using trigonometrical ratios should be used. For example, if the pitch = 32°:

$$\begin{aligned} &\text{then opposite/adjacent} = y/x = \tan 32^\circ \\ &\text{if } y = 1 \text{ then} \quad 1/x = \tan 32^\circ \\ &\text{then} \quad x = 1/\tan 32^\circ \\ &\quad = 1/0.6248 \\ &\quad = 1.6 \end{aligned}$$

Therefore, the rise/run ratio = 1:1.6.

CHECK YOUR UNDERSTANDING

1. Name two ways in which the slope of a roof surface can be specified.
2. What is meant by the following terms as applied to roofing?
   a. Equally pitched
   b. Run of rafter
   c. Rise
   d. Rise/run ratio
   e. Pitch

13.3.3 Pitch—rise in millimetres/metre runs

When it comes to setting up the roof members, it is particularly valuable if the pitch (however it is specified) can be converted to a ratio of rise in millimetres to a metre run (Fig. 13.4).

13.3.3.1 Example 1

Take the right-angled triangle *XYZ*. Side *YZ* is the run and is fixed at one metre (1000 mm). Calculate the rise *XZ* in millimetres if the pitch ratio is 1:2.

Pitch = rise:run
= 1:2
= 1 × 1000/2
= 500 mm/m

Fig. 13.4 Rise/run concept for roof pitch

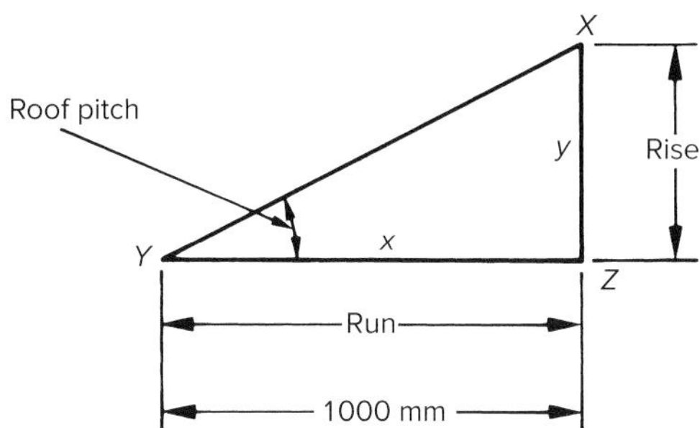

13.3.3.2 Example 2

Calculate the rise *XZ* in millimetres if the pitch ratio is 1:1.6.

Pitch = rise:run
= 1:1.6
= 1 × 1000/1.6
= 625 mm/m

When pitch is expressed in degrees, conversion can be carried out by using trigonometrical methods. For example, if the pitch = 32°:

$$\text{then } y/x \text{ or opposite/adjacent} = \tan 32, \text{ or}$$
$$\text{rise}/1000 = \tan 32$$
$$\therefore \text{rise} = \tan 32 \times 1000$$
$$= 0.6248 \times 1000$$
$$= 624.8, \text{ or } 625 \text{ mm}$$

(Note that answers will be given to the nearest millimetre in future roof calculations.)

For a long time, it has also been common practice to specify pitch as a ratio of rise to span, expressed as a fraction. For example, ¼ pitch indicates that the rise is to be made one-quarter of the span. This pitch can be converted to a rise/run ratio. For example:

- rise 1 unit/span 4 units (if roof equally pitched)
- rise 1 unit/½ span 2 units, or
- rise/run ratio = 1:2

Table 13.1 provides a ready means of converting pitch from degrees to rise/run ratio to mm/m run. Column 1 is the roof pitch expressed in degrees from 1° to 45° in ½° intervals. Column 2 is the pitch in degrees converted to a rise/run ratio and Column 3 is the same pitch converted to a rise in millimetres per metre run.

13.3.3.3 Example 3

Pitch = 27.5° = 1:1.92 = 521 mm/m

13.3.3.4 Example 4

Pitch = 1:2.6 = 21° = 384 mm/m

For a rise/run ratio other than those calculated in Table 13.1, select the closest ratio shown; any error must be less than half a degree, which is a reasonable tolerance. Alternatively, carry out calculations and prepare a more accurate conversion.

A roof can be commonly referred to as:

- *flat*–5° to 1° pitch, or ratios 1:12 to 1:57
- *low pitch*–22.5° to 5° pitch, or ratio 1:2.4 to 1:12
- *traditional*–over 22.5° pitch, or over a ratio of 1:2.4.

WORKPLACE SCENARIO

Chris Nance—owner of Traditional Timber Frames

I guess the thing about roofing is that every part is critical; if you get any element wrong, there's no way of hiding it! I remember working in the United Kingdom on a large oak roof restoration project where I had to scribe in a 4 metre-long underpurlin in 200 × 200 oak. It was part of a very old building and every joint had to be perfectly cut as it was all visible and carrying some very heavy loads. I set up string lines, performed a hundred calculations, worked out every angle imaginable and then cut the joint beautifully—I thought. I was really happy with the joints.

I organised the crane, we slung the timber way up onto this huge roof and lowered it into position only to realise that someone had cut it 100 mm short. Back to the drawing board!

13.4 Ceiling framing

Ceiling framing is closely integrated with roof construction: reference to the other chapters on roof construction may be helpful for a complete understanding. The ceiling framing consists of a number of members.

Table 13.1 Conversion of roof pitch:ratio:rise mm/m

Pitch angle	Pitch ratio	Rise mm/m	Pitch angle	Pitch ratio	Rise mm/m	Pitch angle	Pitch ratio	Rise mm/m
1	1:57.29	17.5	16	1:3.49	287	31	1:1.66	601
1.5	38.19	26	16.5	3.38	296	31.5	1.63	613
2	28.64	35	17	3.27	306	32	1.60	625
2.5	22.90	44	17.5	3.17	315	32.5	1.57	637
3	19.08	52	18	3.08	325	33	1.54	649
3.5	16.35	61	18.5	2.99	335	33.5	1.51	662
4	14.30	70	19	2.90	344	34	1.48	675
4.5	12.71	79	19.5	2.82	354	34.5	1.46	687
5	11.43	87	20	2.75	364	35	1.43	700
5.5	10.39	96	20.5	2.67	374	35.5	1.40	713
6	9.51	105	21	2.61	384	36	1.38	727
6.5	8.78	114	21.5	2.54	394	36.5	1.35	740
7	8.14	123	22	2.48	404	37	1.33	754
7.5	7.60	132	22.5	2.41	414	37.5	1.30	767
8	7.12	141	23	2.36	424	38	1.28	781
8.5	6.69	149	23.5	2.30	435	38.5	1.26	795
9	6.31	158	24	2.25	445	39	1.23	810

Pitch angle	Pitch ratio	Rise mm/m	Pitch angle	Pitch ratio	Rise mm/m	Pitch angle	Pitch ratio	Rise mm/m
9.5	5.98	167	24.5	2.19	456	39.5	1.21	824
10	5.67	176	25	2.14	466	40	1.19	839
10.5	5.40	185	25.5	2.10	477	40.5	1.17	854
11	5.14	194	26	2.05	488	41	1.15	869
11.5	4.92	203	26.5	2.01	499	41.5	1.13	885
12	4.70	213	27	1.96	510	42	1.11	900
12.5	4.51	222	27.5	1.92	521	42.5	1.09	916
13	4.33	231	28	1.88	532	43	1.07	933
13.5	4.17	240	28.5	1.84	543	43.5	1.05	949
14	4.01	249	29	1.80	554	44	1.04	966
14.5	3.87	259	29.5	1.77	566	44.5	1.02	983
15	3.73	268	30	1.73	577	45	1.00	1000
15.5	3.61	277	30.5	1.70	589			

Calculations for this table have been carried out to four decimal places and then abbreviated to two decimal places.

13.4.1 Ceiling framing members

13.4.1.1 Ceiling joists

Ceiling joists serve multiple functions:

- They provide fixing for various types of internal ceiling.
- They tie opposite walls together (Fig. 13.6).
- They form an integral part of the roof framing, completing the couple with the sloping roof rafters.

Fig. 13.5 Ceiling frame—general view

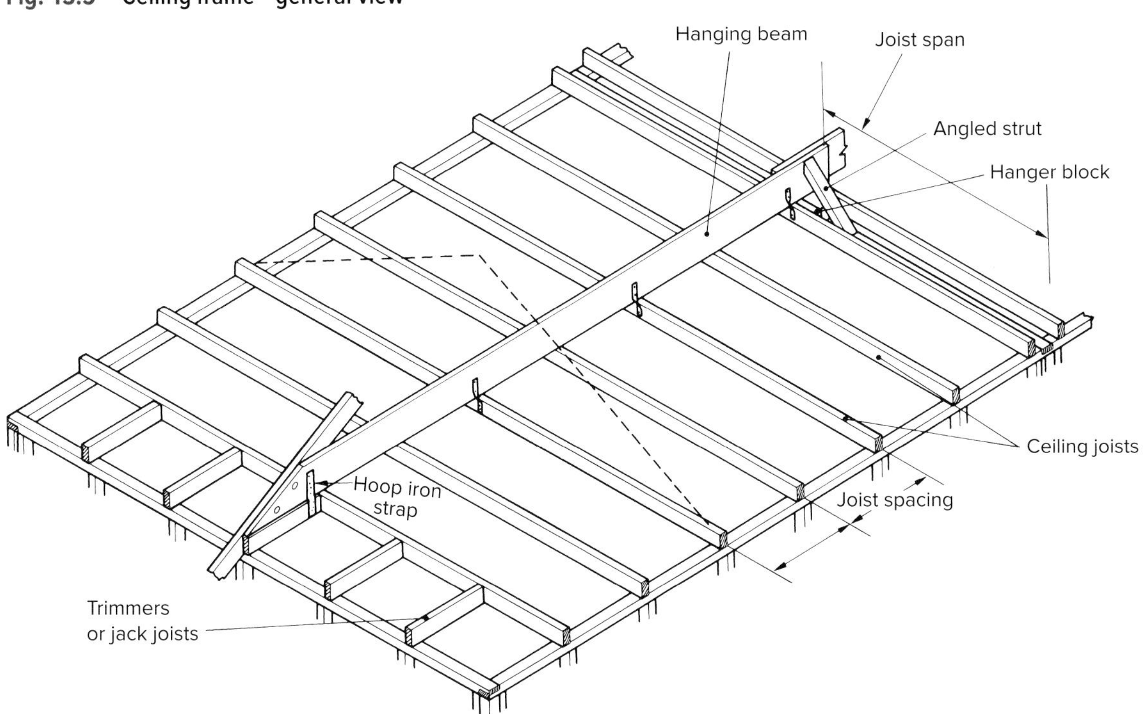

Ceiling joists should run in the same direction as roof **rafters**, as shown in Figure 13.10. They are securely fastened to the heel/foot of the roof rafter. The spacings can vary, but 450 mm or 600 mm between centres are common.

13.4.1.2 Ceiling trimmers

Where the ceiling joist at the end of a building would foul the roof rafters, it is omitted and replaced with *ceiling trimmers,* which run at right angles to the ceiling joists.

13.4.1.3 Hanging beams

Hanging beams are placed on edge over the ceiling joists to prevent them sagging over long spans.

Fig. 13.6 Ceiling framing—concept

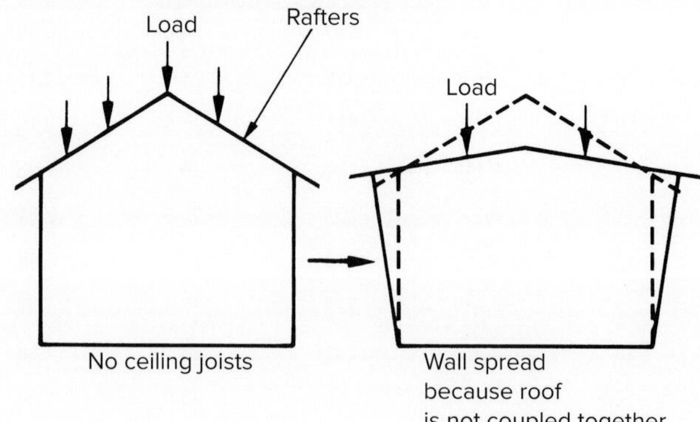

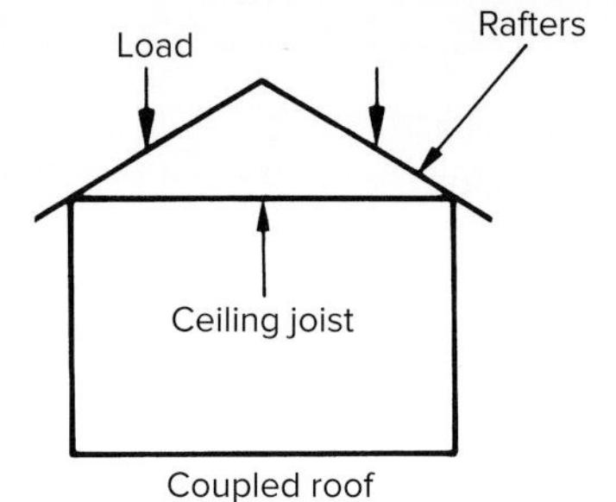

CHECK YOUR UNDERSTANDING

1. What are three functions performed by the ceiling joists?
2. How are the positions to fix the ceiling joists determined? In which direction should ceiling joists be laid?

13.4.1.4 Blocking

Blocking is provided over load-bearing walls to support the ends of hanging beams (Fig. 13.7).

Fig. 13.7 Hanging beam support

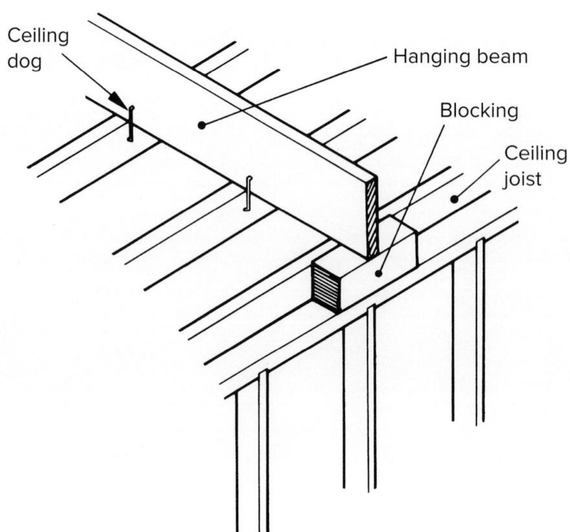

13.4.1.5 Ceiling ties

Ceiling ties, **ceiling joist hangers**, ceiling dogs and wooden cleats are devices used to strengthen the joint between hanging beam and ceiling joist. The section dimensions for ceiling members can be found in AS 1684.

Before the ceiling joists are fixed, the location of the roof rafters must first be marked out on the upper face of the top plate. It is preferable that this be done at an earlier stage when the plates are being set out on the ground. Marking out on the top of the walls after they have been erected can be a slow and precarious procedure. At this stage, note that the ceiling joists are placed beside the positions marked and are reserved for the roof rafters (Fig. 13.8).

Fig. 13.8 Rafter/ceiling joist set-out

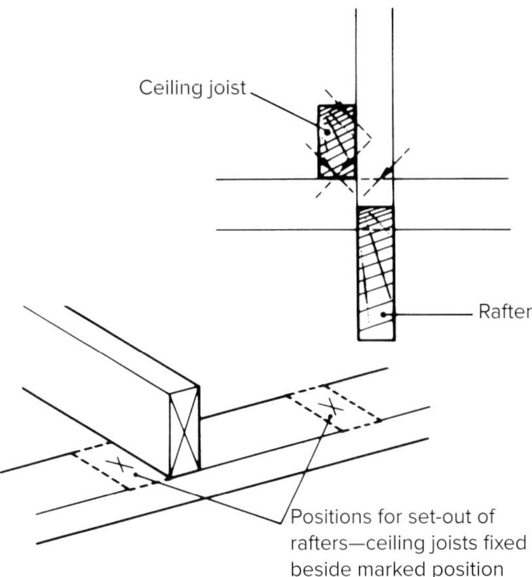

Procedure for laying ceiling joists

1. Stand ceiling joists on end against the wall of the building, approximately where they will be required. Load the roof, laying each ceiling joist on the flat, adjacent to its final location.
2. With a carpenter at each end, stand each joist upright, with the round edge up, and skew nail to the top plates. Where joists are joined in their length, they can be lapped over a wall with the joists of an adjacent room and securely nailed together.
3. Mark the position for the hanging beam and fix the blocking to support the beam at each end (Fig. 13.7).
4. Lay the hanging beam across the ceiling joist–stand it upright and skew nail to blocking.
5. Lay a plank across the ceiling joists beside the hanging beam and, using a short sash cramp or other suitable clamping device, work across each room cramping each joist up to the hanging beam. Ensure that the joists are in line; a string line 90° to the ceiling joists, strung on the underside of ceiling joists and stretched from one wall to the other can be useful to ensure the centre of the joists are in line. Usually, sighting by eye will keep the joists sufficiently straight, but for greater accuracy the position for ceiling joists can be marked on to the hanging beam. Fix a ceiling tie or cleat at each junction on alternate sides of the hanging beam (Fig. 13.9).
6. Cut ceiling trimmers to length where required and fix in position, again beside the positions marked out for the rafters. A word of caution–never walk across the ceiling joists until they have been finally fixed and supported by the hanging beam as a joist could roll or collapse due to a knot or other defect leading to a serious injury.

Fig. 13.9 Hanging beam/rafter supports

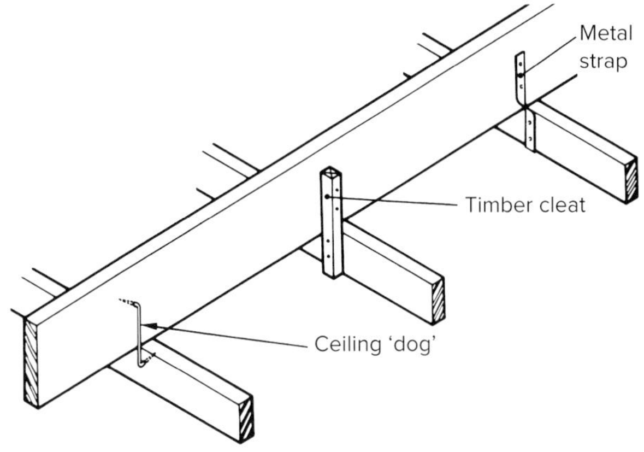

CHECK YOUR UNDERSTANDING

1. What is the purpose of the hanging beams? How are they supported?
2. Where is it necessary to use ceiling trimmers?
3. Describe an important safety precaution that should be observed when fixing ceiling joists.

13.5 The gable roof

The gable roof is a very popular roof form, being comparatively simple in design and readily adaptable to a wide variety of different roof pitches (Fig. 13.10). The gable roof is generally more economical to erect over a plain rectangular floor plan, but it can be adapted to other floor shapes.

The actual part referred to as the **gable** is the vertical triangular section at the end of the roof, from the top plate to the roof rafters. Over many years and in different places it has been traditional to decorate the gable in various distinctive ways; however, the gable can be exposed to the harshest weather conditions and is often finished with materials that will minimise maintenance.

13.5.1 The structural members

13.5.1.1 Common rafters

Common rafters are sloping roof members complete from the wall plate to the ridge. In a gable roof all of the rafters meet this definition, and because of this they can simply be referred to as *rafters*. However, in other roof forms it is necessary to distinguish them from other rafters and give them their full title of **common rafters**. On a plan, the common rafters always lie at right angles to the wall plate on which they are supported. One of the first problems facing the carpenter erecting the gable roof is to determine the length of the common rafters and the bevel required for setting out. This will be discussed in detail later.

Fig. 13.10 Gable roof—general view

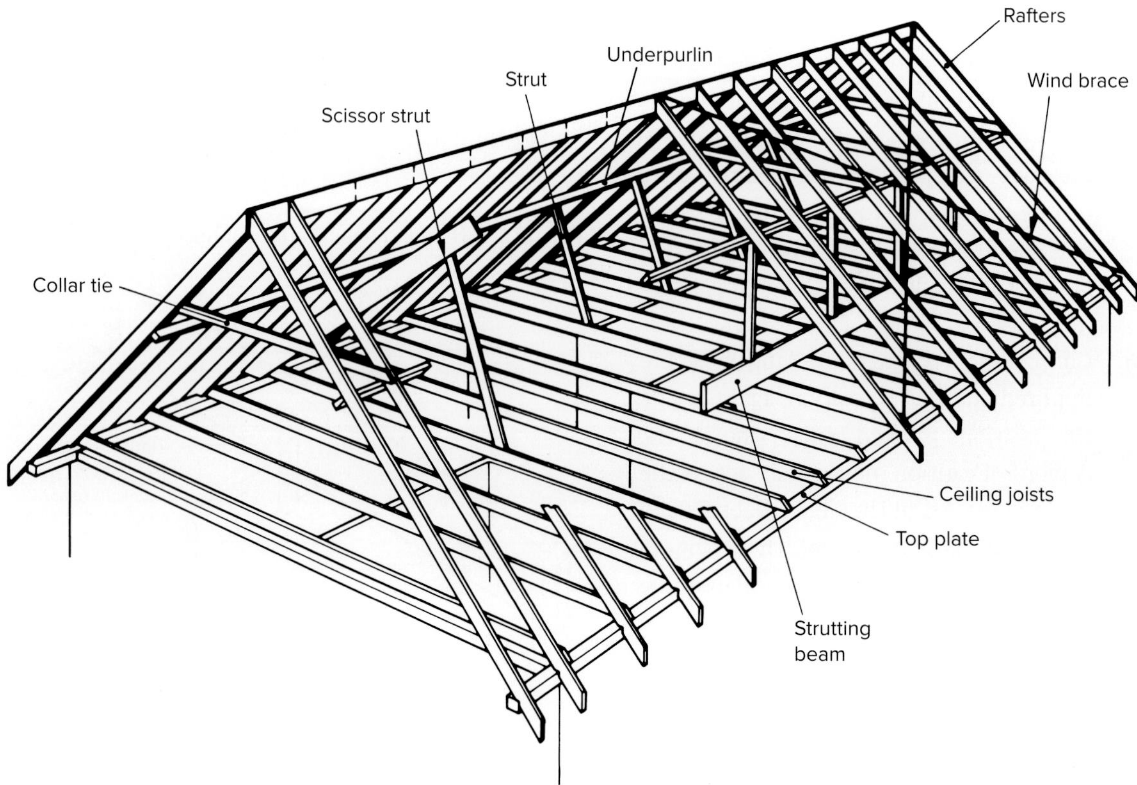

13.5.1.2 Ridge

The ridge runs the length of the roof and provides fixing for the rafters at their apex.

13.5.1.3 Gable studs

Gable studs enclose the gable and provide fixing for external sheeting materials.

13.5.1.4 Underpurlins

Underpurlins run the length of the roof and support the rafters usually midway in their length.

13.5.1.5 Struts/toms/props

Struts support the underpurlins and transfer the load of the roof structure and covering through the rafters and the underpurlins to internal load-bearing walls or suitable strutting beams.

13.5.1.6 Strutting beam

The strutting beam is sometimes used over large rooms to support struts where no internal walls are available. **Strutting beams** can run in any direction and must be supported by load-bearing walls and packed up to maintain a 25 mm clearance over the ceiling joists.

13.5.1.7 Collar ties

Collar ties are usually placed resting on the underpurlins and tie the two sloping roof surfaces together.

13.5.1.8 Wind braces

Wind braces are diagonal braces supported over load-bearing walls, bracing the ridge to prevent the pairs of rafters tending to topple sideways, particularly due to heavy wind loads.

13.6 Setting out rafters

Figure 13.11 shows the position of the rafters set out along the top plate. Usually, a pair of rafters is located at each end of the building and if there is to be an overhanging gable, a pair of rafters is located at the end of the plate to form the gable overhang.

The remaining rafters are set out along the length of the building, not exceeding the maximum spacing. Remember, the rafters must be fixed over the wall studs if the wall framing was designed with this intention. The ceiling joists will be fixed beside these marks.

Fig. 13.11 Rafter set-out

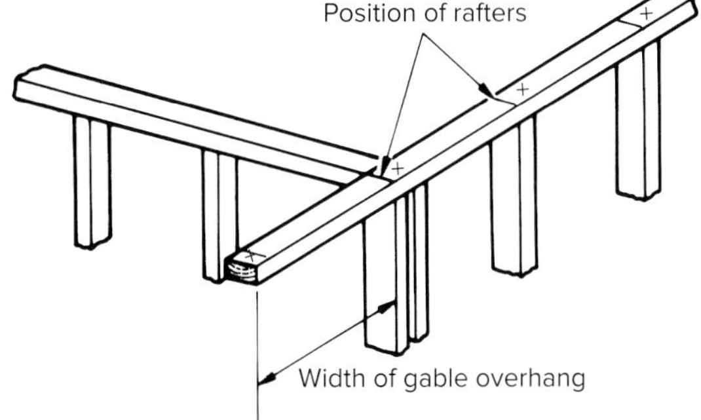

CHECK YOUR UNDERSTANDING

1. Describe the function of the following members in a gable roof:
   a. common rafter
   b. ridge
   c. underpurlin
   d. strutting beam
   e. wind brace.

13.6.1 Setting out the common rafter

In Figure 13.12, the equally pitched roof can be regarded as a right triangular prism resting on a horizontal surface. Angle *BAC* is the pitch of the roof and *AD* is the span. Lines *AB* and *BD* are common rafters represented by a single line.

Isolate the right-angled triangle *ABC*. Then *AC* is the **half span** or run of the common rafter, *BC* is the rise and *AB* is the true length of the common rafter.

Angle *BAC* is the pitch of the roof; it can also be referred to as the **level bevel** for the common rafter. Angle *ABC* is the plumb bevel or cut for the common rafter and is the bevel most commonly used in setting out and cutting the rafters.

As can be seen, if we reproduce triangle *ABC* to the specified information, then the true length of the common rafter and its plumb bevel can be readily obtained. For the purpose of explanation, let us assume that the span of a gable roof–equally pitched–is 8400 mm and the pitch is specified as 1:1.6 or 41°. Therefore, the run (half span) of the rafter is 4200 mm.

Fig. 13.12 Common rafter

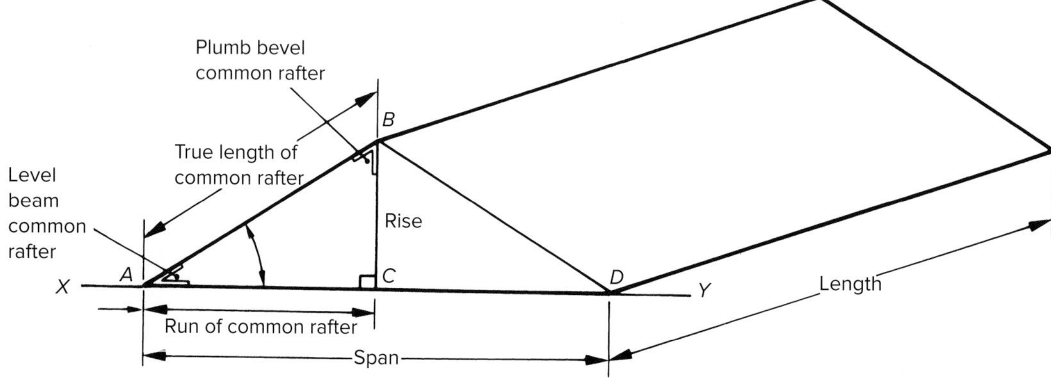

There are two methods for determining rafter length and plumb bevel/cut: by scale drawing and by calculation.

Fig. 13.13 Scale/pitch triangle

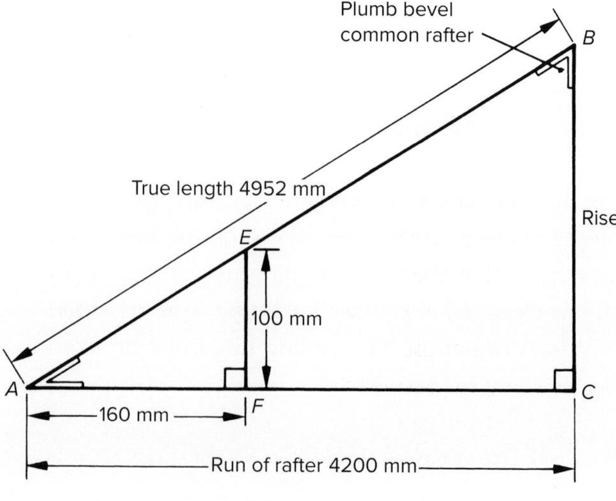

13.6.1.1 By scale drawing

This is done on a suitable plane surface such as paper or plywood. A scale of 1:10 is convenient. A rule can be used if a scale rule is unavailable.

1. Using a scale rule, set off the run of common rafter *AC* equal to 4200 mm. If a scale rule is not available, use a rule and divide each dimension by 10, e.g. 4200/10 = 420 mm.
2. From *A*, set off the pitch of the roof by setting off the vertical and horizontal components as shown. For example, with a ratio of 1:1.6, multiply both components by 100 (100:160).
3. From *C*, set off a right angle to intersect the pitch line at *B*.
4. Measure *AB*, which equals the true length of the common rafter: 4952 mm.

5. Set the sliding bevel to angle *ABC*, which equals the plumb bevel/cut of the common rafter. When setting out to scale, care is essential to ensure accurate results. Set off the pitch with a protractor if the pitch is specified in degrees.

13.6.1.2 By calculation

Run of common rafter = 4200 mm
Roof pitch = 1:1.6
Rise (refer to Table 13.1) = 625 mm/m

Calculate the true length of the common rafter per metre run in Figures 13.14 and 13.15 using Pythagoras' theorem (see Chapter 3).

$$\begin{aligned} XY^2 &= YZ^2 + ZX^2 \\ XY &= \sqrt{YZ^2 + ZX^2} \\ &= \sqrt{1000^2 + 625^2} \\ &= \sqrt{1\,000\,000 + 390\,625} \\ &= \sqrt{1\,390\,625} \\ &= 1179 \text{ mm} \end{aligned}$$

To calculate the length of the common rafter:

$$\begin{aligned} \text{True length of CR} &= \text{CR mm/m run} \times \text{run of rafter in metres} \\ &= 1179 \times 4.2 \\ &= 4952 \text{ mm} \end{aligned}$$

Fig. 13.14 Calculation triangle—common rafter

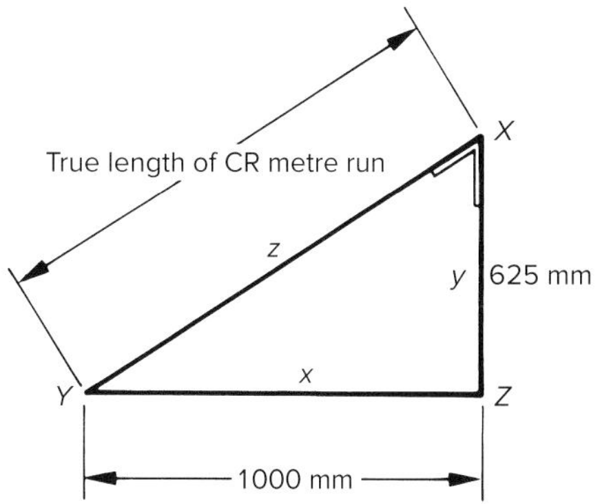

From the set-out drawing (Fig. 13.13), the length of the common rafter, referred to as the *single line length,* and the plumb bevel/cut can be obtained. But roof members have width and thickness, so it is now necessary to consider where this angle and dimension are applied to the common rafter. Also, in most instances, the roof will have an eaves width and this must be provided for in setting out the finished rafter.

Figure 13.16 shows a vertical section through the gable roof. Width, thickness and overhang have been added to the roof members. The triangle *ABC* is also the triangle *ABC* that was isolated and considered in Figure 13.13. Note that *AB* is the single line length of the common rafter and is measured along the top edge of the rafter.

Allowance must be made at the apex of the roof for the ridge to fit between the common rafters. Each rafter is shortened by half the thickness of the ridge measured square off the plumb line. A birdsmouth is cut into the foot of the rafter to obtain a firm bearing and to secure the nailing to the top plate.

The maximum depth of the birdsmouth is one-third the depth of the rafter (Fig. 13.17). In any case, it is more important to leave a constant amount on each rafter from the top plate to the *XY* line; this distance will be referred

Fig. 13.15 Application of common rafter triangle to true length of rafter

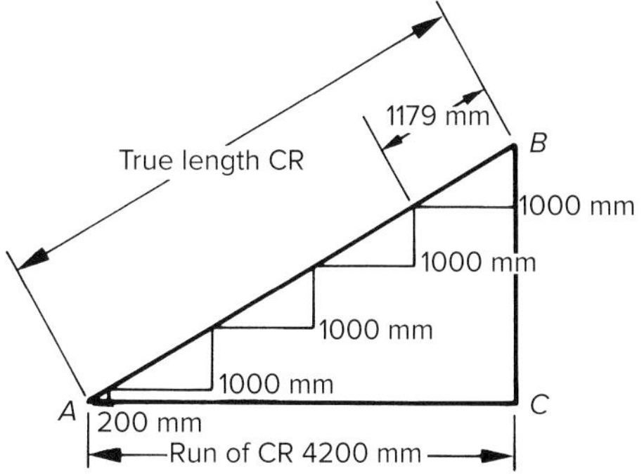

to as *distance* x. No matter what the width of the rafter or pitch of the roof may be, there is no reason for the horizontal cut of the birdsmouth to be greater than the width of the top plate.

The distance *FA* in Figure 13.16 is the eaves width and is a horizontal distance, usually specified by the building designer. The distance *EA*, the eaves overhang, is, of course, a greater distance and is measured along the slope of the rafter.

Fig. 13.16 Vertical section through gable roof

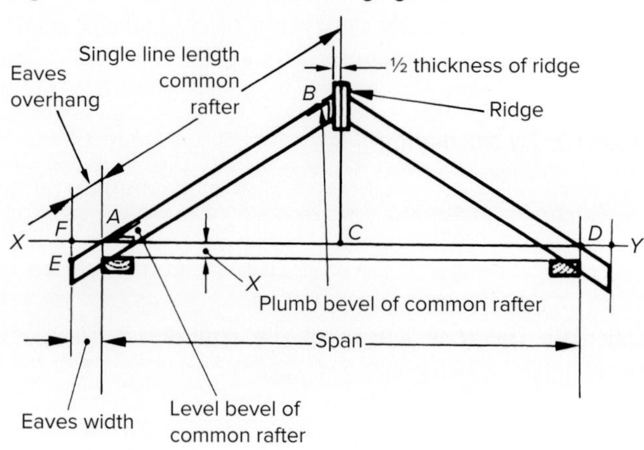

13.6.2 Setting out the pattern rafter

A pattern rafter can now be prepared from which other rafters can be duplicated. The procedure for setting out a pattern rafter is as follows:

Fig. 13.17 Birdsmouth joint depth

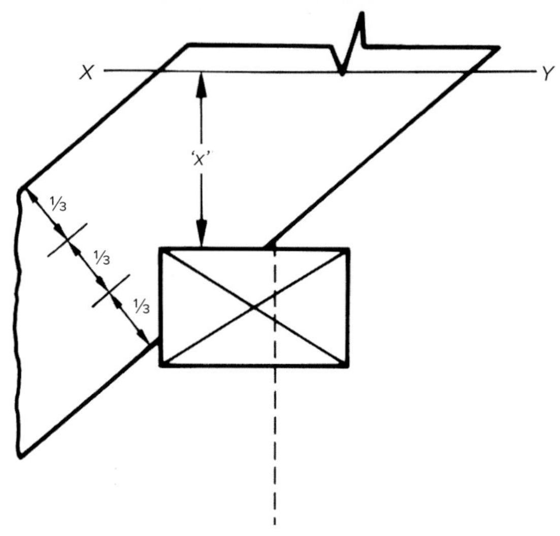

1. Select a common rafter, as straight as practicable, and set a sliding bevel to the plumb bevel of the common rafter.
2. Near the end of the rafter select point *B* and mark the plumb bevel/cut of the common rafter through this point.
3. From point *B* measure off the single line length of the common rafter to point *A* and mark the plumb bevel/cut of the common rafter through this point.
4. From the plumb line through point *B*, measure off at right angles half the thickness of the ridge and again set off the plumb bevel of the common rafter. This is the line at which the rafter will be cut to length.
5. From point *A* measure down the plumb line distance x, and mark off at right angles for the horizontal cut to the birdsmouth.
6. The eaves overhang is set off by measuring square off the plumb line through point *A* a distance equal to the eaves width, and establishing point *E*. Again, mark the plumb bevel of the common rafter through this point. The eaves overhang *EA* can also be calculated readily. For example, if the eaves width is specified at, say, 600 m, then:

$$\begin{aligned}\text{Eaves overhang} &= \text{common rafter mm/m run} \times \text{eaves width (m)}\\ &= 1179 \times 0.6\\ &= 707\text{ mm}\end{aligned}$$

This distance can be measured along the rafter from point *A* to point *E*. The eaves overhang plus an allowance for making the plumb cut is commonly referred to as the tail of the rafter. Quite often the eaves overhang is not marked on the rafter at this stage, but the tail must be left sufficiently long, so that it can be marked and cut off (as described) at a later stage.

At this time, it is no doubt good insurance against possible error to cut a duplicate rafter, and test the pair of rafters for accuracy over the span of the roof. Place a small block of wood, the thickness of the ridge, between the apex of the rafters to allow for the ridge to be fitted between them.

Fig. 13.18 Setting out pattern rafter

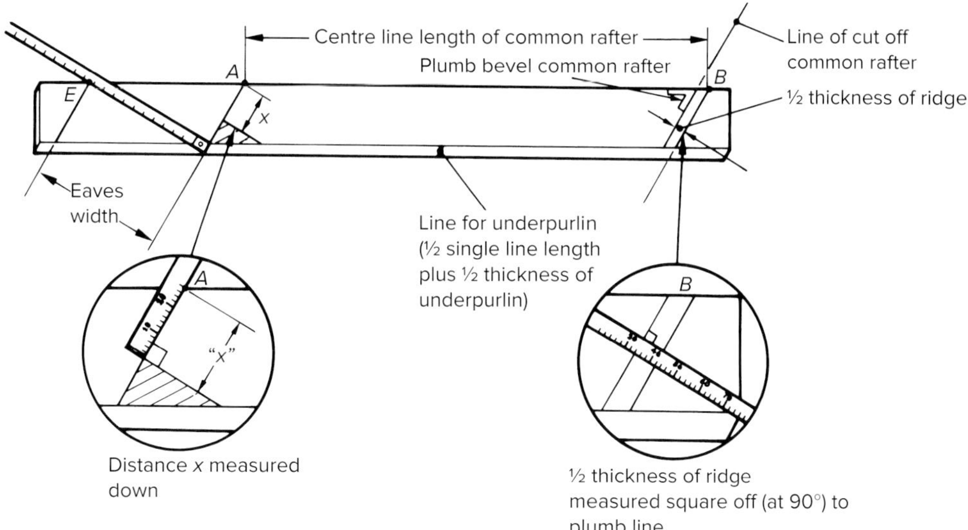

13.6.3 Cutting rafters to pattern

To cut the remaining rafters accurately to the pattern, nail a short cleat across the rafter directly above the birdsmouth and at the top of the rafter. Lay the pattern rafter on each rafter, keeping the round edge up and the cleats hard against the top edge of the rafter. Mark the plumb cut and the birdsmouth. Complete cutting out the remaining rafters.

Fig. 13.19 Cutting rafters to pattern

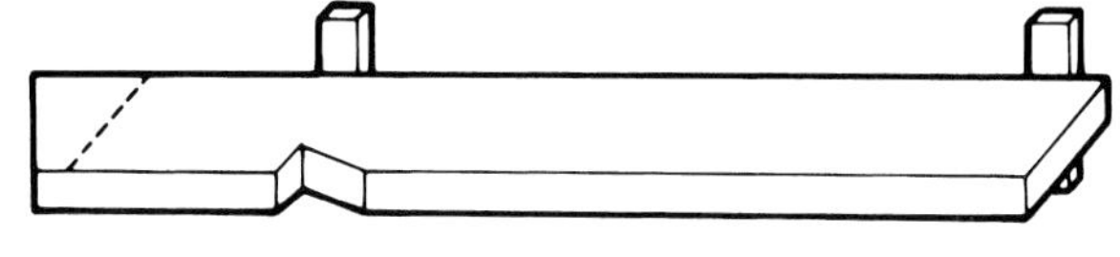

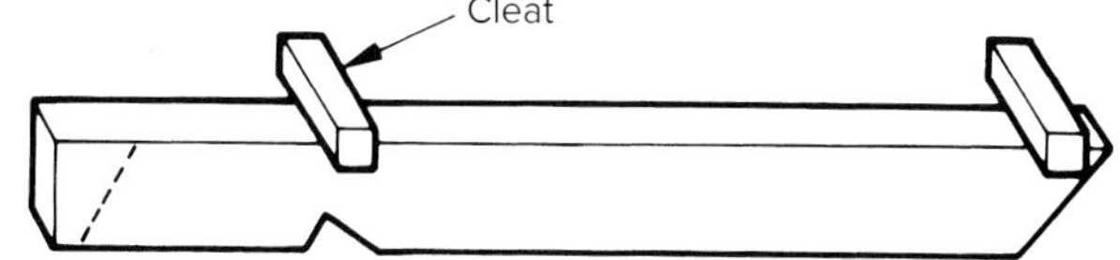

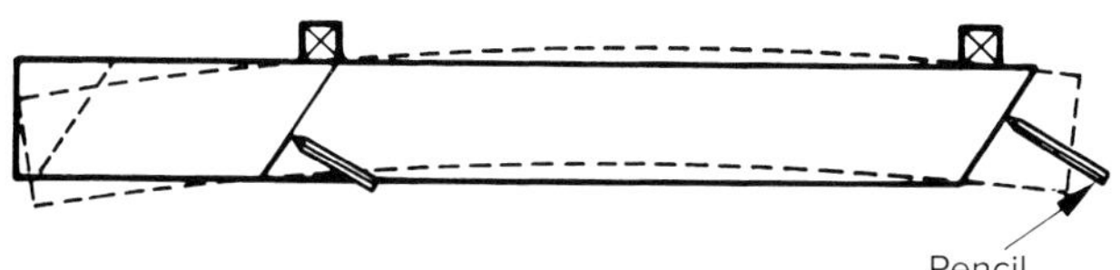

13.6.4 Constructing a rafter boat

The **rafter boat** is another option that can be used when cutting a large number of rafters (Fig. 13.20). It works on a similar principle to a pattern rafter.

> **TIP** The rafter boat is much smaller and lighter than a pattern rafter and can be kept to use on another job with the same roof pitch.

The main difference is that the rafter boat will not give you the rafter length. It will give the plumb cut, the birdsmouth, rafter side cuts and even the eave overhang.

Two pieces of timber are required, generally something light such as pine or plywood. The large piece of timber should be the same depth as the rafter; a cleat of material 19 mm × 70 mm is needed for the top of the boat so it can be positioned and slid along the rafter.

1. First, ensure the timber to be used is straight and free of defects.
2. Identify which edge of the boat will be the top and mark.

3. Transfer the plumb cut to one end of the boat and cut accurately.
4. Locate the position of the birdsmouth at the other end of the boat, mark and cut. It must be identical to the birdsmouth on the rafters.
5. If needed, work out the eave overhang and mark on the boat past the birdsmouth.
6. Centre the cleat along the top of the boat, glue and nail. Ensure the cleat overhang on both sides of the boat are the same.
7. If required, mark and cut the rafter side cuts on the cleat at the plumb cut end of the boat (for a hip roof).
   - **(a)** Use the rafter boat in a similar way to a pattern rafter, select a rafter and ensure the rounds are up.
   - **(b)** Place the boat at the end at which the plumb cut will be and mark the cut using the boat.
   - **(c)** Now the plum cut has been marked, measure along the top of the rafter the desired length. Mark this position.
   - **(d)** Slide the boat down to this new mark and mark out the birdsmouth. To improve the use of the boat, a hole approximately 25 mm can be drilled on the plumb line located on the birdsmouth to make positioning the boat easier.
   - **(e)** If the eave overhang is marked on the boat, mark this while the boat is at the birdsmouth position.

Fig. 13.20 Finished rafter boat

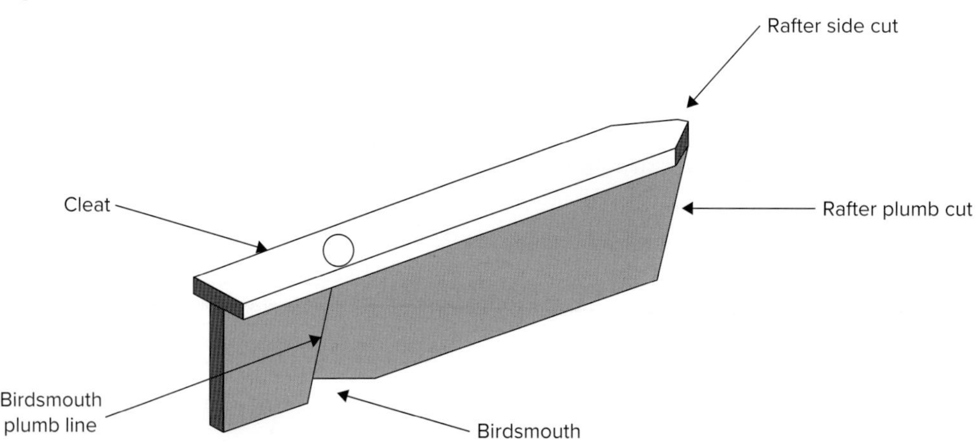

13.7 Preparing the ridge and underpurlins

On the gable roof, the ridge usually has to be joined along its length with a splice joint (Fig. 13.21). It is preferable to make the splice where a pair of rafters will be located, as this ensures that the ridge will be locked in place and kept in alignment.

The ridge is laid out on top of the ceiling joists above the top plate, and the position of the rafters is squared up from the plate on to the ridge and marked (Fig. 13.22).

Fig. 13.21 Joint along length of ridge

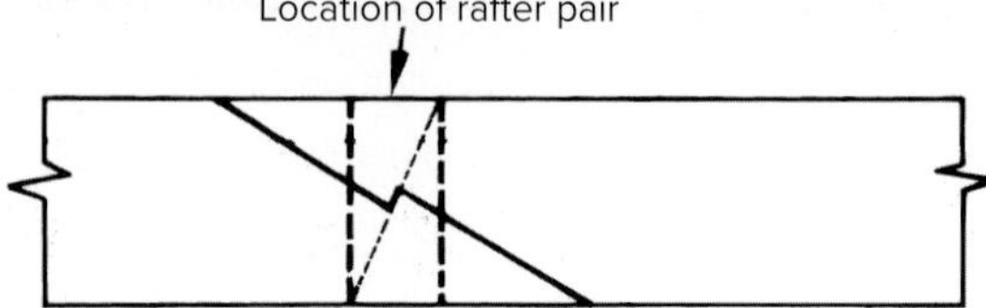

Fig. 13.22 Marking rafters on the ridge

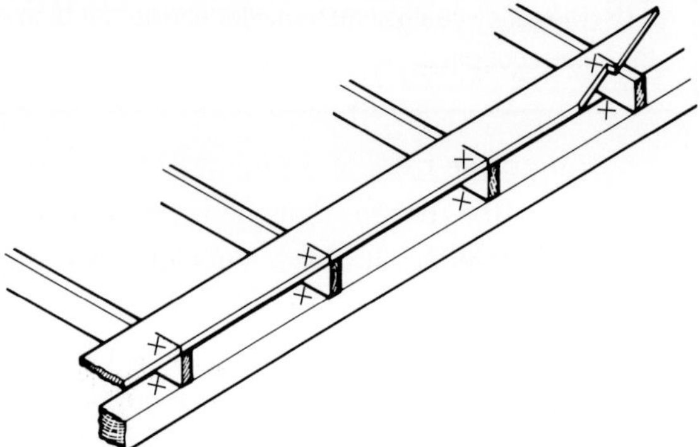

Underpurlins can be prepared on the ground. Where they have to be joined in length, a half-lapped joint is used (Fig. 13.23). It is better to adjust the length of the underpurlins so that the joint can come over a wall and a strut can be fixed underneath the joint.

Fig. 13.23 Joining underpurlins

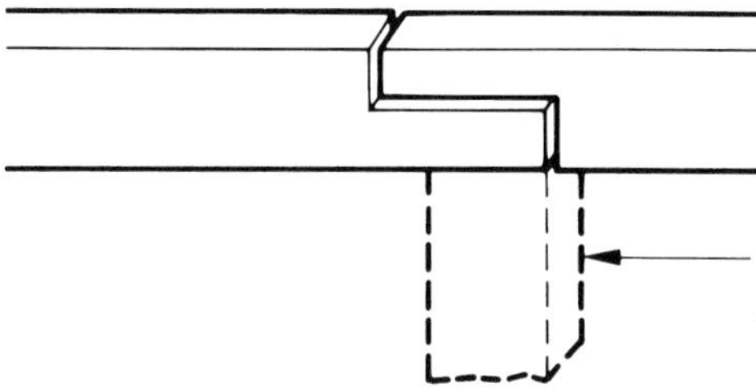

Strut under joint if possible

13.8 Erecting the gable roof

Before erecting the roof, a safe working platform should be provided by either laying planks across and along the length of the ridge or erecting a scaffold. Never commit yourself to working from some insecure or rickety structure, as a fall at this time through the joists can result in some very serious injuries. A catch platform needs to be erected around the perimeter of the building so that, in the case of a fall, you don't fall to the ground. Also, the platform is used to erect the roof.

13.8.1 Procedure

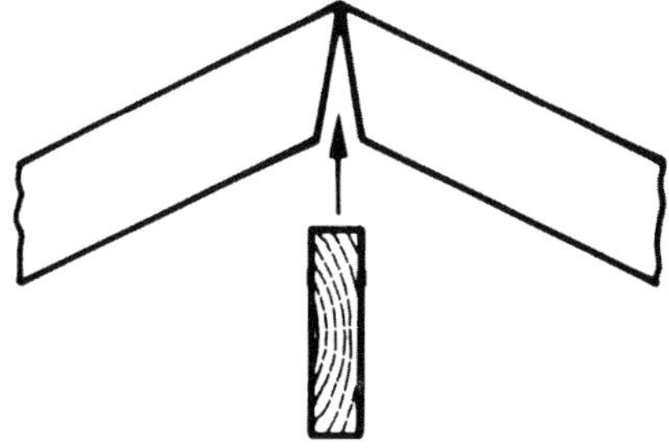

Fig. 13.24 Rafters ready to join ridge

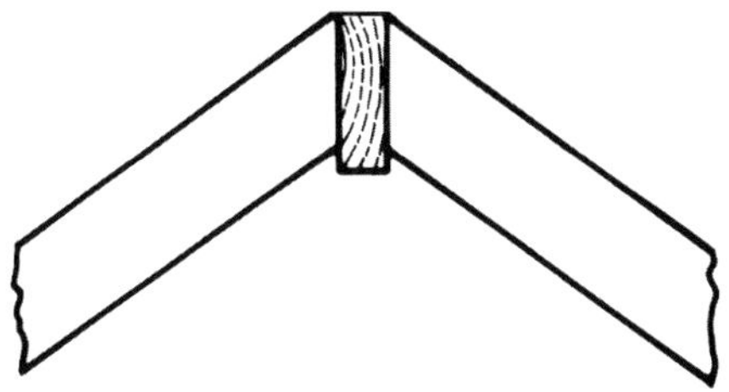

Fig. 13.25 Rafters in place

1. Lay out the ridge along the centre of the building.
2. Load underpurlins onto the roof and lay on the ceiling joists near their final location. If any strutting beams are to be used, also load these onto the ceiling and lay them on their flat face near to the fixing position.
3. Stand the rafters on end along the length of the building beside their fixed position. Pull up the rafters and load the roof with sufficient rafters to set the ridge in place–usually rafters near the end of the building and adjacent to any joint in the ridge.
4. Fix selected pairs of rafters into position, skew nailing through the birdsmouth into the top plate and resting against each other at the apex. Stabilise with a light diagonal brace if necessary.
5. With a carpenter at each end of the ridge, lift it up, separating the rafters so that it can pass between them. Let it down so that it is gripped between the rafters. Taking the weight of the ridge, ease the rafters into place.

   Nailing for the rafter on one side can be carried out directly through the ridge, while the other side is usually secured by skew nailing through the rafter to the ridge. The roof will now be erected to the stage as shown in Figure 13.26.
6. Plumb up the pair of rafters at the end of the roof using the plumb bob or spirit level and straight edge and secure in place with a temporary, but sturdy, brace that will remain in place until permanent wind bracing is provided (Fig. 13.27).
7. Fix the remaining section of the ridge and the rafters as previously described. Remember that underpurlins and strutting have to be cut and fixed to the underside of the rafters, and this can be done much more quickly and easily if the under-roof working area is kept as accessible as possible.

Fig. 13.26 Pairs of rafters in place

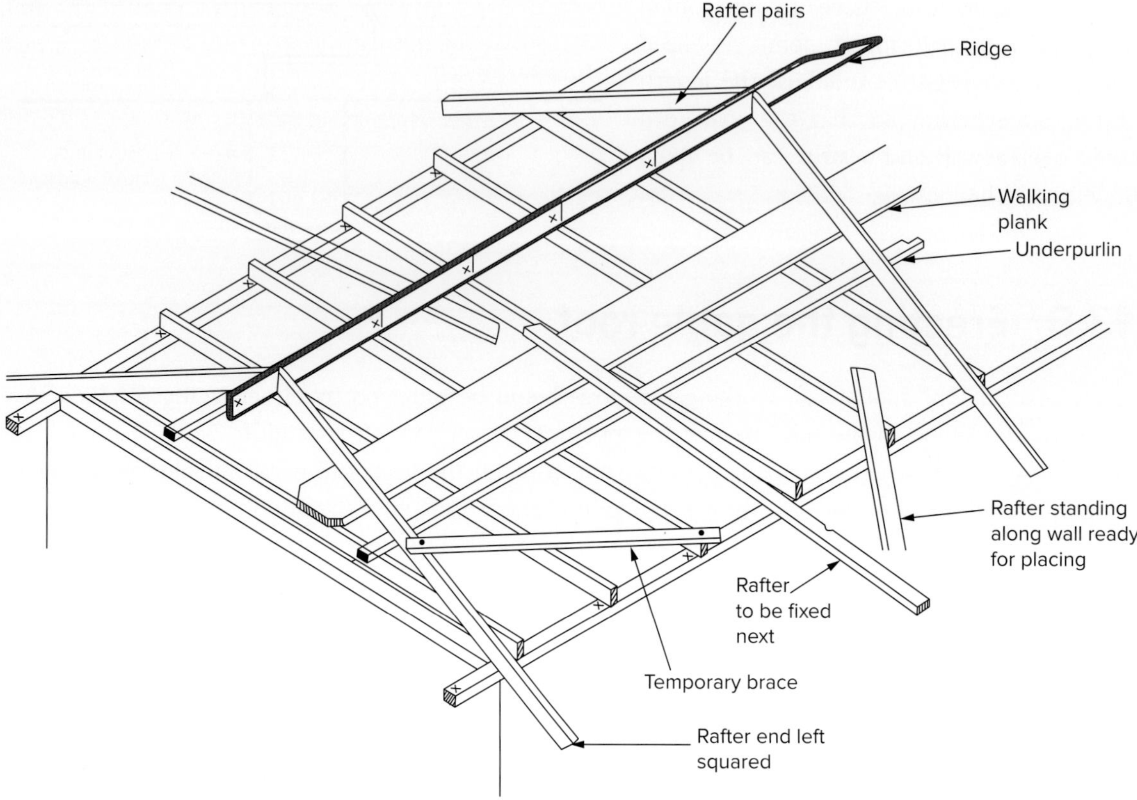

Fig. 13.27 Plumb up of gable end

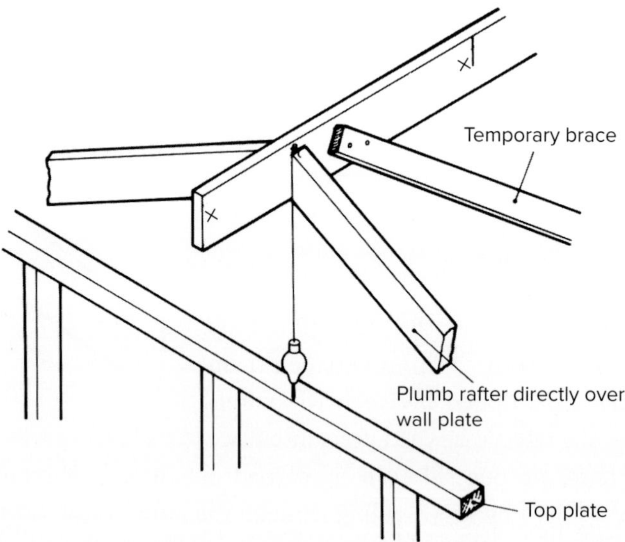

SAFETY TIP At this stage it is suggested that no more rafters than are necessary to keep the ridge and underpurlins in alignment are loaded onto the roof and fixed in position. This is particularly the case if it is a fairly low pitched roof, as you could well save many hours of work and an aching back.

8. Partly drive a nail into the underside of selected rafters on the mark indicating the top edge of the underpurlin. Lift the underpurlins, holding them firmly against the stop provided by the nails, and skew nail the rafters to the underpurlins. A short sash cramp is very useful for pulling the rafters and underpurlins tightly together. Sometimes it is better to fix the underpurlins hollow side up to counteract the rafters that were fixed round side up. This tendency will be much more pronounced in the case of a green hardwood roof.

9. Check that rafters and underpurlins are tightly pulled together, and are now straight enough for strutting. Sighting by eye is usually sufficient. Struts/**props** can be fixed either vertically or at right angles to the common rafter (Fig. 13.28). Struts/props must be supported by a load-bearing wall, not over a doorway, unless the doorway has been specifically designed to take the load.
   (a) Cut the foot of the strut at the plumb bevel of the common rafter and stand beside the underpurlins (Fig. 13.29). Level across from point *A* to point *B* on the strut and from *B* mark out and cut a notch for the underpurlin. The strut will move forward as shown, tightly under the underpurlin. Nail a cleat behind the heel of the strut (Fig. 13.30).
   (b) Stand the strut vertically beside the underpurlin, and level across from point *A* to point *B* on the strut. The angle for the notch on the strut is the plumb bevel of the common rafter, but to allow for any little practical discrepancies, the notch for the underpurlins is probably best marked by paralleling the sides of the underpurlin with the rule as shown (Fig. 13.31).

   Sometimes struts can be more securely and easily fixed if a short sole plate is laid over the ceiling joists on top of a load-bearing wall (Fig. 13.32).

Fig. 13.28 Strutting to support underpurlins

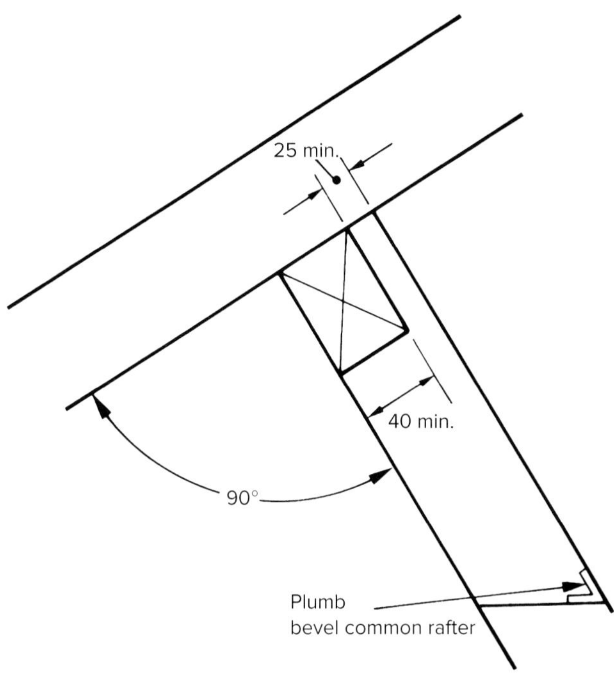

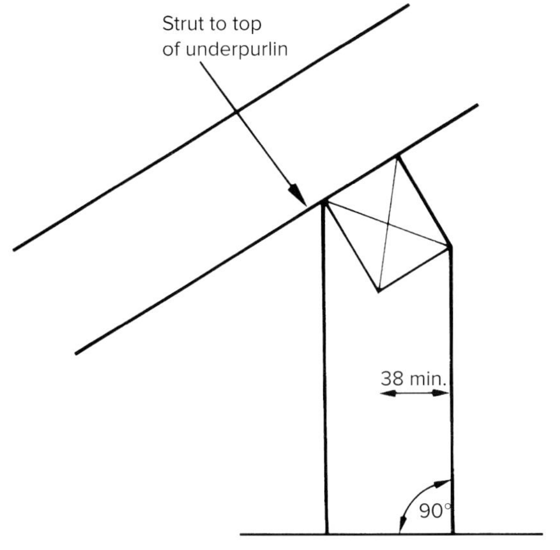

Alternatively, prop gauges are available, which are used to create a quick template for the strut/prop to be cut. The head and foot are adjustable, so any desired angle can be achieved. The tool is also telescopic, so the length of the strut/prop can be ascertained.

Fig. 13.29 Marking strut at right angle to rafters

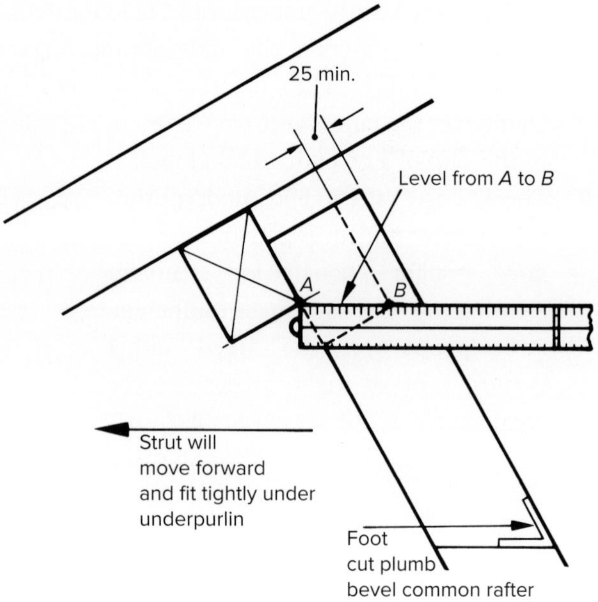

Fig. 13.30 Cleat to foot of strut

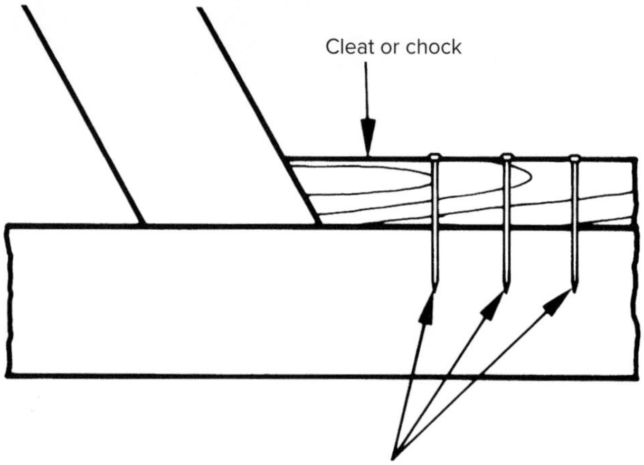

Fig. 13.31 Marking out a vertical strut

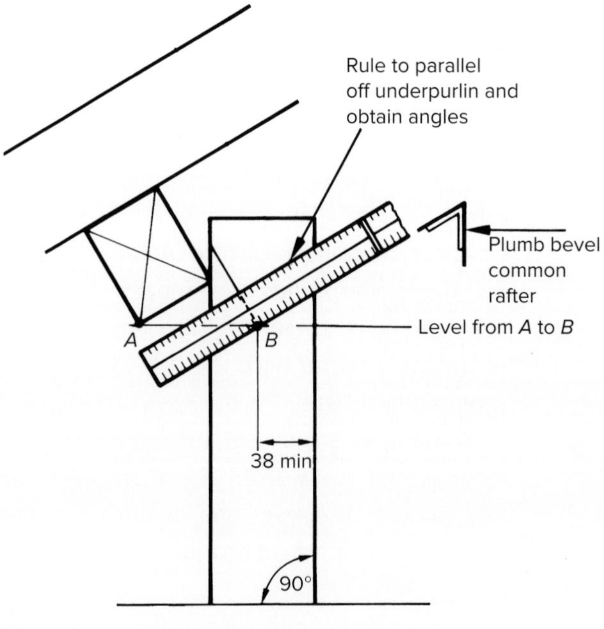

CHECK YOUR UNDERSTANDING

1. How is the ridge marked out for the position of the rafters?

Fig. 13.32 Use of sole plate where strut would land on joist

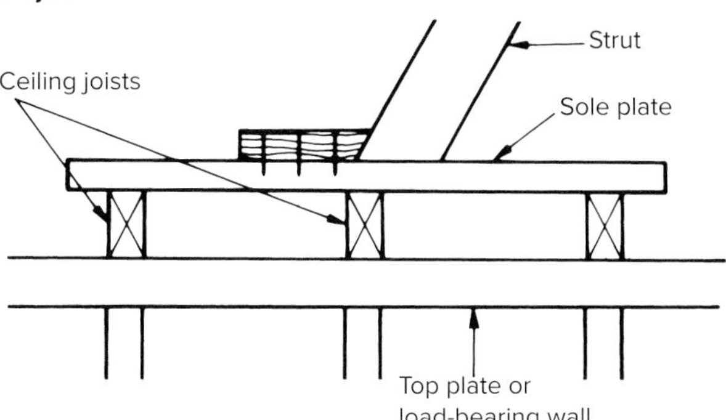

Fig. 13.33 Prop stick

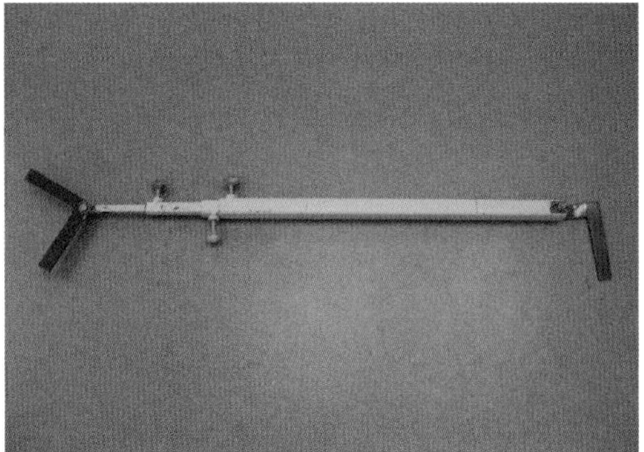

Courtesy of Daniel Bonnici

10. Strutting beams can be used over large rooms where no internal wall is available for supporting struts (Fig. 13.34). Strutting beams can run in any direction across a room, but must be supported by a load-bearing wall and be packed up at least 25 mm above the ceiling joists at both ends. The ceiling joists must *not* support the strutting beam. They should be of seasoned timber; the struts are fixed from the strutting beam to the underside of the underpurlin.
11. Fix two permanent wind braces running in opposite directions. The foot of the brace must be over a load-bearing wall, sloping at an angle of approximately 45°, and scarfed at the top to fit under the ridge. Remember, more rafters and collar ties have to be fixed later, so try to adjust the wind brace so as not to foul other members. Vertical **toms** supporting long sections of ridge can also be fixed at this time. Some carpenters may prefer to fix the rafters and collar ties before the wind braces.
12. Fix the remainder of rafters. No doubt, many would have been fixed as required during the progress of the job.
13. Cut and fix collar ties (Fig. 13.35). Lay a collar tie in position on top of the underpurlins and mark the length and shoulder to fit under the rafters. Prepare collar ties on the ground, cutting the shoulder on one end but leaving the other shoulder to be cut when each collar tie is being fixed in its final position, usually to every second pair of rafters.
14. The lower end of the rafter is the eaves line. The eaves width is usually specified and must be measured along a level line from the face of the wall. Theoretically, the foot of the common rafter can be cut off

Fig. 13.34 Strutting beam

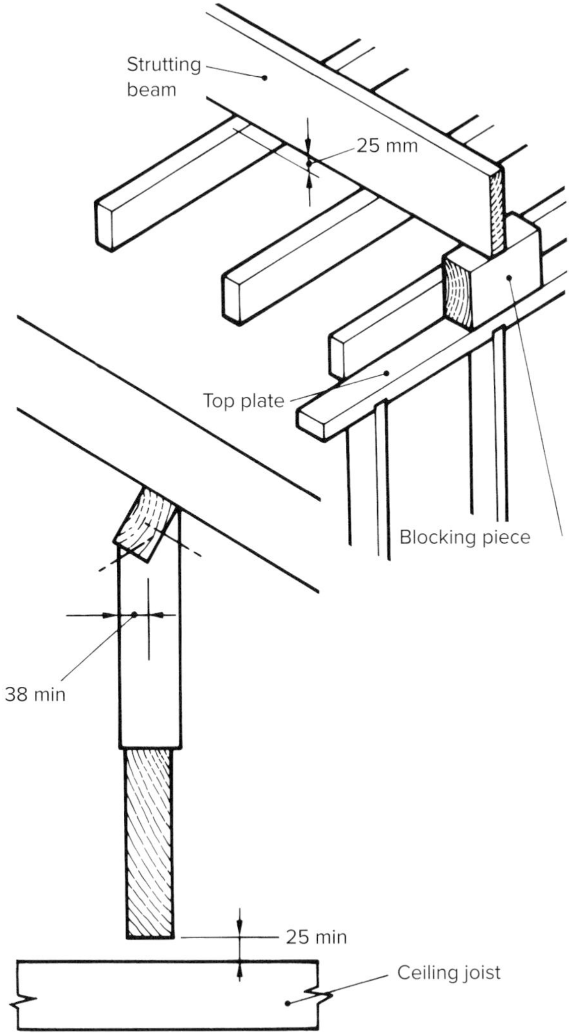

before it is fixed in place. But to ensure a straight fascia line, it is generally agreed that it is much more satisfactory to strike a straight line and cut the rafters to length only after they have been finally fixed and strutted in position.

Measure the eaves width at each end of the wall and mark the plumb line on the rafters. Set up a string line directly over these marks and plumb down from the string line on each rafter, and cut the rafters off to line (Fig. 13.36).

Fig. 13.35 Collar tie

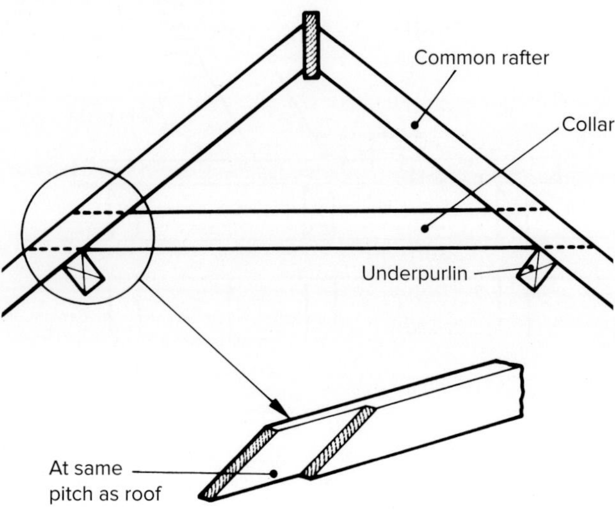

13.9 Roof member sizes

The section size of roof members may vary in different areas, in accordance with local building regulations and established trade practice that have evolved from many years of experience of local conditions. AS 1684 should be consulted to calculate what size of members are required in each situation.

13.9.1 Finishing gable ends

Gable studs are fixed from the top plate and are checked out to fit behind the rafters (Fig. 13.37).

Fig. 13.36 Marking rafters to length

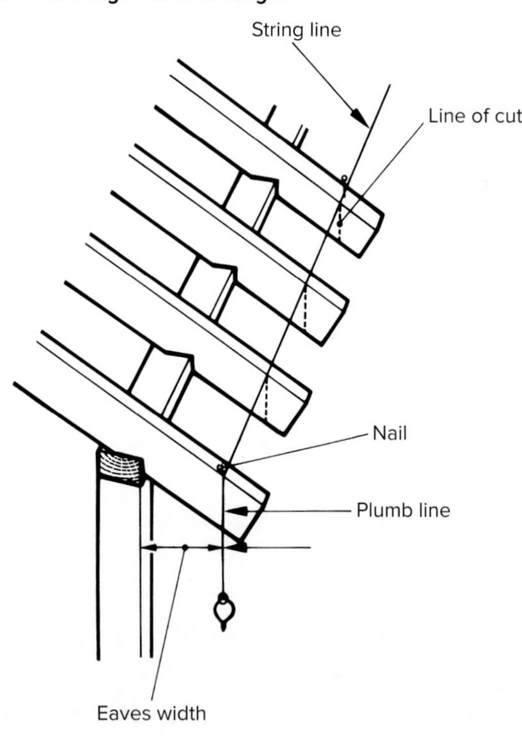

Fig. 13.37 Finishing the gable end

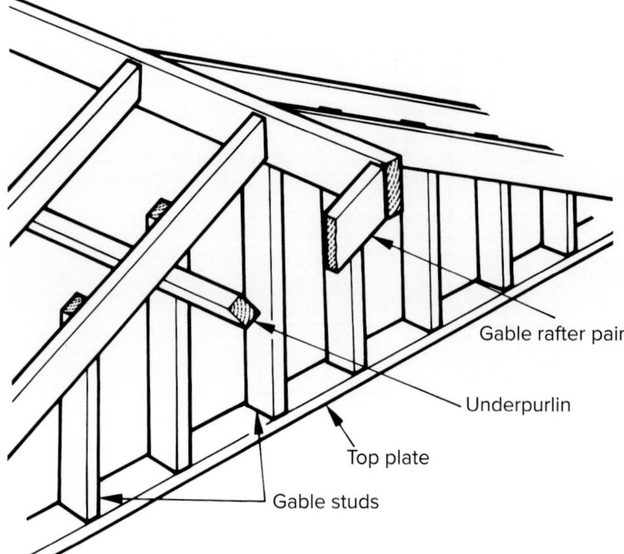

13.9.1.1 Eaves finishes

Eaves can be finished in a number of ways, but some basic requirements should be provided, namely:

- provision for interior room ventilation, where and if required
- fixing for fascia and gutter
- bird proofing
- support for soffit/eave lining if required.

The most commonly adopted eaves finish is probably the *boxed eave* (Fig. 13.38).

Fig. 13.38 Boxed eaves

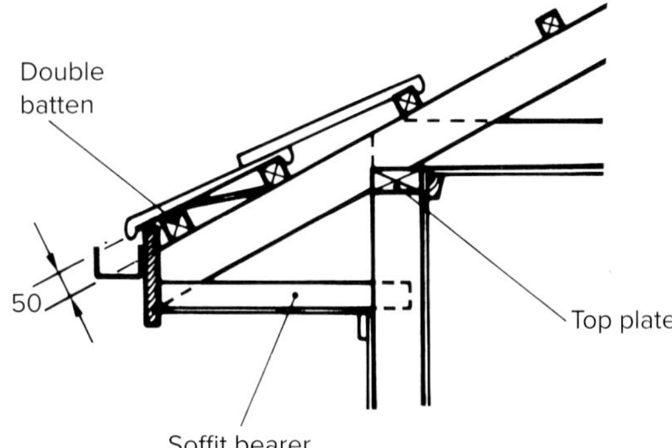

CHECK YOUR UNDERSTANDING

1. How can the tail of the rafter be cut off to a straight line ready for fixing the fascia?

13.10 The skillion roof

Fig. 13.39 Skillion roof frame

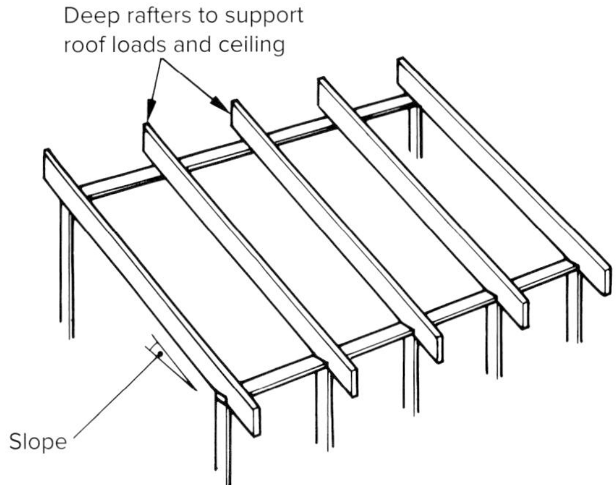

The **skillion** roof, also referred to as a lean-to or a mono-pitch roof, was for a long time thought only suitable for building works of minor importance such as additions, garages and laundries. However, with the introduction of sheet metal roofing materials in long lengths, it has now been adapted successfully to contemporary design in both residential and commercial buildings. The skillion roof can be comparatively quick and simple to install, particularly over irregularly shaped buildings, whereas other roof forms can become very complicated and unattractive.

The function of the rafter and ceiling joist are combined in the one member, which must be sufficiently strong to carry its own mass plus the mass of the roof covering and ceiling linings (Fig. 13.39).

Fig. 13.40 Wind pressure on skillion roof

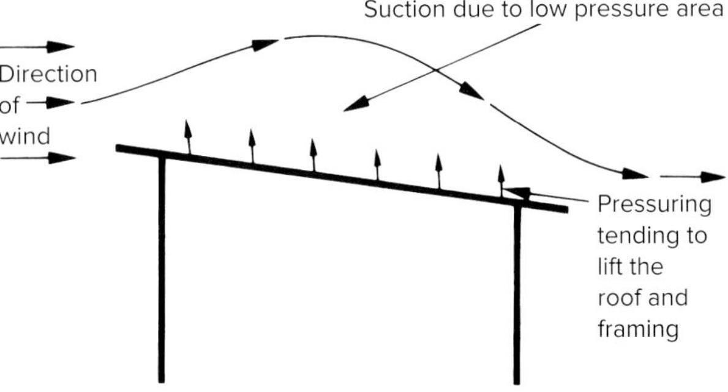

The lower pitched skillion roof (minimum 1:60 = 1°) can be subject to considerable suction due to wind pressure, which tends to lift the roof covering and framing (Fig. 13.40). For this reason, the roof covering must be securely fixed in accordance with the manufacturer's instructions and the rafters skew nailed to plates and tied down to the wall plates with triple-grip connectors.

Setting out the rafters for the skillion roof is usually carried out in a very practical manner. After the walls have been plumbed up, the rafter is laid in position across the walls, which will be of different heights. The location of a birdsmouth (to give the rafter secure fixing to the wall plates) is marked directly from the plates using a small straight edge, for example, a rule and pencil (Fig. 13.41).

Fig. 13.41 Marking out rafter for skillion roof

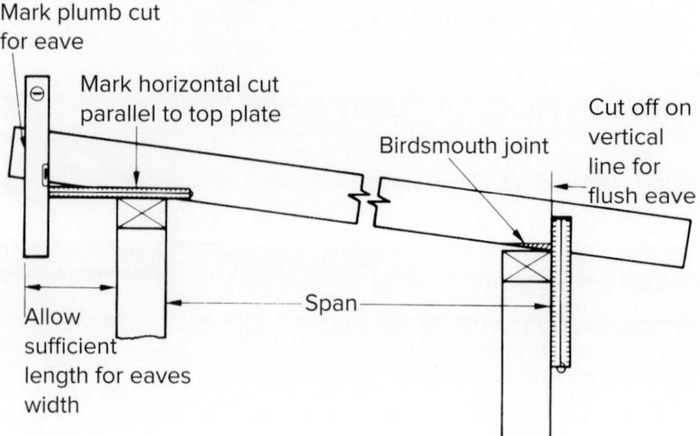

If there is to be an overhanging eave, the width of the eave can be set off along a horizontal line and a plumb line marked on the rafter. It is then cut to length. Alternatively, the rafters can be cut to length after they have been fixed, in the same manner as for the gable roof.

Deep rafters over 200 mm must be blocked to ensure they will not tend to topple sideways when placed under a load. Solid blocking in rows no more than 1500 mm apart is required (Fig. 13.42).

Refer to AS 1684 for roof member sizes.

Fig. 13.42 Strutting for deep rafters: solid strutting

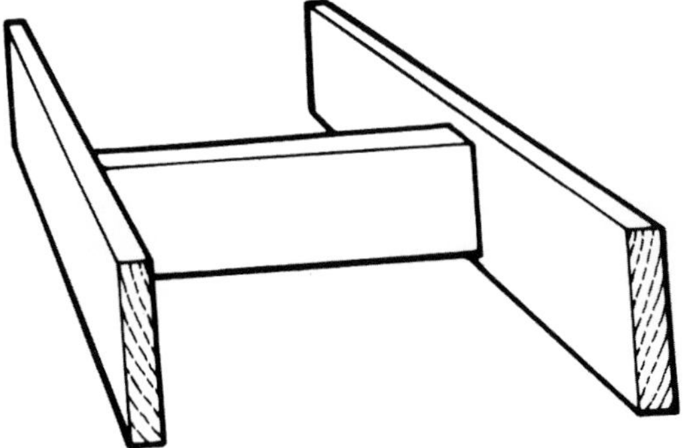

CHECK YOUR UNDERSTANDING

1. How can the rafters of a skillion roof be set out in a practical situation?

Student research

As a group, research:

- how many different types of roofs there are, and briefly describe their major characteristics
- the different roofing apps available, then see if you can use them.

End of chapter activity

Please refer to Chapter 14 for this activity.

AUSTRALIAN BUILDING CODE AND STANDARDS

AS 1684 – Residential Timber-Framed Construction

NCC

Tradie talk—**Sighting the rafter for bow**

ROOFING TRENDS, THEN AND NOW

Roofing has always been the carpenter's special skill. From developing the roof bevels and member lengths, to constructing the roof itself; it's what differentiates them from a joiner or cabinetmaker.

For many years, roofs were constructed from individual members of timber and joined using specialist joints that were constructed by hand. The roofs constructed could vary from simple structures with rafters and ridges that span relatively small distances to ornate truss structures that span large open areas—usually in significant buildings such as churches, halls and palaces. These types of trusses were more than just functional, they were ornate, complex and demonstrated the skill of the artisan; in many cases they were the crowning glory of the building. Below is an example of an ornate truss located in Westminster Hall, United Kingdom.

In Australia, most of the timber-house roofs constructed prior to the 1990s were stick built; that is all the components were cut and manufactured on-site, and then assembled.

Advantages of stick-built roofs

- Easy to construct; just the timber is needed to construct the roof.
- The roof space can be used for storage or converted into rooms if the pitch is high enough.
- It is easy to modify the roof and it can be cut into if an extension is added to the house.
- No need for special metal connectors that require a mechanical press to install.

Disadvantages of stick-built roofs

- Some roof members require intermediate support, which means the walls become load bearing.
- Not suitable for large open spaces; plus, the larger the room or space, the bigger the individual roof components need to be.
- They are slower to erect than trusses.

Hammer beam roof truss–Westminster Hall

Drawing adapted from Joinery and Carpentry, Richard Greenhalgh (editor), Volume IV, 2nd ed, The New Era Publishing Co Ltd, London, 1955

Since the 1990s, the availability of pre-manufactured roof trusses has increased in most areas of Australia, especially areas with large populations like Melbourne and Sydney. This has subsequently increased the amount of truss manufacturers and the speed at which a trussed roof can be erected. Using a trussed roof means that the internal design of the house is no longer required to support the roof, which allows for more flexibility in the layout and larger rooms where possible.

In areas where the availability of trusses is problematic, pitched roofs are often used—especially in high-value builds where the roof space is to be used as part of the dwelling. The construction of a pitched roof has also become easier with the increased availability of manufactured timber products. Manufactured timber members are lighter than solid timber beams, come in most lengths, and are stronger and span further than traditional sawn timber.

Advantages of trussed roofs

- Quick to erect as trusses are pre-manufactured.
- Suitable for large rooms and spaces as the truss can be engineered to suit the size, while still using relatively small individual components.
- Internal house walls are non-load bearing, which means that the room configuration can be easily changed.

Disadvantages of trussed roofs

- Lead-time for truss manufacturer can vary and is dependent on the industry demand.
- Due to the truss design and function the roof space can't be used.
- Access to the site can be an issue, as a semi-trailer delivers the trusses and an on-board crane is used to put the trusses on the wall frames.
- For sites with very poor access, a large mobile crane may need to be hired to place the trusses in the correct position, which adds to the overall cost of the project.

No matter which roof construction method is used, a carpenter must have knowledge of both methods and the techniques required to successfully complete the roof's construction.

Chapter 14

The hip and valley roof

Learning Objectives

LO 14.1 Identify hipped roof components

LO 14.2 Set out the hipped roof

LO 14.3 Develop crown end assembly and shortening distances

LO 14.4 Cut rafters to length

LO 14.5 Erect the hipped roof

LO 14.6 Identify hip and valley roof components

LO 14.7 Set out the minor roof

LO 14.8 Set out valley rafters and creepers

Introduction

The **hipped roof** is pitched on all sides of the building, and if the pitch angle is the same on all sides it is termed *equally pitched*. If an L-shaped building is to be constructed, then the addition of a valley and a minor span to the hipped roof is incorporated into the roof design.

The hipped roof style of roof construction is the most common and allows for a wide variety of roof cladding materials to be used–for example, slate, terracotta or cement tiles, or thatch, tin or timber shingles. As this roof is very prominent and thus becomes part of the overall aesthetic of the building, attractive roofing materials are often used. In this chapter you will learn how to calculate roof member lengths, develop roof cuts, set out the roof members and erect the roof safely for a basic hipped roof and a broken hip and valley roof.

14.1 The hipped roof

Figure 14.1 illustrates the general construction of the hipped roof and the names of the roof members.

Fig. 14.1 General view of a hipped roof

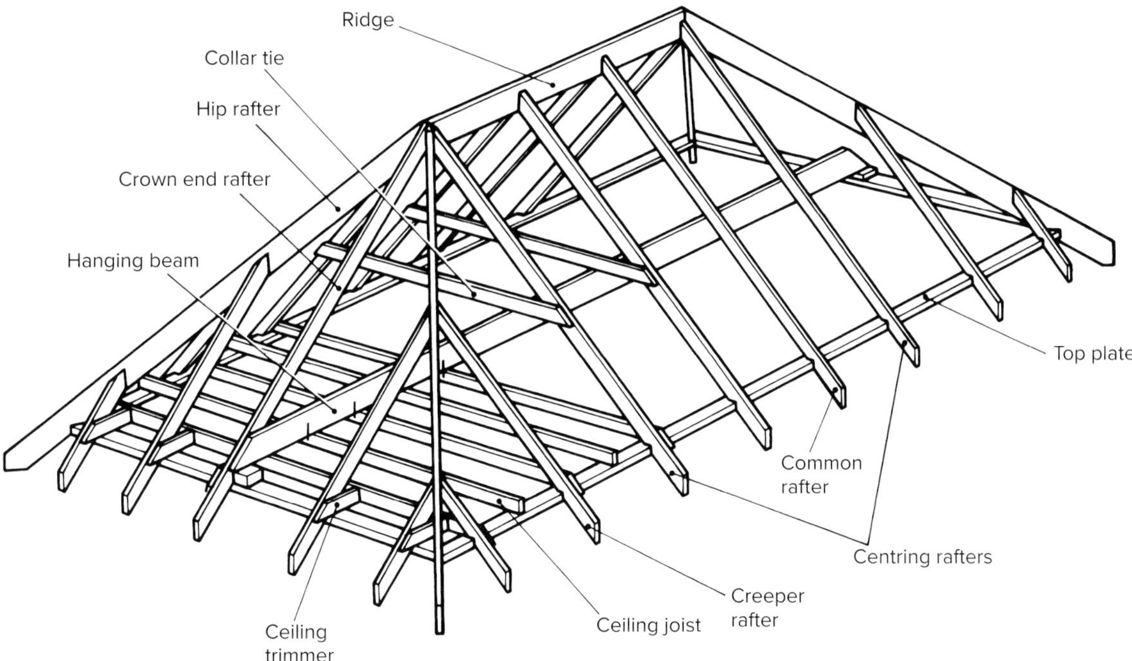

14.1.1 Roof members

14.1.1.1 Common rafter

The common rafter is a rafter complete from wall plate to the ridge, and is similar in all respects to the rafter of the gable roof.

14.1.1.2 Centring rafters

Centring rafters are a particular pair of common rafters. They are located at each end of the ridge and, as will be shown later, form the starting point for the setting out and erection of the roof.

14.1.1.3 Crown end rafter or jack rafter

The crown end rafter or jack rafter is a rafter complete from the wall plate to the end of the ridge. Its single line length is the same as the common rafter, but some small adjustment is necessary to allow for the different thicknesses of the roof members.

14.1.1.4 Hip rafter

The **hip rafter** is a rafter complete from an external corner to the ridge. It forms the junction of two sloping roof surfaces and requires a plumb bevel and an edge bevel. Refer to Table 14.1.

INTERPRETING PLANS

Refer to the floor plan in the Appendix, highlight in different colours the following:

- ridge beam
- hip rafters
- valley rafters
- broken hip.

14.1.1.5 Creeper rafter

The creeper rafter is a rafter running from the wall plate to the hip rafter and varies in length as it meets the hip. It has the same plumb bevel as the common rafter, and the edge bevel will be developed.

14.1.1.6 Underpurlins

The function and strutting of underpurlins are the same as described for the gable roof. Underpurlins butt against the hip rafter at the ends, and require a face and edge bevel.

> **TIP** To calculate the roof member sizes, timber species and stress grades, refer to AS 1684.

CHECK YOUR UNDERSTANDING

1. Describe the function of the following members in the hipped roof:
   a. centring rafter
   b. crown end rafter
   c. hip rafter
   d. creeper rafter.

Fig. 14.2 (a) Common rafter; (b) creeper rafter; and (c) hip rafter

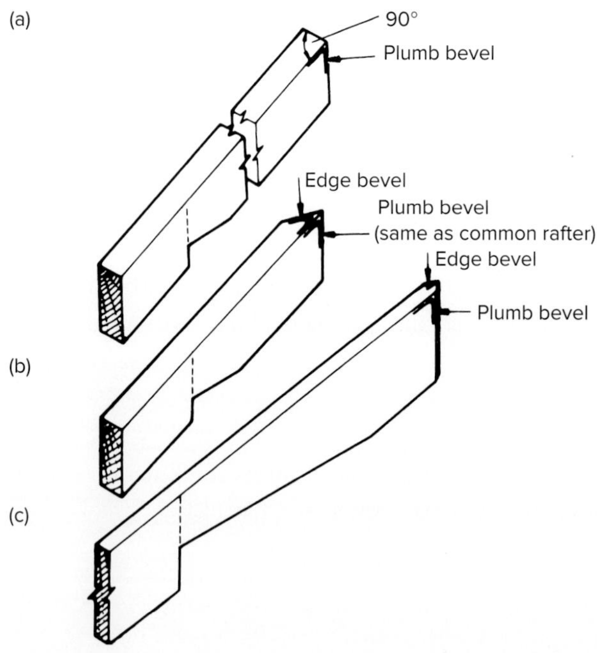

14.2 Setting out the hipped roof

14.2.1 Development of roof bevels

To set out and erect the hipped roof, one of the foremost problems facing the carpenter is to develop the necessary bevels to cut the roof members accurately (Fig. 14.2). These are found by developing a series of right-angled triangles, the included angles of which are applicable to the roof members.

The development and construction of a typical hipped roof is described in the following pages and the general principles demonstrated can be applied to any hipped roof. But for the purpose of explanation, any specific references or dimensions indicated will apply to the drawing and specification for the hipped roof shown in Figure 14.3.

Figure 14.4 shows the hipped roof drawn in single lines (centre lines), resting on a horizontal plane represented by the line *XY*. Refer to the plan and note that as the roof is equally pitched, the hip will bisect the external corners, and if the roof is rectangular on plan, they can be drawn in using a 45° square.

The plumb and level bevels for the common rafter are the only bevels found directly from the drawing: as the vertical section lies in the front vertical plane, the remainder must be developed. As a general guide, note how the *run hip rafter* was rotated around to the vertical plane and projected up to the vertical section and combined with the rise to obtain the plumb bevel hip rafter. The same triangle could also be developed by taking the rise down to the plan and setting off at 90° to the run hip rafter, again developing the plumb bevel hip rafter.

Follow the arrows indicating the direction in which the developments were carried out.

A tabulated list (Table 14.1) can now be prepared and is shown listing the roof bevels and the triangle in which they can be found. In each right-angled triangle, there are two other angles and the last column indicates on which side of the triangle the required bevel is to be found.

Fig. 14.3 Roof specification

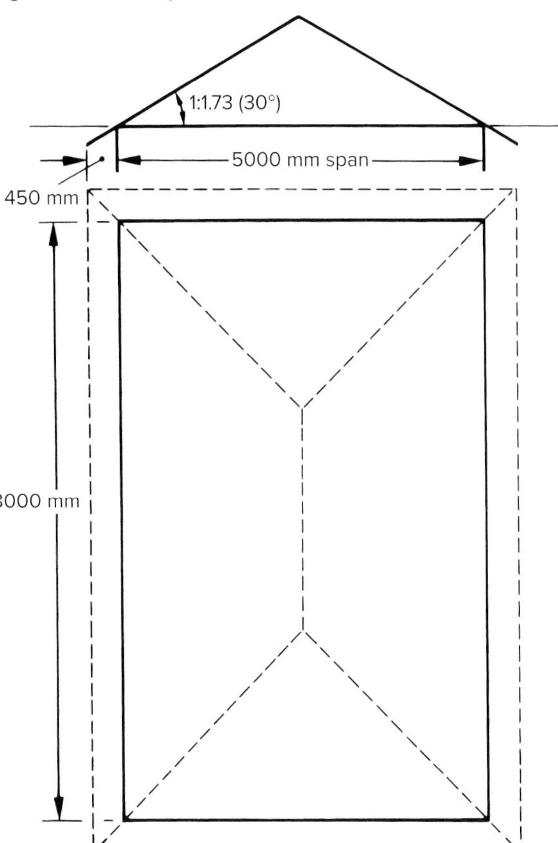

14.2.2 Determining the side cuts of the hip and creeper rafters

The following practical method is required:

1. Using a bevel set to the appropriate angle, plumb a line down one face of the rafter.
2. Measure 90 degrees to the plumb line, the thickness of the material being used for the hip or creeper rafters.
3. Plumb a second line using the original bevel angle at the point marked in step 2.
4. Square a line across the top edge of the timber, connecting to the top of the two plumb lines.
5. Draw a diagonal line from one corner to the other on the top edge of the timber. This diagonal line is the site cut for either the hip or creeper rafters.

Table 14.1 Roof bevels

No.	Bevel	Triangle		Bevel on
1	Level/foot bevel common R	Run common R	Rise	Run common R
2	Plumb bevel/cut common R	Run common R	Rise	Rise
3	Plumb bevel/cut hip R	Run hip R	Rise	Rise

14.2.3 Determining underpurlin bevels

Using a block of timber that is cut square on the end, develop the underpurlin bevels following the method below:

1. Set a bevel to the rafter plumb cut angle and transfer the angle onto the end of the block of timber.
2. Set the bevel to the creeper side cut angle and transfer this angle onto the top of the piece of timber, ensuring this new line connects with the first line.
3. Using a straight edge, join up the lines from steps 1 and 2 on the side of the timber (Fig. 14.6).
4. The required angles for the underpurlin can now be found as indicated in Figure 14.6.

Fig. 14.4 Line representation of rafters

- Span 5000 mm
- Length 8000 mm
- Pitch 1:1.73 (30°)
- Common and creeper rafters 100 mm × 50 mm
- Spacing at 600 centres
- Hips and ridge 175 mm × 32 mm
- Eaves width 450 mm

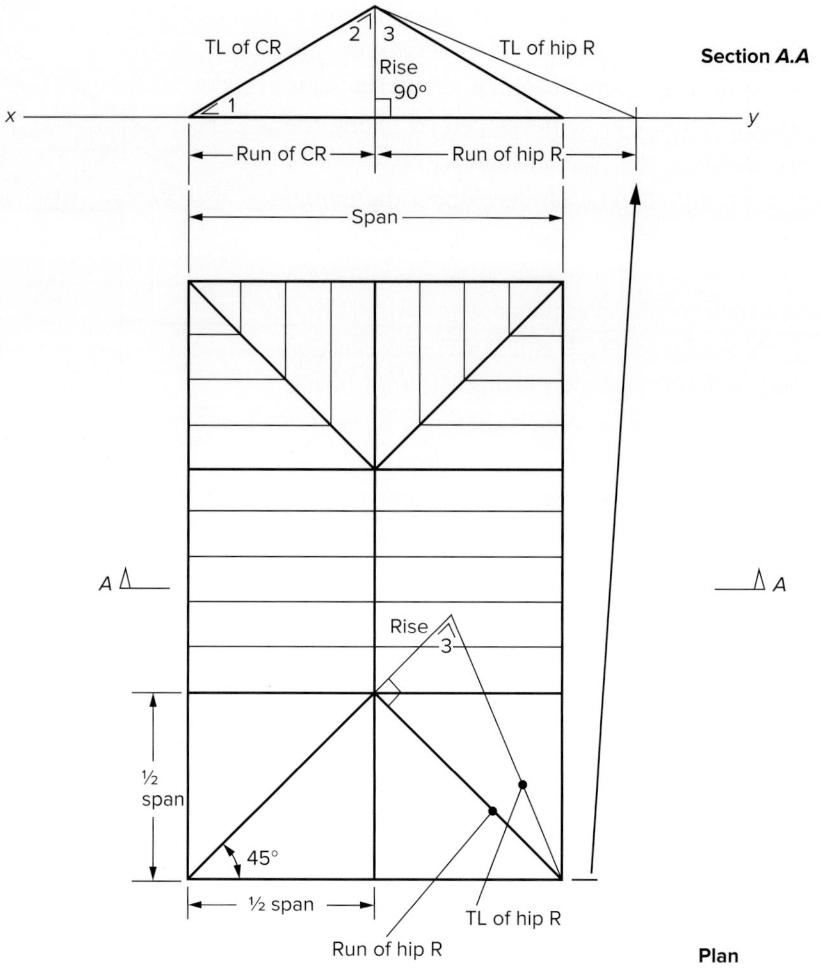

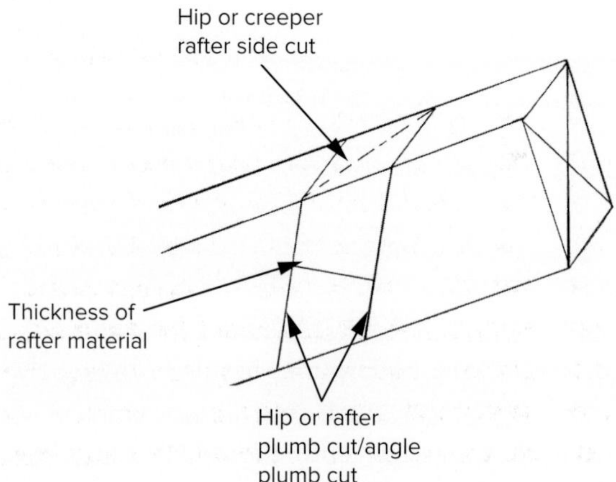

Fig. 14.5 Determining side cuts for the hip and creeper rafters

14.2.4 Determining the length of rafters

Figure 14.7 shows the part plan and vertical section of the hipped roof, with the centre line of the rafters indicated on the plan. Note that the creeper rafters are located by commencing at the centring rafter and crown end rafter, and marking off the rafter spacing back towards the corner. In the process, each pair of creeper rafters diminishes in length, and on the plan each creeper will shorten by a distance equal to the spacing between them.

The true length of the creeper on the centre line is found by projecting its length on the plan up to the vertical section and measuring along the line of the common rafter. Also along this line can be found the difference in the length of the creeper rafters, actually measured along the length of the rafter. This dimension is an important one, and will be used when a pattern rafter is being set out.

The true length of the hip rafter on the centre line is also shown in this drawing.

14.3 Crown end assembly and shortening distances

The centre lines of the common rafter, crown end rafter and the hips all gather together to meet at a common point at the end of the ridge, which is referred to as the *gathering point*. Figure 14.8 shows the details of the most commonly adopted arrangement of the crown end assembly using a double edge cut on the hip rafter.

The common rafters are shortened by half the thickness of the ridge, similar to the gable roof. The crown end rafter that butts against the end of the ridge is shortened by half the thickness of the common rafter, and if it is shorter than the common rafter, as in this example, it will also be fixed slightly below the top of the ridge.

Fig. 14.6 Practical method for obtaining underpurlin cuts

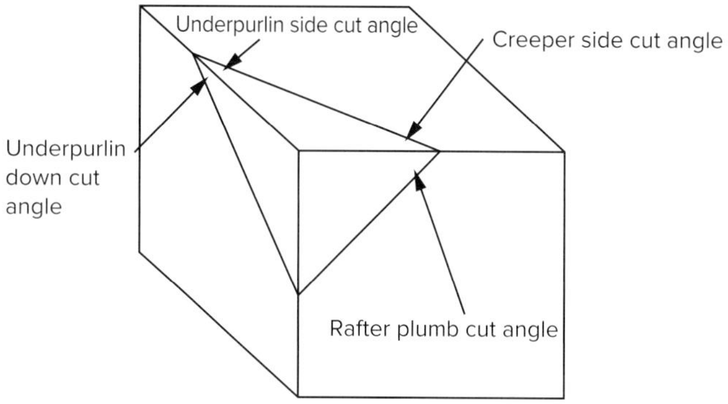

Underpurlin side cut angle
Creeper side cut angle
Underpurlin down cut angle
Rafter plumb cut angle

Fig. 14.7 Rafter spacing—part plan

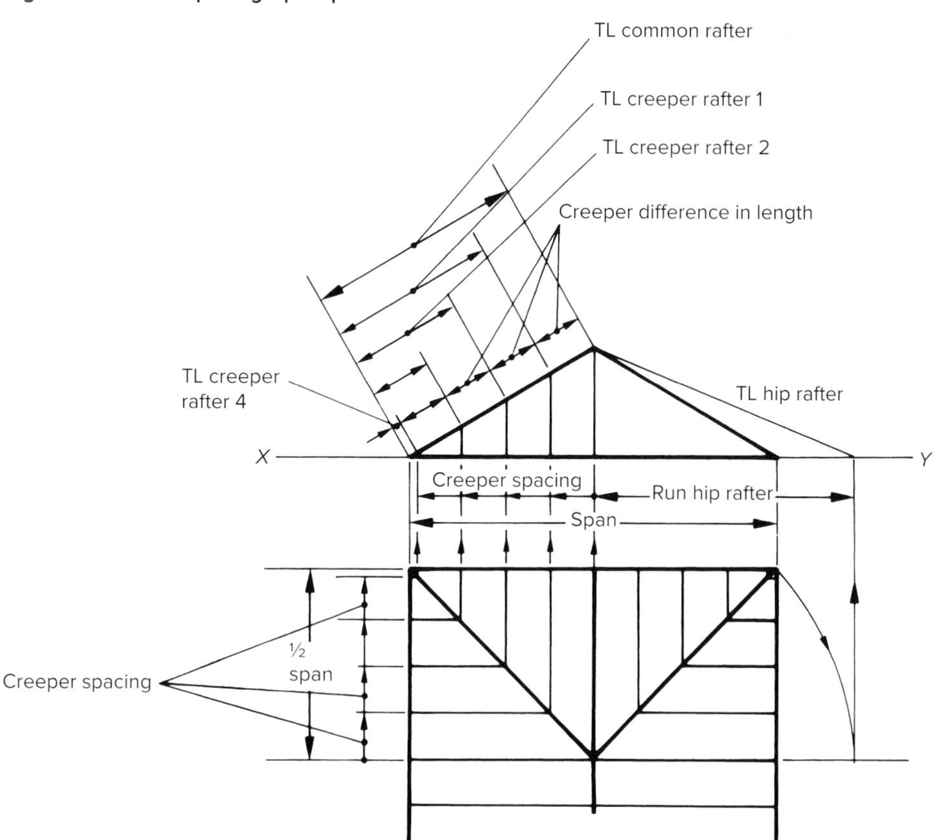

The plan shows how the hip rafter is shortened by measuring along the hip centre line between the end of the hip rafter to the intersection of the centre lines at the crown. This measurement is then subtracted from the length of the hip rafter, square of the hip plumb cut.

TIP Remember that all of these shortening distances must be measured at right angles to the plumb cut.

CHECK YOUR UNDERSTANDING

1. What is the difference in length between adjacent creeper rafters if they are measured at right angles to the plumb cut?

Fig. 14.8 Crown end assembly (point of intersection) and shortening distances

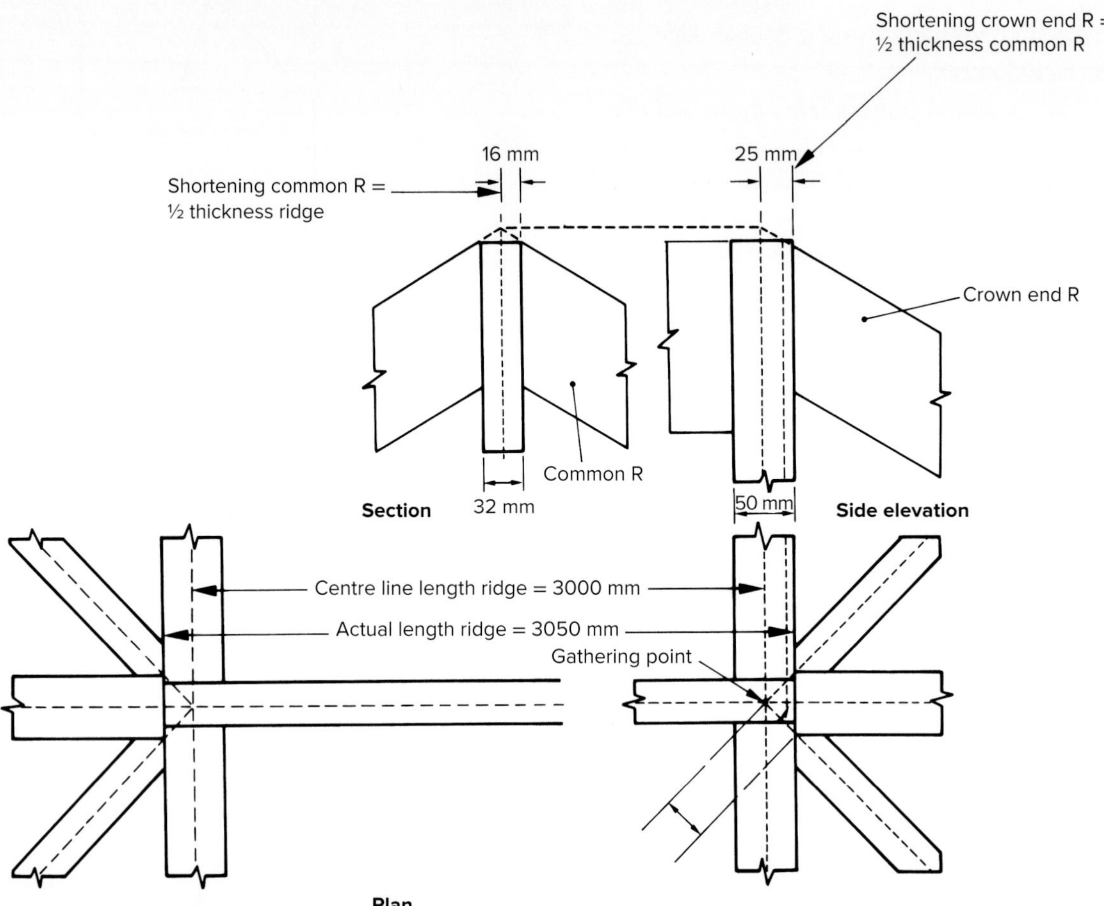

14.3.1 Length of ridge

Figure 14.8 shows how the centre line length of the ridge is measured between the gathering points at each end of the roof, but it must be extended by half the thickness of the common rafter at each end.

The centre line length can be calculated using our example:

centre line length of ridge

= length of roof − span

= 8000 − 5000

= 3000 mm

actual length of ridge

= length of roof − span + thickness of the common rafter

= 8000 − 5000 + 50

= 3050 mm

14.3.2 Setting out wall plates

Figure 14.9 illustrates the set-out on the upper surface of the wall plates for the hipped roof. Commence at each corner and measure back a distance equal to half the span, which gives the centre line for the centring and crown end/jack rafters. The location for these rafters is marked in a distinctive way so that they can be identified readily when the roof is erected. A common practice is to mark them with a double line, crossed diagonally, or to write the letter *J* for jack rafter and *C* for centring rafter at these positions.

Fig. 14.9 Wall plate set-out

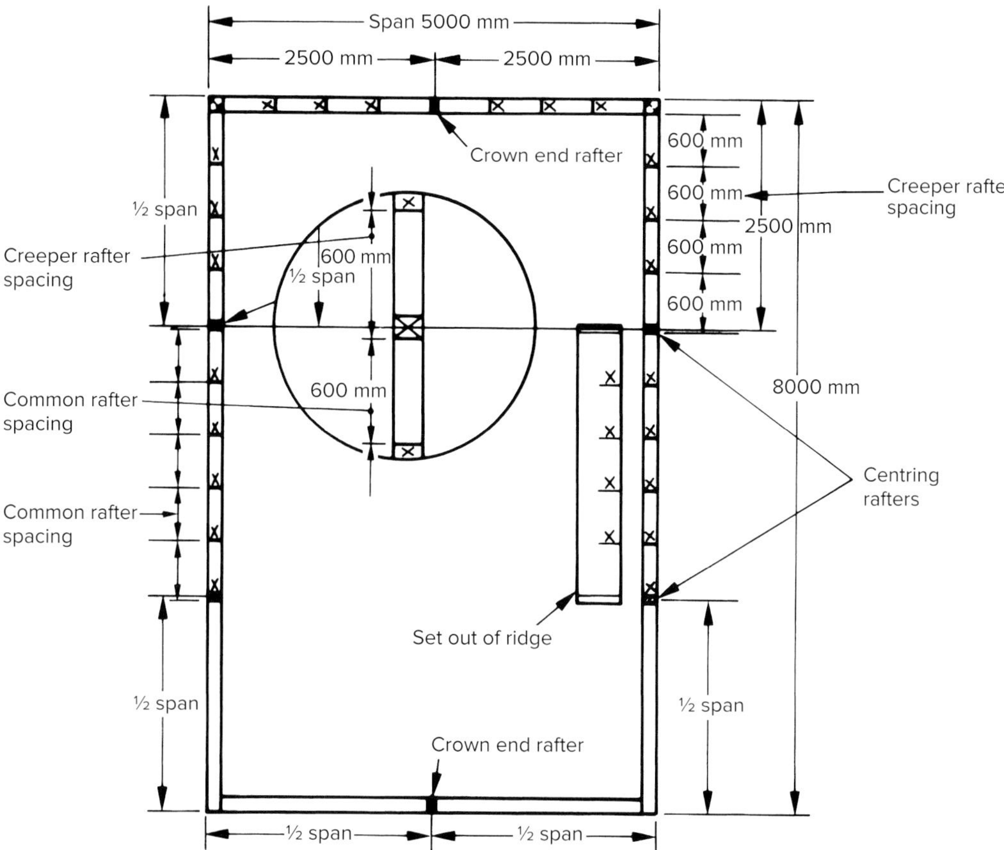

Starting from the side of these marks, furthest from the corner, mark off the spacings for the creeper rafter using the 'in and over' method. It is sufficient to mark the creeper rafter with a single line and a cross indicating the side of the mark on which the rafter will be placed. Between the centring rafters, mark out the positions for the common rafter, not exceeding the maximum spacing.

It is preferable to mark out the ridge at this time by laying it beside the wall plate and transferring the rafter positions onto the face.

CHECK YOUR UNDERSTANDING

1. How can the length of ridge for the hipped roof be calculated?
2. How could you determine the position of the centring rafters and crown end rafters when setting out for the roof members on the top wall plates?

14.3.3 Length of creeper rafter to long point

From the method previously described, the centre line length of the creeper rafters and their difference in length was obtained, but allowance must now be made for the thickness of the hip to be located between the pairs of creeper rafters.

It is usual to measure the creeper rafter length to the long point. Figure 14.10 illustrates the difference between the length of the creeper rafter on the centre line and the length to its long point. This difference will be referred to as the **long point correction** (LP correction) and may be either added to or subtracted from the centre line length.

The LP correction can be calculated from the formula:

$$\text{LP correction} = -\text{creeper rafter shortening} + \text{half thickness of creeper rafter}$$

then

$$\begin{aligned}\text{LP correction} &= -22 + 25 \\ &= +3\ \text{mm}\end{aligned}$$

It will be quickly observed that, in the majority of cases, the LP correction is a very small amount, and no great error would result if it is assumed that the length to long point is the same as the centre line length. Many of the most self-respecting carpenters agree that they could not cut the creepers any more accurately in any case.

As will be shown later, it is quite a simple matter to make the correction. Always be wary of the unusual situation where a correction may have to be calculated or determined as in Figure 14.10.

Fig. 14.10 Length of creeper rafter to long point

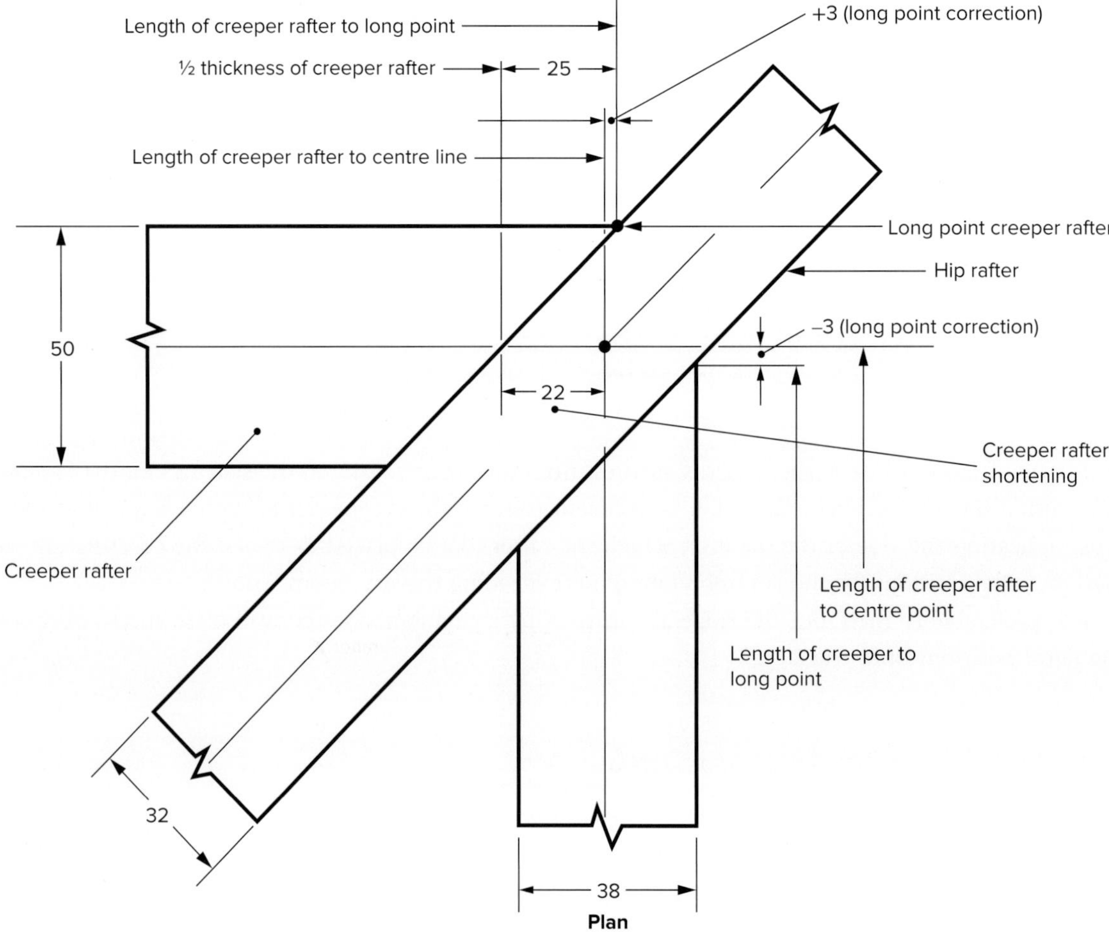

14.3.4 Setting out the pattern rafter

A pattern rafter can be set out on which will be shown the length of all common and creeper rafters (Fig. 14.11). The procedure is as follows:

1. Select a common rafter and set sliding bevels to the plumb bevel for common and creeper rafters and edge bevel creeper rafters.
2. Setting out for the common rafter is a similar procedure to that described for the gable roof (Fig. 13.18), except that the dimensions will be those calculated previously in this chapter.

   distance *AB*, length of common rafter

   = 2887 mm

   FA, eaves overhang

   = 520 mm

   shortening distance of common rafter

   = 16 mm
3. From plumb line through point *B*, again measure off at a right angle half the thickness of the common rafter (25 mm), which gives the length for cutting the crown end rafters.
4. From point *B* measure along the rafter 693 mm, the difference in length between the creeper rafters, to point *K*, and lightly draw a plumb line through this point. This establishes the length of the first creeper rafter to the long point. From this point measure back along the rafter the creeper difference (693 mm) to obtain the length of the remaining creeper rafters to their long point.
5. Mark in the edge bevel of the creeper rafters from their long point. Cut out the birdsmouth and cut to length for the common rafter.

Fig. 14.11 Pattern rafter set-out

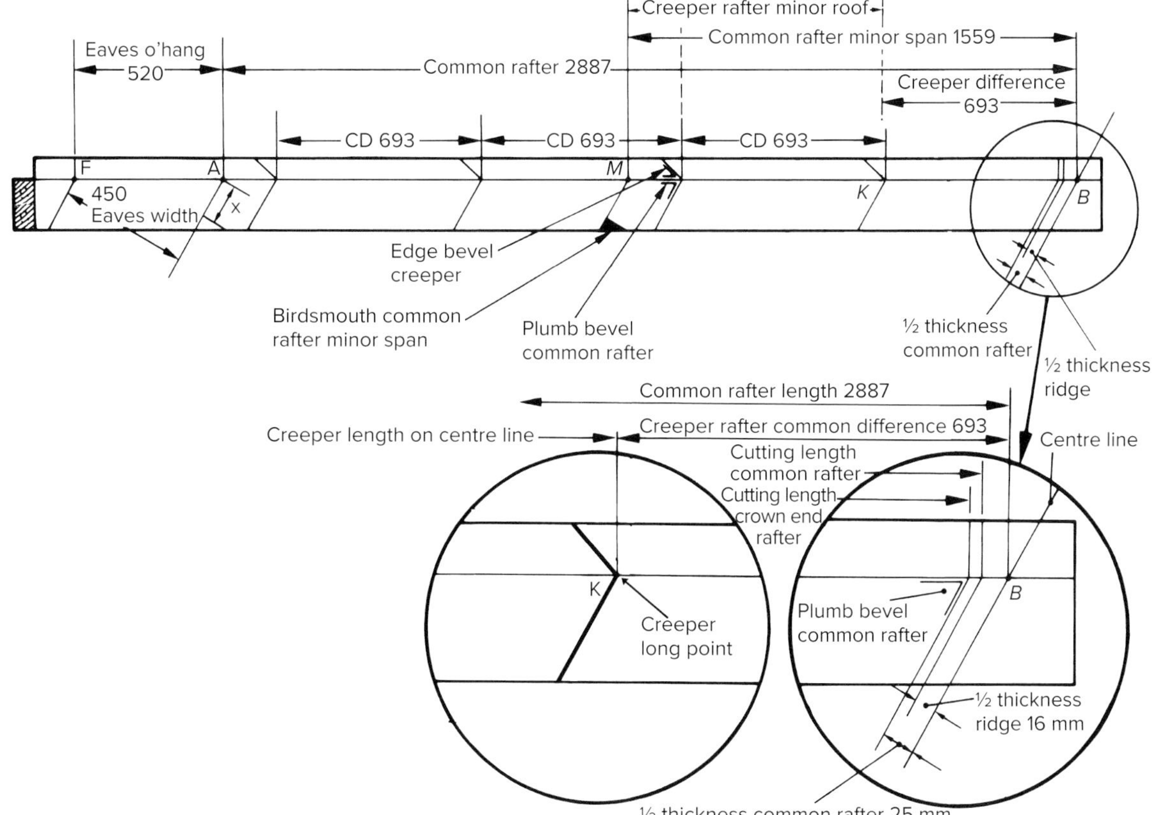

14.4 Cutting rafters to length

Nail two short cleats across the back of the rafter and mark out the common rafters in the same manner as for the gable roof. Select two crown end rafters and mark them to length. Creeper rafters must be set out in pairs for each length required by reversing the direction of the edge cut to give rafters of opposite hand (Fig. 14.12a). The most economical use of timber and the minimum amount of cutting will result if the creepers are cut from each length of timber as in Figure 14.12(b).

When cutting creeper rafters with a power saw (either radial arm saw or powered hand saw), the edge bevel for the creeper rafter is formed by tilting the saw blade to 45°. Figure 14.12(c) shows how the edge bevel developed on the set-out is applied to the square edge of the rafter for cutting with a hand saw.

When a power saw is used, the blade follows the plumb cut; the axis of the blade tilts at right angles to the plumb line, which is the plan angle of the edge bevel or 45° when the roof is rectangular.

Fig. 14.12 A–C Cutting of rafter

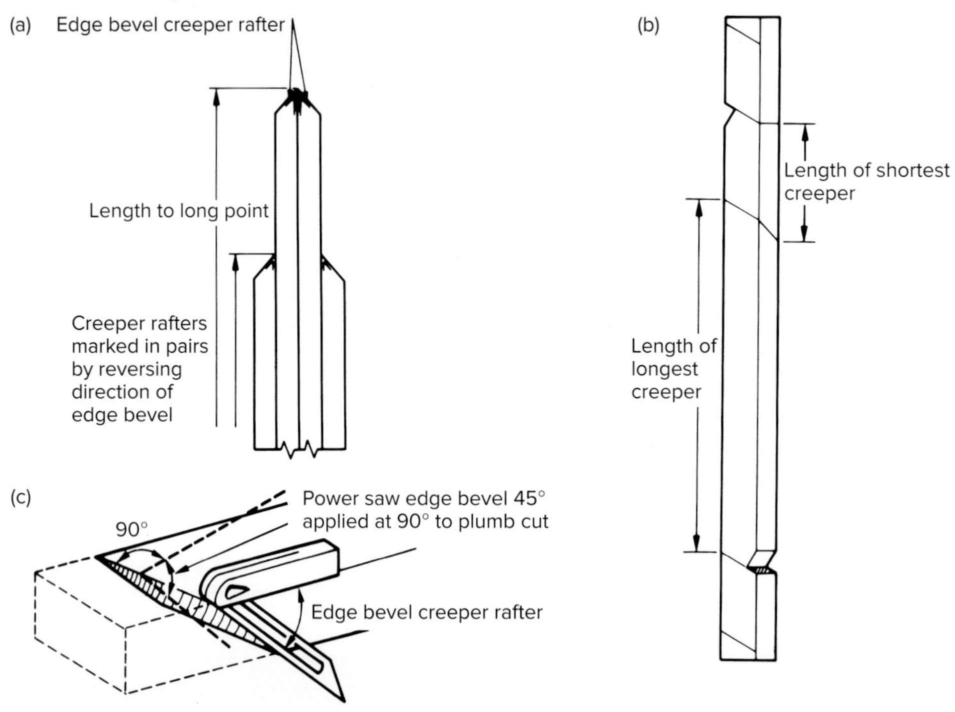

CHECK YOUR UNDERSTANDING

1. To which angle would you tilt the base of the powered circular saw to cut the edge bevel of the creeper rafters for a hip roof rectangular on plan?

14.4.1 Development and cutting of hip rafter

Figure 14.13 illustrates how the hip rafter is developed from the single line set-out. To simplify the fitting of the hip at the birdsmouth, the corner of the plates are cut off at 45° for the thickness of the hip rafter.

Theoretically, it is possible to obtain the true length of the hip by scale drawing or calculation, and to deduct the shortening distances. In practice, however, it is much more satisfactory to measure the length of the hip directly from the roof as it is being erected. Use a steel tape and measure between the points *H* and *J* in Figure 14.15. Proceed to set out the hip in Figure 14.16 as follows:

1. From one end of the hip allow sufficient for the tail, and mark the plumb bevel hip on the face. A practical formula for the tail allowance is shown on the drawing. If necessary, the tail can be cut back to the width of the common rafters.

2. Measure down the plumb line distance *x* and mark off at right angles, giving the horizontal cut for the birdsmouth and establishing point *J*.
3. Measure along hip from point *J* to point *H* on the top edge and square across the edge to the centre line. Mark the plumb line for the hip partly across the face.
4. To obtain the hip edge bevel, refer to the method described earlier.

While the development of the hip edge bevel has been shown in earlier figures, in practice it is simpler to mark the hip directly onto the timber as outlined above.

TIP When cutting the hip edge bevel with a power saw, the correct angle to tilt the saw is 45°, in the same manner as for the creeper rafter (Fig. 14.12c).

CHECK YOUR UNDERSTANDING

1. How would you mark in the edge bevel when setting out the hip rafter?

14.5 Erecting the hipped roof

To erect the hipped roof–or to use the trade term for this operation, *to pitch the hipped roof*–follow these steps:

1. Make preliminary preparations and load the roof in a manner similar to that described for the gable roof.
2. Commence erection with the pairs of centring rafters. Secure these in position and lift the ridge up between the rafters and fix in place.
3. Locate the mark for the crown end/jack rafters and fix these in place by skew nailing them to the wall plates and the end of the ridge.
4. Fix sufficient common rafters in place to keep the ridge in alignment.
5. Hip rafters can now be measured on-site and set out (Fig. 14.16). Cut them to length and fix in position (Fig. 14.17).
6. Lay out pairs of creeper rafters for each hip and commence fixing. Start with a pair near the centre of the hip and, if they are correctly cut and fixed directly opposite each other, the hip should be straight; however, some carpenters may prefer to erect a string line directly over the centre line of the hip to keep it straight as the creeper rafters are fixed.
7. To fix the underpurlins, locate the top edge of the underpurlin on the underside of the rafters and partly drive a few nails to establish the line. Hold a steel tape along this line and extend it to the point

Fig. 14.13 Hip rafter—development

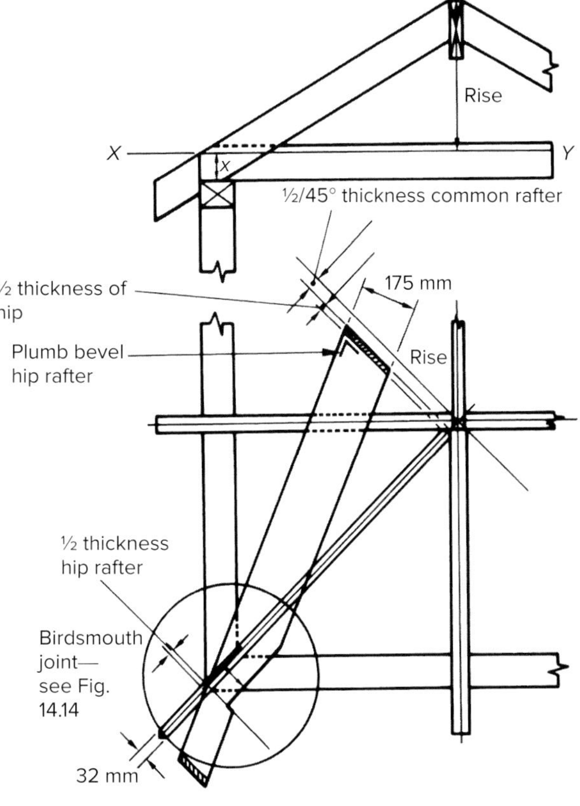

where it strikes the hip at each end. Measure this distance for cutting the underpurlin to length (Fig. 14.18). Apply a face bevel and edge bevel to the underpurlin where it will butt against the hip.

8. Refer to the procedure for fixing underpurlins, strutting, collar ties and any more common rafters to the gable roof and proceed in a similar manner for the hipped roof.

Fig. 14.14 Birdsmouth joint—detail

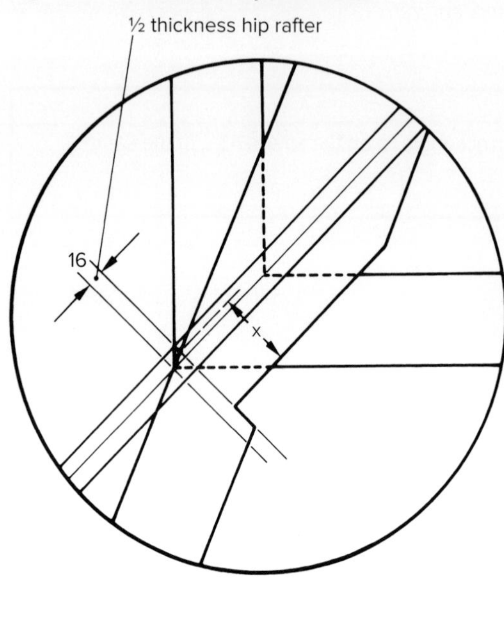

Fig. 14.15 Measuring length of hip rafter with tape

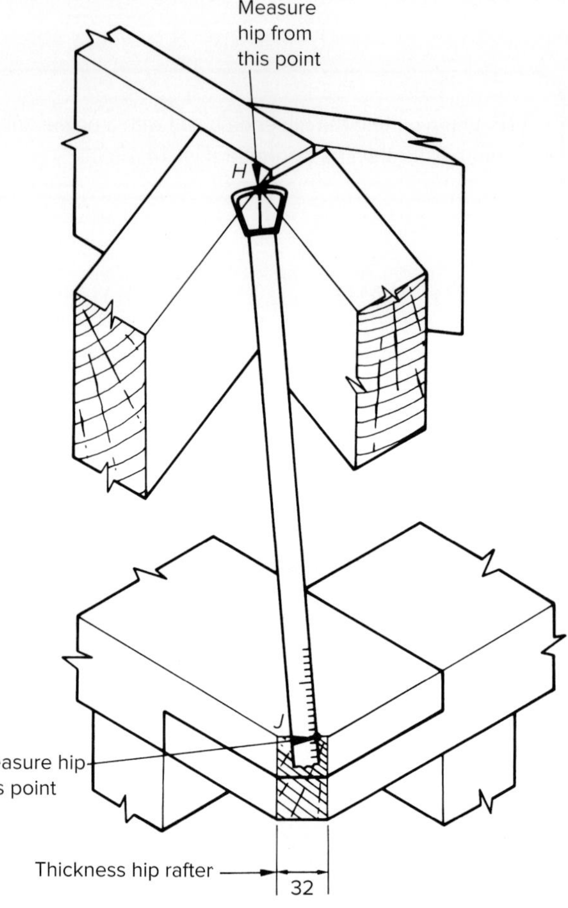

Fig. 14.16 Hip rafter set-out

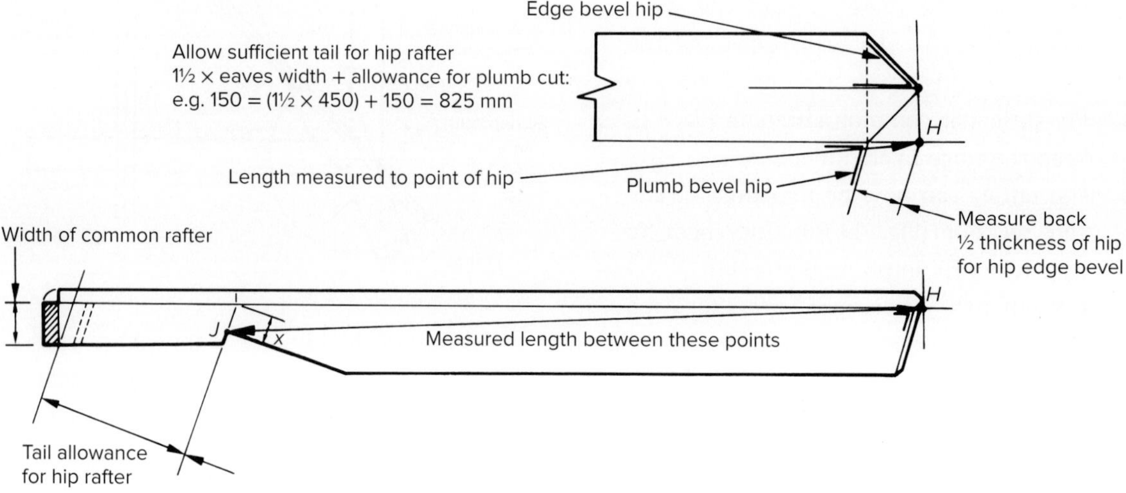

Fig. 14.17 Cutting rafter overhang to length (not to scale)

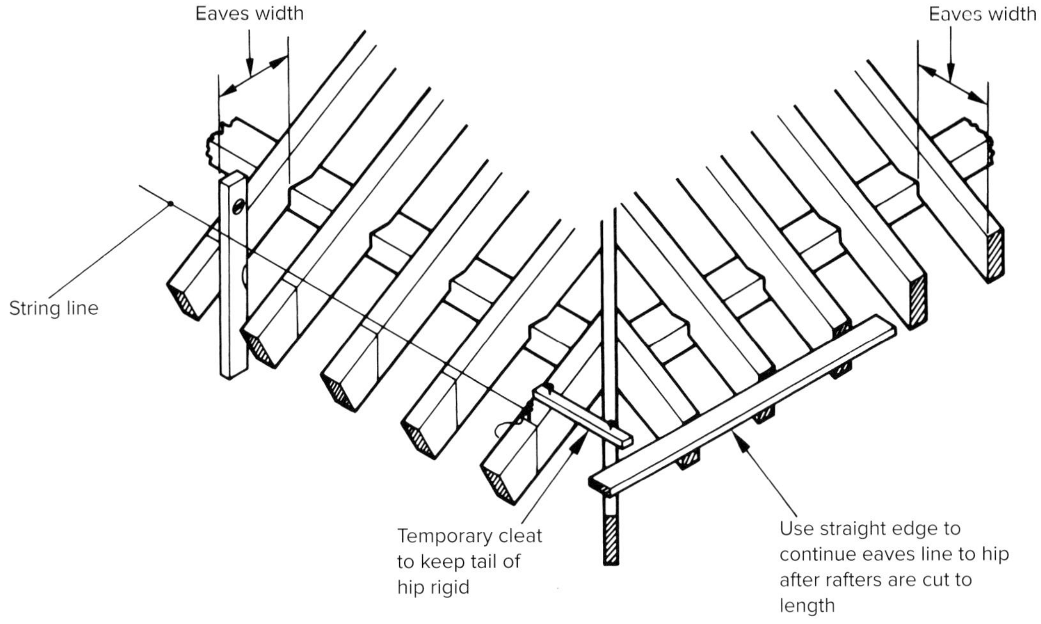

Fig. 14.18 Measured length of underpurlin

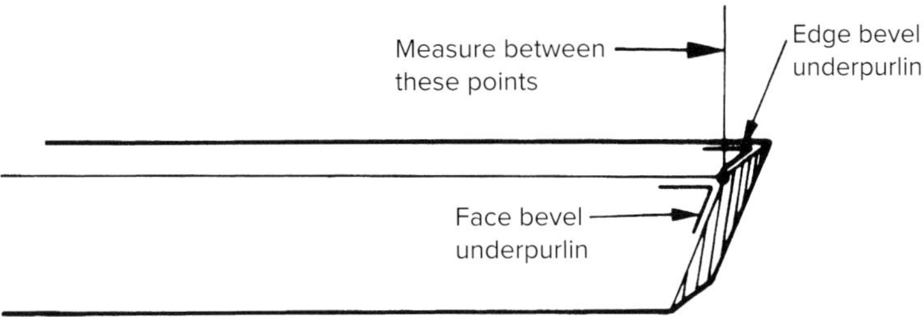

WORKPLACE SCENARIO

Pitching roofs or standing up trusses are dangerous activities. Scaffolding and working platforms are necessary to be safe, to give a good footing and eliminate the risk of falling to the ground. However, as I drive around my home town and see new houses being built, I still see carpenters walking on plates, sometimes on the second storey, with no scaffolding or catch platforms installed. They are walking around on a 90 mm-wide plate, working and carrying—one wrong step and they could fall up to 5 m. It's not worth it; the injuries a person could sustain would be numerous and there would be weeks, if not months, of recovery time. It could also be fatal. The cost of installing the scaffolding is a small price to pay for safety, and it's definitely worth waiting for the scaffold to be installed rather than rushing ahead so you can get on to the next job.

CHECK YOUR UNDERSTANDING

1. Indicate the order in which you would fix the following members when erecting a hipped roof:
   a. hip rafter
   b. crown end rafter
   c. centring rafters
   d. underpurlins
   e. creeper rafters
   f. common rafters.

14.6 The hip and valley roof

A valley is formed in a pitched roof wherever an internal angle is found on the roof plan. A typical example of this is shown in Figure 14.19, where offsets are made, projecting off the main roof. The main roof is usually referred to as the *major roof* and contains the *major span* (S^1) and the *major rise* (R^1). The offset is the *minor roof* and has the *minor span* (S^2) and the *minor rise* (R^2).

In our example (Fig. 14.20), the major and minor spans are unequal and a new set-out must be prepared for the rafters of the minor roof. But if the pitch of the roof remains the same, with the introduction of the valley, no new bevels need to be developed. Figure 14.20 shows how the roof bevels can be developed for a hip and valley roof. The process for development is the same as described earlier.

Fig. 14.19 General view—hip and valley roof

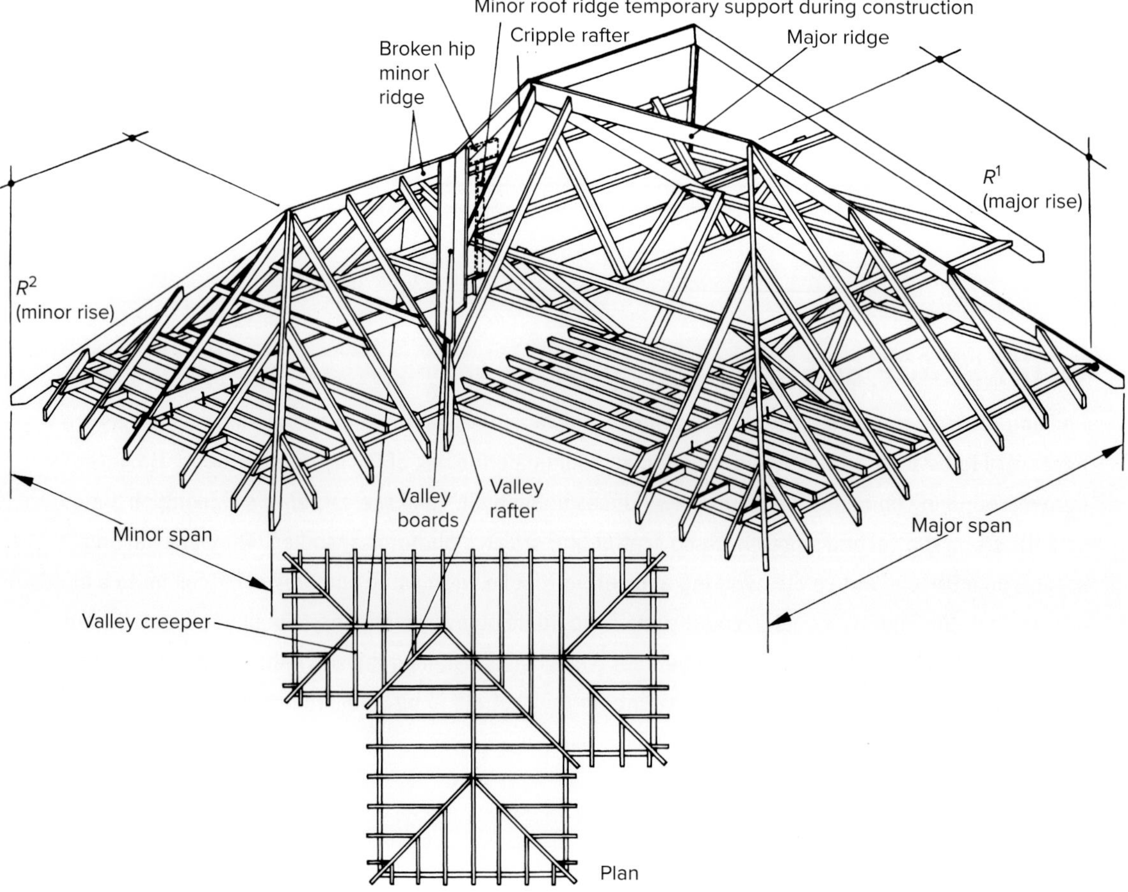

Fig. 14.20 Development of hip and valley roof

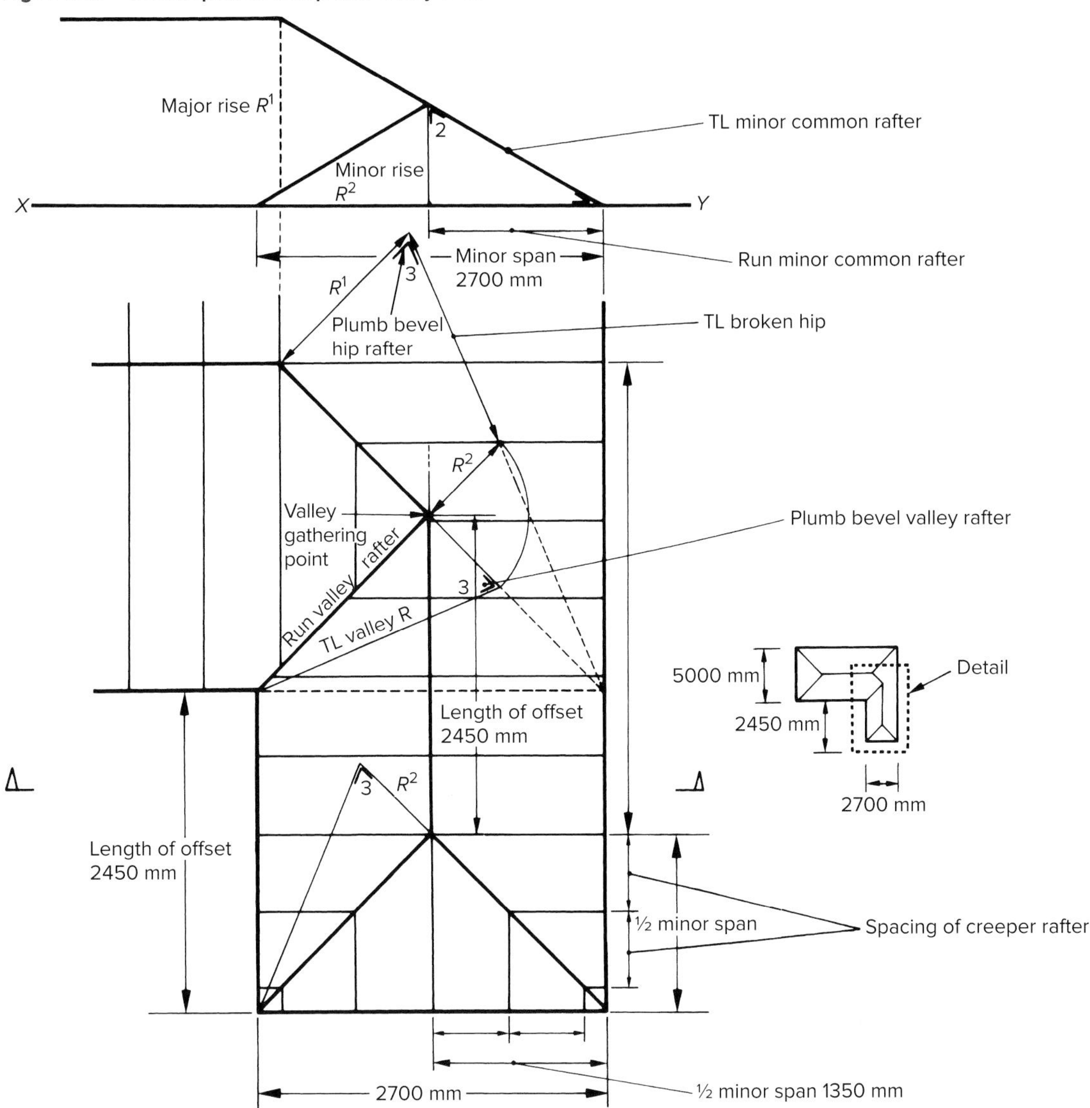

14.6.1 Roof members

The new members now found in the roof are the valley rafter, broken hip rafter, valley creepers and cripple rafter.

14.6.1.1 Valley rafter

The valley rafter runs from an internal corner to the minor ridge and forms the junction of the sloping roof members. On plan it will bisect the internal corner and has a plumb and edge bevel, which are the same as the plumb and edge bevel hip rafter.

14.6.1.2 Broken hip rafter

This rafter joins the main ridge to the minor ridge and is part of a full hip rafter that would form the corner of the main roof before the offset was added; refer to the dotted lines in Figure 14.20.

14.6.1.3 Valley creepers

These are rafters running from a ridge to the valley rafter (Fig. 14.21a) and have the same bevels as other creepers, and, provided the spacing remains constant, the same difference in length.

14.6.1.4 Cripple rafter

The cripple rafter (Fig. 14.21b) runs from the broken hip to the valley rafter (Fig. 14.22), isolating the offset and valley.

Figure 14.22 illustrates the arrangement of the roof members when thickness is added and will be referred to in some of the following explanations.

Fig. 14.21 (a) Valley creeper rafter; and (b) cripple rafter

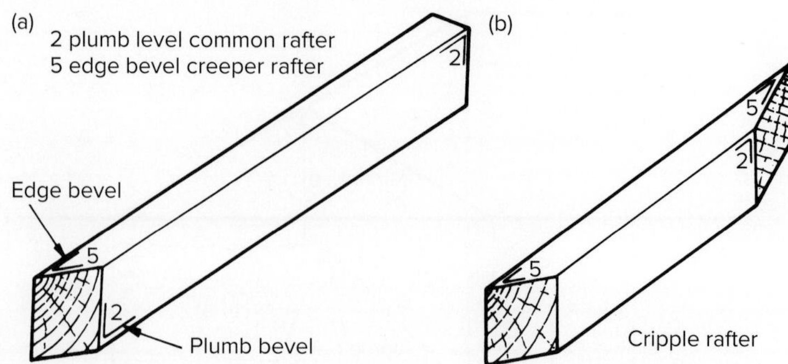

14.6.2 Set-out of wall plates

As shown in Figure 14.24, the wall plates are set out in a manner similar to the main roof. Commence with the centring and crown end rafters at the offset end and mark off common rafters back towards the main roof.

CHECK YOUR UNDERSTANDING

1. Describe briefly the location and function of the following members in the hip and valley roof:
   a. broken hip
   b. cripple rafter
   c. valley creeper
   d. minor ridge.

14.7 Setting out the minor roof

As the wall plates are being set out, it is also an opportune time for setting out the minor ridge. A very strong job will result if this ridge is projected beyond the valley gathering point and fixed to the next rafter (Fig. 14.22). This practice is frequently adopted. Lay the ridge board beside the wall plates and mark the position of the rafters on to the face. The distance from the hip gathering point to the valley gathering point is the same as the length of the offset (Fig. 14.23). Extend the ridge beyond this point to the face of the next rafter in order to determine its length (Fig. 14.24).

14.7.1 Setting out the pattern rafter

Construct the right-angled triangle to scale (Fig. 14.25). The hypotenuse is the true length of the common rafter for the minor roof. This triangle could be superimposed on the previous set-out for the main roof, but for the sake of clarity is kept separate in this example. The calculations for obtaining the true length common rafter (1559 mm) are also shown. To apply this length to the pattern rafter, refer to Figure 14.11.

From point *B* measure back along the rafter 1559 mm to point *M*. Draw a plumb line through this point and mark in a birdsmouth. From *M* to *B* is then the common rafter, and the creepers indicated between those points are the lengths of the creeper rafters for the minor roof.

14.7.2 Pitching the hip and valley roof

Proceed as previously described to the stage illustrated in Figure 14.26. Using a straight edge, check that the minor ridge will line up with the rafters of the main roof and that it is level. A temporary prop may be necessary for holding it firmly in position.

Fig. 14.22 Plan development

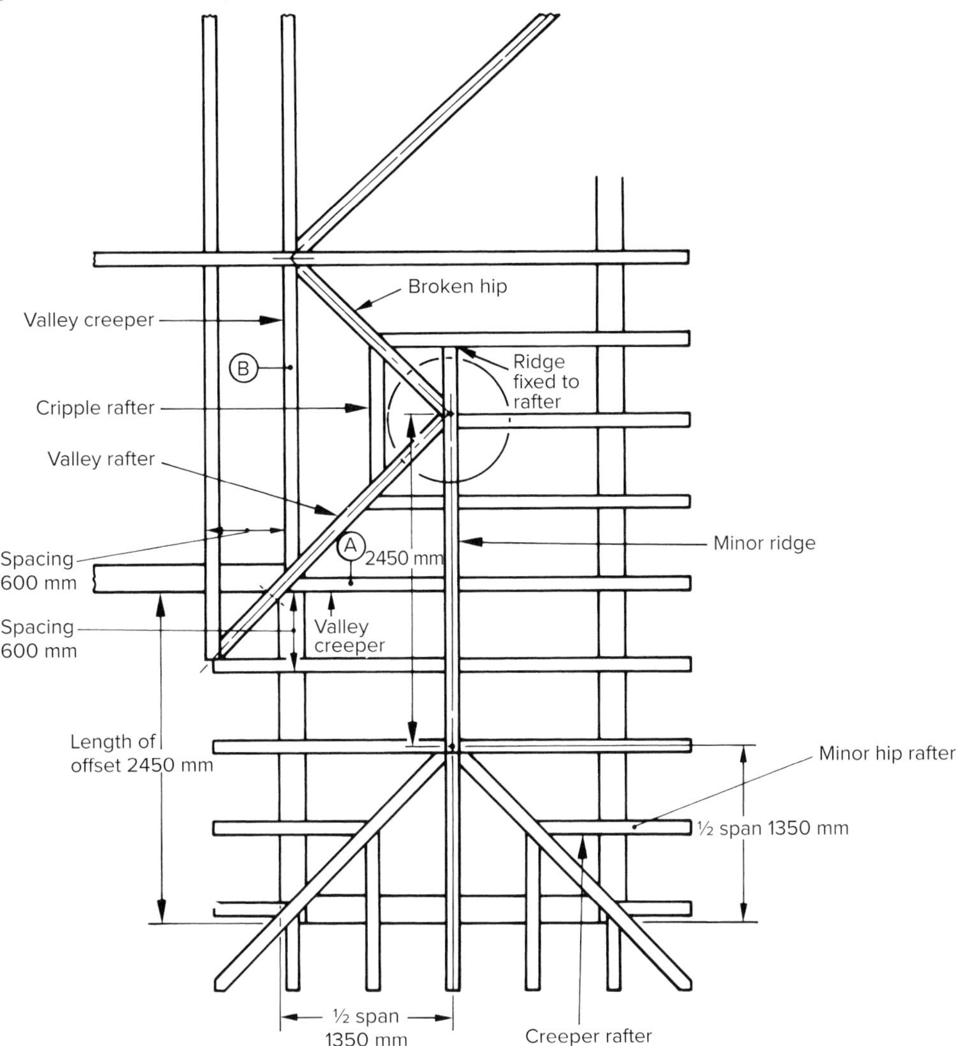

14.8 Setting out valley rafters and creepers

The valley can now be completed as follows:

14.8.1 Gathering point

Locate the valley gathering point and partly drive a nail to indicate its position clearly. The centre line of the broken hip and valley must gather in to this point.

14.8.2 Length of broken hip

Use a steel tape to measure the length of the broken hip along its centre line from the major ridge, point *S*, to a point directly above the centre line where it strikes the face of the minor ridge, point *T* (Fig. 14.27). Use the edge

Fig. 14.23 Valley gathering point

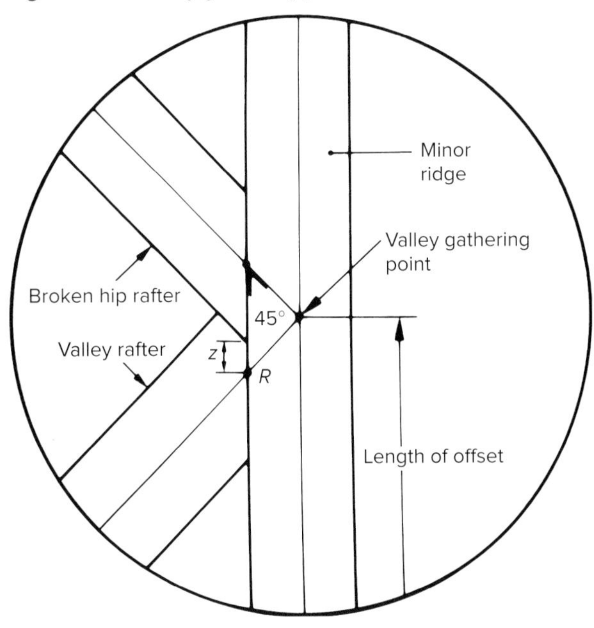

Fig. 14.24 Wall plate set-out for valley

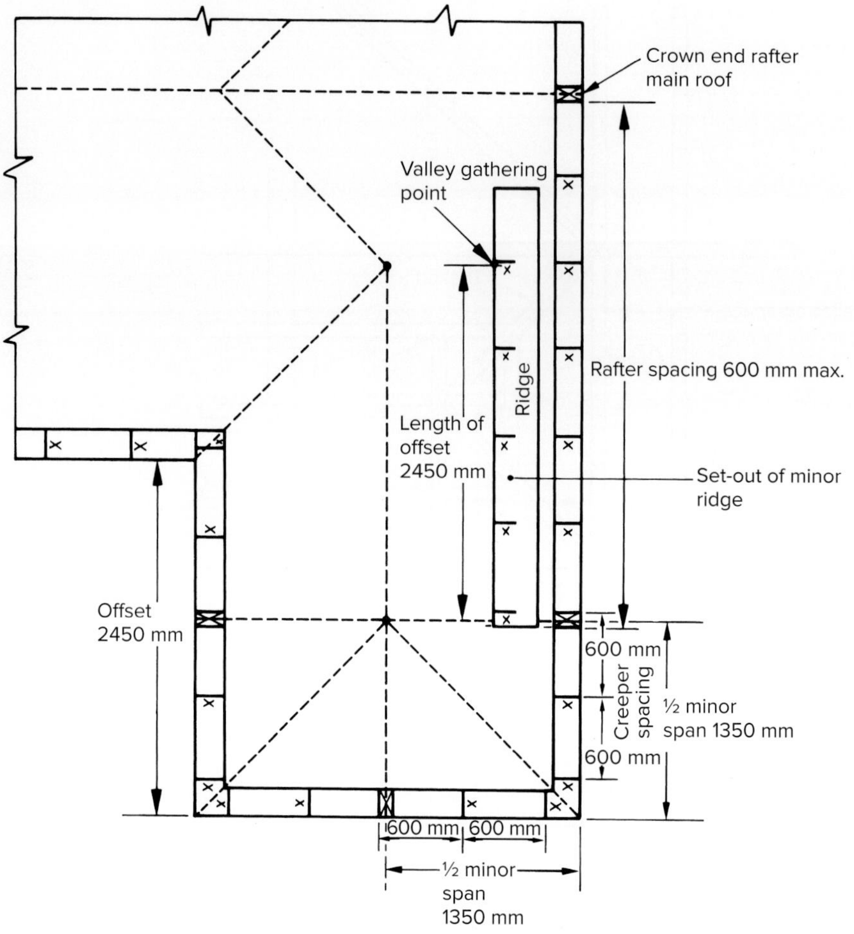

Crown end rafter main roof
Valley gathering point
Ridge
Rafter spacing 600 mm max.
Length of offset 2450 mm
Set-out of minor ridge
Offset 2450 mm
600 mm
Creeper spacing
½ minor span 1350 mm
600 mm
600 mm 600 mm
½ minor span 1350 mm

Fig. 14.25 Minor roof triangle

Plumb bevel creeper rafter
TL common rafter minor span
Creeper difference
693
2
693
Rise z
Pitch level bevel
1
90°
Creeper rafter
600
600
Run common rafter 1350 mm
Common rafter mm/m run = 1155
Common rafter = 1155 × 1.35
= 1559 mm

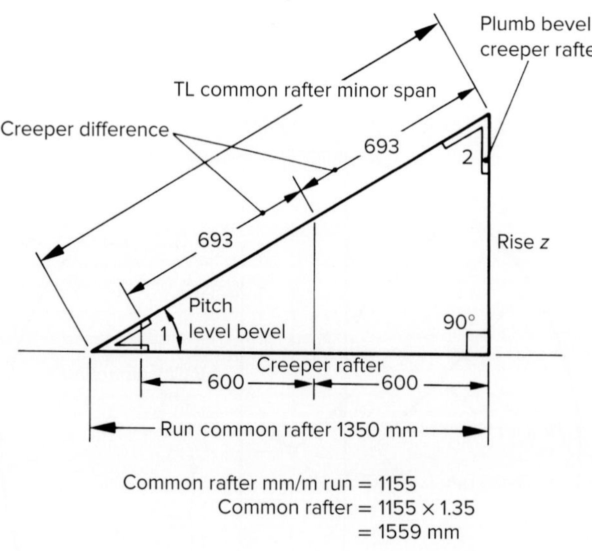

of a rule to project this point accurately on to the tape. Prepare the broken hip and fix it in position, using a short straight edge to align the top of the hip with the opposite edge of the minor ridge (Fig. 14.28).

14.8.3 Length of valley rafter

Again using the steel tape, measure the length of the valley rafter from point *R* to point *V* (Fig. 14.29) and set out the valley (Fig. 14.30).

The set-out is shown with a tail on the valley rafter, but it is common practice to end the valley rafter at point *V* and to allow any common rafters to be located near the internal corner to run through and support the end of the fascia and the valley boards. From one end of the valley allow for a tail and mark a plumb line across the face. From the top edge, measure down distance *x* to locate point *V*, giving the line for the horizontal cut of the birdsmouth. Mark back half the thickness of the valley rafter for the vertical cut of the birdsmouth. From point *V*, measure the length of the valley to the top edge and locate *R* on the centre line; mark the valley edge bevel through this point and continue the plumb bevel across the face.

Depending on the thickness of the members, it may be necessary to make a square cut on the end of the valley where it will butt against the broken hip. Refer to Figure 14.23 detail and note the distance *z*. Measure distance *z* on the roof and transfer this distance to the edge of the valley from the centre line in order to give the position for the square cut. After cutting to detail, fix the valley rafter in position.

Fig. 14.26 Ready for fixing broken hip and valley

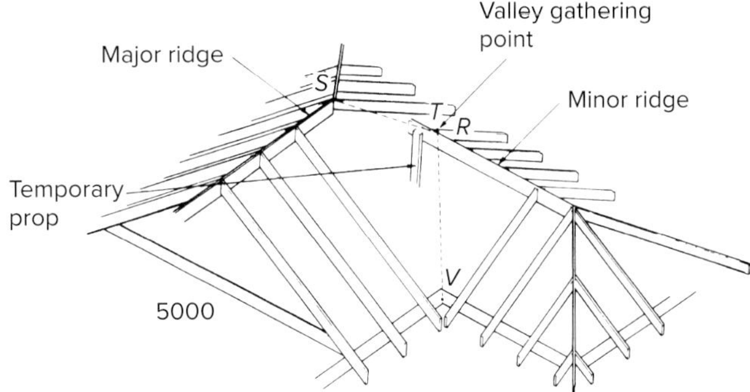

CHECK YOUR UNDERSTANDING

1. In an on-site situation:
   a. How can the length of the valley creeper be obtained?
   b. What is the recommended method for obtaining the length of a hip rafter?
   c. How can the valley gathering point be located in a fully hipped roof? (Refer to the relevant figures if necessary.)

14.8.4 Cutting valley creepers

With the broken hip and valley now in position, the valley creepers can be cut and fixed. Provided the spacing remains the same, the difference in the length of the valley creepers is the same as for the hip creepers–indicated as *A* and *B* in Figure 14.31–and the practical way to find the lengths of the first valley creepers–indicated as *A* and *B* in Figure 14.22–is to mark the rafter spacing from the last common rafter to the long point of the creeper. Using the steel tape, measure from the ridge to this point.

Other creepers will then shorten in length by the creeper difference and the cripple rafters will be shortened by twice the creeper difference. Valley creepers are fixed above the edge of the valley rafter. Use a short straight edge, and the top of the creepers should strike the valley on its centre line (see Fig. 14.32).

Fig. 14.27 Measurement of broken hip

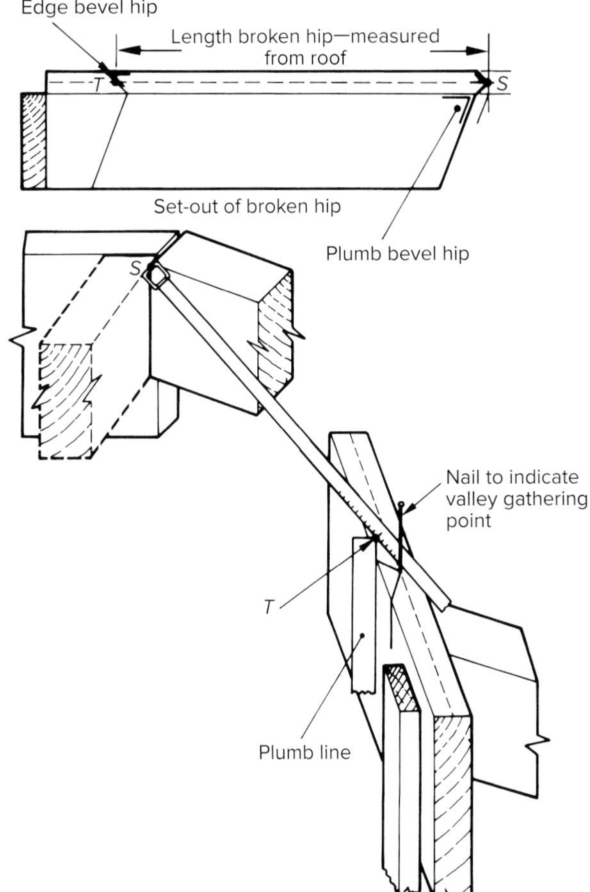

14.8.5 Finishing the hip and valley frame

Proceed to fix any outstanding common or creeper rafters and continue with underpurlins and strutting as previously described. The eaves overhang is marked and cut to length prior to fixing the fascia board. Valley boards are fixed to support the valley gutter and the roof is now ready for the roof plumber. Refer to Figure 14.19.

Fig. 14.28 Fixing broken hip rafter

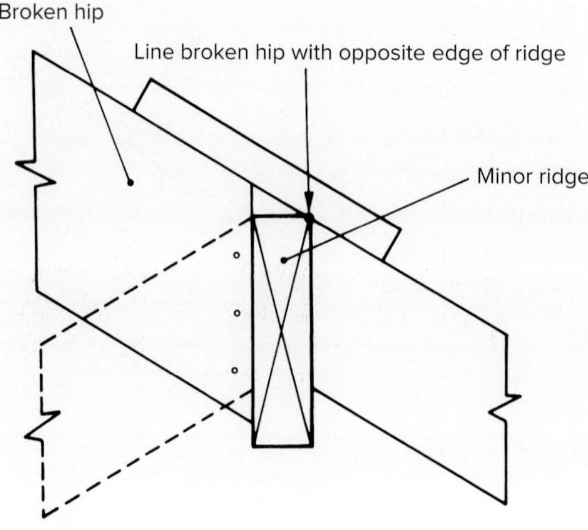

Fig. 14.29 Valley rafter measurement

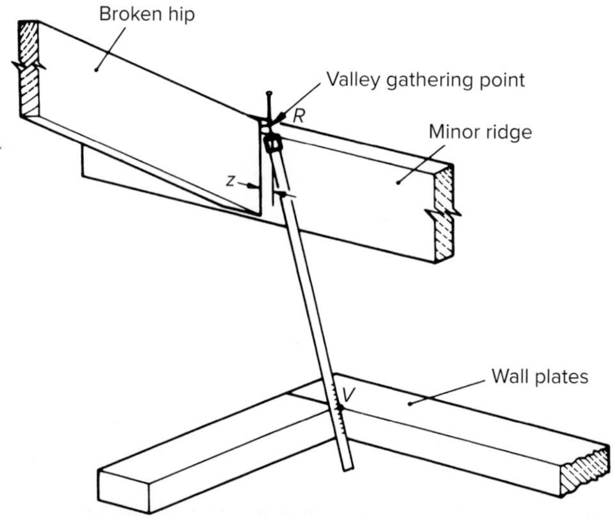

Fig. 14.30 Valley rafter set-out

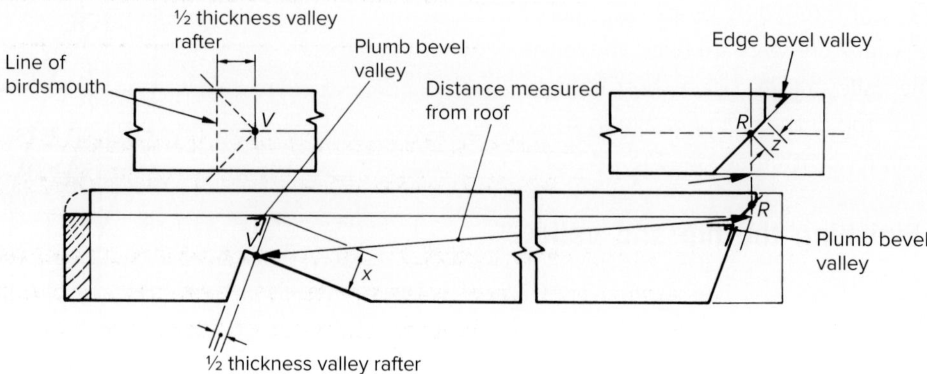

Fig. 14.31 Valley creepers set-out

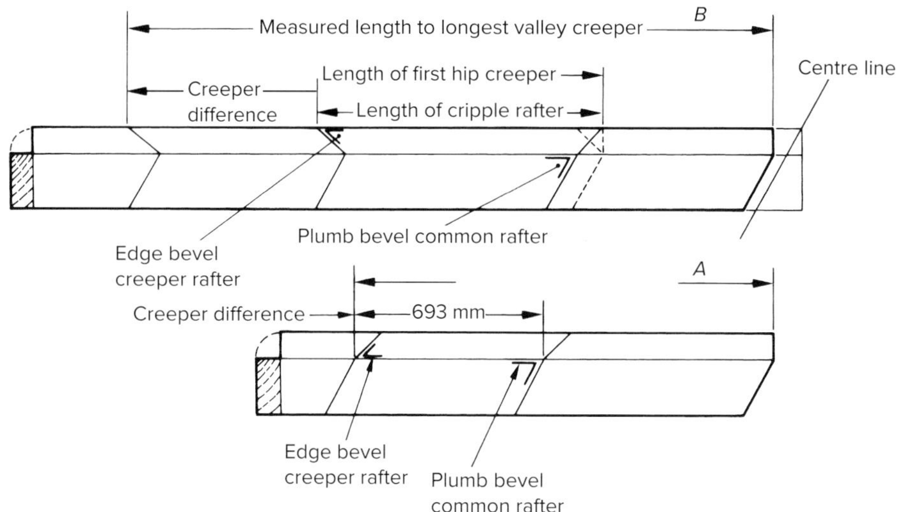

Fig. 14.32 Fixing of valley creeper rafters

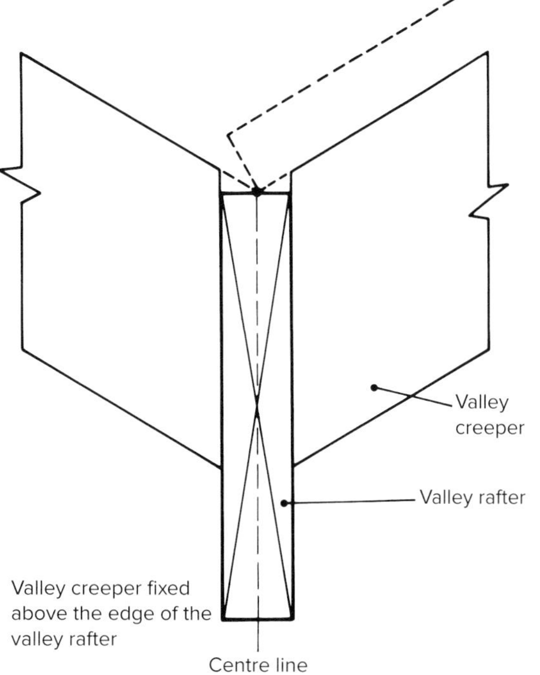

CHECK YOUR UNDERSTANDING

1. Using AS 1684, what size of timber member would you need for a single span rafter 3.4 m long of stress-grade MGP12 with tin roof and collar ties installed?

Student research

Research the different methods of constructing a hip and valley roof and understand where each method is best suited.

End of chapter activity

Activity 1:

Ask your teacher for a half span and pitch of a roof you will be required to build in class. Develop the roof by drawing it to as large a scale as possible and construct a pattern rafter that you will use for the roof construction.

AUSTRALIAN BUILDING CODE AND STANDARDS

AS 1684— Residential Timber-Framed Construction

National Construction Code (NCC)

Tradie talk—**We will be pitching this roof**

Chapter 15

Restricted-height scaffolding and work platforms

Learning Objectives

LO 15.1 Know about licensing and regulation

LO 15.2 Identify types of scaffolding/working platforms

LO 15.3 Understand environmental conditions and scaffolding choice

LO 15.4 Apply scaffolding safety procedures

LO 15.5 Know about scaffolding equipment/components and their application

LO 15.6 Understand procedures for erecting scaffolding

LO 15.7 Inspect working platforms for compliance

LO 15.8 Identify other types of scaffolding

LO 15.9 Know about ladders and their use

Introduction

This chapter deals with some of the most common scaffolding systems you would find in use on an Australian building site. The scaffold and working platforms covered in this chapter are up to a working platform height of 4 metres and therefore no licence is required for this work, as detailed in the Safe Work Australia *General Guide for Scaffolds and Scaffolding Work*.

In this chapter you will learn how to identify and select the appropriate working platforms for a given task and environmental conditions. You will also learn how to identify various scaffolding components and how to erect and safely use working platforms, ladders and scaffolds. In doing so, you will learn about safety and compliance requirements during the erection, use and dismantling of scaffolds, ladders and working platforms.

WORKPLACE SCENARIO

The training you will undertake at trade school is designed to give you an appreciation of how to use scaffolding, ladders and working platforms in a way that is safe and compliant with regulations. You will learn what they should look and feel like when being used, and will be shown how to erect them safely and appropriately for the job you are doing. This training doesn't make you a qualified scaffolder, but it does give you the vital information needed to assess whether the equipment you are being asked to use is safe.

TIP A person who erects, alters or dismantles a scaffold where there is a risk of people or objects falling 4 m or *less* from the platform or structure does *not* require a high-risk work licence.

However, the guide also states that 'a person who erects, alters or dismantles any scaffold must be competent to do the work safely'.

Using any working platform first requires the site to be clean and tidy. It is essential to have clear access so that the scaffold or platform can be erected and accessed safely–in addition, any objects left around the platform pose a danger if someone happens to fall off the platform as they could land on those objects, potentially causing serious injury.

Once the platforms are no longer required, all components need to be stored away appropriately and in an organised manner. This applies especially to scaffolds as these have numerous components, and a tidy and organised storage system ensures the scaffold can be used efficiently the next time it's required.

15.1 Licensing and regulation

If the scaffolding work exceeds 4 m in height, then a licence is required, as detailed in the Safe Work Australia *General Guide for Scaffolds and Scaffolding Work.*

A person undertaking scaffolding work must hold the relevant class of scaffolding high-risk work licence as required by the OHS Regulations. The scaffolding high-risk work licence classes are:

Basic scaffolding licence–required for scaffolding work involving:

- modular or pre-fabricated scaffolds
- cantilevered materials hoists with a maximum working load of 500 kg
- ropes
- gin wheels
- fall arrest systems, including safety nets and static lines
- bracket scaffolds (tank and formwork).

Intermediate scaffolding licence–required for scaffolding work involving:

- cantilevered crane loading platforms
- cantilevered scaffolds
- spur scaffolds
- barrow ramps and sloping platforms
- scaffolding associated with perimeter safety screens and shutters
- mast climbing work platforms
- tube and coupler scaffolds, including tube and coupler covered ways and gantries.

Advanced scaffolding licence–required for scaffolding work involving:

- cantilevered hoists
- hung scaffolds, including scaffolds hung from tubes, wire ropes or chains
- suspended scaffolds.

As scaffolding is a regulated activity in all states of Australia and is considered a high-risk activity, there are numerous Australian standards and codes that cover the production and use of scaffolding systems. These include:

- AS/NZS 1576.3:2015 Scaffolding–Prefabricated and Tube-and-Coupler Scaffolding
- AS 1577:1993 Scaffold Planks
- AS/NZS 4576:1995 Guidelines for Scaffolding
- AS/NZS 1576.1:2019 Scaffolding–General Requirements.

The codes that cover scaffolding can be found at Safe Work Australia or your state's or territory's WorkCover authority. Keep your high-risk activities knowledge current by regularly checking their websites or subscribing to their newsletters and alerts.

15.2 Types of scaffolding/working platforms

The scaffolds or working platforms covered in this chapter fall into three main categories of use, as defined in Safe Work Australia's *General Guide for Scaffolds and Scaffolding Work:*

- *Light duty*–up to 225 kg per platform per bay including a concentrated load of 120 kg. Platforms should be at least two traditional scaffold planks wide–approximately 450 mm. Examples of use include painting, electrical work, many carpentry tasks and other light tasks.
- *Medium duty*–up to 450 kg per platform per bay including a concentrated load of 150 kg. Platforms should be at least four traditional scaffold planks wide–approximately 900 mm. Examples of use include general trades work such as tiling and light steel framing.
- *Heavy duty*–up to 675 kg per platform per bay including a concentrated load of 200 kg. Platforms should be at least 1000 mm wide. This duty of scaffold is needed for concrete blocklaying, bricklaying, concreting, demolition work and most other tasks involving heavy loads or heavy impact forces.

The three types of scaffold covered in the chapter are:

- *quickstage or modular scaffold.* This scaffold is used extensively in the building industry. It can be found on domestic or commercial construction projects. It is very strong and will deal with most building activities. It is very safe and is generally tied into the building being constructed, making it very secure. These scaffolds require a minimum of two people to construct.
- *aluminium mobile scaffold.* This scaffold is very common on construction sites. It is lightweight and moveable, which makes it ideal for jobs such as painting, where it is necessary to be in particular places for only a short period of time before moving on to the next spot. These types of scaffolds don't cope with very heavy loads or too many people on them, so they have their limitations. They require a firm and relatively level surface to make

Fig. 15.1 Quickstage or modular scaffold

Fig. 15.2 Aluminium mobile scaffold

© Baloncici/Shutterstock

Fig. 15.3 Trestles and planks scaffolding

© McGraw Hill Education

it easy to move them around and be safe to use. They can be susceptible to strong winds and have limitations in the height they can be built to. These scaffolds require a minimum of two people to construct.

- *trestles and planks*. These scaffolds are very versatile as they can be made as long or as short as the task requires, take relatively heavy loads (e.g. are used by bricklayers) and can be transported to jobs quite easily. They can be built by a single person on a hard level surface. They are fairly stable, but they can topple if overloaded or used inappropriately. They generally extend to only 1.8 m so they don't require handrails (handrails are required when the potential fall is 2 m or greater).

SAFETY TIP When using a scaffold, ladder or work platform, always investigate the equipment before you use it.

- Can you see any repairs? If so, do they look sound?
- Are all of the components present?
- Has it been well maintained (does it appear to be in good condition)?
- Does it feel stable? Does it rock?
- Is the person erecting it qualified?
- Do you know the equipment's limitations?
- Do you feel safe on the equipment?

If you have any doubts, don't use the equipment.

CHECK YOUR UNDERSTANDING

1. Up to what height can a person construct a scaffold without a licence?

15.3 Environmental conditions and scaffolding choice

There are many factors to take into consideration when choosing the most suitable scaffold to use for a task.

First, determine if the ground or base is suitable to support the scaffold:

- Can the scaffold support the combined weight of the scaffold itself, the people using it and any tools, equipment and materials that will be placed on it?
- Are there any possible voids, pipes, cellars, etc. located in the ground beneath or near where the scaffold is to be set up that could cave in when the scaffold is erected?

> **TIP** If any of these factors are present, then an alternative method may need to be found or work may have to be done to the site to make it suitable to accommodate a scaffold.

Determine the most suitable scaffold for the type of work and the location:

- If the job is small and not too high, a mobile scaffold may be most appropriate.
- If the job is large, is high off the ground and the scaffold will need to accommodate heavy materials and many trades for a long time, then a modular scaffold might be the most appropriate choice.

Remember to keep in mind that each scaffold has its limitations, such as:

- how much weight it can hold
- how high it can go
- how heavy it is
- how easy it is to get the scaffold to the area.

Consider environmental factors:

- Will the scaffold be subject to high winds, rain, etc.? If so, the scaffold will need to be tied into a supporting structure.
- Could the scaffold be subject to potential impacts? If so, an exclusion zone needs to be established to avoid this happening.

Assess the area above for powerlines:

- Are there any overhead powerlines where the scaffold is to be built? If so, Figure 15.4 shows the no-go zone for working near overhead power lines in Victoria. (If you are in another state or territory, check the requirements with your local power supplier/regulator.)

Fig. 15.4 No-go zone around powerlines

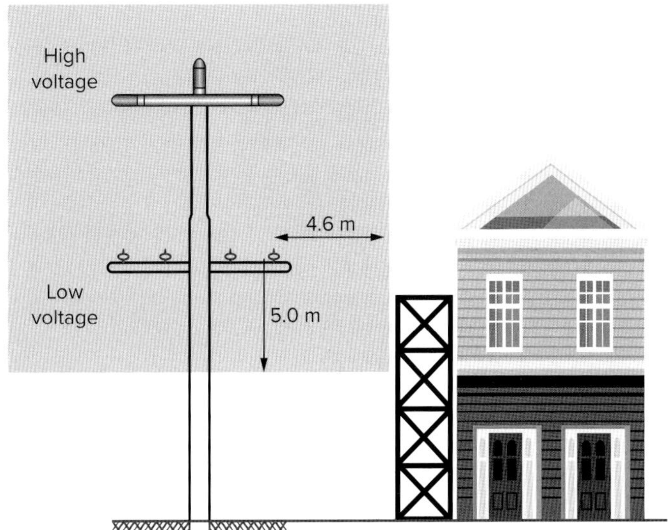

> **TIP** The erection of exclusion zones, fencing, bunting and signs will minimise any of the above incidents occurring.

Assess the area around where the scaffold is to be placed:

- Will the scaffold be located near a road, driveway or anywhere that a powered vehicle, such as a bobcat, forklift or truck could potentially hit the scaffold and make it unsafe to use?
- Will the scaffold be located near a pedestrian walkway, footpath, accessway, etc.? If so, how are people going to be protected from potential items falling from the scaffold or entering the scaffolding zone?
- Will the scaffold be located in an area of a building site where workers will be exposed to potential falling objects?

15.4 Scaffolding safety

Scaffolding work and working at heights are high-risk activities, and accordingly the work health and safety requirements are many. The reasons we use scaffolds are to:

- provide a safe and stable working platform at various heights to support work activities and associated equipment and materials required for the specific task
- reduce the risk of workers falling when working at heights over 2 m by providing handrails, midrails, catch platforms and a fully decked-out work platform
- protect workers and the public below the scaffold by minimising equipment and materials falling from the scaffold by having kickboards and a fully decked-out work platform
- provide easy and safe temporary access to work areas located above ground level.

When using scaffolds there are a number of safety signs that are used, as shown in Figures 15.5–15.7.

These signs let people know what stage of construction the scaffold is at. Signs are only one part of the process to keep both workers and the general public safe. Barriers need to be erected to stop people from walking under the scaffold or getting too close to it just in case, for example, something is dropped. Temporary fencing is ideal if the scaffold is to be left up for an extended period of time.

If the scaffold being used is mobile or consists of trestles and planks, then safety tape/bunting and signs indicating people working overhead will suffice. A person acting as a spotter is also a good idea in certain circumstances.

Fig. 15.5 Incomplete scaffolding sign

Fig. 15.7 A worker's overhead warning sign

Fig. 15.6 Temporary fencing around building site

15.4.1 Safe work method statements for scaffolding work

Erecting a scaffold or working on a scaffold may involve activities defined as high-risk construction work under OHS regulations. High-risk construction work includes any construction work where there is a risk of a person falling more than 2 m. Scaffolding work is defined with a 4 m threshold for licensing purposes. This means that in some cases a high-risk work licence may not be required to erect a scaffold because it is less than 4 m, but there may still be a need for a safe work method statement (SWMS) because it is more than 2 m.

To identify the potential hazards associated with scaffold use and the control measures, an SWMS must be used. An SWMS must be prepared for high-risk construction work before the work starts. The SWMS must:

- identify the type of high-risk construction work being done
- specify the health and safety hazards and risks arising from the work
- describe how the risks will be controlled
- describe how the control measures are to be implemented, monitored and reviewed.

The SWMS must be developed in consultation with workers and their representatives who are carrying out the high-risk construction work.

Further information on high-risk construction work, an SWMS and an SWMS template can be found in the Model Code of Practice–Construction Work (see www.safeworkaustralia.gov.au/sites/swa/about/publications/pages/construction-work).

15.5 Scaffolding equipment/components

Scaffolding equipment and components are shown in Figure 15.8. They can be held in a waist belt so they are easy to access when constructing and dismantling the scaffold.

15.6 Procedures for erecting scaffolding

15.6.1 Trestle scaffolds and planks

Trestle scaffolds should be built on a hard, level surface. If the trestle is to be built on compacted ground or crushed rock, then sole plates need to be used. Each trestle should be in the fully open position. If the trestle scaffold has a height adjustment to allow for the thickness of the plank, it is important to ensure that the pin supplied with the trestle is used. **Don't use a nail or screw**.

The working platform of a trestle scaffold should not be more than 2 m above the ground, as any work platform above 2 m requires fall and edge protection (handrails). If the trestle is being used next to a pit or the edge of a balcony, ensure that the maximum distance a person can fall is not greater than 2 m. Do not piggy-back trestles.

The maximum spacings of trestles should not exceed the maximum spacings given in Table 15.1, or the maximum span of the planks being used as advised by the plank manufacturer.

Table 15.1 Maximum span of solid timber scaffold planks, complying with AS 1577: Scaffolding Planks

Nominal thickness of plank (mm)	Maximum span between putlogs (mm) (point of support)
32 (hardwood only)	1.0
38	1.5
50	2.0
63	2.5

The full width of the trestles being used should be planked. Planks should overhang their end supports by not less than 150 mm or more than 250 mm and where necessary be secured against uplift. Planks may be lapped to form multiple bays of scaffolding, provided there is no obstruction along the full length of the working platform. The working platform should be horizontal.

Work only on the part of the planks between the trestle; don't stand on the overhang of the plank. If the trestle scaffold is more than one bay in length and is being used for bricklaying, blocklaying or similar work, concentrated loads should be placed directly over the trestle.

Scaffold planks can be made of natural timber, aluminium or a man-made timber product such as laminated veneered lumber (LVL).

When constructing the trestle scaffold, inspect the components to make sure the trestles are not damaged or bent and the planks are not cracked, twisted, split, bent or cut through.

Fig. 15.8 Basic scaffolding equipment/tools

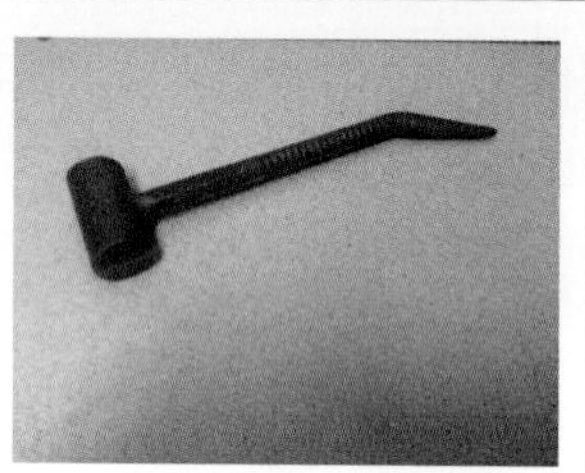	The podger/hammer is used to adjust the height of the jacks (pointy end) and to tap down the captive wedges on a modular scaffold.
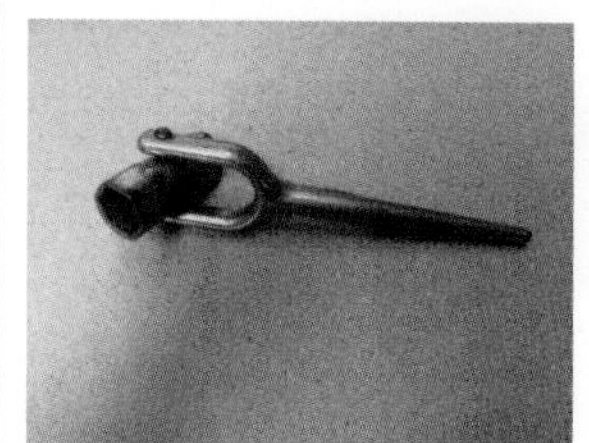	Scaffold keys are mainly used with tube and coupler scaffold (not covered in this chapter), but can also be used with modular scaffold when the scaffold is constructed in a way that couplers need to be used to make the scaffold safe and there is no standard component available. They are also used to tighten the bolts on ladder putlogs and short toeboards.
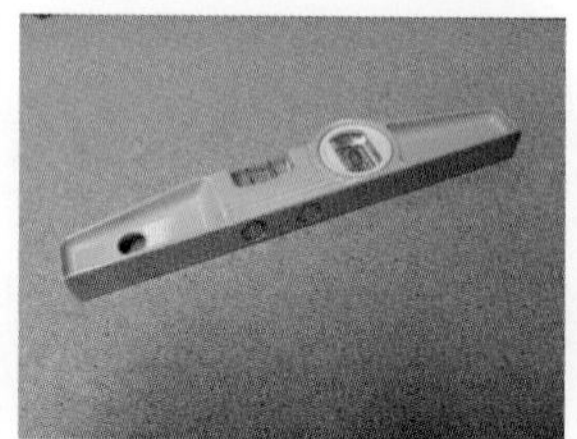	A turbo level is used to level the scaffold and has a magnet on the bottom so that it adheres to steel scaffold.
	A tape measure is used to square the first bay of a modular scaffold by measuring the diagonals of the first bay; it can also be used to work out the required laps and overhangs of the planks when using trestles.
	A shifter is used when the scaffold keys are the wrong size or can't be used in certain situations.

Fig. 15.9 Types of scaffold planks: (a) timber planks; (b) aluminium planks

(a)

(b)

© McGraw Hill Education

Courtesy of Alistair Ford

15.6.1.1 Erecting a trestle scaffold

Below is the sequence for erecting a trestle scaffold.

1. Assess the site on which the trestles are to be used. The area needs to be flat, compact (e.g. concrete) and relatively level. If the area is flat and level but made of dirt, sole plates will need to be used.
2. Unpack the trestles and extend them to the desired height. Ensure that locking pins are present and functioning and that there is a mechanism that will prevent the pins from falling out. As well as the desired height, there is an option of a high hole and a low hole.
3. Space the required number of trestles out for the required distance of scaffold to be used (taking into account the maximum span of the plank being used). For this example, the planks will span a maximum of 1.8 m, so the trestles can be no further apart than 1.8 m.
4. Use planks that are in sound condition only–check that they are not split, cut or damaged. Start placing the planks on the trestles set on the low hole first. Ensure the trestles are fully decked out; if the trestles are designed to take five planks, then use five planks. Ensure that the plank overhang meets the Australian Standard.

Fig. 15.10 Space trestles according to plank length

© McGraw Hill Education

Fig. 15.11 Place planks on trestle

© McGraw Hill Education

5. Now place the planks on the trestles set on the high holes; these planks will also sit on the planks already in place on their ends, causing an overlap to occur. This is normal, but ensure the overlap meets the requirements set out in the Australian Standard.
6. Now the scaffold can be used. However, remember:
   - don't overload the scaffold
   - don't run or jump on the scaffold
   - if the planks are not sitting neatly and wobble, check the trestles are erected properly
   - make sure you can't fall more than 2 m: if the scaffold is set up in an area where a person could fall further than 2 m, then the trestles and planks are the wrong type of scaffold to be using.

Fig. 15.12 The finished scaffold

15.6.2 Mobile scaffold

Mobile scaffolds are generally made from aluminium due to its reduced weight. They are a free-standing scaffold supported on wheels, which vary to suit differing ground conditions. The wheels must be lockable and locked when the scaffold is in use.

Never:

- move the scaffold with a person on it
- climb the outside of a scaffold
- pull the scaffold when moving it as it could tip and fall on you; always push it
- mix components of two different scaffolds
- hit the components of the scaffold with a hammer as they will bend and dent
- force the scaffold components as they could be damaged
- use the scaffold on an uneven surface or a slope
- place near powerlines
- use an incomplete mobile scaffold
- use damaged scaffold components/parts
- use in high winds
- allow the scaffold height to exceed three times the minimum base measurement
- exceed the scaffold's maximum working load.

CHECK YOUR UNDERSTANDING

1. At what height must handrails be installed when working on a trestle or quickstage scaffold?

15.6.3 Modular/quickstage scaffold

This type of scaffold is used widely in the domestic and commercial building sectors. It is designed to take large loads and to have many people working on it. Its height is limited by the height of the building supporting it. It is very stable and heavy and the weight of the scaffold itself must be taken into account when calculating the loads being exerted onto the ground as well as the loads of the materials, equipment and people working on it.

Fig. 15.13 How a correctly erected mobile scaffold should look before use

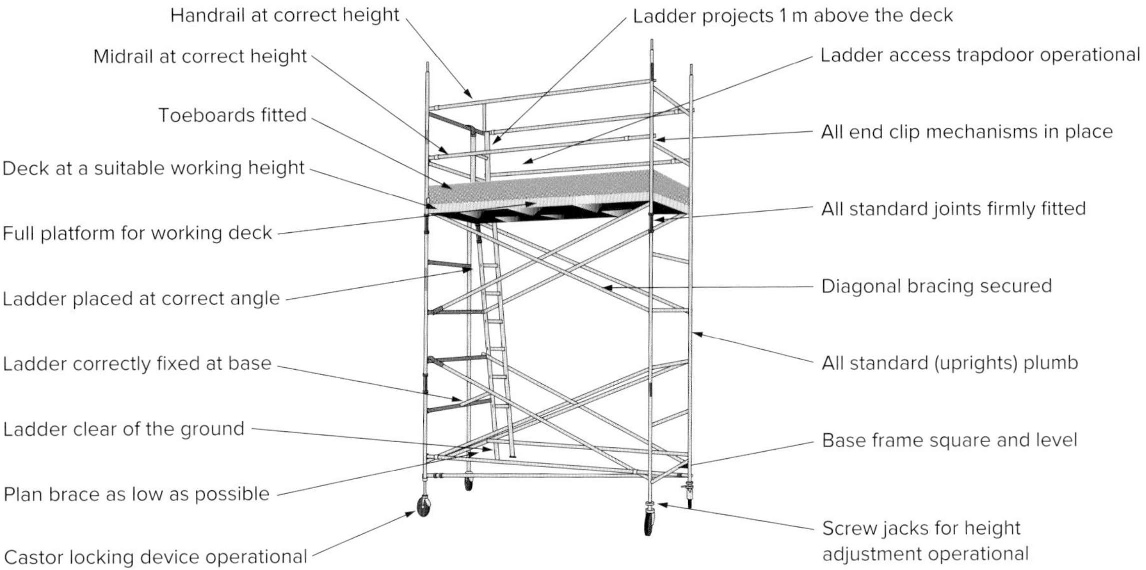

Figure 15.14 shows the major components and where they should be placed.

Fig. 15.14 Components of a modular scaffold

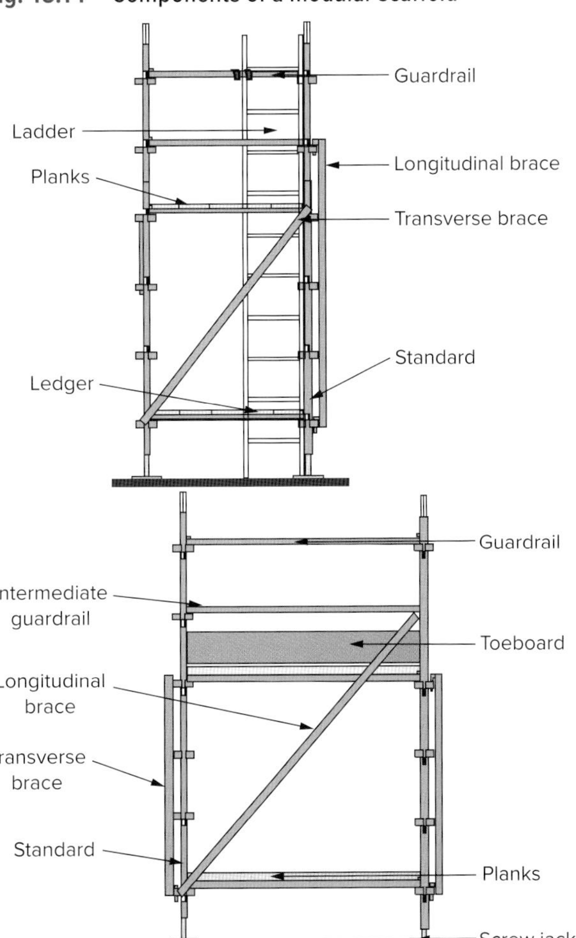

15.6.3.1 Erection procedures for modular scaffold

Inspect the scaffolding components and ensure they are not damaged. Damaged components need to be isolated from the rest of the scaffold and identified as being damaged so that they can be either repaired or destroyed. Also, ensure that components from different scaffolds are not being mixed as not all types of modular scaffold are compatible.

The procedure to erect the modular scaffold is as follows:

1. Adjust the height of the screw jacks so that two- to three-thirds of the shaft is located inside the standard.
2. Choose sole plates to be placed underneath the screw jacks; they will usually be timber, such as damaged trestle planks that can no longer be used safely. The sole plates spread the load from the screw jack over a larger area of the ground on which the scaffold is built. Set out their positions and bed them as level as possible, unless the surface is concrete or another firm surface.

Fig. 15.15 How modular scaffold components fit together

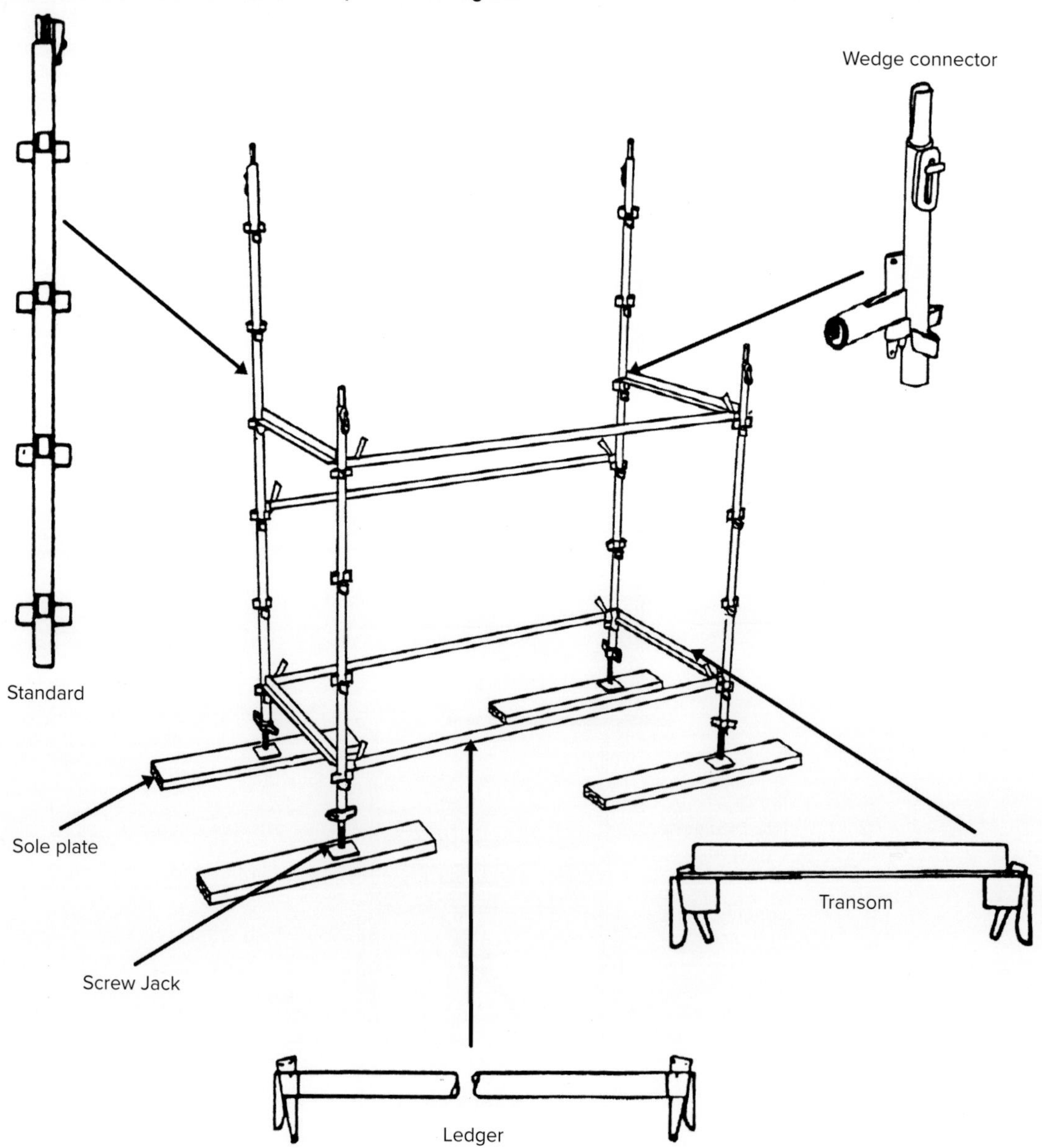

Fig. 15.16 Standard being placed on screw jack

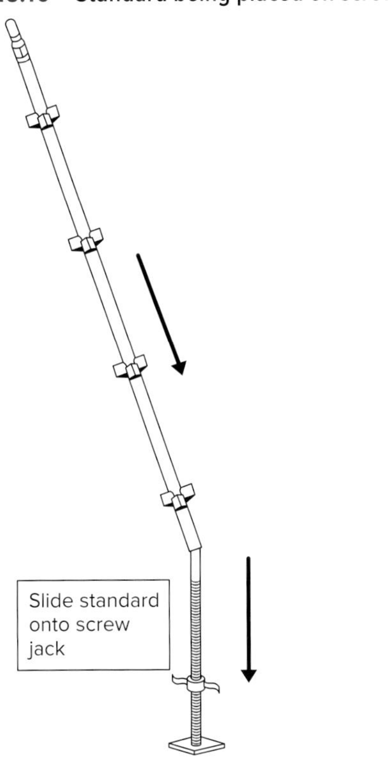

Fig. 15.17 Constructing first end

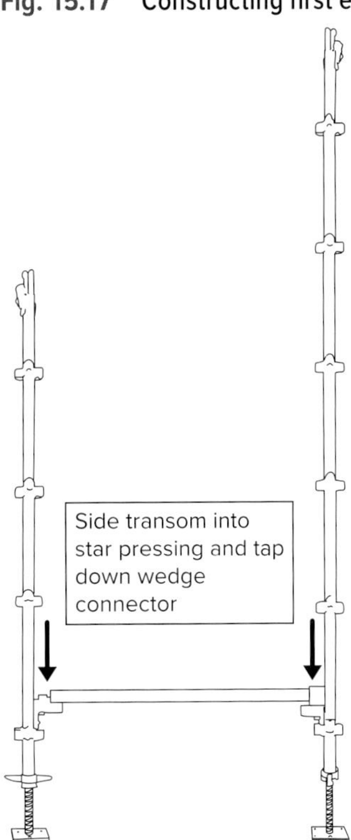

Fig. 15.18 Attach ledgers to the standards

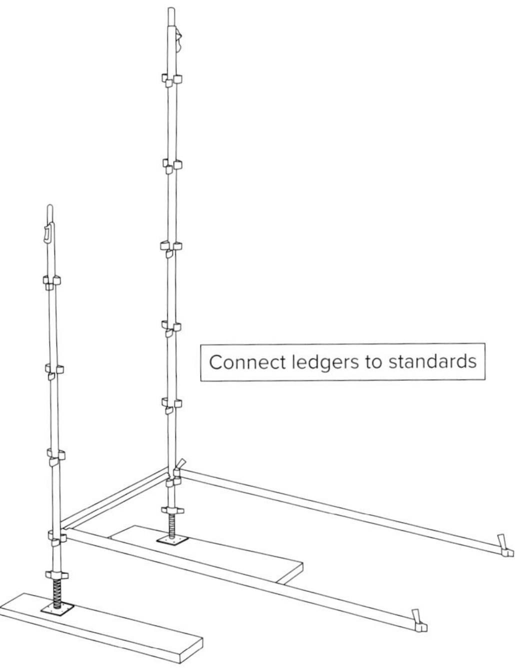

Fig. 15.19 Completed base of first bay

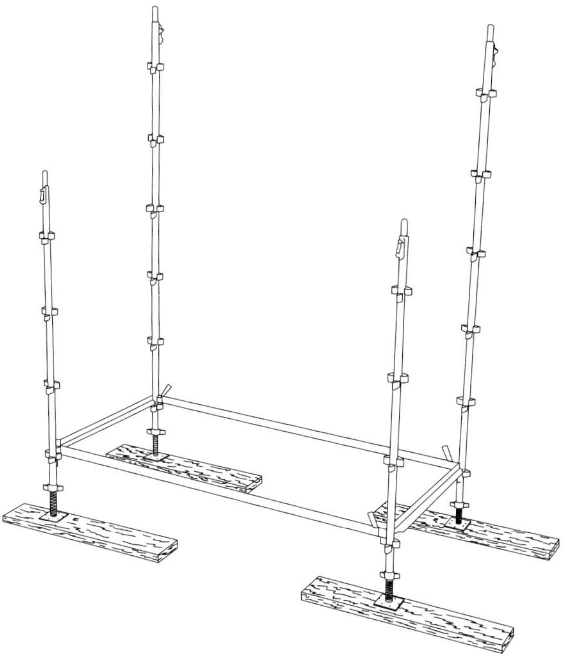

3. Starting at one end of the proposed scaffold, place a pair of screw jacks set to their lowest adjustment on top of the sole plate and slide the end of a standard over the top of each jack.

   Use different height standards on each so that the joins in the standard will be staggered throughout the height, which will increase strength. Insert the transom and fit the wedges to the lowest possible point of connection on the star pressing; do not tighten the wedges at this stage. Ensure that the captive spigots are facing out.
4. With one person holding the two standards, get a second person to attach the ledgers to the standards at the same height as the transom. Do not tighten the wedges.
5. Attach the next two standards to the ends of the ledgers and then fit a transom between them. Do not tighten the wedges at this stage.

   At this point, the scaffold can be levelled. Start at one transom: adjust the jacks to make it level and then move on to the ledgers. Only adjust the jacks that were not adjusted when the first transom was levelled. Once both ledgers have been levelled to the first transom, check the second transom. It should not require any adjustment. If the second transom is not level, check the first transom and then the ledgers and repeat the process. Now the wedges can be tapped down to make the connection rigid.
6. Extend the length of the scaffold, making sure you continue to level the next bay from the previous bay. The first lift can be established, usually 2 m above the first ledger/transom.

   To increase the height of the scaffold, additional standards need to be placed onto the bottom standards and locked in with a captive spigot wedge. A light tap of the hammer will do the trick.
7. Install the galvanised metal planks that fit in between and rest on the transoms. Each subsequent lift should be decked out with planks to enable erection and lifting of members.

Fig. 15.20 Levelling first bay

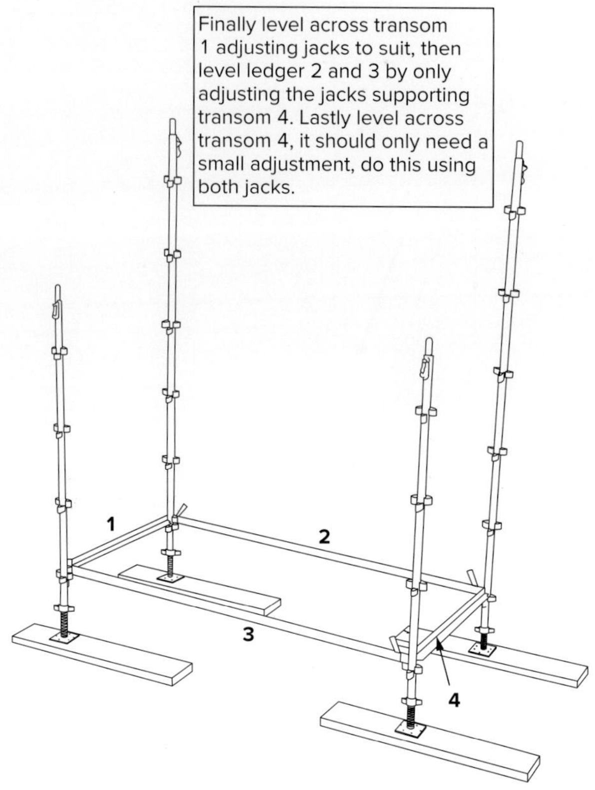

Finally level across transom 1 adjusting jacks to suit, then level ledger 2 and 3 by only adjusting the jacks supporting transom 4. Lastly level across transom 4, it should only need a small adjustment, do this using both jacks.
1
2
3
4

Fig. 15.21 Completed first bay

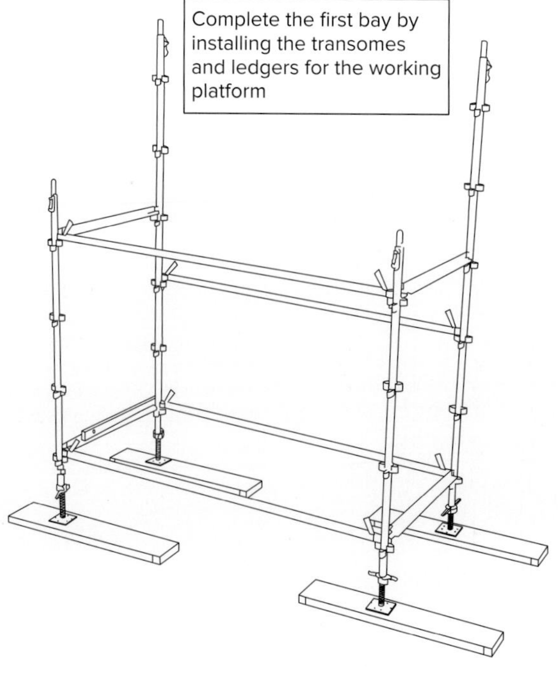

Fig. 15.22 Completed scaffold with toeboards, handrails and midrails

Courtesy of Daniel Bonnici

Fig. 15.23 Access bay should be separate from the working platform

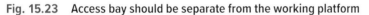

© McGraw Hill Education

When the desired height is reached, fit the longitudinal bracing on the face of the scaffold that is not the working face and the diagonal brace on the ends. Ensure all of the bracing starts off from the same hand; that is, the bottom of the brace has its lowest connection on the left-hand side when you are looking at it. Ensure the wedges are tapped down. Complete the scaffold by fitting toeboards, handrails, midrails, etc. Remember that a separate access bay must be provided as no work is allowed to be conducted on the access bay.

TIP To dismantle the scaffold, reverse the sequence described above. For the correct bracing requirements, please refer to the relevant Australian Standard.

15.7 Inspection and compliance

Scaffolds need to be regularly inspected to ensure they are compliant. Scaffold tags are useful for this task (Fig. 15.24). The tag provides information of when the scaffold was inspected, who inspected it and where the inspection occurred.

For the full details of scaffold inspection and maintenance and an inspection checklist, refer to Safe Work Australia's *Guide to Scaffold Inspection and Maintenance.*

CHECK YOUR UNDERSTANDING

1. What two common materials can scaffold planks be made from?

15.8 Other types of scaffolding

The other types of scaffold in use today include:

- *tube and coupler*–this type of scaffold is made up of lengths of steel tubes and clips/connectors that tie the tubes together (Fig. 15.25)
- *single pole scaffold*–this type of scaffold consists of a single row of standards connected by ledgers (Fig. 15.26). Putlogs are fixed to the ledgers and built into the wall of the building or structure.

Fig. 15.24 Scaffold inspection tags

FRONT TAG

SCAFFOLD TAG

ERECTION AND INSPECTION RECORD

TO BE COMPLETED BY INSPECTOR

LOCATION ______

REF NO. ______

REQUESTED BY ______

BUILT BY ______

DATE ______

INSPECTOR ______

SIGNATURE ______

SCAFFOLD TO BE USED FOR

LIGHT DUTY 225 Kg ☐

MEDIUM DUTY 450 Kg ☐

HEAVY DUTY 675 Kg ☐

THE ABOVE WEIGHTS ARE FOR ANY ONE WORKING PLATFORM DAY AND INCLUDES PEOPLE AND MATERIALS

UNLAWFUL REMOVAL OR INTERFERENCE WITH THIS SIGN COULD MAKE YOU LIABLE TO PROSECUTION AND FINES

BACK TAG

INSPECTION RECORD

AUTHORISED PERSON		
DATE	TIME	SIGNED

LADDER NO. RUNGS: ______

NO. OF STANDARDS: ______

NO. OF LIFTS: ______

M2 BOARDING: ______

STRUCTURE DECOMMISSIONED DATE: ______

Fig. 15.25 Tube and coupler scaffold

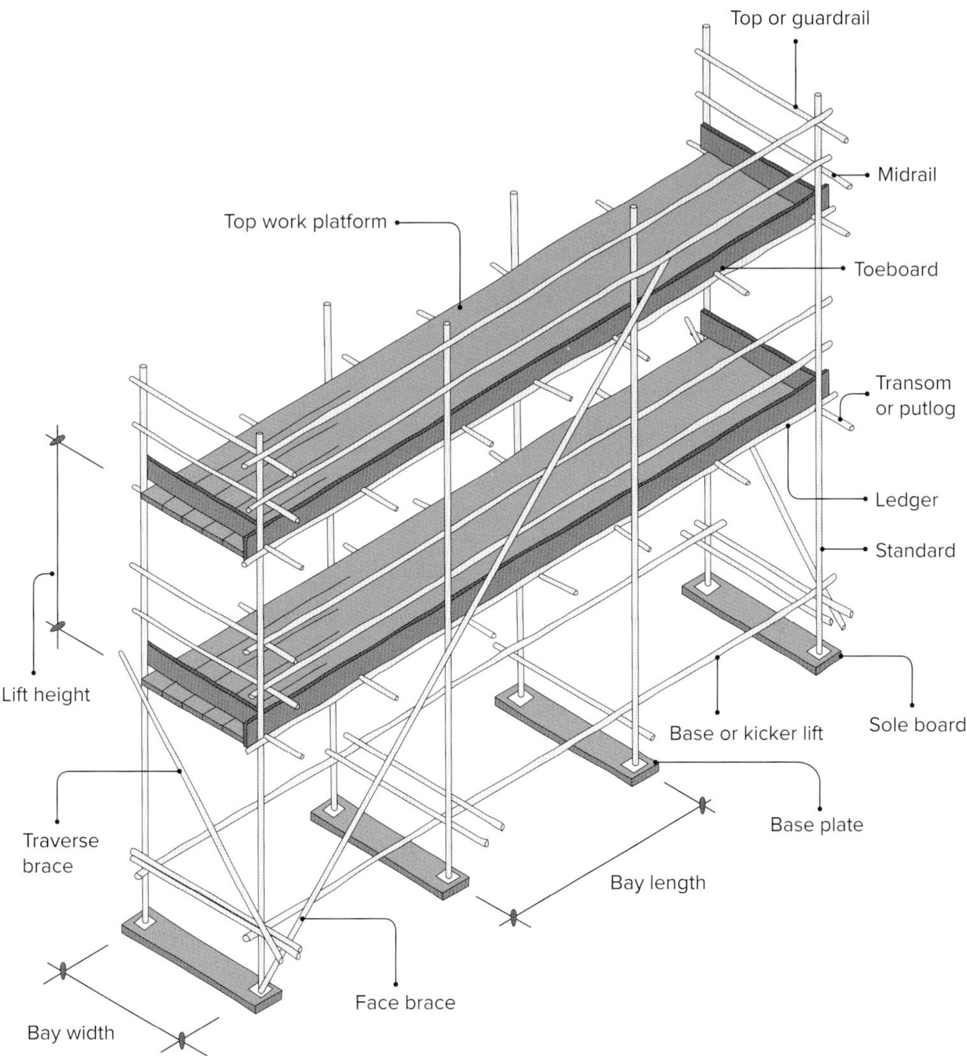

15.9 Ladders

Ladders are an essential and important part of the building industry. They come in various forms.

15.9.1 Stepladders

Ladders are used as a last resort when completing construction tasks and should be used only if there is no other suitable alternative. If completing a large project and access at height is required, more suitable equipment such as scaffolds, trestles, cherry pickers, scissor lifts, etc. should be used. This is because ladders are quite dangerous and are unstable compared to the alternatives.

Fig. 15.26 Single pole scaffold

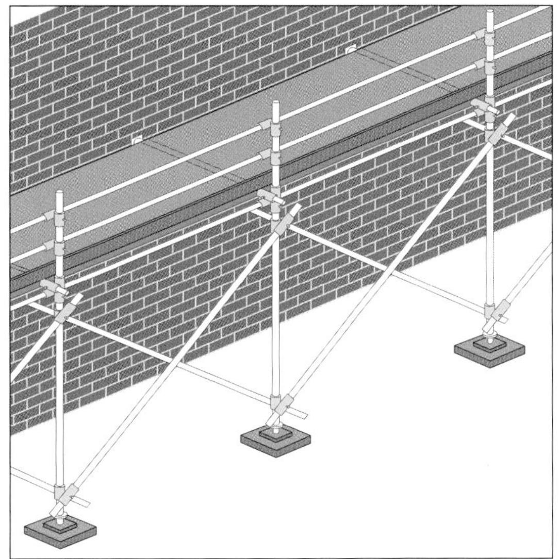

Fig. 15.27 Stepladder

© BK foto/Shutterstock

Fig. 15.28 Platform ladder

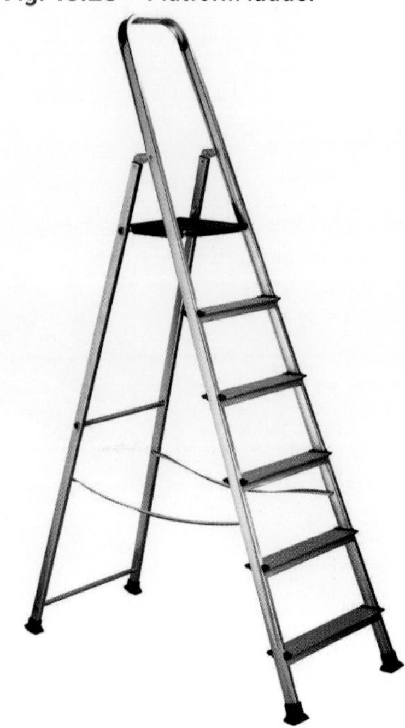

© lynx/iconotec.com/Glow Images

Fig. 15.29 Extension ladders

© Margo Harrison/Shutterstock

As a general rule:

- use ladders only on level and stable ground
- never climb past the second-highest rung on a stepladder
- always try to have someone to foot the ladder you are using
- ensure an extension ladder is at the correct angle when using it–4 to 1
- ensure extension ladders are secured at the top while in use
- make sure extension ladders extend 1 m past the work surface if being used to access a high area such as a roof
- if working around electrical hazards, ensure the ladder is made of non-conductive material such as timber or fibreglass.

For more information about the use of ladders, refer to the National Code of Practice for the Prevention of Falls in General Construction–Safe Work Australia.

CHECK YOUR UNDERSTANDING

1. What part of a stepladder should a person not climb past?

Student research

Check out Safe Work Australia's *General Guide for Scaffolds and Scaffolding Work* and research other types of scaffolds that are used in Australia.

End of chapter activity

Activity 1:

Label the diagram below.

Fig. 15.30 Scaffolding

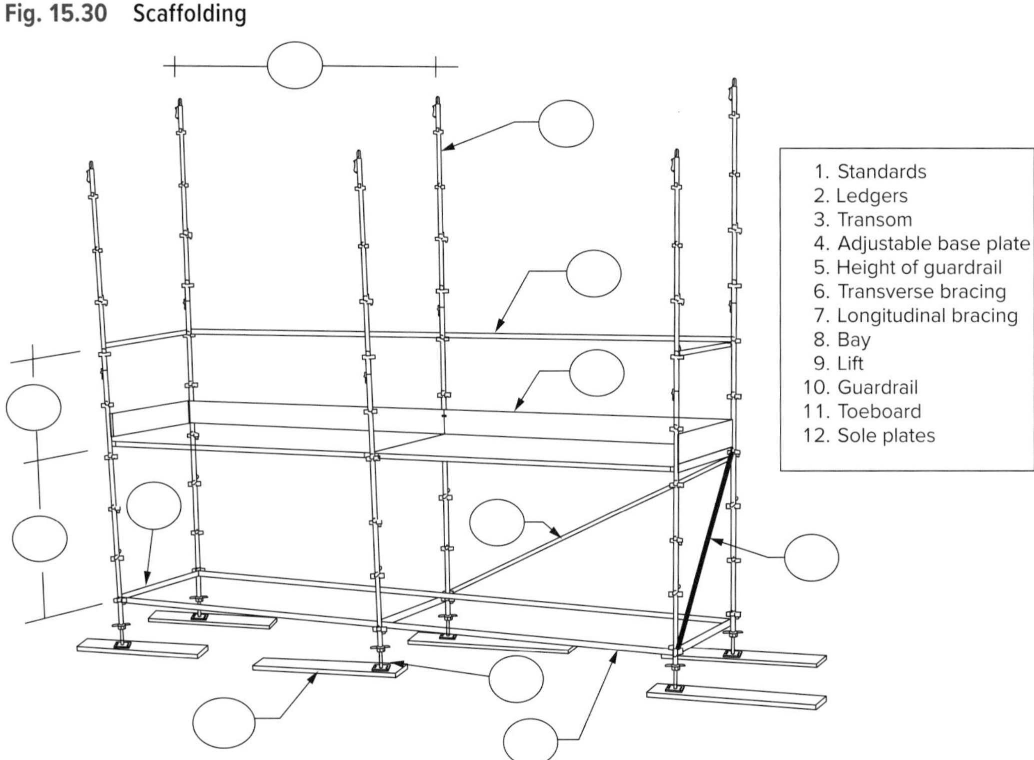

AUSTRALIAN BUILDING CODES AND STANDARDS

Safe Work Australia, National Code of Practice for the Prevention of Falls in General Construction.

AS/NZS 1576.3: 2015 Scaffolding—Prefabricated and Tube-and-Coupler Scaffolding

AS 1577: 1993 Scaffold Planks

AS/NZS 4576: 1995 Guidelines for Scaffolding

AS/NZS 1576.1: 2019 Scaffolding—General Requirements

Tradie talk—Never take a backward step

Chapter 16

External cladding

Learning objective

LO 16.1 Identify and install cladding

Introduction

Cladding is the outer covering applied to the external walls of a framed building. While its main purpose is to protect against elements such as wind, rain, heat and cold, it is also recognised as providing some thermal and acoustic insulation to the building as well as preventing access by intruders. In addition, it contributes to the structural bracing of the building. In this chapter you will learn how to identify various types of cladding, and their installation and fixing requirements. These include sheet material and weatherboarding made of timber, plastic, metal and fibre cement. You will also learn how to calculate the amount of cladding required for a given project.

When commencing cladding work it is important to set up the work area appropriately. Ensure it is safe, that leads are off the ground, there are areas or receptacles for waste and recyclable material, that scaffolds or ladders are well-grounded and the work area is clear of obstacles, etc.

It is good practice to first become familiar with the product that will be used. Read the manufacturer's installation instructions, the SDS (safety data sheet) for the product, taking note of any hazards, and check if the product is recyclable. It is important to familiarise yourself with National Construction Code (NCC) requirements for cladding as any installation needs to be compliant.

Once the cladding has been installed, ensure the worksite has been cleared, that any waste and recyclables have been disposed of appropriately, and that leftover materials, tools and equipment are stored away neatly. If any tools are faulty, this needs to be reported to the supervisor.

CHECK YOUR UNDERSTANDING

1. List the four principle functions of wall cladding.

WORKPLACE SCENARIO

When building in bushfire-prone areas, a dwelling's cladding is an important factor in protecting the house from any fires. All areas that are prone to bushfires now have a BAL (Bushfire Attack Level) rating.

The aim of the residential building standard for bushfire protection is to improve the ability of a building to withstand a bushfire attack. This will provide greater protection for the occupants who may be sheltering inside while the fire front passes. A great deal of scientific modelling has gone into the standard.

The chart in the link below outlines how the baseline data, defined as a BAL, determines the type of construction required. The BAL takes into consideration a number of factors, including the Fire Danger Index, the slope of the land, the types of surrounding vegetation and the proximity to any other building (see www.vba.vic.gov.au/consumers/bushfires).

As a carpenter, it is very important when installing cladding that you follow the manufacturer's instructions and the Australian Standard so that the cladding installation is in line with the provisions for bushfire-prone areas. If these provisions are not adhered to, then the effectiveness of the cladding to protect the occupants and the house from bushfire could be severely compromised.

16.1 Types of cladding

Several materials are commonly used as external wall coverings on timber/metal-framed buildings. Known as **timber siding**, weatherboards or chamferboards, external wall sidings are milled from a number of timber species and to a variety of profiles. Some of the popular profiles are shown in Figure 16.1.

1. **Dressed rusticated siding**–is normally milled from cypress pine, hardwood and pressure-treated radiata pine.
2. **Double log siding**–is milled from cypress pine and pressure-treated radiata pine.
3. **Checked and nosed**–or checked and splayed siding is milled from Western red cedar or Baltic pine.
4. **Sawn splayed siding**–is milled from Western red cedar, Californian redwood or in some areas may be produced as rough splayed from sawn hardwood. **Splayed timber weatherboards** are also milled from Baltic pine and these can be ordered with a dressed face finish. They are available primed with a nominal width of 175 mm in either a square edge or round edge profile.

The nominal sizes and **cover** of the boards also are indicated in Figure 16.1. To calculate the number of linear metres of boards required to cover one square metre, divide 1000 by the cover in millimetres. For example:

$$\begin{aligned} \text{rusticated–nominal width} &= 150\text{, cover } 122\text{ mm} \\ \text{linear metres/m}^2 &= 1000/122 \\ &= 8.2\text{ m} \\ \text{sawn splayed–nominal width} &= 200\text{ mm, cover } 171\text{ mm} \\ \text{linear metres/m}^2 &= 1000/171 \\ &= 5.9\text{ m} \end{aligned}$$

Add an allowance for cutting waste to obtain the total ordering requirements.

Boards are normally finished at external and internal corners by being butt-jointed to corner fillets (weatherboard stops), as shown in Figure 16.2. External corners may sometimes be mitred. Boards should be fixed in long lengths, with the end joints staggered and made over studs. Boards with rebated joints and over 100 mm in width are double nailed at each bearing point.

To apply rebated weatherboards, first fix the internal and external corner stops. Take the first or starter board and cut it to length, taking care not to fit the boards too tightly as this can force the stops out of

Fig. 16.1 Common weatherboard profiles: (a) rusticated nominal 150 × 25; (b) double log cabin 150 × 25; (c) checked and nosed, checked and chamfered; (d) sawn splayed nominal width 200

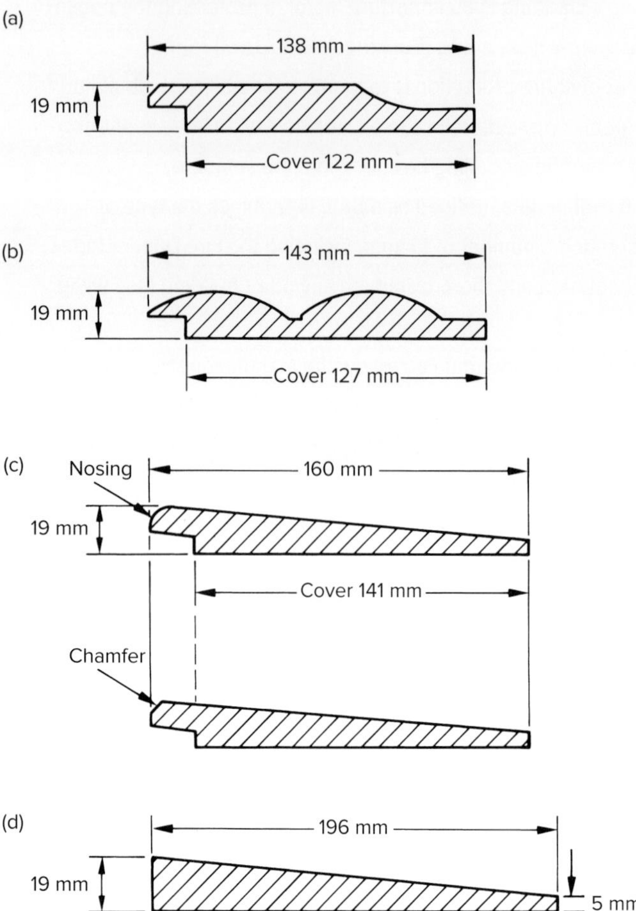

place and open other joints. Tack it lightly into position in line with the bottom edge of the bearer. Permanently nail the two ends and set up a string line to the top edge of the board. Adjust the edge of the board to the string line and permanently nail it into position (Fig. 16.3). Continue fixing subsequent boards, pulling them tightly down on to the rebate and nailing through the face. Sight along the boards occasionally to maintain a straight line.

Splayed weatherboards must be fixed only with hot-dipped galvanised or other non-corrosive nails. First set out the internal and external weatherboard stops, allowing for a minimum lap of at least 25 mm for Baltic pine and 30 mm for hardwood. The stops are marked out evenly to suit the spacing (cover width) of the weatherboards. They are designed so that the boards will fit evenly spaced between the base plinth line and the eaveline (Fig. 16.4).

It is important that stud walls are straightened prior to the installation of any cladding and building paper/vapour barrier. The straightening process is twofold. First, after the walls have been erected, the corner studs and wall junctions are straightened, fixed and plumbed. Then, the top and bottom plates are straightened and plumbed, with the bottom plate being nailed off and the top plate being temporarily braced. This process is covered in Chapter 12 (on wall framing).

Fig. 16.2 Corner fillets for siding

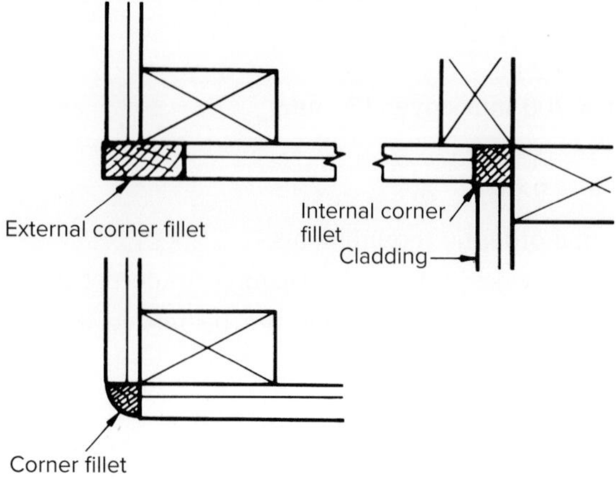

Fig. 16.3 Starting to fix weatherboards

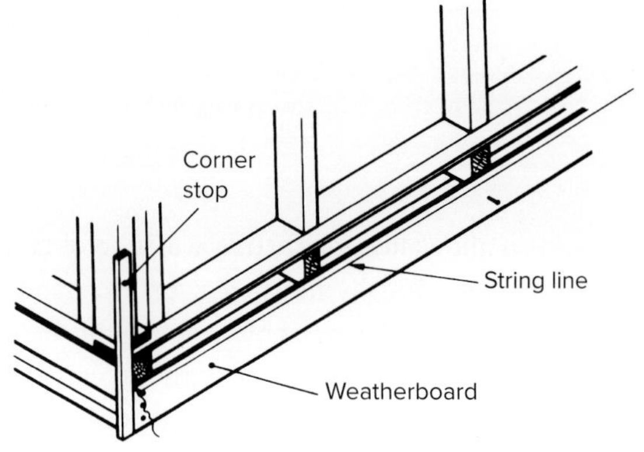

Fig. 16.4 Setting out weatherboard stops

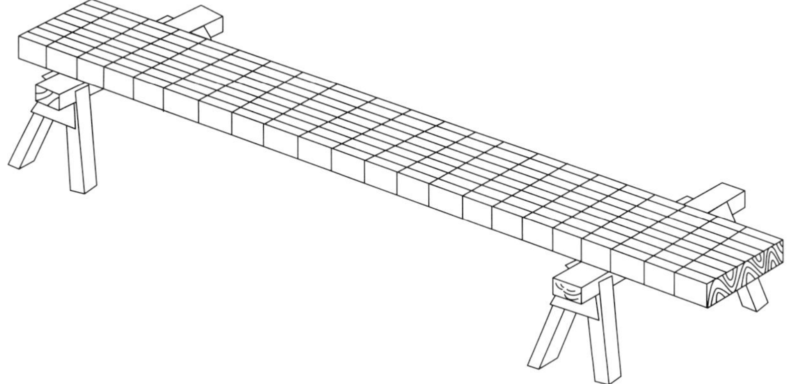

Straightening prior to cladding installation is similar to the process of straightening floor joists before installing floorboards. Using a long straight edge and holding it across the timber studs, identify any high spots. Those high spots on the studs can be planed using an electric plane, ensuring the planed area is feathered with every stroke. Continue this process all over the wall, so as to achieve a consistent flat surface. If any low spots are found, pack them using masonite and again feather out the packing away from the low point on that stud. Once the wall is straightened, a vapour barrier needs to be installed; this will aid with the insulation and weatherproofing of the wall, and help stop moisture from getting into the stud wall.

The stops are now nailed in position at the corner junctions resting on the plinth board. The starter board is packed along the bottom edge to the same thickness as the tapered edge of the boards and is nailed permanently to a straight line. With the first board in place, insert nails into the next lower marking on the weatherboard stops and stretch a line through from end to end. Sight and check the line for any sag and if necessary lift the line in the centre. Nails can now be inserted to the line on every third or fourth stud and act as a rest for the next board while it is being fitted. Boards are scribed and fitted to the corner stops. Joints in the length of a wall are cut square and centred over the studs. Adjacent boards must not be joined on the same stud; all joints must be staggered.

Nailing is done through the face so that the nails will just clear the tip of the board underneath. Continue fixing subsequent boards, checking that joints are neat and not over tight (Fig. 16.5).

Weatherboards, including any with rebates and their ends, should be sealed with primer or some other means, preferably before they are fixed.

TIP Never at any time nail through both boards. Always place the nail so that the board below is narrowly missed.

CHECK YOUR UNDERSTANDING

1. Illustrate with a sketch at least four profiles used in timber siding.
2. Splayed weatherboards with lapped joints must have a minimum lap of how many millimetres?
3. Sketch a neat sectional view of a stud wall with splayed Baltic pine weatherboards attached. Show the position of the plinth board and the bottom three weatherboards nailed correctly in position.
4. What is the recommended fastener for fixing splayed weatherboards to a timber frame?
5. Before fixing weatherboards or chamfer boards to a stud wall, what treatment should be applied to the lap and end joints of those boards?

Fig. 16.5 Fixing splayed weatherboards: (a) to a frame on a timber subfloor system; and (b) to a timber wall frame

(a)

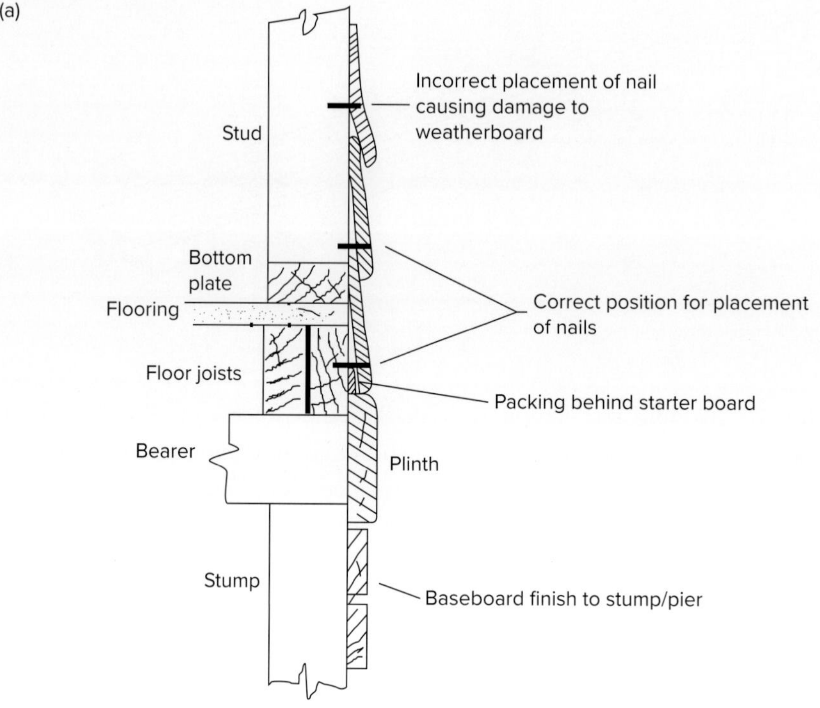

(b)

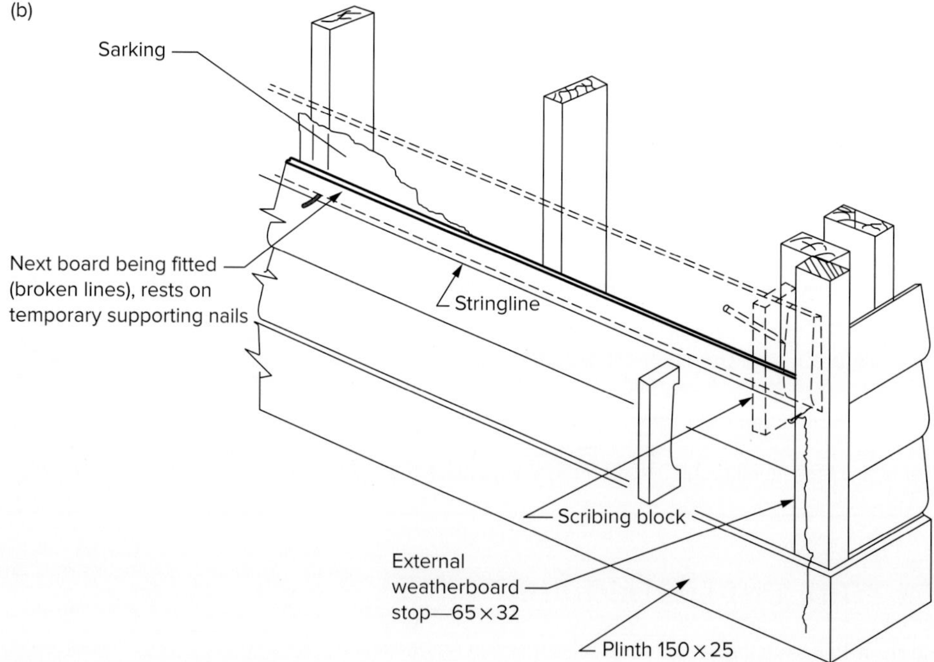

16.1.1 Fibre cement products

Fibre cement products are manufactured from Portland cement, ground sand, cellulose **fibre** and water. They are widely used as external sheeting to timber-framed structures. They will not rot and are fire resistant, comparatively easy to maintain, economical and, if correctly applied, have an almost indefinite life.

16.1.2 Fibre cement flexible sheet

Single-sided building sheet is manufactured in two thicknesses: 4.5 mm and 6.0 mm thick. The sheet width is 900 mm and 1200 mm, and lengths range from 1200 mm to 3600 mm in 300 mm intervals. Like all fibre cement products, all sizes may not be available in all states, so check with local suppliers. The methods of cutting flexible sheet include the following.

16.1.2.1 'Score and snap'

Support the sheet, face side up, on stools and position a straight edge along the line of the cut. Using a tungsten-tipped score and snap knife, score the sheet along the straight edge to about a third of its thickness and snap upwards to achieve a break. Rough edges can be trimmed up with a rasp if necessary.

16.1.2.2 Hand guillotine

This method will produce a clean straight edge. The cut is made on the waste side of the line to allow for the thickness of the blade.

16.1.2.3 Hand sawing

Hand sawing is suitable for making small cuts, but preferably use an old saw.

16.1.2.4 Hole forming

Small rectangular and circular holes can be formed by drilling a series of small holes around the perimeter of the hole and tapping out the waste piece from the face of the sheet. Ensure that the sheet edges are properly supported and trim the hole to shape with a rasp if necessary.

To fix flexible sheets to a frame, use galvanised fibre cement nails of 25 × 2.0 diameter into hardwood, and 30 × 2.0 diameter into softwood. Nails are spaced 200 mm apart and no closer than 10 mm to the edge of sheets and 300 mm apart to the body of the sheets. No nail should be closer than 50 mm to the corners (Fig. 16.10).

Fibre cement flat cover moulds and a range of PVC jointers and angle cover moulds are available for joining flat flexible sheets.

Fig. 16.6 Fibre cement—score and snap

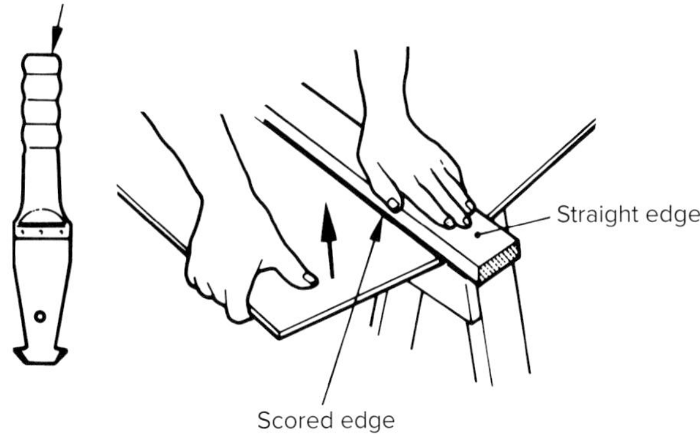

Fig. 16.7 Fibre cement—hand guillotine

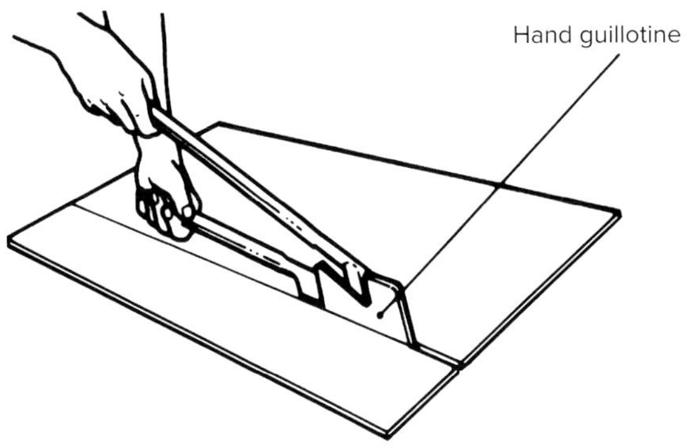

Fig. 16.8 Fibre cement—scoring and hand sawing

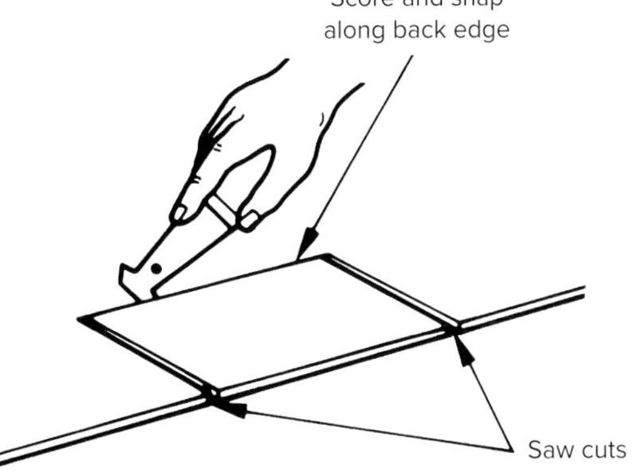

Fig. 16.9 Fibre cement—hole forming

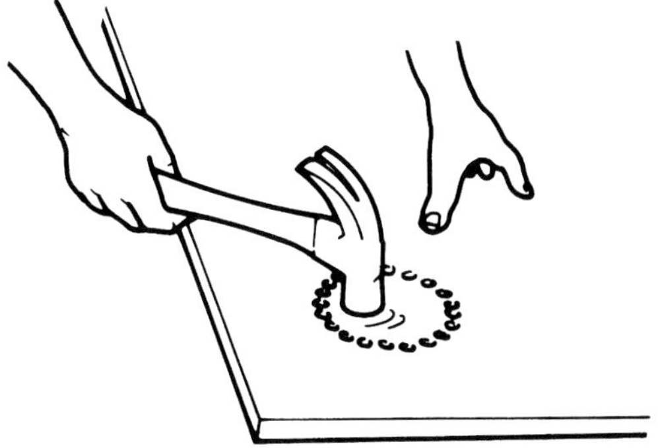

Fig. 16.10 Nailing fibre cement sheets

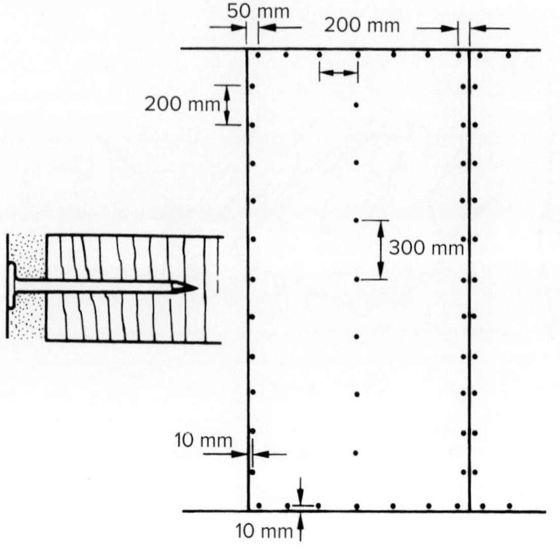

Fig. 16.11 Vertical cover strips and mouldings: (a) vee joint; (b) PVC jointer; (c) cover mould; and (d) PVC accessories

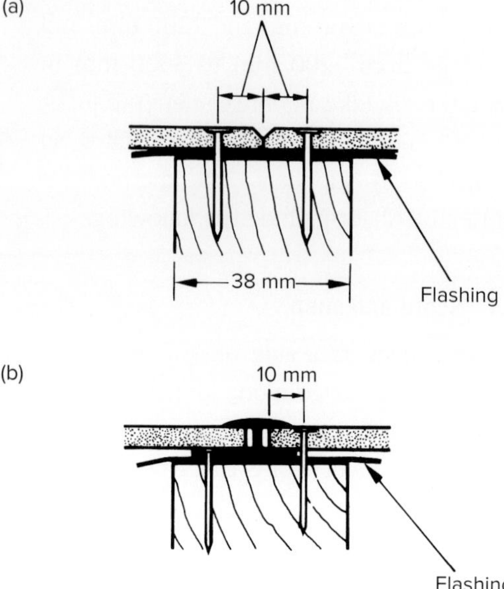

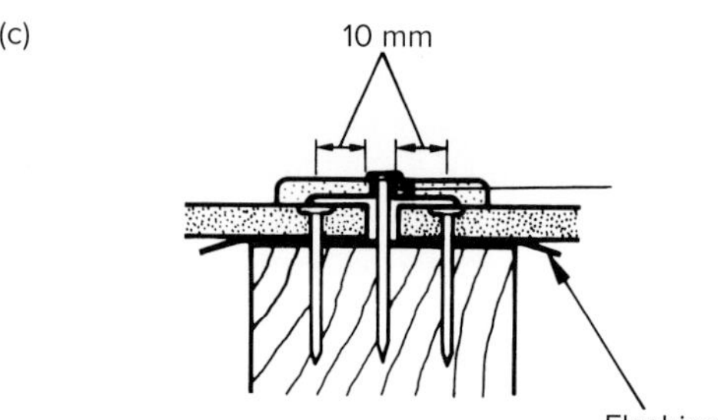

16.1.3 Vertical joints

Wall 'vee' joints (Fig. 16.11) are formed by using a coarse file to form a small chamfer to the edges of the adjoining sheets. Nail both edges at 200 mm apart. Include a strip of aluminium foil flashing between the sheeting and the timber frame.

16.1.3.1 PVC jointer

Nail through the extended leg of the jointer to the framing at 150 mm centres.

16.1.3.2 Fibre cement corner mould

Nail sheets to framing before the cover mould is fixed. Nail clearance holes should be predrilled in the cover mould.

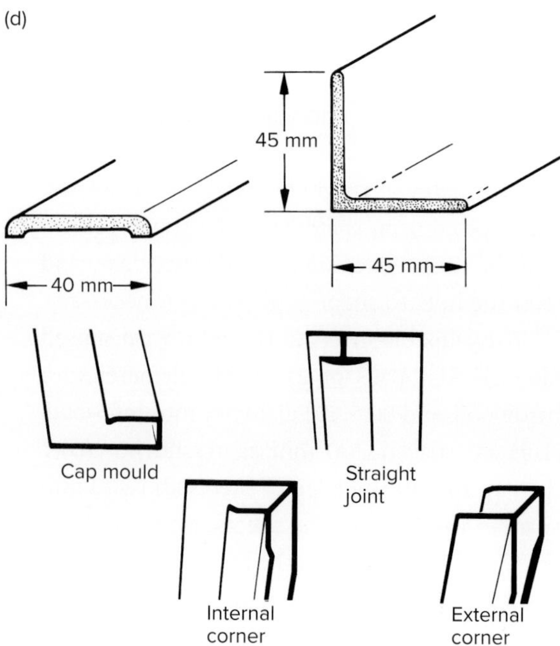

16.1.4 Horizontal joints

When forming horizontal joints, it is essential to include a metal or PVC weather strip. First fix the lower sheet, then apply the weather strip followed by the upper sheet, nailed as recommended (Fig. 16.12).

16.1.5 Fibre cement plank siding

Fibre cement planks are a single-sided building plank, the face of which can be smooth or textured in a variety of wood grains or other patterns. They are used mainly as cladding or part cladding for domestic buildings, particularly in upper floor additions, or as recladding when rejuvenating tired dwellings.

Planks are 7.5 mm thick, and the width can be 170, 180, 230, 300 or 310 mm. Lengths are 3600 mm or 4200 mm. Note again that all sizes may not be available in all states, so check with local suppliers.

Fig. 16.12 Horizontal cover strip

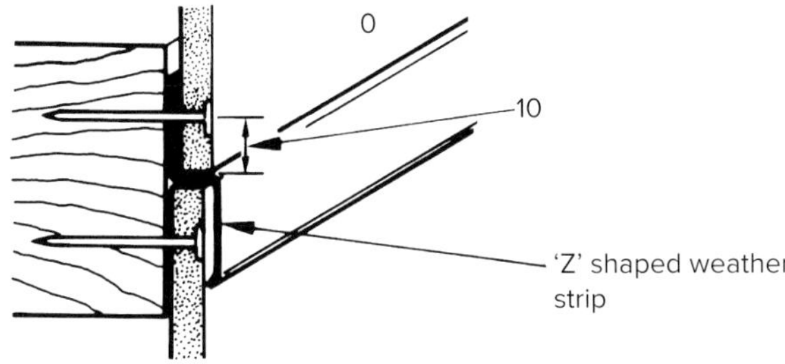

Accessories consist of PVC joiners and preformed metal internal and external corners. Planks are fixed to studs by nailing through both thicknesses of planks with 40 × 2.8 fibre cement nails (Fig. 16.13 and 16.14).

TIP Always refer to the manufacturer's technical notes for the correct fixing details.

16.1.5.1 Fixing procedure

Fix vertical flashings to all external and internal corners. Also fit any sill or head flashings to openings as required. Nail a continuous strip of 40 mm fibre cement cover mould around the perimeter of the building, level with the top of the ant capping.

Locate the height of the top edge of the first plank and level this line around the frame of the building to establish a datum for fixing the first course. Drive a series of nails along this line to act as a guide for fixing the starter plank. Start with the end of the first plank flush with an external corner and against the guide nails, and fix to the framing. Continue fixing the planks around the perimeter of the building, jointing the ends with PVC joiners as necessary and removing the guide nails. Insert preformed metal corners and fix through the hole at the top.

Calculate the overlap of the planks, the minimum overlap being 25 mm. This can be varied to work even courses of planks to the height of the wall. Fabricate two lap gauges. Commence the second course with an offcut so that the vertical joints will be staggered, and use the gauge to maintain the correct lap. Fix subsequent courses by inserting the PVC joiners and metal corners as necessary.

Fig. 16.13 Fibre cement plank accessories

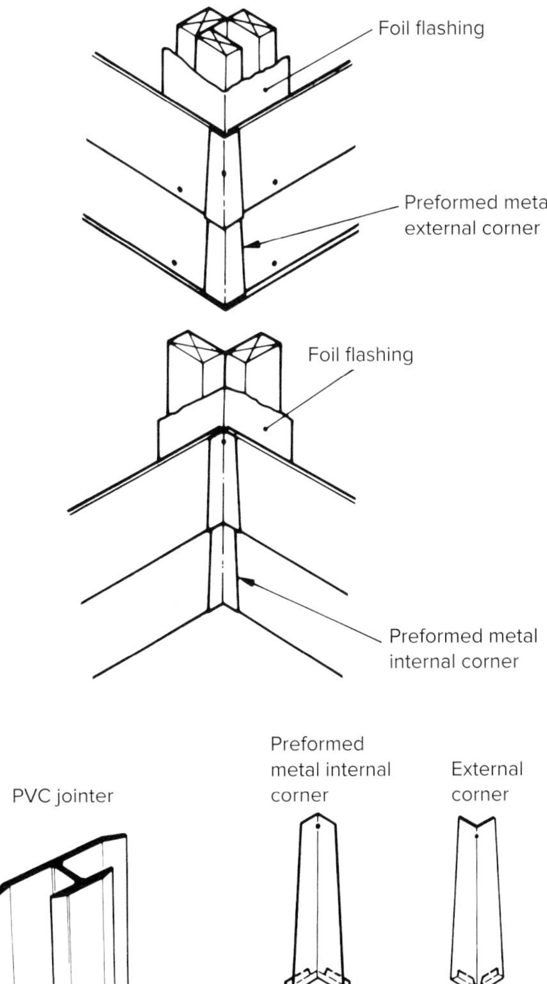

Fig. 16.14 Plank nailing

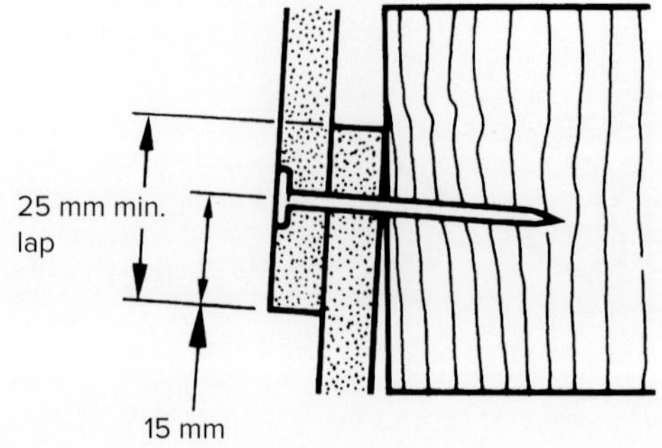

Fig. 16.15 Fixing fibre cement planks

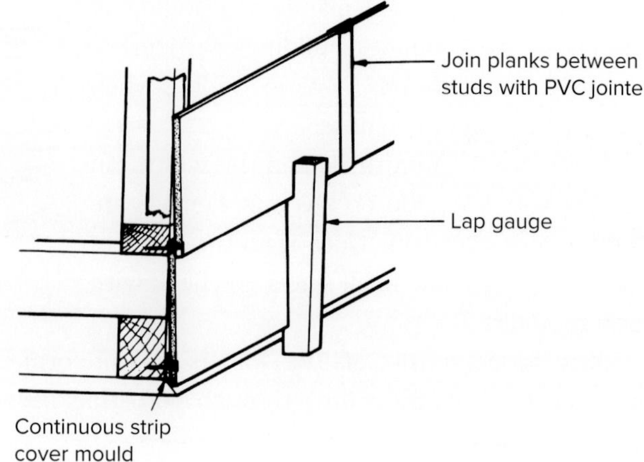

Complete fixing details are available from the manufacturer's information sheets and should be obtained and closely followed when carrying out this work.

TIP An alternative method to using a lap gauge is to position subsequent courses using a storey rod. A storey rod is set out to the required spacing of the boards using a similar procedure that is used for setting out weatherboard stops. Refer to details previously covered in this chapter.

CHECK YOUR UNDERSTANDING

1. Fibre cement flexible sheeting is available in two widths for wall lining. What are those widths?
2. With the aid of a neat sketch, show the nailing requirements for fixing a fibre cement sheet to a timber frame.
3. Describe three effective ways to cut fibre cement flexible sheets.

16.1.6 Hardboard siding

Hardboard is a manufactured timber product, commonly found used as a siding or more often as cladding to upper-storey extensions. The fixing is essentially similar to the fixing of fibre cement products. Again, refer to and check the manufacturer's information sheets for fixing details.

16.1.7 Aluminium siding

Aluminium siding has been available for many years but essentially it is fixed only by licensed applicators. The material is formed from precoloured strip aluminium, either in precut lengths, or as a continuous strip rolled and cut to length on-site. It is installed with insulation backing, usually a foamed plastic. The siding is available in both horizontal and vertical profiles.

16.1.8 Vinyl siding

As with aluminium siding, vinyl siding is generally available only through applicators. The vinyls are available in both thin and thicker grades.

16.1.8.1 Thin vinyls

These are mainly used in the recladding market, and are applied over existing wall claddings. They are less than 2 mm in thickness.

16.1.8.2 Thick vinyls

These sidings are approximately 10 mm thick, and have a board width of around 270 mm with a cover of 250 mm. They are tough faced with a cellular core. The boards are made of PVC powder, with ultraviolet ray inhibitors and fillers added, which is then heated and extruded through a die. The continuous lengths are cut to standard lengths of 5 m, with the join being treated with a preformed joiner strip. The boards interlock when fixed, and can be fixed directly to timber framework.

16.1.9 Plywood cladding

Plywood cladding is a great alternative to other cladding types. It comes in large sheets and in a number styles, including:

- grooved
- textured surface
- flat surface.

Generally, plywood cladding comes in a standard sheet size of 2400 mm × 1200 mm × 12 mm. It is impregnated with a chemical to assist with weatherproofing and inhibit deterioration over time from the elements. (Always refer to the material safety data sheets for the chosen product to establish which chemical has been used in the manufacturing.) Many of the products can be left raw, although it is possible to paint them. Plywood cladding comes with a pre-machined shiplap joint, as can be seen in Figure 16.16.

This shiplap profile weatherproofs the joint and also allows for any movement that can occur in the sheet over time. The installation and fixing detail for plywood cladding is important to keep the waterproof integrity of the product and to accommodate any movement. Figure 16.17 gives an example of the installation and fixing requirements of plywood cladding; however, the manufacturer's specifications for the chosen product should always be consulted prior to use.

Fig. 16.16 Shiplap joint for plywood cladding

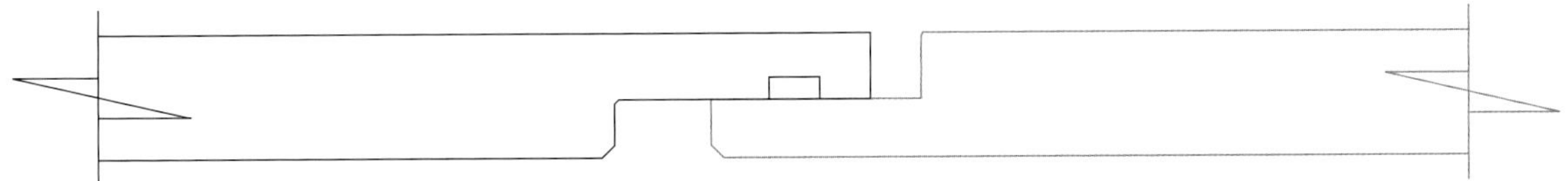

Fig. 16.17 Fixing details for shiplap cladding

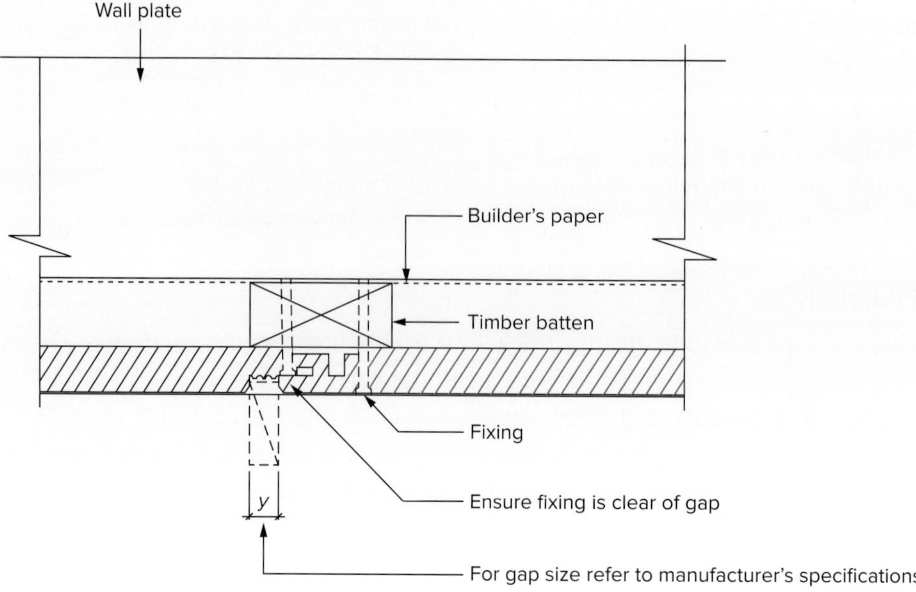

CHECK YOUR UNDERSTANDING

1. Name two styles of plywood cladding.
2. What type of joint does plywood cladding use?

Student research

Investigate what other types of cladding are available.

End of chapter activities

Activity 1:

Identify what type of non-masonry cladding is used on the dwelling shown in the set of plans located in the appendix.

Activity 2:

In how many locations is the non-masonry cladding used on the dwelling in the appendix.

AUSTRALIAN BUILDING CODE AND STANDARDS

National Construction Code (NCC)

AS 1684
Residential Timber-Framed Construction

AS 2904
Damp-Proof Courses and Flashings

Tradie talk—**'Work the cover' for a full board finish**

Chapter 17

Windows and doors

Learning Objectives

LO 17.1 Identify window types

LO 17.2 Know about window manufacture

LO 17.3 Set out rods

LO 17.4 Assemble and glue up the sash

LO 17.5 Know about vertical sliding sash construction and use

LO 17.6 Fit window frames and flashing

LO 17.7 Install glazing

LO 17.8 Identify door construction techniques

Introduction

In this chapter you will learn how to identify different types of doors and windows. You will learn how to construct timber doors and windows and learn about the installation of windows–techniques that require a tradesperson to have a high level of skill. These items can become beautiful adornments to the dwelling as well as serving a practical purpose. Many of the techniques identified in Chapter 7 are used in the construction of these items.

17.1 Window types

Windows are an essential component in residential buildings, their purpose being to admit light and ventilation. Building regulations specify the minimum requirements. Windows are composed of two parts:

1. the window *frame*, which is permanently fixed into the structure of the building
2. the *sashes*, which are the movable parts of the window (in some cases fixed) and are fitted to the frame.

TIP Windows are named according to the way in which the sashes operate within the frame.

17.1.1 Fixed sashes

Fixed sashes are permanently fixed in the frame and admit light only (Fig. 17.1a). All of the sashes to a room cannot be fixed; some must be operable to meet building regulation requirements.

Fig. 17.1 (a) Fixed sashes; and (b) casement sashes

(a)

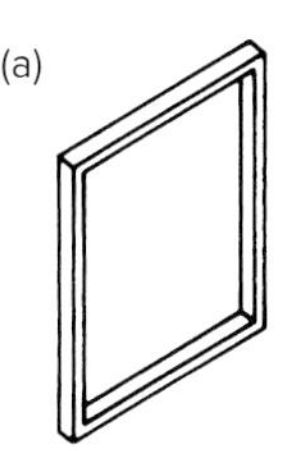

(b)

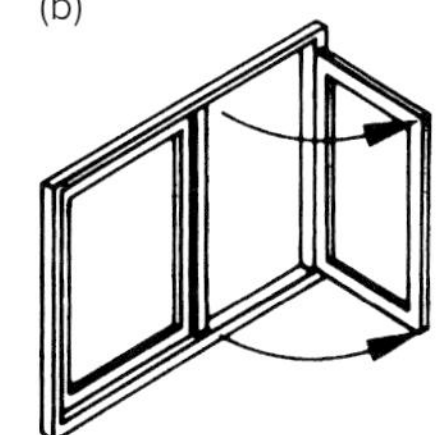

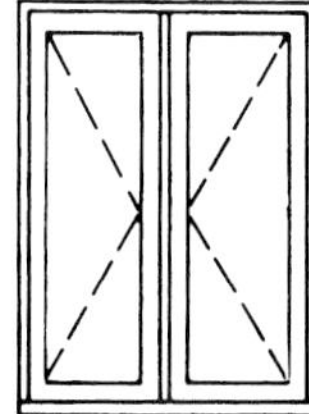

17.1.2 Casement sashes

Casement sashes are either hinged on the side or have a friction stay (Fig. 17.1b). The friction stay allows the sash to pivot near the side and open to its full extent, or lock by friction to any intermediate position (Fig. 17.2).

The advantages of casement sashes are that they will provide maximum ventilation and can be cleaned readily from the inside. On the other hand, ventilation is not so easy to control, insect screens are more difficult to fit, and the sash projecting beyond the face of the wall can sometimes be a dangerous obstruction.

Fig. 17.2 Friction stay

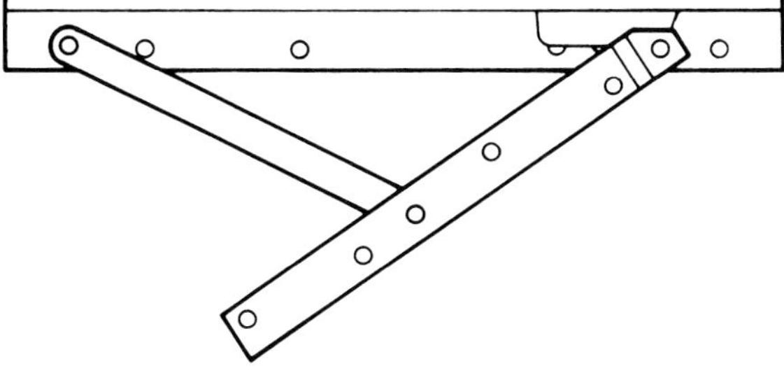

17.1.3 Awning-type sashes

Awning sashes pivot from near the top and can be a single sash or a number of sashes arranged above each other (Fig. 17.3). They are sometimes wrongly named *hopper windows*.

Sashes can be fixed on friction stays; this allows them to open and be held in any position. The manufacturer's instructions should be consulted when selecting a stay so that one long enough to suit the overall height and weight of the sash is chosen.

Single-awning sashes are a popular type of window. They are often fixed with a non-friction-type hinge; an extending chain winder on the sill opens the sash and holds it in position.

Ventilation can be controlled, avoiding direct draughts, and insect screens can be conveniently fitted inside.

Fig. 17.3 Awning sash

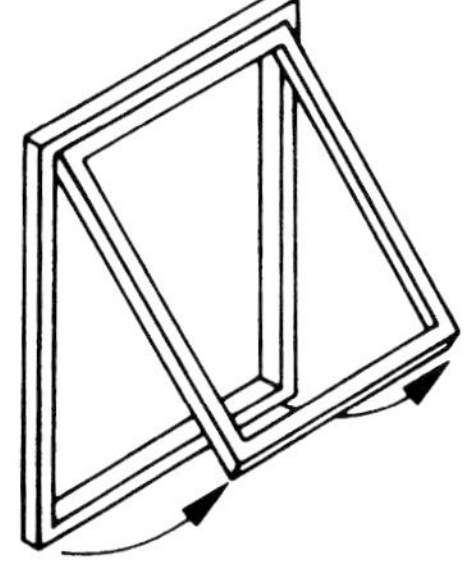

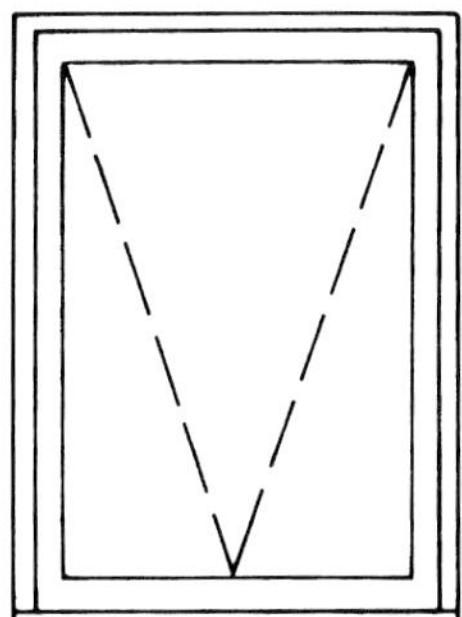

17.1.4 Horizontal sliding sashes

Sliding windows are widely used and an extensive range of stock sizes is available in aluminium windows. The sash slides on metal or plastic rollers along tracks fixed to the sill. The sashes should have a wheel base as wide as possible, and preferably the height of the sash should be no more than two times its width. The sash pull should be set close to the base, where it will not tend to tilt the sash sideways in operation. On a single track, no more than half the window can be opened, and ventilation is not readily controlled. However, as the sash is contained wholly within the frame, insect screens can be conveniently fitted to the outside, and the windows can be quite economical.

Fig. 17.4 Sash winders

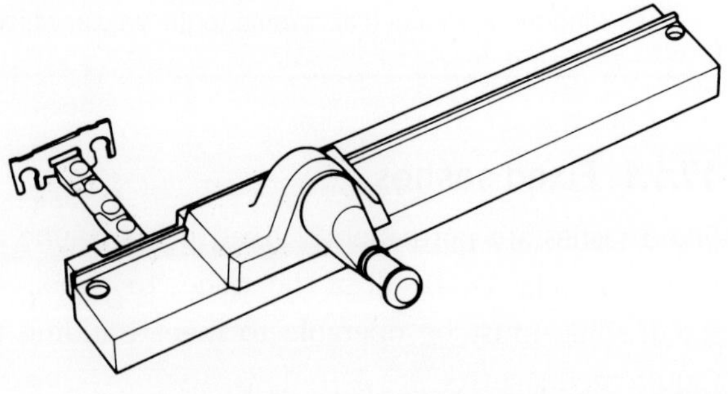

17.1.5 Vertical sliding sashes

Vertical sliding sashes are also sometimes referred to as *box-frame sashes,* or commonly as *double-hung sashes.*

Double-hung windows have long been a very popular type of window; they were originally developed from the box-frame window with sashes hung on cords and counterbalanced by cast-iron weights (Fig. 17.6). Today the timber or aluminium sashes are hung on spiral spring balances.

Fig. 17.5 Horizontal sliding sash

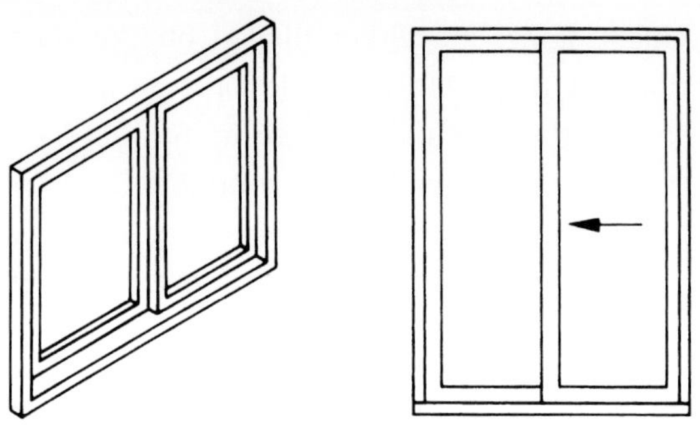

Double-hung sashes have a number of advantages. Ventilation is readily controlled and can be at the top, bottom or both. The sashes are contained within the frame and insect screens can be fitted to the outside. However, only half the window area can be opened at any one time, and poorly fitted sashes can present some problems such as jamming or rattling.

Some double-hung sashes are available that are 'sash-less'; that is, the window glass is held in a counterbalanced suspension system, without being fitted into a timber or metal frame.

Fig. 17.6 Double-hung window

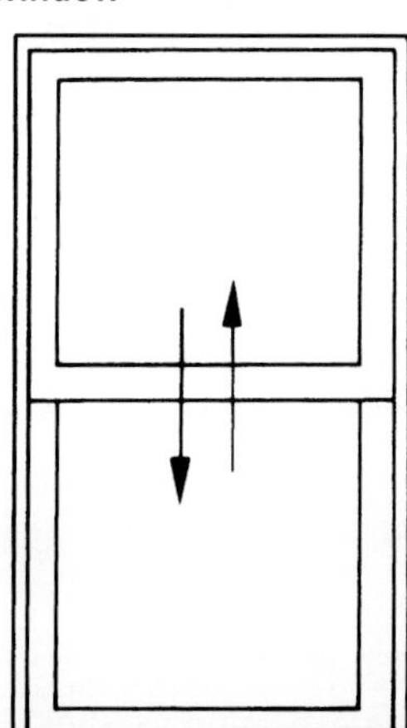

17.1.6 Other window types

Other sashes still seen occasionally, but practically 'extinct' in new construction, are the *hopper-type sash* and the *pivot-hung sash* window (Fig. 17.7). More recently, European-type window systems have started to be seen in Australia. These windows systems are generally made from uPVC and are double glazed. The windows have a high thermal efficiency as well as sound insulation properties. The system can be adapted so that various types of window can be produced from the stock materials, including casement, awning, sliding, tilt and turn, and tilt and slide.

17.2 Window manufacture

Window frames built into the external walls of a building are subject to some of the harshest weather conditions. They must be made from durable materials and constructed in such a way as to resist the penetration of dampness around the sashes. Timber species used include clear Oregon, western red cedar, Tasmanian oak, meranti (Pacific maple) or pressure-treated pine. For sills, which are the most exposed parts, tallowwood, redgum and merbau are often used, and no doubt in some areas other local timbers serve equally well. Window frames vary in construction according to the type of sashes to be fitted.

Fig. 17.7 Pivot-hung sash window

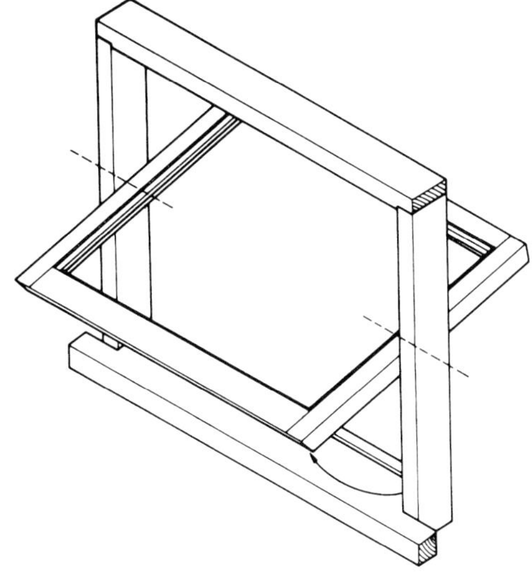

17.2.1 Solid-rebated window frame

A solid rebate is used with casement and awning sashes. Fixed sashes can be fitted; however, it is more usual to set a glass panel directly into the rebate of the frame and secure it with timber fillets or beads.

The basic construction of the solid frame is shown in Figure 17.8. The timber sections are an example of standard metric profiles for milled products available from timber merchants. Manufacturers of mass-produced frames will mill sections to their own details, which also meet the demands of durability and weatherproofing.

Fig. 17.8 Elements of a solid window frame

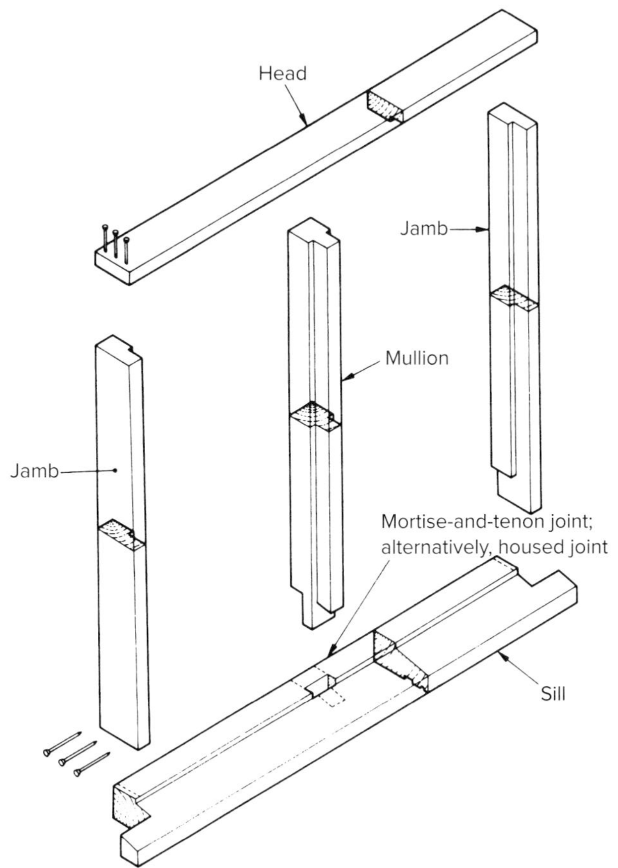

Parts of the solid frame include the *head* and the *stiles*, which are of the same section as single-rebated material. A vertical dividing member between sections is called a *mullion* and may be double-rebated. The bottom member of the frame is the *sill*; this is of special interest as it must resist most of the weather.

Figure 17.9 shows the detail of the sill section, which is described as *sunk, throated* and *weathered*. These are all features that enable the sill to fulfil its function. *Sinking* is the drop in level behind the sash. *Weathering* is the slope on the sill, which varies from 1 in 6 to 1 in 8. *Throating* is the small groove formed at the corner of the sinking; it may not be included in the standard section, but it will increase the weatherproofing of the sill by breaking the capillary action of water tending to seep under the sash, especially when assisted by a stiff breeze.

On the bottom of the sill is a *drip groove* for the purpose of preventing droplets of water forming on the edge of the sill and running underneath. The *milling groove* helps to prevent distortion of the sill by breaking up the wide underside into narrower widths and reducing the overall amount of shrinkage.

Fig. 17.9 Detail of a windowsill

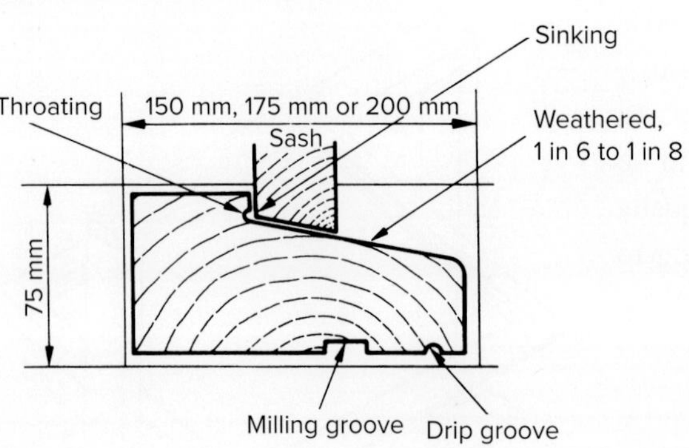

The stiles and mullion are housed into the head of the frame (Fig. 17.8). Note the mortise-and-tenon joint between the mullion and sill. The mortise is made 25 mm wide and lines up on one side with the sinking in the sill. The shoulders on the mullion are bevelled where necessary to butt down tightly onto the sill. The tenon is wedged from underneath the sill. Wedges are driven uniformly to avoid forcing the frame out of square. Alternatively, a housed joint may be used.

Frames are assembled after painting the joints with a timber primer or other equivalent product. They are nailed together with galvanised flat head nails, checked for square by making the diagonals equal, and a temporary brace is fixed to the inside face.

WORKPLACE SCENARIO

When installing windows, it is always good practice to pack the sill of the window. It must be packed level, so start by placing the same size of packer, e.g. 3 mm, on both sides of the opening and then place the spirit level on top of the packers to ensure it's level. If not, increase or decrease the packing on one side until level. (Make sure there is enough room in the opening such that when the window is installed, there is still a gap above the top of the window.)

When the sill of the window is packed, it must be done directly under the stiles to ensure no pressure is put onto the glass or the operating section of the window.

Fig. 17.10 External trim to a solid window frame: (a) timber frame construction; (b) brick veneer; and (c) cavity brick construction

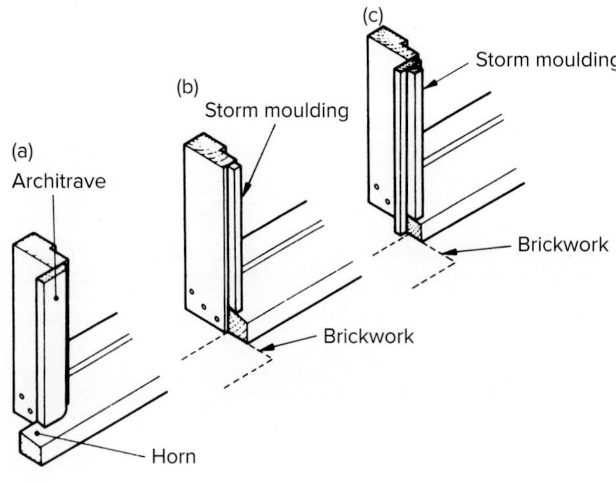

17.2.2 Finishing window frames

The outside is finished with an architrave when used in a timber wall or with storm mouldings in brick-veneer construction, as shown in Figure 17.10.

17.2.3 Sashes

Sashes are made from comparatively slender sections; the names of the members are shown in Figure 17.11.

The procedure for joining the stiles and the top and bottom rails is consistent with the procedure for other framed joinery items. Members dividing the glazing into smaller areas are called *glazing bars* and are either vertical or horizontal; horizontal bars are referred to as *lay bars*.

Each separate pane of glass is a *light*, and a sash can be described by the number of lights and the arrangement of the glazing bars. Examples are given in Figure 17.12.

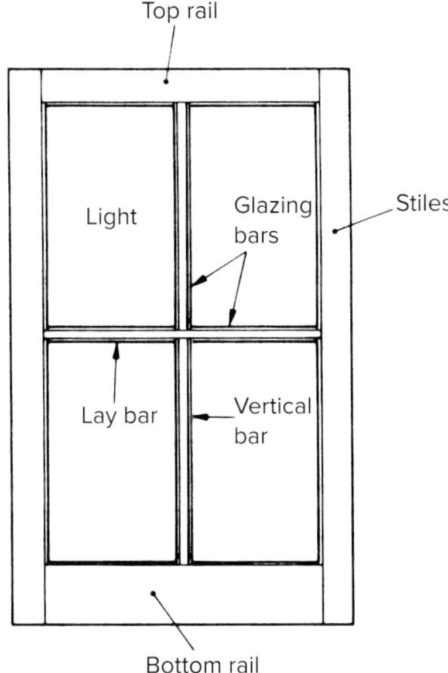

Fig. 17.11 Parts of a sash

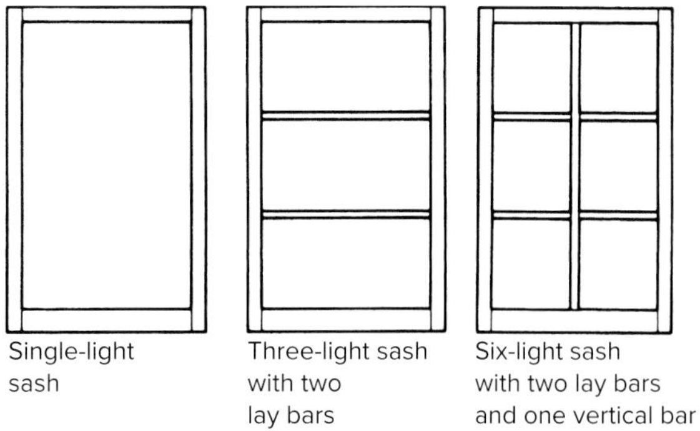

Fig. 17.12 Arrangement of lights and glazing (lay) bars

Sash stock is shaped in a particular manner and the thickness is divided into three approximately equal parts.

The rebate is necessary to hold the glass panels, which are secured in place with putty or glazing beads; it is on the external face of the sash.

The actual amount of daylight seen through the sash is measured to the *daylight line*, and the glass size is measured to the *glass line*, allowing a clearance for expansion and contraction. The difference between these two dimensions is the depth of the rebate. The centre third is the ribbon, on which are located the mortise-and-tenon joints; it is approximately 10 mm wide. The mould on the interior third softens the harsh bulky edge of the frame and can be an *ovolo* mould or, commonly, a *splayed* mould (Fig. 17.13).

Fig. 17.13 Sections of sash stock: (a) ovolo moulding; (b) splayed moulding; and (c) sash bar

(a)

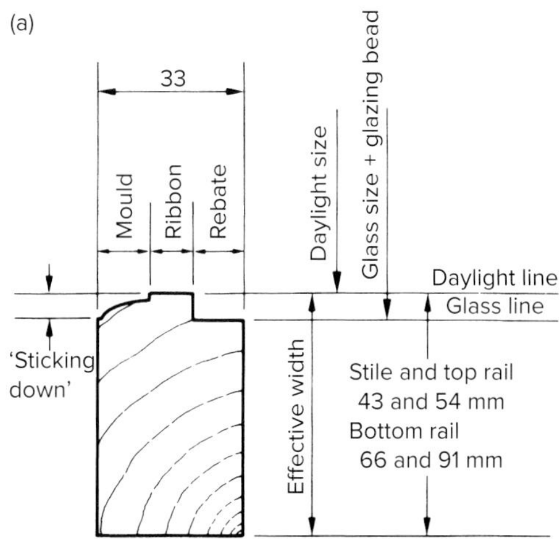

(b)

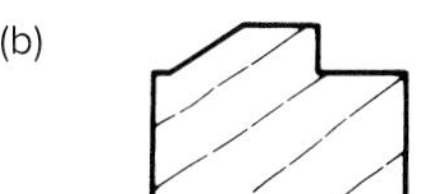

(c)

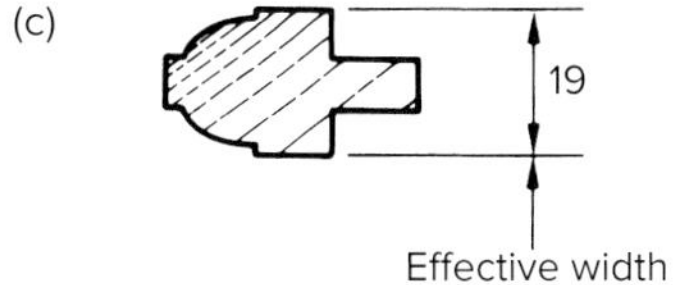

The mould 'sticks down' on the face the same distance as the rebate, making a straight shoulder line on the rails. Note that the effective width is the full-face width of the members, giving the maximum amount of material on which to proportion the mortise-and-tenon joints. Dimensions of members, including sash bars, are given as examples of stock material.

17.2.4 Sash joint

Stiles and rails are joined with mortise-and-tenon joints, modified to the situation and referred to as *sash joints* (Fig. 17.14).

Fig. 17.14 Sash joint

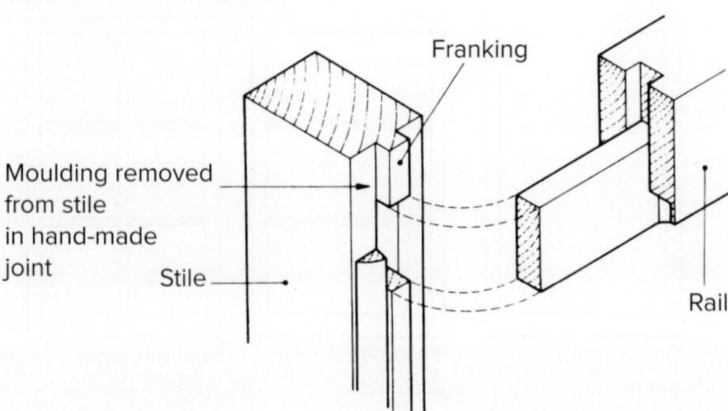

The width of tenon is again two-thirds the effective width of the rail minus one wedge allowance; the thickness is 10 mm. Due to the slenderness of the stile, it is not haunched, but a projection is left on the stile that engages into a corresponding recess in the rail. This type of joint is called a *franked joint* and the actual *franking* is the projection on the stile.

The shoulder on the rail is scribed over the mould on the stile. In a machine-made joint, the scribing is across the full width of the rail (Fig. 17.15). However, if this part of the job is done by hand, it is simpler to remove some of the mould from the stile and scribe only sufficient of the shoulder to make the joint between the moulds.

Where sash bars meet at right angles, there will be a *through bar* and a *cut bar*. The through bar is made in the direction where it will tend to strengthen the sash most effectively. The bars can be stub-tenoned into the stiles and rails. The joint between the bars is shown in Figure 17.16.

The procedure for setting out, using set-out rods, and for making a casement frame and sash, including the glazing bars, will now be described. It is well realised that the bars are not always included in new sashes; so, if single-light sashes are being made up, ignore any reference to glazing bars.

Fig. 17.15 Machine-made sash joint

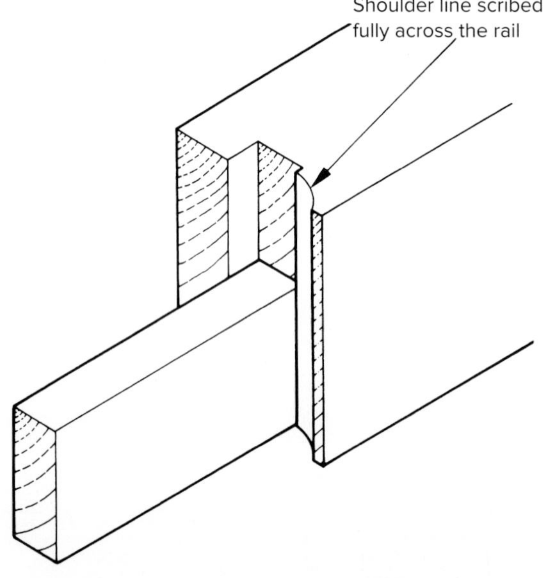

Fig. 17.16 Through and cut glazing bars

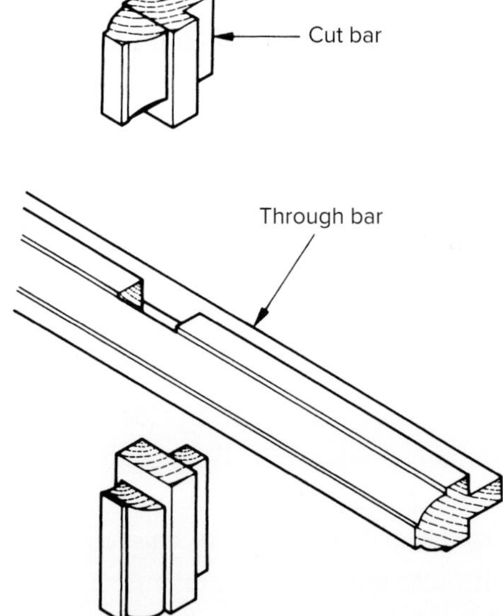

CHECK YOUR UNDERSTANDING

1. List five types of windows and indicate for each type how the sashes operate within the frame.

17.3 Setting out rods

The manufacture of joinery items requires that information be extracted from building plans, correctly interpreted and passed on to the various personnel engaged in the production process. The only adequate way for this information to be prepared and clearly passed on to others is through the preparation of full-size workshop drawings; these drawings are termed *rods*.

An experienced joiner can no doubt make up a piece of simple joinery without the aid of detailed rods, but when he or she becomes part of a production team where the job will pass through different hands for each stage–setting out, preparation of the material, marking off, machining and assembly–before it is completed, a detailed drawing to which all can refer is essential.

The setting out of rods is carried out by a responsible person who must be able to read drawings, be conversant with production methods and have a sound knowledge of building construction so that adequate allowances for manufacture and building in can be provided.

For simple joinery items, such as a rectangular door or window frame, a vertical and a horizontal section will provide all of the necessary information; these can usually be set out full-size on a board, 150 mm to 200 mm wide.

When more complicated shapes such as curves are involved, elevations become necessary and these are set out on a sheet of plywood or hardboard. In rare instances, the only space large enough for a set-out is the floor of the workshop, made available by pushing back the benches.

Table 17.1 Cutting list for sash

Member	Width	Thick	Material	No.	Length	Remarks
Stile	54	33	Oregon	2		Stock ovolo moulded
Top rail	54	33	Oregon	1		Stock ovolo moulded
Bottom rail	91	33	Oregon	1		Stock ovolo moulded
Lay bar	19	33	Oregon	2		Stock ovolo moulded
Vertical bar	19	33	Oregon	1		Stock ovolo moulded

Table 17.2 Cutting list for frame

Member	Width	Thick	Material	No.	Length	Remarks
Sill	142	66	Tallowwood	1		Stock casement sill
Stile	91	41	Oregon	2		Stock rebated
Head	91	41	Oregon	1		Stock rebated
Storm mould	19	19	Oregon	21		Standard quad

The main points when setting out a rod are as follows:

1. Dimensions must be accurate–any discrepancies at this stage will carry on through the job and assume greater proportions the further the job progresses.
2. To assist in the reading of the rod, the sections should always be drawn with the face of the item facing the person setting out, with the bottom member, such as the sill or bottom rail, to his or her right-hand side.
3. Sections are drawn to full size and should show clearly the actual size of the various members and their relationship to each other. Dimensions are not shown except for an overall height and width.
4. The rod should be clearly and easily interpreted. Remove any surplus setting-out lines that could be misleading. Light hatching can be shown on end sections, but remember that the purpose is to make the area easier to identify and not to obliterate the essential information.

   The method of fixing should be clearly indicated–mortises are crossed diagonally and wedge allowances are not shown. The centres of dowel holes are indicated with a cross and circled, with the diameter of the dowel indicated.
5. A cutting list should accompany the rod. This can be regarded as an order for the material required to complete the job. The length of members can be measured directly off the full-size rod and should include any allowances for cutting waste or horns required during manufacture or building in.

Order each item as the number of pieces of the actual length required; do not add them up into long lengths as this could result in many wasted short lengths.

The section sizes of members, if these are finished sizes, are indicated and included in the remarks column. They also draw attention to any other machining necessary, such as rebates or plough grooves indicated on the rod. If making up items from stock material, some of these details can be omitted and the different sections identified by a number or other description.

Copies of the cutting list will probably be required to pass to another part of the workshop for ordering the material, or to go to the office for the job to be costed.

An example of a set-out rod for a solid-frame six-light casement sash is shown in Figure 17.17.

If the frame is to suit a brick veneer or brick cavity wall, the height is measured from underneath the sill to the top of the storm moulding and is a multiple of brick courses. Deduct 10 mm from the height to allow for packing or bedding under the sill. Width is measured over the backs of the storm mouldings and is made to suit the opening in the brickwork.

When bars are set out, it is important that the daylight size to each light is exactly equal. Even with only a small variation, it is possible that every square of glass would have to be cut to a different size–an expensive mistake. The bars are set out by the 'in and over' method.

To set out the rod, straighten one edge and secure the rod to the bench with the straight edge overhanging 10 mm to 15 mm. All gauging is done in pencil from the straight edge, the quickest way being with the rule and forefinger (Fig. 17.18).

Alternatively, a wooden runner attached to the rule and gripped between the thumb and forefinger is an aid to accurate gauging. Square across the rod from the straight edge.

The steps in setting out the rod are illustrated in Figure 17.20 and are described as follows:

1. Mark off the overall height and gauge a line to represent the interior face of the frame. Use short lengths of the stock material, cut from scrap, as templates to mark in the outline of each section.
2. After the sill is marked, continue with the head, first positioning the storm moulding and then locating the head relative to it. Using short lengths of sash stock, mark in the sash top and bottom rails, and space the bars in between (Fig. 17.17).
3. Proportion the mortise-and-tenon joints, mark in and cross the mortises.
4. Follow a similar procedure with the horizontal section, first setting off the width and locating the storm mouldings, followed by the other members.
5. Finally prepare the cutting list.

Fig. 17.17 Set-out rod for casement window sash

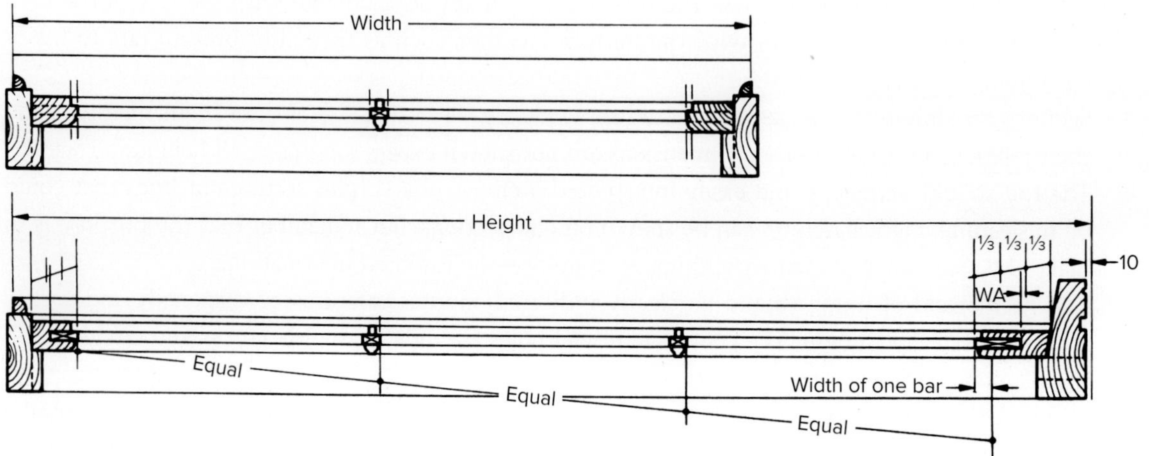

Fig. 17.18 Finger gauging

Fig. 17.19 Using a wooden runner

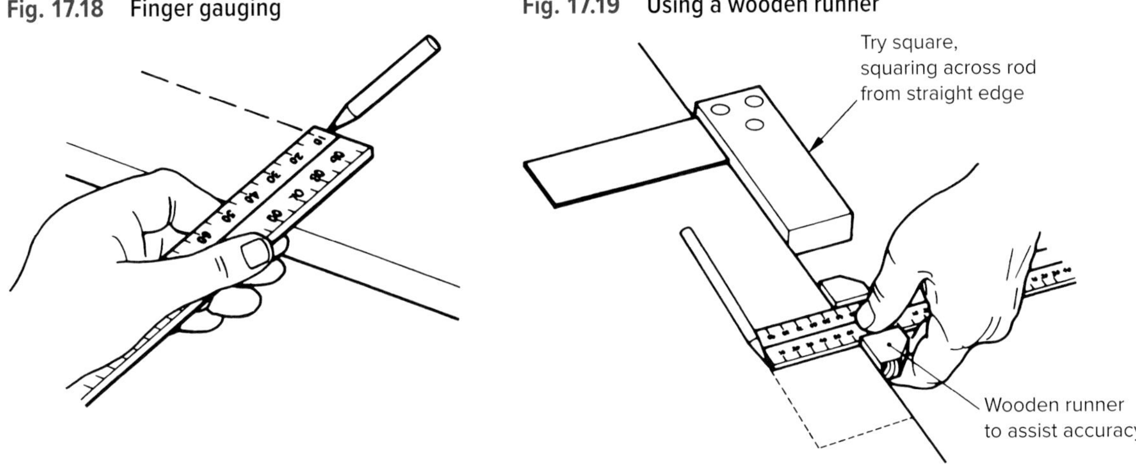

Fig. 17.20 Setting out steps for the rod

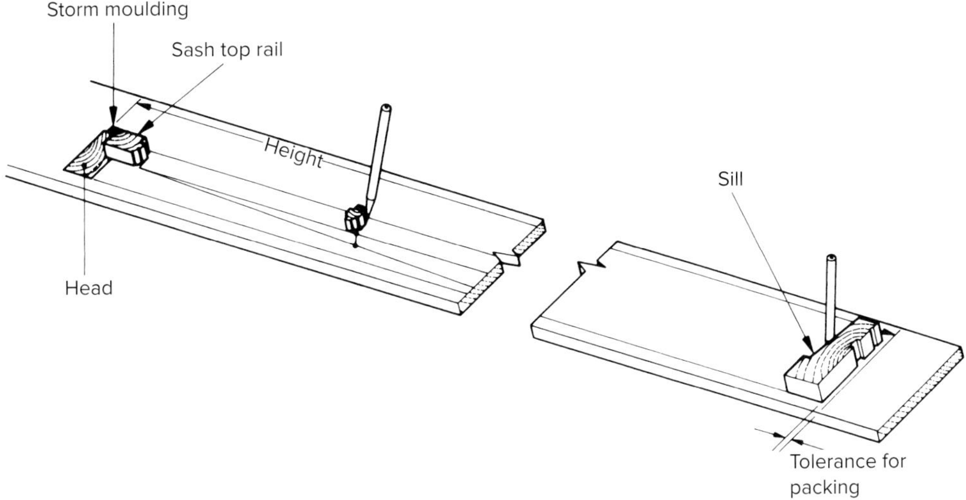

17.3.1 Marking out from the set-out rod

Marking out members of both the frame and the sash is done by laying them face down on the rod and projecting the relevant points directly onto the timber.

17.3.2 Marking out the stiles

Marking out the sash stiles is shown in Figure 17.21.

The procedure is as follows:

1. Lay the stile on the vertical section; the point of the try square can assist in transferring the position of the mortises onto the ribbon of the stile. To square around a section where the sharp corner has been removed by a mould, etc., use a *box square*. This is made up from wood, or folded from metal, and the ends must be square (Fig. 17.22).
2. Always mark out stiles in pairs; hold them together with the face edges together and the face sides opposite (Fig. 17.23). With sash stock, the moulded side is taken as the face side.
3. Square the mortises across both stiles and transfer the position around to the back of the stiles where there are to be through members. The try square can be adapted to square across the stiles by attaching an extension to the face: make a saw cut in a light lath of timber and slide it over the blade of the square.

Fig. 17.21 Marking out the stiles

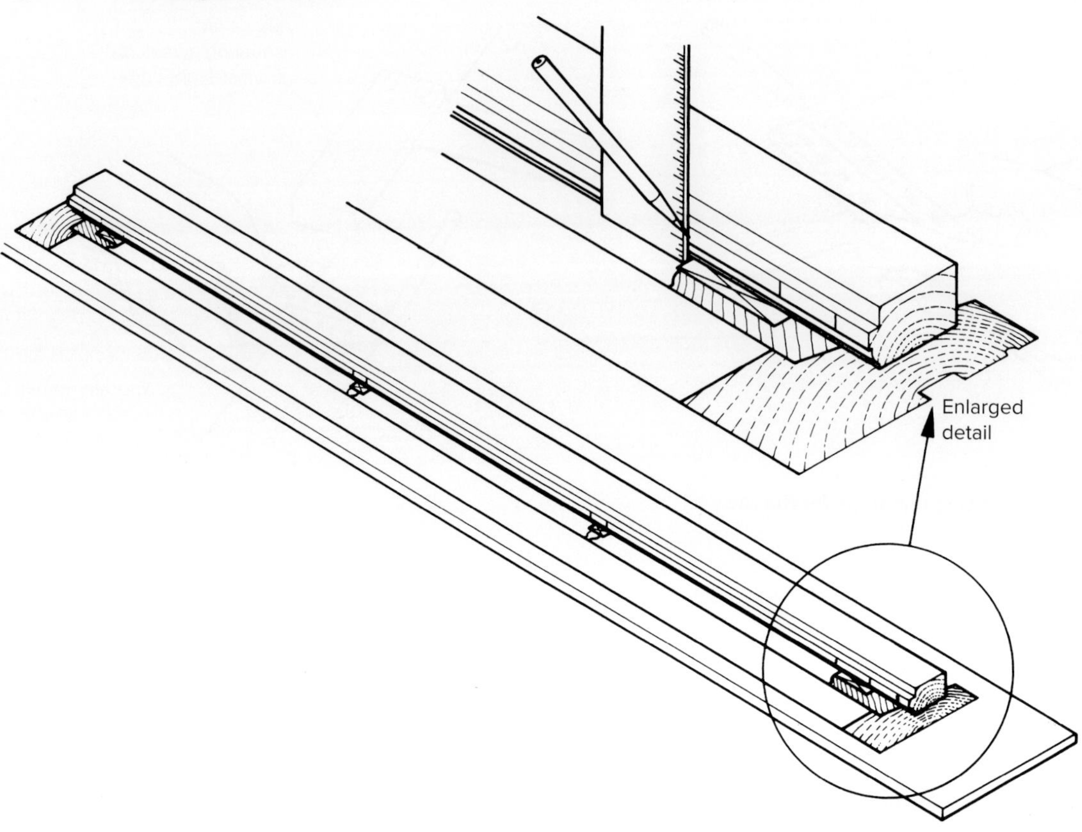

Fig. 17.22 Box square for mouldings

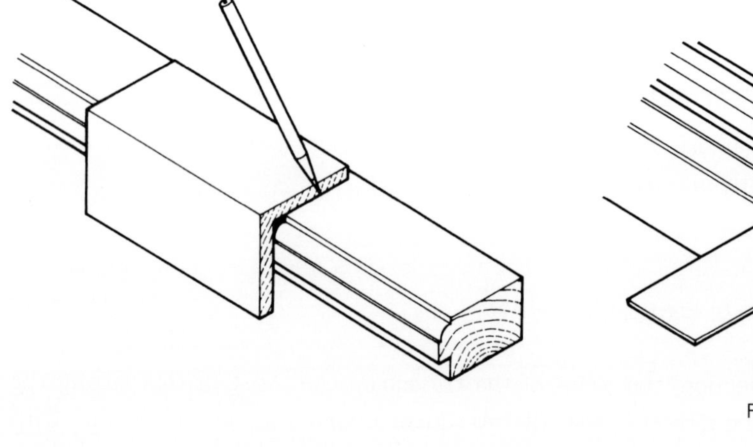

Fig. 17.23 Marking out the stiles in pairs

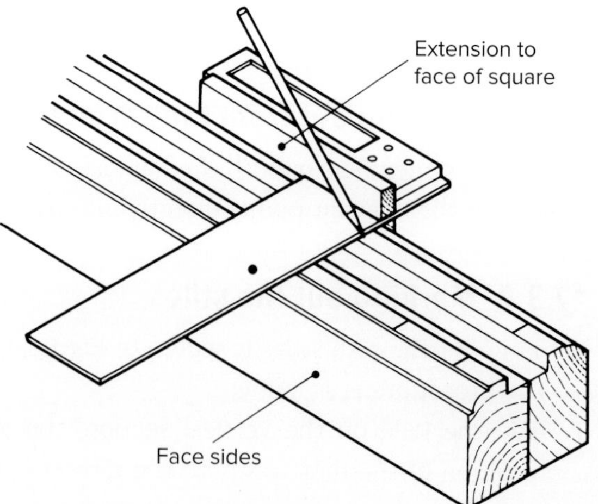

If a number of stiles are to be marked out to the same pattern, proceed as shown in Figure 17.24.

Arrange the stiles in pairs, half of the face sides pointing in one direction, the other half pointing in the other direction, with the pattern stiles to the outside. Square the ends and hold them together with a light sash cramp. Use a short straight edge to mark the mortises across.

Fig. 17.24 Marking out multiple stiles

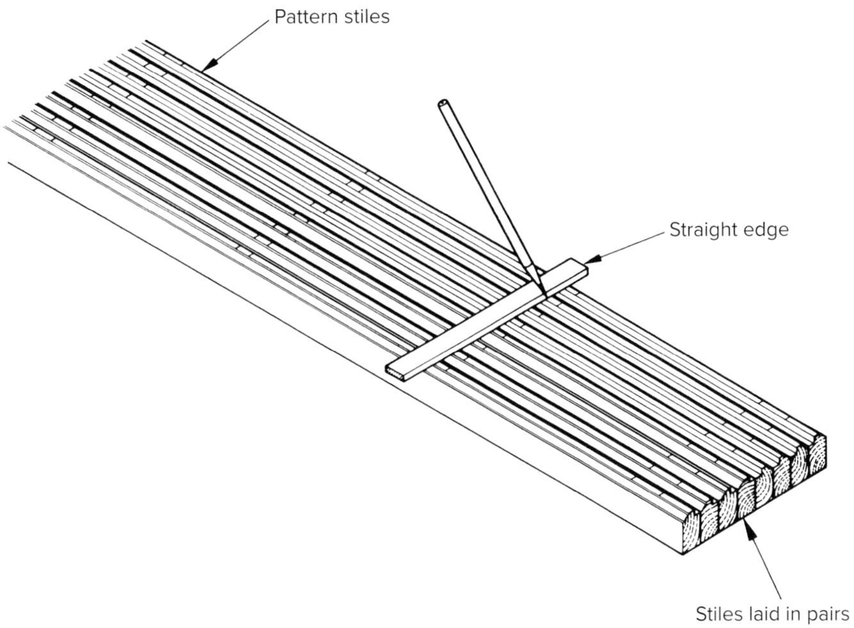

17.3.3 Marking out the rails

Fig. 17.25 Marking out the rails

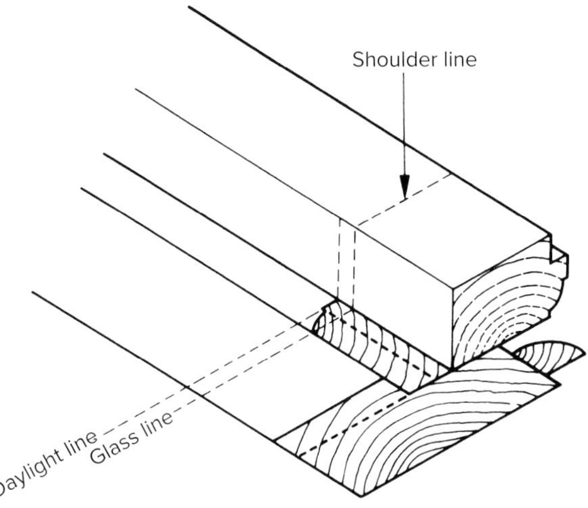

The procedure for marking out the rails is as follows:

1. Lay a rail on the horizontal section and transfer from the rod both the glass line and the daylight line. Square the glass line across the faces; this becomes the shoulder line. Cut back, to the daylight line, the recess for the franking. If there is to be a vertical glazing bar, mark also the position of the mortise.
2. Hold the rails together and mark from the pattern rail to ensure they are all of the same length.
3. Gauge the mortise-and-tenon joints as described previously, lining up one side of the mortise with the rebate.

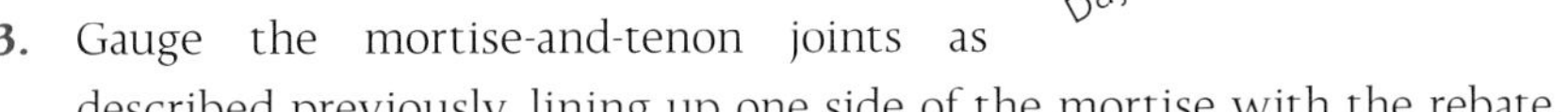

Depending on the type of machining equipment available in the workshop, it may be necessary to set out only one rail in detail to enable the machines to be set up.

17.3.4 Making the sash joints

Proceed by carrying out the following steps. (Some of these steps are more fully described in Chapter 7, 7.1.2 Making the mortise-and-tenon joint.)

1. Chop the mortises.
2. Rip the tenons.
3. Cut the shoulder lines.
4. Cut the tenon to width, including cutting wedges from the waste.

From this stage on, the joint is modified, as illustrated in the following figures. If the joint is to be finished by hand, follow these steps:

5. Cut the recess for franking (Fig. 17.26).
6. Remove a portion of the mould from the stile, being careful not to damage the franking (Fig. 17.27). Leave sufficient of the mould on the rail to scribe over later.
7. To scribe the mould on the rail, first mitre the corner. Use a brass mitre template or use a wooden one which can be made up.

Hold the template across the corner and, using a sharp chisel guided by the face of the template, pare the mould at 45°.

8. Use a scribing gouge to scribe the shoulder line to the contour of the mitre, and remove the shaded portion shown in Figure 17.29. The joint is now ready for fitting.

Other moulded shapes can be undercut using a chisel or suitable gouge. The mould on glazing bars is scribed over the other moulds as shown in Figure 17.30. First, mitre the mould and undercut to the mitred contour using a coping saw.

Fig. 17.26 Recess for franking

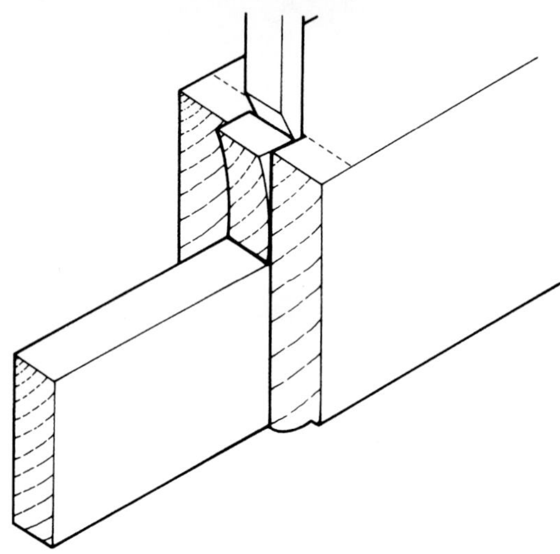

Fig. 17.27 Portion of moulding removed from stile

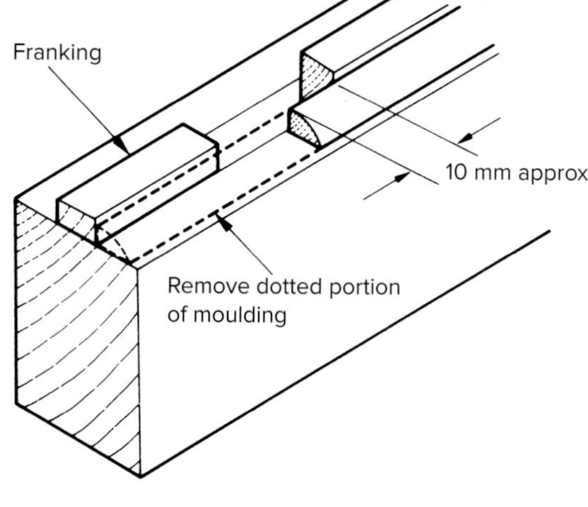

Fig. 17.28 Scribing the moulding

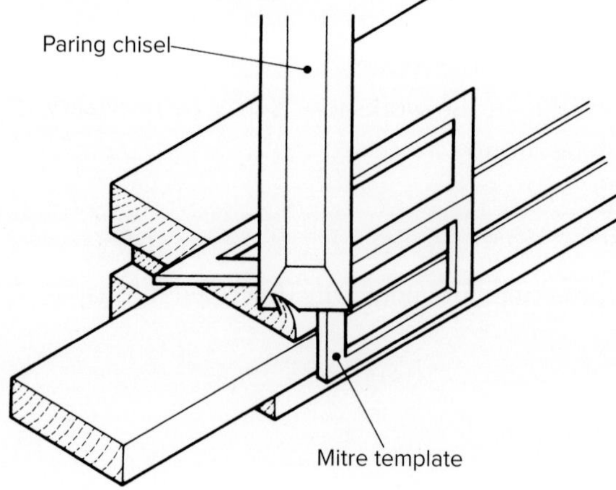

Fig. 17.29 Scribing the shoulder line

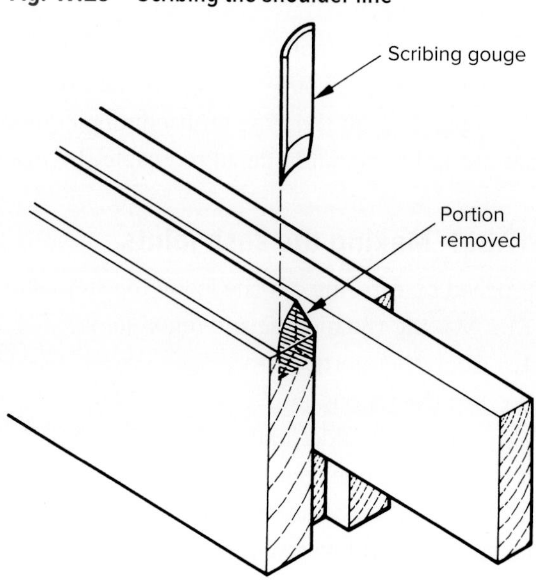

17.4 Assembling and gluing up the sash

Before proceeding to glue up the sash, assemble the necessary equipment and materials. Lay sash cramps out on the gluing-up bench and site across the bars of the cramps to ensure that they are out of 'wind'. Another requirement is a squaring rod, consisting of a light lath of timber, which is at least long enough to measure the diagonals of the frame. At one end of the squaring rod, drive a nail in, or attach a block, to positively locate the end when measuring.

Prepare the adhesive and have ready a suitable application brush. For most joinery, a PVA adhesive is satisfactory unless a more water-resistant adhesive is specified. Wedges and hand tools must also be to hand.

To assemble the sash, proceed as follows:

1. Apply adhesive to the tenons, particularly around the shoulder line and the franking to the stiles. Enter the joints and tap the frame lightly together. The procedure may vary depending on the arrangement of members. In the example shown in Figure 17.31, the assembly would commence with the sash bars, followed by the top and bottom rails and finally the stiles.

Fig. 17.30 Scribing the mould on glazing bars over other moulds

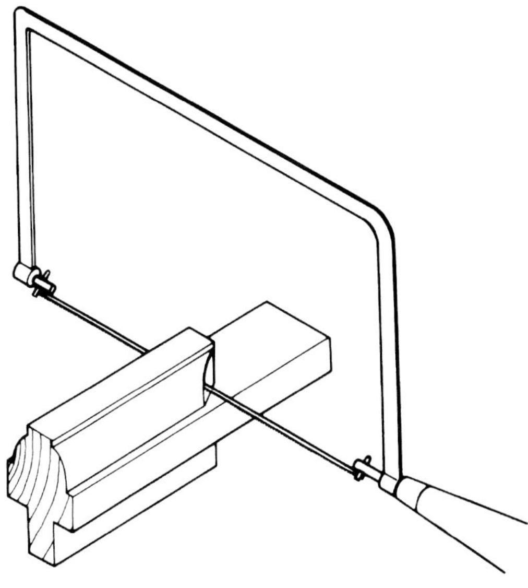

Fig. 17.31 Gluing-up procedure

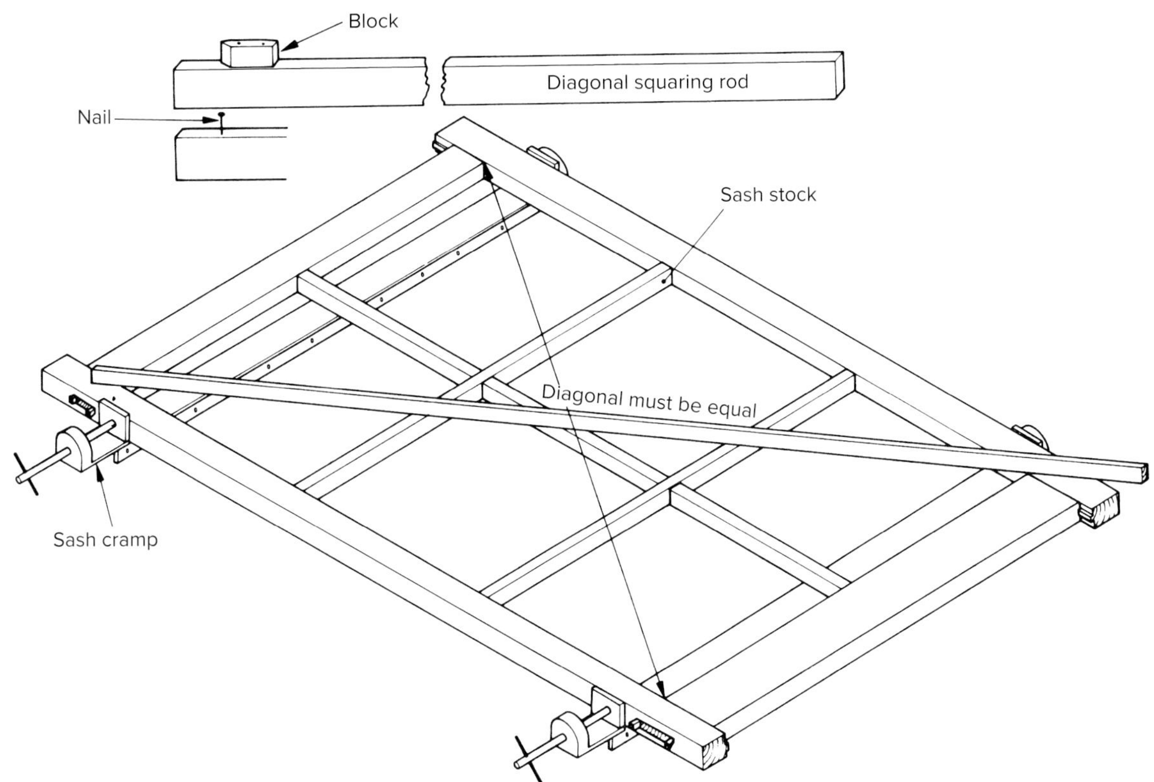

2. Lay the sash in the cramps, keeping it hard down on the bars. Ensure that the cramps are applied square across the sash, and cramp the joints together. The cramp should be positioned so as not to foul the wedges or bow the stiles; positioning across the horns is often the most satisfactory.
3. To square the sash, use the squaring rod to measure the diagonals. These must be made equal. If one diagonal is longer than the other, slacken off the cramps and position them at a slight angle across the sash to pull the long diagonal shorter.
4. Before wedging, it is essential that the sash is square and perfectly flat or, to use the trade term, out of 'wind'. Apply adhesive to the wedges and enter them into the wedge allowance. Commence driving the end wedge first, followed by the other wedges; drive them uniformly so as not to force the sash out of square.
5. Wipe away any surplus adhesive with a cloth that is just damp, remove the sash from the cramps, and stack flat for the adhesive to set ready for the cleaning off.

17.4.1 Cleaning off the sash

Cleaning off the sash by hand requires a smoothing plane, the blade of which is sharp and set to remove a fine shaving. The sash must be laid flat on the benchtop and held securely. Corner blocks, or often a cleat fitting between the horns and attached to the bench, will provide adequate restraint.

Hold the plane as shown in Figure 17.32, with the toe pointing to the outside edge of the stiles or rails. Work around the framing, flushing off the joints by removing any rough spots or other blemishes, and setting-out marks. Take only a fine shaving and follow the direction of the grain. Test the surface for flatness, particularly across the joints, using a straight edge or tilting the sole of the plane.

Finish the surface with an abrasive paper held over a sanding block. First sand across the rails, followed by the stiles and work up to the shoulder line; remove any scratches. If the sanding is carried out with an electric sander, follow a similar procedure.

Fig. 17.32 Cleaning off joints

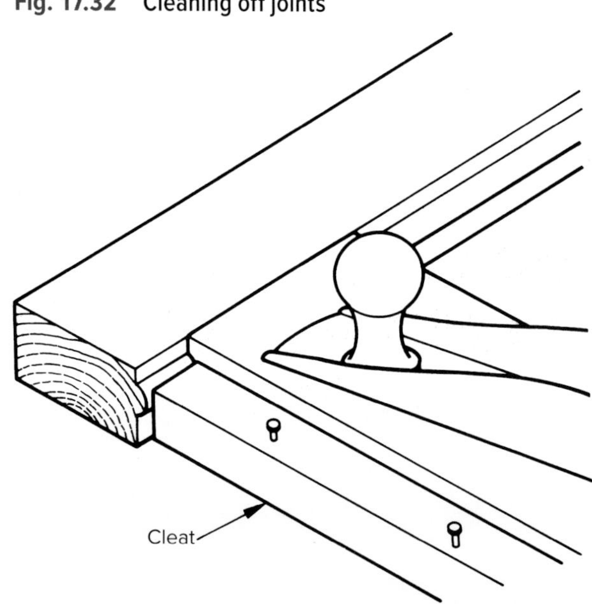

17.5 Vertical sliding sashes

17.5.1 Box frames

Box frames are used with vertical sliding sashes that are counterbalanced with cast-iron or lead weights. The jambs of the frame are made up from comparatively light sections and are formed into a boxed section, which gives them rigidity and provides a space in which the weights can operate.

The frame members that form the box are the *pulley stile*, the *outside lining*, the *inside lining* and, in brick construction, the *back lining*. A *pendulum strip* hangs in the box and separates the weights. Two *sash channels* are formed on the face of the box by the outside lining, the parting bead that separates the sliding sashes and the stop bead. The sashes slide vertically in the sash channels. In addition to the stile and the top and bottom rails, there are the *meeting* rails, which come together when the sashes are closed. Note the difference in the profile of the top and bottom meeting rails and the manner in which they are splayed to come together and seal the junction (Fig. 17.33). The sashes are hung on *sash cords,* which are fixed into a plough groove at the back of the sash stile; they pass up and over axle pulleys

Fig. 17.33 Structure of a box frame

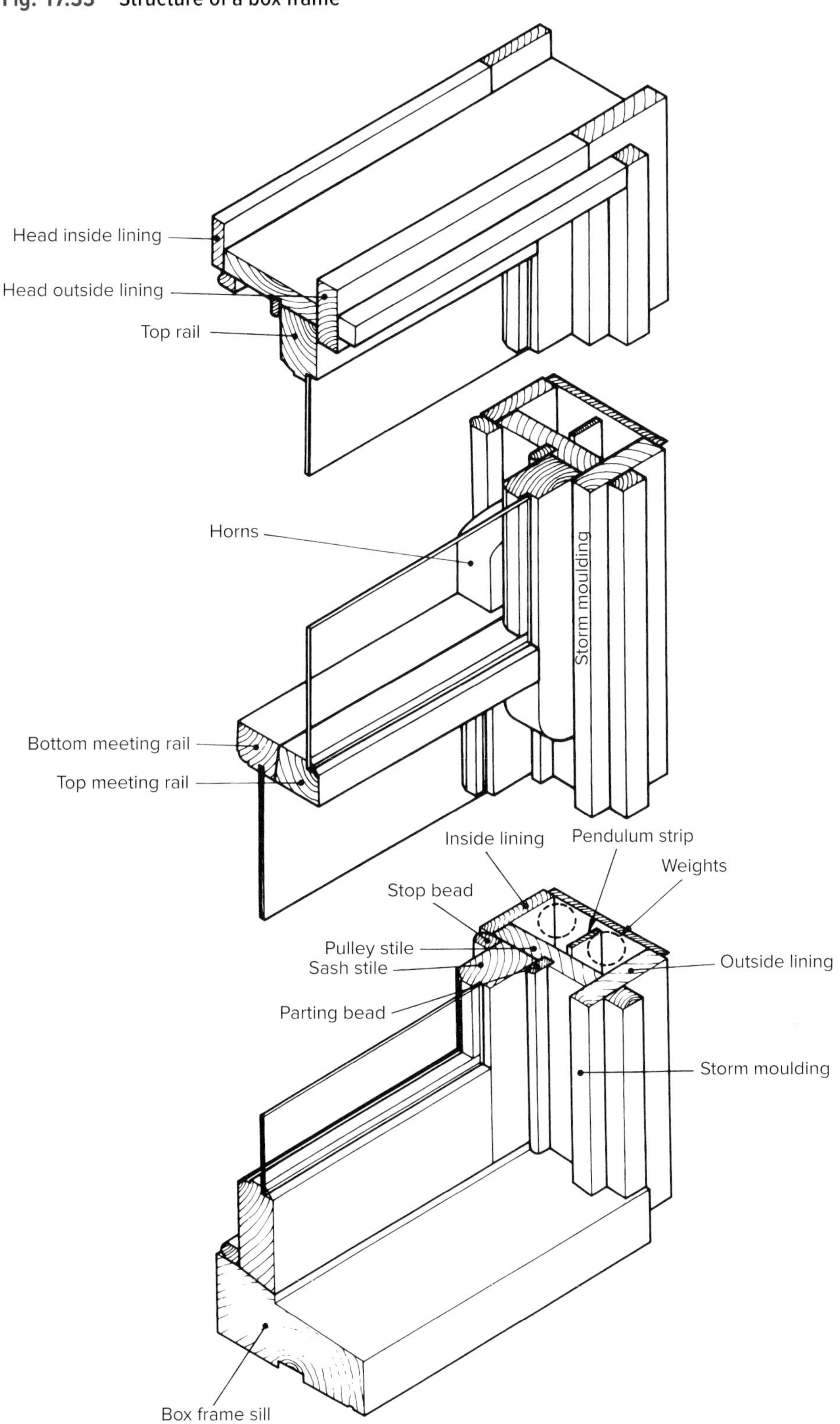

near the top of the frame and are attached to the weights inside the boxed section. Some very old windows may even have a chain instead of cords.

For many years, well-made box-frame windows provided one of the most popular and reliable types of window available. However, in recent years, for various reasons–possibly mainly economic–the box frame has been almost completely superseded by other types of window in new constructions.

CHECK YOUR UNDERSTANDING

1. Draw the profile of a typical casement windowsill; name each part and state its purpose.
2. Illustrate with a sketch what is meant by the description 'a six-light sash with two vertical bars and one lay bar'.
3. What is the purpose of a joiner's rod?
4. What views are shown on the rod for most joinery items?

17.5.2 Repairing box-frame windows

The main interest in box-frame windows, for today's practising carpenter, is their maintenance, installing new sash cords in particular. The following discussion will be limited to the procedure for re-cording a box-frame window. The steps are as follows:

1. **Removing the sashes.** To remove the sashes from the frame, use a suitable chisel to prise away the stop beads and remove them from the frame. Remove the bottom sash. If there is a cord that is not broken, lift the sash to near the top of the frame and cut the cord. The weight will fall the short distance to the bottom of the box.

   Prise out the parting bead–it should not be nailed, but it can be stuck in with paint, and care is necessary to avoid it being broken. Remove the top sash.
2. **Removing the weights.** When the sashes are removed, find the axle pulleys near the top of the frame. Apply a spot of oil to ensure that they are rotating freely.

   On the inside sash channel, near the bottom, is the *pocket piece.* This provides access to the weights. Remove the pocket piece. Move the pendulum strip aside and remove the weights through the pocket. Do not mix the weights as the top and bottom sashes may have slightly different weights–with the bottom sash usually being the heavier.

Fig. 17.34 Axle pulleys and pocket piece

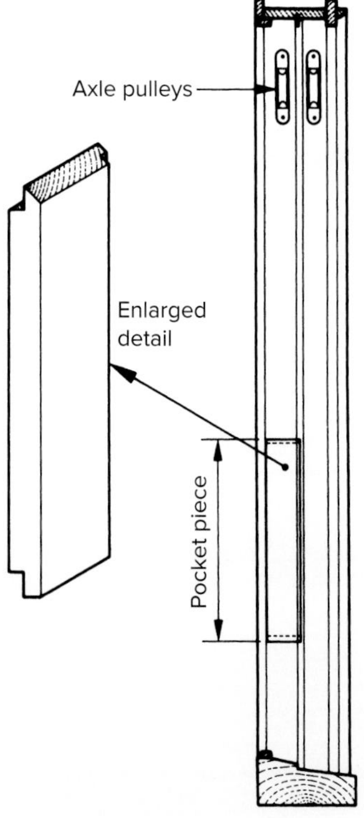

3. **Marking the length of the cords.** To mark the length of the cords, note the plough groove in the back edge of the sash stile in which the sash cord has been fixed. Mark the end of the cord on the stile and remove the old sash cord. Measure the distance from the end of the cord to the bottom of the sash and mark the same distance up from the sill on the pulley side. On the drawing in Figure 17.35, this distance is indicated as *c*. Do this for both sashes–the marks might not necessarily come in line.
4. **Threading new cord** (Fig. 17.36). To thread new sash cords, a 'mouse' is necessary (Fig. 17.37). This consists of a length of string with a small weight on the end, which can be passed over the axle pulleys.

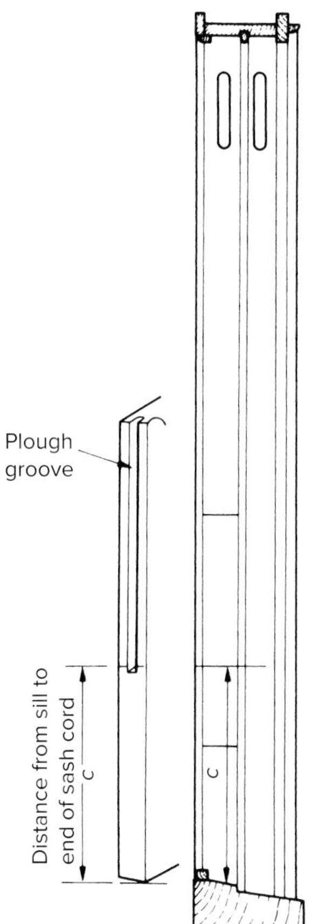

Fig. 17.35 Marking the length of the sash cords

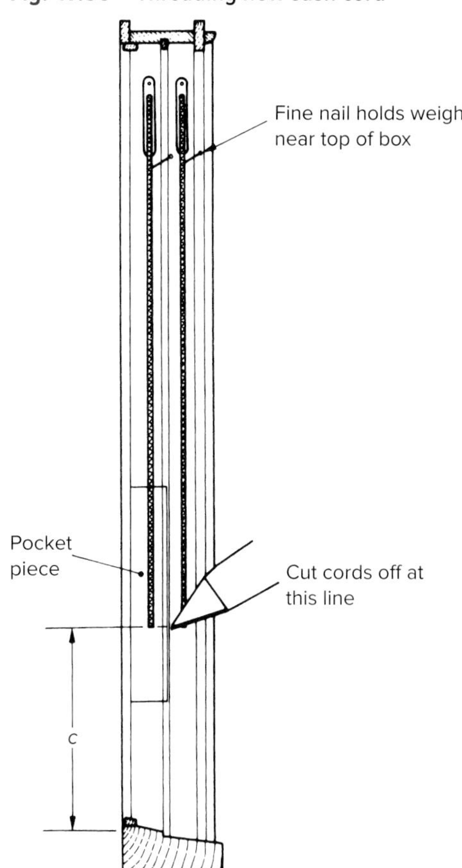

Fig. 17.36 Threading new sash cord

Use a No. 8 sash cord for most applications. Attach the mouse to the end of the cord with a few half hitches. Pass the weighted end over the axle pulley and out through the pocket, followed by the end of the sash cord. Attach the appropriate weight to the ends of the cord and secure by a figure-eight knot on the end of the cord (Fig. 17.38).

Fig. 17.37 A 'mouse'

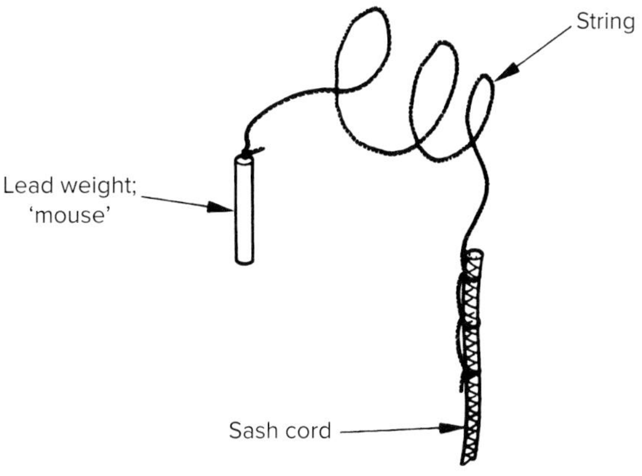

Place the weight back into the box and pull it to the top of its travel. Drop it back about 30 mm from the top and secure it in this position by driving a fine nail through the cord just below the axle pulley (Fig. 17.36). Cut the cord off at the mark on the pulley stile. Repeat this step for each weight and replace the pocket piece.

5. **Attaching the cords to the sashes** (Fig. 17.39). Replace the top sash first. Fit the new sash cords into the plough groove with the end on the mark made previously and secure with no more than three 20 mm clout nails. The top nail must be below the level of the axle pulley when the sash is fully up. Remove the nail holding the weight and locate the sash in its channel. Replace the parting beads, first sanding off any surplus paint that will affect the smooth operation of the sash. Attach the bottom sash to the new cords in a similar manner and locate it in its channel.

Fig. 17.38 Attaching the weight

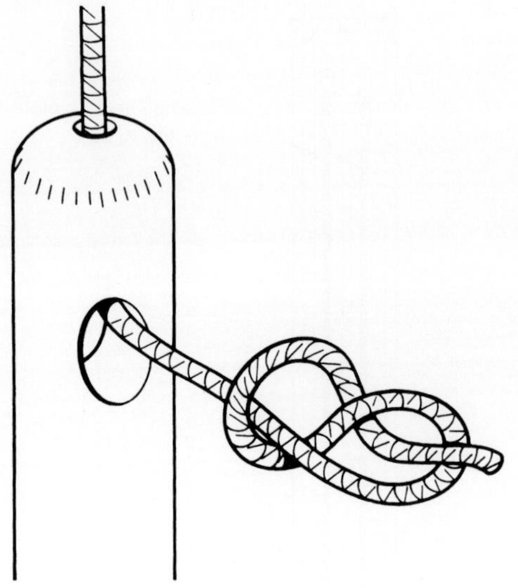

Fig. 17.39 Attaching the cord to the sash

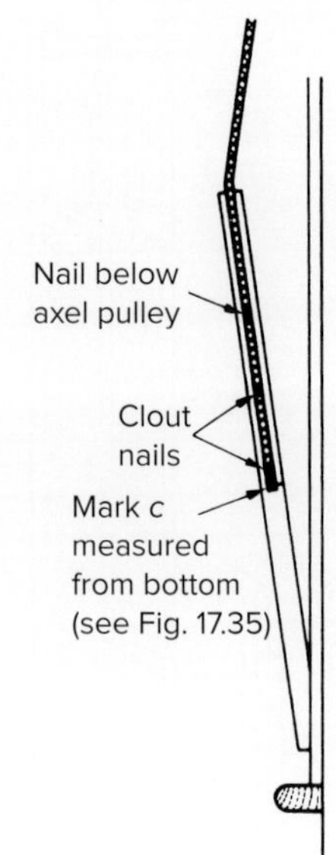

Fig. 17.40 Spiral spring balance

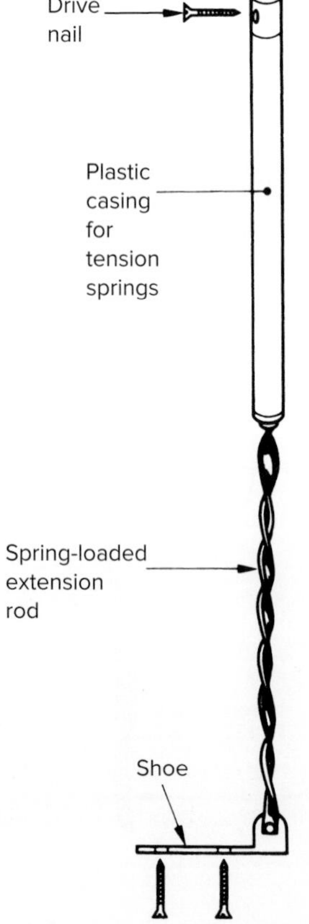

Fig. 17.41 Sash–stile groove for spring balance

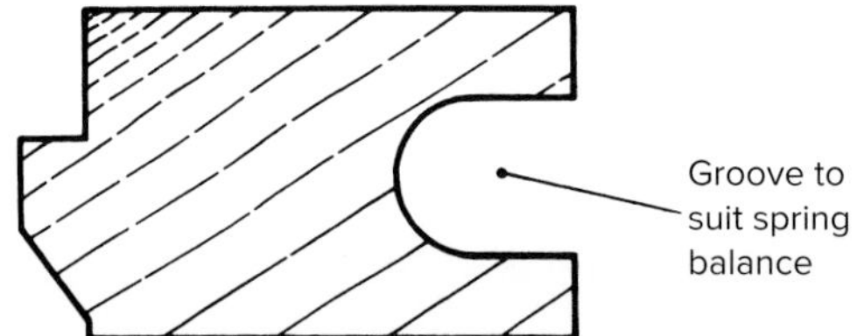

De-nail the stop beads–usually less damage to the bead will result if the heads of the nails are pulled right through with a pair of pincers. Replace the stop beads and renail, allowing the sash a clearance of no more than 2 mm in the channel so that it can operate freely without rattling.

Check the pair of sashes for satisfactory operation.

17.5.3 Spiral spring balances

Spiral spring balances have almost entirely superseded cast-iron weights as a means of counterbalancing vertical sliding sashes in new domestic constructions. The balance in Figure 17.40 consists of a plastic casing in which are housed the tension springs and a twisted extension rod to which the sash is attached. The balance is accommodated in a groove, formed in the back edge of the sash stile (Fig. 17.41).

For trouble-free operation, it is important that the grooves be of the correct size (check with the manufacturer), and it is recommended that they be made after the sashes have been fitted to the frame.

Window frames incorporating spiral spring balances can be adapted to any type of construction. The example in Figure 17.42 shows a frame fitted into timber-framed construction. Note the two sash channels and how the outside lining also functions as an architrave.

17.5.3.1 Set-out for spiral balances

Details of a set-out rod for a double frame and sashes with spiral spring balances to suit the situation in Figure 17.42 are shown in Figure 17.43.

Fig. 17.42 Spiral spring balance in a double-hung window

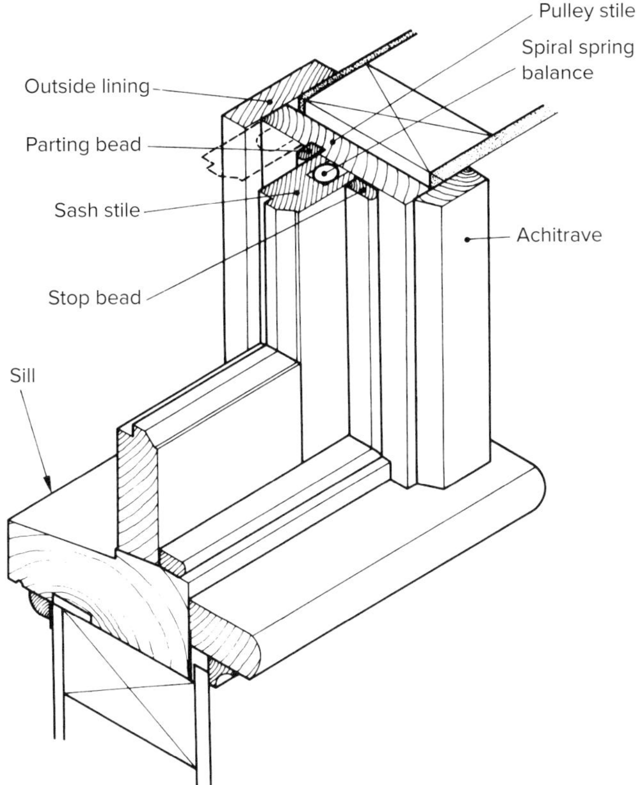

Fig. 17.43 Set-out rod for a double-hung window with spiral spring balances

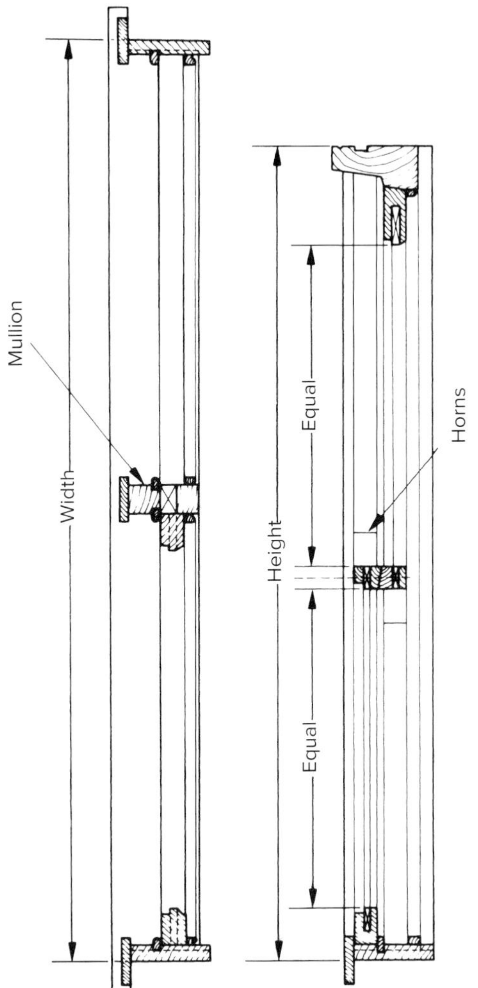

A standard box-frame sill is used, and when grooving the pulley stiles for the parting bead, make the outside width of the sash channel equal to the thickness of the sashes plus 2 mm clearance.

To set out the sashes, mark in the sash channels; bisect the distance between the daylight lines to locate the centre line of the meeting rails, which can now be marked in. If the setting out is correct, there should be equal daylight opening to both the top and bottom sashes. Mark the length of the horns and a common mortise-and-tenon joint on the meeting rails. Note, in the horizontal or width section, how the spring balances allow the use of a narrow mullion, therefore permitting a maximum daylight opening to the frame.

Spring balances are sold in sets and it is necessary to know the overall height of sashes, and their weight after glazing, to select the correct set. It has been found that most problems involving the faulty operation of spiral balances are the result of faulty installation rather than any defect in the balance itself. With each set of balances, detailed installation instructions are included and should be carefully followed for satisfactory results.

CHECK YOUR UNDERSTANDING

1. What information is required on a cutting list?
2. What information must be provided to select the correct spiral balance for a pair of vertical sliding sashes?
3. List five steps carried out when renewing cords on a pair of box-frame sashes.

17.5.4 Window hardware

Window hardware is an integral part of a window's operation.

Window hardware includes:

- latches/locks
- stays
- winders
- handles
- hinges.

The type of hardware to be used on a window is generally determined by the type of window being used as particular hardware may be suited only to a particular window type. The manufacturer's instructions should always be consulted for installation advice. Elements of aesthetics such as finish, colour, design and material type largely vary, so it is essential to consult with the designer or client.

If the windows are new, the hardware will generally already be installed, so the job of a carpenter is to install the completed window to the wall frame opening and then install the architrave once the house is at the lock-up stage.

If the windows are second-hand or already installed, the carpenter's job may be to repair them, replace any broken hardware or even replace all of the hardware if it doesn't suit the client's needs.

Fig. 17.44 Various types of window hardware

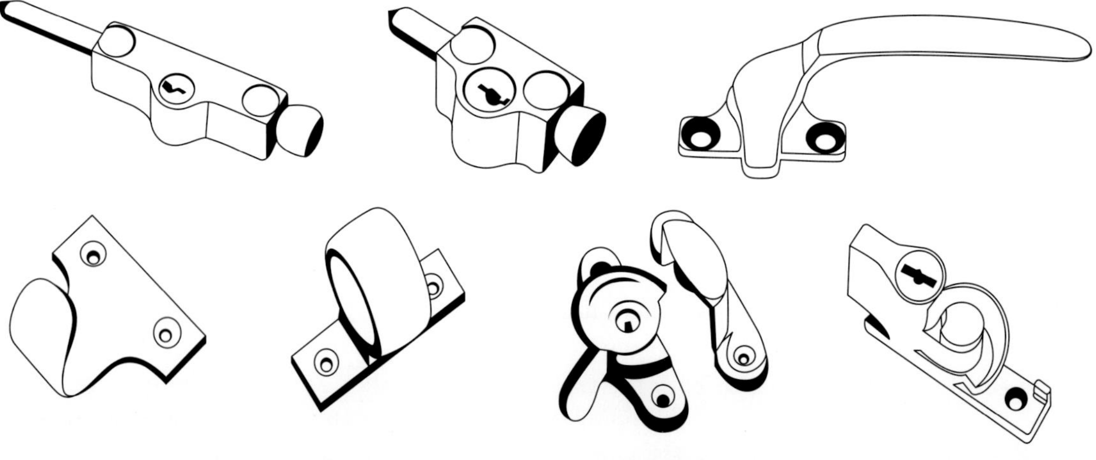

CHECK YOUR UNDERSTANDING

1. Name three types of window hardware.

17.6 Fitting window frames and flashing

Window frames are made up complete with outside architraves and should fit the wall openings, allowing a minimum clearance of 10 mm on the sides and 15 mm between the window head and the wall frame lintel. This is to allow adequate clearance to adjust the frame for levelling and plumbing up. **Flashing** of 1 mm sheet metal or other approved material such as aluminium or plastic must be provided over the head and underneath the sill. Failure to carry out this work adequately could lead to damp patches on the inside walls, a condition that can be very difficult and expensive to rectify.

17.6.1 Sill flashing

The sill flashing consists of a one-piece tray that will turn up at the ends at least 50 mm (Fig. 17.45). The width must be sufficient to turn up the back of the sill at least 25 mm and turn down 25 mm over the face of the external sheeting, whether it is sheet material or timber siding. Care must be taken to fold the flashing without damaging it in order to form a watertight corner.

Fig. 17.45 Windowsill flashing

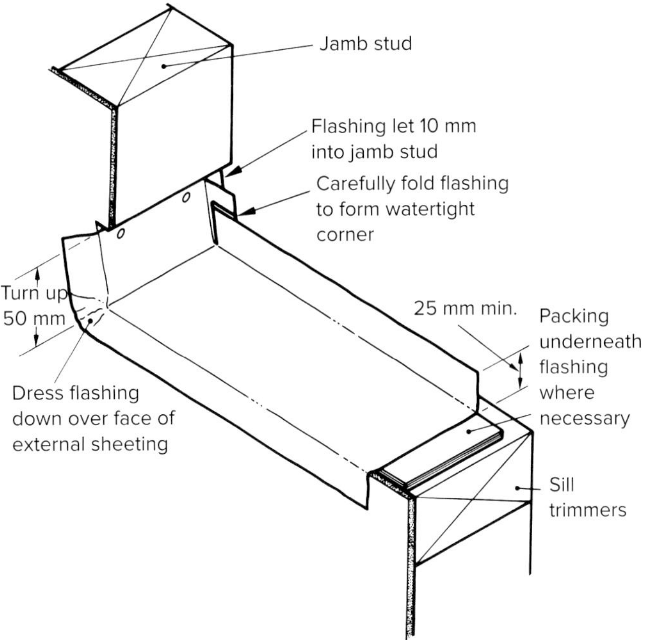

17.6.2 Head flashing

The head flashing is provided across the head of the frame and must turn up under the wall sheeting (Fig. 17.46). It will then cover and turn down over the frame or architrave. In the case of weatherboard wall coverings, the flashing can be inserted into the horizontal joint above the frame and then turned down over the architrave.

Under some circumstances, where adequate overhead protection from the weather is provided, the head flashing can be omitted.

When the head and sill flashings are in place, stand the window frame in the opening provided and first level the sill, placing any sheet material used as packing underneath the sill flashing. Ensure that the sill is straight and adequately supported. Horns left on the sills are used to secure the ends to the wall framing (Fig. 17.47).

Bore a clearance hole through the horn and nail to the framing. Take care that the flashing is not punctured by any nail holes. Plumb the window jambs, pack as necessary and nail through into the jamb stud. Plywood or hardboard offcuts and bituminous felt for making minor adjustments make satisfactory packing.

Fig. 17.46 Head flashing: (a) fibre cement sheeting; and (b) weatherboard siding

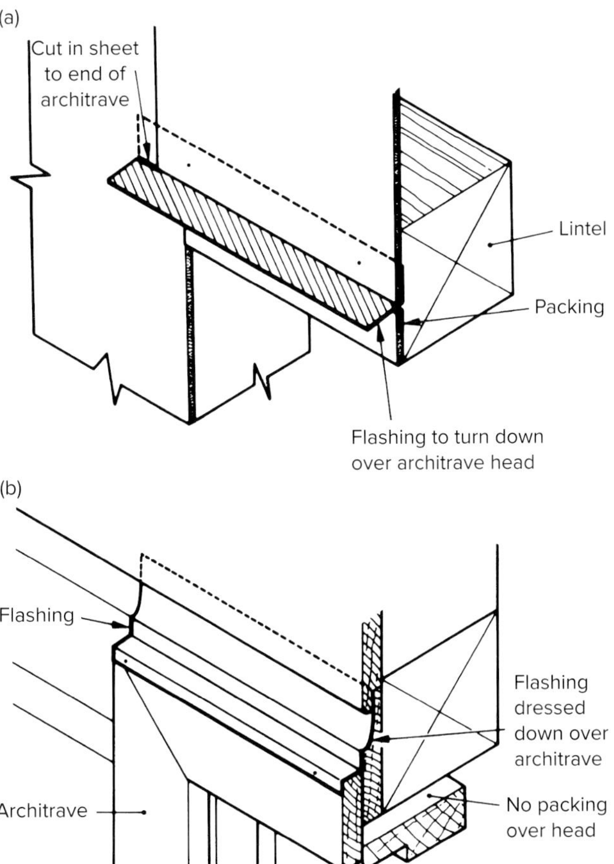

TIP For an in-depth guide to flashing and installing doors and windows, refer to *An Industry Guide to the Correct Installation of Windows and Doors.*

Fig. 17.47 Fixing a windowsill

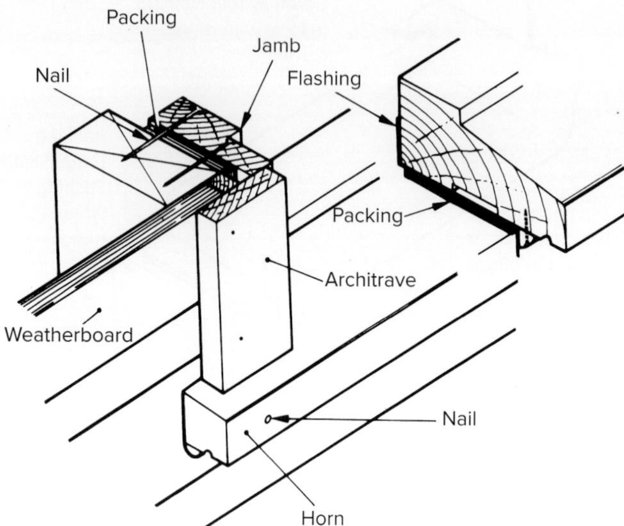

Fig. 17.49 Fixing timber windows

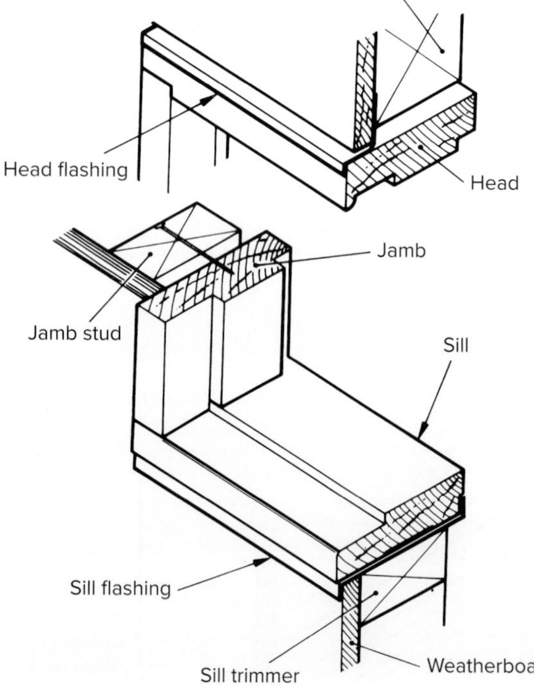

Fig. 17.48 Nailing points for timber-frame window

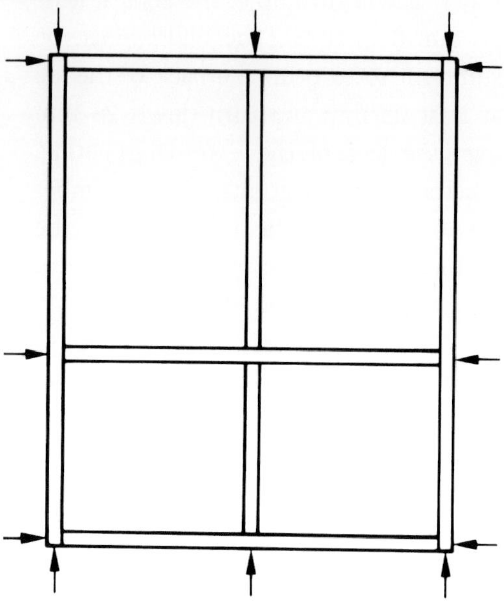

There can be a tendency to open up the joints of the window frame, and it is recommended that, where practical, nailing be done through the framing at the points indicated in Figure 17.48.

The lintel will carry the load over the window opening and no packing should be inserted between the lintel and the window head. Alternative details in fixing window frames are shown in Figure 17.49. This figure is adopted from a popular range of window frames manufactured from Western red cedar, and the application of the head and sill flashing is clearly seen. Note that the ends of the weatherboards butt up to the window frame and that the frame will have to be fixed in place before the weatherboards are applied. Sealing this joint with a bead of **silicone mastic** would be sound building practice.

TIP Always check the manufacturer's installation requirements. Many manufacturers of timber window and door frames will now request that additional side flashings be fitted to the frames on installation.

17.6.3 Aluminium window frames

Aluminium frames from different manufacturers vary considerably in the way they are fixed and waterproofed. Some include a metal flashing, while others rely heavily on the application of a non-hardening sealant to render them watertight. Always obtain the manufacturer's instructions and follow them carefully when fixing aluminium frames.

Flashings to metal window frames should be of a non-reactive approved material usually supplied by the manufacturer on request. A simple and attractive method of fixing aluminium frames into timber-framed construction is to purchase the frames with the timber reveal linings attached.

A typical example of an aluminium window frame with timber reveals is given in Figure 17.50. The frame must be fixed into the wall first and then the external siding is fitted to the metal frame. To fix the frame, stand it in position in the opening. Check that the sill is level, packing where necessary, and nail through the timber reveal into the sill trimmer. Pack the sides of the frame, taking care not to fix the frame out of square and nail through into the jamb studs. Do not pack above the window heads.

Fig. 17.50 Fixing aluminium windows: (a) section through window; and (b) plan of jamb

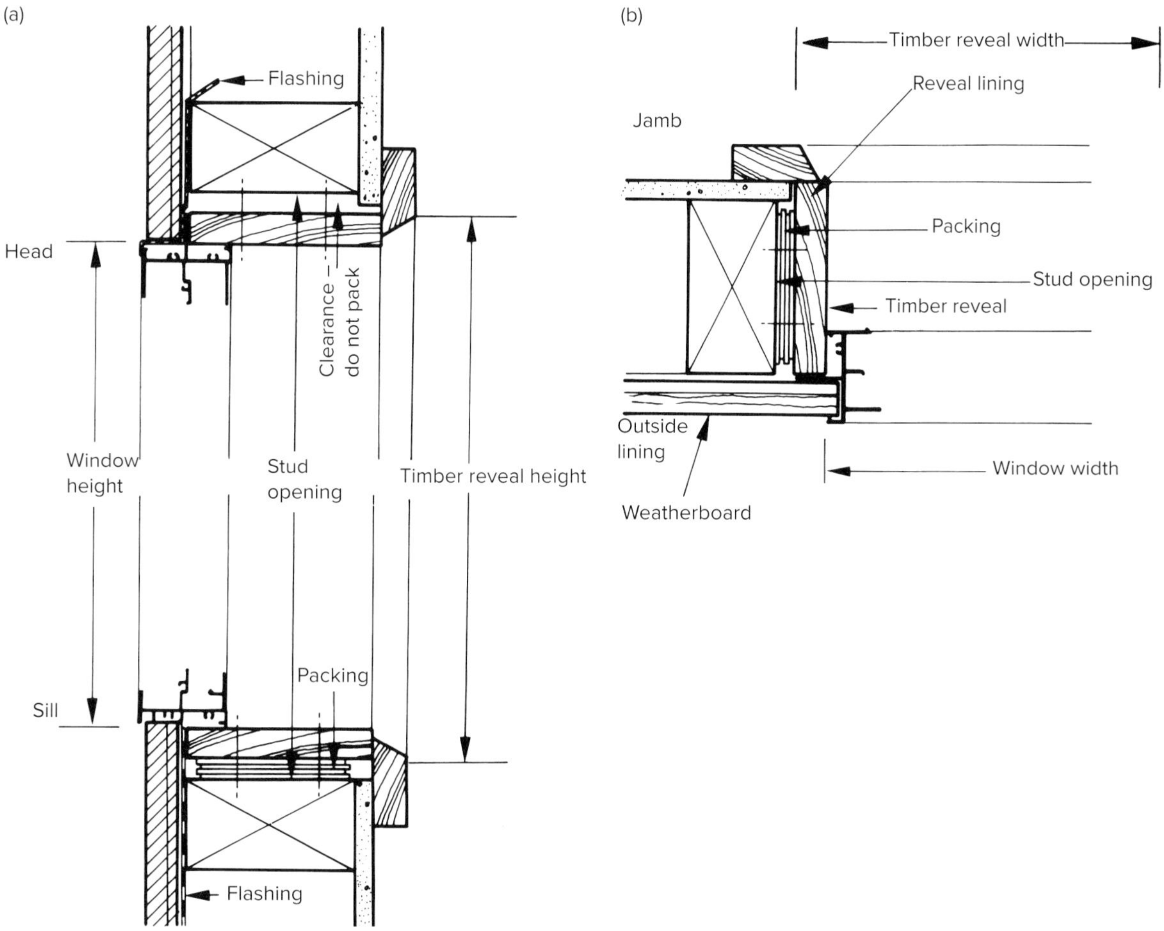

CHECK YOUR UNDERSTANDING

1. Show with a sketch what flashings must be provided for a timber window frame.

17.7 Glazing

Sashes are usually glazed with a flat transparent glass; in bathrooms and WCs, obscure glass in a variety of patterns may be specified.

Most flat transparent glass used today is termed *float glass*. It is manufactured by a process where the raw materials are melted in a furnace and the molten glass moves from the furnace in a continuous ribbon over a bath of molten tin. Here, all irregularities are melted out and both surfaces become flat and parallel with a fire-polished finish.

Float glass is manufactured in thicknesses of 3, 4, 5, 6, 8, 10, 12, 15, 19 and 25 mm. Most domestic glazing uses thicknesses of 3, 4, 5 and 6 mm.

There must always be a proper relationship between size, thickness and use of glass. Where doubt exists, it is better to use a thicker glass. For internal applications, consideration should be given to the degree of use, exposure to traffic and the possibility of accidents; some situations may indicate that a toughened form of glass should be used. In situations such as picture windows, window walls and full-height windows adjacent to doors, authorities may require the use of toughened or laminated safety glass in accordance with AS1288:2006: Glass in Buildings–Selection and Installation.

The glass is bedded and fixed in the sash rebate with linseed-oil putty and metal sprigs (Fig. 17.51). The rebate should be primed with a suitable priming paint (or similar) before glazing. This will seal the timber and prevent the oil from the putty soaking into the timber, causing it to dry out and crack prematurely.

The procedure for glazing sashes using putty is now described.

After the rebate has been primed and the paint is dry, apply a bead of putty to the rebate and press the glass into place, allowing a bed of about 2 mm. Fix the glass with sprigs 300 mm apart, and apply the front putty splayed off to a point 2 mm below the daylight line of the sash (Fig. 17.51).

When it is specified that sashes keep their natural clear finish, the glass may be bedded in clear glazing mastic and then secured on the face with a timber-glazing bead.

Fig. 17.51 Glazing with putty

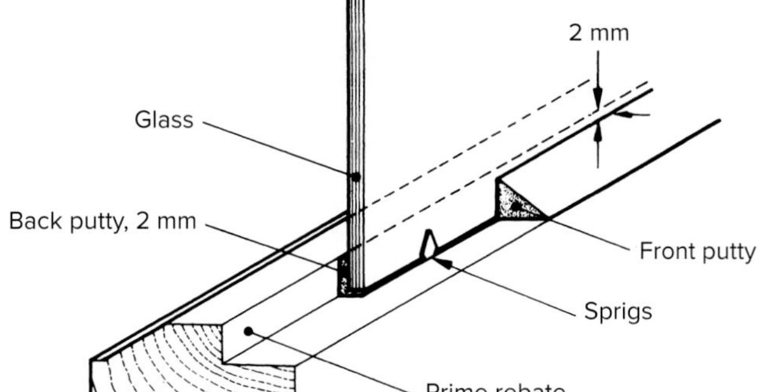

Fig. 17.52 Timber-glazing bead

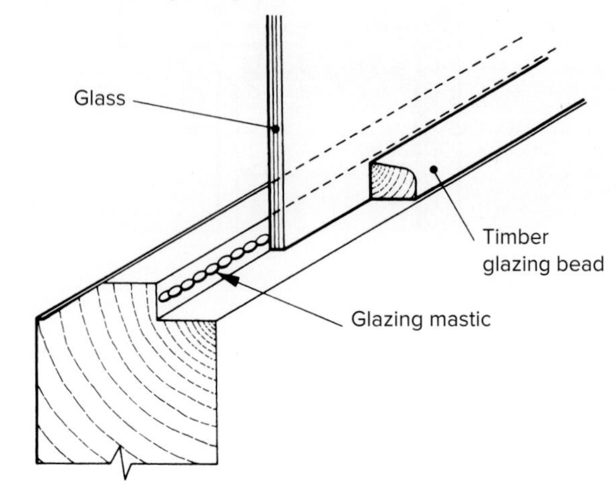

17.8 Door construction

Doors are made in a number of types, each designed to provide satisfactory service in different situations. Two broad divisions can be made:

1. doors suitable for exterior use
2. doors intended only for interior use.

A door designed for exterior use must have a flat face so that water can run off freely. It must be made in such a way that

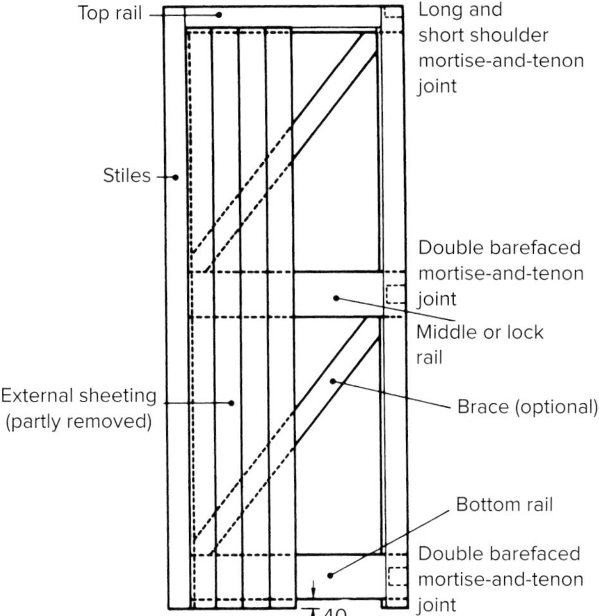

Fig. 17.53 Framed ledge door (sheeting partly removed)

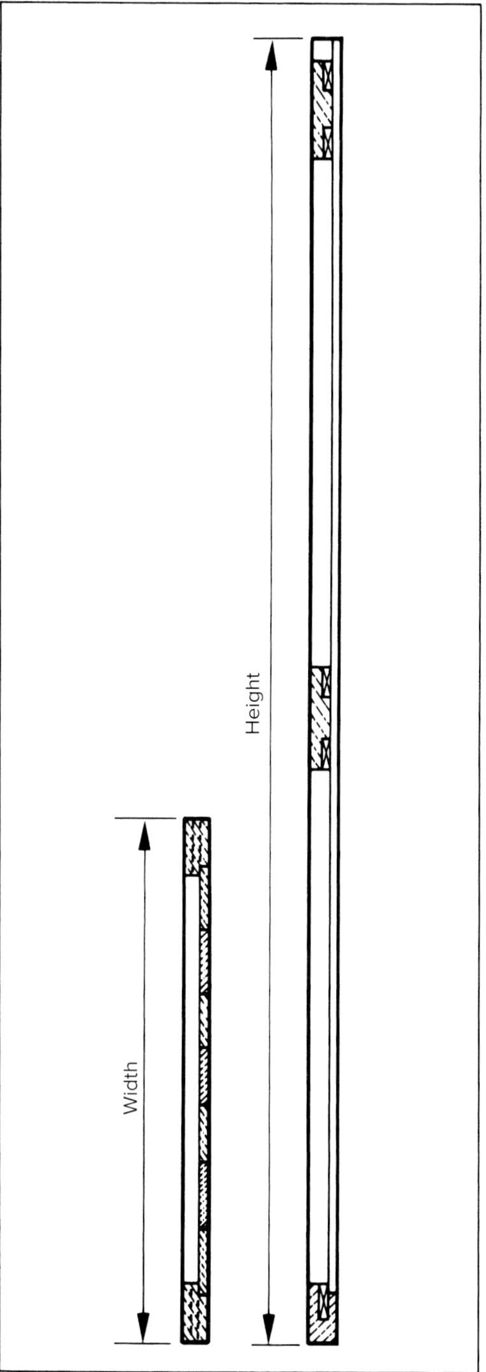

Fig. 17.54 Set-out rod for a framed ledge door

variations in moisture content of the timber can occur without the door becoming distorted.

Doors for interior use can incorporate elaborate panels and moulds to achieve their purpose without fear of moisture and decay penetrating the joints and ruining the door.

17.8.1 Door types

Factory-made flush-panel doors are now the most commonly used door in new constructions. They are suitable for interior use, and also exterior use when bonded with water-resistant adhesives. Other types of fabricated doors, sometimes referred to as *joinery doors,* include these which are now described.

17.8.1.1 Framed ledge door

The framed ledge door is a very strong type of door, suitable for exterior use where it could be subjected to hard wear and tear. The stiles and ledges are mortised and tenoned together to form a rigid frame, which can be further strengthened by the addition of a brace. The face is then sheeted with vertical tongue and grooved boards.

The top rail has the same thickness as the stiles, but the thickness of the ledges is less the thickness of the sheeting. The ledges are joined to the stiles with barefaced mortise-and-tenon joints, as explained earlier. If the ledge is wide, a double barefaced mortise-and-tenon joint is used. The external sheeting is fitted to rebates in the top rail and the stiles and over the ledges to the bottom of the door.

A set-out rod for a framed ledge door, shown in Figure 17.54, illustrates the construction. Details of proportioning the mortise-and-tenon joints are explained earlier in this chapter.

17.8.1.2 Framed and panelled doors

The use of panelled doors is now usually limited to interior situations; however, they may be used as external entrance doors only where adequate protection is provided from the prevailing weather.

The design of panel doors can vary considerably; a basic four-panel door is shown in Figure 17.55.

Note that the stiles and the top rail are of the same width and that the bottom rail is about twice that width. The middle rail, also called the *lock rail,* is about the same width as the bottom rail. Vertical members dividing panels into smaller areas are *muntins.*

The following sections will describe some of the traditional features of panelled doors and how they have been adapted to the mass-produced doors of today.

17.8.1.3 Panels

Framed and panelled doors are largely identified by the arrangement and type of panels used in the design. The panels can be solid timber, plywood, glass or various hardboards. Sections through panels are shown in Figure 17.56.

1. **Flush panels** are for external use and are made from solid timber. A bead joint, where the panel meets the framing, will help to hide any shrinkage that occurs in the width of the panel during seasonal changes. Note that the face of the panel is slightly below the level of framing by approximately 1 mm. This allows the face of the panel to be cleaned off before being assembled into the framing; the framing can then be cleaned off without touching the face of the panel.

Fig. 17.55 Four-panel door

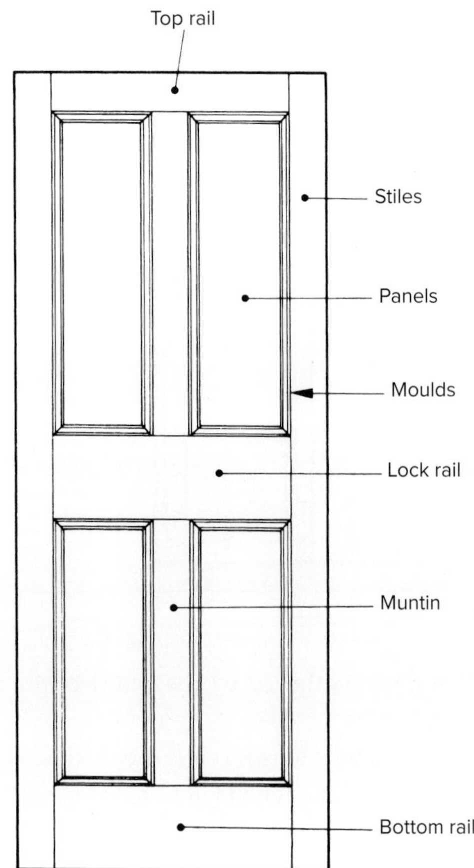

Fig. 17.56 Sections through panels: (a) flush panel; (b) flat-sunk panel; (c) raised panel; (d) laminated raised panel; and (e) panel fixed into rebates

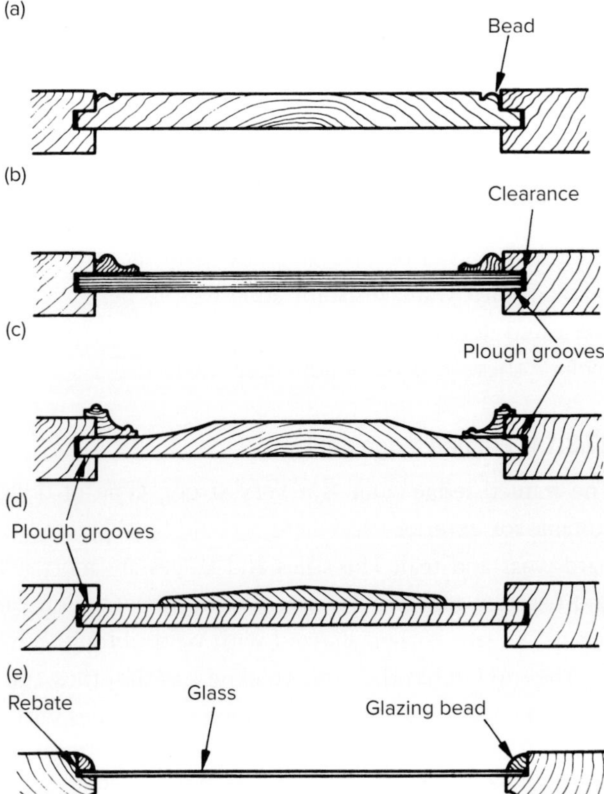

2. **Flat-sunk panels** have a flat face that is sunk below the level of the framing. They may be of plywood, but if they are of solid timber, they must be free to move with changes in moisture content. Any mouldings applied around the panel should be fixed to the framing; if fixed to the panel, they will shrink away as the panel moves.
3. **Raised panels** have the centre of the panel raised and the outside perimeter moulded and may be referred to as the *field*. The contour can be varied, and the face of the panel is again kept just below the level of framing. A simple way of forming a raised panel is to fix a raised centre section to a flat-sunk panel.

Timber panels are usually held in the framing by being fitted to a plough groove that runs around the inside edge. A clearance is allowed in the bottom of the groove.

Using a rebate is another way of fixing panels. Rebates are used particularly for installing glass panels, which, of course, cannot be fitted until the framing is completed. The glass is retained with glazing beads or putty.

17.8.1.4 Mouldings

Mouldings are placed around panels to soften the harsh change in level between the panel and framing, and to improve the appearance of the door. Mouldings can be of two types: they may be 'run in the solid', where the moulding is run directly into the solid timber (Fig. 17.57), or they may be 'planted', where the moulded section is machined as a separate piece in long lengths, and then mitred and fixed around the panels (Fig. 17.58).

Planted mouldings are again divided into two types:

1. **inlay mouldings**, where the profiles can vary but the thickness is such that the top of the moulding will finish just below the level of the framing
2. **bolection mouldings**, where the moulding will rebate over, and finish above, the level of the framing.

Bolection mouldings give a door the appearance of being thicker and more solid than perhaps it really is. In conjunction with raised panels, it makes a most impressive entrance door.

Fig. 17.57 'Run in the solid' mouldings

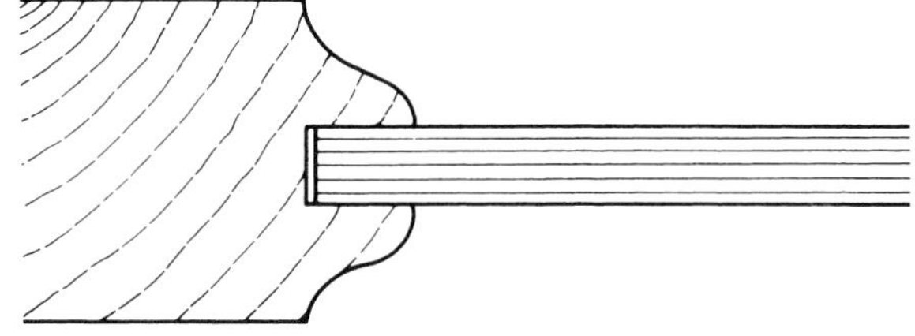

Fig. 17.58 'Planted' mouldings

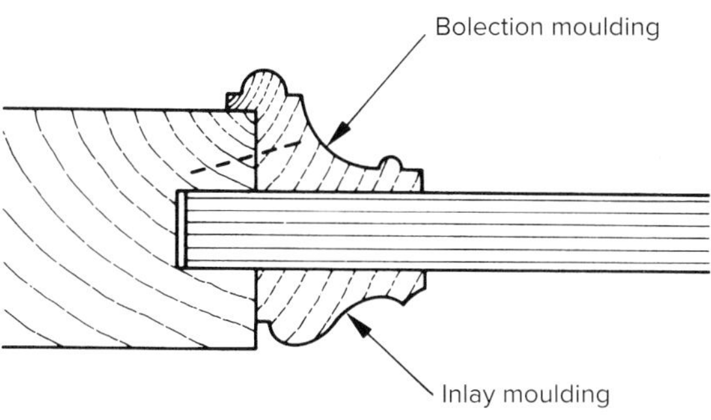

Fig. 17.59 Simulated panelled door

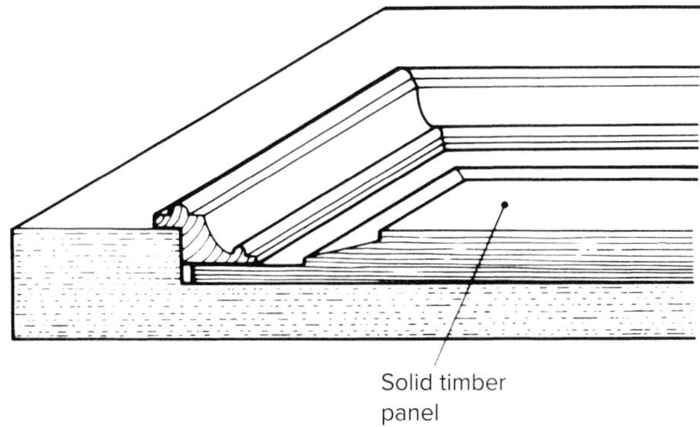

17.8.1.5 Simulated panelled doors

Figure 17.59 and 17.60 show examples of modern doors where the design is routed out of the solid panel. The panel material is usually MDF. The routing may be further enhanced by the addition of planted mouldings.

Fig. 17.60 Simulated panelled door, routed and with a planted moulding

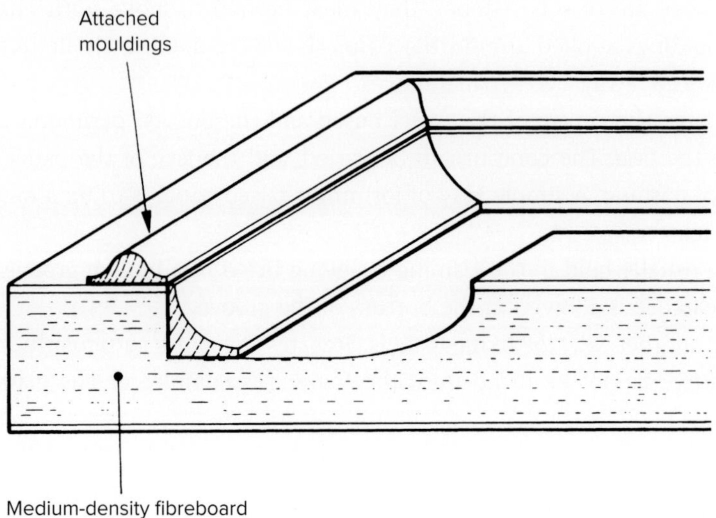

CHECK YOUR UNDERSTANDING

1. List five types of panels that can be found in doors.
2. Draw a horizontal section through a framed and panelled door, incorporating the following features: plough groove, raised panel, muntin, bolection moulding, inlay moulding.

Windows and glazing

Approximately 10 per cent of energy is gained or lost through typical window openings in a house. Heat energy is gained or lost through:

- radiation through the glazing (about two-thirds)
- conduction through glazing bars (if any)
- air leakage around the opening sash and frame (about one-third)
- conduction through the window frame itself.

The actual amount is dependent on the frame material and its sealing in the wall structure. In general, timber frames perform better than metal frames, although some composite material frames (metal with softwood core and PVC and softwood core) outperform timber.

The type of opening sash also determines the amount of air leakage (heat in or heat out). Traditional double-hung windows perform the worst and casement/awning windows the best. Gaps between the frame and wall should be sealed with low-expanding foam and single-sided, preformed and high-performance adhesive tape to provide an airtight bond and vapour-control seal.

Heat is absorbed through the glazing in two ways:

- solar gain directly transmitted through the glass
- energy absorbed by the glass itself and subsequently transferred inwards by convection and radiation.

The type of glass specified for windows is an important decision for lowering heat gain. Glass is available in a range of thicknesses and types. The most common are:

- *low-emissivity glass (low-e)*–glass that has a coating on the outer surface of metal oxide
- *double glazing or insulated glass unit*–two panes of glass in a frame where the gap between the panes is filled with a gas; improves the window performance by reducing the conduction of heat

- *solar control glass*–glass with a high iron content, which reduces its light transmittance and therefore its solar gain
- *tinted glass*–glass with a ceramic non-metallic film or coating applied to the outer surface, which converts solar gain to infrared radiation; similar to low-e glass
- *laminated glass*–this consists of two sheets of glass on either side of a centre layer, which bonds the glass; this glass eliminates approximately 99% of harmful UV rays.

Fig. 17.61 Heat is absorbed through glazing in two ways: direct transmission and energy absorbed and transferred inwards

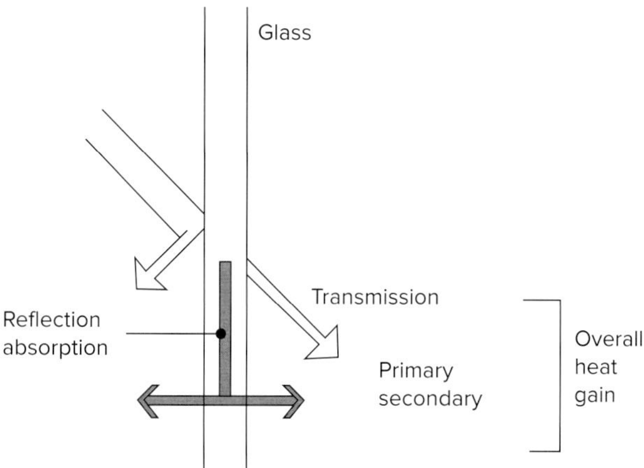

Student research

Investigate the other types of windows that are available for modern buildings. Find out about the range of materials that windows can be made from as an alternative to timber.

End of chapter activity

Activity 1:

Refer to the plans located in the appendix of this publication.

1. Identify how many external windows and doors the building has.
2. Identify how many internal doors (excluding cupboards, etc.) are located in the building, and what size they are.

AUSTRALIAN BUILDING CODES AND STANDARDS

AS2688—Timber and Composite Doors

AS2047—Windows and External Glazed Doors in Buildings

Australian Window Association—*An Industry Guide to the Correct Fixing of Windows and Doors*

AS 4145.2-2008—Locksets and Hardware for Doors and Windows—Mechanical Locksets for Doors and Windows in Buildings

Tradie talk—Make sure the jambs are in wind

Chapter 18

Internal linings and fixings

Learning Objectives

LO 18.1 Use plywood

LO 18.2 Install timber panelling

LO 18.3 Identify and install doorsets

LO 18.4 Hang doors

LO 18.5 Install trims

Introduction

Fixing is the term applied to the interior finishing of a building and includes such things as fitting door frames and hanging doors, applying finishing moulds around door and window frames, fixing built-in cupboards and fitting hardware furniture to doors and windows. (Note: If a fitted floor is used in a timber-framed building, it must first be laid before any fixing out work can be commenced.) This chapter deals with the traditional methods of fixing out a house.

Setting up a good work area is essential for achieving the best results. If possible, one room should be used as a cutting area workstation, to minimise the spread of waste, dust, etc. around the new building. Where possible, the materials you will be using should be located next to the machines, waste vessels should be established, and tools should have a base, instead of being scattered throughout the building–which means they are harder to locate and risk getting lost.

18.1 Plywood

Plywood for interior use will have one side with a face veneer of an attractive timber species. Plywood is commonly available in lengths of 2400 mm or 2440 mm, with a width of 1200 mm to 1220 mm. Occasionally, longer sheets are available.

Plywood can be supported on battens or studs, and spacings should suit the width of the sheet being used. Noggings should be inserted at 600 mm centres for vertical support. Ply sheets can be nailed or screwed.

Refer to the manufacturer's specifications for fixing details. The fixings can be hidden where nail holes are punched, or screws are recessed and filled with a coloured stopping compound matching the veneer colour. Sometimes the fixings are part of the desired look, usually screws, so it is very important to get the spacings and lines of screws right.

Plywood may also be fixed using wallboard adhesive, according to the manufacturer's instructions.

18.2 Timber panelling

Jointed timber panelling is also referred to as *shiplap* and *tongue and grooved panelling.* For internal linings it is normally at least 12 mm thick.

The lining boards selected should be seasoned. Wide boards should be avoided where there are extremes of temperature or humidity, as the timber may tend to cup or otherwise distort, affecting the appearance of the surface. Where wide fluctuations in atmospheric conditions are likely to occur, it may be best to use shiplap rather than tongue and grooved boards because of the easier movement between timber panels.

Tongue and grooved boards may be 'V' jointed or flush-edge jointed as in floorboards. Most popular is the 'V' joint, often with a separate 'V' machined out of the back of the board to give the choice of widths for different effects.

Shiplap boards were originally designed so that one panel overlapped the next, thus hiding a direct joint. Modern milling methods make available shiplap with a tongue and groove jointing system, which allows for secret nailing so that no nails are visible in the finished panelled surface.

Timber panelling is either fixed to the noggings or to battens nailed to the studs when the panels are vertical. Occasionally, they may be installed diagonally or horizontally where the studs themselves must then be used. The nails are normally 40 mm × 1.6 mm bullet head nails, for the smallest possible head size.

Fig. 18.1 Profiles of lining boards

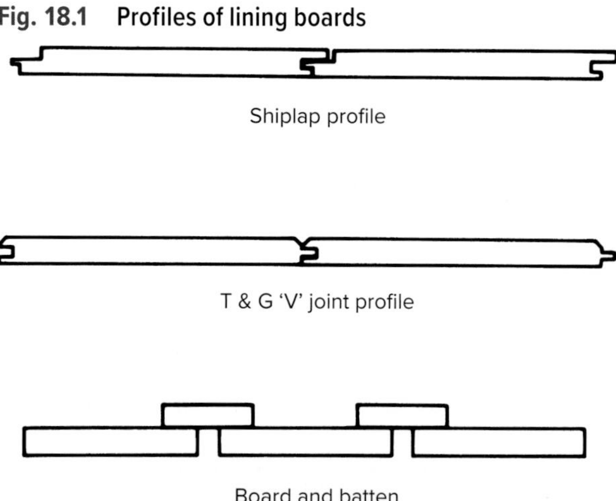

18.2.1 Timber mouldings

Timber mouldings have numerous uses, including:

- to cover junctions where wall linings meet, e.g. the corner of a wall
- to cover a junction where two different wall linings meet, e.g. where dado pine lining meets plasterboard, generally 1.2 metres up from the floor
- as an alternative to cornice at the wall-to-ceiling junction
- to provide decoration or adornment around windows, doors, openings, etc.
- to cover a gap.

There are countless profiles (shapes); traditionally, many of the profiles had specific applications.

Joining mouldings along a run or around an external corner requires a mitred joint to be used. If joining at an internal corner, then a scribed joint is preferable. These joining methods are discussed in more detail later in this chapter.

Fig. 18.2 Examples of common timber mouldings

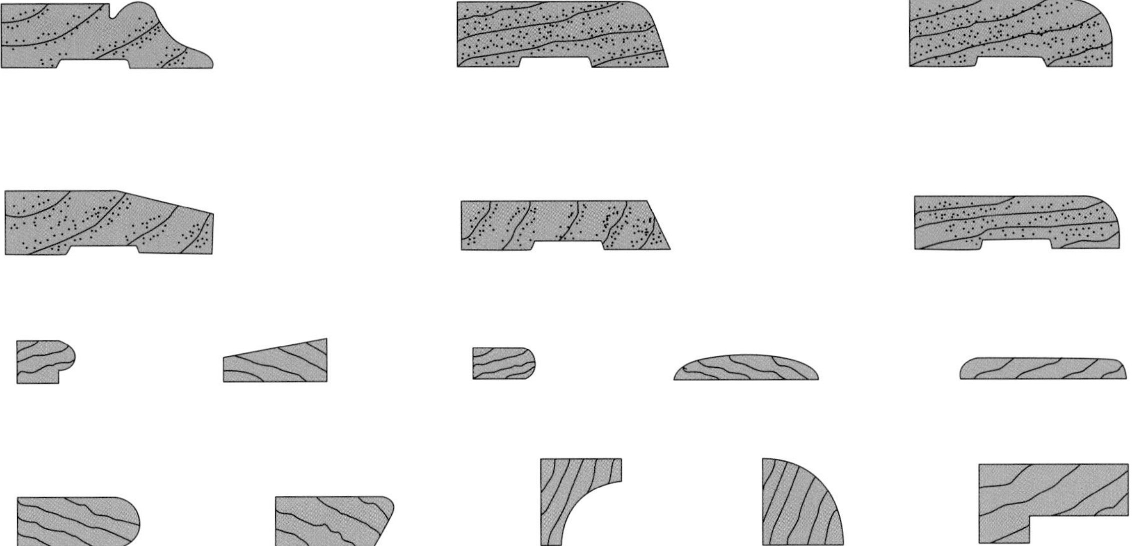

18.3 Doorsets

Doorsets consist of two components:

- the jamb/frame lining, which is accurately fixed into the structural opening and which will support the door
- the door, which is, of course, the movable leaf fitted to the jamb lining.

18.3.1 Standard dimensions of the door opening

Standard opening sizes in a frame for doors are:

- Height–2100, 2400 and 2700 mm
- Width–700, 800, 900 and 1000 mm for single doors
  –1200, 1500, 1800 for double doors.

Standard-sized doors are made to fit into these opening sizes together with their jambs/frames, with room for adjustment for squareness and packing.

The actual standard door sizes are:

- Height–2040, 2340 and 2640 mm
- Width–620, 720, 820 and 920 mm for single doors
  –1120, 1420 and 1720 mm for double doors.

Robe doors are also available in odd sizes such as 2340 mm × 840 mm.

The thickness of doors should not be less than 35 mm for internal doors and 40 mm for external openings. Doors manufactured to these standard sizes are readily available and should be adopted for greatest economy. However, doors can be made to any required size.

18.3.2 Door types

18.3.2.1 Flush doors

Flush doors are widely used in domestic construction. They have a flat, smooth face and are easy to keep clean. Since they are mass produced, they are the most economical.

18.3.2.2 Hollow core doors

Hollow core or honeycomb doors are one of the most common types in use. The stiles and rails are timber with a minimum width of 32 mm, held together at the corners by staples. The core of the door is composed of a cellular infill of expanded cardboard the same thickness as the stiles. Bonded to each side by an adhesive is a sheet of plywood or hardboard with a minimum thickness of 3 mm. A lock block at least 400 mm long can be located on one or both sides depending on the manufacturer; its position is indicated on the outside edge of the stile.

For external use, doors made up with waterproof adhesives are available. This fact should be indicated on the edge of the door.

18.3.2.3 Solid core doors

In solid core doors the core is made of a blockboard infill consisting of individual strips of timber no more than 50 mm wide and all of the same species. Veneers are then bonded to each face: first a cross veneer on each side and finally a face veneer of selected timber, with the grain running in the same direction as the core. Edge strips matching the face veneer are applied to the edges.

18.3.2.4 Medium-density fibreboard core doors

In medium-density fibreboard core doors, the stiles are of timber no less than 42 mm wide. The top and bottom rails are doubled with a minimum width of 84 mm. The infill consists of solid fibreboard. Veneers are bonded to the faces, or the faces can be of a hardboard or plywood no less than 3 mm thick. The face of this type of door is often carved to give a panelled effect and can be most impressive as an entrance door.

Fig. 18.3 Hollow core door

Fig. 18.4 Solid core door

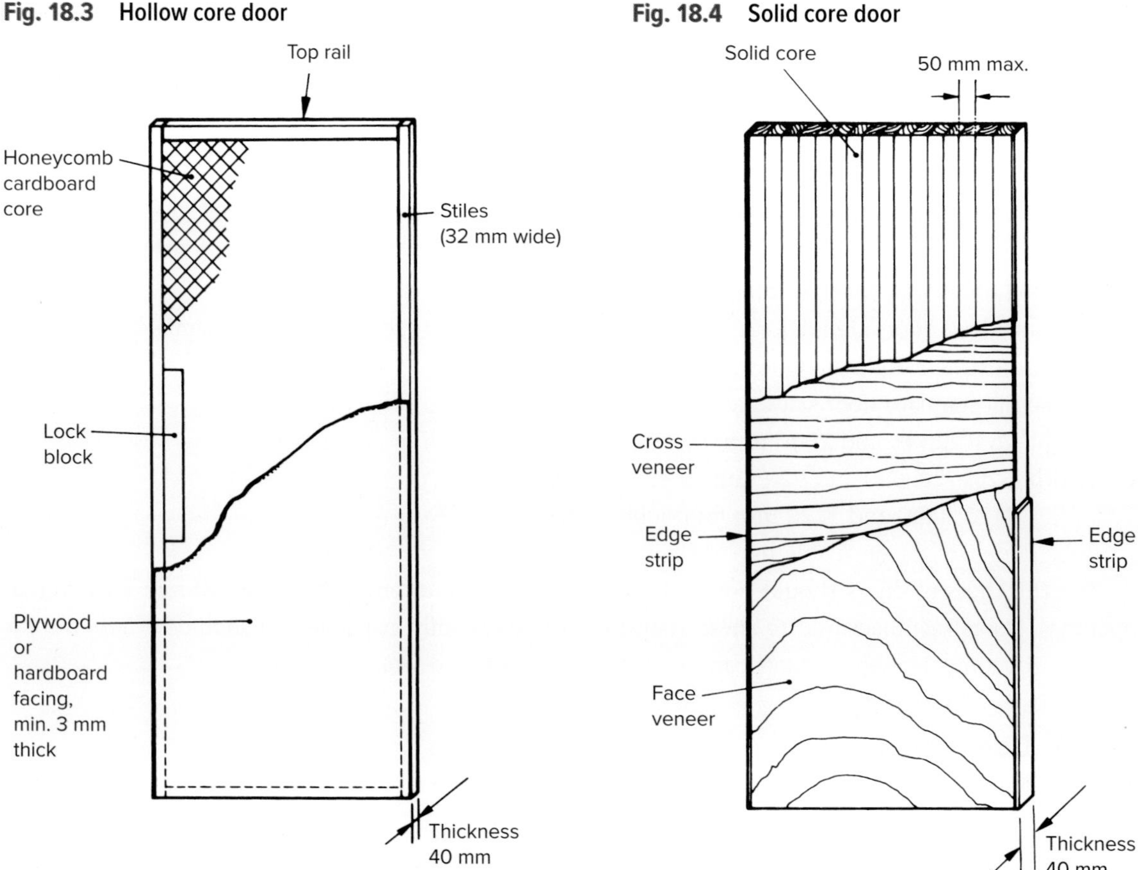

CHECK YOUR UNDERSTANDING

1. What are the standard sizes for single doors?
2. Select one type of flush door and describe its construction.

18.3.3 Jamb linings

Solid jamb linings/door frames come pre-made or in sets that need to be made up. They should have a nominal thickness of at least 38 mm and are rebated 12 mm deep to receive the door, as shown in Figure 18.5. The width of the rebate must be at least equal to the thickness of the door plus 3 mm for clearance. These frames can also come with a timber sill.

Fig. 18.5 Solid jamb lining

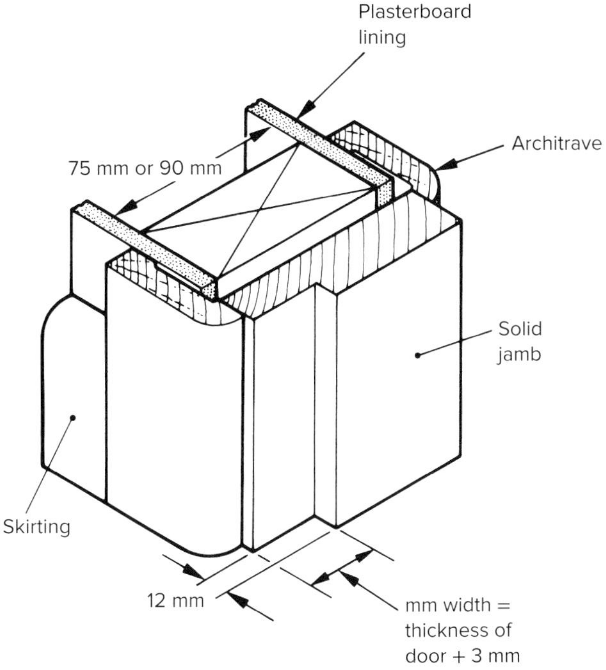

Rectangular door jambs at least 25 mm thick with a separate planted door stop can also be used. In the example shown in Figure 18.6, the jamb lining is fixed and the door is hung before the planted stop is fixed to the line of the door. For security reasons, loose stops are not recommended for external door openings.

Jamb linings are made up into sets: the vertical side members are the *jambs* and the top is the *head*. A set of rebated jamb

Fig. 18.6 Assembled or 'made up' jamb linings

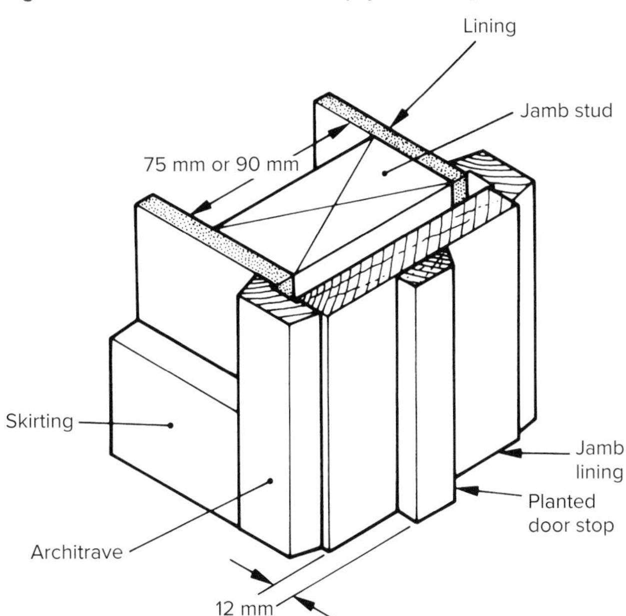

linings is shown made up in Figure 18.7. Note the *brace* and the *spreader* to maintain the correct width at the bottom. Accuracy is essential in setting out jamb linings if the door is to fit with a minimum of effort.

Many doors have very narrow stiles and rails and it is good practice to remove no more than is absolutely necessary during fitting. Allowance should be made at the bottom to allow for floor coverings. For carpets allow at least 20 mm and sometimes even more. The clearance at each side and top should be no more than 3 mm. In our example, 2 mm is allowed.

The length of the stile is set out to the height of the door plus the top and bottom clearances, say 2040 plus 2 plus 20 equals 2062 mm at least. If it is certain that a carpet is not being contemplated, then the bottom clearance can be reduced to 10 mm.

For width, set out the distance between the rebates equal to the width of the door plus a clearance of, say, 4 mm. The stile is housed into the head; see Figures 18.8 and 18.9, which show the setting out of the head.

Mark the distance between the rebates on the face. Take a short piece of jamb material and line the rebate up on this mark, and then mark each side for the width of the housing. Gauge to the depth of the rebate and remove the waste. Cut the spreader to the distance between rebates, assemble the jambs, hold square and attach the brace.

Sets of rectangular jambs with loose stops are made up in much the same manner

Fig. 18.7 Housing of jamb linings

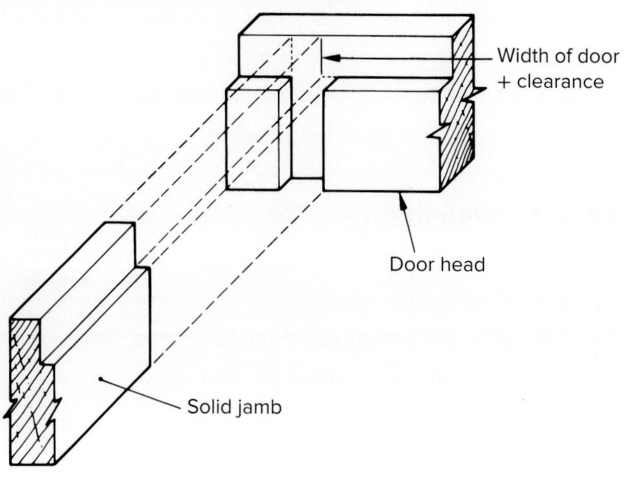

Fig. 18.8 Set-out of head

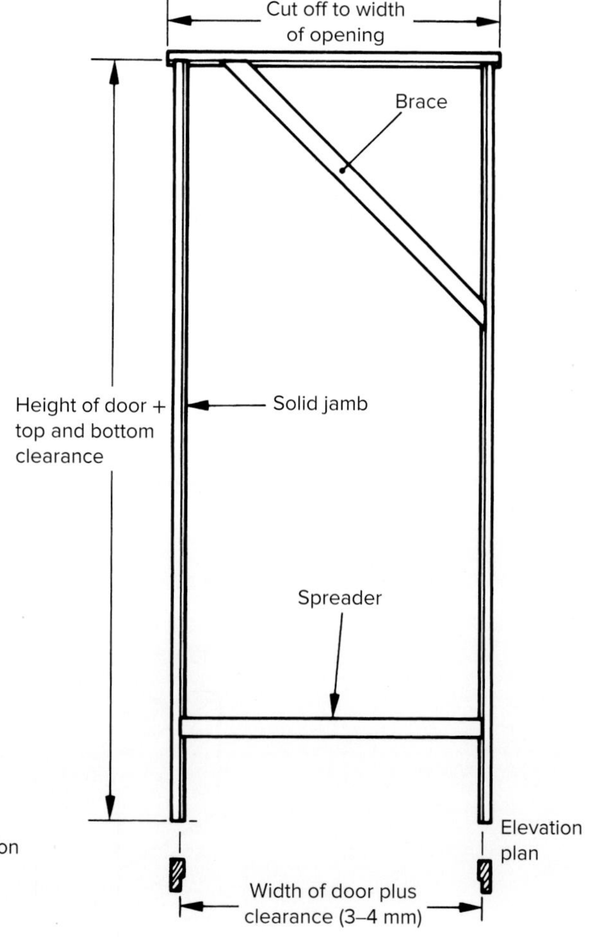

Fig. 18.9 Marking out jamb and head

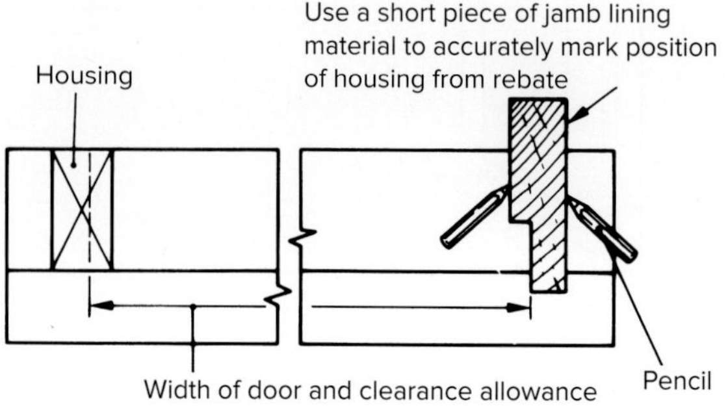

except that the stiles are made longer to house into the head at least 5 mm. The loose stops can be cut to length, temporarily pinned to the jamb and finally fixed after the door has been hung.

18.3.4 Fixing jamb linings

To fix the jamb linings, follow these steps:

18.3.4.1 Step 1

Check the floor across the opening for level. Theoretically it should be level, but due to some of the uncertainties when working with timber, there could be some minor discrepancies. Say the floor at *B* is found to be 5 mm lower than at *A*, then 5 mm must be cut off the stile at *A* so that the head will remain level.

18.3.4.2 Step 2

The jamb head is cut 1 mm shorter than the stud opening; this alleviates packing/wedging the top of the jamb and also allows for equal clearance on both sides of the jambs.

18.3.4.3 Step 3

Stand the linings in the opening and lightly drive a wedge directly over the stile between the jamb and the head trimmer to hold the jambs in place.

Fig. 18.10 Fixing jamb linings

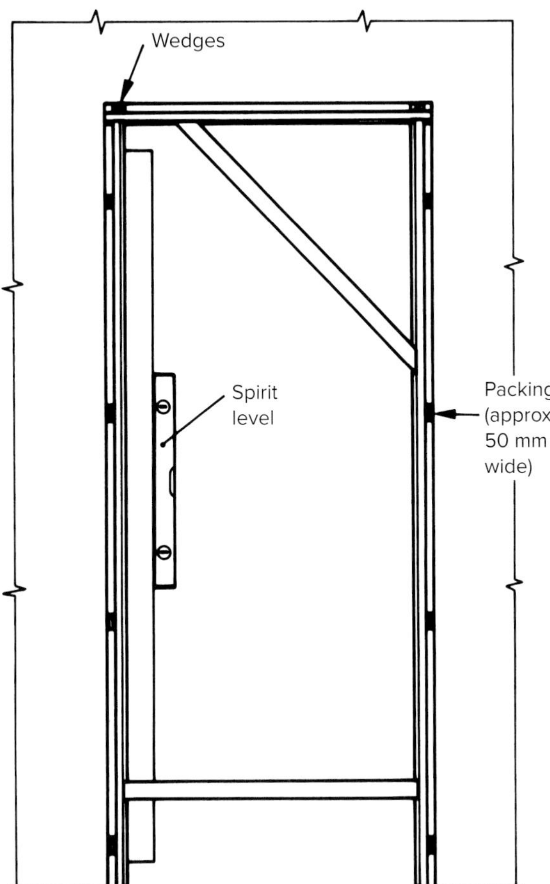

18.3.4.4 Step 4

Using a straight edge and level, check the jamb on one side for plumb in both directions and adjust as necessary. At a point about 150 mm from the bottom, pack the space between the jamb and door stud and secure the jamb to the stud. Suitable packing can consist of strips of plywood, hardboard or bituminous felt for making minor adjustments. Ensure that the nails are below the packing so that the packing does not fall and minor adjustments can be made to the packing if needed. Nails for the fixing of jamb linings are long and of a slender gauge, about 65 mm or 75 mm × 2.8 mm. Use two nails at each fixing point, one through the face and one through the rebate. Continue packing and fixing the jamb lining at the top and at intermediate points, taking care to keep it straight and plumb.

Fig. 18.11 Checking floor for level at door opening

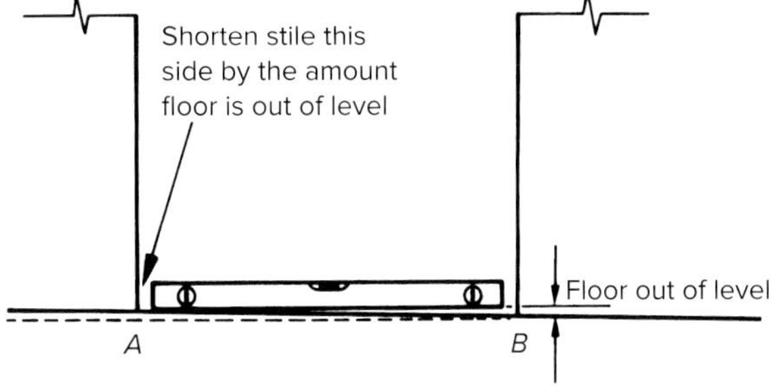

18.3.4.5 Step 5

Test and adjust the second side for straightness and plumb. Pack and secure in place, maintaining the correct width between the rebates. Take care that the two stiles are accurately plumbed laterally and are in 'wind' (i.e. the jambs are parallel in the vertical plane, when sited through opening). Remove spreaders and braces ready for fitting and hanging the door.

CHECK YOUR UNDERSTANDING

1. When making up a set of door jambs, how would you determine the length of the stile?
2. When fixing a set of door jambs, how would you ensure that the head will be level?

18.4 Door hanging

The process of fitting–hinging and applying the lock furniture to doors–is covered by the trade term *hanging doors.* When a door is hung, the sides assume a particular identity and are referred to as the *hinge side,* or *hanging side,* and the *closing side.* Doors are sometimes referred to as *right* or *left hand.*

If you view the door from the hinge side, then the side of the hinges is the *hand* of the door. Where pairs of double rebate doors are specified, the leaf that will open first is indicated as the left or right hand. When fitting doors with a separate lock block on one side, this side must become the closing stile and the hinges will be fitted to the opposite stile.

A *door block* is used on-site to hold the door upright when planing or working on the edges. It is worthwhile making up a door block that will clamp the door and hold it rigid without damaging the face. Figure 18.15 shows a door block using the flat wedging principle that will give satisfactory service for a long time.

Doors in domestic construction are usually hung with steel butt hinges; however, brass butts may sometimes be specified.

Generally speaking, only two hinges are required for the installation of most doors. It is good practice for external doors to have three hinges. Usually external doors are heavier and three hinges provide more security. Very heavy and tall doors may have four hinges, with two hinges being installed close together at the top of the door to support the door weight and stop the door from sagging.

Fig. 18.12 Door swing for right-hand door

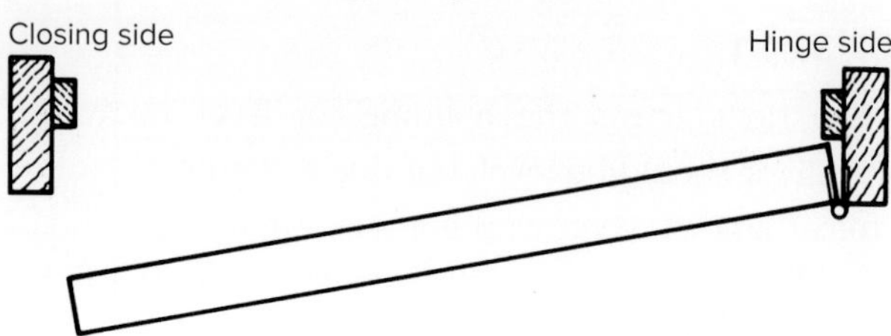

Fig. 18.13 Door swing for left-hand door

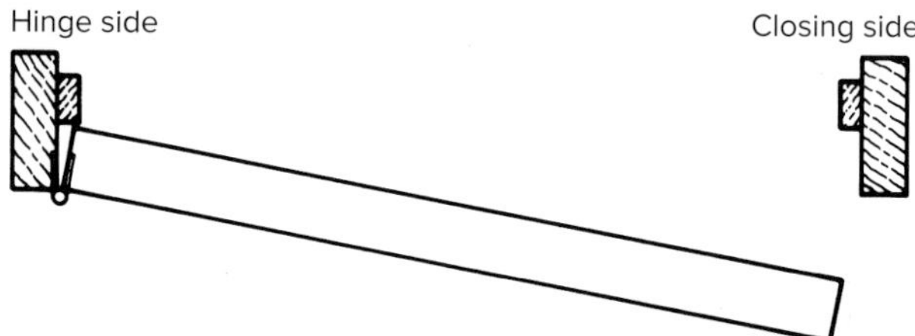

Fig. 18.14 Double door opening

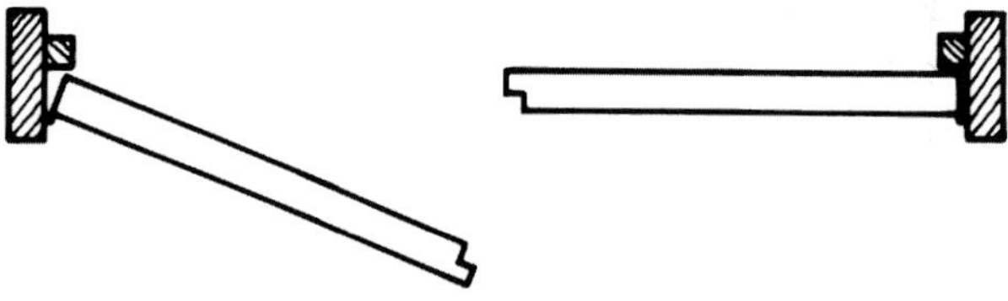

Fig. 18.15 Door block

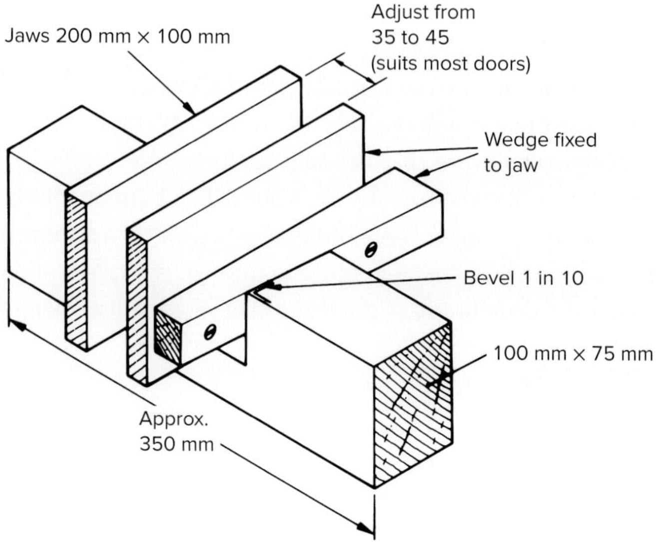

Fig. 18.16 (a) Broad butt hinge; and (b) standard butt hinge

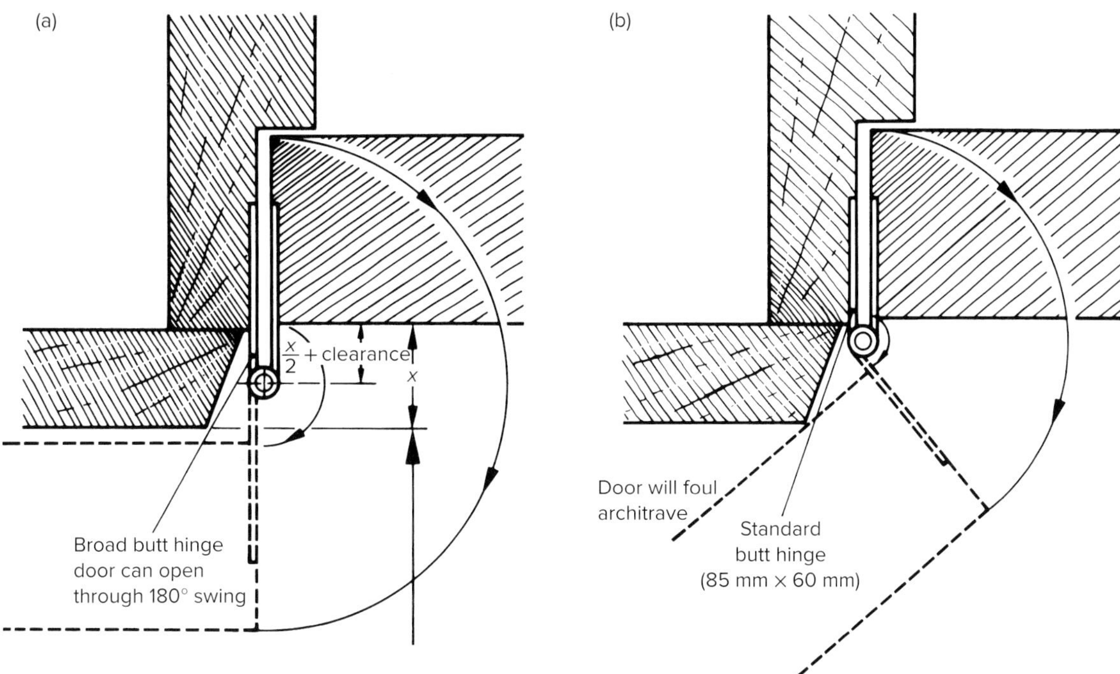

The standard hinge is satisfactory for any domestic doors, but where the door is required to open through 180°, this type of hinge will often foul the architrave and lead to the edge splitting and pulling away from the hinges. The broad butt hinge will overcome this.

Note the distance x from the hinge face of the door to the face of the architrave in Figure 18.16(a). The position of the hinge must now be adjusted so that the distance from the centre of the pin to the face of the door is half of x plus an allowance for clearance, say at least 2 mm, giving a clearance over the architrave of twice that amount.

18.4.1 Slide-in door sets

Sliding doors that disappear into a cavity formed within the thickness of the wall are commonly used as an alternative to hinged doors and permit the maximum use of floor and wall area.

A precision factory-made cavity unit can be obtained and fitted into walls as little as 70 mm thick. Split jambs are on one side of the unit, and concealed behind the jambs are steel angles that keep the sides of the cavity quite rigid. Built-in noggings on each side of the unit provide fixing for any specified wall sheeting, which is applied according to normal practice. The manufacturer's installation details are supplied with each unit and should be followed carefully.

TIP Determining opening dimensions for doors

The dimension of the opening required to install various applications of sliding door pockets is provided. As an example, for a single door 2040 mm × 820 mm that will project 50 mm beyond the split jambs, the wall opening size is as follows:

height (h) = door height + 110 mm measured from top of floor to underneath head

width (w) = 2 × door width + 15 mm

Therefore, opening size, h = 2150 and w = 1655 mm.

Only one jamb lining is required in the door opening and the door running track is fixed to a pelmet across the head of the opening. A removable pelmet permits the door to be installed or removed from the cavity within a few minutes and conceals the running gear. The bottom of the door is fitted with a low-friction rubbing strip and nylon guides that are adjustable and fixed to the split jambs (Fig. 18.18).

Fig. 18.17 Slide-in door sets

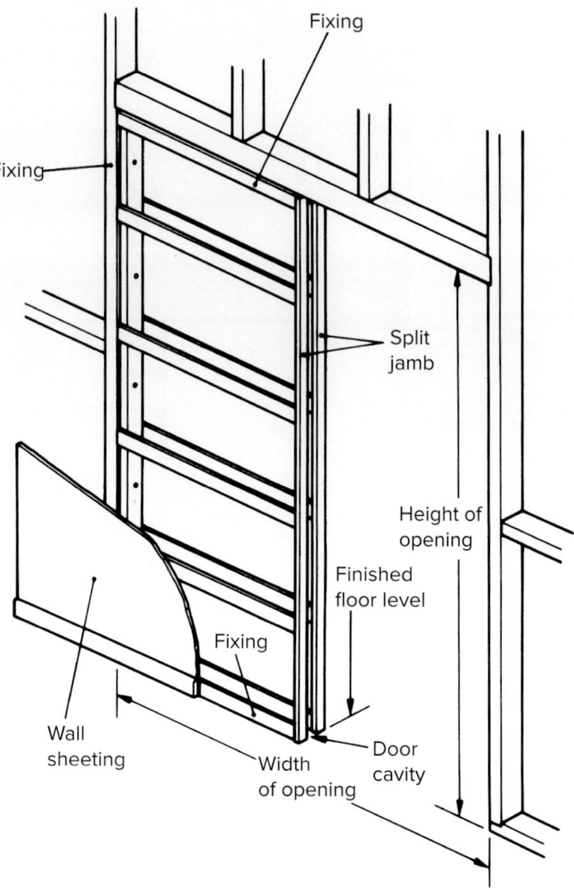

18.4.2 Fitting and hinging of doors

Follow these steps when fitting and hinging doors:

18.4.2.1 Step 1

Stand the door in the opening and note any adjustments that should be made to maintain a 2 mm clearance at both sides and at the top. The object is to reduce individual fitting to a minimum by making up and fitting the jambs accurately. For the average door, the closing stile is bevelled off 2 mm so that the back edge will clear the jamb as it rotates around the hinge pivot point. Very narrow or thick doors may have to be bevelled even more.

18.4.2.2 Step 2

With the door standing in the opening and with a spacer at the top to maintain a clearance, lightly wedge underneath to hold the door in position. Mark the position of the hinges–down 150 mm to 180 mm for the top hinge and up 200 to 250 mm from the bottom, are the usual positions.

Fig. 18.18 Split jambs

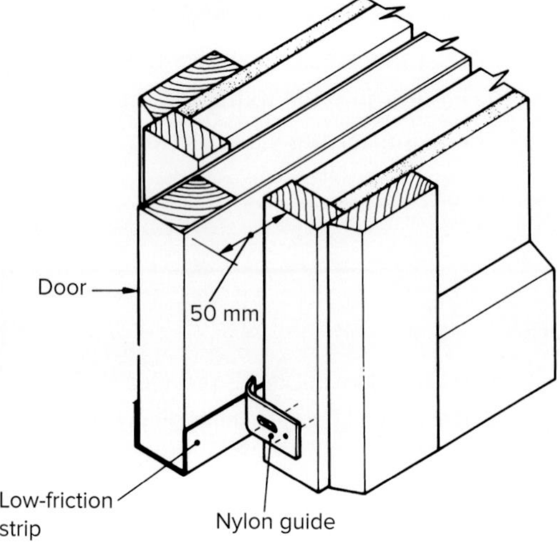

Fig. 18.19 Head and door track

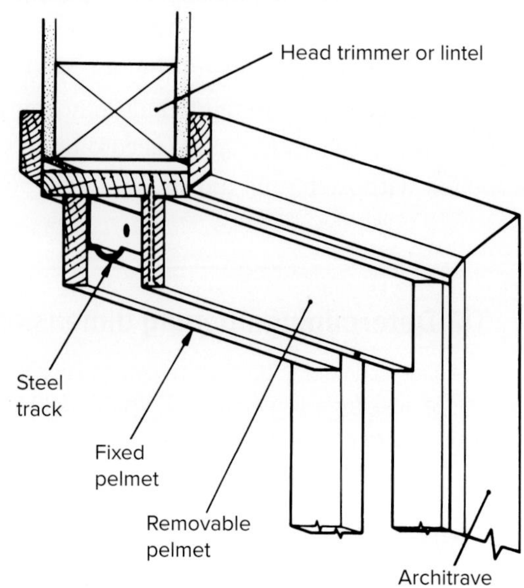

Fig. 18.20 Half and half hinging

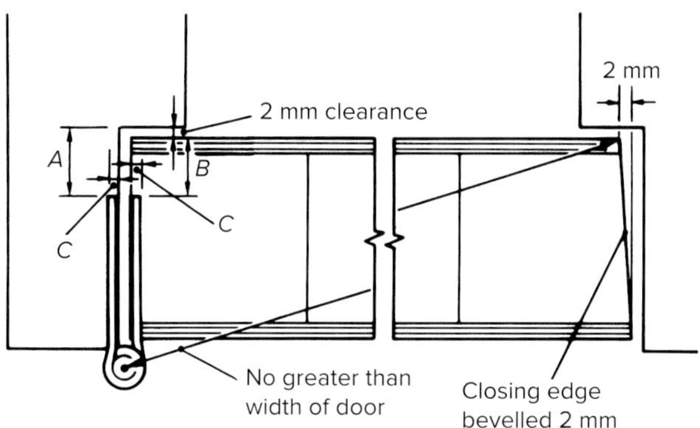

Fig. 18.21 Marking the length of the hinge

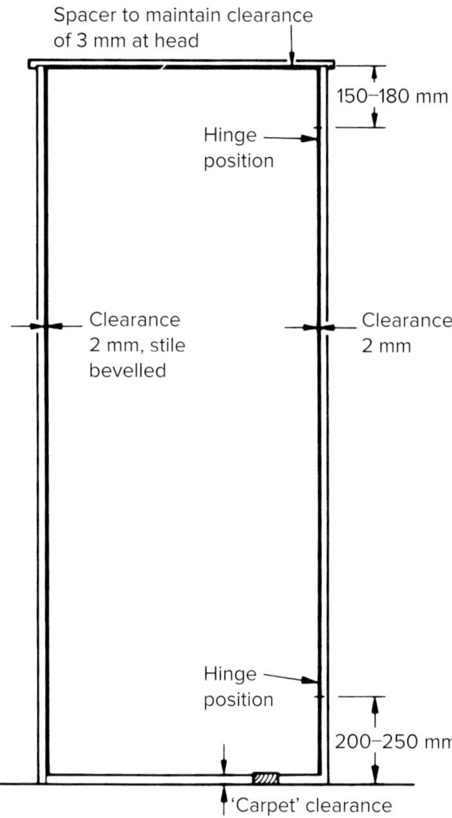

18.4.2.3 Step 3

Doors are usually hung by *half and half hinging,* that is, the flaps of the hinges are let into the door and into the jamb, maintaining the required clearance in between. When hanging a door to a rebated jamb there are three dimensions to be gauged to locate the door on the jamb (Fig. 18.20).

- Distance *A* is from the face of the rebate to the back edge of the hinge.
- *B* is the distance from the closing face of the door to the hinge where the difference between the two is the clearance against the rebate.
- *C* is the depth of the check-out for the thickness of the hinge and is gauged on the face of the door and the edge of the jamb. There is no need to let the back edge of the hinge in any more than just below flush.

18.4.2.4 Step 4

Hold the door in the door block and, using the hinge as a template, mark the length of the hinge. Gauge for distances *B* and *C*. Check out for the hinge and fix the flap of the hinges with two screws.

18.4.2.5 Step 5

Again using a hinge, mark the length of the hinges on the jamb. Gauge *A* from the face of the rebate and *C* on the edge of the jamb. Check out for the hinge.

> **TIP** The checking out of the hinge can either be performed with a chisel or a router. If installing many doors in the one job, a jig can be constructed to use with the router to speed up the checking in process.

18.4.2.6 Step 6

Stand the door beside the opening, locate the hinges in the check-out in the jamb and secure with two screws. Always drill a pilot hole for screws so that they can enter squarely and accurately without damaging the door or jamb.

18.4.2.7 Step 7

Check the operation of the door and make any minor adjustments to maintain a minimum clearance of about 2 mm. The thickness of the blade of a combination square is a handy gauge to check for clearance. Insert the remaining screws and make any minor adjustments to the gauge setting for further use.

In some areas, with the spread of the sub-contract system, door hanging and hinging has become a specialist task; some carpenters do nothing else. By equipping themselves with special tools (often made up by themselves), such as hinge router templates and portable mortise machines for fitting locks, it is remarkable how the work can be sped up. Once the basic requirements are understood, the procedures can also be varied. For example, some carpenters hinge the door to the jamb before it is fixed into the opening and adjust the jambs to suit the door.

1. Cut the jambs to length, set out and trench the head.
2. Prepare the door, bevel off the closing stile.
3. Preferably using a router with a router template, check out for flap of hinges on jamb and door hanging stile. Using loose pin hinges, fix mating flaps to door and stile.
4. Make up jamb linings, check the floor for level and adjust. Stand in the opening and fix the hinge stile, accurately plumbing up in both directions.
5. Using loose pins, hang the door; the closing stile can now be fixed, adjusting clearances to suit the door.

Obviously, whatever procedure is adopted, accuracy in working is essential if it is to produce the expected results and adequate rewards for the carpenter.

Fig. 18.22 Marking and check-out for hinge on door

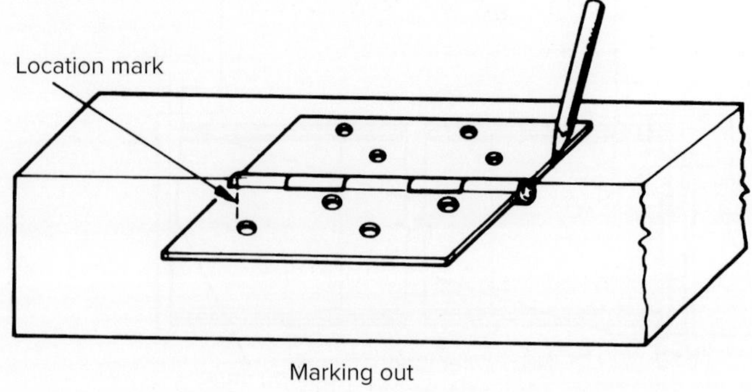

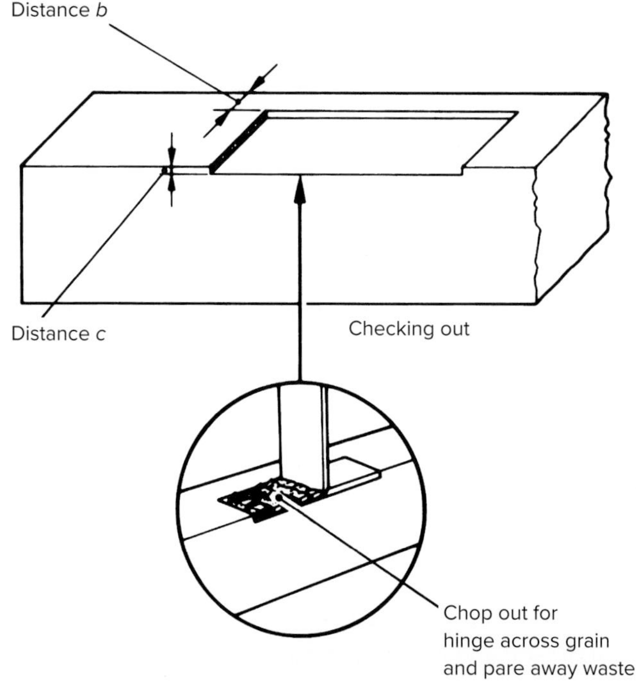

18.4.3 Butt gauge

The butt gauge is a gauge specially designed for use when fitting butt hinges. Gauging for *A*, *B* and *C* can be done with the one gauge, without any need for change. The gauge is adjusted so that when distance

B is set, distance *A* will automatically be approximately 2 mm greater, that is, the allowance for clearance.

18.4.4 Fitting to the jamb with loose stop

When doors are being hinged to a jamb with loose stop, gauging for the width of the hinge check-out is made from the hinge face of the door and the outside edge of the jamb, so that the face of the door will be a little below the level of the jamb. The loose stop is later fixed, allowing a clearance to the door.

18.4.5 Door furniture

Door furniture serves a dual purpose: it enables the door to stay closed, open or locked; plus, it adorns the door, almost like a piece of jewellery, enhancing the door's appearance and the interaction between it and the user.

Types of door furniture include:

- barrel bolts/panic bolts
- rim or mortise locks
- passage sets
- indicator bolts
- closers
- push plates and pull handles.

Additionally, locks require separate door handles, escutcheons, etc. and the style, shape, colour or finish are generally infinite. These items, along with the hardware listed above, are what people see and interact with when using a door.

When installing the door furniture, it is important to read the construction specifications, and to talk to the client or designer to establish the height at which the hardware is to be installed. All locks and passage sets require some sort of drilling of the door, so it is essential to know the correct height at which to install the hardware.

On the next page is the installation sequence of a mortise lock and passage set. Accuracy in marking is essential.

Fig. 18.23 Hinge check-out on jamb

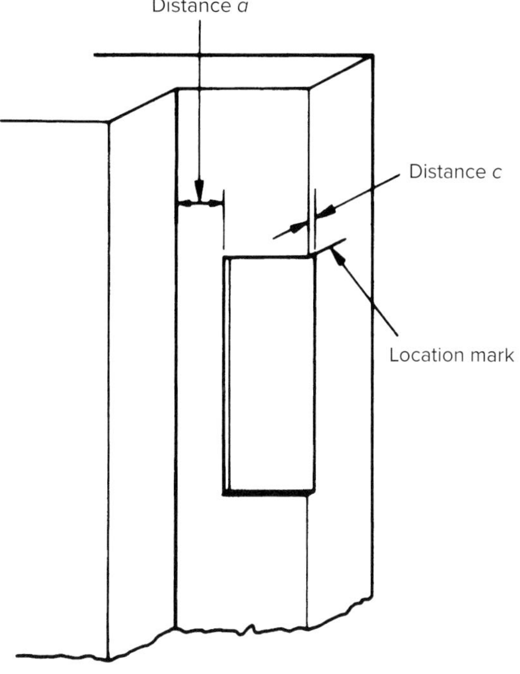

Fig. 18.24 Butt gauge

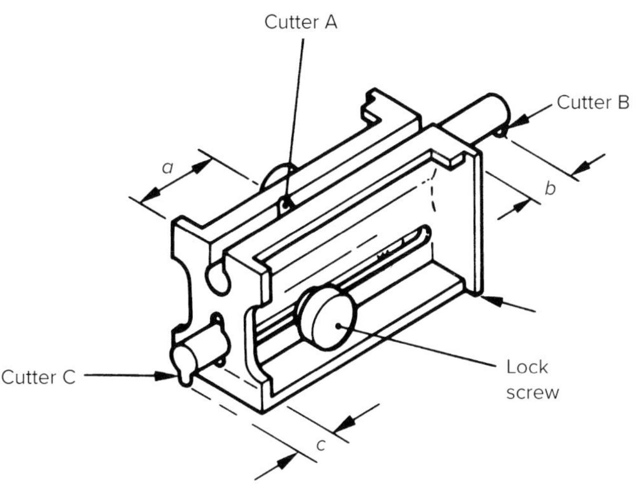

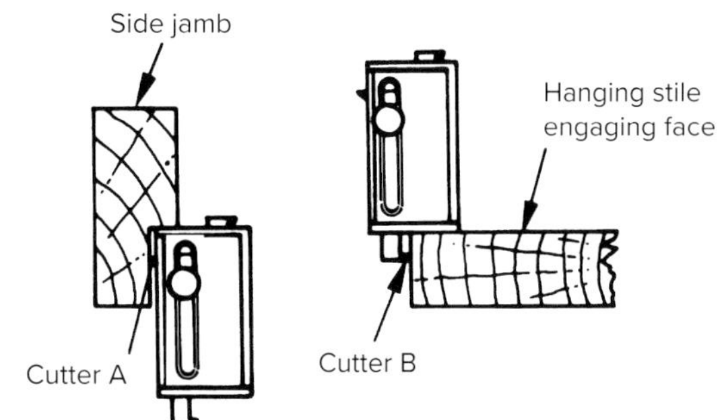

Fig. 18.25 A & B Installation sequence of a mortise lock and passage set

(a)

5400 Series Grade 1 Cylindrical Knob Lockset Installation Instructions

Tools Required

- Chisel & hammer
- #2 phillips screw driver
- Handle removal tool (supplied)
- ⅜" or ½" drill
- 2 ⅛" (54mm) hole saw
- 1" (26mm) boring bit
- 7⁄64" (2.8mm) drill bit

Any retrofit or other field modification to a fire-rated opening can potentially impact the fire rating of the opening, and Yale Locks & Hardware makes no representations or warranties concerning what such impact may be in any specific situation. When retrofitting any portion of an existing fire-rated opening, or specifying and installing a new fire-rated opening, please consult with a code specialist or local code official (Authority Having Jurisdiction) to ensure compliance with all applicable codes and ratings.

1 Mark Door

Mark line across edge of door for centerline of lock. Fold template over edge of door, centering on horizontal line. Mark centers of holes at proper backset.

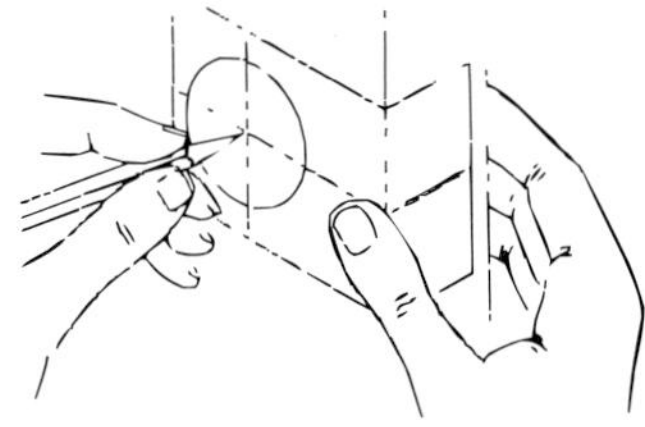

2 Drill Door

A. Drill 2–⅛" (54mm) hole thru the door. Cut Notches as shown on the template.

B. Drill 1" (25mm) hole in edge of door. Cut out for latch front 1–⅛" (29mm) wide × 2–¼" (57mm) high × 5⁄32" (4mm) deep.

C. Drill (2) 7⁄64" (2.8mm) holes for latch assembly

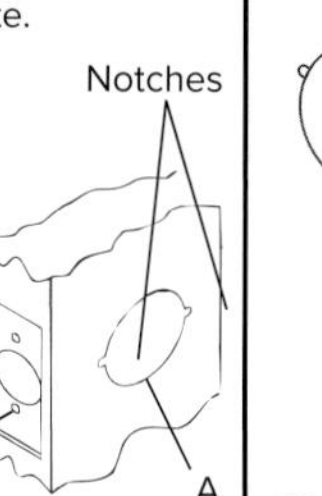

3 Install Latch Unit

Install latch assembly with screws provided.

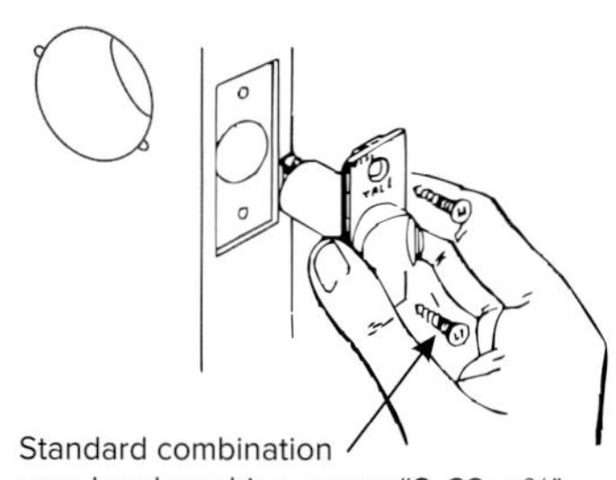

4 Remove Inside Knob

To remove knob, depress retainer with small end of spanner wrench, through small hole "A" in trim cap. Keep spanner wrench perpendicular with spindle. Pull knob off. Remove inside rose by unscrewing counterclockwise. For cylinder knob see step 8 and 9.

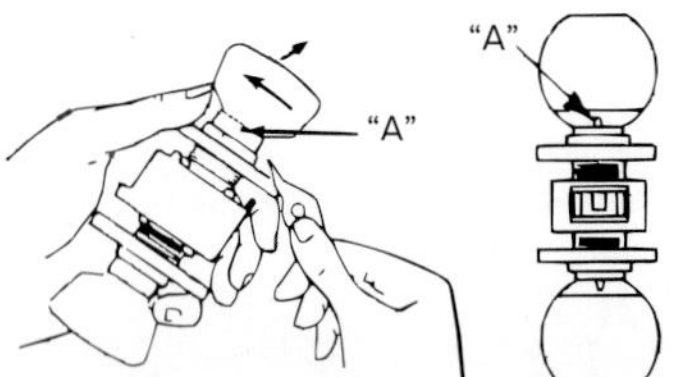

5 Adjust for Door Thickness *If necessary*

(Lock is packed preadjusted for 1–¾" (44mm) doors.)

Rotate the outside rose nut assembly until proper dimension is reached for required door thickness. Measure from the end of the pin to the lock flange.

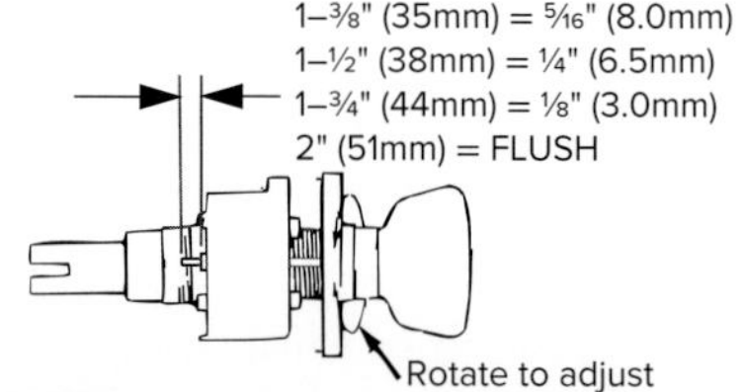

6 Install Lock

With lock case in place, insert lock assembly into 2–⅛" (54mm) hole, making sure that lock case hooks the retainer legs and retractor engages the bolt tail.

Caution: Do Not Force

If Lockbody does not engage latch easily, check door prep for errors.

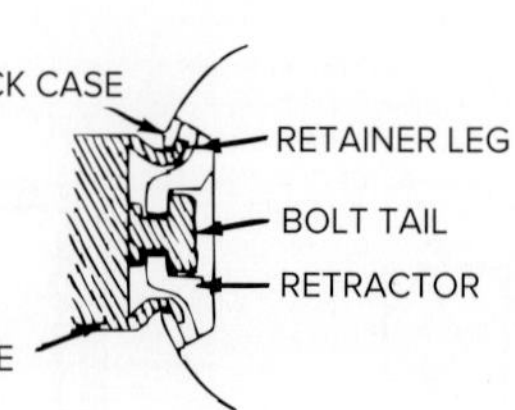

7 Attach Inside Rose

Slide inside rose over inside spindle and sleeve and screw rose into place. Tighten securely with spanner wrench supplied. Catch small lug "B" of spanner wrench in small hole "C" in rose and turn clockwise until tight.

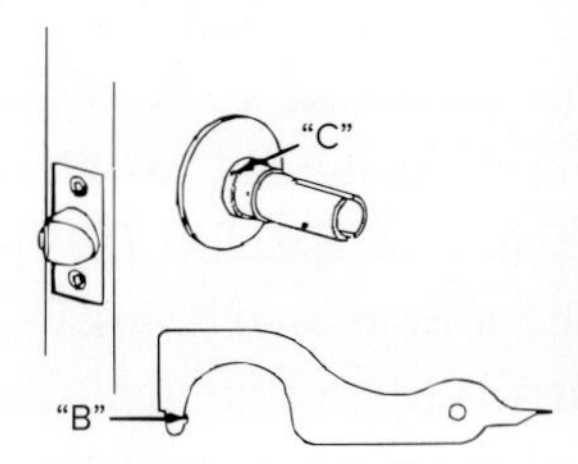

80-9150-0003-010 (08-12)

ASSA ABLOY

ASSA ABLOY Australia Pty Ltd

(b)

3770 SERIES MORTICE LOCK

MOUNTING INSTRUCTIONS FOR TIMBER BOOKS

P/No. 377$_{0-1}$21 01$_1$ 6

HAND OF DOOR ACCORDING TO ENGLISH PRACTICE

DOOR OPENING IN	DOOR OPENING OUT
INSIDE RIGHT HAND LOCK / LEFT HAND LOCK OUTSIDE	INSIDE RIGHT HAND LOCK / LEFT HAND LOCK OUTSIDE

- **Determine hand of lock required from chart above.**
- Ensure only holes required by handling, function and furniture are cut into door.
- Disregard instructions referring to features not required.
- The lock is supplied in passage mode. The hold back and anti-lock functions are de-activated.

1 LOCK SETTING

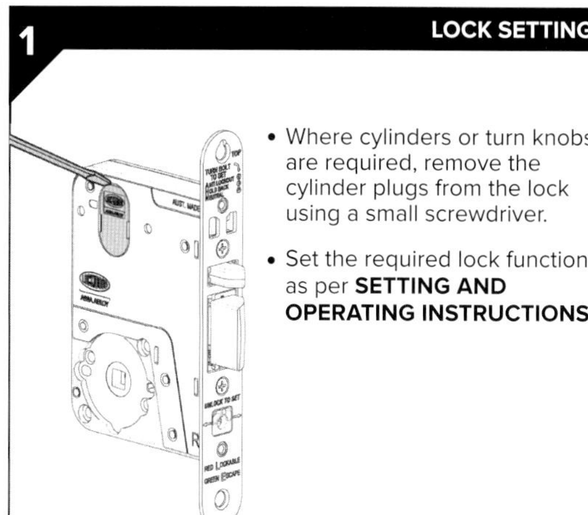

- Where cylinders or turn knobs are required, remove the cylinder plugs from the lock using a small screwdriver.
- Set the required lock function as per **SETTING AND OPERATING INSTRUCTIONS.**

2A LOCKING ADAPTOR INSTALLATION

- Rotate the locking adaptor so one leg is pointing towards the locking bar.
- With the locking bar in the unlocked (up) position; fit the locking adaptor into the lock. Ensure the cam is engaged within the locking bar.
- Secure by pressing in the cylinder retainer pin.

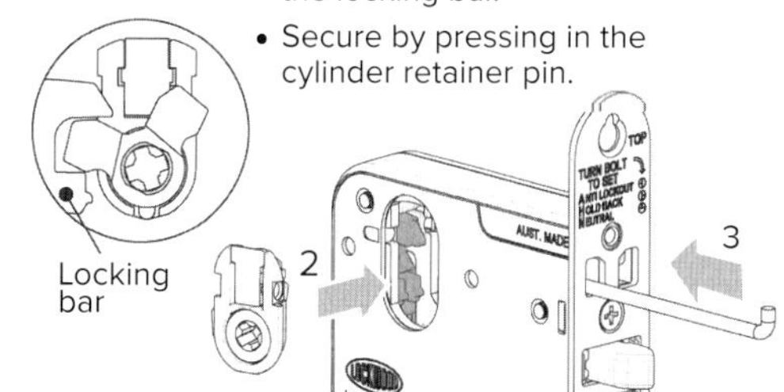

2B LATCHING ADAPTOR INSTALLATION

- Depress the latch bolt and fit the latching adaptor.
- Secure by pressing in the cylinder retaining pin.
- **Note: The face plate must be fitted to ensure correct function.**

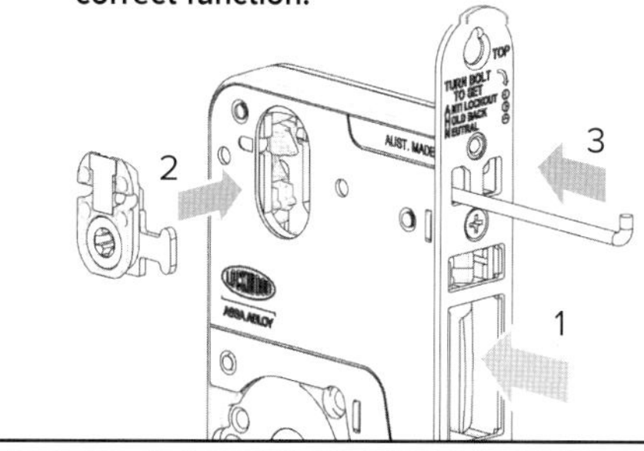

3 DOOR PREPARATION

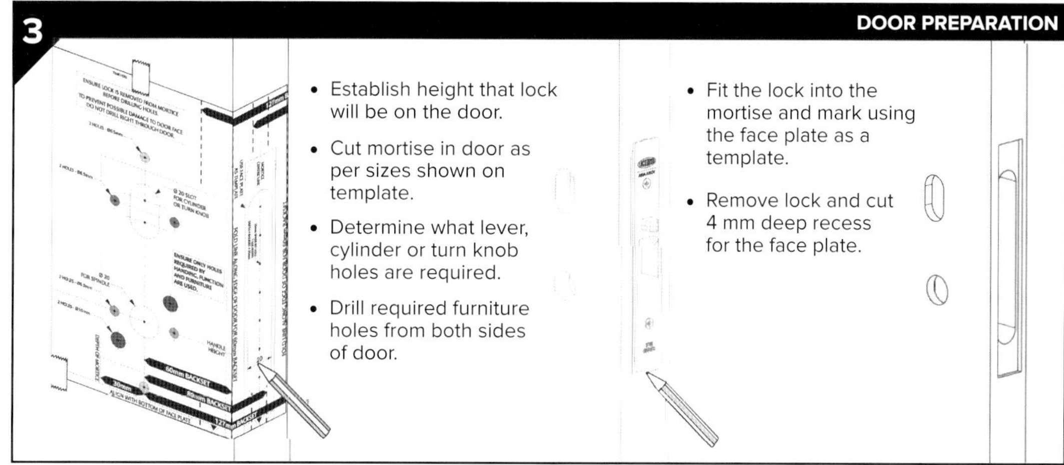

- Establish height that lock will be on the door.
- Cut mortise in door as per sizes shown on template.
- Determine what lever, cylinder or turn knob holes are required.
- Drill required furniture holes from both sides of door.
- Fit the lock into the mortise and mark using the face plate as a template.
- Remove lock and cut 4 mm deep recess for the face plate.

(continued)

3770 SERIES MORTICE LOCK

MOUNTING INSTRUCTIONS FOR TIMBER BOOKS

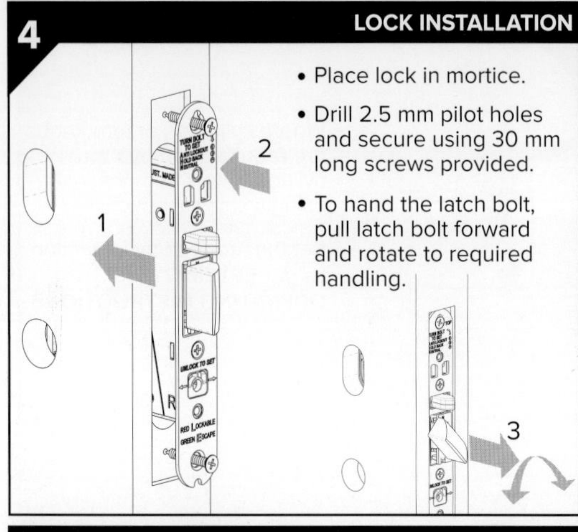

4
LOCK INSTALLATION
• Place lock in mortice.
• Drill 2.5 mm pilot holes and secure using 30 mm long screws provided.
• To hand the latch bolt, pull latch bolt forward and rotate to required handling.
1
2
3

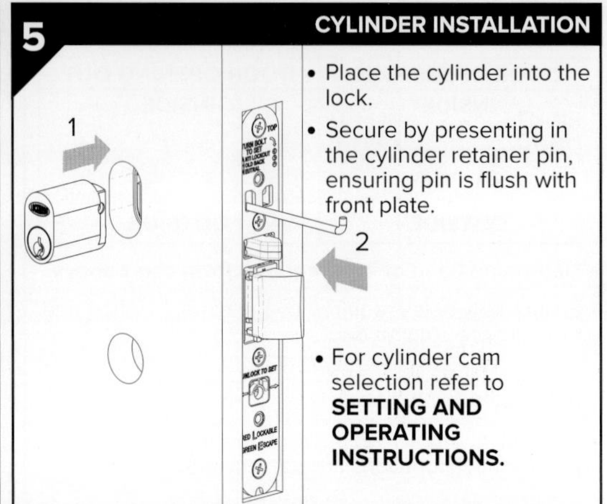

5
CYLINDER INSTALLATION
• Place the cylinder into the lock.
• Secure by presenting in the cylinder retainer pin, ensuring pin is flush with front plate.
• For cylinder cam selection refer to SETTING AND OPERATING INSTRUCTIONS.
1
2

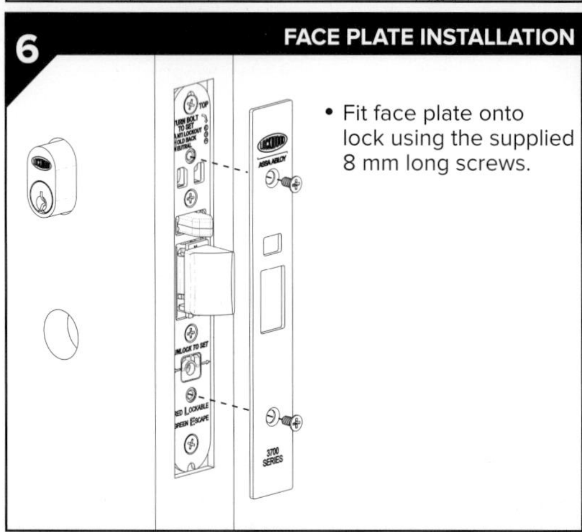

6
FACE PLATE INSTALLATION
• Fit face plate onto lock using the supplied 8 mm long screws.

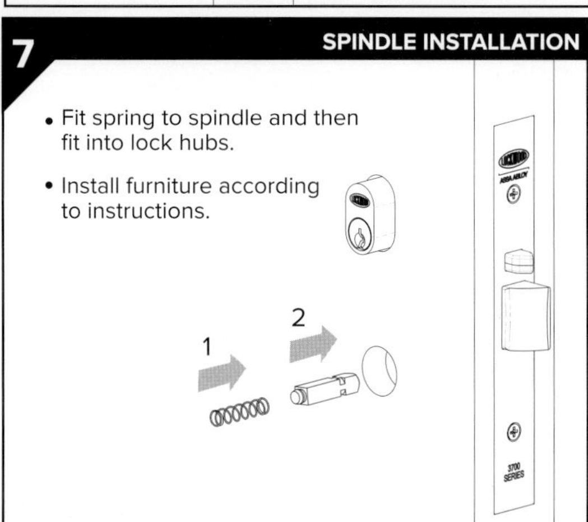

7
SPINDLE INSTALLATION
• Fit spring to spindle and then fit into lock hubs.
• Install furniture according to instructions.
1
2

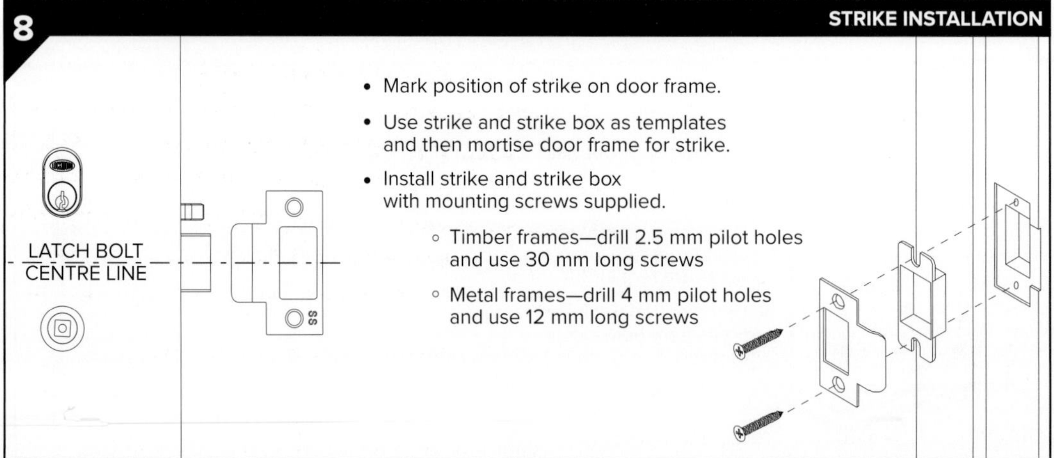

8
STRIKE INSTALLATION
• Mark position of strike on door frame.
• Use strike and strike box as templates and then mortise door frame for strike.
• Install strike and strike box with mounting screws supplied.
◦ Timber frames—drill 2.5 mm pilot holes and use 30 mm long screws
◦ Metal frames—drill 4 mm pilot holes and use 12 mm long screws
LATCH BOLT
CENTRE LINE

ASSA ABLOY Australia Pty Ltd

CHECK YOUR UNDERSTANDING

1. List three types of door furniture.
2. Indicate in a sketch what distances must be measured from the top and bottom of a jamb so that the hinges can be correctly installed.

18.5 Trims

18.5.1 Architraves

Architraves are mouldings used to frame the outline of doors and windows and to cover the joint between the frame and the structural members. Architraves are commonly 50 mm to 75 mm in width and 15 mm to 25 mm in thickness. They can be a variety of moulded profiles; some of the most used profiles are shown in Figure 18.26.

Note how some architraves may have a wide milling groove taken out of the back face. This makes them much easier to fit, as the groove will bridge any irregularities in the level of the wall sheeting while the edges will fit down tightly on the wall and frame.

The corners of architraves are mitred.

When fitting architraves to an opening, cut the side pieces and fix in position, allowing a *quirk*, approximately 3 mm, at the edge of the jamb. Cut and fit the head piece and fix the bottom edge to the jamb head. Flush the mitred joint and secure with a nail through the top edge only.

In the two top corners of the jamb mark the 3 mm quirk on the head and the stile. This intersection will give you the length of each piece of architrave to the short point of the mitre.

It is good practice to remove the arris (sharp edge) off the moulding prior to installing the architraves. This can be achieved using either a sharp block plane, which you can keep in your nail bag, or some abrasive paper (sandpaper). Select a paper with a grit of about 180 to avoid too rough a finish. Removing the arris from timber ensures that any finish to be applied, such as paint, stain, lacquer, etc., will

Fig. 18.26 Architrave profiles

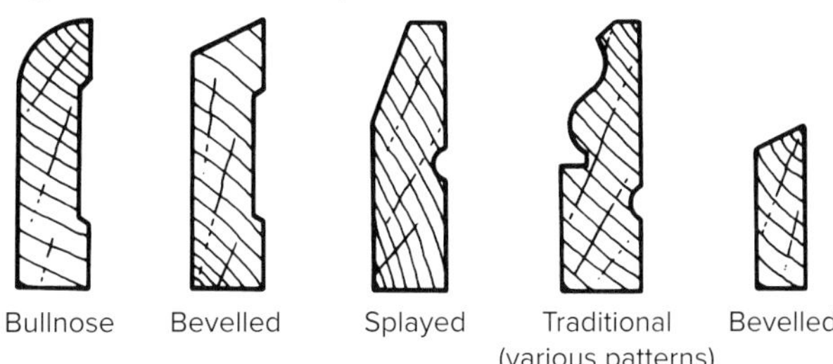

Fig. 18.27 Architrave mitre joint

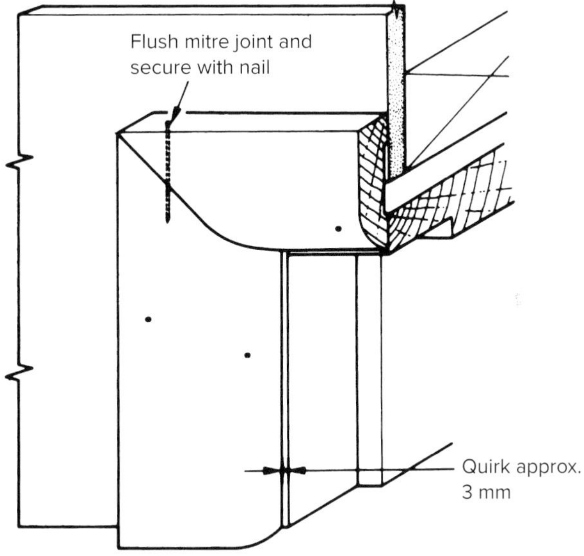

Fig. 18.28 Procedure for fixing architraves

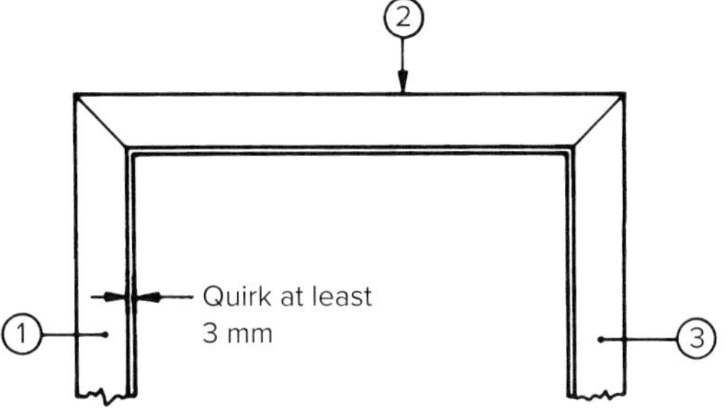

stick (nothing sticks to a sharp edge). Removing the arris prior to installation also makes the task much easier. This step should be carried out prior to the installation of any finished timber work.

1. Starting from the left side, cut the mitre on the architrave and then cut to length and fix.
2. Cut the left-hand mitre on the head piece of architrave and fit. When this mitre fits well, cut the second mitre and fix.
3. Hold the right-side architrave on the jamb and mark off the length of the architrave. Cut the mitre and offer it up to see if the mitre joint fits well. There is scope for some adjustment on this cut, but only a small amount.
4. Once fitting well, fix and remove the arris off the back of the architrave with some sandpaper.

CHECK YOUR UNDERSTANDING

1. In what sequence should architraves around a door be installed?

WORKPLACE SCENARIO

It's always good practice to cut one piece of architrave at a time. Therefore, you are dealing with only one mitre at a time. I like to start from the left. If you want to, instead of measuring the length of the architrave, place the architrave where it's going to be installed and mark its length directly off the quirk line, which you would have marked at the top two corners of the jamb. Cut the architrave and install. Next, start with the head. Cut the left-hand mitre and offer it up to the first piece. At this point, the mitre can be adjusted so that it fits properly. Repeat the process, working on one cut at a time. If you don't do this, and precut your architraves, adjusting the mitres will be hard, as removing anything from them will decrease the length of the architrave, thus making it too short—meaning it won't sit on the floor.

18.5.2 Skirting

Skirting is located at the base of the internal walls and covers the joint between the wall and the floor, protecting the wall from damage.

Moulded profiles for skirtings are similar to those for architraves except that skirtings can range in size from under 75 mm to 100 mm wide; in the restoration of old buildings, skirtings 150 mm to 200 mm wide, and occasionally more, may be used.

When fitting skirtings, scribed joints are used for internal corners, and external corners are mitred.

Fig. 18.29 Skirting and skirting blocking

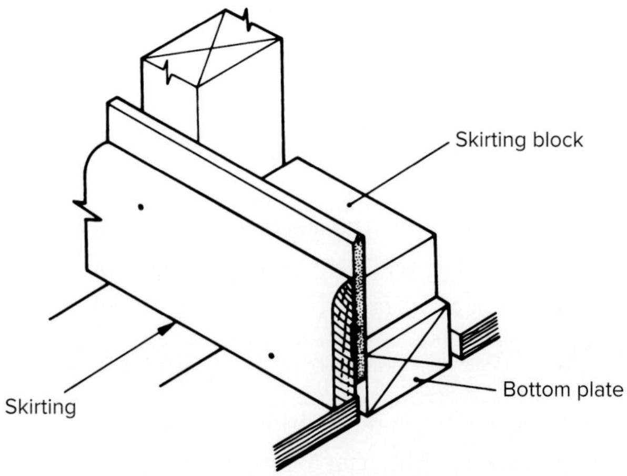

18.5.2.1 Scribed joint

To make the scribed joint, the first piece *A* is cut off square and fixed in position. The second piece *B* is mitred at half the angle of intersection, usually 90°, as in this example, making the mitre angle 45°. The contour of the mould is then cut with a coping saw at the angle of intersection, in this case 90°.

The advantage of the scribed joint for internal corners is that piece *B* can be cut slightly long (say 2 mm, depending on length) and sprung into place, cramping the joint together. Thus, the joint will not be forced apart when nailed to the frame. Joints can at times open up due mainly to shrinkage of the structural framing and should be so arranged that a person will not be looking straight into the open joint from the most probable viewing positions.

Generally, the skirting opposite the door is fixed first and then work continues back towards the door, finally butting against the back of the architrave. A *pinch rod* can be used to measure the length of skirting quickly. This consists of two light laths of timber that are extended to measure the required lengths. Pinch them together and mark across them to ensure there is no movement as they are laid along the skirting and the length is marked from the rod.

Fig. 18.30 (a) Mitred joint; and (b) scribed joint

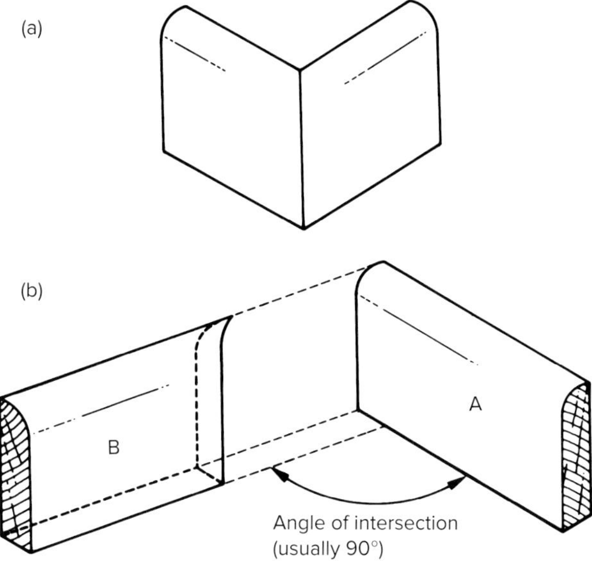

18.5.3 Fixing trim to window openings

Timber window frames are finished at the back of the sill with a window nosing and apron mould. Window nosings are milled from stock 50 mm or 75 mm × 38 mm and, when cut to length, are fitted around the window stud, and the mould returned on the ends. The nosing is drilled and fixed to the back of the sill using nails, which are so positioned that they will penetrate the window flashing close to its top edge only.

The apron mould covers the joint below the nosing and can be made up of various sections–bullnose, bevelled or scotia, for example. Great care must be exercised when nailing the apron mould to ensure no nails are allowed to penetrate the sill flashing.

The sides and head of the window opening are finished with an architrave in a manner similar to door openings, but depending on the thickness of the wall and the window frame used, an elbow lining may be necessary to cover the reveal from the frame to the face of the interior lining.

Fig. 18.31 A & B Making a scribed joint

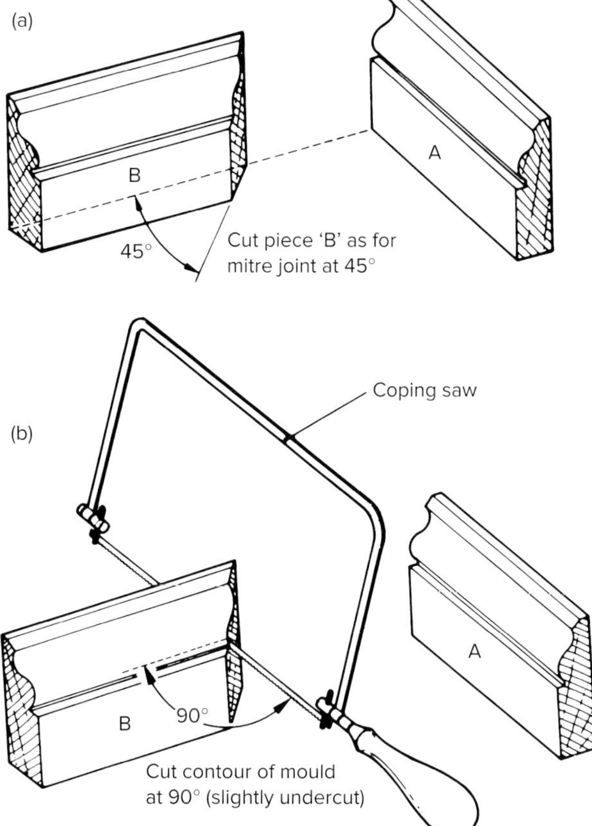

Fig. 18.32 Order of fixing skirting to a room

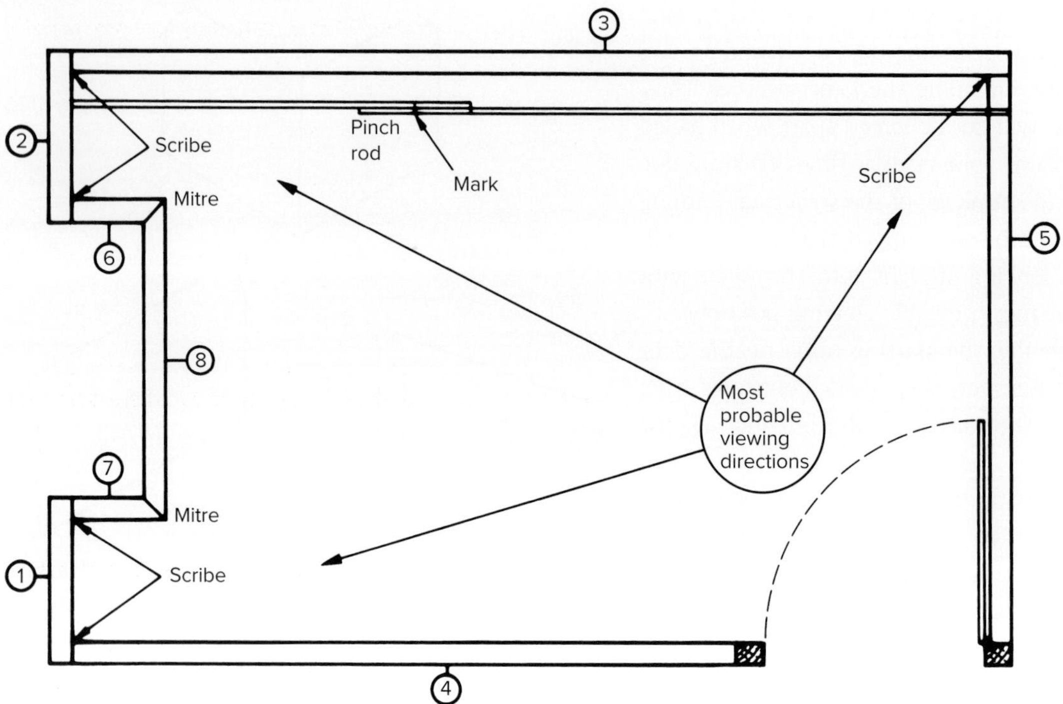

Fig. 18.33 Fixing of internal trim to window openings: (a) timber; and (b) aluminium

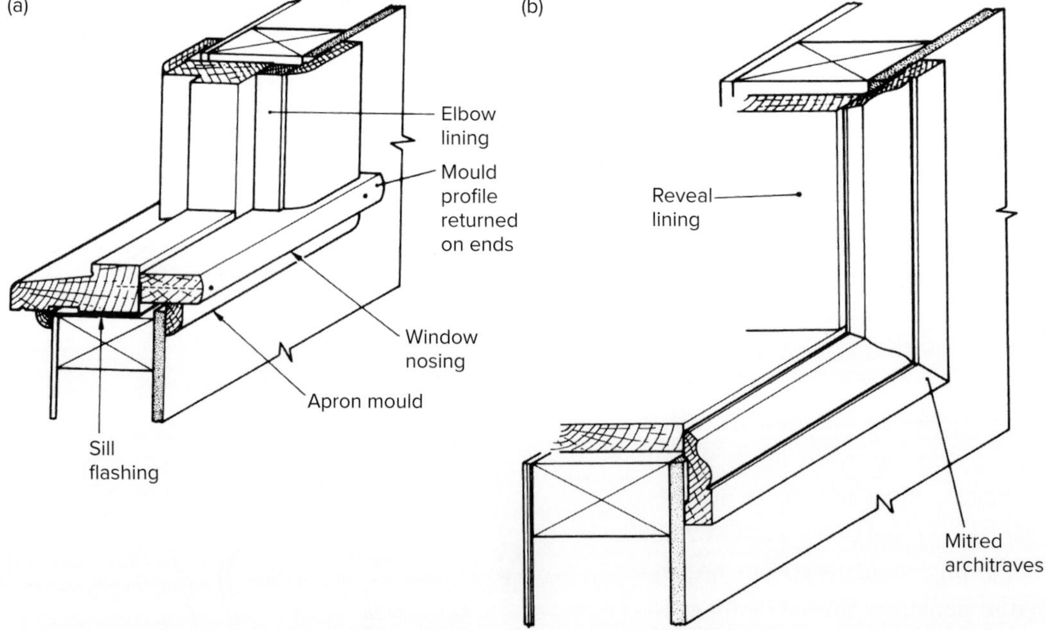

CHECK YOUR UNDERSTANDING

1. What general procedure should be adopted when fixing skirting boards to a room? Give an example.
2. Describe the joint you would use to join a skirting board at an internal corner.
3. What is the function of the following members when used to finish around a window opening?
   a. Architrave
   b. Nosing

18.5.4 Exterior door openings

Timber-framed buildings use a timber threshold to external doorways. This is milled from stock usually 175 mm × 38 mm of equivalent durable timber.

The threshold is made weather resistant by fixing it with a slope to the outside of, say 5 mm, which allows a strip of hardboard to be used as packing to the inner edge.

The jambs are housed to the ends of the threshold, which is also cut around the face of the exterior sheeting to form a horn. The front edge of the threshold can be fixed by nailing through the horn, but be sure to drill a hole with ample clearance to prevent splitting this close to the end: only the end of the jamb and housing should be painted before assembly.

> **TIP** Thresholds can be manufactured with rebates to stop wind and rain from entering the house. There is also a wide range of weather strips available to make doors more draft and rain resistant.

Fig. 18.34 External door openings

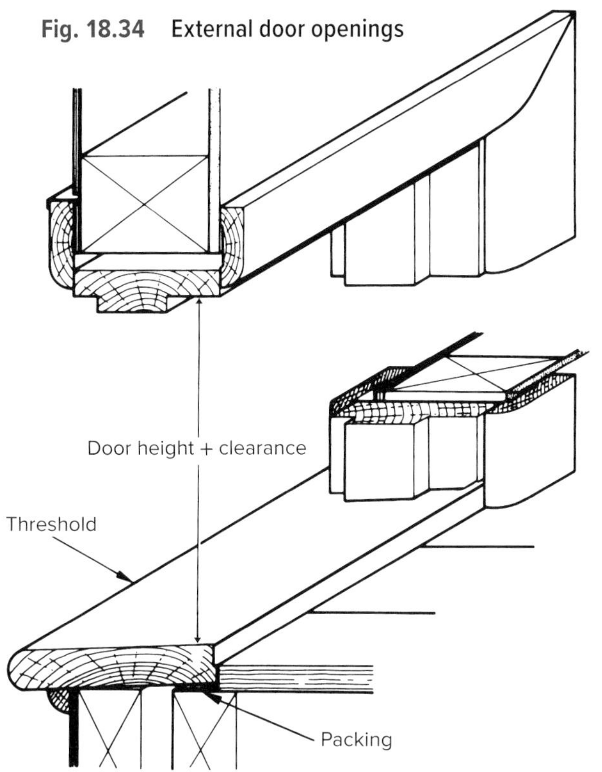

Fig. 18.35 External jamb lining housed into threshold

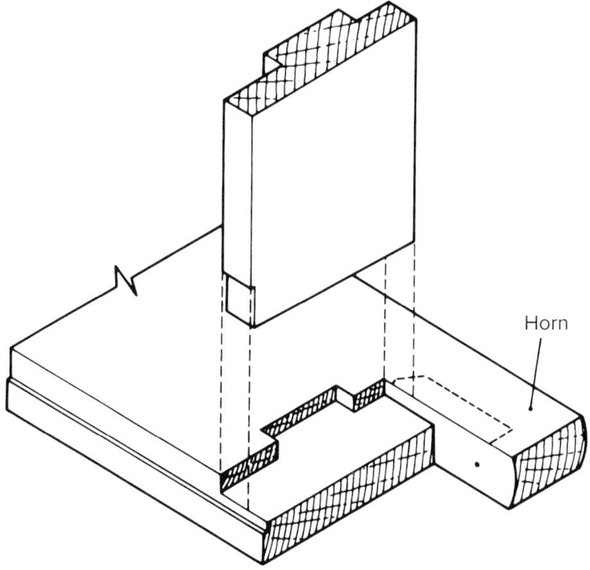

Student research

Research how many different types of hinges there are and their applications. Then practise letting in a butt hinge on a scrap piece of timber.

End of chapter activity

Activity 1:

Research what materials other than those mentioned can be used to line the walls of a home.

AUSTRALIAN BUILDING CODE AND STANDARDS

AS 1684—Residential Timber-Framed Construction

NCC

AS 4145—Locksets and Hardware for Doors and Windows—Mechanical Locksets for Doors and Windows in Buildings

AS 2688:2017—Timber and Composite Doors

AS/NZS 2272:2006—Plywood—Marine

Tradie talk—**Always scribe an internal mitre**

WORLDSKILLS AUSTRALIA

WorldSkills Australia (WSA) aims to develop and foster the skills of young Australians. WorldSkills' purpose is to promote and build a skills culture by inspiring young people, celebrating skills excellence and providing them with an opportunity to showcase their trade and skills talent. We achieve this goal through competitions held on a regional, national and international level.

Courtesy of WorldSkills Australia

WorldSkills Regional Competitions (VETiS)

WorldSkills Australia Regional VETiS Competitions are open to secondary school students undertaking a VET subject. The competitions are a series of work-simulated projects based on industry standard frameworks, which are designed and administrated by practising teachers and industry professionals.

VETiS medallists from State and Regional Competitions are eligible to compete against their peers from around the country in 14 VETiS categories offered at the WorldSkills Australia National Championship.

Competitions take place over a two-year cycle, with Regional Competitions leading into the National Championships the following year.

Talk to your VET teacher or Careers Advisor about competing.

Why compete?

- Dramatically boost your skills and performance at work and school.
- Showcase your potential to current and future employers.
- Stand out from the crowd and create prestige for your school.
- Get access to invaluable free training from world-class experts.
- Opportunity to travel interstate and overseas.
- Challenge your skills against others in your field.

Courtesy of WorldSkills Australia

Skills competition areas

VETiS Automotive Services
VETiS Bricklaying
VETiS Business Services
VETiS Commercial Cookery
VETiS Construction
VETiS Electrotechnology
VETiS Food & Beverage
VETiS Hairdressing
VETiS Information Technology
VETiS Metals & Engineering
VETiS Plumbing
VETiS Primary Industries
VETiS Retail
VETiS Tourism

Chapter 19

Formwork construction and basic concreting

Learning Objectives

LO 19.1 Understand OHS considerations for concreting

LO 19.2 Know about concrete formwork

LO 19.3 Identify types of reinforcement materials

LO 19.4 Understand reinforced concrete footings

LO 19.5 Know about concrete materials

LO 19.6 Know the factors affecting concrete quality

LO 19.7 Know batching proportions for concrete

LO 19.8 Mix and place concrete

LO 19.9 Cure concrete

LO 19.10 Know the compressive strength of concrete

Introduction

Concrete is one of the most important structural components used in residential construction today. Together with timber and steel framing, structural steel and masonry, it is an important part of the structural fabric that supports the final external and internal finish. Most homes are built on reinforced concrete footing systems, while concrete is also used in many of the available building finishes such as concrete driveways, concrete masonry, paving and exposed concrete floors, and walls and ceilings.

Concreting is also a recognised trade under the Australian Qualifications Framework and is available as a Certificate III in Concreting. Closely allied trade qualifications include the Certificate III in Formwork/Falsework and the Certificate III in Steel Fixing.

Fig. 19.1 Steel stumps supporting a house—set on concrete pads and backfilled to ground level with more concrete

Courtesy of Alister Ford

Fig. 19.2 Basic formwork for a footpath, with reinforcing mesh cut and fitted; bar chairs will be installed before the pour begins

Courtesy of Alister Ford

Form workers create temporary moulds known as formwork into which concrete is poured to become a designed concrete structure. Steel fixers position and secure steel bars or steel mesh to reinforce concrete structures.

When concrete is poured with reinforcing steel, it is called reinforced concrete; this is used for strip footings, slabs on the ground such as driveways and house slabs, beams, columns and floors.

Concrete that has no reinforcement is simply called plain concrete. It is used for simple poured piers or blob footings supporting brick piers, columns and posts.

In this chapter you will learn about the occupational health and safety (OHS) aspects of working with formwork, reinforcement materials, dry concrete, wet concrete and the required tools and equipment.

You will also learn about the basics of formwork construction in the building industry using different types of formwork materials and methods of construction.

In addition, you will learn about the steel reinforcement sheets, bars and accessories that are placed in concrete for strengthening purposes, the materials and additives that make up the concrete mix and the ratios that need to be applied when adding these materials so that the correct consistency and strength is achieved.

Finally, you will learn how to correctly place, work and finish the concrete, then how to remove, clean and store the formwork.

WORKPLACE SCENARIO

In 2008 I was teaching apprentice carpenters at the University of Ballarat TAFE division and one of my classes was due to complete the concreting unit, which is part of their apprenticeship training.

I was trying to think up a few ways to make a week of trade school learning about formwork, reinforcement, pouring and finishing concrete a bit more interesting and to find a real-time concreting example that the apprentices would not generally see on a domestic or commercial building site.

A major construction job being undertaken in the Central Highlands at this time was the Waubra wind farm, which is situated about 35 km from Ballarat. A total of 128 wind turbines were being constructed, with towers measuring 71.5 m high and blades approximately 40 m in length. These wind turbines would need a very substantial foundation and footing to serve this purpose over many years of service.

I contacted the principal construction company, introduced myself and explained that I wanted to show the apprentices some concrete work from preparation to finish of the likes that they had not seen before. The construction company representative was fantastic. She made all of the arrangements for us to visit the site and all we had to do was arrive at the main construction site offices at the given time on the day of our visit and we would be assigned a personal guide for the day.

Upon arrival, the 10 apprentices and myself were given a full site and company induction, before we went out into the field. All of the personal protective equipment was supplied: hi-vis reflective vests, safety glasses and hard hats. Then we got back into the minibus and followed our guide's four-wheel-drive dual cab out to the first tower site.

This site had been excavated and the workers were in the process of setting up the reinforcing steel in the excavation in readiness for the concrete pour in a few days' time. The excavation measured 15 m long × 15 m wide × 1.5 m deep. With the pedestal for the tower included on top of this huge block of concrete-to-be, this equated to approximately 360 cubic metres of concrete needed to construct the base.

The scope of this job meant that the two local concrete companies contracted to supply the concrete for these tower bases had actually set up their own concrete batching plants on-site rather than mixing in Ballarat and having a constant convoy of trucks continuously doing the 70 km round trip.

Inside each concrete foundation there was 32 tonnes of steel reinforcing that needed to be manually placed and tied or welded into position. Some of the steel bars, which were 70 mm in diameter and 15 m long, needed one worker per metre to carry and place them in the correct position.

Some of the apprentices started doing some calculations among themselves. Pretty soon the whole group was involved. Six cubic metres of concrete per truckload meant 60 full trucks would be required for a 360 cubic metre pour; 2400 kg per cubic metre equated to 864 000 kg or 864 tonnes of concrete. Add 32 tonnes of steel reinforcement and this meant the weight of each foundation would be 896 tonnes.

Considering a concrete slab for an average-sized house may be between 30 and 40 cubic metres of concrete, this would certainly be the biggest concrete pour they had ever seen. Out came the calculators again. Somewhere in the vicinity of 25 house slabs in each foundation multiplied by 128 towers equated to something like 3200 house slabs—enough houses to create a whole new town.

The second site we visited was a concrete pour in process. A steady stream of concrete trucks in and out of the site were placing the concrete into the concrete pumping truck chute and then placing it via the pump and pipe into the set excavation.

It felt like a truckload of concrete made absolutely no impression, considering the size of the hole in the ground that needed to be filled. The time to complete it would take anywhere from six to eight hours, with the workers rotating roles frequently for OHS reasons. Up to 10 truckloads per hour placed on a good weather day.

After watching the concrete pouring process, we headed back to the main site offices for an official debrief and question/answer session with the company representative, which we had to call to a close because of the amount of questions the apprentices kept asking. Seeing construction of this magnitude was certainly an eye-opener. It made a huge impression on these carpentry apprentices, and the field trip is often mentioned whenever I speak to any of them when our paths cross.

19.1 OHS considerations

19.1.1 Hazard and risk control

Depending on the scope and location of the concreting work, a variety of hazards may be encountered. To meet their OHS obligations, the person conducting the business or undertaking (PCBU) is required to conduct a risk assessment of the job. The risk assessment may be carried out in the form of a job safety analysis (JSA) or a safe work method statement (SWMS).

The location and nature of the concrete pour will dictate the importance of barricades, signage, traffic and pedestrian management and personal protective equipment (PPE) such as hi-vis clothing, hard hats, work boots, work pants, gloves, eyewear and hearing protection. The scope of the job will dictate the type of mechanical aids required for the pour. These may include pumps, chutes or kibbles, all of which introduce new hazards to the job.

Fig. 19.3 Concrete workers wearing correct PPE: hi-vis jackets, footwear and hard hats

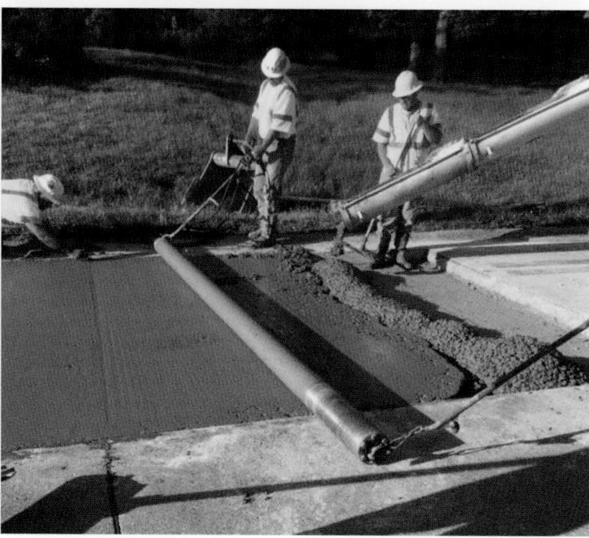

Courtesy of Riviera Concrete

An additional requirement under the OHS Act is a material safety data sheet (MSDS) for all hazardous materials that are in use. This information should be made available to all workers on-site. The MSDS is a document that describes the properties and uses of a substance. These include chemical and physical properties, health hazard information, precautions for use, first aid and safe handling information.

The MSDS can be sourced on the internet by visiting the manufacturer's website.

19.1.2 Personal protection equipment

Common personal protective equipment (PPE) that may be required include:

- protective clothing
- safety vest
- gloves
- rubber boots
- safety glasses
- steel-capped work boots
- hard hat or protective sun hat.

19.1.3 Working with dry concrete

Concrete dust contains particles of crystalline silica, which, if small enough, can be inhaled into the lungs. The dust can be associated with the handling of dry cement powder or through dust generated by cutting, drilling or grinding. Exposure to this dust over a long period may result in the lungs being affected by damage and scarring, causing shortness of breath and silicosis.

Control of these hazards may require extraction systems, wet saws, wet sweeping, vacuums and/or dust masks and respirators.

19.1.4 Working with wet concrete

Contact with wet concrete can cause irritation to the skin and eyes, resulting in skin conditions such as cement burns, cement dermatitis, dry skin and irritation to the eyes. Contact with wet cement should be managed through the correct use of PPE. Any contact with skin or eyes should be rinsed with clean water. Barrier creams can be useful for long-term exposure issues.

19.1.5 Site preparation

Before concrete can be poured, some important preparation tasks need to be undertaken:

- The slab must be set out according to a plan or known requirements.
- Topsoil must be removed and the foundation must be excavated to an established design level and must be firm enough to be able to support the slab to be laid.
- Road base may be required to raise the foundation to the design level of the underside of the slab.
- A waterproof membrane or underlay may be required to prevent water being drawn into the finished slab. This is particularly important to combat rising ground salts.
- Sand blinding may be required to protect the membrane.
- Tools required to place and finish the concrete must be available.
- The quantity of concrete must be estimated and ordered at least one day before the work is to be carried out.

TIP Have buckets of water available so that concreting tools can be cleaned regularly or dipped in water as the concrete is being poured and finished.

19.2 Concrete formwork

19.2.1 What is formwork?

Formwork is the temporary mould used to encase fresh plastic concrete and create the desired shape. It represents a large part of the cost of any concrete structure. The erection of concrete formwork requires special skills, and on many routine jobs, the design and satisfactory performance of the formwork becomes the responsibility of the site manager or supervisor. The specifications for the construction of any formwork, either from the formwork manufacturer, supplier or from the plans of the job, need to be closely followed as any shortcuts taken or incorrect modifications to the formwork could have disastrous results. If the formwork fails, there will be a significant risk of injury to any workers in the vicinity. There will also be significant cost through concrete loss, formwork repair or replacement, plant and equipment requirements, time and labour.

Basic formwork materials can include:

Timber:

- strong
- very economical
- comes in many different sizes
- easily fixed in place and easily removed
- lightweight
- re-usable for a time.

Plywood or formply:

- strong
- lightweight
- can be cut to different sizes or shapes
- comes in different thicknesses
- re-usable for a time.

Steel:
- excellent for repeated use
- strong
- excellent finishing quality
- can be expensive to buy or replace.

Aluminium:
- excellent for repeated use
- strong
- lightweight
- easy to assemble and fix.

Plastic:
- excellent for repeated use
- extremely light
- easy to handle and assemble.

Particleboard:
- re-usable for a time
- can be cut to different sizes or shapes
- comes in different thicknesses
- must be properly treated to restrict moisture absorption.

Formwork is covered by AS 3610.1: 2010 Formwork for Concrete.

At the time of writing, persons engaged in the erection of formwork require no special qualifications; however, a national qualification, the Certificate III in Formwork/Falsework, is available under the Australian Qualifications Framework. Concreters employed on a single dwelling residential construction can normally install the required formwork and reinforcing steel.

19.2.2 Formwork guide

Good formwork should be:
- accurate in size, shape and direction
- strong
- rigid and propped or pegged, braced and tied in all directions to prevent any bulging or displacement of forms
- tight-jointed to prevent the escape of the cement paste
- simple to install and dismantle and easy to handle
- cleaned immediately and checked for any damage upon stripping and dismantling. All nails and screws should be removed, and the formwork components neatly stacked ready for re-use or transport to the next job. Any detected faults need to be reported and then repaired or have the piece of formwork replaced. Arriving at the next job site with damaged, unusable formwork components will hinder the timelines of the job and cost unnecessary money in time and productivity.

To ensure that the formwork strips away easily from the semi-hardened concrete, the inside surfaces are coated with a form release agent that can be applied by brush, roller or spray equipment. This is best done before the formwork is erected or before any steel is placed in the forms, as the release agent must *not* come into contact with the steel reinforcement. Contact would destroy the bond between the steel and concrete, which is an essential factor if the reinforced concrete is to achieve its design strength.

Fig. 19.4 Rebated slab, cantilevered formwork

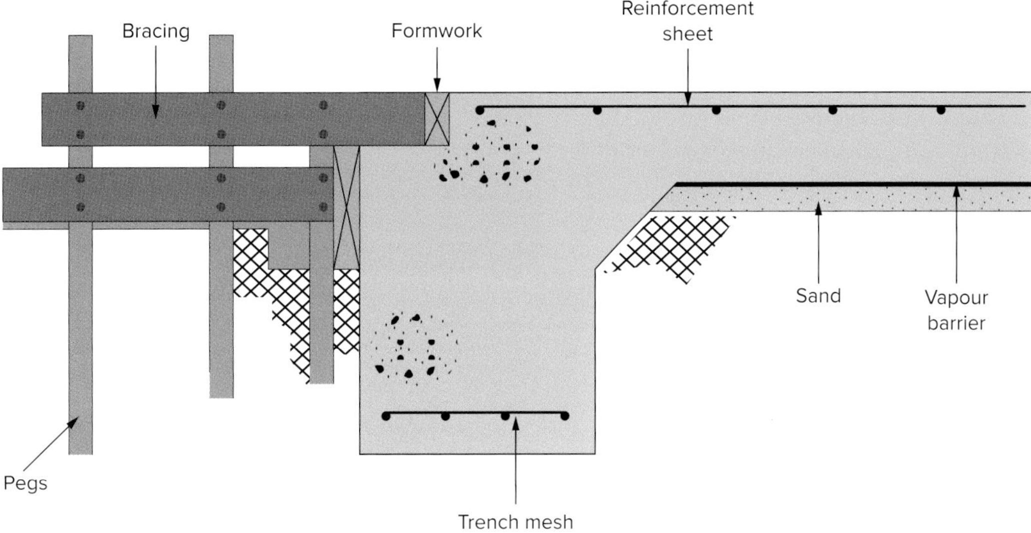

Fig. 19.5 Rebated slab, foam block out

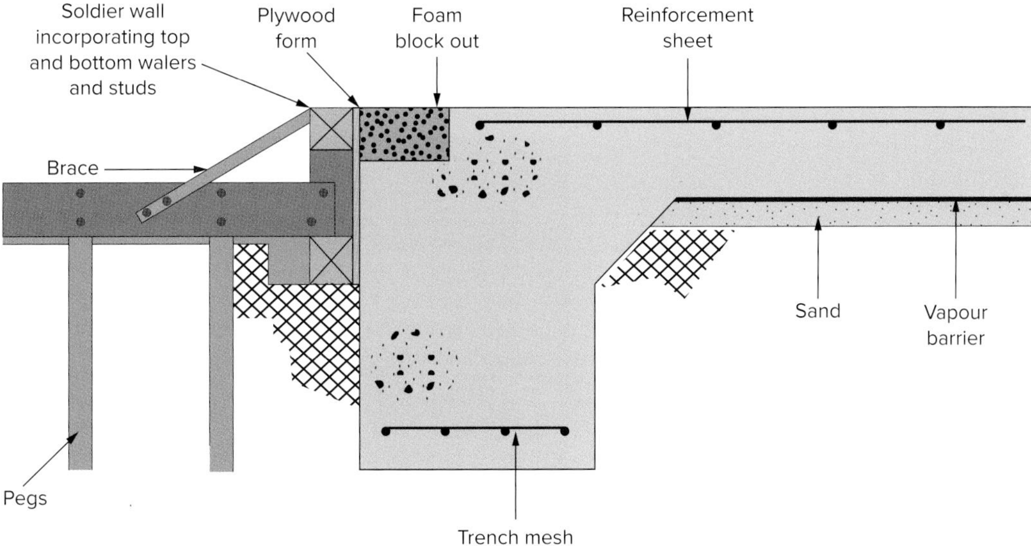

19.2.3 Formwork for curved edges

To form a curved edge, use a flexible form material such as strips of formply. Where a sharp curve is required, it is necessary to place a series of shallow saw cuts almost halfway through the form material to allow it to be bent. The sharper the curve, the more closely spaced the saw cuts must be. The cut surface is placed on the outside of the formwork.

19.2.4 Tilt panel construction

Tilt panels are now a very popular form of construction in the commercial sector. The panels can be manufactured on the building site or at the manufacturer's factory then transported to the building site.

Tilt panels effectively make up the walls of a building. If they are made or 'cast' on the building site, this can be done on the existing concrete slab that will be the floor of the completed building or in cast beds constructed and placed specifically around the building site to enable the minimum distance of travel when the time comes to erect the panels.

The formwork for each panel in the building can be custom-designed and made as some panels will need block outs inserted in them for window and door openings. Any other inserts for footing, wall, roof attachments and lifting points are fixed in place. After the formwork is completed, the steel reinforcing sheets and deform bars to strengthen the wall panels are added and tied, and the concrete poured and finished.

When the concrete has set and hardened, another tilt panel can be poured on top. A bond breaker is placed in between the panels so that they easily lift away from the panel below. A stack of tilt panels can now be produced in the one cast.

Fig. 19.6 A & B Simple timber and bracket formwork for a rebated concrete slab

(a)

(b)

Courtesy of Alister Ford

Fig. 19.7 Curved formwork for a large tank base using flexible form, steel pegs and spreaders

Courtesy of Alister Ford

19.3 Reinforcing steel

19.3.1 Steel mesh or fabric

Mesh is manufactured in flat sheets with bars up to 12 mm in diameter. The sheets are typically 6 metres by 2.4 metres. The fabric consists of a reinforcing bar welded in either a square or rectangular grid.

- Square mesh comprises bars of the same thickness, spaced 200 mm or 300 mm apart in both directions and is coded as SL82, which means square mesh, 8 mm main wire at 200 centres.
- Rectangular mesh comprises longitudinal wires spaced at 100 mm centres, transverse at 200 mm.

19.3.1.1 Trench mesh reinforcement

Trench mesh reinforcement consists of wires welded together to form narrow sheets ranging from 200 mm wide to 600 mm for footings. Common trench mesh strip widths are 200 mm (3 wires), 300 mm (4 wires) and 400 mm (5 wires). Trench mesh is coded by the number and size of the main longitudinal wires. For example, 3-L10TM means the mesh has three main wires, each 10 mm in diameter, spaced at 100 mm apart.

Fig. 19.8 A & B Square reinforcement mesh

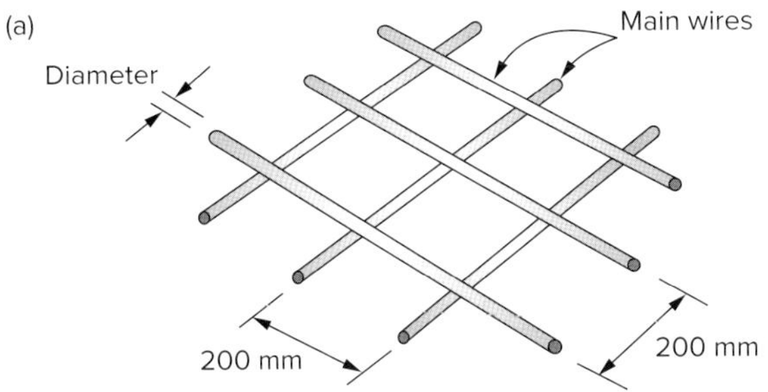

Courtesy of Alister Ford

19.3.2 Steel bars

19.3.2.1 Deformed bar

A deformed bar, or N-bar, is used as the main structural reinforcement for suspended concrete floors, walls, beams and columns and for footing systems.

An N-bar has a yield stress of 500 megapascals (MPa) and is the most commonly specified reinforcing steel for residential construction. An N-bar is sometimes mistakenly referred to as a Y-bar, which has not been manufactured since 1988.

Fig. 19.9 Rectangular mesh

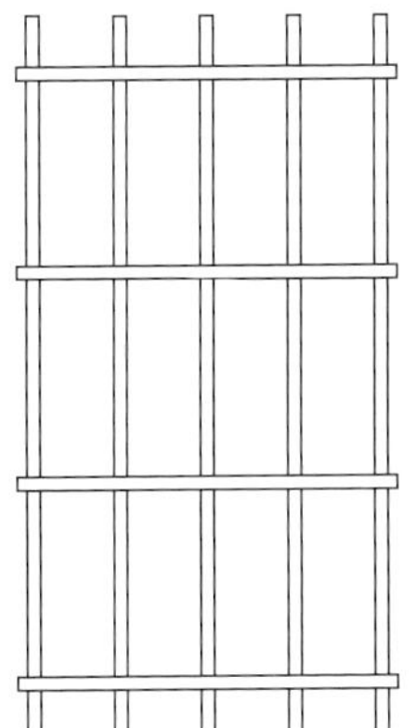

19.3.2.2 Round bar

Round bar is commonly available in 6.5 mm and 10 mm diameter steel in 6 m lengths. It is used primarily for fitments (e.g. stirrups, spacers) to separate the top and bottom layers of trench mesh in footings and the fabrication of mesh reinforcement cages for columns and beams.

19.3.3 Reinforced concrete

Concrete is strong in compression but weak in tension, so steel rods are placed in concrete to carry the tensile stresses. The combination of steel and concrete is known as reinforced concrete.

19.4 Reinforced concrete footings

Reinforced concrete footings generally have two layers of trench mesh or N-bars and are formed into a cage with stirrups or spacers out of round bar or N-bar. The cage is supported underneath by bar chairs spaced at 500 mm centres maximum. The top of the footing should be finished no higher than 150 mm below the existing ground height to ensure that the footing and first course of masonry are well covered. The reinforcing steel should be covered with a minimum of 40 mm of concrete and as the concrete is poured it should be vibrated to ensure the concrete achieves full strength and provides full protection for the reinforcing steel against corrosion.

Fig. 19.10 A & B Trench mesh reinforcement

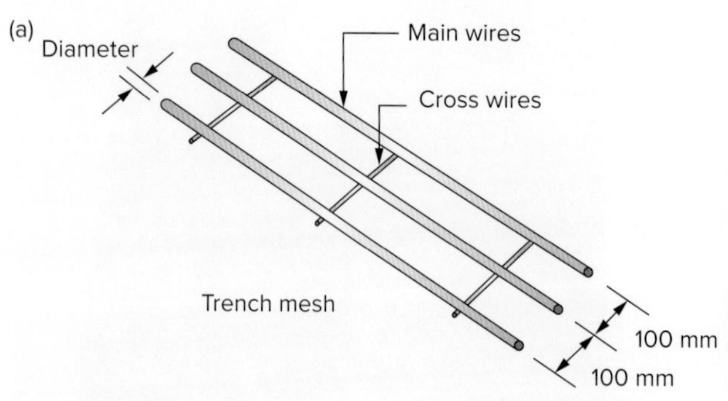

(b)

Courtesy of Alister Ford

Fig. 19.11 Reinforcing cage out of deformed bar

© Luckylife298/Shutterstock

Fig. 19.12 An elevation view of a round bar ligature or spacer

Concrete strip footings have two layers of steel to resist the downward load of the building they support and to resist any upward movement of the foundation that the footings are placed on. Such movement can occur when there are extended periods of rain and the ground swells.

Extended periods of dry weather may also cause clay foundations to shrink. The double layer of steel reinforcement in the concrete supports the footing beam between piers that are drilled down to a more stable foundation.

1. Support length of trench mesh or reinforcing bars on two saw horses or work platform.
2. Loop spacers onto trench mesh or reinforcing bars.
3. Tie off spacers to mesh or bars at spacings up to 900 mm between spacers and 100 mm between bars using annealed wire.
4. Place bottom layer of mesh or bars onto spacers and tie them off.
5. Run a length of wire diagonally from top to bottom on both sides of the cage as a brace.

Lapping of reinforcement

(See Fig. 19.17–19.21.)

- Bars must be lapped at least 500 mm (Fig 19.17).
- Trench mesh must be lapped at least 500 mm when lapping lengths and by the width of the trench mesh at corners and junctions. L-bars to external corner of L-shaped corners (Fig 19.17).
- Mesh sheets must be lapped on their sides by the two outermost edge wires (Figs 19.18 and 19.20).
- The ends of mesh sheets must also be lapped by at least two main wires at the end of the sheet regardless of any overhanging longitudinal wire (Fig. 19.19).
- Two mesh sheets can be joined with a section of sheeting that laps and splices the two sheets together with at least two main wires (Fig. 19.21).

Fig. 19.13 Simple beam

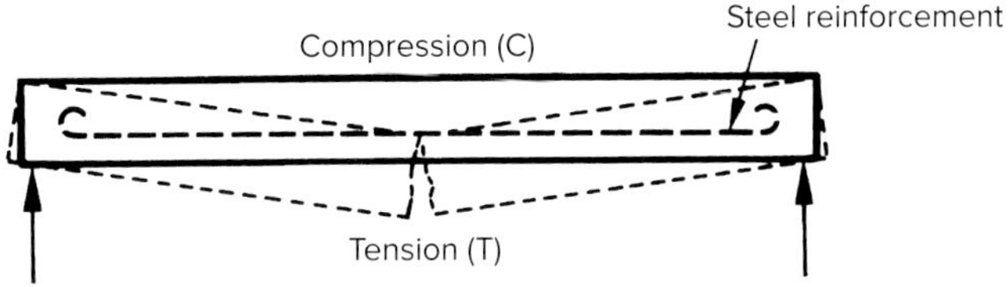

Fig. 19.14 Continuous beam

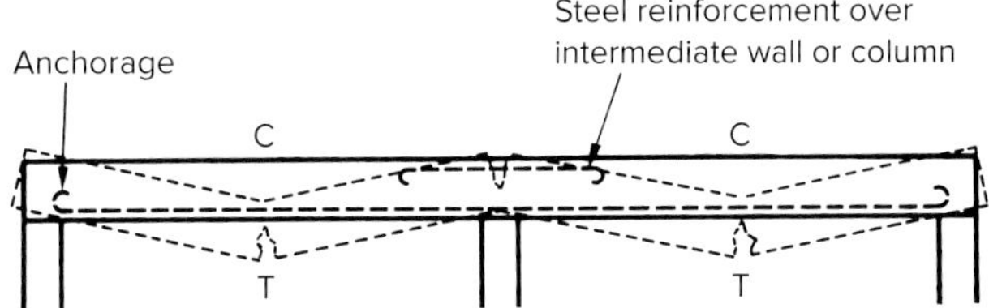

Fig. 19.15 Cantilever beam

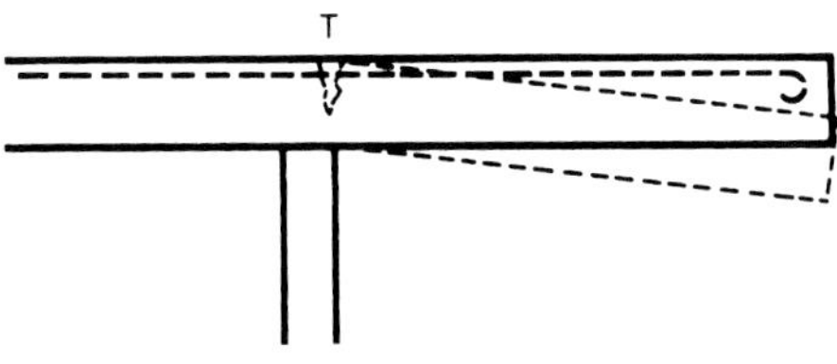

Fig. 19.16 Reinforcement cage 4-L10TM: (a) two layers with R8 cogged ties at 900 centres (b) Fixing reinforcement cages

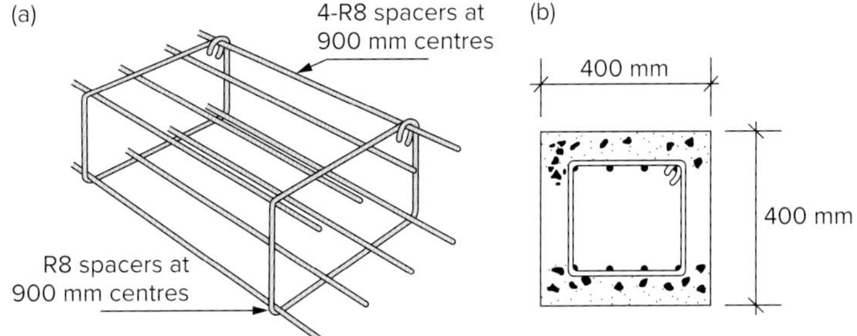

Fig. 19.17 Lapping of reinforcement

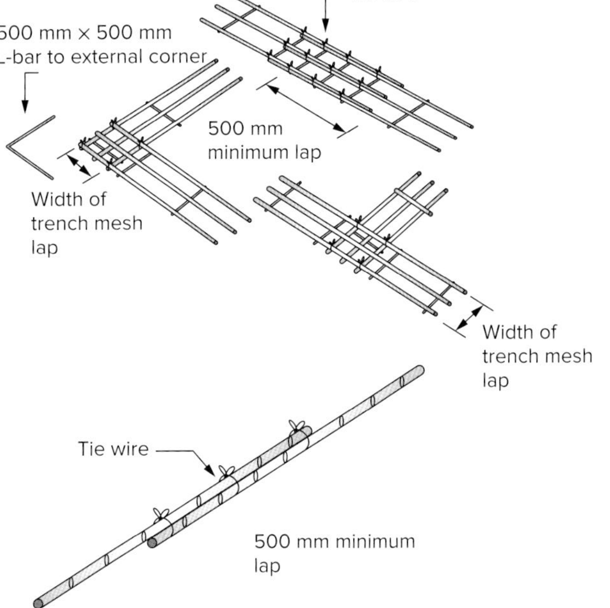

> **CHECK YOUR UNDERSTANDING**
>
> 1. List the three main types of mesh reinforcement.
> 2. What is the purpose of placing steel rods in reinforced concrete? Provide two examples.
> 3. Why is steel reinforcing added to concrete?

19.5 Concrete materials

19.5.1 What is concrete?

Concrete is an artificial rock and is a combination of cement, fine aggregate, coarse aggregate and water. The raw materials are mixed to form a plastic material, which can then be poured or moulded into the required shape to harden by a chemical reaction. This chemical reaction is one of hydration, which specifically involves the action of water and cement. The resultant dense mass will be capable of carrying great loads.

Producing quality concrete requires an understanding of the process and careful control of all stages of production, from the selection of materials through to mixing, pouring, vibrating, levelling, finishing off and, finally, curing. If any of these steps is either missing or not carried out effectively, the desired results will not be attained.

Fig. 19.18 Two examples of lapping ends of two sheets

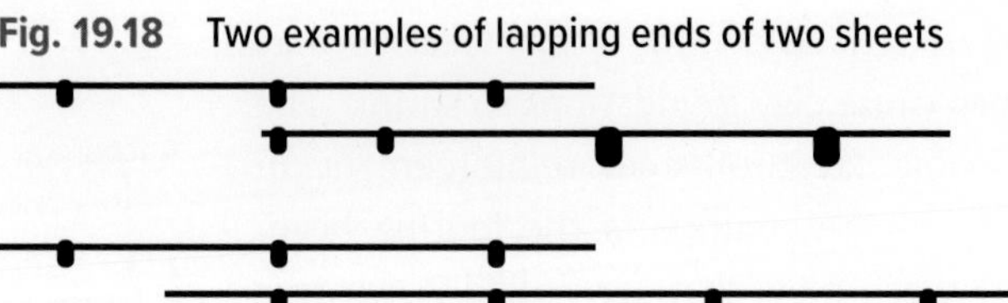

Fig. 19.19 Lapping sides of mesh sheets by at least two main edge wires

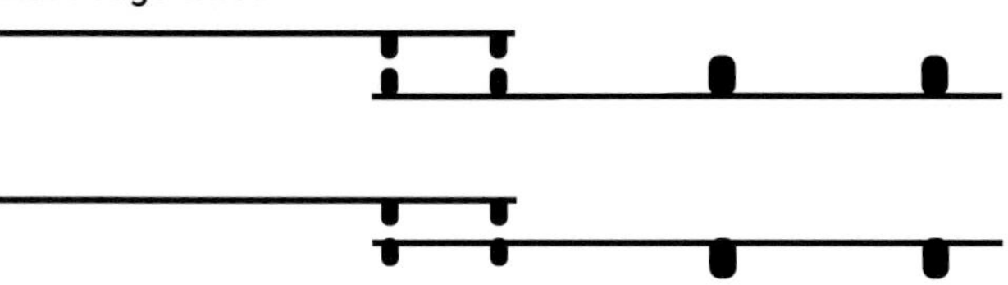

Fig. 19.20 Lapping two sheets of square mesh

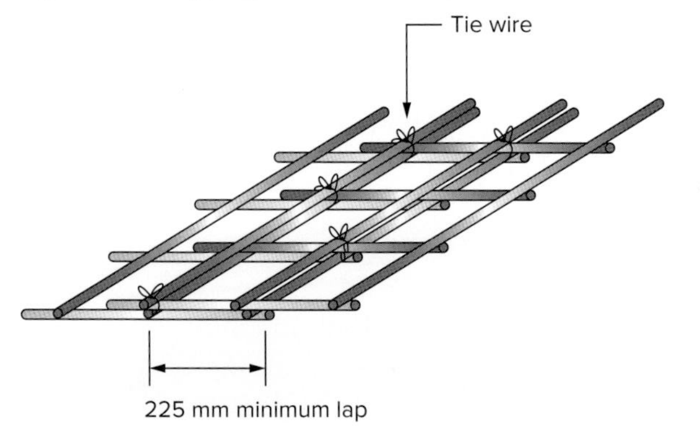

Fig. 19.21 Lap splice of two similar sheets

19.5.2 Cement

Portland cement is so named because of its close resemblance to natural Portland stone in Britain. It is manufactured from suitable shale, limestone and smaller amounts of gypsum to control the setting times. The shale and the limestone are crushed separately and then mixed and burnt at a very high temperature of around 2000°C. The resulting clinker is crushed to form a fine powder. This powder is then bagged (packed in paper bags with a thin plastic liner) or transported in bulk containers.

$$1 \text{ bag of cement} = 20 \text{ kg}$$

$$108 \text{ bags of cement} = 1 \text{ m}^3$$

$$1 \text{ bag of cement} = 0.009 \text{ m}^3$$

Portland cement must comply with AS 3972: 2010 General Purpose and Blended Cements, which classifies cement used in Australia as:

- *General purpose cement (Type GP).* This ordinary Portland cement is the most commonly used cement for the manufacture of concrete, which is used for the majority of building construction for both concrete and mortars. Off-white cement is also a Type GP cement and is used where design requirements demand an off-white colour finish.

- *Blended cement (Type GB).* This is used in all types of building and construction work. Early rates of strength gain may be lower than those of Type GP, and curing may be more critical for full-strength development.
- *General purpose limestone cement (Type GL).* This cement has up to 7.5% of additional lime replacing the cement and has been developed to reduce the amount of cement manufacture in an effort to reduce carbon dioxide emissions in the manufacture of cement.

19.5.3 Special purpose cement

19.5.3.1 High early strength (HES)

High early strength (HES) cement gains its strength a little faster than Type GP, even though it sets in around the same time. However, its eventual strength will be little different from ordinary cement. Due to its rapid gaining of strength, HES cement also generates more heat in reacting, and so mixes made with it are more prone to cracking in uncontrolled conditions and when in large masses.

19.5.3.2 Low heat (LH)

During setting and early hardening, cement generates a considerable amount of heat, which is called *heat of hydration.* In large masses of concrete, this can cause damage. LH cement does not produce the same levels of heat as ordinary cement. For this reason, LH cement is useful in mass concrete to minimise the amount of heat generated. The gain in strength of LH cement is slower than ordinary cement, but the ultimate strength is the same.

19.5.3.3 Sulphate resisting (SR)

Sulphate resisting (SR) cement is similar to ordinary cement, with a greater resistance to sulphate attack, such as may occur where there is a high-salt content in the soil, or where there are high concentrations of industrial waste or in sewer applications. Sulphate attack causes the concrete to expand, which in turn causes cracking and softening of the concrete.

19.5.4 Aggregates

Aggregates for concrete should consist of clean, hard particles free from organic matter or other impurities that could affect the strength of the concrete.

19.5.4.1 Coarse aggregate

Coarse aggregate is usually produced from crushed granite, quartzite, dolerite or basalt, and is often known as 'blue metal' or crushed river gravel. The particles in a concrete aggregate should be graded in size so that the voids formed between the larger particles will be filled by smaller particles. For reinforced concrete, the maximum nominal size of aggregate is 20 mm, but up to half the total weight may be 10 mm gauge with a small proportion being even smaller in size. Common sizes are 7, 10, 14, 20, 30 and 40 mm.

19.5.4.2 Fine aggregate

Fine aggregate can pass through a 5 mm sieve and usually consists of natural sand, although it can also be produced from crushed stone. The sand must be clean and sharp and free of salt or organic matter. Rubbing a sample between the hands can be a rough test of its cleanliness: if the hands remain clean, the sand is probably satisfactory. Enquiries as to its source may indicate whether further testing is warranted. Sand should also be screened for particle size as it should not contain more than 10% by volume of fine dust.

19.5.5 Water

Water for concrete and mortar should be clean and free of chemical impurities. Generally, water that is fit for drinking should be satisfactory. Avoid bore water as it may be high in sulphate or chloride salts.

19.5.6 Additives

Additives or admixtures are chemical-based and can be added to the concrete mix before or during the concrete mixing process.

There are five main classes of chemical admixtures:

- accelerating
- retarding
- water-reducing
- air-entraining
- plasticisers (super plasticisers).

Different admixtures will have varying reactions with the concrete, depending on the required modifications to the finished, hardened concrete.

Most commonly, admixtures can be used to accelerate the curing time of concrete, especially in colder climates where it can take a long time for the concrete to set. The concrete strength can be achieved much more quickly.

They can also work in the exact opposite, as a retarding agent–meaning that the curing time can be decelerated. This is especially useful in warmer climates, where the concrete can lose its workability very quickly.

Water-reducing additives can be used to reduce the amount of water by up to 10% required for the mixture. These allow the concrete mix to reach the required slump without using as much water, which means the concrete can reach a higher strength upon curing and hardening than concrete without the additive.

If an even higher water content reduction is required, a plasticiser or high-range water reducer can be used. These can reduce water content by as much as up to 30% and still create a high slump flowing concrete mix. The drawback with plasticisers is that they are effective for only a short period of time and the concrete can start to lose its workability in as little a time as 30 minutes. For this reason, plasticisers are generally added when the concrete arrives at the job site.

Sometimes, the mass of the finished concrete will need to be reduced and the workability improved. Air-entraining admixtures are used to form minute air bubbles in the concrete to achieve this. Some speciality admixtures are also used for varying applications such as concrete colouring, bonding, damp proofing, reinforcement corrosion inhibition, concrete shrinkage reduction and enhanced workability.

19.6 Factors affecting concrete quality

There are a number of factors that can affect the final quality of the concrete, the main four being:

- water/cement ratio
- cement to aggregate ratio
- compaction
- curing.

19.6.1 Water/cement ratio

After the concrete aggregates have been selected and graded to form a dense mass, they must be bonded together. This is done with the cement paste, which consists of cement and water. The strength of the cement paste is a critical factor affecting the final quality of the concrete.

The water/cement (W/C) ratio is the ratio between the amount of water and the amount of cement present in the concrete at the time of mixing and is expressed as the weight of water divided by the weight of cement. For example, if one cubic metre of concrete contains 150 kg of water and 300 kg cement, then the W/C ratio is equal to 150 divided by 300, or 0.5.

In some circumstances, the W/C ratio may be more conveniently expressed as litres of water per bag of cement. For example, one litre of water weighs 1 kg. One bag of cement weighs 20 kg. In the example above, the number of litres of water per bag of cement equals 0.5 × 20 = 10 litres.

19.6.2 Cement-to-aggregate ratio

The strength of the concrete can also be improved by increasing the proportion of cement to aggregate (shown in Table 19.1). A typical mix for high-strength concrete is 1:2:3: one part cement to five parts aggregate. This mix will make a dense, watertight finished product. Where general purpose concrete is required, a mix of 1:2.5:4, or one part cement to six-and-a-half parts aggregate, will provide concrete with lower compressive strength.

19.6.3 Compaction

By eliminating the air trapped in the concrete, its strength is significantly increased. This action allows the concrete to achieve its design strength. Compaction also provides good adhesion to reinforcing as well as ensuring that the concrete completely fills the forms, providing good off-form appearance.

Figure 19.22 represents the loss of strength as a result of air voids in the concrete. Air voids of 5% will cause a 30% reduction in the relative strength of the concrete.

The most common way to achieve compaction is through the use of vibration; this process reduces the friction between the aggregate, improving workability without the need to add additional water.

Compaction can be achieved through the use of:

- immersion (poker) vibrators
- surface vibrators (vibrating screeds)
- form (external) vibration
- tamping
- rodding.

Fig. 19.22 The impact of air voids on concrete strength

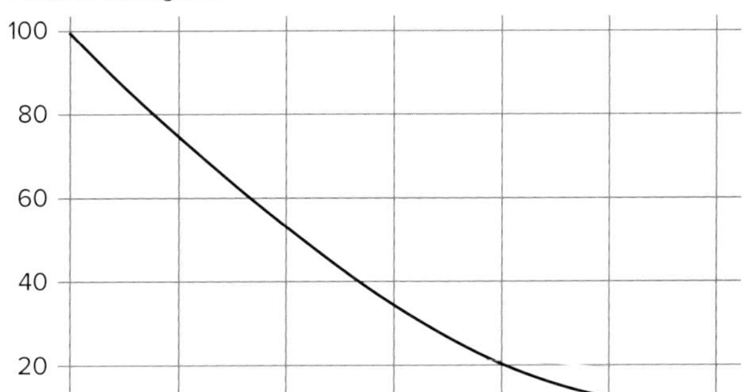

19.7 Proportions for concrete

Concrete raw materials are proportioned by batching, either by mass or by volume.

19.7.1 Mass batching

Concrete may still be batched on remote construction sites where ready-mix concrete is not available. In those cases, responsibility for the strength and quality of the concrete is entirely with the site contractor producing the concrete.

Fig. 19.23 Mobile site batching plant

© Baloncici/Shutterstock

For large construction sites in major cities, a site concrete batching plant is often set up to reduce the need for heavily laden concrete trucks driving through traffic-choked streets and to ensure quality and availability of supply. The raw materials are delivered outside peak traffic hours in bulk loads and the freshly batched concrete is delivered directly to where it is required on-site by pumps.

19.7.1.1 Ready-mixed concrete

Ready-mixed concrete is the most convenient and economical way of obtaining high-quality concrete that has been accurately proportioned by mass (weight) through a batching plant. Ready-mixed concrete is available in most areas. It is ordered by the cubic metre in increments of 0.2 m^3. Trucks are available that can deliver up to 7 m^3; in some areas, mini-trucks will deliver loads of as little as 0.2 m^3. The concreters must be ready to accept the concrete as it is delivered.

TIP When ordering ready-mixed concrete, indicate the strength of concrete required in MPa, as it is usually delivered with a slump not exceeding 80 mm and any departure from this amount must be specified when ordering.

Water should never be added to ready-mixed concrete to increase its workability unless the W/C ratio is maintained by also adding cement in the same proportion. The practice of 'wetting up' by adding water to the concrete in a concrete truck changes the W/C ratio and the design strength of the concrete, potentially resulting in concrete that has a lower MPa than required. If concrete with a lower slump is required because a concrete pump is being used, the concrete should be ordered with a higher slump.

19.7.2 Volume batching

Volume batching can be rather inaccurate due mainly to the bulking of materials, particularly sand, which when damp can bulk up to as much as 25%.

For the most accurate results, materials should be accurately proportioned by volume using consistently sized containers such as 10–15 L buckets, which can be picked up manually and emptied into a concrete mixer.

TIP A filled 20 L container is too heavy for one person to lift repeatedly and may cause injury.

Site mixing concrete using a concrete mixer

1. Position mixer close to materials and stabilise the mixer to prevent it from moving.
2. Ensure all electrical leads are off the ground and elevated above the heads of workers. Connect electrical supply lead to an earth leakage detector and connect mixer lead to the earth leakage detector.
3. Wear relevant PPE. Avoid contact with your skin and concrete or cement powder and keep your hair, hands and body parts away from any moving parts. Don't fill buckets with more than 20 kg of material at any time (10–15 L of aggregate is approximately 20 kg).
4. Place all of the required water into the mixer for a full batch of concrete after the previous batch has been emptied.
5. Place one part of sand in the mixer and allow this to mix for a short period of time. This will clean the mixer.
6. Slowly add all of the cement required. Adding the cement now will allow the powder to thoroughly mix with the water while minimising the amount of dust created. Stand upwind from the mixer or use a protective dust mask.
7. Slowly add the remaining materials, alternating between fine and coarse aggregate.
8. If additional water is required, add a proportional amount of cement also based on the required W/C ratio.
9. Don't overfill the mixer! Smaller batches are better mixed and easier to handle in a wheelbarrow.

For volume-mixed concrete, the proportions shown in Table 19.1 can be used. Concrete should be mixed for at least three minutes and only clean water added sufficient to produce a slump not exceeding 80 mm.

19.7.3 Quantities of dry materials

In order to manufacture one cubic metre of wet concrete, approximately 1.45 m^3 of dry materials will be required. A large amount of the fine aggregate will fill the voids between the coarse aggregate, with the cement filling some of the voids between the sand.

Fig. 19.24 Coarse and fine aggregate

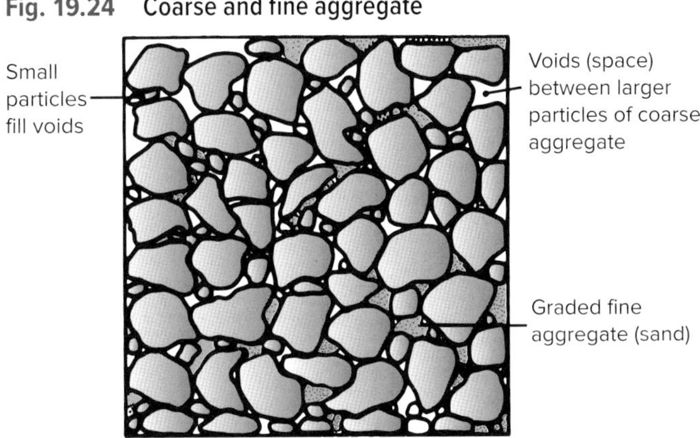

Table 19.1 Volume-mixed concrete proportions

Material	High strength	General use	Large mass
Cement (Portland)	1	1	1
Fine aggregate (sharp sand)	2	2.5	3
Coarse aggregate (20 mm)	3	4	5

19.7.3.1 Concrete mix estimating

To calculate dry mix quantities for high-strength concrete, we need to convert all of the units to m^3. This means that although the cement is sold by weight (20 kg), we need to convert it to volume.

$$1\ m^3 \text{ of cement} = 108 \text{ bags } (20\ kg) = 2160\ kg$$

$$1 \text{ bag} = 0.009\ m^3$$

$$1.45\ m^3 \text{ of dry mix} = 1\ m^3 \text{ finished concrete}$$

The proportion for this batch of concrete, for 1 m^3, will be:

$$\begin{aligned} \text{cement (1 part)} &\quad 0.242\ m^3 \div 0.009 = 27\ \text{bags} \\ \text{sand (2 parts)} &\quad 0.483\ m^3 \\ \text{aggregate (3 parts)} &\quad 0.725\ m^3 \\ 6\ \text{parts} &= 1.45\ m^3 \\ 1\ \text{part} &= 1.45\ m^3 \div 6 = 0.242\ m^3 \end{aligned}$$

TIP Cement has a limited 'shelf life'; contact with air and moisture will cause the cement powder to begin setting. When ordering dry mix concrete, order only enough for the immediate task. If you do need to store the dry mix, ensure that it is kept off the ground and under cover in a dry location.

CHECK YOUR UNDERSTANDING

1. How does adding water to wet concrete affect its strength?
2. How many 20 kg bags of cement would you need to make 1.7 m^3 of finished concrete using a 1:2:3 ratio mix?
3. Explain why batching by mass is preferable to volume batching in the production of high-quality concrete.

19.8 Mixing and placing concrete

19.8.1 Transport and placing

After concrete has been mixed, it must be placed before it starts to stiffen. Generally, it will take four hours from when concrete is placed to when it initially sets. There is no appreciable stiffening for approximately 30 minutes, after which stiffening will accelerate at a rate depending on the concrete mix, the temperature of the concrete and weather conditions. High temperature accelerates stiffening and must be taken into account when determining how many people will be needed. In many cases, if the concrete must be moved by wheelbarrow over any distance, it may be more economical to hire a concrete pump.

TIP There are approximately 15 wheelbarrows of concrete for each cubic metre of concrete.

19.8.1.1 Segregation

Segregation in concrete refers to the separation of the coarse aggregate from the cement paste. This can occur during transporting and placing if unsuitable methods are used. This results in non-uniform concrete with weak and porous honeycomb patches.

Concrete should be proportioned in such a way as to give a cohesive concrete; harsh, overly wet mixes will segregate more readily. On smaller projects, avoid excessive vibration of the concrete by transporting it in

Fig. 19.25 Site delivery chute

rubber-tyred barrows or dump trucks over smooth, prepared runways. Deposit the concrete directly over the final location.

When placing concrete, it should be shovelled into place and not allowed to flow over long distances. It must be well compacted by mechanical vibration but avoid over-vibration, which may also lead to segregation. In addition, concrete should not be dropped from a great height as this also leads to segregation. Where the drop exceeds 2 m, a drop chute may be necessary.

Immediately after placing, a screed (straight edge) is used to level off the concrete.

19.8.2 Ordering concrete

When ordering concrete, you must provide the following information:

- the name of the customer, contact mobile/phone and method of payment–normally the concrete supplier will need credit card details if you don't have an account
- the address and location of the site with any special instructions (e.g. 'number 90B Hill Street, near the corner of Hill Street and View Road, up the driveway and around the back; contact number 0230 910 245')
- the quantity in cubic metres (m^3) in increments of 0.2 m^3; orders less than 3 m^3 attract additional charges to cover delivery costs
- slump; for example, if a concrete pump is being used it is better to order concrete with a slump of 120 mm
- setting time required–concrete can be ordered with a guaranteed setting time for an additional charge (e.g. if the weather is forecast to be hot, concrete with a guaranteed setting time of four hours or more can provide extra time to place and finish the concrete)
- compressive strength in MPa (e.g. 25 MPa)
- day, date and time of delivery–allow time for workers to arrive and prepare, for formwork and steel to be checked; for the concrete pump to arrive and set up; and traffic control to be set up
- rate of delivery or how many cubic metres per hour–for example, you can request two consecutive 6 m^3 truck loads, 45 minutes apart with a 'message'. When the second load is discharged from the truck, the driver will ask 'What's the message?'. Whereby, a quick check will show that another 1.5 m^3 is required, so the 'message' is 1.5 m^3. Without the advice at the time of order that additional concrete may be needed you may have to wait until a concrete truck is available.

Charges apply to take excess concrete away, so it is best to attempt to accurately estimate the concrete order. Always have a plan to use or get rid of excess concrete (e.g. pour a small concrete pathway or slab; spread the concrete over the site or in one location so that it can be easily broken up before it sets and used later as fill; place it behind a retaining wall; fill any low areas that can be covered with soil).

19.8.3 Level control

When pouring slabs and footings, the finished height is critical to the successful completion of the project. The height is given as a reduced level (RL). This is usually given as a vertical height above or below a fixed known point, also known as a bench mark.

When slabs are being laid, height control can be achieved by setting the top edge of any formwork to the finished RL required. This can be achieved through the use of a laser level, an automatic level or, at a pinch, a water level.

Of all of these methods, the laser level is the most productive as it can be achieved by one person and has the added benefit of being able to provide spot readings over the entire slab surface as the pour is being conducted. This is particularly important when large areas are being laid, avoiding high or low spots in the slab.

When pouring footings, the finished height at the top of the footing should be fixed, by working back from the finished floor level stated on the plans. Before the pour commences, the finished height of the concrete can be marked by pushing a dowel into the side of the trench at the height required.

Fig. 19.26 Laser level monitoring depth of trench

© TFoxFoto/Shutterstock

> **TIP** On a sloping site, a stepped footing may be required. When pouring a stepped footing, the height of each step should correspond to brick courses. This would eliminate the need for the bricklayer to lay a course of split bricks.

19.8.4 Example: Pouring and finishing a simple reinforced concrete driveway

19.8.4.1 Planning procedure

1. Establish the finished height of the concrete and all grades and falls.
2. Estimate the concrete and reinforcing steel required.
3. Install edge formwork.
4. Install waterproof membrane if required. Sand blinding may be required to prevent damage to the membrane.
5. Establish how many control joints or expansion/contraction joints are required and their spacing. Generally, these are up to 6 metres apart.
6. Install mesh reinforcement with bar chairs but only if the driveway can be easily accessed by a concrete truck or if a concrete pump is being used.
7. If you plan to drive the truck on the steel, do not install the bar chairs. The steel mesh must be lifted off the ground by at least 50 mm to provide 'cover' as the concrete is discharged using steel hooks or the hook on a concrete rake.
8. Order at least 25 MPa concrete for improved durability and workability.
9. Avoid pouring concrete during either extremely hot weather or when rain is forecast.
10. Ensure that you have enough workers to place the concrete and finish it within four hours of the first load arriving.
11. Do not attempt to pour concrete 'uphill' on steep driveways. Any minor grade can turn an easy job into a hard job if you have to drag and **screed** concrete 'uphill'. Consider using a concrete pump!
12. Finish concrete with a broom, cove trowel or wood float finish to improve traction.

> **TIP** When colouring concrete, it is safer to order coloured concrete because the oxide used to colour the surface contains a hardener that is particularly hazardous. There is little cost difference.

19.8.4.2 Preparation of foundation

1. Remove all grass and topsoil to achieve a hard foundation for the driveway.
2. Fill any low spots and compact the fill by tamping with a piece of timber or with a mechanical compactor.
3. Compacted road base may need to be placed on top of the foundation or sub-grade to reduce the amount of concrete required and to provide the required level of the foundation.

19.8.4.3 Placing concrete

1. The concrete should be placed where it is needed, working from one end and across the driveway, spreading it to an even specified depth or slightly more by using a concrete rake or shovel.
2. As the concrete is spread, it should be compacted by tamping or using a mechanical vibrator to maintain durability and strength.
3. Screeding should commence immediately with another person assisting the person/people screeding by 'dragging' excess concrete away and adding additional concrete behind the screed as required. Screeding is progressed methodically from one end to the other.
4. The concrete supplier allows 15 minutes per cubic metre for the concrete to be placed. Use that time to place the concrete where and when it is needed. That means the truck driver must understand when and where to discharge concrete and when to stop. Instructions must be clear with no indecisiveness.
5. Between loads of concrete, use the time to complete placement and screeding; collect and clean tools; and check the formwork for any movement.
6. Start to apply the initial edging finish as soon as possible. Unless the edging is started while the concrete is still plastic, it will be more difficult to finish the edging after the initial set.

Fig. 19.27 Road base ready for final compaction before pouring

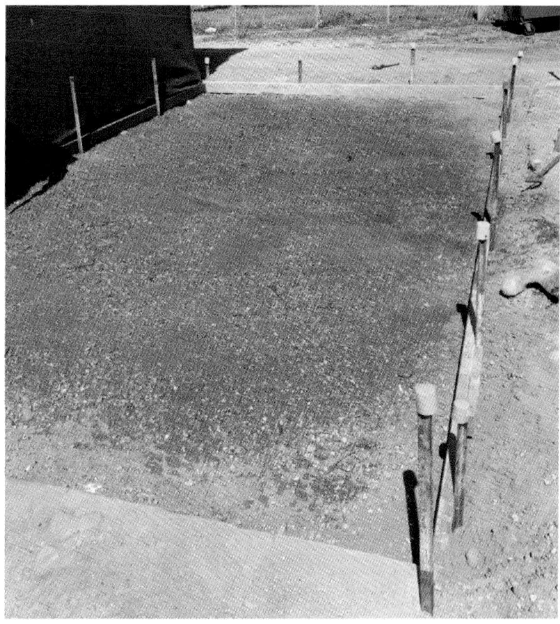

Courtesy of Alister Ford

Fig. 19.28 Screeding concrete

Courtesy of Riviera Concrete

Fig. 19.29 Vibrating screed

© Mahlebashieva/Shutterstock

Fig. 19.30 A & B Bull float

(a)

© Readin Wilson/Shutterstock

(b)

Courtesy of Riviera Concrete

Fig. 19.31 Edging tool

Courtesy of Alister Ford

19.8.4.4 Initial finish to concrete

1. Screed concrete to a consistent grade or level with no shallow spots or humps.
2. Initial edging finish is completed while concrete is wet to set up the aggregate along the edge form. A bull float can be applied over initial edge finish.
3. A bull float is applied when the screeding is completed. This will provide the initial finish. At this stage, the concrete is still too wet to apply the final finish.
4. The final finishing can commence once the concrete is firm but still plastic and all excess water that bleeds to the surface has evaporated.

19.8.4.5 Final finish to concrete in two steps

1. While the concrete is still plastic but no excess water is on the surface, apply a hand or power trowel finish to prepare the surface for the final finish. Use hand trowels, such as a steel trowel, wood float or bull float.
2. Trowel edging for a second time and then apply final finish.

Fig. 19.32 Powered helicopter trowel

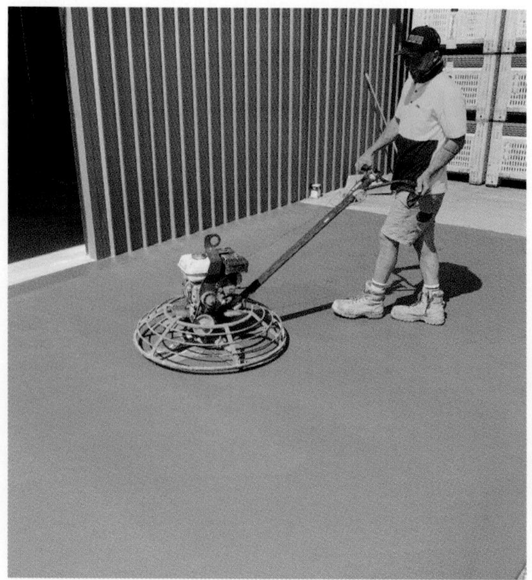

Courtesy of Alister Ford

TIP Edging may be trowelled a third and final time if a different final finish is required for the edging.

19.8.4.6 Suitable slip-resistant finishes for a concrete driveway

Suitable slip-resistant finishes for a concrete driveway are:

- *Cove trowel finish.* Applied in same way as other hand trowels but leaves a raised or rippled surface finish. Edges may be finished with a smooth edge trowel or left as a cove finish.
- *Wood float finish.* Applied in the same way as other hand trowels, but leaves a slightly sandy or slightly rough surface, which is ideal for pathways and driveways. The edges may be finished with a smooth edge trowel or left as a wood trowel finish.
- *Broom finish.* Applied with a special long-handled concrete finishing broom. The technique requires the broom to be placed gently on a far edge just off the concrete. It is then drawn steadily back from the far edge towards the person applying the finish across the full width of the slab. The edges may be finished with a different trowel or left with a broom finish.

Fig. 19.33 (a) Cove trowel finish for non-slip pedestrian access; (b) marking the break for finishing; (c) smooth finish abutting cove trowel finish

(a)

(b)

19.8.4.7 Control joints

Control joints are used to prevent irregular or random cracking occurring after the concrete has set. They are either formed with a jointer when the concrete is wet or can be cut with a concrete saw between 4 and 12 hours after the concrete is poured. Control joints should be between 25 and 30 mm deep for a 100 mm slab and are spaced between 3 and 6 metres apart, depending on the size of the steel mesh bars. The larger the diameter of the steel bars, the greater the spacing between control joints.

(c)

Courtesy of Alister Ford

19.8.4.8 Construction joints

Construction joints separate large sections of concrete and allow the concrete to expand or contract. The dowels, in this case, slide inside an outer casing on the other side of the plastic construction joint, which has a 'honeycomb' construction to allow contraction and expansion of the concrete. The dowels maintain the alignment of the two connecting slabs while allowing them to move independently.

Fig. 19.34 Broom finish

© TFoxFoto/Shutterstock

Fig. 19.35 Sawn control joint

Courtesy of Alister Ford

19.8.5 Environmentally friendly waste management

Substantial fines are issued for contaminating stormwater channels with concrete waste. A waste management system should be established prior to work commencing. Tools and equipment should not be washed into the gutter. A washdown area should be established on-site where the run-off will not enter the stormwater system.

Excess concrete should be poured into manageable piles on a plastic sheet. When dry, they can be broken up, recycled and used as road base or fill or disposed of in a rubbish skip as a last resort.

Ready-mix concrete suppliers offer environmentally friendly waste management. This involves 'take back' of excess concrete delivered to site. This concrete is taken back to the batching plant and put through a process to recover the aggregates and waste water. This is not a free service and can be costly. This cost can be avoided by accurate quantity estimating.

Fig. 19.36 Dowelled construction joint placed every 18 metres

Courtesy of Riviera Concrete

Fig. 19.37 Dowelled construction joint between two separate concrete slabs

Courtesy of Riviera Concrete

Fig. 19.38 Finished foam expansion joint

Courtesy of Alister Ford

Fig. 19.39 Excess and clean demolished concrete piled for recycling

Courtesy of Alister Ford

Fig. 19.40 Recycling by machine

Courtesy of Alister Ford

Thermal mass of building materials

Thermal mass is the ability of a material to absorb heat energy. A lot of heat energy is required to change the temperature of high-density materials such as concrete, bricks and tiles. They are, therefore, said to have high thermal mass. Lightweight materials such as timber have low thermal mass.

In climates that have large fluctuations between day and night-time temperatures, high thermal mass can act as a thermal shock absorber. The building will heat up slowly during the day and hold that heat into the night, thereby providing a more constant temperature inside the house. This process can be seen in places such as Coober Pedy where a lot of the population live underground. Despite extreme temperatures on the surface, the houses stay at a fairly constant 24 degrees.

It is not feasible to build all houses underground; however, the same principles can be applied by using high thermal mass concrete structures.

Fig. 19.41 Recycled concrete as road base

Courtesy of Alister Ford

Table 19.2 Volumetric heat capacity of common building materials

Material	Thermal mass Kj/M^3
Water	4186
Concrete	2060
Sandstone	1800
Rammed earth	1673
FC sheets	1530
Brick	1360
Earth wall	1300
AAC (Hebel)	550

CHECK YOUR UNDERSTANDING

1. What is the maximum height that concrete may be dropped in order to avoid segregation of concrete?
2. What is the difference between control joints and construction joints?
3. When placing a steep concrete driveway, what type of surface finish would you use? Why have you selected this finish?
4. List at least four precautions that must be taken to avoid the segregation of concrete during transporting and placing.

19.9 Curing

When concrete begins to harden, the process of gaining strength will continue for some time provided that favourable conditions are maintained. This process is termed *curing* and two conditions are required: (1) the presence of moisture and (2) a favourable temperature.

Excessive evaporation of water from the concrete can cause the hydration process to stop.

On-site curing methods can include:

- spraying curing compounds over the surface, forming a liquid membrane that limits the evaporation of moisture
- covering the concrete with hessian or a layer of sand and keeping this moist by watering periodically
- covering the concrete with building paper or plastic sheeting to prevent loss of moisture.

TIP The curing process should continue for at least seven days. Concrete reaches 60% of its strength after seven days, and its final working strength when it is more than 28 days old.

19.10 Compressive strength of concrete

The compressive strength of concrete is a measure of its ability to resist crushing. The standard test for compressive strength is made when the concrete is 28 days old. It is measured in megapascals (MPa).

Test specimens are prepared when the concrete is being poured. These are cured under controlled conditions and at 28 days are subjected to a test procedure by crushing them to destruction to determine their compressive strength.

Standard grades of structural concrete are 15, 20, 25, 30, 40, 45 and 50 MPa. For domestic work, concrete of 20 to 32 MPa will meet most specifications. Concrete is designed to reach a specified strength by determining the W/C ratio. Examples of different concrete strengths and the required W/C ratios are shown in Table 19.3.

Fig. 19.42 Compressive strength test cylinder

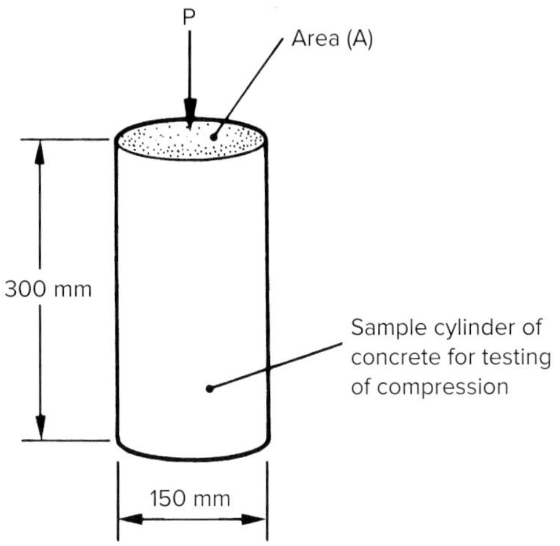

Table 19.3 Required W/C ratios

Compressive strength (MPa)	W/C ratio in kg	W/C ratio in litres per 20 kg bag of cement
15	0.7	14
20	0.6	12
25	0.55	11

TIP An allowance must be made in the amount of added water in order to compensate for the amount of water that may be present in the aggregates already, particularly damp or wet sand.

Concrete must be made sufficiently workable so that it will flow around any reinforcement and into the furthest corners of the formwork. When additional water is added to improve workability, an additional and proportional amount of cement must also be added to maintain the design strength.

19.10.1 Slump test

The workability of concrete is measured by the slump test and is made using a standard cone and tamping rod.

1. Stand the cone on a level surface and hold it firmly by placing feet on the foot rests.
2. Fill the cone with concrete in three layers of 100 mm each, tamping each layer evenly over its area 25 times with the tamping rod.

3. Strike level at the top. Lift the cone carefully but firmly straight up and then allow the concrete to subside.
4. Invert the cone and place it beside the sample.
5. Using the tamping rod and a rule, measure the slump, which is the difference in height between the cone and the concrete.

The recommended slump for differing applications and types of construction is shown in Table 19.4.

Fig. 19.43 Slump test equipment

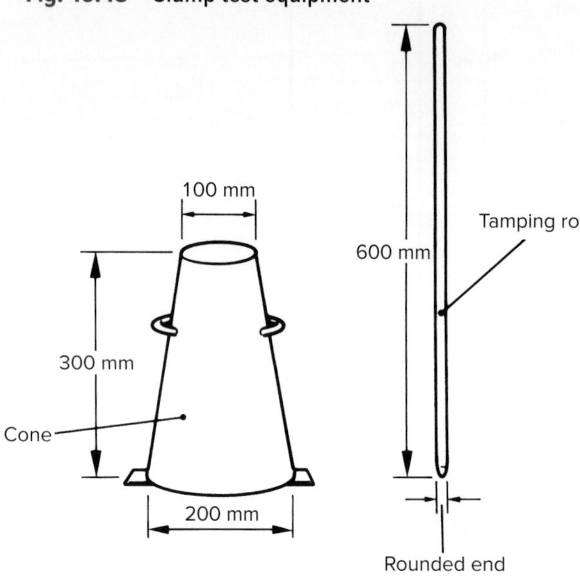

Fig. 19.44 Measuring concrete slump

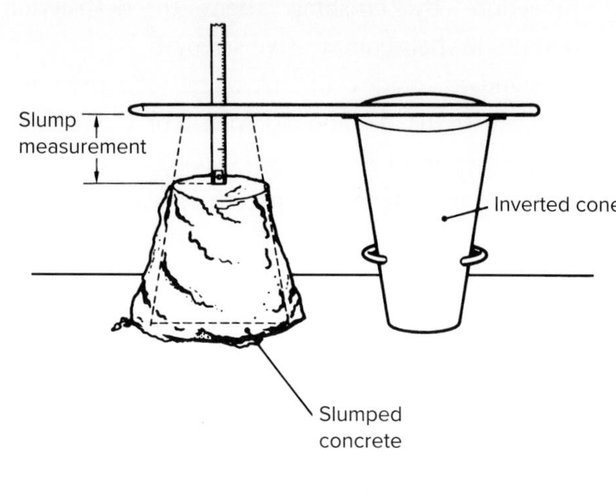

Table 19.4 Recommended slumps

Type of construction	Recommended slump (mm)	
	Minimum	Maximum
Pavements and slabs	50	80
Beams	50	100
Reinforced footings	50	100
Columns	50	100
Reinforced thin walls	80	120
Core fill for blocks	230	250

CHECK YOUR UNDERSTANDING

1. What are the four main components of concrete?
2. What are the benefits of compaction?
3. List six requirements of good formwork.
4. What items of PPE would you consider using when pouring a slab outdoors?
5. Why is it necessary to 'cure' concrete?

Student research

Search online for a MSDS for cement. Calculate the concrete and steel required for a reinforced concrete driveway.

1. Using the house plan provided in the appendix, estimate the amount of concrete that should be ordered to pour the driveway from the front of the garage to the front boundary. Assume that the dimensions are 5.8 m wide × 8.0 m long. Allow 5% for waste for the concrete. Reinforcing steel will be standard sheet sizes using SL82 mesh and the driveway is 120 mm in thickness. Order concrete up to the nearest 0.2 of a cubic metre above your answer.

The following websites will provide detailed information in areas that go beyond the scope of this chapter:

- Cement Concrete & Aggregates Australia (CCAA): www.concrete.net.au
- Concrete Institute of Australia: www.concreteinstitute.com.au.

Search online for 'laying concrete' to see more detail on concrete finishing.

End of chapter activities

Activity 1:

Using your drawing tools and equipment, neatly draw to scale two elevations of alternative methods of setting up the formwork to create a rebated slab for a brick veneer house. See drawings earlier in the chapter.

Specifications:

- scale 1:10
- timber and plywood formwork, braces and pegs; foam block outs
- footing size 400 mm wide × 500 mm deep
- slab thickness 150 mm
- reinforcing, trench mesh, sand blinding and vapour barrier to be shown in correct positions
- sand blinding 50 mm thick
- sheet reinforcement SL-82
- trench mesh 4-L10TM
- show correct hatching for earth, concrete and foam block out.

Activity 2:

Fill out a safe work method statement (SWMS) for the construction of only one of the rebated slab formwork systems that you have just drawn. The slab will be on a flat building site and everything is prepared and ready for the formwork to be constructed.

AUSTRALIAN BUILDING CODE AND STANDARDS

AS 3610.1: 2010
Formwork for Concrete

AS 3610.1: 2018
Formwork for Concrete. Part 1: Specifications

AS 4671: 2001
Steel Reinforcing Materials

National Code of Practice: Pre-cast, Tilt-up and Concrete Elements in Building Construction. February 2008

WorkSafe Victoria Industry Standard: Precast and Tilt-up Concrete for Buildings

Tradie talk—Too wet, too weak

Chapter 20

Carry out demolition

Learning Objectives

LO 20.1 Understand the safety aspects of demolition

LO 20.2 Plan for demolition work

LO 20.3 Prepare for demolition work

LO 20.4 Carry out demolition of basic timber structures

LO 20.5 Carry out demolition of brickwork

LO 20.6 Identify re-usable and recyclable materials

LO 20.7 Carry out demolition site clean-up

Introduction

Building and construction workers will at some stage be involved in the demolition of building structures. These structures will all be different in terms of their size, building materials and construction methods–making every demolition job unique in its own way.

It is important to note that even small-scale demolition projects may be difficult and expose workers to a wide range of hazards and risks. These hazards and risks need to be identified, then controlled or removed before the work commences.

Workers become qualified demolition practitioners by completing the required high-level demolition training. These are separate qualifications under the Australian Qualifications Framework that address demolition work.

Some construction workers specialise in large-scale demolition work, including the removal of multistory concrete, steel or timber-framed buildings or structures.

In this chapter you will learn about the safety aspects of demolition work, including your own personal safety and the safety of your fellow workers and members of the general public. You will also learn how to plan for, prepare and carry out the demolition of basic timber structures and brickwork, as well as how to identify and deal with potentially hazardous building materials and situations.

Other topics covered are the paperwork and permits required before the job commences, personal protective equipment (PPE) and how to fit and use it correctly, plus the tools and equipment required for the job. You will also learn how to identify and reclaim used building materials for re-use and recycling, followed by the clean-up procedures for the construction site and the correct disposal of demolition waste materials.

20.1 Demolition safety

Workplace safety is the most important aspect of any job, as you cannot undo a workplace accident. When carrying out demolition work, you must be aware of any dangers that exist for that particular job and proceed in a manner that protects you and your workmates at all times.

Some important points to remember:

- Always have safe work method statements (SWMS) and demolition plans in place before starting work.
- Make sure you correctly select and fit any PPE (Fig. 20.1).
- Use safe manual handling procedures.
- De-nail as work proceeds.
- Do not work under any structure that is being demolished.
- Follow the Safe Work Australia Code of Practice for Demolition Work–October 2018.
- Follow the WorkSafe Victoria Demolition Compliance Code–Edition 1, May 2018.

Fig. 20.1 PPE for demolition

WORKPLACE SCENARIO

Early in my apprenticeship, I was involved in the demolition of several brick tank stands on the top floor of a hospital boiler house. There were some old double timber doors on this floor and a steel gantry with a big steel pulley on the end installed above the doors. This was for getting water-pumping equipment in and out.

We had to get the bricks down from this fourth-story level somehow, so we sourced a length of thick rope that was probably a hundred metres or so long and rigged up some slings to hold the wheelbarrow level. We put the rope through the pulley then dropped it back down and tied it to the back of the hospital groundman's tractor. We then put the wheelbarrow in the slings and drove the tractor up the road.

This lifted the wheelbarrow up to the fourth floor with no problems. We loaded the barrow with bricks, the tractor reversed, bringing the wheelbarrow back down to ground level, and we went down and unloaded and stacked the bricks. A lot more barrowloads later, and all of the bricks were neatly stacked ready to be removed from the site. This method would not be acceptable under any circumstances by today's standards, and in hindsight I think we were very lucky to get the job completed without any incidents.

Fig. 20.2 Codes of Practice

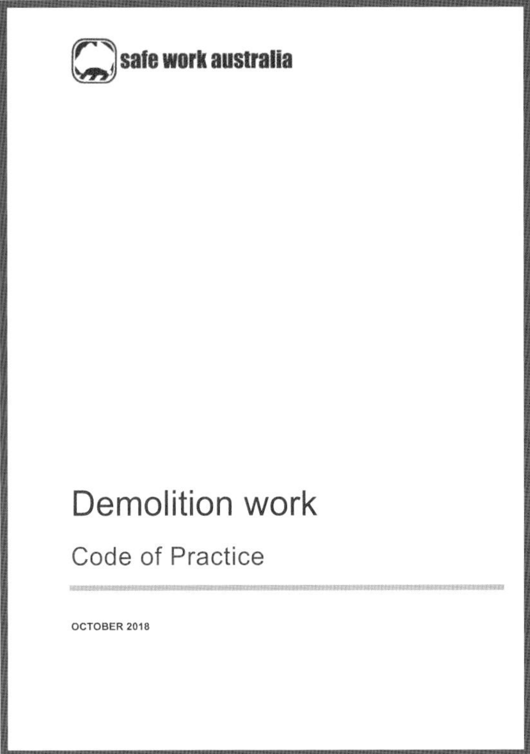

Courtesy of Safe Work Australia CC BY-NC 4.0 https://creativecommons.org/licenses/by-nc/4.0/

20.1.1 Hazardous materials and chemicals

Hazardous materials and chemicals can be present in all types of buildings. Structures built many years ago are more likely to contain such materials as they were seen at the time to be hard-wearing. However, very little, if anything, was known then of the dangers that such materials would pose in later years. The most common hazardous materials are listed below.

20.1.1.1 Asbestos

What is asbestos? Asbestos is made up of naturally occurring minerals in our environment. When these minerals are mined, refined and then bonded, asbestos products can be manufactured. The substance was used in construction for many purposes because it was cheap, easy to use and very durable. However, it is a known health hazard and is associated with diseases such as mesothelioma and asbestosis.

Asbestos can be divided into two types: non-friable and friable.

- *Non-friable:* This type of asbestos is dry. It cannot be pulverised, crumbled or reduced to powder by hand pressure.
- *Friable:* This type of asbestos breaks, crumbles or turns to dust when handled.

Who can remove asbestos? This is one of the most commonly asked questions by apprentices and pre-apprentices because it is common on even the smallest demolition task to come across some type of asbestos containing building material.

There are three different asbestos removal levels: Class A licence; Class B licence; and no licence required.

- *Class A licence:* permits the removal of any amount of friable or non-friable asbestos, asbestos-containing material or asbestos-containing dust.

- *Class B licence:* allows the removal of any amount of non-friable asbestos, non-friable asbestos-containing material or non-friable asbestos-containing dust.
- *No asbestos removal licence required:* in these cases it is permissible to remove up to 10 m^2 of non-friable asbestos or non-friable asbestos-containing material. Asbestos-contaminated dust that is associated with the removal of less than 10 m^2 of non-friable asbestos or non-friable asbestos-containing material is considered to be only a minor contamination.

The total time for asbestos removal should not exceed one hour in any seven-day period (this time is cumulative for work carried out by all workers over a seven-day period). The workers must have been trained correctly in the safe removal of asbestos.

Examples of where you may encounter asbestos in demolition work include: external wall cladding on houses or buildings, corrugated sheets on roofs or fences, internal linings on walls and ceilings, particularly in bathrooms and laundries, spouting, guttering and downpipes, eave linings, splashbacks in wet areas, floor and wall tile underlay, carpet underlay, vinyl floor and wall linings (in both sheet and tile form), electrical meter box backing boards, pipe lagging (wrapped around water pipes, particularly hot water pipes), dust (in roof spaces and wall cavities where the covering is asbestos sheeting).

TIP A laundry or bathroom sized 1.8 m × 2.4 m with a 2.4 m ceiling height, a single door and a window has around 20 m^2 of wall and ceiling linings. Thus, most bathrooms and laundries will exceed the limit of 10 m^2 for non-specialist removal of asbestos and a licensed asbestos removal contractor must be used. The ceiling of a typical laundry or bathroom up to 2.4 m × 2.4 m will be around 6 m^2.

Bonded asbestos cement sheeting that is damaged and friable must be removed only by a licensed asbestos removal contractor.

20.1.1.2 Lead

Lead can be present in various construction materials and applications, such as:

- plumbing fittings, solders, sheets, pipes and flashings
- leadlight windows and glass
- lead-based paint.

If it is suspected that lead-based paint may be present, a test must be carried out to confirm its presence. Refer to *AS 4361.1-1995, Guide to lead paint management–industrial applications,* for more information and guidance.

Fig. 20.3 Specialised tools allow whole sheets of asbestos sheeting to be removed without breakage

© Logtnest/Shutterstock

When carrying out demolition that contains lead materials, some basic precautions can be taken:

- Continually clean the work area during the demolition process.
- As far as possible, minimise lead dust and fumes.
- Correctly select and fit your personal protective equipment (PPE).
- Thoroughly clean the work area upon completion of the task.
- Maintain good personal hygiene levels.

20.1.1.3 Synthetic mineral fibres

These are most commonly known as wall and ceiling insulation batts.

- Removal needs to be carried out carefully to avoid disturbing the fibres as they can cause skin and eye irritation.
- The fibres can also become airborne and may be inhaled into the lungs.
- Damping down can suppress the fibres and dust during removal.

20.1.1.4 Rogue services

It is vitally important to be aware of another hazard known as 'rogue services' when carrying out demolition work. These are services that have not been identified when all of the checks have been made and the reports finalised.

Rogue services most commonly occur in older buildings where renovations have been carried out, sometimes several times over many years. Most commonly, electrical cable or wiring is terminated in a wall cavity or roof space and capped off, leaving it attached to the switchboard and still live. This can also be the case with gas and water services being capped and left as a branch from another known pipe that is connected to the mains.

Demolition saws and other demolition tools have been known to hit rogue services, sometimes with disastrous consequences.

Fig. 20.4 **A near miss that could have been deadly: these wires were disconnected, taped and pushed back into the brickwork cavity during a previous renovation. The demolition saw was only millimetres away from cutting through them. A licensed electrician was immediately brought in to fix the situation**

Courtesy of Alister Ford

CHECK YOUR UNDERSTANDING

1. List the items of PPE that would be required when undertaking demolition work.
2. What is the difference between bonded asbestos cement and friable asbestos?
3. What is the maximum amount of bonded asbestos cement sheeting that can be removed by a person who is not a licensed asbestos removalist?

20.2 Planning for demolition work

20.2.1 Plan well before any work takes place

The written plan will vary for demolition work–from very basic plans for small, uncomplicated work through to very detailed and comprehensive plans for large-scale demolition. But no matter the size or scope of the demolition, the plan must be easy to understand, effective and followed by all workers involved in the demolition job. Local government authorities have specific requirements relating to how demolition work should be carried out.

Before any demolition plan is developed or work is started, the local authorities should be consulted. Building plans should be lodged and approved, inspections should be conducted and permits to work obtained. Failure to do so can lead to heavy penalties being imposed.

Demolition is considered to be a high-risk work activity under current legislation. It is therefore a requirement to produce a safe work method statement (SWMS) to identify any hazards, and the risks created by these hazards, for anyone involved in the demolition task as well as the general public.

The SWMS will also detail how the identified hazards and risks will be controlled or eliminated, and who will be responsible for these actions. The SWMS will be used in conjunction with the demolition plan to ensure a safe, well-functioning work site.

Items that could be included in the demolition plan include:

- the address where the demolition will take place
- entry and exit points to the site
- the type of structure to be demolished
- the size of the structure
- protection of adjoining properties
- required barricading, fencing or hoardings
- road or footpath closures and detours for any closures
- site plans, building plans and building specifications
- reports, which may include engineering, asbestos or environmental details
- location of services, whether underground, overhead or installed in the structure to be demolished
- methods of demolition for the specific job
- the sequence of demolition and proposed timelines
- designated storage areas or bin placement for disposable materials, hazardous materials, recyclable materials or re-usable materials
- emergency plans, which could include: emergency services contact details (000), first-aid kits and first-aid areas in case of personal injury, rescue equipment in case of structure collapse, firefighting equipment, evacuation procedures, evacuation assembly points.

20.3 Preparing for demolition work

Once the planning for the demolition job has been completed, you can begin to prepare for the work using the plan as a guide. In other words, once all of the building plans have been submitted and approved, all permits have been obtained, testing for hazardous materials or substances has been completed and identified hazards have been professionally removed, preparation for the demolition work can now begin.

> **SAFETY TIP** The correct personal protective equipment will need to be selected, fitted and used correctly.

An important point to remember is that you may not be the worker carrying out a specific task, but you may still be required to use the same PPE as that person because of close proximity.

Let's look at some of the specific PPE that you may be required to wear or use on a demolition task.

- *Eye protection:* safety glasses or a full-face shield. These should be clear with no damage, scratches or scuffs to impair your vision. Note: if you are working outside and some of the demolition is at eye level or above, safety sunglasses may be a good option to reduce glare and give better vision.
- *Hearing protection:* ear muffs or ear plugs. Should be clean with no damage.

- *Head protection:* hard hat, hair nets and caps. All should be adjusted correctly and fitted so they won't fall off.
- *Hand protection:* leather gloves, fabric gloves, rubber gloves.
- *Foot protection:* steel-capped leather work boots.
- *Body protection:* overalls, coveralls, long-sleeve hi-vis work shirts, long work pants, hi-vis work jackets, disposable coveralls.
- *Respiratory protection (breathing, lungs):* disposable respirators, non-disposable respirators with changeable filters.

All of your PPE should be selected and used in accordance with the relevant Australian Standard for that particular item.

An example of respiratory protection could be: P1 respirators, which filter hazardous dusts and asbestos particles; P2 respirators, which also filter smoke particles. The carbon in the mask shown in Figure 20.5 reduces the nuisance levels of organic fumes but will not filter hazardous fumes.

- Remember, your PPE is your own. Look after it. Replace anything that gets damaged immediately. Do not wait until the next job.
- For good health and hygiene, wear only your own PPE. Do not borrow from or loan your PPE to anyone else.
- Prepare access and egress points from the site.
- Implement any footpath or road closures and detours.
- Erect and position signage, barricading, hoardings or barriers.
- Set up protection for neighbouring properties.
- Set up areas for recyclable or re-usable or waste materials. These could be large bins or skips, trailers, pallets or areas on the ground for stacking and later removal.

Also select and check the correct tools and equipment for the job.

Fig. 20.5 P2 respirator with valve and carbon filter

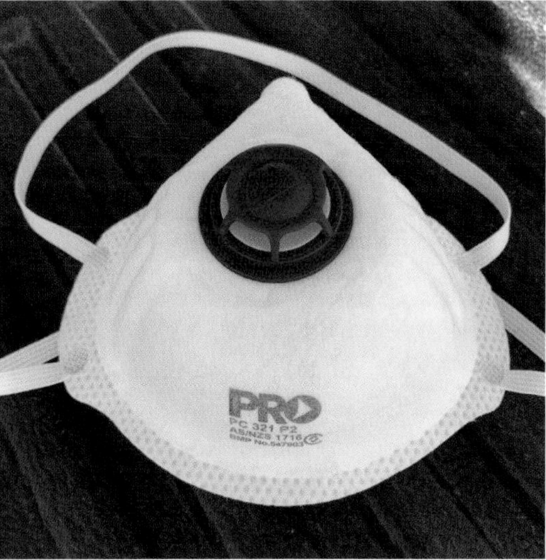

Courtesy of Alister Ford

Fig. 20.6 Sheet hoardings fixed onto the scaffolding for the building facade demolition process and rebuild

Courtesy of Alister Ford

20.4 Demolition of basic timber structures

20.4.1 Typical small-scale demolition projects

In residential construction, demolition is not on the same scale as the demolition of high-rise office or industrial buildings. Many residential renovation projects require the demolition of small-scale structures such as laundries or bathrooms, the removal of internal load-bearing or non-load-bearing walls to create open-plan living areas, the removal of external load-bearing wall sections so that an existing house can be extended and, occasionally, the demolition of a complete home.

Other residential projects can involve demolishing out-buildings such as verandahs, pergolas, patios, garages, tool sheds and workshops. Small tasks may also involve the demolition of fencing.

An important point to remember that will help you with any demolition task is 'Do not rush in and start to demolish'. Stand back and look at the structure that you are about to demolish and think about how it was constructed.

- Where did the builder start?
- What steps did they take to carry out the job?
- How did they fix things in place?

Demolition should be conducted in the opposite order to the construction. Remember: 'The last piece in is the first piece out.' This will create a safer work environment for you and your fellow workmates, and will minimise the risk of collapse or structural failure, which more often than not leads to workers being injured, sometimes critically.

20.4.1.1 Bathroom and laundry renovations

As it is classed as a domestic wet area, a bathroom or laundry in a house built before 1990 will most likely include asbestos cement sheeting as wall or ceiling lining, and perhaps tile or vinyl underlay on the floor. The vinyl tiles or sheeting used in these areas may also contain asbestos.

In full cavity (or double) brick homes, asbestos sheeting was used for bathroom and laundry ceilings only. In brick veneer or timber-framed houses with cladding, asbestos sheeting was used as both an internal wall lining and as a ceiling lining for bathrooms and laundries.

As discussed in the earlier section on demolition safety, asbestos cement sheeting that covers more than 10 m^2 must be removed by a licensed asbestos removalist. Besides the potential presence of airborne asbestos particles, when asbestos cement sheet ceilings or any ceiling linings are removed, abundance of dust is an issue. Filthy, black, sticky dust that contains many types of hazardous materials, including the remains of dead animals and bird excrement, must be handled with care.

The best method is to use an industrial exhaust fan with a length of ducting to remove all of the dust and move it outside into a collection point so that it can be safely and easily removed later (see Fig. 20.7). Alternatively, the top of the ceiling can be cleaned with an industrial vacuum cleaner before the ceiling is removed.

Another risk that construction workers and renovators face is the unknown condition and location of electrical wiring and the gas and water services. This should be established before demolition commences. Extreme care should be taken while demolition takes place in case rogue wires and water pipes exist in wet areas from previous renovation work.

Bathrooms are often located in an area that can be accessed only by walking through parts of the home. Large amounts of demolished materials must be moved manually from the bathroom or laundry and out through the house and into the specified area, depending on the plan for the demolished materials.

Fig. 20.7 Industrial exhaust system for the removal of dust

Courtesy of Alister Ford

Before work commences, the access walkway to the bathroom should be determined and plastic sheeting should be placed over floors, walls and ceiling to prevent damage to floor coverings and painted surfaces.

Entries into other parts of the home must be sealed to prevent dust travelling throughout the home. This can be done by closing the doors to other rooms and hanging plastic drop sheets from the top of the door-frame architraves and fitting them down onto the floor to prevent any dust penetration whatsoever. Placing protective covers over all exposed surfaces will result in a safe and efficient demolition that doesn't impact on the rest of the home. This also makes it easier and quicker to clean up after the work is completed.

20.4.1.1.1 Demolition plan for a bathroom renovation

Before demolition commences:

- inspect the ceiling space above the bathroom and identify any obstacles, fixed structures, and water, electrical and gas services
- inspect the bathroom and identify all services that have to be removed or relocated
- identify any potential asbestos and plan for this to be removed separately; asbestos removal should be conducted as a separate process and not in conjunction with the demolition of other structures forming the bathroom
- arrange for services to be relocated or removed before the demolition begins
- organise the correct signage and barrier tape
- organise waste, recyclable and re-usable storage areas for removed material.

During demolition:

- wear appropriate PPE
- install protective coverings to all surfaces in areas to be used as a walkway
- if the bathroom door or window is being retained, it will also need to be protected
- seal doors that open into the work area with tape to prevent any dust entering other parts of the home; secure these doors so they cannot be opened
- install signs and barrier tape (Fig. 20.8)
- shut down electrical, gas and water services at the main tap or switch and attach safety tags
- remove or temporarily move and shut down and secure all services–electrical, water and drainage, and gas
- isolate the bathroom work area from all services by shutting off power and capping other services to the bathroom
- identify and remove all prime-cost items, taps and other fixtures and fittings
- install a vacuum exhaust fan for dust removal and seal around exhaust ducting where it passes through the bathroom window
- remove the ceiling in accordance with the demolition plan in the next section
- remove wall tiles and the wall lining if it is to be removed
- remove floor tiles using appropriate demolition tools.

Fig. 20.8 Asbestos removal warning sign and barrier tape

Courtesy of Alister Ford

20.4.1.2 Demolition plan for removal of ceiling linings

- Turn power off at the mains and attach safety tags.
- Inspect the roof framing, supporting ceiling battens and the ceiling lining before commencing any demolition.
- Identify any insulation, obstacles or services that may have to be removed or moved temporarily.
- Identify any fixed structures that will interfere with the removal of the ceiling.
- Wear PPE: gloves, P1 or P2 mask or respirator, coverall clothing, safety glasses.
- Work from a raised work platform 600 to 900 mm above the floor placed inside the bathroom or laundry.
- Don't demolish the ceiling from inside the roof space.
- Ensure there are two workers on the job, to assist each other.
- Remove all dust from the top of the ceiling in the roof space using a class H vacuum cleaner.
- Cut ceiling lining into manageable pieces, e.g. 1.2 m × 1.2 m (four pieces for a 2.4 m × 2.4 m ceiling).
- Consider all identified obstacles, services and structures as the ceiling is cut.
- Select the right tools, which may include a cordless reciprocating saw, a cordless 100 mm grinder with cutting disc or a hand saw.
- Remove each piece separately but don't work underneath the piece being removed as you remove it.
- Remove each section from the demolition area as it is taken down.

TIP Vacuum cleaners used to remove asbestos fibres and friable asbestos must comply with Australian Standards and must display the high hazard (H) class symbol (Fig. 20.9) and contain warning information about dust in the vacuum cleaner collection system.

Fig. 20.9 (a) H-class label; (b) H-class vacuum cleaner with HEPA filter

(a)

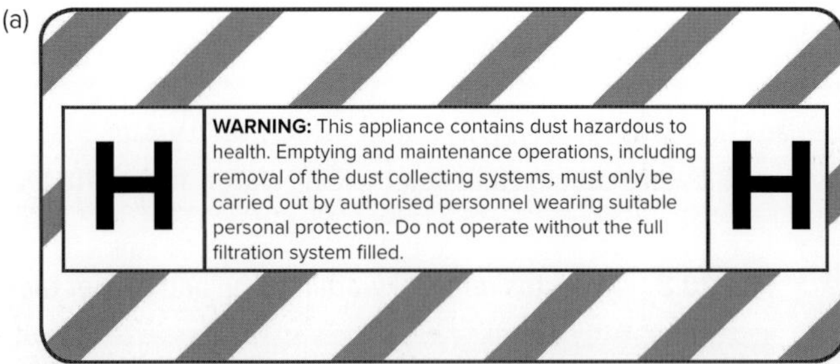
H
WARNING: This appliance contains dust hazardous to health. Emptying and maintenance operations, including removal of the dust collecting systems, must only be carried out by authorised personnel wearing suitable personal protection. Do not operate without the full filtration system filled.
H

(b)

Numatic

Courtesy of Alister Ford

20.4.1.3 Removal of external cladding and roofing

Removing external cladding and roofing requires the use of suitable scaffolding or work platforms. You may also be required to use a fall-prevention device such as safety harnesses attached to designated anchorage points. If the roofing or cladding is an asbestos-based product, it must be removed by a licensed asbestos removalist.

Insulation in wall cavities and roof spaces may also have to be removed.

Type P1 or P2 respirators or disposable masks must be used to prevent the inhalation of insulation products such as fibreglass and mineral wool. PPE such as safety gloves, safety glasses and disposable protective overalls should also be used to minimise skin irritation. Old insulation collects dust, which may contain unknown elements such as decayed dead animals and insects, and must be assumed to be hazardous.

Steel sheet roofing and clay and cement tiles can be recycled for re-use or conversion to other products. External cladding made from **fibre cement** (not asbestos cement) can be recycled, and timber cladding can be re-used if it is in good condition, otherwise it can be used as landfill.

Fig. 20.10 Asbestos disposable plastic bags

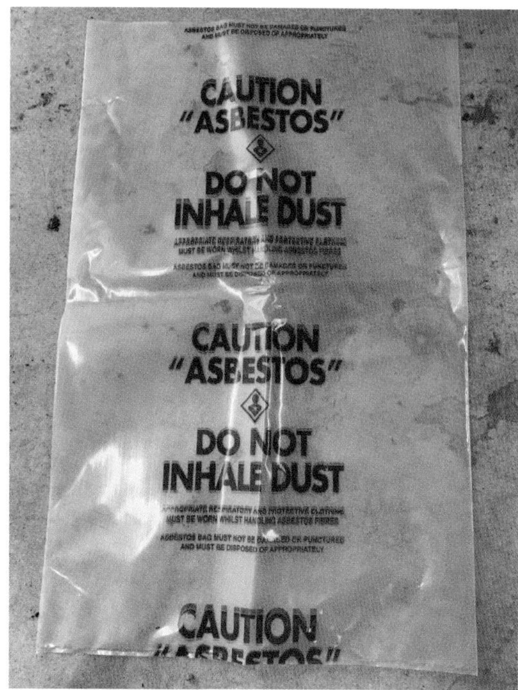

Courtesy of Alister Ford

20.4.1.4 Removing load-bearing and non-load-bearing timber-frame walls

Internal walls may be removed to create larger open-plan living areas and improve access. These walls may be either **load-bearing** or **non-load-bearing**. Once the load that a wall was carrying has been directed to other points of the structure, the techniques for removal are similar for both.

It is not always a simple process to remove load-bearing walls–in fact, it can be very complicated. Many factors need to be considered and addressed with regard to the structural integrity of the building. Failure to do so can lead to some disastrous outcomes.

Redistributing the load that the soon-to-be-demolished wall was carrying is a vital part of the demolition process. No structural part of the wall can be removed until this has been done or at least until the load has been temporarily held up with props and toms, to be completed at a later stage. The redistribution of the load must be carefully planned and approved to ensure the new members that will take the load comprise the correct materials and are of adequate size and strength.

The load must also be redistributed through new point load positions that are deemed to be adequate to transfer the new load, as well as any previous load that this point was carrying. It is often necessary to add strength to remaining walls by way of adding extra studs into the frame to create new load points.

If the subfloor in the structure is of conventional bearer and joist design, it may also be necessary to add more timbers to the subfloor structure. The load will then be transferred through the remaining walls, onto the subflooring system and then to the footings of the structure. The risk of structural failure and accidents is lower when removing non-load-bearing walls.

Internal load-bearing timber-frame walls are designed to support many individual items in ceilings, conventionally pitched roofs and the upper story of two-story houses. These items can include:

In ceilings:

- ceiling joists
- ceiling trimmers
- hanging beams
- counter beams
- combined strutting/hanging beams
- combined strutting/counter beams
- ceiling lining materials

- ceiling insulation
- light fittings, ceiling fans, exhaust fans, surround-sound systems
- electrical cabling, gas and water pipes, air-conditioning ductwork
- hot water services.

In conventionally pitched roofs:

- common, jack, creeper, hip, valley rafters
- ridge boards
- underpurlins
- collar ties
- Barrup trusses
- struts
- strutting beams
- roof battens
- roof coverings
- television or communication aerials or dishes
- air-conditioning units.

Fig. 20.11 Temporary propping of floor joists in upper floor—note services

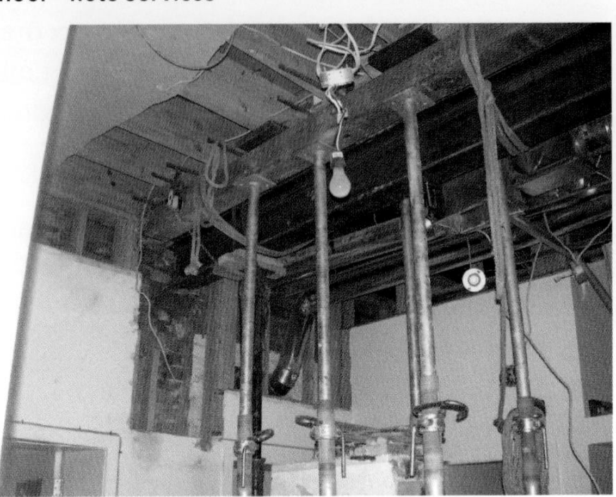

Viktor Ginić, B Design Engineering Services Inc.

In two-story constructions:

- upper-story bearer, joist and flooring systems
- upper-story deep joist and flooring systems
- in-floor cabling and piping
- upper-story floor coverings and underlays
- upper-story wall framing, insulation and linings
- ceiling and roof system as described above
- upper-story dead load–bathtubs (full), shower units, built-in cabinetry
- upper-story live load–moveable furniture, people, wardrobe and cupboard contents, paintings and wall hangings.

If the roof is constructed with a truss system, all ceiling loads and roof loads are transferred to the external walls. This allows *internal* non-load-bearing timber-frame walls (or masonry walls) to be removed without the need to support the roof framing temporarily. However, if the removal is for the purpose of creating a larger opening in an *external* wall, then the roof trusses must be temporarily supported.

When timber walls or roof framing is damaged by termites, the remaining structures that would normally be supported by the damaged framework must be temporarily propped while the damaged framework is removed and replaced.

20.4.1.5 Demolition and replacement of timber decking

Homes built in the 1970s or 1980s and earlier often used unsuitable timber such as Oregon or Canada pine for external cantilevered joists. Timber decking was often a mixture of hardwoods with varying grades of durability. Decking was commonly used in highly-exposed situations or areas close to a marine environment. Bright steel nails were used to nail into untreated softwood floor joists through the hardwood decking. The steel nails would corrode and allow water to enter into the timber joists, causing them to rot over time.

Cantilevered balconies add another dimension to structural damage, with the real possibility of collapse as floor joists decay and finally fail, often resulting in sudden collapse when heavily loaded with people.

For the renovator or repairer, the first task in demolishing and replacing timber decking is to remove all of the structurally damaged materials. Working at height on a structurally unsound framework of timber floor joists that are cantilevered over steeply sloping land is a hazardous situation so a safe work platform must be used.

Sometimes cantilevered joists that run back for several metres as support under internal flooring must be removed and replaced. If the floor framing joists below the upper floor can be easily accessed, the replacement of cantilevered floor joists is fairly simple, but if the lower floor area is another habitable room, then part of the ceiling lining below must be removed.

Fig. 20.12 A–C Conversion of a two-door garage into a one-door garage: (a) previous two-door garage; (b) brickwork removed and temporary props; and (c) new single garage door installed

(a)

(b)

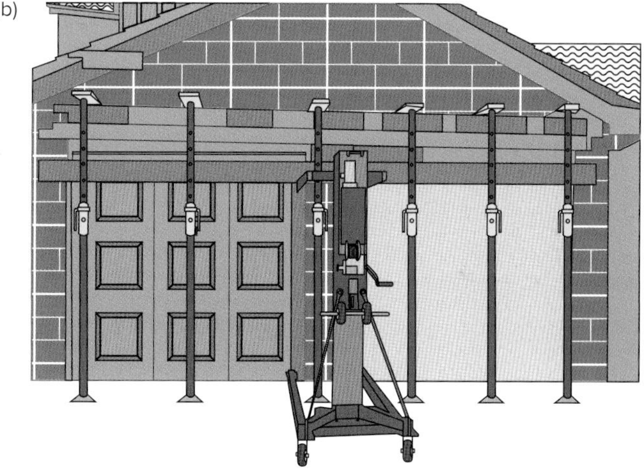

(c)

20.5 Demolition of brickwork

20.5.1 Demolition of brickwork to enlarge openings and install new doors and windows

The removal of existing masonry to form an opening for a door or window requires the use of temporary props. As masonry is removed, the work above that was previously supported must be re-supported.

The procedure involves the removal of one or two bricks to create small openings that continue across the full width of the wall plus the insertion of **needles** through these openings in the wall (Fig. 20.12 a-c). The needles are then propped on each side and held up tight against the masonry above the small opening. Needles may be up to 1200 mm apart or further, subject to the strength of the mortar that was used originally to construct the brickwork.

Brickwork between the needles is removed, starting at the top of the proposed opening. As the brickwork is removed, an opening larger than the size of the door or window to be installed is created (Fig. 20.13a). The brickwork is **toothed** on the sides of the opening as bricks are removed to allow bricks to be re-laid alongside the newly installed window or door (Fig. 20.13b).

Fig. 20.13 **(a) Brickwork removed and toothed; (b) door frame installed and bricks re-laid**

(a)

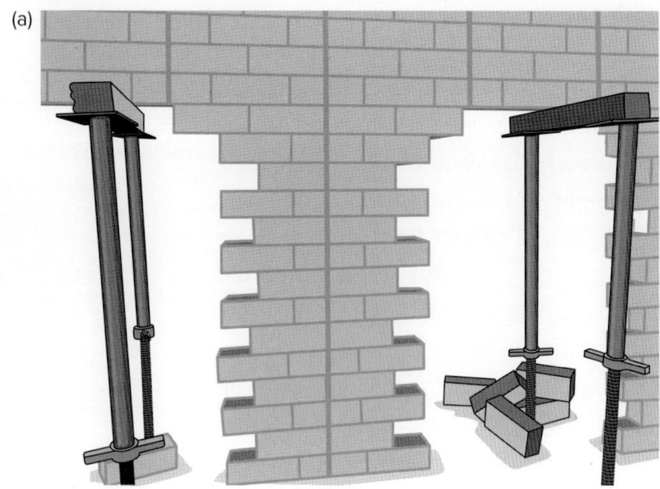

(b)

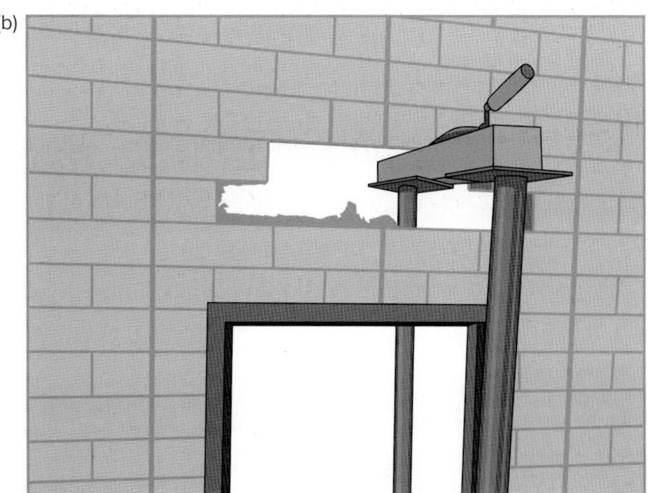

Once the brickwork has been re-laid up to the needle, the temporary props and the needles can be removed. However, before removing these props, the new lintel must be temporarily supported until the brickwork has set for at least seven days. The brickwork should be cured with occasional wetting to facilitate hydration.

Alternatively, an opening only slightly larger than the window or door to be installed can be cut out with a demolition saw. Normally this would be done only where the brickwork is not seen and is covered by solid plaster.

Masonry walls are also often removed to open up connecting rooms and create one large room. Any structures supported by these walls should be temporarily propped until a new steel beam or lintel is installed over the new opening to support any masonry above the opening.

Any surplus bricks remaining after the opening is completed can be re-used or recycled for road base materials or decorative gravel.

CHECK YOUR UNDERSTANDING

1. Briefly describe a simple process to support brickwork when enlarging or creating an opening.
2. How would you undertake the installation of an opening in a non-load-bearing timber-frame wall? Take into account services such as water, gas and electricity.
3. Briefly describe how you would create a large opening in a load-bearing timber-frame wall.

20.6 Re-usable and recyclable materials

20.6.1 Reduce, re-use or recycle if you can

Construction waste is generated in considerable quantities on building sites. Finding a sustainable solution has become an urgent environmental problem. The crucial strategy to managing waste is to reduce, re-use or recycle materials.

One of the most important tasks during demolition is to sort demolished materials into separate categories.

Always remember the 'three Rs' of waste minimisation:

- reduce
- re-use
- recycle.

20.6.1.1 Reduce

We need to minimise the current amount of construction waste. This can be done through accurate ordering of materials and good site management. Excess packaging should be avoided, and suitable dry storage facilities provided on-site. Often the amount of new construction materials spilled or thrown away exceeds demolition waste: up to 20% of construction material on a commercial site may never be used.

20.6.1.2 Re-use

It is often more sustainable and cost-effective to re-use construction material rather than purchasing new ones. Re-using materials is frequently more energy-efficient because it eliminates the additional processes and energy involved in recycling. Timber, roofing tiles, bricks and blocks of low cement mortar are often re-used and are equal to, and sometimes better than, new.

Re-useable building materials can be sold or given away. Some money can be made by selling any re-usable materials. If something cannot be sold, it can often be given away. Second-hand building materials such as timber framing, timber flooring, timber lining and cladding, windows and window frames, doors and door frames, and cast-iron products are often highly sought after by home renovators, builders and furniture makers. These items sometimes cost more to buy than a new product to be used for the same purpose.

TIP Re-usable timber must be de-nailed before resale as any personal injuries that result from nails could result in litigation. There is also the risk of causing costly damage to woodworking machinery. Always use the correct PPE when handling materials from a demolition job.

As any demolition job proceeds, it is important to de-nail any timber as you go. This makes it much easier to stack, taking up less space as the job proceeds, which in the long run saves time and money. One way to check if the timber available for re-use has been properly de-nailed is to run a metal detector over the timber. It is quite easy to miss a nail–sometimes nails will snap off, leaving part of the nail in the timber unseen. Most second-hand timber yards will check any timber they purchase this way and if they detect any nails still present, they may lower the price substantially.

The process used to de-nail any re-usable timber is also very important. Timber can be ruined for its next purpose if it is not de-nailed correctly.

Timber-framing material will most probably be used for timber framing again. The process of de-nailing timber framing is a straightforward one.

- Straighten the nail at the back of the timber to its original position by hitting with your hammer or pulling it up straight with the claw of the hammer or a pinch bar.
- Place the timber on a solid support and hit the nail back through the timber with your hammer.
- Turn the timber back over, place a block under your hammer or pinch bar for extra leverage and pull the nail out.
- Throw the nail into a box or bucket for later disposal.

Fig. 20.14 A–E The correct way to de-nail timber framing

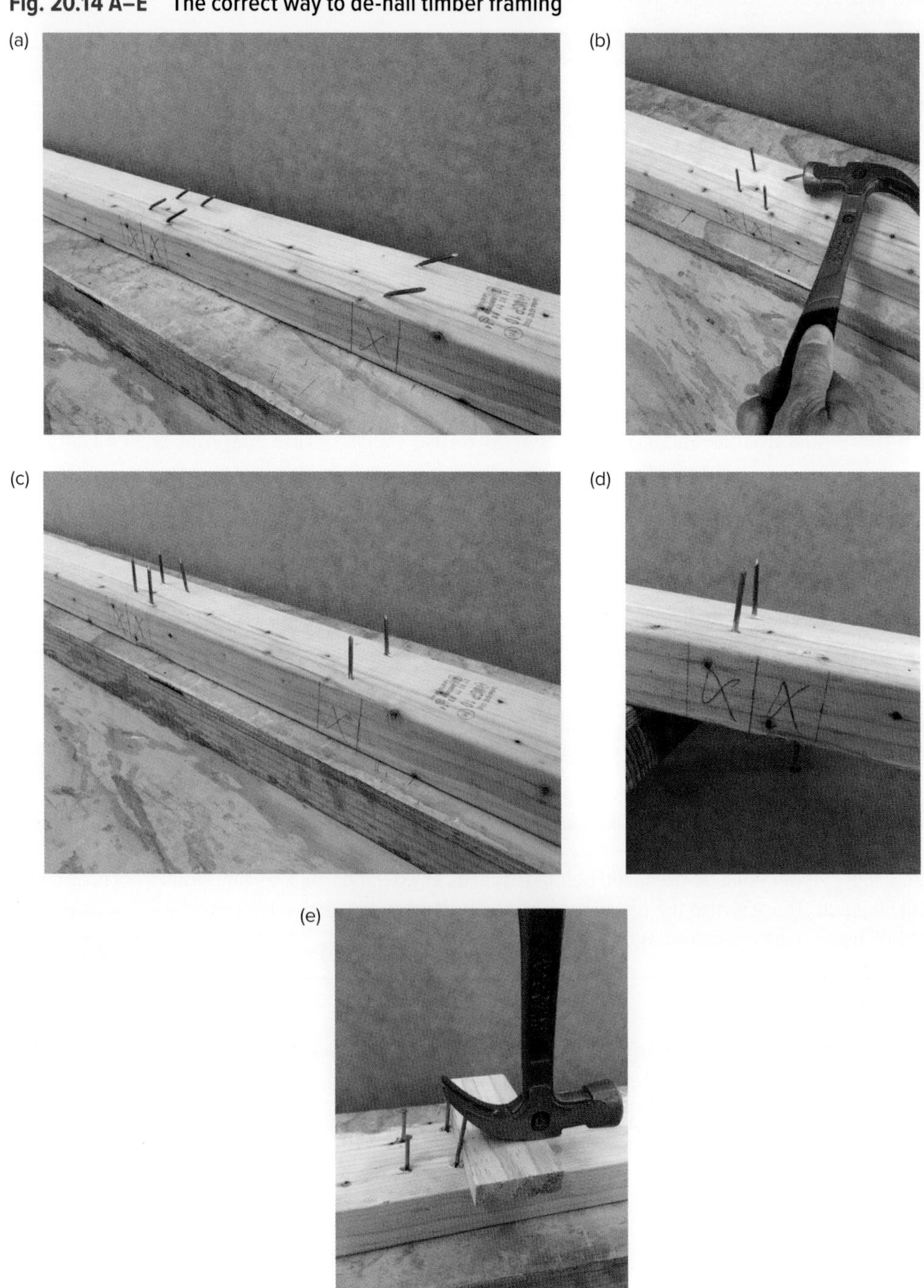

Courtesy of Alister Ford

It can be very costly to buy timber flooring and lining boards second-hand, so it is well worth the time and effort to remove them and de-nail them properly. De-nailing is a bit different here from normal framing timber because we do not want to damage the face edge of the board by 'exploding' the head of the nail back through the timber.

- Once the board has been carefully removed, lay it on a flat surface and bend the nail to about 45 degrees to the back surface of the board.
- Place a thin strip of timber under your pinch bar and place the claw of the pinch bar under the bent nail.
- Hit the pinch bar onto the nail until you feel the bar has securely grabbed the nail.
- Pull the nail through the timber, removing it from the back surface and leaving the face surface nice and clean with no nail removal damage.
- Throw the nail into a box or bucket for later disposal.

Fig. 20.15 A & B Timber flooring damaged by incorrect de-nailing process

(a)

(b)

Courtesy of Alister Ford

20.6.1.3 Recycle

Often, in reality, recycling in the building industry is actually 'down-cycling'. This is when a material is reduced to a lower-level use such as hardcore fill or landscaping bulk. Some recycling processes now produce materials that have equal properties to new, or natural, materials–for example, synthetic wood made from recycled rubber, plastic and sawdust (but at an energy cost).

Recyclable materials can often be given away at no cost to recycling yards. (Sometimes, however, there is a small fee for concrete and brick products, but it is far less than standard tip fees.)

Hard materials such as cement render, concrete, brickwork and tiles are recycled by being crushed down into different aggregate sizes to be used as road base or **hardcore** fill on new construction projects. Hard materials may also be suitable as clean fill on-site or may be used by others who require clean fill on vacant land to raise it above a flood level.

CHECK YOUR UNDERSTANDING

1. Explain why it is important to recycle demolished materials.
2. What products can be manufactured from recycled concrete?

20.7 Demolition site clean-up

When a job has been completed and all re-usable and recyclable materials have been removed, the demolition site will need to be cleaned up and made good for the next phase of the construction process.

You may have a skip, waste-disposal bin, small tip truck, trailer or ute on-site to collect all of the disposable waste, ready for transport to the local tip or landfill site.

The whole demolition site needs to be inspected for rubbish; areas to be checked include:

- the place of demolition
- the designated paths from the demolition to the re-usable, recyclable and waste material areas
- the ground around these areas
- the entry and exit points of the job site
- the roadway leading to and from the site.

Rubbish may include:

- nails
- screws
- bolts
- timber
- tiles
- plaster
- glass
- rubble
- tin
- insulation
- anything that has unknowingly been dropped or been blown away during the transport of the demolished materials.

All waste must be properly secured for transport as heavy fines can be applied for unsecured loads. The slightest sign of asbestos cement materials mixed with landfill waste will result in all demolition materials being declared asbestos-contaminated and the cost of tipping will escalate dramatically. Asbestos-based materials must be wrapped in plastic and clearly labelled as 'hazardous asbestos material' and taken to specially designated landfill sites.

Once the demolition site has been thoroughly cleaned up and all materials disposed of, you will need to thoroughly check and pack away all of your tools and equipment. The nature of demolition work, with all sorts of building materials being pulled down, means that it can be quite easy to damage tools and equipment without realising it.

All hand and power tools should be checked and cleaned. Any faults such as damaged leads, guards, blades, casings or handles will need to be reported so they can be repaired.

Remember: If any tool or piece of equipment gets damaged while you are using it, stop using it immediately, isolate it and report it. You are not going to be in trouble for creating a safe work environment. Never try to hide anything away. The next person who picks up and uses the damaged tool or piece of equipment may get badly injured.

Fig. 20.16 (a) Reclaimed materials stacked and banded ready for removal from the demolition site; (b) Demolished brickwork being cleaned and stacked on pallets ready for removal from the demolition site for re-use

(a)

Courtesy of Alister Ford

(b)

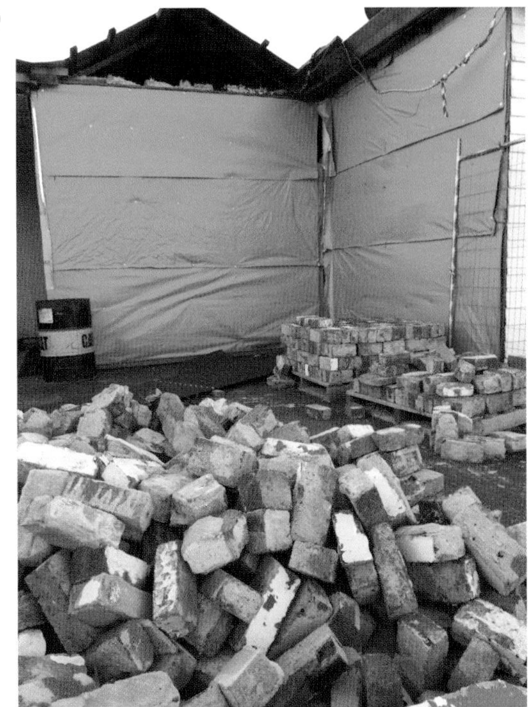

Student research

- Visit Safe Work Australia: www.safeworkaustralia.gov.au, or any other state workers' safety authority (e.g. WorkSafe Victoria) and locate and review the Code of Practice for the Safe Removal of Asbestos and the Code of Practice for Demolition Work, or similar.
- Search YouTube using the search term 'asbestos removal' and review the process of removing asbestos.
- Search YouTube using the search terms 'house demolition' and 'building demolition', and compare the different processes.

End of chapter activities

1. Activity 1

Create a basic plan for the demolition of a small out-building such as an old free-standing garage.

Specifications:

- 7 m long × 4.5 m wide × 2.7 m high
- 1.2 m from the side boundary fence. The residence on the neighbouring property is 1.2 m from the same fence
- unseasoned hardwood wall frames externally clad horizontally with hardwood weatherboards (nailed and painted)
- unseasoned hardwood skillion roof frame externally clad with corrugated iron (nailed)
- 125 mm thick concrete slab, reinforcement unknown
- no internal linings
- 2 × 1.5 m wide × 2 m high ledge and braced timber entry doors for vehicles
- 1 × 0.8 m wide × 2 m high ledge and braced timber side door
- 1 × 2 m wide × 1.2 m high fixed pane timber-framed side window
- 1 × double fluorescent light fixed to the roof
- 2 × double power points fixed to wall frames
- 1 × garden tap plumbed and attached to the external side wall.

2. Activity 2

Fill out the SWMS template (Fig. 20.17) for the task of demolishing the same garage for which the demolition plan has been formulated.

3. Activity 3

Create a tools and equipment list for the small demolition task in Activity 1.

The list should include three sections:

1. personal protective equipment (PPE)
2. required tools
3. required plant and equipment.

Fig. 20.17 Work Safe SWMS template

APPENDIX A—SAMPLE SWMS TEMPLATE FOR HIGH RISK CONSTRUCTION WORK (HRCW)

DUTIES: 1) A SWMS **must** be prepared if proposed works involve any of the HRCW activities listed below and that work poses a risk to the health and safety of any person. 2) Affected employees and their HSRs **must** be consulted in the preparation of the SWMS. **3)** Once a SWMS has been developed and implemented, the HRCW to which it relates **must** be performed in accordance with the SWMS. **4)** Duty holders (builder and sub-contractor) **must** stop the HRCW immediately or as soon as it is safe to do so if the SWMS is not being complied with; the HRCW **must not** resume until the SWMS is complied with or reviewed and revised as necessary. **5)** The SWMS **must** be reviewed and if necessary, revised whenever the HRCW changes, or after any incident that occurs during HRCW, or if there is any indication that risk control measures are not adequately controlling the risks. **6)** An employer **must** retain a copy of the SWMS for the duration of the HRCW.

Direct employer:		**Principal contractor (PC)** (Name and contact details)	
Work supervisor: (Name and contact details)		**Date SWMS provided to PC:**	
Work activity: (Job description)		**Workplace and works location:**	

High risk construction work:	☐ Where there is a risk of a person falling more than two metres.	☐ On or adjacent to roadways or railways used by road or rail traffic.	☐ In, over or adjacent to water or other liquids where there is a risk of drowning.
	☐ At workplaces where there is any movement of powered mobile plant.	☐ Structural alterations that require temporary support to prevent collapse.	☐ In an area where there are artificial extremes of temperature.
	☐ On or near energised electrical installations or services.	☐ Involving a trench or shaft if the excavated depth is more than 1·5 metres.	☐ On or near pressurised gas distribution mains or piping.
	☐ Involving demolition.	☐ Involving a confined space.	☐ On or near chemical, fuel or refrigerant lines.
	☐ Involving tilt-up or precast concrete.	☐ On telecommunications towers.	☐ Involving diving.
	☐ Involving removal or likely disturbance of asbestos (note: preparation of an asbestos control plan is taken to be preparation of a SWMS).	☐ In an area that may have a contaminated or flammable atmosphere.	☐ Involving the use of explosives.
			☐ Involving a tunnel.

Person responsible for ensuring compliance with SWMS:		**Date SWMS received:**	
What measures are in place to ensure compliance with the SWMS? (e.g. direct supervision, regular spot checks)			
Person responsible for reviewing SWMS control measures (e.g. PC's representative):		**Date SWMS received by reviewer:**	
How will the SWMS control measures be reviewed?			
Review date:		**Reviewer's signature:**	

Selecting risk controls:

Any risk to health or safety must be **eliminated**, or if that is not reasonably practicable, **reduced** so far as is reasonably practicable by:

- implementing any mandated controls specified by law (e.g. the *OHS Regulations 2017*)
- substituting a new activity, procedure, plant, process or substance (e.g. scaffold in preference to ladders)
- isolating persons from the hazard (e.g. fence off areas for mobile plant operation)
- using engineering controls (e.g. guard rails, trench shields)—or a combination of the above.

If any risk to health or safety remains, it must be reduced by using:

- administration controls (e.g. activity specific safety training, work instructions, warning signs)
- PPE such as respiratory protection, hard hats, high-visibility clothing—or a combination of the above.

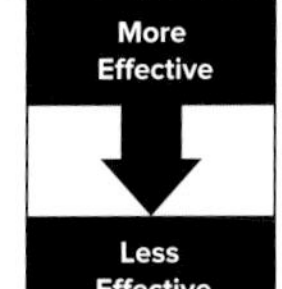

What are the tasks involved?	What are the hazards and risks?	What are the risk control measures?
List the work tasks in a logical order.	What aspects of the work could harm workers or the public?	Describe what will be done to make the activity as safe as possible?

Name of Worker	Signature	Date	Name of Worker	Signature	Date	Name of Worker	Signature	Date

4. Activity 4—Practical

Correctly fit PPE, including eye protection, hearing protection, head protection, respiratory protection and hand protection.

5. Activity 5—Practical

Undertake the correct removal of nails from timber framing and flooring as described in the chapter.

6. Activity 6—Practical

Demolish a small timber structure, observing all of the requirements as listed in the chapter.

AUSTRALIAN BUILDING CODES AND STANDARDS

Safe Work Australia. Demolition Work Code of Practice—October 2018.

WorkSafe Victoria. Compliance Code, Demolition—Edition 1, May 2018. Safe Work Australia.

How to Safely Remove Asbestos Code of Practice—October 2018.

WorkSafe Victoria. Guidance Notes, Removing Asbestos—October 2010.

Tradie talk—The last piece in is the first piece out

Appendix

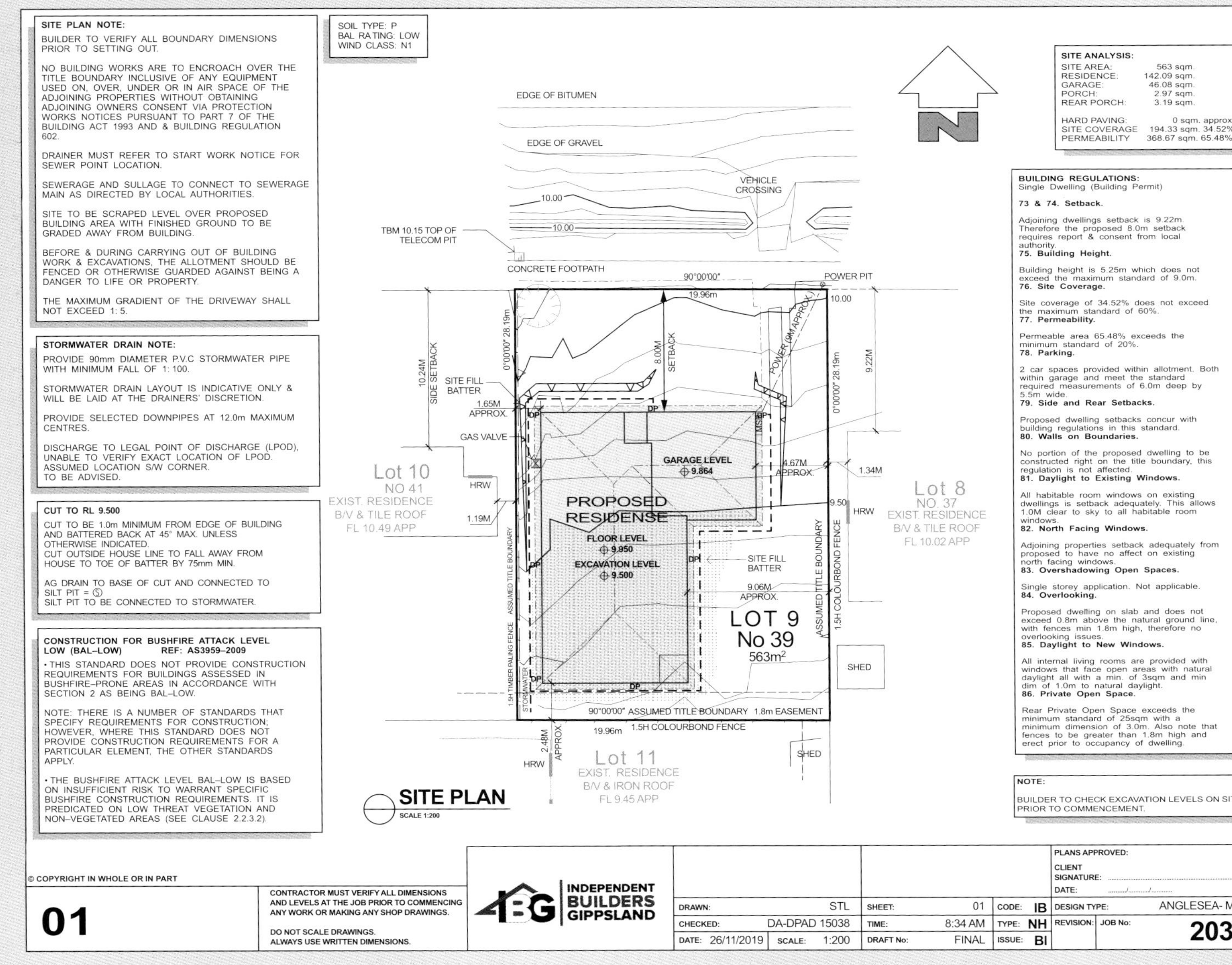

Figure 1
Independent Builders Gippsland

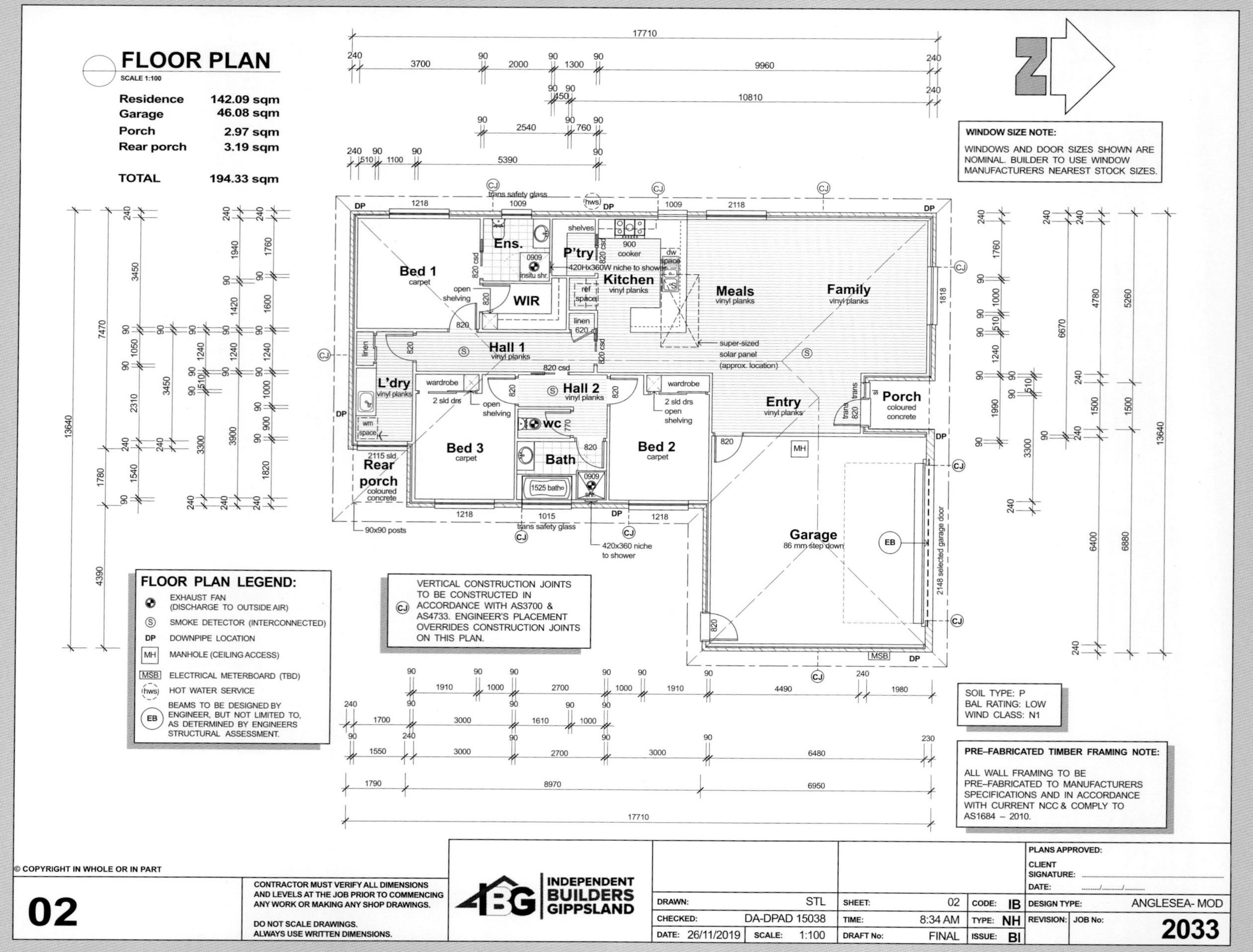

FLOOR PLAN
SCALE 1:100
Residence 142.09 sqm
Garage 46.08 sqm
Porch 2.97 sqm
Rear porch 3.19 sqm
TOTAL 194.33 sqm
WINDOW SIZE NOTE:
WINDOWS AND DOOR SIZES SHOWN ARE NOMINAL. BUILDER TO USE WINDOW MANUFACTURERS NEAREST STOCK SIZES.
Bed 1
Ens.
WIR
P'try
Kitchen
Meals
Family
Hall 1
L'dry
Hall 2
WC
Bath
Bed 3
Bed 2
Entry
Porch
Rear porch
Garage
86 mm step down
2148 selected garage door
super-sized solar panel (approx. location)
FLOOR PLAN LEGEND:
EXHAUST FAN (DISCHARGE TO OUTSIDE AIR)
SMOKE DETECTOR (INTERCONNECTED)
DP DOWNPIPE LOCATION
MH MANHOLE (CEILING ACCESS)
MSB ELECTRICAL METERBOARD (TBD)
hws HOT WATER SERVICE
EB BEAMS TO BE DESIGNED BY ENGINEER, BUT NOT LIMITED TO, AS DETERMINED BY ENGINEERS STRUCTURAL ASSESSMENT.
VERTICAL CONSTRUCTION JOINTS TO BE CONSTRUCTED IN ACCORDANCE WITH AS3700 & AS4733. ENGINEER'S PLACEMENT OVERRIDES CONSTRUCTION JOINTS ON THIS PLAN.
SOIL TYPE: P
BAL RATING: LOW
WIND CLASS: N1
PRE-FABRICATED TIMBER FRAMING NOTE:
ALL WALL FRAMING TO BE PRE-FABRICATED TO MANUFACTURERS SPECIFICATIONS AND IN ACCORDANCE WITH CURRENT NCC & COMPLY TO AS1684 – 2010.
© COPYRIGHT IN WHOLE OR IN PART
02
CONTRACTOR MUST VERIFY ALL DIMENSIONS AND LEVELS AT THE JOB PRIOR TO COMMENCING ANY WORK OR MAKING ANY SHOP DRAWINGS.
DO NOT SCALE DRAWINGS. ALWAYS USE WRITTEN DIMENSIONS.
INDEPENDENT BUILDERS GIPPSLAND
DRAWN: STL
CHECKED: DA-DPAD 15038
DATE: 26/11/2019
SCALE: 1:100
SHEET: 02
TIME: 8:34 AM
DRAFT No: FINAL
CODE: IB
TYPE: NH
ISSUE: BI
PLANS APPROVED:
CLIENT SIGNATURE:
DATE:
DESIGN TYPE: ANGLESEA- MOD
REVISION:
JOB No: 2033

Figure 2
Independent Builders Gippsland

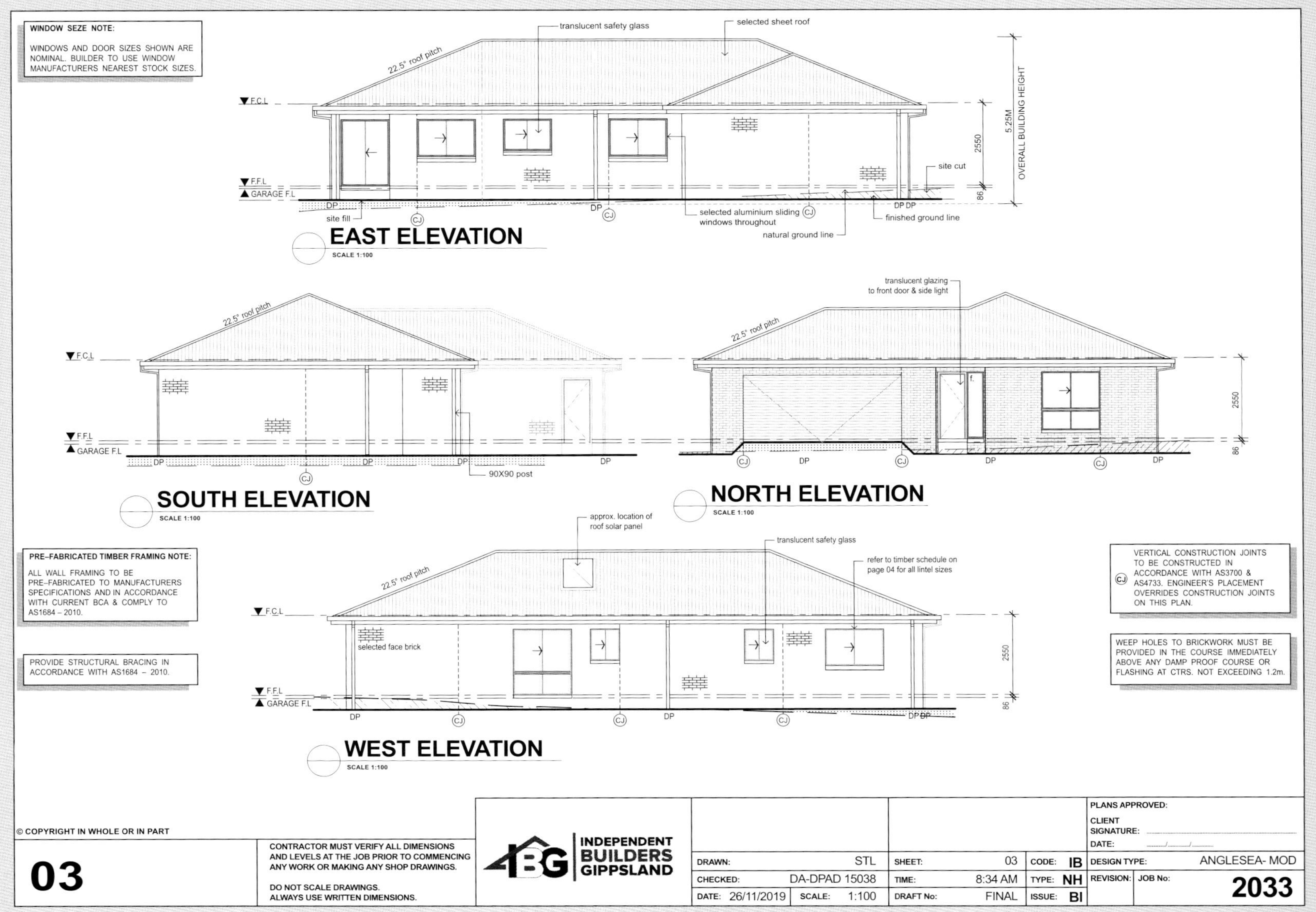

Figure 3
Independent Builders Gippsland

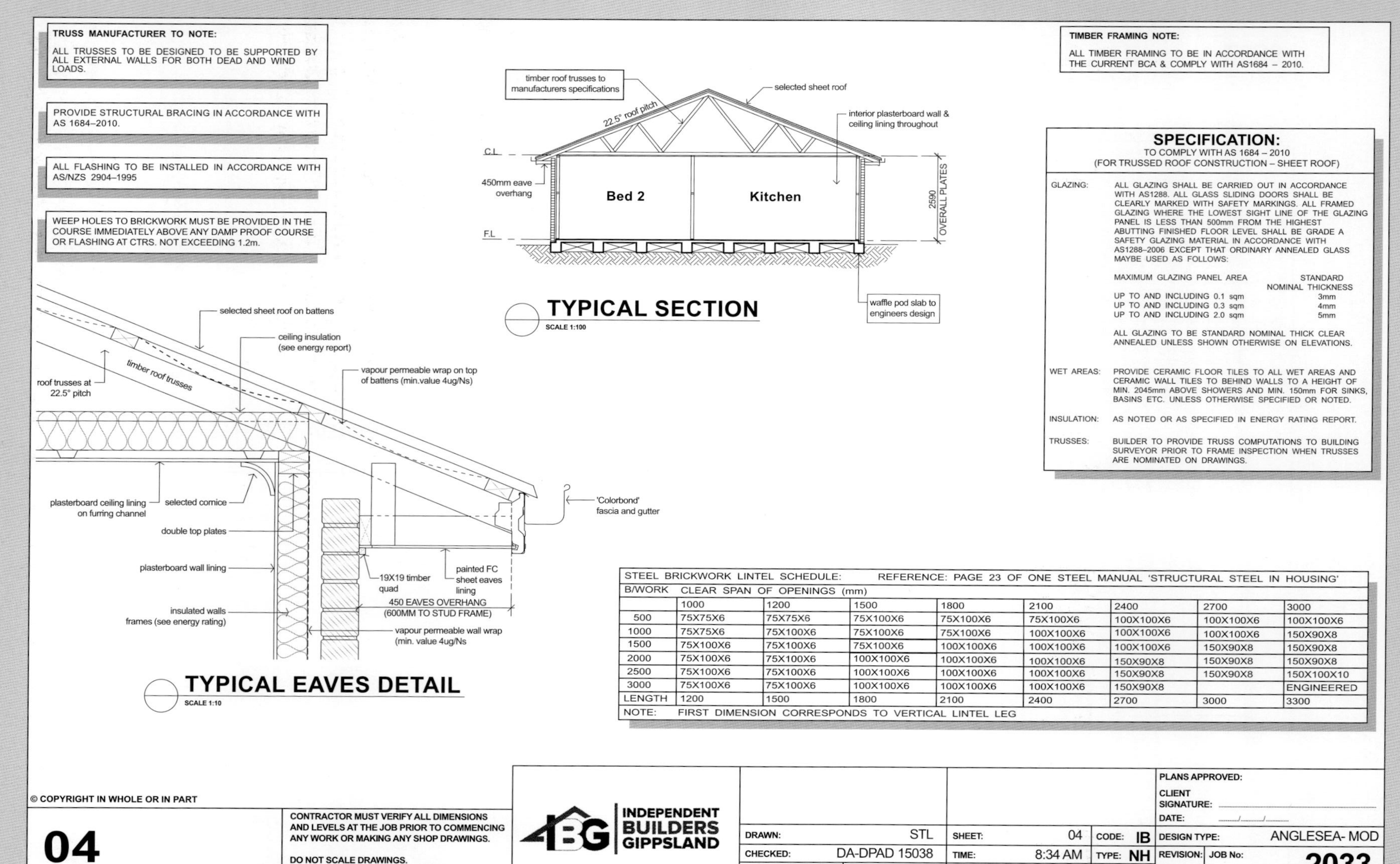

TRUSS MANUFACTURER TO NOTE:
ALL TRUSSES TO BE DESIGNED TO BE SUPPORTED BY ALL EXTERNAL WALLS FOR BOTH DEAD AND WIND LOADS.
PROVIDE STRUCTURAL BRACING IN ACCORDANCE WITH AS 1684–2010.
ALL FLASHING TO BE INSTALLED IN ACCORDANCE WITH AS/NZS 2904–1995
WEEP HOLES TO BRICKWORK MUST BE PROVIDED IN THE COURSE IMMEDIATELY ABOVE ANY DAMP PROOF COURSE OR FLASHING AT CTRS. NOT EXCEEDING 1.2m.
timber roof trusses to manufacturers specifications
22.5° roof pitch
selected sheet roof
interior plasterboard wall & ceiling lining throughout
C.L.
450mm eave overhang
Bed 2
Kitchen
F.L.
2590
OVERALL PLATES
waffle pod slab to engineers design
TYPICAL SECTION
SCALE 1:100
TIMBER FRAMING NOTE:
ALL TIMBER FRAMING TO BE IN ACCORDANCE WITH THE CURRENT BCA & COMPLY WITH AS1684 – 2010.
SPECIFICATION:
TO COMPLY WITH AS 1684 – 2010
(FOR TRUSSED ROOF CONSTRUCTION – SHEET ROOF)
GLAZING: ALL GLAZING SHALL BE CARRIED OUT IN ACCORDANCE WITH AS1288. ALL GLASS SLIDING DOORS SHALL BE CLEARLY MARKED WITH SAFETY MARKINGS. ALL FRAMED GLAZING WHERE THE LOWEST SIGHT LINE OF THE GLAZING PANEL IS LESS THAN 500mm FROM THE HIGHEST ABUTTING FINISHED FLOOR LEVEL SHALL BE GRADE A SAFETY GLAZING MATERIAL IN ACCORDANCE WITH AS1288–2006 EXCEPT THAT ORDINARY ANNEALED GLASS MAYBE USED AS FOLLOWS:
MAXIMUM GLAZING PANEL AREA — STANDARD NOMINAL THICKNESS
UP TO AND INCLUDING 0.1 sqm — 3mm
UP TO AND INCLUDING 0.3 sqm — 4mm
UP TO AND INCLUDING 2.0 sqm — 5mm
ALL GLAZING TO BE STANDARD NOMINAL THICK CLEAR ANNEALED UNLESS SHOWN OTHERWISE ON ELEVATIONS.
WET AREAS: PROVIDE CERAMIC FLOOR TILES TO ALL WET AREAS AND CERAMIC WALL TILES TO BEHIND WALLS TO A HEIGHT OF MIN. 2045mm ABOVE SHOWERS AND MIN. 150mm FOR SINKS, BASINS ETC. UNLESS OTHERWISE SPECIFIED OR NOTED.
INSULATION: AS NOTED OR AS SPECIFIED IN ENERGY RATING REPORT.
TRUSSES: BUILDER TO PROVIDE TRUSS COMPUTATIONS TO BUILDING SURVEYOR PRIOR TO FRAME INSPECTION WHEN TRUSSES ARE NOMINATED ON DRAWINGS.
selected sheet roof on battens
ceiling insulation (see energy report)
vapour permeable wrap on top of battens (min.value 4ug/Ns)
timber roof trusses
roof trusses at 22.5° pitch
plasterboard ceiling lining on furring channel
selected cornice
double top plates
plasterboard wall lining
insulated walls frames (see energy rating)
'Colorbond' fascia and gutter
19X19 timber quad
painted FC sheet eaves lining
450 EAVES OVERHANG (600MM TO STUD FRAME)
vapour permeable wall wrap (min. value 4ug/Ns
TYPICAL EAVES DETAIL
SCALE 1:10

STEEL BRICKWORK LINTEL SCHEDULE: REFERENCE: PAGE 23 OF ONE STEEL MANUAL 'STRUCTURAL STEEL IN HOUSING'

B/WORK	CLEAR SPAN OF OPENINGS (mm)							
	1000	1200	1500	1800	2100	2400	2700	3000
500	75X75X6	75X75X6	75X100X6	75X100X6	75X100X6	100X100X6	100X100X6	100X100X6
1000	75X75X6	75X100X6	75X100X6	75X100X6	100X100X6	100X100X6	100X100X6	150X90X8
1500	75X100X6	75X100X6	75X100X6	100X100X6	100X100X6	100X100X6	150X90X8	150X90X8
2000	75X100X6	75X100X6	100X100X6	100X100X6	100X100X6	150X90X8	150X90X8	150X90X8
2500	75X100X6	75X100X6	100X100X6	100X100X6	100X100X6	150X90X8	150X90X8	150X100X10
3000	75X100X6	75X100X6	100X100X6	100X100X6	100X100X6	150X90X8		ENGINEERED
LENGTH	1200	1500	1800	2100	2400	2700	3000	3300
NOTE:	FIRST DIMENSION CORRESPONDS TO VERTICAL LINTEL LEG							

Figure 4
Independent Builders Gippsland

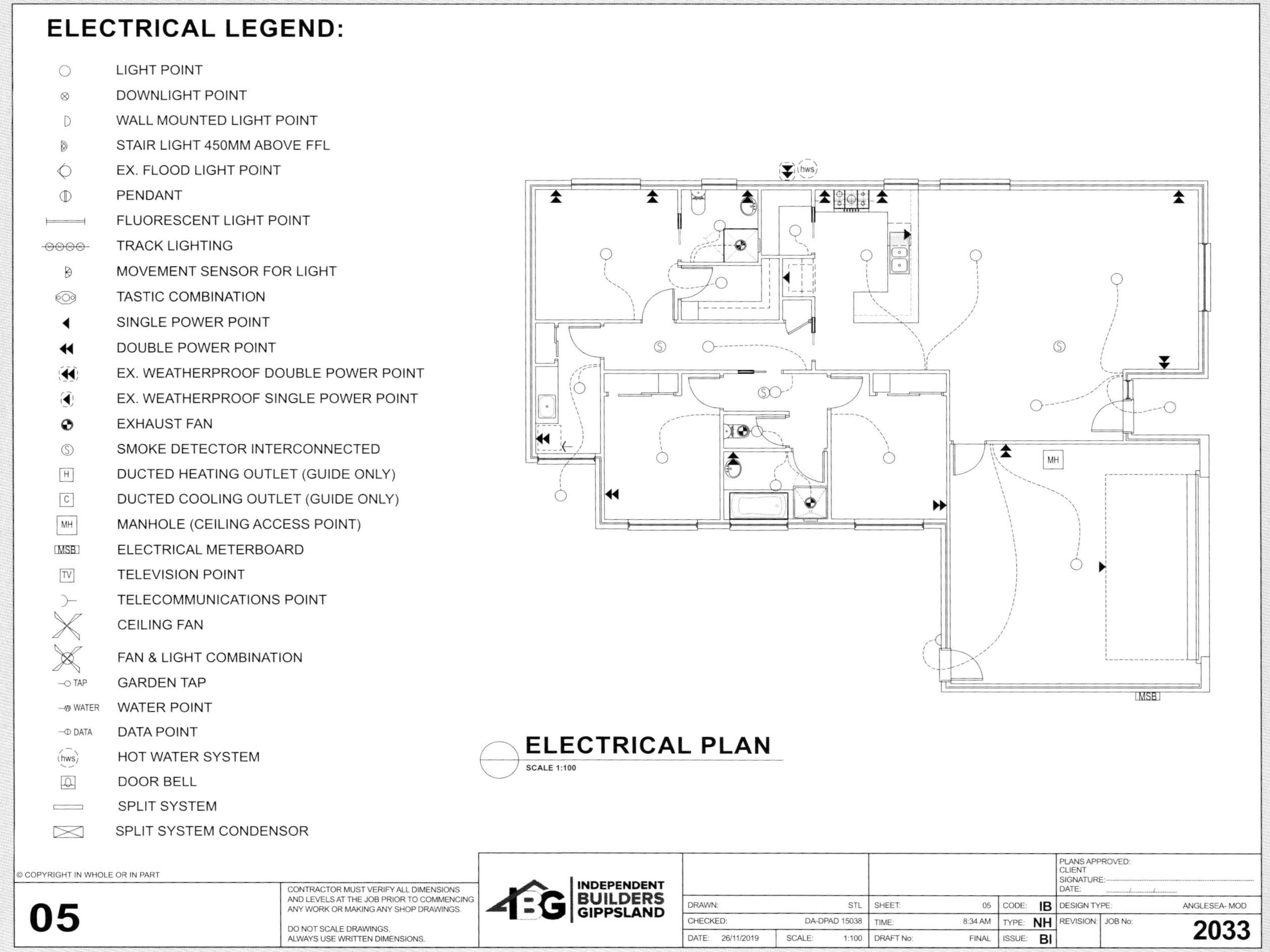

ELECTRICAL LEGEND:

- LIGHT POINT
- DOWNLIGHT POINT
- WALL MOUNTED LIGHT POINT
- STAIR LIGHT 450MM ABOVE FFL
- EX. FLOOD LIGHT POINT
- PENDANT
- FLUORESCENT LIGHT POINT
- TRACK LIGHTING
- MOVEMENT SENSOR FOR LIGHT
- TASTIC COMBINATION
- SINGLE POWER POINT
- DOUBLE POWER POINT
- EX. WEATHERPROOF DOUBLE POWER POINT
- EX. WEATHERPROOF SINGLE POWER POINT
- EXHAUST FAN
- SMOKE DETECTOR INTERCONNECTED
- H DUCTED HEATING OUTLET (GUIDE ONLY)
- C DUCTED COOLING OUTLET (GUIDE ONLY)
- MH MANHOLE (CEILING ACCESS POINT)
- MSB ELECTRICAL METERBOARD
- TV TELEVISION POINT
- TELECOMMUNICATIONS POINT
- CEILING FAN
- FAN & LIGHT COMBINATION
- TAP GARDEN TAP
- WATER WATER POINT
- DATA DATA POINT
- hws HOT WATER SYSTEM
- DOOR BELL
- SPLIT SYSTEM
- SPLIT SYSTEM CONDENSOR

hws

MH

MSB

ELECTRICAL PLAN

SCALE 1:100

05

CONTRACTOR MUST VERIFY ALL DIMENSIONS AND LEVELS AT THE JOB PRIOR TO COMMENCING ANY WORK OR MAKING ANY SHOP DRAWINGS.

DO NOT SCALE DRAWINGS. ALWAYS USE WRITTEN DIMENSIONS.

IBG INDEPENDENT BUILDERS GIPPSLAND

DRAWN: STL	SHEET: 05	CODE: IB	DESIGN TYPE: ANGLESEA- MOD
CHECKED: DA-DPAD 15038	TIME: 8:34 AM	TYPE: NH	REVISION: JOB No: 2033
DATE: 26/11/2019 SCALE: 1:100	DRAFT No: FINAL	ISSUE: BI	

PLANS APPROVED:
CLIENT
SIGNATURE:
DATE: / /

Figure 5
Independent Builders Gippsland

GENERAL NCC NOTES

ALL MATERIALS AND WORK PRACTICES SHALL COMPLY WITH, BUT NOT LIMITED TO THE BUILDING REGULATIONS 2018, NATIONAL CONSTRUCTION CODE SERIES 2019 BUILDING CODE OF AUSTRALIA VOL 2 AND ALL RELEVANT CURRENT AUSTRALIAN STANDARDS (AS AMENDED) REFERRED TO THEREIN.

UNLESS OTHERWISE SPECIFIED, THE TERM BCA SHALL REFER TO NATIONAL CONSTRUCTION CODE SERIES 2019 BUILDING CODE OF AUSTRALIA VOLUME 2.

ALL MATERIALS AND CONSTRUCTION PRACTICE SHALL MEET THE PERFORMANCE REQUIREMENTS OF THE BCA. WHERE A PERFORMANCE SOLUTION IS PROPOSED THEN, PRIOR TO IMPLEMENTATION OR INSTALLATION, IT FIRST MUST BE ASSESSED AND APPROVED BY THE RELEVANT BUILDING SURVEYOR AS MEETING THE PERFORMANCE REQUIREMENTS OF THE BCA.

ALL WORKS SHALL COMPLY TO THE FOLLOWING AUSTRALIAN STANDARDS;

- AS 1288 - GLASS IN BUILDINGS: SELECTION AND INSTALLATION
- AS 1562 - DESIGN AND INSTALLATION OF SHEET ROOF AND WALL CLADDING
- AS 1684 - NATIONAL TIMBER FRAMING CODE
- AS 1860 - INSTALLATION OF PARTICLEBOARD FLOORING
- AS 2049 - ROOF TILES
- AS 2050 - FIXING OF ROOF TILES
- AS 2870 - RESIDENTIAL SLABS AND FOOTINGS
- AS 2904 - DAMP PROOF COURSES AND FLASHINGS
- AS 3600 - CONCRETE STRUCTURES
- AS 3660.1 - CODE OF PRACTICE FOR PHYSICAL BARRIERS USED IN THE PROTECTION OF BUILDINGS
- AGAINST SUBTERRANEAN TERMITES
- AS 3700 - MASONRY IN BUILDINGS
- AS 3786 - SMOKE ALARMS
- AS 4055 - WIND LOADINGS FOR HOUSING
- AS 4100 - STEEL STRUCTURES

GLAZING:
GLAZING, INCLUDING SAFETY GLAZING, SHALL BE INSTALLED TO A SIZE, TYPE AND THICKNESS SO AS TO COMPLY WITH:

- BCA PART 3.6 FOR CLASS 1 AND 10 BUILDINGS WITHIN A DESIGN WIND SPEED OF NOT MORE THAN N3; AND
- BCA VOL 1 PART B1.4 FOR CLASS 2 AND 9 BUILDINGS.

WINDOW SIZES NOMINATED ARE NOMINAL ONLY. ACTUAL SIZE MAY VARY ACCORDING TO MANUFACTURER. WINDOWS TO BE FLASHED ALL AROUND.

WET AREAS :
WATERPROOFING OF WET AREAS, BEING BATHROOMS, SHOWERS, SHOWER ROOMS, LAUNDRIES, SANITARY COMPARTMENTS AND THE LIKE SHALL BE PROVIDED IN ACCORDANCE WITH AS 3740-2010: WATERPROOFING OF DOMESTIC WET AREAS

PROVIDE IMPERVIOUS FLOOR & WALL FINISHES TO ALL WET AREAS IN ACCORDANCE WITH BCA PART 3.8.1.2.

ENERGY RATING REPORT:
THESE DRAWINGS SHALL BE READ IN CONJUNCTION WITH ANY HOUSE ENERGY RATING (HERS) REPORT AND SHALL BE CONSTRUCTED IN ACCORDANCE WITH THE STAMPED PLANS ENDORSED BY THE ACCREDITED THERMAL PERFORMANCE ASSESSOR WITHOUT ALTERATION.

STAIRS & BARRIERS:
STEP SIZES (OTHER THAN FOR SPIRAL STAIRS) TO BE:

- RISERS (R) 190MM MAXIMUM AND 115MM MINIMUM
- GOING (G) 355MM MAXIMUM AND 240MM MINIMUM
- 2R + 1G = 700MM MAXIMUM AND 550MM MINIMUM
- WITH LESS THAN 125MM GAP BETWEEN OPEN TREADS

ALL TREADS, LANDINGS AND THE LIKE TO HAVE A SLIP-RESISTANCE CLASSIFICATION OF P3 OR R10 FOR DRY SURFACE CONDITIONS AND P4 OR R11 FOR WET SURFACE CONDITIONS, OR A NOSING STRIP WITH A SLIP-RESISTANCE CLASSIFICATION OF P3 FOR DRY SURFACE CONDITIONS AND P4 FOR WET SURFACE CONDITIONS.

PROVIDE BARRIERS WHERE CHANGE IN LEVEL EXCEEDS 1000MM ABOVE THE SURFACE BENEATH LANDINGS, RAMPS AND/OR TREADS. BARRIERS (OTHER THAN TENSIONED WIRE BARRIERS) TO BE:

- 1000MM MIN. ABOVE FINISHED SURFACE LEVEL OF BALCONIES, LANDINGS OR THE LIKE, AND
- 865MM MIN. ABOVE FINISHED SURFACE LEVEL OF STAIR NOSING OR RAMP, AND
- VERTICAL WITH LESS THAN 125MM GAP BETWEEN, AND
- ANY HORIZONTAL ELEMENT WITHIN THE BARRIER BETWEEN 150MM AND 760MM ABOVE THE FLOOR MUST NOT FACILITATE CLIMBING WHERE CHANGES IN LEVEL EXCEEDS 4000MM ABOVE THE SURFACE BENEATH LANDINGS, RAMPS AND/OR TREADS.

WIRE BARRIER CONSTRUCTION TO COMPLY WITH NCC 2019 BCA PART 3.9.2.3 FOR CLASS 1 AND 10 BUILDINGS AND NCC 2019 BCA VOLUME 1 PART D2.16 FOR OTHER CLASSES OF BUILDINGS.

TOP OF HAND RAILS TO BE MINIMUM 865MM VERTICALLY ABOVE STAIR NOSING AND FLOOR SURFACE OF RAMPS.

TERMITE PROTECTION:
WHERE THE BUILDING (EXCLUDES A DETACHED CLASS 10) IS LOCATED IN A TERMITE PRONE AREA THE BUILDING IS TO BE PROVIDED WITH A TERMITE MANAGEMENT SYSTEM.

CONCRETE STUMPS:

- UP TO 1400MM LONG TO BE 100MM X 100MM(1 NO. H.D. WIRE)
- 1401MM TO 1800MM LONG TO BE 100MM X 100MM(2 NO. H.D. WIRES)
- 1801MM TO 3000MM LONG TO BE 125MM X 125MM(2 NO. H.D. WIRES)

100MM X 100MM STUMPS EXCEEDING 1200MM ABOVE GROUND LEVEL TO BE BRACED WHERE NO PERIMETER BASE BRICKWORK PROVIDED.

TOILET DOORS:
TOILET DOORS ARE TO BE FITTED WITH REMOVABLE HINGES, OR ARE TO SWING OUT, OR BE SLIDING WHERE THE HINGE SIDE OF DOORWAY IS WITHIN 1200MM OF THE PAN.

SMOKE ALARMS:
SMOKE ALARMS TO BE INSTALLED IN ACCORDANCE WITH B.C.A PART 3.7.2. SMOKE ALARMS MUST BE CONNECTED DIRECTLY TO MAINS POWERS AND INTERCONNECTED

MECHANICAL VENTILATION:
MECHANICAL VENTILATION MUST BE DUCTED TO THE EXHAUST TO THE OUTSIDE OF THE BUILDING TO COMPLY WITH PART 3.8.5 OF THE BCA & AS1668.2

MASONRY:
BUILDINGS IN MARINE OR OTHER EXPOSURE ENVIRONMENTS SHALL HAVE MASONRY UNITS, MORTAR AND ALL BUILT-IN COMPONENTS AND THE LIKE COMPLYING WITH THE DURABILITY REQUIREMENTS OF TABLE 4.1 OF AS 4773.1-2010 'MASONRY IN SMALL BUILDINGS' PART 1: DESIGN.

STORMWATER:
ALL STORMWATER TO BE TAKEN TO THE LEGAL POINT OF DISCHARGE TO THE RELEVANT AUTHORITIES APPROVAL.

CLASS 6 UPVC STORMWATER LINE LAID TO A MINIMUM GRADE OF 1:100 AND CONNECTED TO THE LEGAL POINT OF STORMWATER DISCHARGE. PROVIDE INSPECTION OPENINGS AT 9000MM C/C AND AT EACH CHANGE OF DIRECTION. THE COVER TO UNDERGROUND STORMWATER DRAINS SHALL BE NOT LESS THAN:

- 100MM UNDER SOIL
- 50MM UNDER PAVED OR CONCRETE AREAS
- 100MM UNDER UNREINFORCED CONCRETE OR PAVED DRIVEWAYS
- 75MM UNDER REINFORCED CONCRETE DRIVEWAYS

THESE DRAWINGS SHALL BE READ IN CONJUNCTION WITH ALL RELEVANT STRUCTURAL AND ALL OTHER CONSULTANTS' DRAWINGS/DETAILS AND WITH ANY OTHER WRITTEN INSTRUCTIONS ISSUED IN THE COURSE OF THE CONTRACT.

BAL (BUSHFIRE ATTACK LEVEL):
SITE BUSHFIRE ATTACK ASSESSMENT (SIMPLIFIED METHOD) REFERENCE DOCUMENT 'AS 3959-2009 CONSTRUCTION OF BUILDINGS IN BUSH FIRE PRONE AREAS'. CLASSIFICATION = BUSHFIRE ATTACK LEVEL TO BE DETERMINED BY BUSHFIRE ASSESSMENT REPORT

DESIGN GUST WIND SPEED/WIND CLASSIFICATION:
BUILDING TIE-DOWNS TO BE PROVIDED IN ACCORDANCE WITH AS1684-2015 FOR AN ASSUMED DESIGN GUST WIND SPEED/WIND CLASSIFICATION TO BE CONFIRMED ON SITE BY RELEVANT BUILDING SURVEYOR AT FIRST INSPECTION. REFER TO AS1684 FOR CONSTRUCTION REQUIREMENTS.

CORROSION PROTECTION OF BUILT-IN STRUCTURAL MEMBERS:
PROVIDE CORROSION PROTECTION OF BUILT-IN STRUCTURAL STEEL MEMBERS SUCH AS STEEL LINTELS, SHELF ANGLES, CONNECTORS, ACCESSORIES (OTHER THAN WALL TIES) IN ACCORDANCE WITH TABLE 4.1 OF AS4773.1-2015 MASONRY IN SMALL BUILDINGS PART 1: DESIGN SUITABLE FOR THE ENVIRONMENT CLASSIFICATION.

CORROSION PROTECTION FOR SHEET ROOFING:
PROVIDE CORROSION PROTECTION FOR SHEET ROOFING IN ACCORDANCE WITH BCA TABLE 3.5.1.1A SUITABLE FOR THE ENVIRONMENT CLASSIFICATION OF

SITE CLASSIFICATION:
REFER TO SOIL REPORT PROVIDED BY LICENSED SOIL ENGINEER FOR SITE CLASSIFICATION.

SITE PLAN MEASUREMENTS IN METRES - ALL OTHER MEASUREMENTS IN MILLIMETRES UNLESS NOTED OTHERWISE.

FIGURED DIMENSIONS TAKE PRECEDENCE OVER SCALED DIMENSIONS.

THE BUILDER SHALL TAKE ALL STEPS NECESSARY TO ENSURE THE STABILITY AND GENERAL WATER TIGHTNESS OF ALL NEW AND/OR EXISTING STRUCTURES DURING ALL WORKS.

THE BUILDER AND SUBCONTRACTORS SHALL CHECK AND VERIFY ALL DIMENSIONS, SETBACKS, LEVELS AND SPECIFICATIONS AND ALL OTHER RELEVANT DOCUMENTATION PRIOR TO THE COMMENCEMENT OF ANY WORKS. REPORT ALL DISCREPANCIES TO THIS OFFICE FOR CLARIFICATION.

INSTALLATION OF ALL SERVICES SHALL COMPLY WITH THE RESPECTIVE SUPPLY AUTHORITY REQUIREMENTS.

THE BUILDER AND SUBCONTRACTOR SHALL ENSURE THAT ALL STORMWATER DRAINS, SEWER PIPES AND THE LIKE ARE LOCATED AT A SUFFICIENT DISTANCE FROM ANY BUILDINGS FOOTING AND/OR SLAB EDGE BEAMS SO AS TO PREVENT GENERAL MOISTURE PENETRATION, DAMPNESS, WEAKENING AND UNDERMINING OF ANY BUILDING AND ITS FOOTING SYSTEM.

THESE PLANS HAVE BEEN PREPARED FOR THE EXCLUSIVE USE BY THE CLIENT OF 'SKETCH BUILDING DESIGN' ('THE DESIGNER') FOR THE PURPOSE EXPRESSLY NOTIFIED TO THE DESIGNER. ANY OTHER PERSON WHO USES OR RELIES ON THESE PLANS WITHOUT THE DESIGNER'S WRITTEN CONSENT DOES SO AT THEIR OWN RISK AND NO RESPONSIBILITY IS ACCEPTED BY THE DESIGNER FOR SUCH USE AND/OR RELIANCE.

THE CLIENT AND/OR THE CLIENT'S BUILDER SHALL NOT MODIFY OR AMEND THE PLANS WITHOUT THE KNOWLEDGE AND CONSENT OF 'SKETCH BUILDING DESIGN' EXCEPT WHERE A REGISTERED BUILDING SURVEYOR MAKES MINOR NECESSARY CHANGES TO FACILITATE THE BUILDING PERMIT APPLICATION AND THAT SUCH CHANGES ARE PROMPTLY REPORTED BACK TO 'SKETCH BUILDING DESIGN'

THE APPROVAL BY THIS OFFICE OF A SUBSTITUTE MATERIAL, WORK PRACTICE, VARIATION OR THE LIKE IS NOT AN AUTHORISATION FOR ITS USE OR A CONTRACT VARIATION. ALL VARIATIONS MUST BE ACCEPTED BY ALL PARTIES TO THE AGREEMENT AND WHERE APPLICABLE THE RELEVANT BUILDING SURVEYOR PRIOR TO IMPLEMENTING ANY VARIATION.

07

CONTRACTOR MUST VERIFY ALL DIMENSIONS AND LEVELS AT THE JOB PRIOR TO COMMENCING ANY WORK OR MAKING ANY SHOP DRAWINGS.

DO NOT SCALE DRAWINGS.
ALWAYS USE WRITTEN DIMENSIONS.

IBG INDEPENDENT BUILDERS GIPPSLAND

DRAWN:	STL	SHEET:	07	CODE:	IB	DESIGN TYPE:	ANGLESEA- MOD
CHECKED:	DA-DPAD 15038	TIME:	8:34 AM	TYPE:	NH	REVISION:	JOB No: 2033
DATE: 26/11/2019	SCALE: N/A	DRAFT No:	FINAL	ISSUE:	BI		

PLANS APPROVED:
CLIENT SIGNATURE: ____________
DATE: ____/____/____

Figure 6
Independent Builders Gippsland

CONSTRUCTION FOR BUSHFIRE ATTACK LEVEL LOW (BAL-LOW) REF: AS3959-2009

- THIS STANDARD DOES NOT PROVIDE CONSTRUCTION REQUIREMENTS FOR BUILDINGS ASSESSED IN BUSHFIRE-PRONE AREAS IN ACCORDANCE WITH SECTION 2 AS BEING BAL-LOW.

NOTE: THERE IS A NUMBER OF STANDARDS THAT SPECIFY REQUIREMENTS FOR CONSTRUCTION; HOWEVER, WHERE THIS STANDARD DOES NOT PROVIDE CONSTRUCTION REQUIREMENTS FOR A PARTICULAR ELEMENT, THE OTHER STANDARDS APPLY.

- THE BUSHFIRE ATTACK LEVEL BAL-LOW IS BASED ON INSUFFICIENT RISK TO WARRANT SPECIFIC BUSHFIRE CONSTRUCTION REQUIREMENTS. IT IS PREDICATED ON LOW THREAT VEGETATION AND NON-VEGETATED AREAS (SEE CLAUSE 2.2.3.2).

08

CONTRACTOR MUST VERIFY ALL DIMENSIONS AND LEVELS AT THE JOB PRIOR TO COMMENCING ANY WORK OR MAKING ANY SHOP DRAWINGS.

DO NOT SCALE DRAWINGS.
ALWAYS USE WRITTEN DIMENSIONS.

IBG INDEPENDENT BUILDERS GIPPSLAND

DRAWN: STL		SHEET: 08	CODE: IB	DESIGN TYPE: ANGLESEA- MOD	
CHECKED: DA-DPAD 15038		TIME: 8:34 AM	TYPE: NH	REVISION:	JOB No: 2033
DATE: 26/11/2019	SCALE: N/A	DRAFT No: FINAL	ISSUE: BI		

PLANS APPROVED:
CLIENT SIGNATURE: ....
DATE: ..../......./.......

Figure 7
Independent Builders Gippsland

Glossary

1/3 fixing method a method of fixing plasterboard ceiling sheets where extra fixings are placed at 1/3 spacings across a plasterboard sheet together with adhesive spaced between these fixings

A

abutments the jambs of the wall or pier supporting the arch

access bay a bay of scaffold that gives access to the scaffold work platform by means of ladders or stairs

aluminium oxide synthetic abrasive made into stones, wheels or abrasive paper for sharpening tools

angle of repose the angle to the horizontal at which a given soil type will rest in situ or when stockpiled, and not crumble or otherwise fall away

asbestosis a disease of the lungs caused by an accumulation of asbestos in the lungs that reduces a person's ability to breathe

Australian Standard approved standard for material, equipment, technique or procedure as set down by the Standards Association of Australia (SAA)

autoclaved aerated concrete (AAC) lightweight concrete manufactured with a mix of sand, lime and cement with water and aluminium paste which causes a chemical reaction to form gas bubbles. The liquid mix is placed in forms and, as the gas bubbles expand and increase, the mixture rises creating a lightweight concrete product. Common applications include blocks and reinforced panels for building construction

awning sash top pivoted sash

B

backblocking technique applied to reinforce the joints between sheets of plaster by fixing smaller pieces of plasterboard behind the finished surface but over joints between plasterboard sheets

backpropping after formwork has been dismantled, all concrete previously supported by the formwork is stripped and temporary props are installed underneath it to allow work to continue on the new floor above

barefaced mortise-and-tenon joint a mortise and tenon that has only one shoulder

base material the material into which the fastener shaft is driven

battering to form the face, side or wall of an excavation to an angle, usually less than the natural angle of repose, to prevent earth slippage

bay the horizontal distance/length between any two adjacent standards

belt sander a powered machine where an abrasive belt revolves over rollers

benching the horizontal stepping of the face, side or wall of an excavation

bevel an angle formed between two straight lines meeting at an angle other than 90°

bit an interchangeable cutting tool inserted into chuck of drilling machine

blinding layer a 50 mm layer of coarse sand (washed river sand) laid over the top of filling or road base under a concrete slab that allows water to drain through and also prevents puncturing of the waterproof slab underlay

blob footings plain concrete footings poured into a square or round hole. Often called pad footings and used to support brick piers or timber or steel posts that support the floor framing for a suspended timber floor; load-bearing concrete elements

blockboard *see* coreboard

blocking short lengths of timber used to support ends of hanging beams, etc.

bolection moulding moulding that finishes higher than the framing

bond beam a course or courses of blockwork that are core filled with concrete poured around reinforcing steel to form a reinforced concrete beam. The beam may support roof framing and may also tie a series of vertical core filled blocks together to form a concrete framed structure

bonded asbestos cement asbestos fibres bonded together with cement and formed into sheeting and other products including sheet roofing and cladding

bond strength the strength of the bond of mortar onto bricks and between bricks and the ability of mortar to hold masonry units together

bore diameter of centre hole in circular saw blade

bottom plate a horizontal member forming the bottom side of wall framing

building drawings the site plan showing the location and dimensions of the building site; the floor plan or view looking down on the building at floor level; four external elevations for each aspect and cross sections through the building that indicate internal details or specific construction requirements

building line a line established by the local council which is the minimum distance that must be maintained from the building to the site boundary

bullnose step the curve to the first riser in a flight of stairs

butter a term used to describe how a trowel or a similar tool is used to apply an adhesive or mortar coating in a swiping motion to materials to be laid or fixed in place

butt hinge hinge commonly used for fitted doors

butt joint a joint between two pieces of timber in which the two members abut without lapping two flat surfaces brought into contact and secured with nails, screws or adhesives

C

cantilevered where a structural element (e.g. a beam) is anchored and supported at one end and unsupported at the other

carcase the basic unit of furniture and fitments

carcinogenic cancer causing

casement sash side-hinged sashes

ceiling joist structural member which binds the wall and roof framing together and carries the mass of the ceiling sheeting

ceiling joist hangers metal product designed to connect the ceiling joist to the hanging beam

ceiling ties devices used to strengthen the joint between hanging beams and ceiling joists

centring rafters pair of rafters located at the end of the ridge in a hipped roof running at 90° to the ridge of the wall plates

chuck the part of a drilling machine that grips the drill bit

cladding external covering to the walls of a framed construction

cleaning eyes every third or fourth brick is left out at ground level or on top of the DPC/flashing for a raft slab to facilitate cleaning of the cavity in cavity wall construction. This is undertaken regularly as work progresses using a hose, timber battens, trowels or chains to remove excess mortar that has squeezed out from masonry units as they are laid. The hose is used to flush out the excess mortar and to clean the cavity

cleats small pieces of timber used to join pieces of timber

clenching when nails longer than what is required are used to secure two or more pieces of timber together and the protruding point is bent over and hit into the timber parallel to the grain, thus pulling the timber together tightly

codes of practice practical guides to achieving the health, safety and welfare standards required under OHS legislation. They represent the lower tier of the OHS hierarchy which includes the Act, the Regulation and codes of practice. They must be approved as a code of practice and are admissible in court proceedings. Compliance with an approved code of practice is considered to demonstrate compliance with OHS legislation

collar a circular steel disc that clamps each side of a saw blade

collar tie a roof framing member that ties the two rafters together, usually halfway between the ridge and the wall plate

common rafter a rafter that is continuous from the fascia to the ridge

compressive strength in concrete measured in MPas, the ability of concrete to withstand pressure. The greater the amount of cement in concrete, the greater its ability to withstand pressure. Concrete is naturally strong in compression

compressive strength of masonry measured in MPas, the ability of masonry units and mortar to withstand pressure

concentrated load a force applied at a single point on a beam or structure

construction work defined in the Model Regulations as any work carried out in connection with the construction, alteration, conversion, fitting out, commissioning, renovation, repair, maintenance, refurbishment, demolition, decommissioning or dismantling of a structure

continuous span a beam supported at three or more points

contour line a line drawn on a site plan joining points of the same ground level

conventional fixing method the conventional method of fixing plasterboard sheets is to fix around the perimeter of the plasterboard sheet with screws or nails while the internal area of the board is laid against adhesive placed onto metal or timber framing that is to be 'plastered or lined'

coreboard plywood type with central core of timber strips bonded together

countersink a tapered recess, cut around a pilot hole for a screw, to receive the head of the screw

cover the actual distance a board will cover in width after milling; notably in flooring, weatherboards and lining boards

crane chaser or dogman a worker who places slings in position around loads of materials and then attaches the slings to a crane hook ready to lift the material

creeper cuts the bricks cut into the extrados of an arch

creeper rafter a rafter running from wall plate to hip rafter

cross cutting cutting timber across the grain

crown the upper half of the arch

crown end assembly junction of ridge, hip rafter, crown end rafter and centring rafters

crown end rafter a rafter running from the end of the ridge to the wall plate

curing of concrete maintaining sufficient moisture in the concrete to allow the process of hydration to be completed

cutters blades and bits used with power tools

cutting an open cut excavation having a single wall or face

D

damp-proof coursing (DPC) a layer of impervious material including lead, copper, stainless steel, aluminium, embossed polyethylene or chemicals added to mortar. Often placed between brick or masonry courses to prevent rising dampness or used vertically to create a damp-proof barrier between an external damp wall and an internal dry wall where there is no cavity

datum a horizontal plane of known height to which the elevation of different points can be referred

daylight line the edge of daylight seen through a window

dead load permanent load due to weight of materials

deformed bar round steel bars with lugs, or deformations, rolled into the surface of the bar during manufacturing. These deformations create a mechanical bond between the concrete and steel. Deformed steel bar is designated as N-bar and has a yield strength of 500 MPa

diagonal brace oblique framing member securing wall framing laterally in vertical position

direct load situation where roof rafters are placed directly over wall studs

door (flush) smooth-faced door

door (hollow core) a door with infill of expanded cardboard

door (solid core) a door with blockboard infill

doorset door plus frame (jamb)

double-hung sash vertically sliding sashes

double log cabin pattern of timber siding

double mortise-and-tenon a joint used where a rail is much wider than the stile

dovetailed tee halved joint a halved joint where one half is dovetailed into the other

dowel borer horizontal boring machine for dowelling

dowelled joints framing and carcase joints using dowels fitting into mating holes

dowelling jig a metal guide to boring dowel holes

draw boring preparing holes for draw pin, where one is slightly offset to draw a joint together

draw pin a wooden peg used to draw a joint together after draw boring

dressed rusticated pattern of timber siding

drop stripping a high-risk practice that does not comply with the formwork code of practice. Formwork is partially dismantled and then dropped as a single item rather than a section at a time

dumpy level an optical levelling instrument

durability a measure of the ability of mortar or concrete to resist the long-term effects of weather and environment on their surface

duty of care a reasonable and responsible person is expected to act cautiously and with foresight and logic while being attentive to the potential for damage or injury to others. To act otherwise would be considered negligent

E

easement a section of the building site that must be left clear of construction so that services such as sewer and stormwater drains can be provided

eave soffit *see* soffit

elevations scaled drawings showing what a building looks like when viewed from one side

end grain a cross-cut section of timber, e.g. at right angles to the grain

engineers detailed drawings drawings that convey specific information on how a building is to be constructed so that it is structurally sound, e.g. reinforced concrete strip footings or a raft slab with details of reinforcement dimensions and location of reinforcement in concrete

excavation to remove soil, earth, rock or rubble to create a hollow, hole, pit, trench, dam, pool, pond or similar

exclusion zone an area from which all non-essential persons are excluded from excavation work (also called a no go zone)

exposure grade bricks bricks that are capable of resisting salt attack

extrados the external circumference of an arch

F

face an exposed sloping or vertical surface resulting from the excavation of material

face work first quality masonry work–bricks or blocks–that is finished with a clean 'face' free of mortar smears and the joints are properly finished, e.g. round iron finish. Face work masonry is the final finish. There are no surface coatings and the masonry is seen. The face side of a brick or block is the side that is laid level and plumb

false deck the extension of a formwork platform of deck beyond the formwork that will actually support poured concrete and is used as a scaffolding or work platform

fanlight a sash placed over a door opening

FCL finished ceiling level

feather a tongue used to join timber

fence guide for running tool or material along an edge

FFL finished floor level

fibre wood tissue mainly for support of wood

fibre cement the modern alternative to asbestos cement. Cellulose fibre is combined with cement to produce a range of fibre cement products including compressed sheet flooring, cladding and roofing, and circular hollow columns for use as formwork and drainage pipes

fish plate a plate of steel or timber nailed or bolted over a butt joint in timber, used to extend timber in length

flashing a layer of impervious material including lead, copper, stainless steel, aluminium, zincalume or embossed polyethylene. Used to prevent the ingress of rain by covering or 'capping' a gap or joint between two building elements; to prevent moisture crossing between walls in a cavity wall and to redirect water to an outside wall to prevent entry into a wall cavity. Often placed above and below openings in walls, such as windows and doors

floor (fitted) flooring fitted between interior wall frames

floor (platform) a construction system where all frames are fixed on top of the flooring

footing a structural building element that distributes the load of the building through the foundation. Located at the base of a wall or pier, it is usually a concrete strip or pad, supporting the mass of the building

foundation the ground on which a building is built and that supports the mass of the building

fox wedges wedges concealed in a stump mortise-and-tenon joint

framed and panelled door a frame-type door with panels fitted to grooves or rebates

framed ledge door a framed door sheeted with vertical boards

franking projection on stile in a sash joint

friable asbestos asbestos in the form of small particles or that can be easily crumbled, pulverised or reduced to small particles by hand pressure, e.g. asbestos lagging, sprayed insulation, millboard, felt, woven asbestos matting and bonded asbestos that has been damaged

furring channels U-shaped sections of light gauge steel that are fixed to the underside of a floor so that suspended plasterboard ceilings can be fixed to them and supported

G

gable vertical triangular area formed at the end of a roof between the top plate and two sloping roof surfaces

gable stud stud enclosing the gable

galvanised a sacrificial zinc coating used on steel to prevent rusting

glazing fitting of glass to sashes

glazing bar bar dividing glazing in sashes into small sections

going the horizontal distance covered by a step

going of flight the total horizontal distance covered by a flight of stairs

green blocks concrete masonry blocks are made by placing wet concrete into a mould. After the initial set the concrete block is released from the mould and cured in a heated chamber. Freshly moulded blocks are given the term 'green'. The same term applies to masonry walls after they have been laid and the mortar has not properly set. Wet, partly set mortar or concrete has a dark green appearance

green hardwood timber that has been milled and sold without being seasoned/dried. The timber has high moisture content

H

half-lapped joint a joint in woodwork where two members are halved in thickness and joined by lapping

half span the run of a rafter (half the width of the building)

hammer drill a portable drilling machine which exerts hammer blows to the head of the drill as it rotates for drilling masonry materials

handrail bolt a heavy duty joiner of timber in length

hanging beam a beam placed on edge over ceiling joists to prevent them sagging over long spans

hanging door fixing a door and its hardware into its frame

hardboard timber sheet product manufactured of compressed wood fibre

hardcore a hard sub-base material around 40 mm diameter with a small amount of fines that is compacted in layers as a sub-base for less coarse base material such as road base

harmonisation a process applied by the state, territory and federal governments to provide consistency in the implementation of legislation across different state and territory and commonwealth jurisdictions. With OHS this has meant the development of model legislation by the Commonwealth to be adopted by the states and territories

haunch the portion of tenon left near shoulder to prevent rail from twisting in stile; part of arch from springing line to crown

haunched mortise-and-tenon framing joint to ends of stile

head top horizontal member of door jamb

head trimmer (head) beam supporting load over a window or door opening in wall framing

high-risk construction work hazardous construction work that has the potential to harm the health and safety of people or to damage plant and equipment. Specific activities that are defined in the regulations require a relevant SWMS

hipped roof a roof shape where the sloping surfaces are pitched on all sides of the building

hip rafter a rafter that runs from the end of the ridge to an external corner and forms the junction of two sloping roof surfaces

hopper sash bottom-pivoted sashes

horizontal line a line at a tangent to the curvature of the earth's surface, e.g. at any given point it is parallel with the surface of still water

horn waste left on stile while joint is being worked

hydration in concrete a chemical reaction caused by the addition of water to cement. The cement compounds form chemical bonds with water molecules and become hydrates or hydration products

I

inlay moulding moulding that finishes below the level of the framing

intrados or soffit the internal circumference or underside of an arch

J

jack stud a short stud above and below a window or door opening

jamb lining milled to fit into door opening and rebated to accept door

jamb stud the stud supporting the ends of lintels at door and window openings

JSA job safety analysis

K

kerf a cut made by a saw

key the central voussoir in masonry arches; it is often decorated in a distinctive way

knockdown fitting a clamping device for assembly of carcase framing

L

level bevel an angle used on rafters; equals pitch of roof

lift the vertical distance between levels of a scaffold at which a platform can be constructed

lightweight truss common timber domestic-type truss

lintel a structural member or beam carrying loads over an opening

live load load to the use of an area

load bearing structural elements such as walls that support other structural or non-structural building elements

long and short shoulder mortise-and-tenon a joint where the shoulders are of different lengths

long point correction allowance when measuring length of creeper rafters for thickness of the rafter

low angle of incidence a reference to the low angle of a light source that lights a finished surface area, such as painted plasterboard. The lower the angle of incidence, the more likely that shadows will highlight surface imperfections

M

machine gauged structural members that have been passed through a machine to reduce them to a uniform width and thickness

mandatory construction induction training general safety training that must be provided to construction workers before they commence work in the construction industry

manual excavation an excavation created by the use of basic hand tools

masonry bricks, blocks or panels made from natural stone or man-made products, such as concrete, burnt clay or cement, and fibre stabilised clay (mud bricks) that are bound together by mortar

masonry veneer a building system that incorporates an external loadbearing frame of timber or steel with an external cladding of masonry that is tied to the frame for stability

MDF board *see* medium-density fibreboard

mechanical excavation an excavation created by the use of a machine such as a scraper, backhoe, bulldozer, excavator or similar

medium-density fibreboard a manufactured timber board made from very fine timber particles that are bonded together with glue and compressed into square sheets

megapascals (MPa) a measurement of the pressure that can be withstood by concrete before it fails. It is determined by a calculation using the mass of a load or building element, the rate of acceleration of gravity and the area the pressure is applied to. MPa is also used to measure internal pressure, stress and tensile strength

mesothelioma a cancer caused by asbestos fibres that become embedded in the lining between the lungs and the diaphragm

mitre saw circular saw which can rotate on a swivel base for cutting timber at angles up to 45°

mortise-and-tenon joint framing joint for stiles and rails

mortise board a holding device on a bench for timber being mortised or chopped

mortise gauge a marking tool with two spurs to mark out mortises

mouldings timber ornamentation

mullion vertical divide between sash of window

muntin vertical frame members dividing door panels into smaller sections

mutually recognised where states and territories agree to recognise licenses and training delivered by different states

N

nap a reference to the length of the pile on a paint roller

needles used as horizontal support members in vertical shoring to provide temporary support for materials that are removed and replaced with a new element

nogging horizontal blocking fitted between studs to hold studs straight and for fixing linings

non-load bearing non-structural elements that only support their own mass, e.g. internal partition walls

nosing the front edge of the tread

O

open excavation an excavation where the width is equal to or greater than the depth

orbital sander a powered machine where a sheet of abrasive paper attached to a platen revolves in a circular motion

overlaid door a cabinet door fitted over carcase

overlay hinge hinges for overlaid doors

P

panelling fixing of plywood sheets to walls

particleboard manufactured material formed by bonding together flakes of wood and pressing them into a dense sheet

PCBU person conducting business or undertaking

piano hinge a continuous hinge

pilot hole a hole drilled to receive a nail or screw

pit a hole or cavity in the earth

plain concrete concrete that has no reinforcement; commonly used to support downward or compressive loads only

plan view a drawing that shows the details of a house or building when viewed from above

plasterboard a layer of set gypsum or plaster between two layers of heavy duty paper to form a board which is used to line wall and ceiling framing

platform construction building construction where the floor is fitted with wall framing applied over finished 'platform'

pneumatic nailer a machine tool for driving nails

pneumoconiosis an occupational lung disease that is very restrictive, caused by the inhalation of dust

post hole a hole for erecting posts in

pozidrive a type of slot formed in the head of screws providing a very positive grip for the screwdriver

practical completion a point in time when a building is ready for occupation, although there may still be minor defects to be rectified

principal contractor a person who has management or control of a workplace where construction work is undertaken. Under OHS legislation they have specific responsibilities.

Additional responsibilities are applied once the value of the construction work reaches $250 000

profile frame set up on a building site on which is indicated the position of the building

progressive collapse the effect of formwork collapsing from one point through to another with the mass of collapsing concrete, reinforcing and wet concrete causing damage to adjacent formwork structures and causing those to also fail

prop *see* underpurlin

R

rafter a sloping roof member supporting the roof covering

rafter boat a jig which generally contains the rafter plumb cut, birdsmouth and rafter side cut of the roof being constructed. It is used in place of a pattern rafter

rail horizontal member connecting stiles

random loading where roof rafters are spaced without reference to location of the wall studs and exceed the limits of direct loading

reactive soils soils or foundations that respond to different moisture levels by expanding and contracting. Reactive clay soils are common in Australia, requiring double layered reinforcement in concrete and the use of piling systems that transfer the loads placed on footing systems down to a stable foundation

reinforced concrete when reinforcing steel is placed inside concrete, the tensile strength qualities of the steel and the bonding of concrete to the steel result in a structural element that has the compressive strength of concrete and the tensile strength of steel

reinforced concrete footings generally have a top and bottom layer of reinforcing steel to withstand tensile stress on the bottom of the footing caused by building loads or upward forces caused by soils that expand when affected by groundwater causing tensile stress on the top of the footing

reinforced concrete masonry contains embedded reinforcing steel that imparts the strength and characteristics of reinforced concrete. In the case of hollow masonry blocks that are core-filled, the blockwork acts as a formwork

ripping sawing timber in the direction of the grain

rise the vertical height of the arch from the centre of the springing line to the underside of the crown. Also, the vertical distance covered by a step

rise of flight the total vertical distance covered by a flight of stairs

rise/run ratio the ratio that exists between the rise of a roof rafter and its run

riser the board forming the vertical face of the step

riser shutter the board retaining the concrete for the rise of the stair

roof load width (RLW) used as an indicator of the loading carried by some roof members as well as load-bearing wall frame members and supporting substructure

roof pitch the angle formed between a sloping roof surface and a horizontal line

roof rise the height of a roof from its base to the apex

roof span the distance overall between the wall plates supporting the roof

rough-in 'the rough-in' of gas, water, drainage and electrical services is the initial installation of these services before the final fittings are connected and the services are commissioned

router a powered machine mainly used for cutting grooves and trimming edges

run the horizontal distance covered by a sloping roof member

S

safe operating procedures (SOP) a sequence of logical steps that describe how to safely operate an item of plant or how to use a specific hand or power operated tool

safety tags tags attached to switches or valves as a warning not to open or turn on a service, such as electricity, or to start up an item of equipment. This is done to ensure the safety of any workers required to work in areas where such services or equipment would be a hazard if they were made operational or reconnected. Normally used to isolate or shutdown services and equipment during maintenance repairs or when hazardous work such as demolition is undertaken

safe work method statement (SWMS) a document that identifies the risks associated with high-risk work and that develops a step-by-step process that mitigates or removes the risk altogether

salt attack damage that occurs in areas where ground salts in solution are drawn into masonry in contact with the ground by capillary action. The solution evaporates on the masonry surface leaving salt crystals growing below which damage either the mortar or the masonry units, causing spalling or erosion

sarking a reflective foil laminate (RFL) that contributes to radiant and convective insulation and also provides a condensation barrier

sash cramp a cramp for cramping wide joinery items

sash joint a modified mortise-and-tenon joint for stiles and rails

saw kerfing a method of bending timber by a series of close saw cuts

sawn splayed a pattern of timber siding

screw gauge a number indicating the diameter of the shank of a woodscrew

scribed joint a joint where mouldings intersect, where one member is fixed and the other mitred at half the angle of intersection and then cut out by coping saw to the contour formed

sealant group of non-hardening materials used to seal joints against penetration of moisture

shaft lock a lever operated locking device which, when operated, prevents the spindle of a tool from rotating

shiplap timber milled for one member to overlap the next

shoring the provision of support for excavated face(s) to prevent the movement of soil and ground collapse

shoulder line the line on rail where it meets the face of the stile

shutter normally a reference to the 'plywood' used to form the internal mould and finish of formed concrete

siding *see* cladding

silicon mastic a waterproof adhesive type sealant used for sealing off gaps and thus preventing water penetration

sill trimmer the horizontal member supporting the window frame in a wall frame

simulated panel door solid door where design is routed into panel

single leaf construction a single load-bearing masonry wall with no cavity to prevent moisture entering a building. The wall supports the roof framing and must be waterproofed with an external coating to prevent internal dampness

single span a beam supported at two points only

site safety audit a methodical process of identifying and listing all potential work hazards in a workplace and then identifying appropriate risk management strategies that have been or need to be applied

site safety management plan a management plan developed and administered by the head contractor on a construction site where the contracted work is valued at $250 000 or more

site-specific induction training safety training provided to all workers when first entering a new worksite. This training identifies specific site hazards, site rules and site amenities

skewbacks the immediate sloped masonry that supports the bricks of a segmental arch

skew nail a nail driven at an oblique angle through a piece of timber and into the receiving piece. Usually in opposite pairs

skillion a roof shape consisting of a single sloping surface

skirting moulding to cover the joint between floor and wall lining

sleeper pier a pier supporting floor framing at points not attached to dwarf walls

slip-tongue joint a joint used for joining segments in curved work

slope of roof the angle formed between the slope of the roof surface and the horizontal line

slump test a test to measure the water content and workability of concrete

soffit underside of eaves

soffit decking formwork in contact with the underside of the stair and landing

solid plaster a 10 mm coating of cement mortar placed onto masonry units and finished with a smooth sandy texture ready for painting. Surface may also be finished with a thin 2 mm coat of plaster leaving a smooth texture-free finish

spall the process of layers of masonry being damaged by the growth of salt crystals below the surface of masonry units. The growing salt crystals cause small layers of masonry to be fractured and forced off the face of the masonry unit

spalling the process of layers of masonry being damaged by the growth of salt crystals below the surface of masonry units; the growing salt crystals cause small layers of masonry to be fractured and forced off the face of the masonry unit

span the horizontal distance between the abutments

specifications a document that details exactly what materials and systems are to be used to construct a building and includes both structural elements such as the footings and non-structural building elements such as a prescribed paint system. Specifications take precedence over building drawings if there is conflicting information

splayed weatherboards a pattern of timber siding

springing line the horizontal line from which the arch commences; the points at each end of the spring line are referred to as the springing points

starter bars steel reinforcing bars that are left protruding above finished concrete or core-filled concrete masonry. They allow different phases of work to be completed at different times. As a new section of work is started the starter bars overlap and connect with the steel reinforcement in the new section to structurally connect two different phases of concrete or reinforced concrete masonry

stile vertical side member of a frame

stopped housed joint a joint where the housing is stopped short at face edge

straight edge a length of timber that is made straight and parallel, and used during levelling operations or setting up a number of points to a straight line

stress grade visual or mechanical grading of whole lengths of timber into working stress categories, based on resistance to bending

stretcher bond bonding in masonry is the lap of one block or brick over another to strengthen the wall. When the lap is equal to half the length of the masonry units being used it is termed 'stretcher bond' or 'half bond'

stringer a member laid over the top of the formwork which supports brackets that prevent the riser shutter bulging in the centre

string shutter provides fixing at each end for the riser shutter and retains the concrete at the sides of the stair

strutting beam a beam used over large rooms to support struts where no internal walls are available

stud a vertical member in wall framing

stump a vertical member of timber, pre-cast concrete or steel sunk into the ground for the support of a floor or wall framework

subfloor a loadbearing timber or metal structure which the flooring material is secured to. The subfloor sits upon the stumps or piers

substrate underlying layer that a surface finish, such as paint, plasterboard or ceramic tiles, is applied or fixed to

SWMS *see* safe work method statement

T

tensile stress in concrete the response of concrete when loads are applied causing it to bend and resulting in a tendency to split or tear along the convex surface of the concrete element. Plain concrete has a low tensile strength

tension bracing bracing, usually galvanised hoop iron. It is effective in tension only and must be fixed in two directions

termite barrier the traditional termite barrier for Australian construction systems has been galvanised steel or zincalume sheeting placed as a continuous layer between timber framing and supporting subfloor masonry or posts. Today they include granular granite or recycled glass, chemically impregnated sheeting, stainless steel micromesh sheeting and the use of termite resistant structural building elements

timber siding boards specially milled to function as an effective wall cladding to timber-framed buildings

tom *see* underpurlin

tongue and groove jointing system for timber boards where one side is grooved to allow the insertion of a tongue from the adjacent board

tongued and trenched joint a joint where a tongue on one piece engages a trench in the other

toolbox meeting short informal meetings used to convey important information to workers on a construction site. Safety requirements for new work activities are generally the focus but work systems updates may also be topics for discussion

toothed (brickwork) refers to the practice of removing bricks or leaving bricks out as work progresses so that the brickwork steps in and out vertically in line with the bond or overlap of bricks. Forms a vertical 'tooth-like' appearance

top plate horizontal member forming top side of wall framing

traffic control plan for the construction industry, a traffic control plan documents the steps to be taken to minimise the risk of motor vehicle related accidents and interruption of traffic flow adjacent to where building or construction work is being undertaken

transom a horizontal member dividing doors or sashes in window or door frames

traversing planing diagonally across a board

trench an excavation having a length that exceeds its width, measured at the bottom

twin mortise-and-tenon a joint used where rails and stile are very thick

U

underpurlin roof member running the length of the roof supporting rafters

unreinforced concrete masonry masonry that has no embedded reinforcing steel to increase the tensile strength of the masonry but may have embedded joint reinforcement to stop joint cracking

V

vermin wire a layer of 12 mm steel mesh that is fixed to the bottom plate of wall framing at 200 mm centres and turned down onto a course of masonry in masonry veneer construction. The mesh prevents the entry of rats and other vermin into the wall cavity from the subfloor area and through into the roof space

vertical line a line at right angles to a horizontal line

vibrating screed comprises a length of straight steel or aluminium with a mechanical vibrator mounted on top that is drawn along by the operator. Levels and compacts the concrete

voussoirs the arch bricks forming the arch; voussoirs are supported by mutual pressure from their neighbours; the joints in voussoirs should always be normal to the curve of the arch

W

waffle raft slab a two-way reinforced concrete slab system with integral beams in two directions beneath it which forms a slab with a waffle-like appearance underneath. Polystyrene or plastic forms are placed directly on a prepared and levelled foundation with the slab underlay on top of the foundation

waterproof membrane or slab underlay a continuous layer of polyethylene sheeting placed under a reinforced concrete slab to prevent ground moisture and ground salts in solution rising into the slab. The underlay is lapped 300 mm and taped when lengths are joined. High impact underlay should always be used to prevent puncturing of the membrane

weatherboard *see* timber siding

wedge allowance space provided for wedges, as in mortise-and-tenon joints

weepholes unfilled perpendicular joints or gaps left between masonry units to allow moisture to drain from a cavity where flashing crosses the cavity. They also provide cavity ventilation. Proprietary inserts have a mesh cover to prevent entry of insects and vermin. In bushfire areas the mesh is stainless steel to prevent entry of flying embers or sparks

wet area(s) a room in a building which houses water-using fixtures or appliances

wind brace diagonal members bracing ridge to prevent rafters toppling sideways due to wind load

wind load load due to wind action

work activity training training that must be provided by an employer to ensure that employees are able to undertake work, of which they have no prior experience, safely and competently

work method statement a document that identifies the risks associated with high-risk work and that develops a step-by-step process that mitigates or removes the risk altogether

Index

F

G

H

I

J

K

L

M

N

O

P

Q

R

S

T

U

V

W

Worksheet 1

Apply OHS requirements, policies and procedures in the construction industry

Date: ______________________ Grade: Satisfactory/Unsatisfactory

Name: ______________________ Student number: ______________________

Course code: ______________________ Course name: ______________________

Unit code: ______________________ Unit name: ______________________

1 Name the federal legislation that addresses workplace health and safety in Australia.

2 Which training is mandatory for all construction workers to undertake before they commence work on a construction worksite?

3 Which four important points would an employer or site manager cover during a site-specific induction training session?

4 Name the type of meeting that often occurs informally on-site, so that workers are given an opportunity to participate and have a say on any work health and safety issues?

5 A blue sign on a white background with black lettering gives you what type of instruction?

6 List five actions that could be classed as forms of bullying on a construction site.

7 What is the cause of most fatalities in the construction industry?

8 List five safety precautions to be aware of while working on mobile scaffolding.

9 Which item of PPE must be worn if working adjacent to or in an area where work is being conducted overhead?

10 List four types of injuries that could occur from manual handling.

11 Describe the colouring of the lettering, symbol and triangle on a hazard warning sign.

12 Homes built before which year should be considered to contain hazardous asbestos cement until confirmed otherwise?

13 List three building materials that you may use that would require you to wear either a P1 or P2 respirator.

14 Name the document/book where you could get information about extremely hot or cold workplace conditions, and any personal safety requirements for these conditions.

15 In an emergency evacuation of a construction site, who would take charge of the situation?

16 Why must any injuries or near misses be reported and documented, following the correct procedure?

17 Define the meaning of construction work according to the OHS Regulations 2011.

18 Codes of practice provide guidelines for meeting the requirements of which occupational health and safety documents?

19 How would the identification of workplace hazards commence on a construction site?

20 Using the risk assessment matrix in Figure 1.28, determine the level of risk if the likelihood of an accident or incident was possible and the consequences were major.

21 List the five recommended steps when creating a JSA.

Worksheet 2

Conduct workplace communication

Date: ______________________ Grade: Satisfactory/Unsatisfactory

Name: ______________________ Student number: ______________________

Course code: ______________________ Course name: ______________________

Unit code: ______________________ Unit name: ______________________

1 List at least four types of electronic communication that may be used in the construction industry.

2 What do you think might be the most used form of communication on construction sites?

3 How could you make certain that any instructions you have passed on have been understood correctly by your workmate?

4 List four types of signs you may see on a construction site.

5 What is the name for a short on-site meeting to discuss items such as safety issues, new materials, work plans or instructions or other work-related announcements?

6 What type of communication can be used to overcome differences in language, occupations and levels of work experience?

7 What might be the main form of communication if you are working a long distance from someone, such as a surveyor, but still in sight of each other?

8 What would you do if the hand signals that you were receiving from a coworker were unclear and hard to understand?

9 If an accident were to happen on-site, what type of reporting might need to be done if you were a witness?

10 How might you check if all of the items listed on the delivery docket are there when the delivery arrives?

11 If any of the delivery items are missing or damaged, what steps would you take?

12 List three ways in which you can contribute to or participate in a work meeting.

13 Generally, what is the main reason that workplace disputes occur?

14 If a visitor arrives at the building site asking for directions to the correct person or office, how might you go about communicating with and assisting them?

15 What may be the outcome if you do not fill out your time sheet correctly?

16 What would you do if you were unclear about any information given to you in work plans, schedules or instructions?

17 How could you acknowledge the good work practices being carried out by other members of your team?

18 How could you send a visual message to all workers on the construction site that personal safety and personal protective equipment (PPE) are very important things to be aware of?

19 Would it be good work practice to remove your hearing protection if you are in a noisy area to talk to a work colleague? Explain your answer.

20 Who is responsible for making sure that communication has been delivered correctly and understood?

Worksheet 3

Carry out measurements and calculations

Date: ______________________ Grade: Satisfactory/Unsatisfactory

Name: ______________________ Student number: ______________________

Course code: ______________________ Course name: ______________________

Unit code: ______________________ Unit name: ______________________

1 Which document lists the windows and details of their sizes, widths and locations?

2 What document needs to be completed prior to starting any high-risk activity?

3 List four pieces of equipment you can use to measure distance.

4 For the site plan below, list two methods that you could use to measure the four sides of the property's boundary.

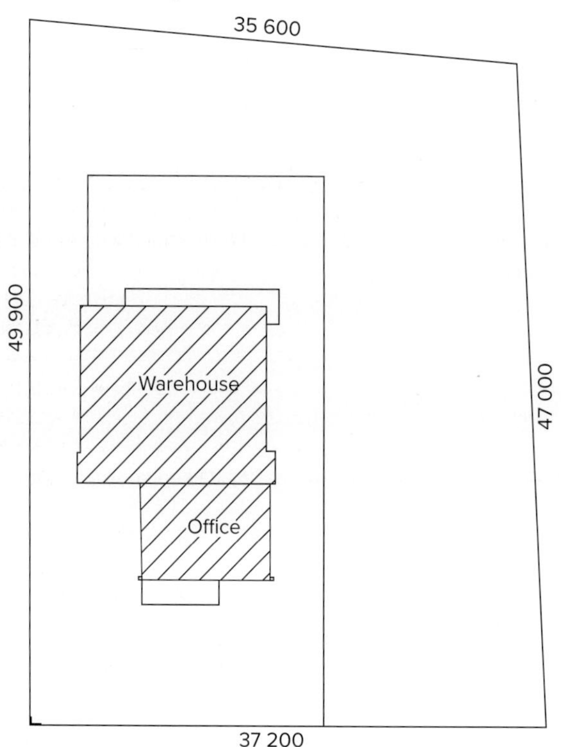

5 Using a rule, draw a line that measures 157 mm.

6 Referring to the drawing below, calculate the area of wall 3, excluding the window opening.

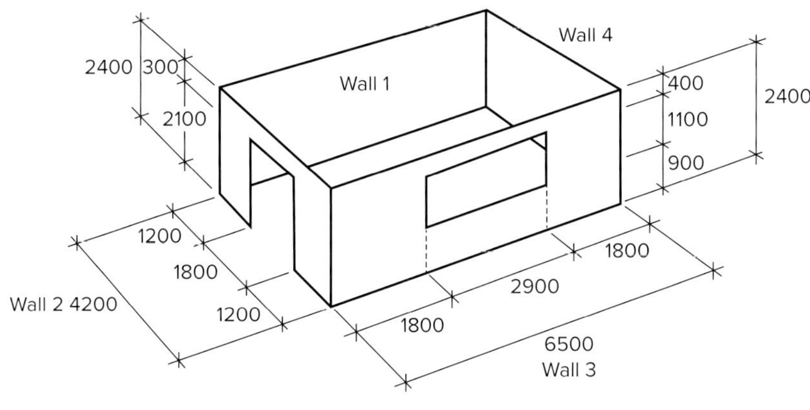

7 Using the above drawing, calculate the volume of the building.

8 Calculate the volume of concrete required for the concrete slab below. The slab thickness is 180 mm.

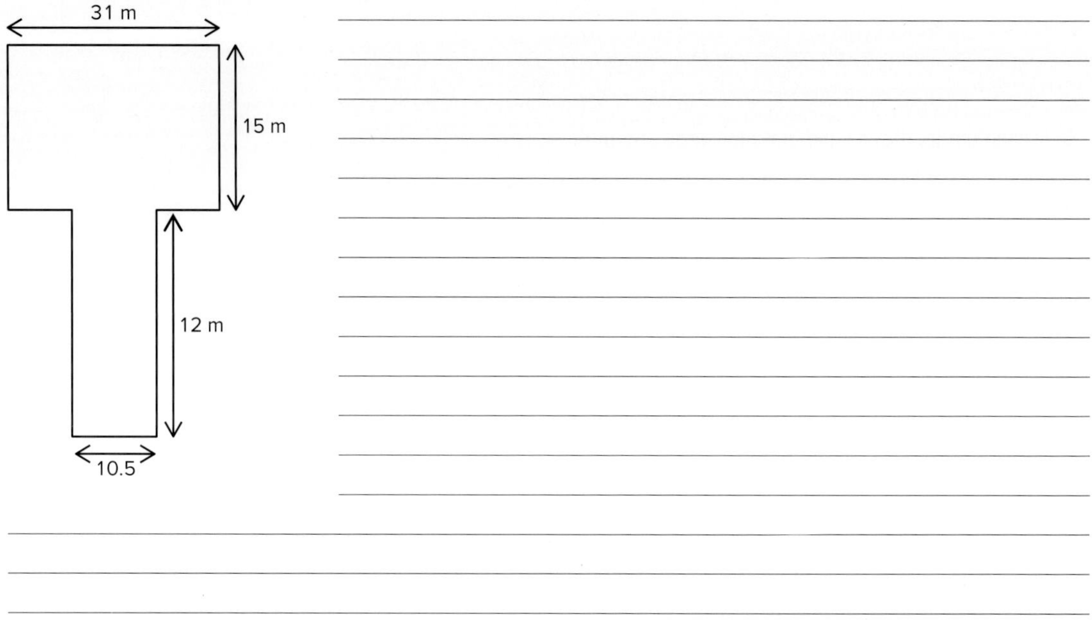

9 Depth of concrete to be 120 mm.

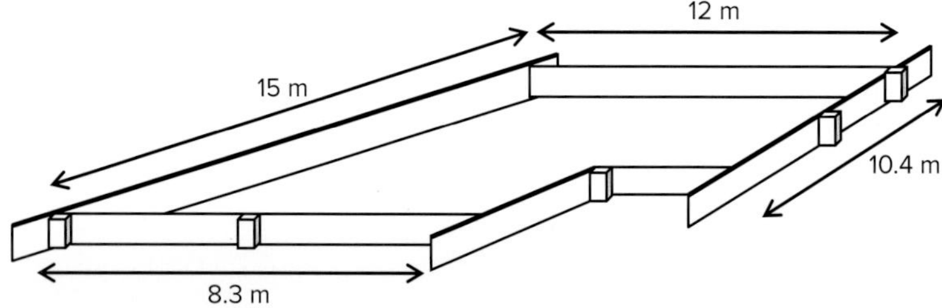

From the above diagram, calculate the following.

(a) Calculate the total length of timber required for the formwork—omit pegs. Allow an extra 150 mm for each overlap.

(b) Calculate the surface area of the floor.

(c) Calculate the volume of the proposed concrete floor.

(d) Estimate the amount of ready-mix concrete to order (order to the nearest m^2).

10 Write out the following formulas: volume of a cube; area of a circle; volume of a cylinder.

11 For the drawing below, calculate the amount of material required and its length.

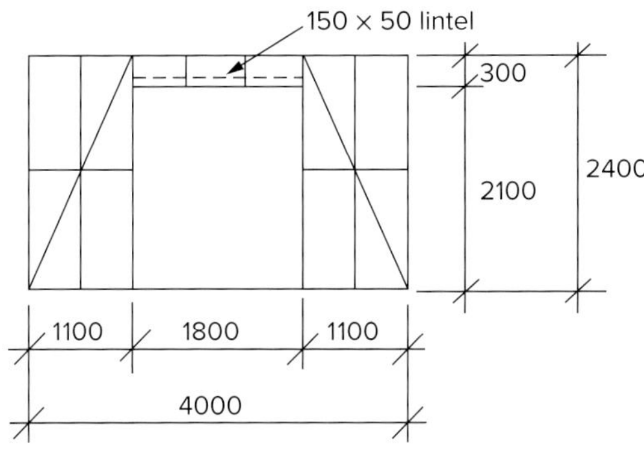

Member	Quantity	Length
Plates		
Studs		
Noggings		
Head trimmers		
Lintels		
Braces		

Worksheet 4

Interpret and apply basic plans and drawings

Date: ______________________ Grade: Satisfactory/Unsatisfactory

Name: ______________________ Student number: ______________________

Course code: ______________________ Course name: ______________________

Unit code: ______________________ Unit name: ______________________

1 Name the two basic levels of drawings that are used for the design and construction of a building and give a brief description of each one.

2 Identify the abbreviations for the following building terms:

(a) concrete

(b) smoke alarm

(c) Australian Standard

(d) rolled steel joist

(e) drainage

(f) switchboard

(g) water closet

(h) brickwork

(i) plasterboard

(j) finished floor level

3 Sketch two symbols for each of the following building materials:

(a) brickwork

(b) cut stone masonry

(c) stud walls

(d) insulation

(e) timber.

4 Name the process of drawing buildings or parts of buildings to a proportionate size so they can be fitted on to paper.

5 How are the elevations of a four-sided building named?

6 What are the three main types of 'view' of a building?

7 What are the three most common scales used for site plans?

8 A comprehensive list of all common building abbreviations is available in which Australian Standard?

9 What is the main purpose of using symbols and abbreviations on building plans?

10 Without a scale rule, what formula could you use to convert the measurements on your preliminary sketches to the correct scale length or dimension?

11 When producing plans electronically, what does the acronym CAD stand for?

12 List four items that you may find on a site plan.

13 Which plan would you refer to for the dimensions of all of the rooms in the planned building?

14 Why do amendments need to be signed off by both the builder and the client?

15 List three changes that may be made that require an amendment to the plans and specifications.

16 If an amendment is made to the plans and specifications on a large-scale construction job, which document might you be directed to for information regarding the amendment?

17 If working from a plan drawn at a scale of 1:20 and with dimension lines measured at 190 mm apart, what would be the true measurement of the finished item?

18 A site plan needs to be drawn at a scale of 1:200 for a block of land measuring 18.6 m × 35.8 m. What will the drawn size of the block on the plan be when completed?

19 Why must plan drawing tools and equipment be treated the same as building tools of the trade?

20 Why is it important that plans and specifications are drawn and written clearly and accurately?

Worksheet 5

Work effectively and sustainably in the construction industry

Date: ______________________ Grade: Satisfactory/Unsatisfactory

Name: ______________________ Student number: ______________________

Course code: ______________________ Course name: ______________________

Unit code: ______________________ Unit name: ______________________

1 Approximately how many people are employed in the Australian construction industry?

2 The construction industry is regarded as Australia's largest non-service industry: true or false?

3 Explain the difference between direct and indirect employment in the construction industry.

4 Give a brief description of what may be included in your contract of employment, to be observed by both yourself and your employer.

5 The *Occupational Health and Safety Act 2011* states that employers must provide for the safety and wellbeing of all employees in a way that is free of risk. List the five main points that this statement refers to.

6 If you were to observe a fellow employee acting in a drug- or alcohol-affected manner at work, what steps might you take?

7 If you felt that you or a workmate were being bullied, harassed or discriminated against in any way while working in the construction industry, what steps might you take to address the situation?

8 List five factors that could affect construction tasks and completion timelines.

9 Which chart is an excellent tool for planning and scheduling the activities that will take place on the construction site so that the job runs smoothly to the given timelines and any delays can be identified and addressed?

10 What are the benefits of being organised with your workload time frames?

11 What is generally the main contributing factor towards disharmony within a work team, and what can be the outcome of such disharmony?

12 Explain the advantages of keeping an up-to-date work journal for your own professional development and your contribution to day-to-day activities on the work site.

13 List six specific reporting and recording systems that companies may use for stock and resource control.

14 Explain why it is necessary that each of the records referred to in Question 13 is well maintained and kept current. Address each document individually.

15 Give a brief definition of the term *sustainable building.*

16 List the most important energy measures for achieving low operating energy in a building.

17 List and summarise six storage options for construction materials, to reduce material wastage due to loss or damage.

18 Across Australia, state and local laws require that a building site is clean and ordered to protect both the environment and the workers on-site. What can be the consequences if these laws are not adhered to?

19 If any environmental breaches occur or potential breaches are detected, what must immediately happen?

Worksheet 6

Identify and handle carpentry tools and equipment

Date: ______________________ Grade: Satisfactory/Unsatisfactory

Name: ______________________ Student number: ______________________

Course code: ______________________ Course name: ______________________

Unit code: ______________________ Unit name: ______________________

1 From the list below, highlight three things you would check prior to using a power saw.

- The brand of the saw
- The test tag—is it current?
- Is the guard operating correctly?
- The model of the saw
- Is the tool lead free of nicks and cuts?
- Is the blade installed correctly?

2 Fill in the blanks in the following sentences.

(a) I use a ______________________ to mark a line 90 degrees to the edge of the timber.

(b) A ______________________ is a power tool used to create moulds and shapes on the edges of timber.

(c) When I push on a ______________________ or ______________________ it cuts into timber.

(d) A ______________________ is a power tool used to cut curves.

3 Answer the following questions.

(a) What two tasks can be performed using a carpenter's hammer?

____

____

____

____

(b) Describe the use of a power planer.

____

____

____

____

____

(c) Describe how to use a chisel, when checking out a hinge.

____

____

____

____

____

____

(d) Which electric power tool would be used to drill a hole?

4 List all of the PPE that would be used when cutting treated pine with a power saw.

5 List four of the pre-operational checks required prior to using a belt sander.

6 Describe the step-by-step process of dressing a piece of timber using a hand plane. Include the hazards and the control measures.

7 Read through the following steps taken when using a jig saw, then correctly order and number them from 1–10.

Obtain the tool from the store and inspect it to ensure it is tagged and safe to use.

Complete the cut. Allow the tool to stop running prior to removing the tool from the timber.

Secure the timber to be cut firmly using clamps or similar to a bench or saw stool.

Check the tool blade is sharp and straight and select the appropriate speed and function according to the cut to be made.

Clean up any mess.

Plug the tool into a socket or lead.

Ensure the work area is clear of debris and other hazards.

Put on appropriate PPE (ear/eye protection and dust mask).

Place the base of the tool on the timber and start cutting, ensuring the blade is clear of the timber initially and the tool has come to full speed before commencing the cut.

Put the tool back in the store.

8 Before using any power tool you should check to make sure the equipment is in good working order. If you find the power tool not in good working order, then it should be withdrawn from service.

True or false?

9 Answer the questions below.

(a) Which on-site piece of plant would you use to provide the power to run a power saw?

(b) Which on-site piece of plant is used to supply air to a nail gun?

(c) Name two types of load-shifting equipment.

10 Answer the following questions.

(a) What piece of plant would you use to move crushed rock on-site?

(b) Explain the correct and safe procedure for using this piece of equipment to move crushed rock.

11 List the checks you would carry out before using an air compressor.

12 What should be checked prior to using a generator?

13 Once you have completed using a power saw, what should you do?

Worksheet 7

Perform basic setting out

Date: __________ Grade: Satisfactory/Unsatisfactory

Name: __________ Student number: __________

Course code: __________ Course name: __________

Unit code: __________ Unit name: __________

1 Which drawing would you look at to establish where a house was going to be situated on a block?

2 List four hazards that could be encountered when setting out a house.

3 What code deals with setting out on a building site?

4 Label the diagram below:

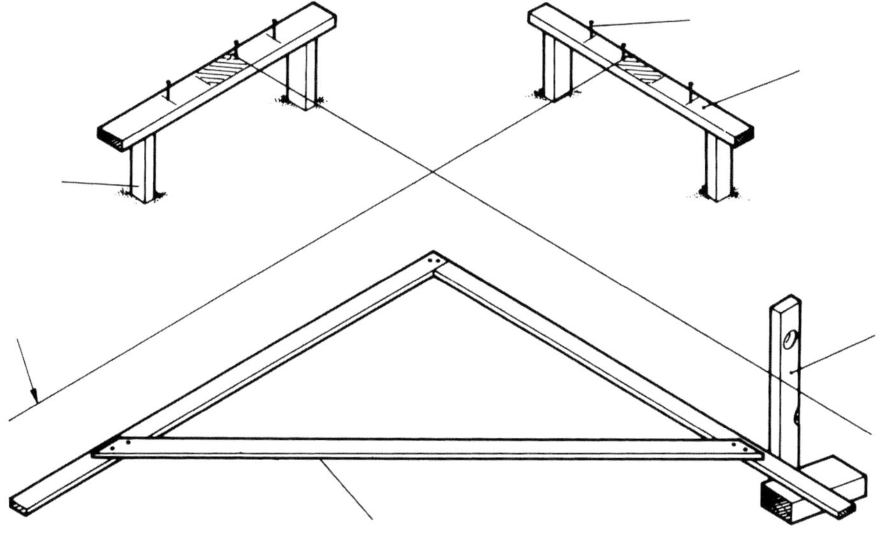

5 Setting out is generally an outdoor activity. What environmental hazard would you encounter during summer and how would you deal with it?

6 List six pieces of equipment and materials used in setting out.

7 Describe the sequence of setting out a square building.

8 How should waste and remaining material be dealt with after completing the setting out activity?

9 How should tools and equipment be dealt with after the setting out has been completed?

Worksheet 8

Apply basic levelling procedures

Date: ______________________ Grade: Satisfactory/Unsatisfactory

Name: ______________________ Student number: ______________________

Course code: ______________________ Course name: ______________________

Unit code: ______________________ Unit name: ______________________

1 When setting out a building, why is it important to ensure the hurdles are all at the same level?

2 What document is completed prior to a construction task, identifying the hazards that are present or could be experienced?

3 What test do you perform on a dumpy level to check that it is reading accurately?

4 List two ways in which people could communicate with each other if they are using a dumpy level and there is a large distance between the person taking the readings and the person holding the staff.

5 What term is used to identify the required height of the new building floor?

6 How should waste and remaining material be dealt with after completing the levelling activity?

7 How should tools and equipment be dealt with after levelling has been completed?

Worksheet 9

Construct basic sub-floor

Date: ______________________ Grade: Satisfactory/Unsatisfactory

Name: ______________________ Student number: ______________________

Course code: ______________________ Course name: ______________________

Unit code: ______________________ Unit name: ______________________

1 Fill in the missing words. What are the titles of the two drawings below?

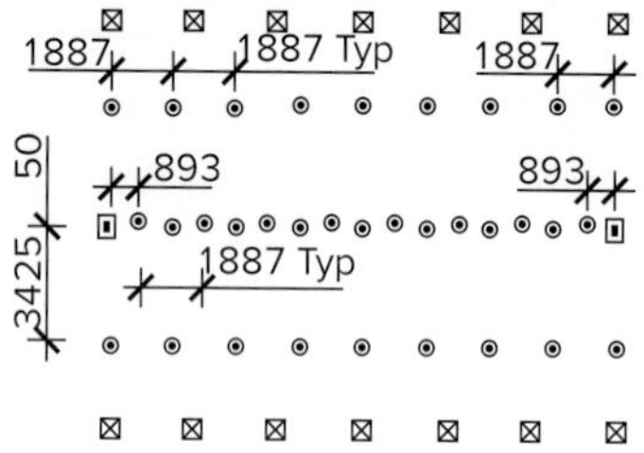

1887 1887 Typ 1887

50 893 893

3425 1887 Typ

Legend:

⊙ 350 DIA × 150 DEEP 20Mpa PAD FOOTING

◘ 600 SQ × 700 DEEP 20Mpa PAD FOOTING

⊠ 500 SQ × 500 DEEP 20Mpa PAD FOOTING

__________ Plan

Bearer

Joist

Ant cap

Stump

150

500

350 DIA

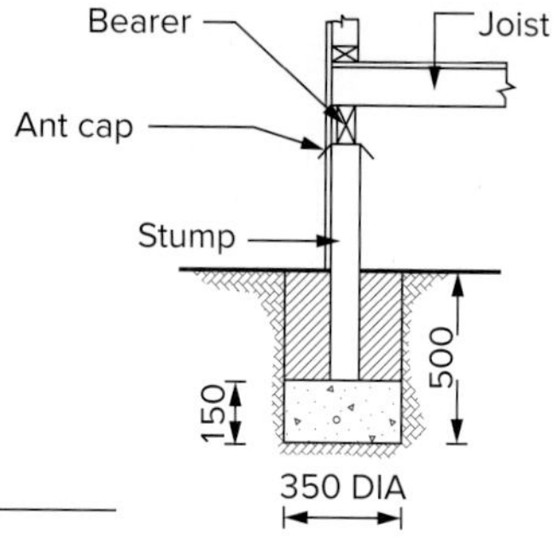

Footing __________

2 List three hazards you may encounter when digging holes for stumps.

____

____

____

____

____

____

____

3 Which Australian Standard deals with timber subfloor construction?

____

____

4 What should happen to the waste timber once the subfloor framing has been completed?

____

____

____

____

____

____

5 Label the following drawing:

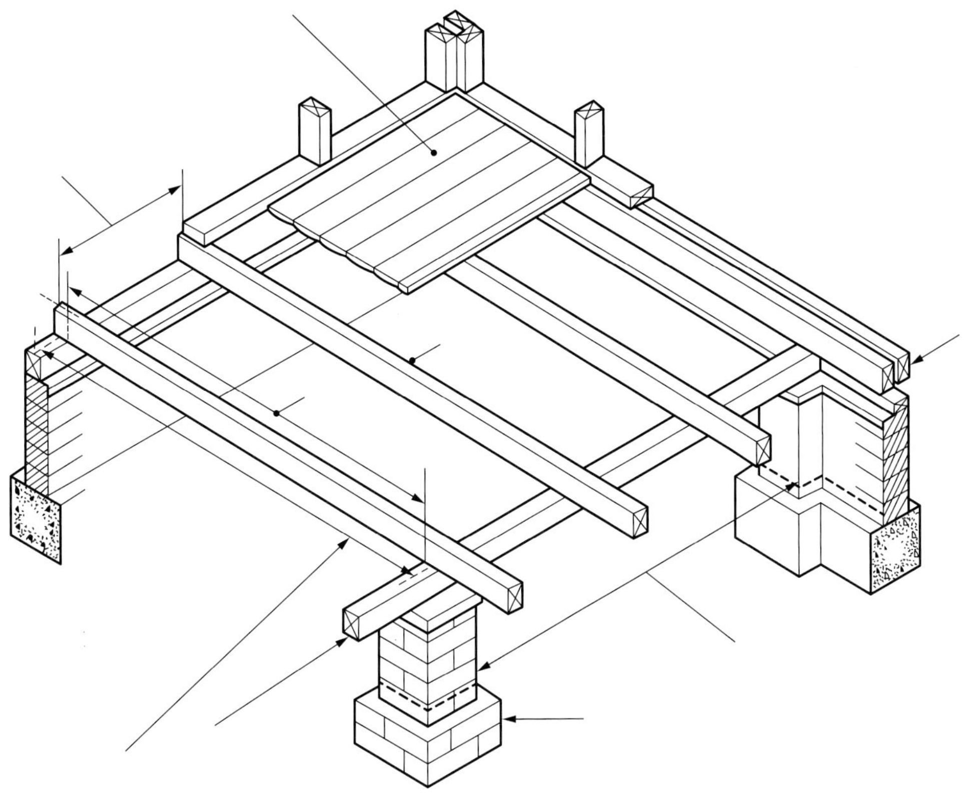

6 What PPE would be used when cutting and constructing a timber subfloor?

____

____

7 What part of a set of drawings would you use to calculate the amount of bearer material required to build the subfloor?

____

____

8 List six major tools you would use to construct a timber subfloor frame.

____

____

____

____

____

____

____

9 Describe the on-site process of setting out and installing timber stumps.

10 What method would you use to ensure the floor joists are flat and true prior to the flooring being installed?

11 How should waste and remaining material be dealt with after completing the construction of a timber subfloor frame?

12 How should tools and equipment be dealt with after the subfloor frame has been constructed?

Worksheet 10

Construct basic wall frames

Date: ______________________ Grade: Satisfactory/Unsatisfactory

Name: ______________________ Student number: ______________________

Course code: ______________________ Course name: ______________________

Unit code: ______________________ Unit name: ______________________

1 Which drawing could you use to see how a timber wall frame should be built?

2 Correct lifting technique is important when standing up wall frames. Describe the correct technique.

3 Which Australian Standard deals with timber wall framing?

4 What should happen to any waste timber once the framing has been completed?

5 Label the wall frame components on the diagram:

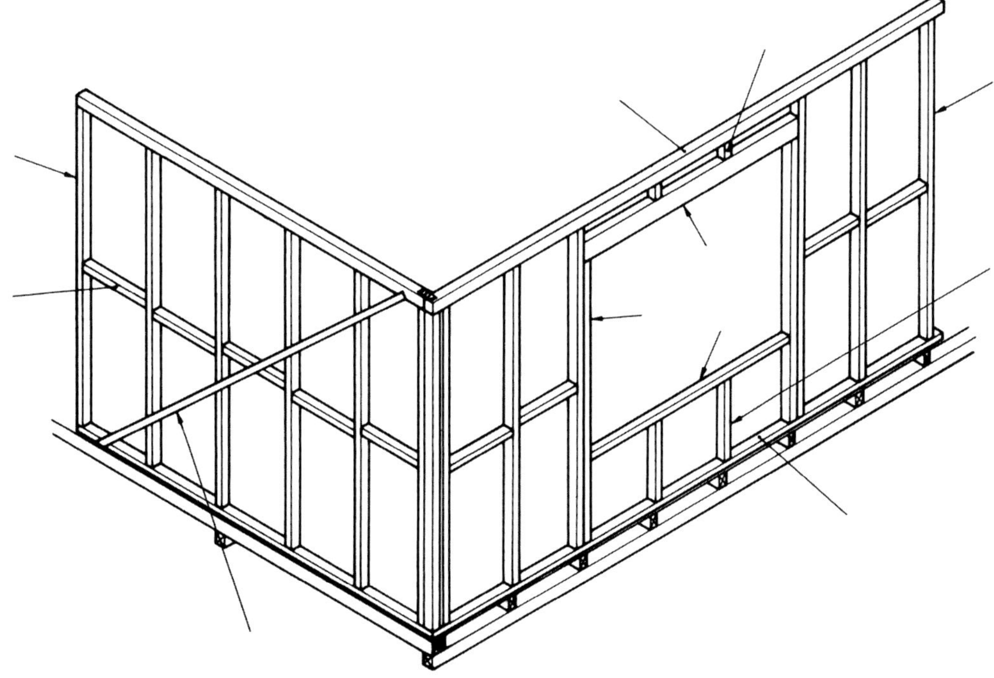

6 What PPE would be used when cutting and constructing a timber wall frame?

7 What part of a set of drawings would you use to find out how many walls need to be constructed?

8 List six major tools you would use to construct a timber wall frame.

9 Describe the sequence for setting out wall plates.

10 Describe the sequence for assembling wall frames.

11 How should waste and remaining material be dealt with after completing the construction of wall frames?

12 How should tools and equipment be dealt with after the wall frames have been constructed?

Worksheet 11

Construct a basic roof frame

Date: ______________________ Grade: Satisfactory/Unsatisfactory

Name: ______________________ Student number: ______________________

Course code: ______________________ Course name: ______________________

Unit code: ______________________ Unit name: ______________________

1 Which drawings in a set of plans would show you how the roof members connect to each other?

____

____

2 List four hazards you would encounter when erecting a timber roof on-site.

____

____

____

____

____

____

3 Which Australian Standard deals with timber wall framing?

____

____

4 What should happen to waste timber once the framing has been completed?

____

____

5 Label the following drawing:

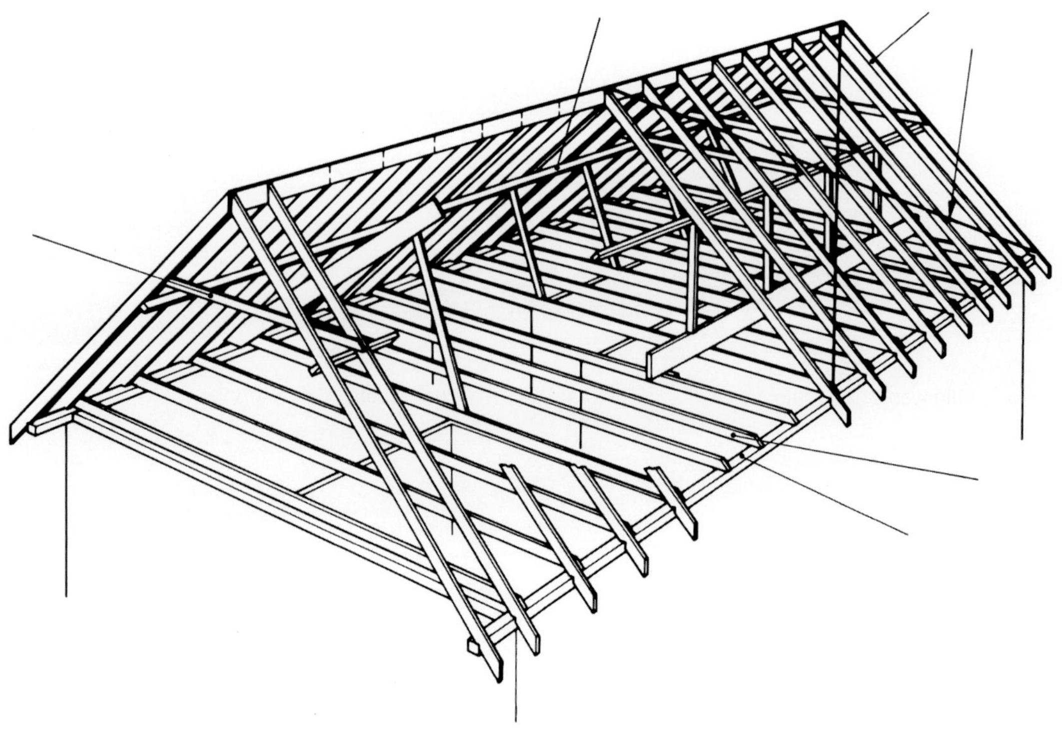

6 What PPE would be used when cutting a timber roof frame?

7 What part of a set of drawings would you use to work out the plan length of the hip rafters and ridge?

8 List six major tools you would use to construct a timber wall frame.

9 Name two methods that can be used to calculate roof bevels and rafter lengths.

10 List the sequence for erecting a hipped roof.

11 How should waste and remaining materials be dealt with after completing the construction of wall frames?

Worksheet 12

Erect and safely use working platforms

Date: ______________________ Grade: Satisfactory/Unsatisfactory

Name: ______________________ Student number: ______________________

Course code: ______________________ Course name: ______________________

Unit code: ______________________ Unit name: ______________________

1 When erecting a mobile aluminium scaffold, what documents should be obtained and studied before erecting and using the scaffold?

____

____

____

2 Which Australian Standard deals with OHS requirements relating to the erection and use of working platforms and scaffolds?

____

____

____

3 What should happen to scaffold components that have been damaged?

____

____

____

____

4 Label the following diagram:

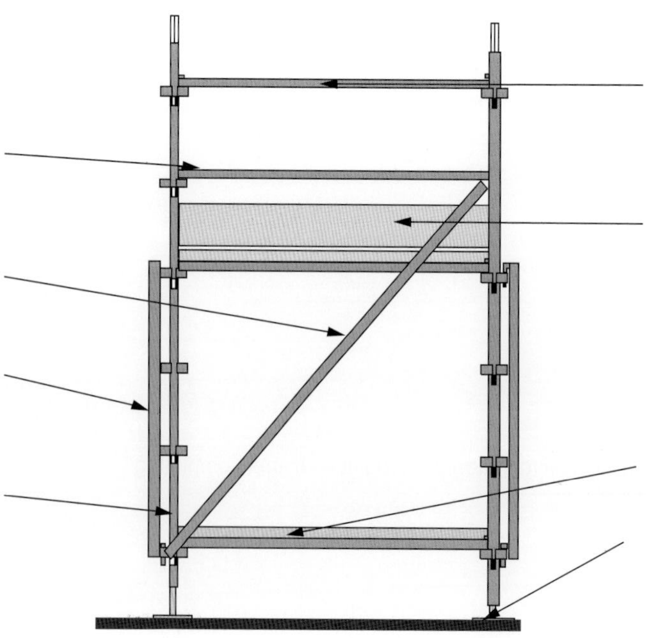

5 You are asked to install weatherboards onto the second storey of a house. An extension ladder is the most appropriate type of platform to use to install the weatherboards: true or false?

6 What PPE would you use when erecting scaffolding?

7 Label the following tools.

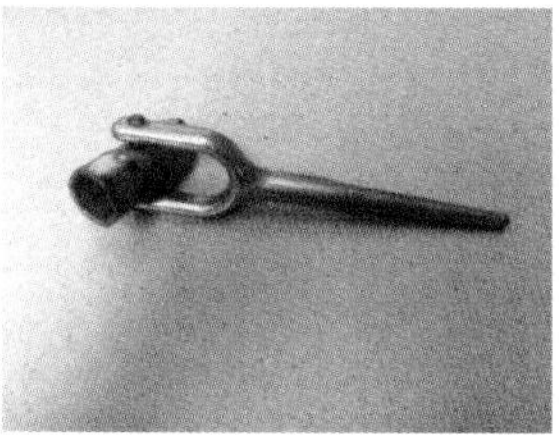

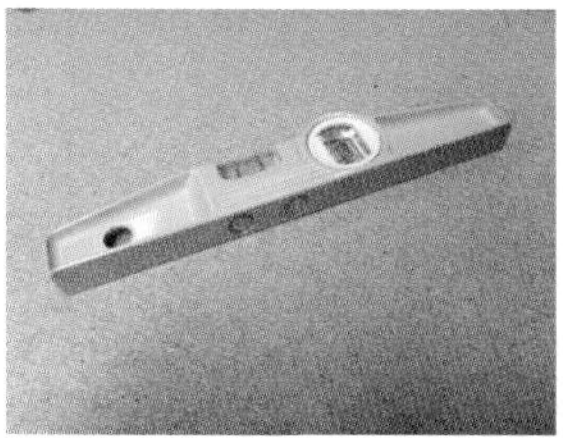

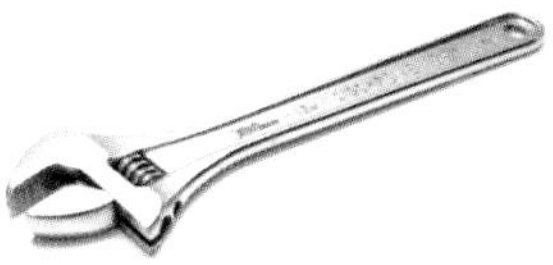

8 Number the drawings below, showing the erection sequence for a modular scaffold, in the correct order.

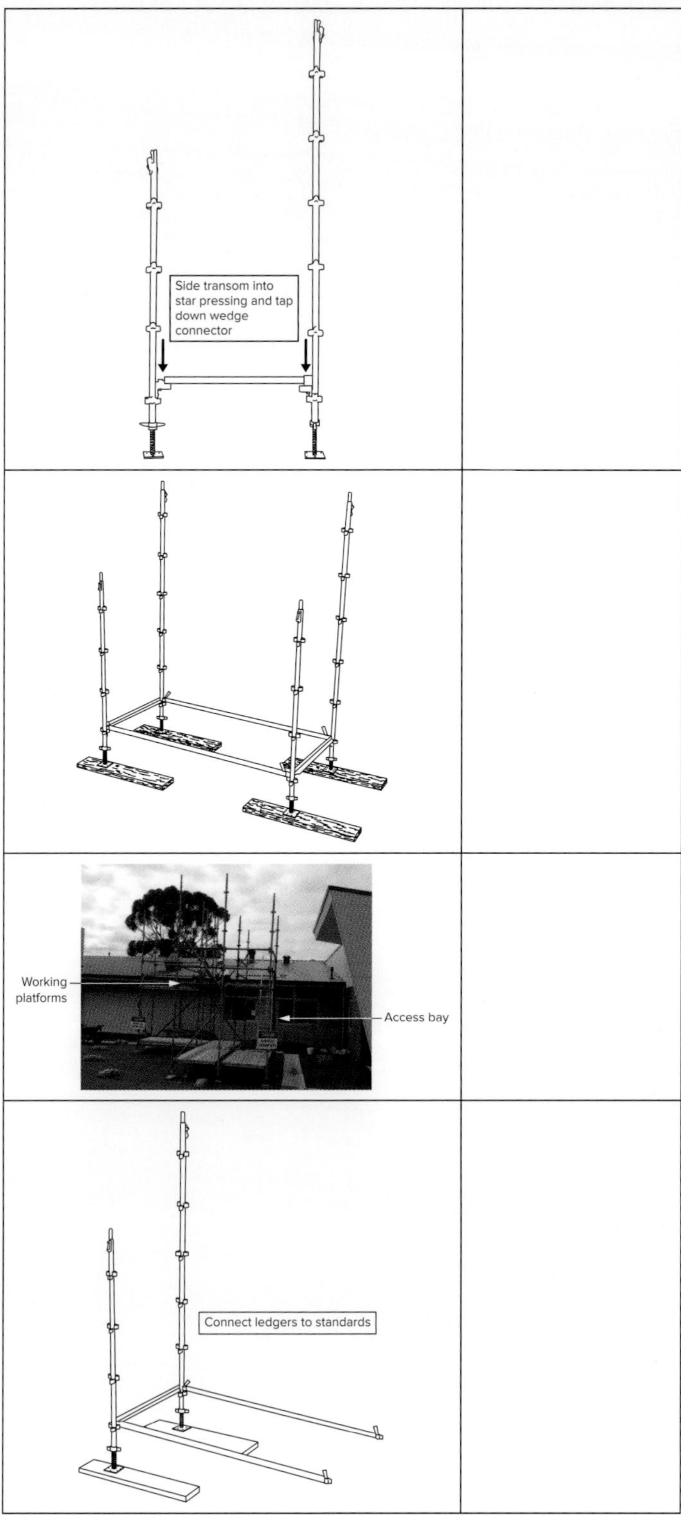

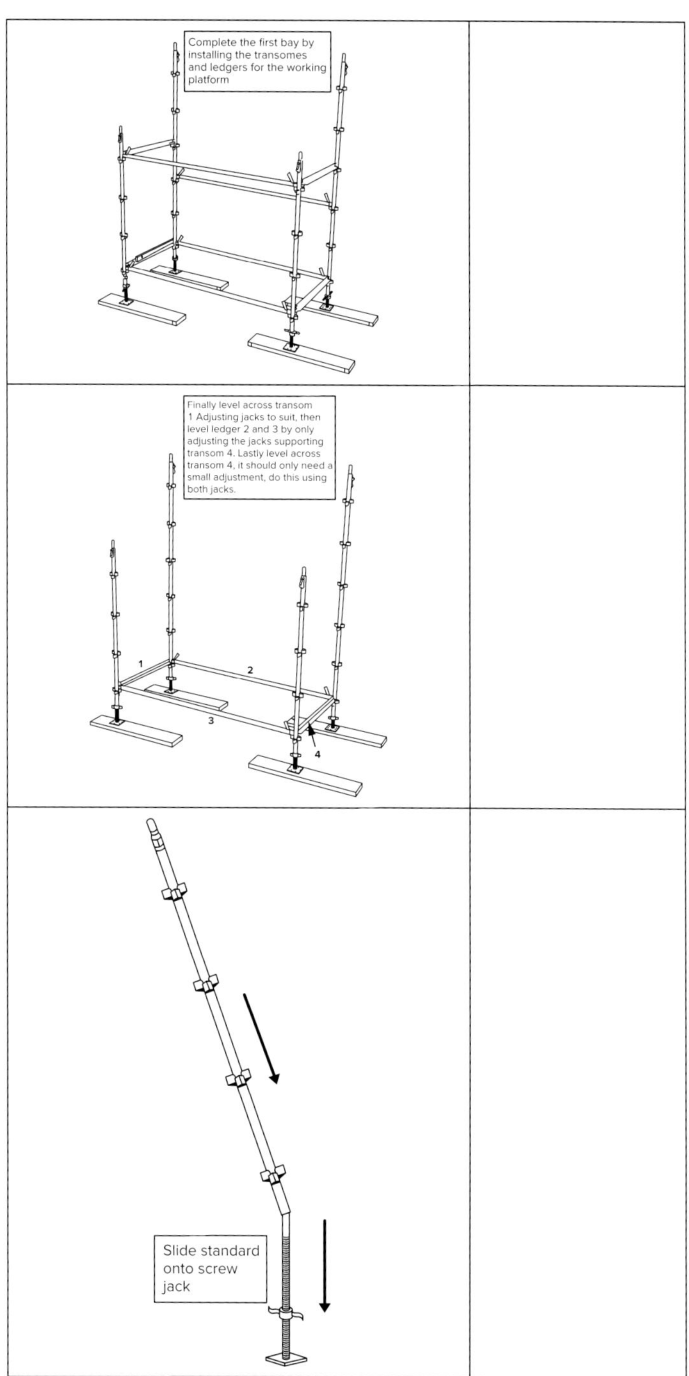
Complete the first bay by installing the transomes and ledgers for the working platform
Finally level across transom 1 Adjusting jacks to suit, then level ledger 2 and 3 by only adjusting the jacks supporting transom 4. Lastly level across transom 4, it should only need a small adjustment, do this using both jacks.
1
2
3
4
Slide standard onto screw jack

9 How should waste and remaining material be dealt with after erecting, using and dismantling working platforms and ladders?

10 How should scaffold components, tools and equipment be dealt with after they have been dismantled?

Worksheet 13

Install basic external cladding

Date: ______________________ Grade: Satisfactory/Unsatisfactory

Name: ______________________ Student number: ______________________

Course code: ______________________ Course name: ______________________

Unit code: ______________________ Unit name: ______________________

1 What document would you consult to find out about the fixing requirements for the cladding you are about to install?

2 Which document(s) would you refer to in order to identify any potential hazards associated with the use of a new cladding product?

3 Insert the missing words: When setting up a work area when installing cladding, an area or receptacle(s) should be established to deal with the ______________________ from the cladding. Where possible, it should be ______________________ and not disposed of in landfill.

4 Match each word on the left with its meaning on the right:

Weatherboard stop	**A. Shape of the weatherboard**
Profile	**B. Piece of vertical timber attached to the corner of the frame**
Effective cover	**C. The actual amount of building each weatherboard covers**

5 What PPE would you use when installing untreated Baltic pine weatherboards?

6 What three things need to be obtained prior to calculating the amount of cladding required for a job?

7 Name one source of information where you can find out what tools are required to cut and fix a cladding product.

8 What tools would be used to check a timber wall is straight and true?

9 Which Australian Standard covers the blocking and nogging requirements for a timber wall frame?

10 What tools would be used to plumb and straighten the corner of a wall frame?

11 Draw the layout of external corner wall studs to suit weatherboard cladding.

12 What two methods of fixing could be used to attach a timber weatherboard stop to a timber frame?

13 The top of a window and door frame requires flashing prior to the cladding being installed: true or false?

14 Which document would be used to determine the cover of a weatherboard?

15 Sketch a method to set out weatherboard stops.

16 How should waste and any remaining material be dealt with after completing the installation of cladding?

17 How should tools and equipment be dealt with once clearing-up has been done after cladding has been installed?

Worksheet 14

Install basic window and door frames

Date: ______________________ Grade: Satisfactory/Unsatisfactory

Name: ______________________ Student number: ______________________

Course code: ______________________ Course name: ______________________

Unit code: ______________________ Unit name: ______________________

1 Which drawing would be used to locate the positions of the windows and doors on a new house?

2 When installing window frames, what OHS issues or hazards could be encountered? List three.

3 Which code deals with the installation of windows and doors?

4 If using aluminium flashing to flash a window head, how should the waste flashing be dealt with?

5 What do the following words mean?

(a) Still

(b) Head

(c) Stiles

(d) Jambs

6 When installing door jambs using a nail gun, what four pieces of PPE would you be required to wear?

7 What part of a house plan would you use to find out the dimensions of the windows to be installed and the identification number to locate the window on the plan?

8 List four tools that would be used to install a window or door.

9 How should waste and remaining materials be dealt with after completing the installation of windows and door frames?

10 How should tools and equipment be dealt with after the windows and doors have been completed?

Worksheet 15

Install interior fixings

Date: ______________________ Grade: Satisfactory/Unsatisfactory

Name: ______________________ Student number: ______________________

Course code: ______________________ Course name: ______________________

Unit code: ______________________ Unit name: ______________________

1 Which drawing would be used to calculate the total amount of skirting required for a new house?

2 When cutting skirting with a drop saw, what OHS issues or hazards are created? List three.

3 Which Australian Standard deals with the manufacture of doors?

4 What do the following words mean?

(a) Scribed joint

(b) Architrave

(c) Skirting

5 Referring to Question 2, what four pieces of PPE would be used to minimise the potential risk?

6 What part of a working drawing would be used to determine the amount of fixing materials the project required for skirting and architraves?

7 List four tools that would be used for interior fixing.

8 List the correct order when installing jamb linings.

9 Label the drawing with the correct installation sequence for skirting.

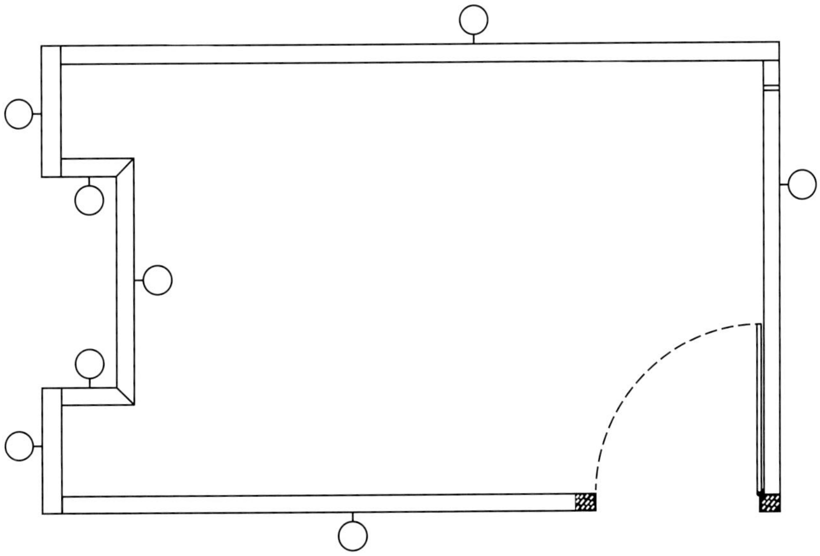

10 How should tools and equipment be dealt with after the fixing has been completed?

Worksheet 16

Construct basic formwork for concreting

Date: ______________________ Grade: Satisfactory/Unsatisfactory

Name: ______________________ Student number: ______________________

Course code: ______________________ Course name: ______________________

Unit code: ______________________ Unit name: ______________________

1 List four different materials from which formwork could be constructed.

2 List five items of PPE that may be required while constructing formwork.

3 List four common uses in the building industry for reinforced concrete.

4 List two situations where unreinforced concrete may be used.

5 Which official forms must be used to document all of the hazards and risks associated with the construction of formwork?

6 What type of health problems could occur if you are not properly protected and come into contact with wet cement?

7 What could be some of the outcomes if formwork is not properly braced and fails during the pour?

8 What are some of the advantages of using formply to construct formwork?

9 Which Australian Standard covers the construction of formwork?

10 What needs to be done prior to the pour to ensure that the formwork will strip away easily from the semi-hardened concrete?

11 How can formwork be treated to achieve sharp curves in the finished concrete?

12 Typically, what size would a large sheet of steel reinforcing be?

13 What may be installed directly below the concrete slab to stop moisture rising up through the foundations and entering the concrete?

14 To which plan would you refer to find the correct position for the slab on the block of land?

15 Which plans would you refer to, to determine the finished height of the slab above the ground and the depth of any excavations?

16 Describe what should be done to formwork as you dismantle it and strip it away from the poured concrete.

17 What signage or barricading may need to be in place before concrete trucks begin delivering concrete to the site?

18 What items would you use to lift sheet reinforcement to the required height according to the plans and specifications?

19 After the formwork has been constructed and before putting the reinforcement in place, ready to pour the concrete, what must you do to make sure there will be no contamination to the concrete after pouring?

20 If using a concrete pump, what is the maximum height from which you could drop concrete to avoid segregation?

21 On a smaller pour, name the two most common ways to distribute the concrete into the formwork structure.

22 How can concrete be placed into the formwork and be finished accurately to the required height?

23 After the concrete has been placed and moved into position by shovelling and vibrating, what is the process called that levels concrete to the required height with a long straight edge?

24 List three finishes that may be applied to concrete that will provide a slip-resistant surface.

Worksheet 17

Carry out basic demolition of timber structures

Date: ______________________ Grade: Satisfactory/Unsatisfactory

Name: ______________________ Student number: ______________________

Course code: ______________________ Course name: ______________________

Unit code: ______________________ Unit name: ______________________

1 Because demolition is classed as high-risk work, which document must be written before any work commences?

2 Name the Australian Standard that deals with lead paint management.

3 List three types of hazardous materials that you may find on a demolition task.

4 What is the maximum amount of non-friable asbestos someone can remove without having a removal licence?

5 List five building items that may contain asbestos.

6 Rogue services are most commonly found in what type of building?

7 Who could you consult for information before writing a demolition plan?

8 The emergency section of your demolition plan could include what type of information?

9 List seven items of PPE that you may need to carry out demolition.

10 What type of areas need to be set up for materials as the demolition job proceeds?

11 List four hand tools, four power tools and four items of equipment that could be used to carry out demolition.

12 List three things that an internal wall might hold up that would class it as load-bearing.

13 List four important points to consider when you stand back, look at and consider the best way to go about the demolition task.

14 If doing a bathroom renovation, what might you need to do to the floors, walls and doors to create a safe route for the removal of any demolished materials?

15 If removing a load-bearing wall, what must you do to the carried load before demolition?

16 If a roof is constructed with a truss system, which walls carry the load of the roof?

17 If demolishing and then replacing termite-damaged load-bearing wall framing, how would you support the roof load while you replace the damaged framework?

18 What are the 'three Rs' of demolition waste minimisation?

19 Why would you de-nail any timber as the demolition job progresses?

20 Materials such as cement render, concrete, brickwork and tiles can be crushed down into different aggregate sizes. List two things that this aggregate can be used for.

21 List three ways in which you could transport demolition rubbish to the tip or landfill site.

22 Why must you ensure that any loads of rubbish being transported are properly secured?

23 If any tools or equipment that you are using for demolition get damaged, what should you do?